# The Atmosphe

**Citations**

Please use the following example for citations:

Prinn R.G. (2003) Ozone, hydroxyl radical, and oxidative capacity, pp. 1–19. In *The Atmosphere* (ed. R.F. Keeling) Vol. 4 *Treatise on Geochemistry* (eds. H.D. Holland and K.K. Turekian), Elsevier–Pergamon, Oxford.

**Cover photo:** A line of thunderstorms is backlit by the setting sun in this photograph taken from the National Scientific Balloon Facility in Palestine, Texas. (Photograph provided by Darin Toohey)

# The Atmosphere

*Edited by*

## R. F. Keeling
*University of California, CA, USA*

## TREATISE ON GEOCHEMISTRY
### Volume 4

*Executive Editors*

## H. D. Holland
*Harvard University, Cambridge, MA, USA*

and

## K. K. Turekian
*Yale University, New Haven, CT, USA*

ELSEVIER

AMSTERDAM – BOSTON – HEIDELBERG – LONDON – NEW YORK – OXFORD

PARIS – SAN DIEGO – SAN FRANCISCO – SINGAPORE – SYDNEY – TOKYO

Elsevier
The Boulevard, Langford Lane, Kidlington, Oxford OX5 1GB, UK
Radarweg 29, PO Box 211, 1000 AE Amsterdam, The Netherlands

First edition 2006

**British Library Cataloguing in Publication Data**
A catalogue record for this book is available from the British Library

**Library of Congress Cataloging-in-Publication Data**
A catalog record for this book is available from the Library of Congress

ISBN-13: 978-0-08-045091-9
ISBN-10: 0-08-045091-9

The following chapters are US Government works in the public domain and not subject to copyright:
    Biomass Burning: The Cycling of Gases and Particulates from the Biosphere to the Atmosphere

For information on all Elsevier publications
visit our website at books.elsevier.com

Printed and bound in the United Kingdom
Transferred to Digital Print 2011

# DEDICATED
## TO

**HANS OESCHGER**
**(1927–1998)**

**CHRISTIAN JUNGE**
**(1912–1996)**

# Contents

# Executive Editors' Foreword

H. D. Holland

*Harvard University, Cambridge, MA, USA*

and

K. K. Turekian

*Yale University, New Haven, CT, USA*

Geochemistry has deep roots. Its beginnings can be traced back to antiquity, but many of the discoveries that are basic to the science were made between 1800 and 1910. The periodic table of elements was assembled, radioactivity was discovered, and the thermodynamics of heterogeneous systems was developed. The solar spectrum was used to determine the composition of the Sun. This information, together with chemical analyses of meteorites, provided an entry to a larger view of the universe.

During the first half of the twentieth century, a large number of scientists used a variety of methods to determine the major-element composition of the Earth's crust, and the geochemistries of many of the minor elements were defined by V. M. Goldschmidt and his associates using the then new technique of emission spectrography. V. I. Vernadsky founded biogeochemistry. The crystal structures of most minerals were determined by X-ray diffraction techniques. Isotope geochemistry was born, and age determinations based on radiometric techniques began to define the absolute geologic timescale. The intense scientific efforts during World War II yielded new analytical tools and a group of people who trained a new generation of geochemists at a number of universities. But the field grew slowly. In the 1950s, a few journals were able to report all of the important developments in trace-element geochemistry, isotopic geochronometry, the exploration of paleoclimatology and biogeochemistry with light stable isotopes, and studies of phase equilibria. At the meetings of the American Geophysical Union, geochemical sessions were few, none were concurrent, and they all ranged across the entire field.

Since then the developments in instrumentation and the increases in computing power have been spectacular. The education of geochemists has been broadened beyond the old, rather narrowly defined areas. Atmospheric and marine geochemistry have become integrated into solid Earth geochemistry; cosmochemistry and biogeochemistry have contributed greatly to our understanding of the history of our planet. The study of Earth has evolved into "Earth System Science," whose progress since the 1940s has been truly dramatic.

Major ocean expeditions have shown how and how fast the oceans mix; they have demonstrated the connections between the biologic pump, marine biology, physical oceanography, and marine sedimentation. The discovery of hydrothermal vents has shown how oceanography is related to economic geology. It has revealed formerly unknown oceanic biotas, and has clarified the factors that today control, and in the past have controlled the composition of seawater.

Seafloor spreading, continental drift and plate tectonics have permeated geochemistry. We finally understand the fate of sediments and oceanic crust in subduction zones, their burial and their exhumation. New experimental techniques at temperatures and pressures of the deep Earth interior have clarified the three-dimensional structure of the mantle and the generation of magmas.

Moon rocks, the treasure trove of photographs of the planets and their moons, and the successful search for planets in other solar systems have all revolutionized our understanding of Earth and the universe in which we are embedded.

Geochemistry has also been propelled into the arena of local, regional, and global anthropogenic problems. The discovery of the ozone hole came as a great, unpleasant surprise, an object lesson for optimists and a source of major new insights into the photochemistry and dynamics of the atmosphere. The rise of the $CO_2$ content of the atmosphere due to the burning of fossil fuels and deforestation has been and will continue to be at the center of the global change controversy, and will yield new insights into the coupling of atmospheric chemistry to the biosphere, the crust, and the oceans.

The rush of scientific progress in geochemistry since World War II has been matched by organizational innovations. The first issue of *Geochimica et Cosmochimica Acta* appeared in June 1950. The Geochemical Society was founded in 1955 and adopted *Geochimica et Cosmochimica Acta* as its official publication in 1957. The International Association of Geochemistry and Cosmochemistry was founded in 1966, and its journal, *Applied Geochemistry*, began publication in 1986. *Chemical Geology* became the journal of the European Association for Geochemistry.

The Goldschmidt Conferences were inaugurated in 1991 and have become large international meetings. Geochemistry has become a major force in the Geological Society of America and in the American Geophysical Union. Needless to say, medals and other awards now recognize outstanding achievements in geochemistry in a number of scientific societies.

During the phenomenal growth of the science since the end of World War II an admirable number of books on various aspects of geochemistry were published. Of these only three attempted to cover the whole field. The excellent *Geochemistry* by K. Rankama and Th.G. Sahama was published in 1950. V. M. Goldschmidt's book with the same title was started by the author in the 1940s. Sadly, his health suffered during the German occupation of his native Norway, and he died in England before the book was completed. Alex Muir and several of Goldschmidt's friends wrote the missing chapters of this classic volume, which was finally published in 1954.

Between 1969 and 1978 K. H. Wedepohl together with a board of editors (C. W. Correns, D. M. Shaw, K. K. Turekian and J. Zeman) and a large number of individual authors assembled the *Handbook of Geochemistry*. This and the other two major works on geochemistry begin with integrating chapters followed by chapters devoted to the geochemistry of one or a small group of elements. All three are now out of date, because major innovations in instrumentation and the expansion of the number of practitioners in the field have produced valuable sets of high-quality data, which have led to many new insights into fundamental geochemical problems.

At the Goldschmidt Conference at Harvard in 1999, Elsevier proposed to the Executive Editors that it was time to prepare a new, reasonably comprehensive, integrated summary of geochemistry. We decided to approach our task somewhat differently from our predecessors. We divided geochemistry into nine parts. As shown below, each part was assigned a volume, and a distinguished editor was chosen for each volume. A tenth volume was reserved for a comprehensive index:

(i) *Meteorites, Comets, and Planets*: Andrew M. Davis

(ii) *Geochemistry of the Mantle and Core*: Richard Carlson

(iii) *The Earth's Crust*: Roberta L. Rudnick

(iv) *Atmospheric Geochemistry*: Ralph F. Keeling

(v) *Freshwater Geochemistry, Weathering, and Soils*: James I. Drever

(vi) *The Oceans and Marine Geochemistry*: Harry Elderfield

(vii) *Sediments, Diagenesis, and Sedimentary Rocks*: Fred T. Mackenzie

(viii) *Biogeochemistry*: William H. Schlesinger

(ix) *Environmental Geochemistry*: Barbara Sherwood Lollar

(x) *Indexes*

The editor of each volume was asked to assemble a group of authors to write a series of chapters that together summarize the part of the field covered by the volume. The volume editors and chapter authors joined the team enthusiastically. Altogether there are 155 chapters and 9 introductory essays in the Treatise. Naming the work proved to be somewhat problematic. It is clearly not meant to be an encyclopedia. The titles *Comprehensive Geochemistry* and *Handbook of Geochemistry* were finally abandoned in favor of *Treatise on Geochemistry*.

The major features of the Treatise were shaped at a meeting in Edinburgh during a conference on Earth System Processes sponsored by the Geological Society of America and the Geological Society of London in June 2001. The fact that the Treatise is being published in 2003 is due to a great deal of hard work on the part of the editors, the authors, Mabel Peterson (the Managing Editor), Angela Greenwell (the former Head of Major Reference Works), Diana Calvert (Developmental Editor, Major Reference Works),

Bob Donaldson (Developmental Manager), Jerome Michalczyk and Rob Webb (Production Editors), and Friso Veenstra (Senior Publishing Editor). We extend our warm thanks to all of them. May their efforts be rewarded by a distinguished journey for the Treatise.

Finally, we would like to express our thanks to J. Laurence Kulp, our advisor as graduate students at Columbia University. He introduced us to the excitement of doing science and convinced us that all of the sciences are really subdivisions of geochemistry.

# Contributors to Volume 4

W. S. Broecker
Columbia University, Palisades, NY, USA

P. R. Buseck
Arizona State University, Tempe, AZ, USA

P. J. Crutzen
University of California, San Diego, CA, USA and
Max-Planck-Institut für Chemie, Mainz, Germany

W. C. Graustein
Yale University, New Haven, CT, USA

J. Jouzel
Institut Pierre Simon Laplace, Saclay, France

J. S. Levine
NASA Langley Research Center, Hampton, VA, USA

R. O. Pepin
University of Minnesota, Minneapolis, MN, USA

D. Porcelli
University of Oxford, UK

R. G. Prinn
Massachusetts Institute of Technology, Cambridge, MA, USA

W. S. Reeburgh
University of California Irvine, CA, USA

S. E. Schwartz
Brookhaven National Laboratory, Upton, NY, USA

M. H. Thiemens
University of California, San Diego, CA, USA

K. K. Turekian
Yale University, New Haven, CT, USA

R. von Glasow
University of California, San Diego, CA, USA

D. Yakir
Weizmann Institute of Science, Rehovot, Israel

# Volume Editor's Introduction

R. F. Keeling

*University of California, San Diego, La Jolla, CA, USA*

The atmosphere is a crucial geochemical reservoir of the Earth. Although the mass of the atmosphere is dwarfed by that of the oceans and solid Earth, the atmosphere plays a more direct role than these other reservoirs in controlling the Earth's climate via the absorption and scattering of sunlight and infrared radiation. The atmosphere also provides the conditions essential for sustaining life, such as supplying $O_2$, $CO_2$, moisture, and many nutrients. As humans, it is not hard to appreciate the importance of the atmosphere—the atmosphere is the one geochemical reservoir in which we reside. The atmosphere, in spite of its low mass, is also a dominant reservoir of many volatile chemical species, for example, $N_2$, $O_2$, and $CH_4$, and even for less volatile species the atmosphere can be the principal reservoir for production, transport, or destruction of the species.

To a geochemist (or biogeochemist) the study of the atmosphere has generally centered on specific themes, depending on the timescale of interest. On the billion-year timescale, a key question is how the Earth acquired its volatile elements and how and when these elements were degassed from the Earth's interior. On the somewhat shorter $\sim 100$ Ma timescale, a key question is how the abundance of the major biogeochemical gases, for example, $O_2$ and $CO_2$ are controlled, and the relation of these gases to the evolution of life and to global tectonic change. Shorter still are the time-scales associated with glacial–interglacial cycles and with recent human-induced climate changes. The key questions here concern the relationship between changing atmospheric composition and changes in climate and other life-sustaining resources, such as water, nutrients, and sunlight.

As a field, "atmospheric geochemistry" spans several distinct methods of enquiry. The most general method involves direct measurement of the chemical properties of the modern atmosphere, such as greenhouse gases, aerosols, or isotope ratios. Another method involves using tracers of atmospheric origin as tools in the study of other systems or reservoirs. One example of this approach includes the use of $^{14}C$ as a dating tool or as a tracer of ocean circulation. Another example is the use of isotopic signatures of atmospheric origin (e.g., photochemically produced excess $^{17}O$ in $O_3$) for tracing chemical pathways in the atmosphere or elsewhere. A third method involves reconstructing the past composition of the atmosphere from chemical signatures retained in ice cores, soils, or in marine sediments.

The twelve chapters of this volume, written by experts in their fields, provide a broad overview of these themes and methods of enquiry. The first chapter, by Prinn, reviews the controls on atmospheric OH and $O_3$ abundance. These chemicals play an essential role by "cleansing" the atmosphere of the emissions of greenhouse gases and various noxious compounds that otherwise would build up in the atmosphere to intolerable levels. The next chapter, by Von Glasow and Crutzen, provides a synthesis of tropospheric halogen chemistry, and includes a review of the main halogen reactions, their interaction with other elements. The chapter also outlines the importance of halogen chemistry to ozone in the Arctic, to the chemistry of the marine boundary layer, and to the chemistry downwind of salt lakes and salt pans.

The next three chapters emphasize subjects of central importance to the theme of human-induced climate change. Reeburgh (Chapter 4.03) reviews the subject of the global biogeochemisty of methane, an important greenhouse gas that has increased at a rapid pace since the beginning of the Industrial Revolution. This chapter discusses the processes controlling the recent rise as well as the record of past variations as reconstructed from ancient air trapped in ice cores. Buseck and Schwartz (Chapter 4.04) provide a broad overview of tropospheric aerosols, which are increasingly recognized as an important determinant of regional and global climate through their effect on atmospheric radiation. This chapter reviews recent progress in the understanding of aerosol sources, properties, distributions, and climate impacts. Levine (Chapter 4.05) reviews the subject of biomass burning, with discussions on the geographical distribution of burning, on gaseous and particulate emissions, on the impacts of burning on atmospheric chemistry—particularly effects on tropospheric and stratospheric ozone—and on the scattering and absorption of sunlight.

Applications of stable isotopes are emphasized in the following three chapters. Thiemens (Chapter 4.06) reviews the subject of non-mass-dependent isotopic effects and their applications, which include the study of $SO_2$ oxidation in the atmosphere, global biological productivity, $NO_3$ biogeochemistry, and the origin of atmospheric $O_2$. Yakir (Chapter 4.07) discusses the subject of the controls on the isotopic composition of atmospheric $CO_2$. The exchange of $CO_2$ with the land and the oceans produces unique signatures in the $^{13}C/^{12}C$ and $^{18}O/^{18}O$ ratio of $CO_2$, which can be observed in space and time and exploited to identify exchange rates and mechanisms. Jouzel (Chapter 4.08) discusses the dominant controls on the isotopic composition (i.e., D/H and $^{18}O/^{16}O$ ratios) of water in the Earth's atmosphere, in precipitation, and in the Earth's ice caps. An important theme of this chapter is the use of isotope ratios in ice cores as a paleotemperature proxy.

The next two chapters discuss radioisotopes. Broecker (Chapter 4.09) reviews the geochemistry of $^{14}C$ in relation to changes in the Earth's magnetic field, solar activity, the circulation of the oceans, and the fate of anthropogenic $CO_2$. Turekian and Graustein (Chapter 4.10) discuss the production and fates of other natural radionuclides in the atmosphere, such as the radiogenic nuclides $^{222}Rn$ and its daughters (especially, $^{210}Pb$, $^{210}Po$, and $^{210}Bi$) and the cosmogenic nuclides $^{10}Be$, $^{7}Be$, $^{32}P$, and $^{31}P$. The chapter highlights the application of these nuclides as tracers of atmospheric mixing and of aerosol scavenging.

The final two chapters concern the ultimate origin of volatile elements in the atmosphere. The first of these, by Porcelli and Turekian

(Chapter 4.11), discusses the degassing history of the Earth, drawing largely on evidence based on mantle inert-gas abundances and discussing implications for other elements, such as carbon, nitrogen, and hydrogen. The chapter highlights the strengthening evidence that most of the degassing of volatile elements from the Earth's interior to the atmosphere must have occurred early in Earth's history. The chapter also points to major unresolved questions, such as how to reconcile the heat and helium budgets of the Earth's interior. The final chapter by Porcelli and Pepin (Chapter 4.12) reviews the question of how the Earth accumulated volatile elements from the parental solar nebula during the process of early planet formation and the question of what controls the differences in volatile elements between the Earth, Venus, and Mars.

Because the geochemistry of the atmosphere is such a rich subject, this volume does not attempt an exhaustive treatment of all possible topics. Some atmospheric themes, such as the controls on $O_2$ and $CO_2$, are discussed in parallel volumes of this Treatise. This volume also does not attempt to provide a comprehensive treatment of atmospheric photochemistry, which is a vast field in its own right. In soliciting chapters for this volume, the decision was made to concentrate on subjects pertaining to atmospheric evolution, budgets, composition, and structure, consistent with a "geochemical" as opposed to a more pure "chemical" perspective. The decision was also made to focus on emergent fields that could most benefit from up-to-date synthesis.

This volume is dedicated to two pioneering atmospheric geochemists, Hans Oeschger and Christian Junge, who were founding fathers of their respective fields of ice-core geochemistry and atmospheric chemistry. The monumental achievements of Oeschger and his group in Bern, such as the recovery of past atmospheric composition from bubbles trapped in the cores, and the discovery of abrupt climate change during the last glacial period (now known as Dansgarrd–Oeschger events), are fundamental to our understanding of the controls on global climate and its sensitivity to human activities. Junge was among the first to take a global approach to the study of the atmospheric chemistry and he played a major role in building the community of researchers in this field. His pioneering work covered many areas, including aerosol chemistry and size distribution, sulfate oxidation, and distributions of trace gases and aerosols.

# 4.01

# Ozone, Hydroxyl Radical, and Oxidative Capacity

R. G. Prinn

*Massachusetts Institute of Technology, Cambridge, MA, USA*

## 4.01.1 INTRODUCTION

The atmosphere is a chemically complex and dynamic system interacting in significant ways with the oceans, land, and living organisms. A key process proceeding in the atmosphere is oxidation of a wide variety of relatively reduced chemical compounds produced largely by the biosphere. These compounds include hydrocarbons (RH), carbon monoxide (CO), sulfur dioxide ($SO_2$), nitrogen oxides ($NO_x$), and ammonia ($NH_3$) among others. They also include gases associated with advanced technologies such as hydrofluoro carbons and hydrochlorofluorocarbons. A summary of the composition of the atmosphere at the start of the twenty-first century is given in Table 1. Due to their key role, oxidation reactions are sometimes referred to as nature's atmospheric "cleansing" process, and the overall rate of this process as the "oxidation capacity" of the atmosphere. Without this efficient cleansing process, the levels of many emitted gases could rise so high that they would radically change the chemical nature of our atmosphere and biosphere and, through the greenhouse effect, our climate.

Oxidation became an important atmospheric reaction on Earth once molecular oxygen ($O_2$) from photosynthesis had reached sufficiently high levels. This $O_2$ could then photodissociate in the atmosphere to give oxygen atoms, which combine with $O_2$ to form ozone ($O_3$). When $O_3$ absorbs UV light at wavelengths less than 310 nm, it produces excited oxygen atoms ($O(^1D)$) which can attack water vapor to produce the hydroxyl free radical (OH). It is the hydroxyl radical that, above all, defines the oxidative capacity of our $O_2$- and

1

**Table 1**   Gaseous chemical composition of the atmosphere (1 ppt $= 10^{-12}$, 1 ppb $= 10^{-9}$, 1 ppm $= 10^{-6}$).

| Constituent | Chemical formula | Mole fraction in dry air | Major sources |
|---|---|---|---|
| Nitrogen | $N_2$ | 78.084% | Biological |
| Oxygen | $O_2$ | 20.948% | Biological |
| Argon | Ar | 0.934% | Inert |
| Carbon dioxide | $CO_2$ | 360 ppm | Combustion, ocean, biosphere |
| Neon | Ne | 18.18 ppm | Inert |
| Helium | He | 5.24 ppm | Inert |
| Methane | $CH_4$ | 1.7 ppm | Biogenic, anthropogenic |
| Hydrogen | $H_2$ | 0.55 ppm | Biogenic, anthropogenic, photochemical |
| Nitrous oxide | $N_2O$ | 0.31 ppm | Biogenic, anthropogenic |
| Carbon monoxide | CO | 50–200 ppb | Photochemical, anthropogenic |
| Ozone (troposphere) | $O_3$ | 10–500 ppb | Photochemical |
| Ozone (stratosphere) | $O_3$ | 0.5–10 ppm | Photochemical |
| NMHC | $C_xH_y$ | 5–20 ppb | Biogenic, anthropogenic |
| Chlorofluorocarbon 12 | $CF_2Cl_2$ | 540 ppt | Anthropogenic |
| Chlorofluorocarbon 11 | $CFCl_3$ | 265 ppt | Anthropogenic |
| Methylchloroform | $CH_3CCl_3$ | 65 ppt | Anthropogenic |
| Carbon tetrachloride | $CCl_4$ | 98 ppt | Anthropogenic |
| Nitrogen oxides | $NO_x$ | 10 ppt–1 ppm | Soils, lightning, anthropogenic |
| Ammonia | $NH_3$ | 10 ppt–1 ppb | Biogenic |
| Hydroxyl radical | OH | 0.05 ppt | Photochemical |
| Hydroperoxyl radical | $HO_2$ | 2 ppt | Photochemical |
| Hydrogen peroxide | $H_2O_2$ | 0.1–10 ppb | Photochemical |
| Formaldehyde | $CH_2O$ | 0.1–1 ppb | Photochemical |
| Sulfur dioxide | $SO_2$ | 10 ppt–1 ppb | Photochemical, volcanic, anthropogenic |
| Dimethyl sulfide | $CH_3SCH_3$ | 10–100 ppt | Biogenic |
| Carbon disulfide | $CS_2$ | 1–300 ppt | Biogenic, anthropogenic |
| Carbonyl sulfide | OCS | 500 ppt | Biogenic, volcanic, anthropogenic |
| Hydrogen sulfide | $H_2S$ | 5–500 ppt | Biogenic, volcanic |

Source: Brasseur *et al.* (1999) and Prinn *et al.* (2000).

$H_2O$-rich atmosphere (Levy, 1971; Chameides and Walker, 1973; Ehhalt, 1999; Lelieveld *et al.*, 1999; Prather *et al.*, 2001). As of early 2000s, its global average concentration is only $\sim 10^6$ radicals cm$^{-3}$ or $\sim 6$ parts in $10^{14}$ by mole in the troposphere (Prinn *et al.*, 2001). But its influence is enormous and its life cycle complex. For example, it reacts with CO, usually within $\sim 1$ s, to produce $CO_2$ (and also a hydrogen atom, which quickly combines with $O_2$ to form hydroperoxy free radicals, $HO_2$). Another key player is nitric oxide (NO), which can be produced by lightning, combustion, and nitrogen microbes, and undoubtedly became present at significant levels in the Earth's atmosphere as soon as $O_2$ and $N_2$ became the dominant atmospheric constituents. NO can react with either $O_3$ or $HO_2$ to form $NO_2$. The reaction of NO with $HO_2$ is special because it regenerates OH giving two major sources ($O(^1D) + H_2O \rightarrow 2OH$ and $NO + HO_2 \rightarrow NO_2 + OH$) for this key cleansing radical. The $NO_2$ is also important because it photodissociates even in violet wavelengths to produce oxygen atoms and thus $O_3$. Hence, there are also two major sources of $O_3$ (photodissociation of both $O_2$ and $NO_2$).

**Table 2**   Global turnover of tropospheric trace gases and the fraction removed by reaction with OH according to Ehhalt (1999). The mean global OH concentration was taken as $1 \times 10^6$ cm$^{-3}$ (Prinn *et al.*, 1995) (1 Tg $= 10^{12}$g).

| Trace gas | Global emission rate (Tg yr$^{-1}$) | Removal by OH (%) | Removal by OH (Tg yr$^{-1}$) |
|---|---|---|---|
| CO | 2,800 | 85 | 2,380 |
| $CH_4$ | 530 | 90 | 477 |
| $C_2H_6$ | 20 | 90 | 18 |
| Isoprene | 570 | 90 | 513 |
| Terpenes | 140 | 50 | 70 |
| $NO_2$ | 150 | 50 | 75 |
| $SO_2$ | 300 | 30 | 90 |
| $(CH_3)_2S$ | 30 | 90 | 27 |

The importance of OH as a removal mechanism for atmospheric trace gases is illustrated in Table 2, which shows the global emissions of each gas and the approximate percentage of each of these emitted gases which is destroyed by reaction with OH (Ehhalt, 1999).

Evidently, OH removes a total of $\sim 3.65 \times 10^{15}$ g of the listed gases each year, or an amount equal to the total mass of the atmosphere every $1.4 \times 10^6$ yr. Without OH our atmosphere would have a very different composition, as it would be dominated by those gases which are presently trace gases.

Because both UV radiation fluxes and water vapor concentrations are largest in the tropics and in the southern hemisphere, the levels of OH generally have their maxima in the lower troposphere in these regions and seasons (Figure 1). Also, because much $O_3$ is produced in and exported from polluted areas, its concentrations generally have maxima in the northern hemisphere mid-latitude summer (Figure 1).

Even from the very brief discussion above, it is clear that the levels of OH in the atmosphere are not expected to remain steady but to change rapidly on a wide variety of space and timescales. In the lower atmosphere, OH sources can be decreased or turned off by lowering UV radiation (night time, winter, increasing cloudiness, thickening stratospheric $O_3$ layer), lowering NO emissions, and lowering $H_2O$ (e.g., in a cooling climate). Its sinks can be increased by increasing emissions of reduced gases (CO, RH, $SO_2$, etc.). Hence, OH levels are expected to have changed

substantially over geological and even recent times. But because natural or anthropogenic combustion of biomass or fossil fuel is a primary source of gases driving the oxidation processes, and because CO and NO are both products of combustion with (usually) opposite effects on OH levels, the system is not as unstable as it could otherwise be.

Because $O_3$ is a powerful infrared absorber and emitter (a "greenhouse" gas), and because many other greenhouse gases, notably $CH_4$, are destroyed principally by reaction with OH, there are significant connections between atmospheric oxidation processes and climate. Added to these greenhouse gas related connections is the fact that the first step in the production of aerosols from gaseous $SO_2$, $NO_2$, and hydrocarbons is almost always the reaction with OH. Aerosols play a key role in climate through absorption and/or reflection of solar radiation.

These oxidation processes also dominate in the chemistry of air pollution, where the occurrence of harmful levels of $O_3$ and acidic aerosols depends on the relative and absolute levels of urban emissions of NO, CO, RH, and $SO_2$. In the marine and polar troposphere, reactive compounds of chlorine, bromine, and iodine provide a small but significant additional pathway for oxidation besides OH, as discussed by von Glasow and Crutzen (see Chapter 4.02).

This chapter reviews the possible evolution of oxidizing capacity over geologic time, the fundamental chemical reactions involved, and the influences of meteorology and human activity on atmospheric oxidation. The measurement of oxidation rates (i.e., oxidative capacity), and the complex interactive models of chemistry, transport, and human and other biospheric activities that are now being developed to help understand atmospheric oxidation as a key process in the Earth system are also discussed.

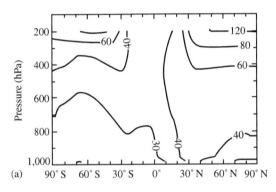

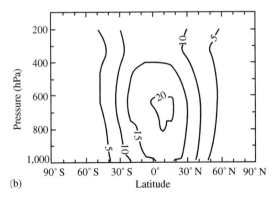

**Figure 1** (a) Ozone mole fractions (in ppb) and (b) hydroxyl radical concentrations (in $10^5$ radicals cm$^{-3}$) as functions of latitude and pressure (in hPa) from a model of the atmosphere in the 1990s (source Wang and Jacob, 1998).

## 4.01.2 EVOLUTION OF OXIDIZING CAPABILITY

A discussion of the evolution of $O_3$ and OH over geologic time needs to be placed in the context of the evolution of the Earth system (see Table 3). The Earth accreted between 4.5 and 4.6 billion years (Gyr) before present (BP). Enormous inputs of kinetic energy, ice, and hydrated minerals during accretion would have led to a massive $H_2O$ (steam) atmosphere with extremely high surface temperatures (Lange and Ahrens, 1982). This earliest atmosphere may also have been partly or largely removed by the subsequent Moon-forming giant impact. Due to the apparent obliteration of rocks much older than 4 Gyr on Earth, very little is known directly

**Table 3** Schematic showing time in billions of years before present, major eras and associated biospheric and geospheric evolutionary events, and hypothesized atmospheric composition. Gases are listed in order of their hypothesized concentrations and those whose sources and/or sinks are dominated by biological processes are denoted with an asterisk. Note the transition from an abiotic to an essentially biologically controlled atmosphere over geologic time.

| Time (Gyr BP) | Major evolutionary events | Hypothesized atmospheric compositions |
|---|---|---|
| | Earth accretes and massive steam atmosphere forms | $[H_2O, CO_2, CO, N_2, H_2]$ |
| | Surface cools, volcanism begins? | |
| | Ocean forms, continental weathering, and carbonate precipitation begin | $[CO_2, CO, N_2, H_2O, H_2]$ |
| 4 | Archean era begins, oldest rocks | |
| | Late heavy bombardment | |
| | Sediment subduction and volatile recycling begins? | $[N_2, CO_2, CO, H_2O, H_2, CH_4, HCN, NO_x]$ |
| | First living organisms evolve, prokaryotes | |
| 3 | First protosynthetic organisms evolve: cyanobacteria, stromatolites | $[N_2, H_2O, CO_2, *O_2, CH_4]$ |
| | Proterozoic era begins: scarce fossils | |
| 2 | Aerobic biosphere with nitrogen and methane microbes evolves | $[*N_2, *O_2, H_2O, *CO_2, *CH_4, *N_2O]$ |
| 1 | Eucaryotes evolve | |
| | Metazoans appear | |
| | Invertebrates appear | |
| | Paleozoic era begins, ubiquitous fossils | |
| | Fishes appear | $[*N_2, *O_2, H_2O, *CO_2, *CH_4, *N_2O]$ |
| | Land plants appear | |
| | Amphibians appear | |
| | Mesozoic (reptilian) era begins | |
| | Fission of Pangean supercontinent | |
| | Cretaceous–Tertiary collision | |
| | Cenozoic (mammalian) era begins | |
| 0 | Present | $[*N_2, *O_2, H_2O, *CO_2, *CH_4, *N_2O, *CF_2Cl_2, etc.]$ |

about the Earth's earliest atmosphere. It is hypothesized that, as the Earth cooled, the massive $H_2O$ atmosphere condensed forming oceans and a much less massive $CO_2/CO/N_2/H_2O/H_2$ atmosphere then evolved.

The next big evolutionary step, involving the aqueous weathering of rocks and removal of $CO_2$ as carbonates, had occurred ~3.8 Gyr ago (Holland, 1984). Thus, the early Archaean atmosphere was probably largely dominated by $N_2$ with smaller amounts of $CO_2$, CO, $H_2O$, $H_2$, $CH_4$, HCN, and $NO_x$ (from lightning). The evolution of the first life forms (procaryotes), and specifically photosynthetic cyanobacteria, would have allowed the slow accumulation of biologically produced $O_2$ in the atmosphere in the late Archaean period. Similarly, the early evolution of methane and nitrogen bacteria along with the evolution of the aerobic biosphere led, apparently early in the Proterozoic, to an $N_2/O_2/H_2O/CO_2/CH_4/N_2O$ atmosphere with all of these gases, with the exception of $H_2O$, being largely controlled by biological sources and biological or atmospheric chemical sinks.

### 4.01.2.1 Prebiotic Atmosphere

Before the evolution of the photosynthetic source of $O_2$, the atmosphere was relatively chemically neutral or even reducing. Its reduction state would have depended significantly on the quantities of reduced gases produced from volcanic sources (e.g., sources in which primordial iron, sulfur, carbon, and organic compounds were thermally oxidized by $H_2O$ and $CO_2$ producing $H_2$, CO, and $CH_4$ (Prinn, 1982; Lewis and Prinn, 1984)). In these prebiotic atmospheres, small amounts of $O_2$ can be produced from $H_2O$ photodissociation (Levine, 1985), and small quantities of NO can be formed by shock heating, by lightning, and large collisions (Chameides and Walker, 1981; Prinn and Fegley, 1987; Fegley *et al.*, 1986).

However, it is reasonable to conclude that the evolution of the powerful oxidizing processes involving $O_3$ and OH in the present atmosphere had to await the evolution of photosynthetic $O_2$. Before that, atmospheric chemical transformations were probably driven by simple photodissociation and dissolution reactions and by episodic processes like lightning and collisions. The transformation from reducing/neutral to oxidizing atmosphere was itself a long-term competition between abiotic sources of reduced gases, anaerobic microbial processes, and finally the aerobic biological processes which came to dominate the biosphere. Once the land biosphere evolved, the burning of vegetation ignited by lightning provided an additional source of trace gases (see Chapter 4.05).

## 4.01.2.2 Pre-industrial Atmosphere

We expect that OH and $O_3$ changed even under the approximately constant levels of atmospheric $O_2$ in the Cenozoic era. For example, the glacial–interglacial cycles should have significantly changed atmospheric $H_2O$ (inferred to be low in ice ages) and $CH_4$ (observed to be low in ice ages). But these particular changes would have partially offsetting effects on OH (by lowering both OH production and OH removal rates in ice ages). And unfortunately, the lack of direct information about the abundance of the key short-lived gases (CO, $NO_x$, nonmethane hydrocarbons (NMHC), and stratospheric $O_3$) in paleoenvironments make it difficult to be quantitative about past OH changes (Thompson, 1992). One observational record over the past millennium which is useful in this respect is the analysis of hydrogen peroxide ($H_2O_2$) in Greenland permafrost, which indicates stable $H_2O_2$ from the years 1300 to 1700 followed by a 50% increase through to 1989 (however, most of the increase has occurred since 1980 and is not seen in Antarctic permafrost (Sigg and Neftel, 1991; Anklin and Bales, 1997)).

Increases in $H_2O_2$ during industrialization could signal a decrease in OH due to net conversion of $HO_2$ to $H_2O_2$ rather than recycling of $HO_2$ back to OH. Staffelbach *et al.* (1991) have used $CH_2O$ measurements in Greenland ice cores to suggest that pre-industrial OH levels were 30% higher than the levels in the late twentieth century. Indeed, a variety of model studies which assume low pre-industrial emissions of CO, $NO_x$, and NMHC compared to the levels in the twentieth century suggest that the OH levels in the years 1200–1800 were ~10–30% higher than the levels at the end of the twentieth century (Thompson, 1992). Ozone levels in the 1880s can be inferred from iodometric measurements (Volz and Kley, 1988; Marenco *et al.*, 1994) indicating mole fractions as low as $(7-12) \times 10^{-9}$, which are lower by a factor of 3 or more from those observed in the 1990s. Such low levels of $O_3$ are difficult to reconcile with CO measurements in ice cores (Haan *et al.*, 1996)

indicating CO mole fractions of $\sim 60 \times 10^{-9}$ (Antarctica) and $\sim 90 \times 10^{-9}$ (Greenland), which are not much less than those observed today. As we will discuss in Section 4.01.3.1, oxidation of each CO molecule ought to produce one $O_3$ molecule, so that presuming OH levels in the 1880s to be similar to those in the 1990s implies similar $O_3$ production rates from CO.

## 4.01.3 FUNDAMENTAL REACTIONS

### 4.01.3.1 Troposphere

As we noted in Section 4.01.1, the ability of the troposphere to chemically transform and remove trace gases depends on complex chemistry driven by the relatively small flux of energetic solar UV radiation that penetrates through the stratospheric $O_3$ layer (Levy, 1971; Chameides and Walker, 1973; Crutzen, 1979; Ehhalt *et al.*, 1991; Logan *et al.*, 1981; Ehhalt, 1999; Crutzen and Zimmerman, 1991). This chemistry is also driven by emissions of NO, CO, and hydrocarbons and leads to the production of $O_3$, which is one of the important indicators of the oxidizing power of the atmosphere. But the most important oxidizer is the hydroxyl free radical (OH), and a key measure of the capacity of the atmosphere to oxidize trace gases injected into it is the local concentration of hydroxyl radicals.

Figure 2 summarizes, with much simplification, the chemistry involved in production and removal of $O_3$ and OH (Prinn, 1994). This chemistry is remarkable because it involves compounds that also influence climate, air pollution, and acid rain.

The greenhouse gases involved here include $H_2O$, $CH_4$, and $O_3$. Primary pollutants emitted mainly as a result of human activity (including fossil-fuel combustion, biomass burning, and land-use) include RH, CO, and nitrogen oxides (NO, $NO_2$). Reactive free radicals or atoms are in two categories: the very reactive species, such as $O(^1D)$ and OH, and the less reactive ones, such as $HO_2$, $O(^3P)$, NO, and $NO_2$.

The atmospheric cleansing role of OH is very evident in Figure 2. When OH reacts with an RH, the latter is oxidized in a series of steps mostly to CO. These steps consume OH but may also produce $HO_2$. In turn, OH oxidizes CO to $CO_2$, and it also oxidizes the gases $NO_2$ and $SO_2$ to nitric ($HNO_3$) and sulfuric ($H_2SO_4$) acids, respectively. The primary OH source involves water vapor which reacts with the very reactive singlet oxygen atom, $O(^1D)$, that comes from photodissociation of $O_3$ by solar UV radiation at wavelengths less than 310 nm. Typically, OH within a second of its formation oxidizes other compounds either by donating oxygen or by

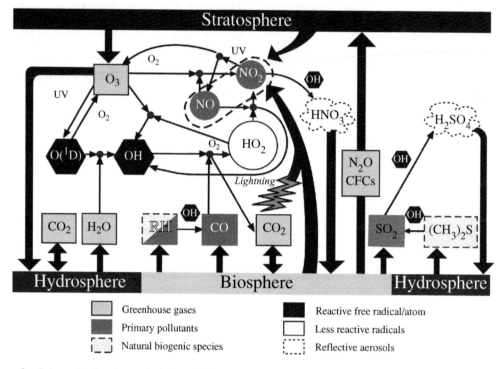

**Figure 2** Schematic showing principle oxidation processes in the troposphere in $NO_x$-rich air (after Prinn, 1994). In $NO_x$-poor air (e.g., remote marine air), recycling of $HO_2$ to OH is achieved by reactions of $O_3$ with $HO_2$ or by conversion of $2HO_2$ to $H_2O_2$ followed by photodissociation of $H_2O_2$. In a more complete schematic, nonmethane hydrocarbons (RH) would also react with OH to form acids, aldehydes and ketones in addition to CO.

removing hydrogen leaving an H atom or organic free radical (R). Then R and H attach rapidly to molecular oxygen to form hydroperoxy radicals ($HO_2$) or organoperoxy radicals ($RO_2$), which are relatively unreactive. If there is no efficient way to rapidly recycle $HO_2$ back to OH, then levels of OH are kept low.

Adding the nitrogen oxides (NO and $NO_2$), which have many human-related sources, significantly changes this picture. Specifically, NO reacts with the $HO_2$ to form $NO_2$, and reform OH. UV radiation then decomposes $NO_2$ at wavelengths less than 430 nm to form $O_3$ and reform NO, so the nitrogen oxides are not consumed in this reaction. In a study of relatively polluted air in Germany, Ehhalt (1999) estimated that the production rate of OH by this secondary path involving NO is 5.3 times faster than the above primary path involving $O(^1D)$. Note that the reaction of NO with $HO_2$ does not act as a sink for the so-called odd hydrogen (the sum of the H, OH, and $HO_2$ concentrations). Rather, it determines the ratio of OH to $HO_2$. In the Ehhalt (1999) study, the $[HO_2]:[OH]$ ratio was 44:1. This is due to, and indicative of, the much greater reactivity of OH compared to $HO_2$. To summarize, the principle catalyzed reactions creating OH and $O_3$ in the troposphere are shown in Equations (1)–(3), (4)–(9), and (10)–(15):

$$O_3 + UV \rightarrow O_2 + O(^1D) \quad (1)$$

$$O(^1D) + H_2O \rightarrow 2OH \quad (2)$$

$$\text{Net effect}: O_3 + H_2O \rightarrow O_2 + 2OH \quad (3)$$

and

$$OH + CO \rightarrow H + CO_2 \quad (4)$$

$$H + O_2 \rightarrow HO_2 \quad (5)$$

$$HO_2 + NO \rightarrow OH + NO_2 \quad (6)$$

$$NO_2 + UV \rightarrow NO + O \quad (7)$$

$$O + O_2 \rightarrow O_3 \quad (8)$$

$$\text{Net effect}: CO + 2O_2 \rightarrow CO_2 + O_3 \quad (9)$$

and

$$OH + RH \rightarrow R + H_2O \quad (10)$$

$$R + O_2 \rightarrow RO_2 \quad (11)$$

$$RO_2 + NO \rightarrow RO + NO_2 \quad (12)$$

$$NO_2 + UV \rightarrow NO + O \quad (13)$$

$$O + O_2 \rightarrow O_3 \quad (14)$$

---

Net effect: $OH + RH + 2O_2 \rightarrow RO + H_2O + O_3$

$$(15)$$

In theoretical models, the global production of tropospheric $O_3$ by the above pathway beginning with CO is typically about twice that beginning with RH. For most environments, it is these overall catalytic processes that pump the majority of the OH and $O_3$ into the system. If the concentration of $NO_2$ becomes too high, however, its reaction with OH to form $HNO_3$ ultimately limits the OH concentration (see Section 4.01.4). The global impact of the above tropospheric OH production mechanisms has been summarized in Table 2. Hough and Derwent (1990) have estimated the global production rate of tropospheric $O_3$ by the above reactions to be 2,440 Tg yr$^{-1}$ in pre-industrial times and 5,130 Tg yr$^{-1}$ in the 1980s with the increase caused by man-made $NO_x$, hydrocarbons, and CO.

An update to the above 1980s calculation (Ehhalt, 1999) gives a production rate of 4,580 Tg yr$^{-1}$. A comparison of several tropospheric $O_3$ models showed a chemical production rate range for $O_3$ of 3,425–4,550 Tg yr$^{-1}$ (Lelieveld *et al.*, 1999). Since the troposphere contains ~350 Tg of $O_3$ (Ehhalt, 1999), the chemical replenishment time for $O_3$ (defined as the tropospheric content divided by the tropospheric chemical production rate) is 28–37 d. This is much longer than the replenishment time for OH, ~1 s, and indicates the enormous reactivity of OH relative to $O_3$.

How stable is the oxidizing capability of the troposphere? How might it have changed over geologic time even after oxygen reached near today's levels? If emissions of gases that react with OH such as $CH_4$, CO, and $SO_2$ are increasing then, keeping everything else constant, OH levels should decrease. Conversely, increasing $NO_x$-emissions from combustion (of biomass in the past (see Chapter 4.05) augmented by fossil fuel today) should increase tropospheric $O_3$ (and thus the primary source of OH), as well as increase the recycling rate of $HO_2$ to OH (the secondary source of OH).

Also, if the temperatures of oceans are increasing, we expect increased water vapor in the lower troposphere. Because water vapor is part of the primary source of OH, climate warming also increases OH. This increase could be lowered or raised due to changes in cloud cover accompanying the warming leading to more or less reflection of UV back toward space. Rising temperature also increases the rate of reaction of $CH_4$ with OH, thus lowering the lifetimes of both chemicals. The opposite conclusions apply if trace gas emissions increase or the climate cools. Finally, decreasing stratospheric $O_3$ can also increase tropospheric OH.

### 4.01.3.2 Stratosphere

The $O_3$ in the stratosphere is maintained by a distinctively different set of chemical reactions than the troposphere. The flow diagram in Figure 2 shows how the driving chemicals like water vapor, chlorofluorocarbons, and nitrous oxide, are transported up from the troposphere to the stratosphere, where they produce chlorine-, nitrogen-, and hydrogen-carrying free radicals, which destroy $O_3$. There is an interesting paradox here: the nitrogen oxide free radicals, which are responsible for destroying a significant fraction of the $O_3$ in the stratosphere, are the same free radicals that are producing $O_3$ in the troposphere through the chemistry outlined earlier (Figure 3).

When the ratio of nitrogen oxide free radicals to $O_3$ is low, as it is in the stratosphere, the net effect is for $O_3$ to be destroyed overall, while if this ratio is high, as it is in much of the troposphere, then $O_3$ is produced overall (Crutzen, 1979).

The stratosphere is chemically very active due to its high $O_3$ and UV levels. But the densities and usually temperatures are much lower than in the lower troposphere so that only a small fraction of the destruction of the gases listed in Table 2 occurs in the stratosphere. The stratosphere contains ~90% of the world's $O_3$ and exports ~400–846 Tg yr$^{-1}$ of $O_3$ to the troposphere to augment the chemical production discussed in Section 4.01.3.1 (Lelieveld *et al.*, 1999).

There is also an important link between stratospheric chemistry and climate. The precursor gases for the destructive free radicals in the stratosphere are themselves greenhouse gases, as is the stratospheric $O_3$ itself. Therefore, while increases in the concentrations of the source gases will increase the radiative forcing of warming, these increases will, at the same time, lead to decreases in stratospheric $O_3$, thus lowering the radiative forcing. Once again, as in the troposphere, there are important feedbacks

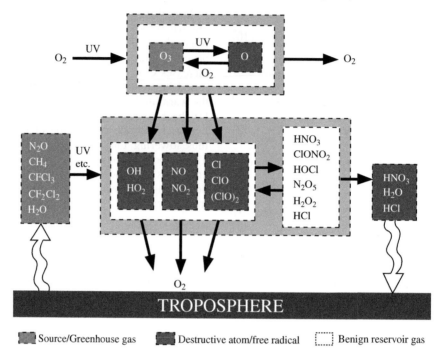

**Figure 3**  Schematic showing major chemical processes forming and removing $O_3$ in the stratosphere (after Prinn, 1994). UV photodissociaton of largely natural ($N_2O$, $CH_4$, $H_2O$) and exclusively man-made ($CFCl_3$, $CF_2Cl_2$) source gases leads to reactive $HO_x$, $NO_x$, and $ClO_x$ free radicals which catalytically destroy $O_3$. The catalysts reversibly form reservoir compounds ($HNO_3$, etc.), some of which live long enough to be transported down to the troposphere.

between chemical processes and climate through the various greenhouse gases. It is not just man-made chemicals that can change the stratospheric $O_3$. Changes in natural emissions of $N_2O$ and $CH_4$, and changes in climate that alter $H_2O$ flows into the stratosphere, undoubtedly led to changes in the thickness of the $O_3$ layer over geologic time even after $O_2$ levels reached levels near those in the early 2000s.

There is also a strong link between the thickness of the stratospheric $O_3$ layer and concentrations of OH in the troposphere. Because $O(^1D)$ can only be produced from $O_3$ at wavelengths less than 310 nm, and the current $O_3$ layer effectively absorbs all incoming UV at less than 290 nm, there is only a very narrow window of 290–310 nm radiation that is driving the major oxidation processes in the troposphere. Decrease in the amount of stratospheric $O_3$, which has occurred since about 1980, can therefore increase the tropospheric OH by increasing the flux of radiation less than 310 nm which reaches the troposphere to produce $O(^1D)$ (Madronich and Granier, 1992). By the same arguments, if the stratospheric $O_3$ layer was very much thicker in the past, it would have removed a great deal of the "cleansing" capacity of the Earth's troposphere leading to much higher levels of gases like CO and $CH_4$.

## 4.01.4  METEOROLOGICAL INFLUENCES

The oxidative capability of the atmosphere is not simply a function of chemistry. Convective storms can carry short-lived trace chemicals from the planetary boundary layer (the first few hundred to few thousand meters) to the middle and upper troposphere in only a few to several hours. This can influence the chemistry of these upper layers in significant ways by delivering, e.g., reactive hydrocarbons to high altitudes. Conversely, the occurrence of very stable conditions in the boundary layer can effectively trap chemicals near the surface for many days, leading to polluted air. Larger-scale circulations serve to carry gases around latitude circles on timescales of a few weeks, between the hemispheres on timescales of a year, and between the troposphere and stratosphere on timescales of a few years.

A convenient measure of the stability of a trace chemical is its local chemical lifetime, which is defined as its local concentration (e.g., in $mol\,L^{-1}$) divided by its local removal rate (e.g., in $mol\,L^{-1}\,s^{-1}$). Concentrations of chemicals whose lifetimes (due to chemical destruction or deposition) are comparable to, or longer than, the above transport times will be profoundly affected by transport. As a simple example, in a steady (constant concentration) state with a horizontal wind $u$, the concentration $[i]$ of chemical $i$ whose lifetime is $t_i$ decreases downwind of a source region (located at horizontal position $x = 0$) according to

$$[i](x) = [i](0)\exp\left[\frac{-x}{(ut_i)}\right]$$

Specifically, the concentration of $i$ decreases by a factor of e over the distance scale $ut_i$. Note also that, if the transport time $x/u$ is much less than $t_i$, the concentration remains essentially constant and the species can be considered to be almost inert. For oxidation by OH of $NO_2$ and $SO_2$ to form the acids $HNO_3$ and $H_2SO_4$, the time $t_i$ is typically 3 d ($2.6 \times 10^5$ s), so the e-folding distance $ut_i$ is only ~800 km for a typical $u$ of $3 \text{ m s}^{-1}$. Alternatively, for the oxidation of CO by OH to form $CO_2$, $t_i = 3$ months ($7.9 \times 10^6$ s), so $ut_i = 2.4 \times 10^4$ km for the same $u$. In a simple way this illustrates why $NO_2$ and $SO_2$ are local or regional pollutants, whereas CO is a hemispheric pollutant.

In the above example, if the chemical lifetime $t_i$ is much less than the transport time $x/u$ then we could ignore transport and simply consider chemical sources and sinks to determine concentrations. However, for the chemicals controlling oxidation reactions, care must be taken in defining $t_i$. Specifically, sometimes members of a "chemical family" are converted from one to another on very short timescales relative to transport, whereas the total population of the family is produced and removed on longer timescales.

For example, in the family composed of NO and $NO_2$ (called the $NO_x$ family) the reactions shown in Equations (16)–(18) rapidly interconvert NO and $NO_2$ on timescales much less than typical transport times and do so without affecting the total $[NO_x] = [NO] + [NO_2]$.

$$NO + O_3 \rightarrow NO_2 + O_2 \quad (16)$$

$$NO_2 + h\nu \rightarrow NO + O \quad (17)$$

$$O + O_2 + M \rightarrow O_3 + M \quad (18)$$

$$NO_2 + OH + M \rightarrow HNO_3 + M \quad (19)$$

However, the removal of $NO_x$ by the reaction shown in Equation (19) is much slower (e.g., ~3 d). Thus, we can use the chemical steady-state approximation (which equates chemical production to chemical loss only) to define the ratio of NO to $NO_2$. We must however take full account of transport in calculating $[NO_x]$. Similar arguments hold for the "odd oxygen" ($[O_x] = [O] + [O_3]$) and "odd hydrogen" ($[HO_x] = [H] + [OH] + [HO_2]$) families.

Besides being a vigorous vertical transport mechanism, moist convection also directly affects the oxidizing reactions in the atmosphere. First, lightning and thundershocks serve to thermally convert atmospheric $N_2$ and $O_2$ to 2NO, leading to a very important natural source of $NO_x$. Second, the formation of cloud droplets is aided by (and consumes) aerosols (as cloud condensation nuclei), and the cloud droplets then serve as sinks for soluble species like $H_2O_2$, $HNO_3$, $NO_x$, $SO_2$, and $H_2SO_4$. For some oxidation reactions, e.g., oxidation of $SO_2$ to $H_2SO_4$, this dissolution leads to an additional nongaseous pathway for oxidation generally considered comparable to the gas-phase processes for $SO_2$. Finally, the formation of convective clouds, and particularly the extensive anvils, leads to reflection of UV radiation enhancing the photo-oxidation processes above the cloud and inhibiting them below (e.g., Wang and Prinn, 2000).

## 4.01.5 HUMAN INFLUENCES

### 4.01.5.1 Industrial Revolution

The advent of industrialization since 1850 has led to significant increases in the global emissions of $NO_x$, CO, NMHC, and $CH_4$ (Wang and Jacob, 1998; Prather *et al.*, 2001). Biomass burning associated with human land use provides an additional source of these trace gases (see Chapter 4.05). Also, the observed temperature increase of $0.6 \pm 0.2 \,°C$ in this time period (Folland *et al.*, 2001) should have led to increases in tropospheric $H_2O$ and possibly changes in global cloud cover (of unknown sign). Anthropogenic aerosols, both scattering and absorbing, have also increased due to human activity (Penner *et al.*, 2001). These aerosol increases should decrease UV fluxes below the cloud tops due to both their direct optical effects (absorbing or reflecting) and their indirect effects as cloud condensation nuclei, which should increase the reflectivity of water clouds. Finally, depletion of stratospheric $O_3$, which has occurred due to man-made halocarbons, will increase tropospheric UV irradiation (Madronich and Granier, 1992).

All of these changes should have affected $O_3$ and OH concentrations, sometimes offsetting and sometimes augmenting each other, as discussed earlier. Table 4 summarizes the expected changes in OH and $O_3$ relative to pre-industrial times. Where quantitative estimates are not available, only the signs of expected changes are indicated. Note in particular that for OH the positive effects of $NO_x$ increases have been more than offset by the negative effects of CO, NMHC, and $CH_4$ increases, so the net change in OH is a modest 10% decrease (Wang and Jacob, 1998; see also Levy *et al.*, 1997).

**Table 4** Effects of anthropogenically driven or observed changes in trace gas emissions, concentrations, or meteorological variables between pre-industrial and present times on the oxidizing capability of the atmosphere (expressed where available as percentage changes ($\Delta$) in [OH] or [$O_3$] from pre-industrial to present).

|  | *Present/pre-industrial* | $\Delta[OH]/[OH]$ | $\Delta[O_3]/[O_3]$ |
|---|---|---|---|
| $NO_x$[a] | 4.7 | +36% | +30% |
| $CO$[a] | 3.2 | −32%[b] | +10%[b] |
| NMHC[a] | 1.6 | | |
| $CH_4$[a] | 2.1 | −17% | +13% |
| Combined[a] | | −10%[c] | +63%[c] |
| $H_2O(T)$ | >1 | >0 | >0 |
| Clouds[d] | >1, <1 | <0, >0 | <0, >0 |
| Scat. aero[e] | >1 | <0 | <0 |
| Abs. aero[f] | >1 | <0 | <0 |
| Strat. $O_3$[g] | <1 | >2% | >0 |

[a] Wang and Jacob (1998). Individual effect of each specific chemical computed in sensitivity runs with pre-industrial conditions except for the specific chemical which is at present levels. [b] CO and NMHC effects combined. [c] CO, NMHC, $CH_4$, and $NO_x$ effects combined. Total effect not equal to sum of individual effects due to interactions. [d] Cloud cover changes unknown. Effects shown for either an increase or a decrease. [e] Specifically aerosols with high single scattering albedos (e.g., sulfates) which reflect UV radiation back to space. [f] Specifically aerosols with low single scattering albedos (e.g., black carbon) which absorb UV radiation. [g] Madronich and Granier (1992) and Krol *et al.* (1998).

**Table 5** Percentages by which either emissions of $NO_x$, CO, NMHC (plus oxidized NMHC), and $CH_4$, or concentrations of tropospheric OH, or column amounts of tropospheric $O_3$, are projected to change between 2000 and 2100 for various IPCC scenarios.

| *Scenario* | $NO_x$ (%) | CO (%) | NMHC (%) | $CH_4$ (%) | OH (%) | $O_3$ (%)[a] |
|---|---|---|---|---|---|---|
| A1B | +26 | +90 | +37 | −11 | −10 | +7.7 |
| A1T | −12 | +137 | −9.2 | −15 | −18 | +5.5 |
| A1FI | +243 | +193 | +198 | +128 | −14 | +47 |
| A2 | +241 | +165 | +143 | +175 | −12 | +46 |
| B1 | −42 | −59 | −38 | −27 | +5 | −8.6 |
| B2 | +91 | +128 | +21 | +85 | −16 | +23 |
| A1p | +27 | +138 | +15 | −13 | −18 | +11 |
| A2p | +238 | +140 | +133 | +163 | −12 | +46 |
| B1p | +3.4 | −8.3 | +2.0 | +9.2 | −3 | +2.6 |
| B2p | +86 | +100 | +13 | +46 | −11 | +19 |
| IS92a | +124 | +64 | +125 | +95 | −11 | +29 |

Source: Prather *et al.* (2001).
[a] Ozone changes decreased by 25% to account for modeling errors as recommended by IPCC.

Since combustion is a common, and (except for $CH_4$) a dominant source for all these gases affecting OH, then the offsetting increases of $NO_x$, CO, and hydrocarbons mean that OH levels are not very sensitive overall to past increases in combustion. Note, however, that if future air pollution controls lower $NO_x$ emissions more than CO and hydrocarbon emissions, then OH decreases should result.

For $O_3$, in contrast to OH, the effects of $NO_x$, CO, and RH emission increases are additive and in sum are calculated to have caused ~63% increase in $O_3$ (Wang and Jacob, 1998). The actual increase may be even larger, since observations (albeit not very accurate) of $O_3$ during the 1800s (Marenco *et al.*, 1994) suggest even lower values than those computed by Wang and Jacob (1998).

### 4.01.5.2 Future Projections

The importance of oxidation processes in climate led the Intergovernmental Panel on Climate Change (IPCC) to investigate the effects on $O_3$ and OH of 11 IPCC scenarios for future anthropogenic emissions of the relevant $O_3$ and OH precursors (Prather *et al.*, 2001). In all of the scenarios, except one, OH was projected to decrease between 2000 and 2100 by 3–18%, while $O_3$ was projected to increase by 2.6–46% (Table 5). The smallest decrease in OH and smallest increase in $O_3$ was seen as expected in the scenario in which emissions of $NO_x$, CO, NMHCs, and $CH_4$ were projected to change by only +3.4%, −8.3%, +2.0%, and +9.2%, respectively, between 2000 and 2100. Also as expected, the greatest OH decreases were seen in two

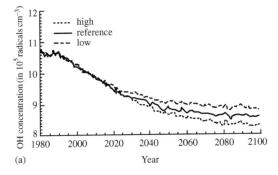

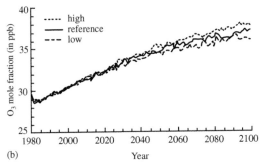

(a)

(b)

**Figure 4** (a) Predicted future tropospheric average concentrations of OH and (b) mole fractions of $O_3$ for three scenarios of future emissions of $O_3$ and OH precursors (as well as greenhouse gases) predicted in an economics model (Prinn *et al.*, 1999). The natural system model receiving these emissions includes simultaneous calculations of atmospheric chemistry and climate (Wang and Prinn, 1999).

scenarios in which small increases or even decreases in $NO_x$ emissions (OH source) were accompanied by relatively very large increases in CO emissions (OH sink). The greatest $O_3$ increases were seen in two scenarios where emissions of all four $O_3$ precursors increased very substantially. Finally, the one anomalous scenario (in which OH increased by 5% and $O_3$ decreased by 8.6%) involved significant decreases in emissions of all four precursors (which is very unlikely).

The above results make it very clear that forecasts of the future oxidation capacity of the atmosphere depend critically on the assumed emissions. The IPCC did not assign probabilities to its emission scenarios but it is apparent that some of these scenarios are highly improbable for oxidant precursor emissions. Integrated assessment models, which couple global economic and technological development models with natural Earth system models provide an alternative approach to the IPCC scenario approach with the added advantage that objective estimates of individual model uncertainties can be combined with Monte Carlo approaches to provide more objective ways of defining means and errors in

forecasts. We will use here one such model (Prinn *et al.*, 1999) that integrates a computable general equilibrium economics model with a coupled chemistry and climate model, and an ecosystem model. The economics model automatically accounts for the strong correlations between emissions of air pollutants (gases and aerosols) and greenhouse gases due to their common production processes (e.g., combustion of coal, oil, gasoline, gas, biomass).

Results from this model for (i) a high-probability reference emissions forecast and (ii) a lower-probability high and low emissions forecasts are shown in Figure 4 (Wang and Prinn, 1999). Emissions driving these projections are summarized in Prinn *et al.* (1999); reference anthropogenic CO, $NO_x$, and $CH_4$ emissions increase by ~18%, ~95%, and ~33%, respectively, between 2000 and 2100 in these projections. Evidently, these emission projections from the economics model (which consistently models all of the relevant industrial and agricultural sectors producing these gases) lead to projections of significant depletion of OH and enhancement of $O_3$ between 2000 and 2100 with magnitudes comparable to the larger of the changes predicted from the IPCC scenarios (Table 5).

## 4.01.6 MEASURING OXIDATION RATES

Given its importance as the primary oxidizing chemical in the atmosphere, a great deal of effort has been given to measuring OH concentrations and trends. Two approaches, direct and indirect, have been taken to address this measurement need.

### 4.01.6.1 Direct Measurement

The direct accurate measurement of local OH concentrations has been one of the major technical challenges in atmospheric chemistry since the early 1980s. This goal was first achieved in the stratosphere (e.g., Stimpfle and Anderson, 1988), but the troposphere proved more difficult (Crosley, 1995). Nevertheless, early long-baseline absorption methods for OH were adequate to test some basic theory (e.g., Poppe *et al.*, 1994). Current successful direct methods include differential optical absorption near-UV spectroscopy with long baselines (e.g., Mount, 1992; Dorn *et al.*, 1995; Brandenburger *et al.*, 1998), laser-induced fluorescence after expansion of air samples (e.g., Hard *et al.*, 1984, 1995; Holland *et al.*, 1995), and a variety of chemical conversion techniques (Felton *et al.*, 1990; Chen and Mopper, 2000; Tanner *et al.*, 1997).

The long-baseline spectroscopic techniques require path lengths in air of 3–20 km, which can be achieved by a single pass through air (e.g., Mount and Harder (1995) used a 10.3 km path to achieve a sensitivity of $5 \times 10^5$ radicals cm$^{-3}$), or multiple reflections in long cells (e.g., Dorn *et al.* (1995) used 144 passes through a 20 m cell to observe six OH absorption lines around 308 nm). The laser-induced fluorescence methods use either 282 nm or 308 nm photons to excite OH, which then subsequently emits at 308–310 nm (e.g., Brune *et al.* (1995); Holland *et al.* (1995) use 308 nm laser light and achieve detection limits of $(1-3) \times 10^5$ radicals cm$^{-3}$).

The common chemical conversion techniques use a flow reactor in which ambient OH reacts with isotopically labeled $SO_2$ or CO to yield observable products (e.g., Felton *et al.* (1990) used $^{14}CO$ with radioactive $^{14}CO_2$ detection, while Eisele and Tanner (1991) used $^{34}SO_2$ with $H_2{}^{34}SO_4$ detection by ion-assisted mass spectrometry).

These various direct OH measurement techniques enable critical tests of current theoretical models of fast photochemistry. This is achieved by simultaneously measuring NO, $NO_2$, CO, $H_2O$, $O_3$, RH, and other relevant trace species, UV fluxes, the frequency of photodissociation of $O_3$ to produce $O(^1D)$, and the concentrations of OH (and sometimes $HO_2$). A model incorporating the best estimates of the relevant kinetic rate constants and absorption coefficients is then used, along with the trace gas and photodissociation rate measurements, to predict OH concentrations for comparison with observations. Examples of results from the Mauna Loa Observatory Photochemistry Experiment (MLOPEX) and the Photochemistry of Plant-emitted Compounds and OH Radicals in North Eastern Germany Experiment (POPCORN) are shown in Figures 5 and 6, respectively.

The 1992 MLOPEX experiment was able to sample both free tropospheric air (associated with downslope winds at this high mountain station) and boundary layer air (upslope winds). It is evident from Figure 5 that agreement between observed (using the $^{34}SO_2$ method) and calculated OH concentrations was always good in the free troposphere, but in the summer the calculated boundary layer OH was about twice that observed (Eisele *et al.*, 1996). Since the measurement accuracy was argued to be much better than factor of 2, this suggests a missing OH sink in their model, which they suggest might be due to unmeasured hydrocarbons from vegetation or undetected oxidation products of anthropogenic compounds.

The 1994 POPCORN experiment in rural Germany compared OH measurements using laser-induced fluorescence to calculations in a regional air chemistry model. Figure 6(a) shows the measured OH concentrations as a function

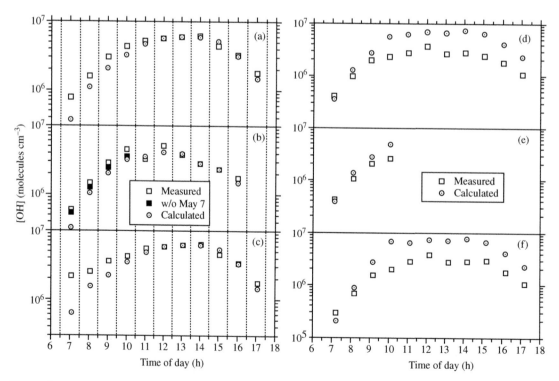

**Figure 5** Hourly average measured and calculated OH concentrations during spring (a)–(c) and summer (d)–(f) of 1992 in MLOPEX: (a), (d) all time; (b), (e) free tropospheric air only; (c), (f) boundary layer air only (after Eisele *et al.*, 1996). May 7 measurements were anomalous.

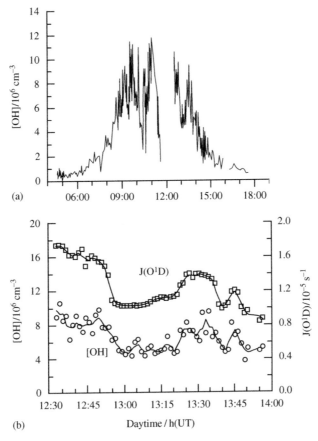

**Figure 6** (a) Diurnal variation of OH concentrations during POPCORN (August 16, 1994) and (b) higher time resolution version of (a) showing fast response of OH concentrations to $J(O^1D)$ variations (solid lines are three-point running averages) (after Hofzumahaus *et al.*, 1996).

of time. Note the expected strong diurnal variation in OH due to the daily cycle in UV radiation. Also shown in Figure 6(b) is a demonstration of the expected strong correlation between the measured frequency of photoproduction of $O(^1D)$ from $O_3(J(O^1D))$ and the measured OH (see Section 4.01.3.1). The strong high-frequency variations in OH evident in Figure 6(a) are real and are due to strongly correlated variations in solar UV fluxes caused by transient cloud cover. The agreement between observed and calculated OH was quite good in POPCORN but, as in MLOPEX, there was also a tendency for the model to overpredict OH (Hofzumahaus *et al.*, 1996; Ehhalt, 1999). Part of the discrepancy may be attributable to the combined effects in the modeling of rate constant errors and measurement errors in the observed RH, CO, NO, $NO_2$, $O_3$, and $H_2O$ used in the model. Poppe *et al.* (1995) have estimated this modeling error to be ±30%.

Overall, considering the difficulties in both the measurements and the models, the agreements between observation and theory are encouraging at least for the relatively clean and clear air environments investigated in these experiments.

The power and great importance of the *in situ* direct measurement is that in conjunction with simultaneous *in situ* observations of the other components of the fast photo-oxidation cycles, they provide a fundamental test of the photochemical theory. However, the very short lifetime and enormous temporal and spatial variability of OH, already emphasized in Section 4.01.1, make it impossible practically to use these direct techniques to determine regional to global scale OH concentrations and trends. It is these large-scale averages that are essential to understanding the regional to global chemical cycles of all of the long-lived (more than a few weeks) trace gases in the atmosphere which react with OH as their primary sink. For this purpose, an indirect integrating method is needed.

### 4.01.6.2 Indirect Measurement

The large-scale concentrations and long-term trends in OH can in principle be measured indirectly using global measurements of trace gases whose emissions are well known and whose primary sink is OH. The best trace gas for this purpose is the industrial chemical 1,1,1-trichloro-ethane (methylchloroform, $CH_3CCl_3$). First, there are accurate long-term measurements of $CH_3CCl_3$ beginning in 1978 in the ALE/GAGE/AGAGE network (Prinn *et al.*, 1983, 2000, 2001) and beginning in 1992 in the NOAA/CMDL network (Montzka *et al.*, 2000). Second, methylchloroform has fairly simple end uses as a solvent, and voluntary chemical industry reports since 1970, along with the national reporting procedures under the Montreal Protocol in more recent years, have produced very accurate emissions estimates for this chemical (McCulloch and Midgley, 2001).

Other gases which are useful OH indicators include $^{14}CO$, which is produced primarily by cosmic rays (Volz *et al.*, 1981; Mak *et al.*, 1994; Quay *et al.*, 2000). While the accuracy of the $^{14}CO$ production estimates, and especially the frequency and spatial coverage of its measurements, do not match those for $CH_3CCl_3$, its lifetime (2 months) is much shorter than for $CH_3CCl_3$ (4.9 yr), so it provides estimates of average concentrations of OH that are more regional than $CH_3CCl_3$. Another useful gas is the industrial chemical chlorodi-fluoromethane (HCFC-22, $CHClF_2$). It yields OH concentrations similar to those derived from $CH_3CCl_3$ but with less accuracy due to greater uncertainties in emissions and less extensive measurements (Miller *et al.*, 1998). The industrial gases $CH_2FCF_3$ (HFC-134a), $CH_3CCl_2F$ (HCFC-141b), and $CH_3CClF_2$ (HCFC-142b) are potentially useful OH estimators but the accuracy of their emission estimates needs improvement (Simmonds *et al.*, 1998; Huang and Prinn, 2002).

The most powerful methods for determining OH concentrations from trace gas data involve solution of an inverse problem in which the observables are expressed as Lagrangian line integrals, and the unknown OH concentrations (expressed as functions of space and time) are the integrands. The inverse problem of interest consists of determining an "optimal" estimate in the Bayesian sense of the unknown OH concentrations from imperfect concentration measurements of say $CH_3CCl_3$ over space and time. The unknown OH concentrations are arrayed in a "state" vector $x^t$ and the $CH_3CCl_3$ measurement errors are arrayed in a "noise" vector. Approximating the line integral by a summation leads to the observed $CH_3CCl_3$ concentrations being expressed as the noise vector plus a matrix of "partial derivatives" ($H$) multiplied by $x^t$. $H$ expresses the sensitivity of the chemical transport model (CTM) $CH_3CCl_3$ concentrations to changes in the state vector elements (i.e., to changes in OH). Given the discrete time series nature of $CH_3CCl_3$ measurements, it is

convenient (but not essential) to solve for $x^t$ using a discrete recursive optimal linear filter such as the discrete Kalman filter (DKF). The DKF has the specific useful property that it provides an objective assessment of the uncertainty in estimates of $x^t$ as each $CH_3CCl_3$ measurement is used and thus of the usefulness of each measurement. Application of the DKF requires a CTM to compute $H$.

While the derivation of the measurement equation uses Lagrangian concepts, $H$ can be equally well derived using an Eulerian CTM. Several intuitive concepts exist regarding the important effects of observational errors and CTM errors on the value of the observations in improving and lowering the errors in the OH estimates (Prinn, 2000). (The latter paper is contained in a teaching monograph designed to introduce researchers to these inverse techniques applied to the biogeochemical cycles.) Application of optimal linear filtering requires careful attention to both the physics of the problem expressed in the "measurement" and "system" (or "state-space") equations or models, and the sources and nature of the errors in the observations and CTMs. This approach was the one adopted for analysis of $CH_3CCl_3$ data by the ALE/GAGE/AGAGE scientists from the inception of the experiment.

The use of $CH_3CCl_3$ has established that the global weighted average OH concentration in the troposphere is, over the 1978–2000 period, $\sim 10^6$ radicals cm$^{-3}$ (Prinn *et al.*, 1987, 1992, 1995, 2001; Krol *et al.*, 1998; Montzka *et al.*, 2000). The weighting factor for this average is formally the rate constant $k$ (which is temperature dependent) of the reaction of OH with methyl-chloroform, multiplied by the $CH_3CCl_3$ concentration, $[CH_3CCl_3]$. Practically, this means that the average is weighted toward the tropical lower troposphere. A similar average concentration is derived from $^{14}CO$ (Quay *et al.*, 2000), although the weighting here is not temperature dependent.

While the average OH concentration appears to be fairly well defined by this method, the temporal trends in OH are more difficult to discern since they require long-term measurements, and very accurate calibrations, model transports, and above all, a very accurate time series of $CH_3CCl_3$ emissions (Prinn *et al.*, 1992). In the first attempt at trend determination, Prinn *et al.* (1992) derived an OH trend of $1.0 \pm 0.8\%$ yr$^{-1}$ between 1978 and 1990. Due primarily to a subsequent recalibration of the ALE/GAGE/AGAGE $CH_3CCl_3$ measurements, the estimated 1978–1994 trend was lowered to $0.0 \pm 0.2\%$ yr$^{-1}$ (Prinn *et al.*, 1995). In a subsequent reanalysis of the same measurements (but not with the same emissions),

Krol *et al.* (1998) derived a trend of $0.46 \pm 0.6\%$ yr$^{-1}$. The reasons for this difference, including differences in models, data processing, emissions, and inverse methods, have been intensely debated (Prinn and Huang, 2001; Krol *et al.*, 2001). However, when the Prinn *et al.* (1995) method is used with the same emissions as Krol *et al.* (1998), it yields a trend of $0.3\%$ yr$^{-1}$, in reasonable agreement with their value (Prinn and Huang, 2001).

In the latest analysis applied to the entire 1978–2000 ALE/GAGE/AGAGE CH$_3$CCl$_3$ measurements, Prinn *et al.* (2001) showed that global OH levels grew by $15 \pm 22\%$ between 1979 and 1989. But, the growth rate was decreasing at a statistically significant rate of $0.23 \pm 0.18\%$ yr$^{-1}$, so that OH began slowly declining after 1989 to reach levels in 2000 that were $10 \pm 24\%$ lower than the 1979 values (see Figure 7). Overall, the weighted global average OH concentration was $[9.4 \pm 1.3] \times 10^5$ radicals cm$^{-3}$ with concentrations $14 \pm 35\%$ lower in the northern hemisphere than in the southern hemisphere. The Prinn *et al.* (2001) analysis included, in addition to the measurement errors, a full Monte Carlo treatment of model errors (transport, chemistry), emission uncertainties, and absolute calibration errors with the last two being the dominant contributors to OH trend errors.

Because these substantial variations in OH were not expected from current theoretical models, these results have generated much attention and scrutiny. Recognizing the importance of their assumed emissions estimates in their calculations, Prinn *et al.* (2001) also estimated the emissions required to provide a zero trend in OH. These required emissions differ by many standard deviations from the best industry estimates particularly for 1996–2000 (McCulloch and Midgley, 2001). They are also at odds with estimates of European and Australian emissions obtained from measurements of polluted air from these continents reaching the ALE/GAGE/AGAGE stations (Prinn *et al.*, 2001). Indeed, if these required zero-trend emissions are actually the correct ones, then the phase-out of CH$_3$CCl$_3$ consumption reported by the party nations to the Montreal Protocol must seriously be in error. This topic is expected to be studied intensively in the future. The good news is that as the emissions of CH$_3$CCl$_3$ reduce to essentially zero (as they almost are in the early 2000s), the influence of these emission errors on OH determinations becomes negligible.

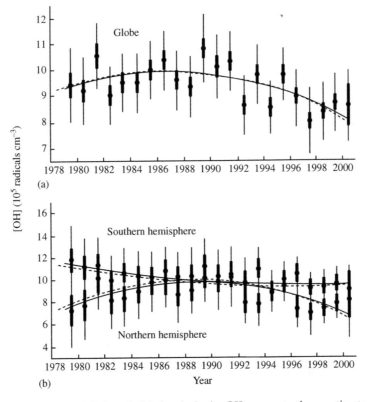

**Figure 7** (a) Annual average global and (b) hemispheric OH concentrations estimated from CH$_3$CCl$_3$ measurements (after Prinn *et al.*, 2001). Uncertainties (one sigma) due to measurement errors are indicated by thick error bars and uncertainties due to both measurement, emission, and model errors are shown by thin error bars. Also shown are polynomial fits to these annual values (solid lines), and polynomials describing OH variations which were estimated directly from CH$_3$CCl$_3$ measurements (dotted lines).

Presuming that they are correct, the above positive and negative OH trends are not simply explained by the measured trends in trace gases involved in OH chemistry. There is no clear negative or positive global-scale trend in the major OH precursor gases $O_3$ and $NO_x$ between 1978 and 1999, while levels of the dominant OH sink CO decreased, rather than increased, in the northern hemisphere (Prather *et al.*, 2001). At the same time, concentrations of another OH sink, $CH_4$, have risen significantly during 1978–1999. Prinn *et al.* (2001) have suggested that increases in tropical and subtropical countries of hydrocarbons and $SO_2$ that react with OH, along with a decrease of $NO_x$ emissions in developed countries, could be involved.

Also, a growth in levels of anthropogenic aerosols could increase heterogeneous OH destruction. Such a growth in aerosols (both absorbing and reflecting) could also lower UV fluxes, through direct and indirect reflection and absorption, thus lowering OH. Neglect of these aerosol effects, as well as urban-scale chemistry, which can remove $NO_x$ locally, may also be the reason why the higher OH levels inferred in the southern hemisphere are not simulated in many of the models in use in the early 2000s.

It is clear that further study on possible causes of OH variations of the type suggested from the $CH_3CCl_3$ analysis is necessary. Toward this end, the lack of long-term global-covering measurement of $O_3$, $NO_x$, $SO_2$, hydrocarbons, and aerosols is a very serious impediment to our understanding of OH chemistry.

## 4.01.7 ATMOSPHERIC MODELS AND OBSERVATIONS

A major approach for testing our knowledge of atmospheric oxidation processes has involved a careful comparison between trace gas and free radical observations and the predictions from three-dimensional (3D) CTMs that incorporate the latest understanding of homogeneous and heterogeneous fast photochemistry. While these 3D CTMs simulate qualitatively many features of the observations, there are discrepancies that point out the need for additional research. Some illustrative examples are briefly discussed here.

Berntsen *et al.* (2000) noted that their 3D CTMs indicate increases in upper tropospheric $O_3$ in the 1980s, whereas observations show a leveling off or a decrease. They point to the possible role of unmodeled stratospheric $O_3$ changes which might explain this discrepancy. Houweling *et al.* (2000) compared their 3D CTM simulations of $CH_4$ and $CH_3CCl_3$ observations providing a test of their $CH_4$ emission assumptions and OH

concentrations. They concluded that further work is needed on simulating seasonal $CH_4$ emissions from wetlands and rice paddies. Spivakovsky *et al.* (2000) extensively tested their OH calculations using $CH_3CCl_3$, $CHClF_2$, and other trace gas data and concluded that their 3D CTM gave good agreement except in the two tropical winters (possible OH underestimates) and southern extratropics (possible OH overestimate). They noted the greater usefulness of $CH_3CCl_3$ for regional OH tests, as its emissions are becoming negligible.

Models have also been tested using paleodata as noted in Sections 4.01.2.2 and 4.01.5.1. Wang and Jacob (1998) noted their significant overestimation of $O_3$ levels in the 1800s and discussed the possible effects of incorrectly modeled surface deposition and troposphere–stratosphere exchange in producing this discrepancy. Mesoscale 3D models, which couple the dynamics of convection with cloud microphysics, and fast photochemistry have also been tested against observations; e.g., Wang and Prinn (2000) concluded that $NO_x$ produced from lightning can deplete $O_3$ in the tropical Pacific upper troposphere as is often observed, but there were difficulties in simulating the observed high $O_3$ concentrations in the middle troposphere in these regions.

Underpinning these tests of our understanding are a large number of observational experiments using surface, aircraft, balloon and satellite platforms, and ranging from short-term campaigns to multidecadal measurement networks. Lelieveld *et al.* (1999) and Ehhalt (1999) have reviewed some of the relevant experiments focused on tropospheric oxidants. They noted the advances made in direct $NO_x$ and $HO_x$ observations for testing photochemical models. Thompson *et al.* (2003) have reviewed the gaps in tropospheric $O_3$ observations and have presented an important new data set for the southern hemisphere tropics that will provide new tests of chemical models.

## 4.01.8 CONCLUSIONS

Oxidation processes have played a major role in the evolution of Earth's atmosphere. Observations of trace gases and free radicals in the atmosphere in 1978–2003, and of chemicals in ice cores recording the composition of past atmospheres, are providing fundamental information about these processes. Also, basic laboratory studies of chemical kinetics, while not reviewed here, have played an essential role in defining mechanisms and rates. Models have been developed for fast photochemistry and for coupled chemical and transport processes that encapsulate current laboratory and theoretical understanding and help explain some of these atmospheric observations.

But there are important discrepancies between models and observations for OH and $O_3$ (both locally and globally) that still need to be resolved.

Looking into the twenty-first century, it is clear that there is urgent need to gain more observations to improve our understanding. We need much more extensive global 3D distributions for many trace gas species ($O_3$, CO, $NO_x$, reactive hydrocarbons, $H_2O_2$, etc.) and many more measurements of key reactive free radicals (OH, $HO_2$, etc.) to improve and ultimately validate models of the atmosphere. More attention needs to be given, in both modeling and observational programs, to the role of aerosols and clouds in governing UV fluxes and heterogeneous chemical reactions. Refinement of existing methods for determining long-term trends in OH, along with development of new independent approaches is clearly important.

Innovative methods for discerning the composition of past atmospheres are needed, since such information both tests current understanding and helps elucidate the process of atmospheric evolution. Finally, human and natural biospheric activities serve as the primary sources of both oxidant precursors and oxidant sink molecules, and the past, present, and possible future trends in these sources need to be much better quantified if we are to fully understand the oxidation processes in our atmosphere.

## REFERENCES

Anklin M. and Bales R. C. (1997) Recent increases in $H_2O_2$ concentrations at Summit, Greenland. *J. Geophys. Res.* **102**, 19099–19104.

Berntsen T. K., Myhre G., Stordal F., and Isaksen I. S. A. (2000) Time evolution of tropospheric ozone and its radiative forcing. *J. Geophys. Res.* **105**, 8915–8930.

Brandenburger U., Brauers T., Dorn H.-P., Hausmann M., and Ehhalt D. H. (1998) In-situ measurement of tropospheric hydroxyl radicals by folded long-path laser absorption during the field campaign POPCORN in 1994. *J. Atmos. Chem.* **31**, 181–204.

Brasseur G. P., Orlando J. J., and Tyndall G. S. (1999) *Atmospheric Chemistry and Global Change.* Oxford University Press, Oxford.

Brune W. H., Stevens P. S., and Mather J. H. (1995) Measuring OH and $HO_2$ in the troposphere by laser-induced fluorescence at low pressure. *J. Atmos. Sci.* **52**, 3328–3336.

Chameides W. L. and Walker J. C. G. (1973) A photochemical theory of tropospheric ozone. *J. Geophys. Res.* **78**, 8751–8760.

Chameides W. L. and Walker J. C. G. (1981) Rates of fixation by lightning of carbon and nitrogen in possible primitive atmospheres. *Origins of Life* **11**, 291–302.

Chen X. and Mopper K. (2000) Determination of tropospheric hydroxyl radical by liquidphase scrubbing and HPLC: preliminary results. *J. Atmos. Chem.* **36**, 81–105.

Crosley D. R. (1995) The measurement of OH and $HO_2$ in the atmosphere. *J. Atmos. Sci.* **52**, 3299–3314.

Crutzen P. J. (1979) The role of NO and $NO_2$ in the chemistry of the troposphere and stratosphere. *Ann. Rev. Earth Planet. Sci.* **7**, 443–472.

Crutzen P. J. and Zimmerman P. H. (1991) The changing photochemistry of the trophosphere. *Tellus* **43AB**, 136–151.

Dorn H.-P., Brandenburger U., Brauers T., and Hausmann M. (1995) A new *in situ* long-path absorption instrument for the measurement of tropospheric OH radicals. *J. Atmos. Sci.* **52**, 3373–3380.

Ehhalt D. H. (1999) Gas phase chemistry of the troposphere. In *Global Aspects of Atmospheric Chemistry*, Topics in Physical Chemistry (ed. R. Zellner). Springer, Darmstadt, vol. 6, pp. 21–109.

Ehhalt D. H., Dorn H., and Poppe D. (1991) The chemistry of the hydroxyl radical in the troposphere. *Proc. Roy. Soc. Edinburgh* **97**, 17–34.

Eisele F. L. and Tanner D. J. (1991) Ion-assisted tropospheric OH measurements. *J. Geophys. Res.* **96**, 9295–9308.

Eisele F. L., Tanner D. H., Cantrell C. A., and Calvert J. G. (1996) Measurements and steady state calculations of OH concentrations at Mauna Loa Observatory. *J. Geophys. Res.* **101**, 14665–14679.

Fegley B., Prinn R. G., Hartman H., and Watkins G. H. (1986) Chemical effects of large impacts on the Earth's primitive atmosphere. *Nature* **319**, 305–308.

Felton C. C., Sheppard J. C., and Campbell M. J. (1990) The radiochemical hydroxyl radical measurement method. *Environ. Sci. Technol.* **24**, 1841–1847.

Folland C. K., Karl T. R., Christy J. R., Clarke R. A., Gruza G. V., Jouzel J., Mann M. E., Oerlemans J., Salinger M. J., and Wang S. W. (2001) Observed climate variability and change. In *Climate Change 2001: The Scientific Basis* (ed. J. T. Houghton), Cambridge University Press, New York, pp. 99–181.

Haan D., Martinerie P., and Raynaud D. (1996) Ice core data of atmospheric carbon monoxide over Antarctica and Greenland during the last 200 years. *Geophys. Res. Lett.* **23**, 2235–2238.

Hard T. M., O'Brien R. J., Chan C. Y., and Mehrabzadeh A. A. (1984) Tropospheric free radical determination by FAGE. *Environ. Sci. Technol.* **18**, 768–777.

Hard T. M., George L. A., and O'Brien R. J. (1995) FAGE Determination of Tropospheric HO and $HO_2$. *J. Atmos. Sci.* **52**, 3354–3372.

Hofzumahaus A., Aschmutat U., Hebling M., Holland F., and Ehhalt D. (1996) The measurement of tropospheric OH radicals by laser-induced fluorescence spectroscopy during the POPCORN field campaign. *Geophys. Res. Lett.* **23**, 2541–2544.

Holland F., Hessling M., and Hofzumahaus A. (1995) *In-situ* measurement of tropospheric OH radicals by laser-induced fluorescence. A description of the KFA instrument. *J. Atmos. Sci.* **52**, 3393–3401.

Holland H. D. (1984) *The Chemical Evolution of the Atmosphere and Oceans.* Princeton University Press, Princeton.

Hough A. M. and Derwent R. G. (1990) Changes in the global concentration of tropospheric ozone due to human activities. *Nature* **344**, 645–648.

Houweling S., Dentener F., and Lelieveld J. (2000) The modeling of tropospheric methane: how well can point measurements be reproduced by a global model? *J. Geophys. Res.* **105**, 8981–9002.

Huang J. and Prinn R. (2002) Critical evaluation of emissions for potential gases for OH estimation. *J. Geophys. Res.* **107**, D24, 4784, doi: 10.1029/2002 JD002394.

Krol M., van Leeuwen P. J., and Lelieveld J. (1998) Global OH trend inferred from methyl chloroform measurements. *J. Geophys. Res.* **103**, 10697–10711.

Krol M., van Leeuwen P. J., and Lelieveld J. (2001) Reply to comment by Prinn and Huang (2001) on Krol *et al.* (1998) *J. Geophys. Res.* **106**, 23158–23168.

Lange M. A. and Ahrens T. J. (1982) The evolution of an impact-generated atmosphere. *Icarus* **51**, 96–120.

Lelieveld J., Thompson A. M., Diab R. D., Hov O., Kley D., Logan J. A., Nielson O. J., Stockwell W. R., and Zhou X. (1999) Tropospheric ozone and related processes.

In *Scientific Assessment of Ozone Depletion: 1998* (ed. D. Albritton). World Meteorological Organization, Geneva, pp. 8.1–8.42.

Levine J. S. (1985) The photochemistry of the early atmosphere. In *The Photochemistry of Atmospheres* (ed. J. S. Levine). Academic Press, Orlando, pp. 3–38.

Levy H. (1971) Normal atmosphere: large radical and formaldehyde concentrations predicted. *Science* **173**, 141–143.

Levy H., Kasibhatla P. S., Moxim W. J., Klonecki A., Hirsch A., Oltmans S., and Chameides W. L. (1997) The global impact of human activity on tropospheric ozone. *Geophys. Res. Lett.* **24**, 791–794.

Lewis J. and Prinn R. (1984) *Planets and their Atmospheres: Origin and Evolution.* Academic Press, Orlando.

Logan J., Prather M., Wofsy S., and McElroy M. (1981) Tropospheric chemistry: a global perspective. *J. Geophys. Res.* **86**, 7210–7254.

Madronich S. and Granier C. (1992) Impact of recent total ozone changes on tropospheric ozone photodissociation, hydroxyl radicals, and methane trends. *Geophys. Res. Lett.* **19**, 465–467.

Mak J. E., Brenninkmeijer C. A. M., and Tamaresis J. (1994) Atmospheric $^{14}CO$ observations and their use for estimating carbon monoxide removal rates. *J. Geophys. Res.* **99**, 22915–22922.

Marenco A., Gouget H., Nedelec P., and Pages J. P. (1994) Evidence of long-term increase in tropospheric ozone from Pic du Midi data series, consequences and positive radiative forcing. *J. Geophys. Res.* **99**, 16617–16632.

McCulloch A. and Midgley P. (2001) The history of methyl chloroform emissions: 1951–2000. *Atmos. Environ.* **35**, 5311–5319.

Miller B. R., Weiss R. F., Prinn R. G., Huang J., and Fraser P. J. (1998) Atmospheric trend and lifetime of chlorodifluoromethane (HCFC-22) and the global tropospheric OH concentration. *J. Geophys. Res.* **103**, 13237–13248.

Montzka S. A., Spivakovsky C. M., Butler J. H., Elkins J. W., Lock L. T., and Mondecl D. J. (2000) New observational constraints on atmospheric hydroxyl on global and hemispheric scales. *Science* **288**, 500–503.

Mount G. H. (1992) The measurement of tropospheric OH by long-path absorption: 1. Instrumentation. *J. Geophys. Res.* **97**, 2427–2444.

Mount G. H. and Harder J. W. (1995) The measurement of tropospheric trace gases at Fritz Peak Observatory by long-path absorption: OH and ancillary gases. *J. Atmos. Sci.* **52**, 3342–3353.

Penner J., Andreae M., Annegarn H., Barrie L., Feichter J., Hegg D., Jayaraman A., Leaitch R., Murphy D., Nganga J., and Pitari G. (2001) Aerosols: their direct and indirect effects. In *Climate Change 2001: The Scientific Basis* (ed. J. T. Houghton). Cambridge University Press, New York, pp. 289–348.

Poppe D., Zimmerman J., Bauer R., Brauers T., Bruning D., Callies J., Dorn H. P., Hofzumahaus A., Johnen F. J., Khedim A., Koch H., Koppman R., London H., Muller K. P., Neuroth R., Plass-Dulmer C., Platt U., Rohrer F., Roth E. P., Rudolph J., Schmidt U., Wallasch M., and Ehhalt D. H. (1994) Comparison of measured OH concentrations with model calculations. *J. Geophys. Res.* **99**, 16633–16642.

Poppe D., Zimmerman J., and Dorn H. P. (1995) Field data and model calculations for the hydroxyl radical. *J. Atmos. Sci.* **52**, 3402–3407.

Prather M., Ehhalt D., Dentener F., Derwent R., Dlugokencky E., Holland E., Isaksen I., Katima J., Kirchhoff V., Matson P., Midgley P., and Wang M. (2001) Atmospheric chemistry and greenhouse gases. In *Climate Change 2001: The Scientific Basis* (ed. J. T. Houghton). Cambridge University Press, New York, pp. 239–288.

Prinn R. (1982) Origin and evolution of planetary atmospheres: an introduction to the problem. *Planet. Space Sci.* **30**, 741–753.

Prinn R. (1994) The interactive atmosphere: global atmospheric–biospheric chemistry. *Ambio* **23**, 50–61.

Prinn R. and Fegley B. (1987) Bolide impacts, acid rain, and biospheric traumas at the Cretaceous–Tertiary boundary. *Earth Planet. Sci. Lett.* **83**, 1–15.

Prinn R. and Huang J. (2001) Comment on Krol et al. (1998). *J. Geophys. Res.* **106**, 23151–23157.

Prinn R., Rasmussen R., Simmonds P., Alyea F., Cunnold D., Lane B., Cardelino C., and Crawford A. (1983) The atmospheric lifetime experiment 5. Results for $CH_3CCl_3$ based on three years of data. *J. Geophys. Res.* **88**, 8415–8426.

Prinn R., Cunnold D., Simmonds P., Alyea F., Boldi R., Crawford A., Fraser P., Gutzler D., Hartley D., Rosen R., and Rasmussen R. (1992) Global average concentration and trend for hydroxyl radicals deduced from ALE/GAGE trichloroethane (methyl chloroform) data for 1978–1990. *J. Geophys. Res.* **97**, 2445–2461.

Prinn R., Jacoby H., Sokolov A., Wang C., Xiao X., Yang Z., Eckhaus R., Stone P., Ellerman D., Melillo J., Fitzmaurice J., Kicklighter D., Holian G., and Liu Y. (1999) Integrated global system model for climate policy assessment: feedbacks and sensitivity studies. *Climat. Change* **41**, 469–546.

Prinn R., Weiss R., Fraser P., Simmonds P., Cunnold D., Alyea F., O'Doherty S., Salameh P., Miller B., Huang J., Wang R., Hartley D., Harth C., Steele L., Sturrock G., Midgley P., and McColloch A. (2000) A history of chemically and radiatively important gases in air deduced from ALE/GAGE/AGAGE. *J. Geophys. Res.* **105**, 17751–17792.

Prinn R. G. (2000) Measurement equation for trace chemicals in fluids and solution of its inverse. In *Inverse Methods in Global Biogeochemical Cycles Geophysical Monograph* (ed. P. Kasibhatla). American Geophysical Union, Washington, DC, vol. 114, pp. 3–18.

Prinn R. G., Cunnold D. M., Rasmussen R., Simmonds P. G., Alyea F. N., Crawford A., Fraser P. J., and Rosen R. (1987) Atmospheric trends in methylchloroform and the global average for the hydroxyl radical. *Science* **238**, 945–950.

Prinn R. G., Weiss R. F., Miller B. R., Huang J., Alyea F. N., Cunnold D. M., Fraser P. J., Hartley D. E., and Simmonds P. G. (1995) Atmospheric trends and lifetime of $CH_3CCL_3$ and global OH concentrations. *Science* **269**, 187–192.

Prinn R. G., Huang J., Weiss R. F., Cunnold D. M., Fraser P. J., Simmonds P. G., McCulloch A., Harth C., Salameh P., O'Doherty S., Wang R. H. J., Porter L., and Miller B. R. (2001) Evidence for substantial variations of atmospheric hydroxyl radicals in the past two decades. *Science* **292**, 1882–1888.

Quay P., King S., White D., Brockington M., Plotkin B., Gammon R., Gerst S., and Stutsman J. (2000) Atmospheric $^{14}CO$: a tracer of OH concentration and mixing rates. *J. Geophys. Res.* **105**, 15147–15166.

Sigg A. and Neftel A. (1991) Evidence for a 50% increase in $H_2O_2$ over the past 200 years from a Greenland ice core. *Nature* **351**, 557–559.

Simmonds P. G., O'Doherty S., Huang J., Prinn R., Derwent R., Ryall D., Nickless G., and Cunnold D. (1998) Calculated trends and the atmospheric abundance of 1,1,1,2-tetrafluoroethane, 1,1-dichloro-1-fluoroethane, and 1-chloro-1, 1-difluoroethane using automated *in situ* gas chromatography-mass spectrometry measurements recorded at Mace Head, Ireland, from October 1994 to March 1997. *J. Geophys. Res.* **103**, 16029–16037.

Spivakovsky C. M., Logan J. A., Montzka S. A., Balkanski Y. J., Foreman-Fowler M., Jones D. B. A., Horowitz L. W., Fusco A. C., Brenninkmeijer C. A. M., Prather M. J., Wofsy S. C., and McElroy M. B. (2000) Three dimensional climatological distribution of tropospheric OH: update and evaluation. *J. Geophys. Res.* **105**, 8931–8980.

Staffelbach T. A., Neftel A., Stauffer B., and Jacob D. J. (1991) Formaldehyde in polar ice cores: a possibility to characterize the atmospheric sink of methane in the past? *Nature* **349**, 603–605.

Stimpfle R. M. and Anderson J. G. (1988) *In situ* detection of OH in the lower stratosphere with a balloon borne high repetition rate laser system. *Geophys. Res. Lett.* **15**, 1503–1506.

Tanner D. J., Jefferson A., and Eisele F. L. (1997) Selected ion chemical ionization mass spectrometric measurement of OH. *J. Geophys. Res.* **102**, 6415–6425.

Thompson A. M. (1992) The oxidizing capacity of the Earth's atmosphere: probable past and future changes. *Science* **256**, 1157–1165.

Thompson A. M., Witte J. C., Mc Peters R. D., Oltmans S. J., Schmidlin F. J., Logan J. A., Fujiwara M., Kirchoff V., Posny F., Coetzee G., Hoegger B., Kawakami S., Ogawa T., Johnson B., Vomel H., and Labow G. (2003) The 1998–2000 SHADOZ Tropical Ozone Climatology: comparison with TOMS and ground-based measurements. *J. Geophys. Rev.* **108**, D2, 8238, doi: 10.1029/2001 JD000967.

Volz A. and Kley D. (1988) Evaluation of the Montsouris Series of ozone measurements made in the nineteenth century. *Nature* **332**, 240–242.

Volz A., Ehhalt D. H., and Derwent R. G. (1981) Seasonal and latitudinal variation of $^{14}$CO and the tropospheric concentration of OH radicals. *J. Geophys. Res.* **86**, 5163–5171.

Wang C. and Prinn R. G. (1999) Impact of emissions, chemistry and climate on atmospheric carbon monoxide: 100-year predictions from a global chemistry model. *Chemosph. Global Change* **1**, 73–81.

Wang C. and Prinn R. G. (2000) On the roles of deep convective clouds in tropospheric chemistry. *J. Geophys. Res.* **105**, 22269–22297.

Wang Y. and Jacob D. J. (1998) Anthropogenic forcing on tropospheric ozone and OH since pre-industrial times. *J. Geophys. Res.* **103**, 31123–31135.

# 4.02
# Tropospheric Halogen Chemistry

R. von Glasow

*University of California, San Diego, CA, USA*

and

P. J. Crutzen

*University of California, San Diego, CA, USA and*
*Max-Planck-Institut für Chemie, Mainz, Germany*

## 4.02.1 INTRODUCTION

Halogens are very reactive chemicals that are known to play an important role in anthropogenic stratospheric ozone depletion chemistry, first recognized by Molina and Rowland (1974). However, they also affect the chemistry of the troposphere. They are of special interest because they are involved in many reaction cycles that can affect the oxidation power of the atmosphere indirectly by influencing the main oxidants $O_3$ and its photolysis product OH and directly, e.g., by reactions of the Cl radical with hydrocarbons (e.g., $CH_4$).

Already by the middle of the nineteenth century, Marchand (1852) reported the presence of bromine and iodine in rain and other natural waters. He also mentions the benefits of iodine in drinking water through the prevention of goitres and cretinism. In a prophetic monograph *"Air and Rain: The Beginnings of a Chemical Climatology,"* Smith (1872) describes measurements of chloride in rain water, which he states to originate partly from the oceans by a process that he compares with the bursting of "soap bubbles" which produces "small vehicles" that transfer small spray droplets of seawater to the air. From deviations of the sulfate-to-chloride ratio in coastal rain compared to seawater, Smith concluded that chemical processes occur once the particles are airborne.

For almost a century thereafter, however, atmospheric halogens received little attention. One exception was the work by Cauer (1939), who reported that iodine pollution has been significant in Western and Central Europe due to the inefficient burning of seaweed, causing mean gas phase atmospheric concentrations as high as or greater than 0.5 $\mu$g m$^{-3}$. In his classical textbook *Air Chemistry and Radioactivity*, Junge (1963) devoted less than three pages to halogen gas phase chemistry, discussing chlorine and iodine. As reviewed by Eriksson (1959a,b), the main atmospheric source of halogens is sea salt, derived from the bursting of bubbles of air which are produced by ocean waves and other processes. Early work by Cauer (1951) had shown that Cl/Na and Cl/Mg ratios were lower in air than in seawater, indicating loss of chlorine by "acid displacement" from sea salt by the strong acids, $H_2SO_4$ (Eriksson, 1959a,b) and $HNO_3$ (Robbins et al., 1959). Already the first measurements of bromine in aerosols by Duce et al. (1963) showed that bromine, like chlorine, was lost from the sea salt particles, whereas iodine was strongly enriched (Duce et al., 1965). Research since the early 1980s has shown that photochemical processes are actively involved.

Interest in the chemistry of atmospheric halogens took a steep upward surge after it was postulated that the release of industrially produced halocarbons, in particular the chlorofluorocarbons (CFCs), $CFCl_3$, and $CF_2Cl_2$, could cause severe depletions in stratospheric ozone (Molina and Rowland, 1974) by the reactions involving the CFC photolytic product radicals, Cl and ClO, as catalysts. The first stratospheric measurements of ClO did indeed show its presence in significant quantities in the stratosphere so that by the end of the 1970s USA, Canada, and the Scandinavian countries issued laws against the use of CFC gases as propellants in spray cans. In the mid-1980s the springtime stratospheric ozone hole over Antarctica was discovered by Farman et al. (1985), involving heterogeneous reactions on polar stratospheric clouds that lead to chlorine activation (Solomon et al., 1986). Ten years later, in 1996, a complete phaseout of the production of the CFCs and a number of other chlorine- or bromine-containing chemicals came into effect for all nations in the developed world. In this contribution we will, however, concentrate on the impact of reactive chlorine, bromine, and iodine on tropospheric ozone chemistry.

Halogens have the potential to be important in many facets of tropospheric chemistry. A multitude of gas phase reactions and gas–particle interactions occur that include coupling with the sulfur cycle and reactions with hydrocarbons. Loss of ozone by catalytic reactions involving halogen radicals lowers the concentrations of the hydroxyl radical OH and thus the oxidation power of the atmosphere. Figure 1 shows these and other relevant halogen-related processes schematically. The sum of particulate and gaseous halogen concentrations maximize in the marine troposphere. Important for our climate—via feedback with cloud microphysics mainly in the large regions of marine stratocumulus—are links between halogen chemistry and the sulfur cycle. $HOBr_{aq}$ and $HOCl_{aq}$ can increase the liquid phase oxidation of S(IV) to S(VI), while BrO can decrease the most important *in situ* source for $SO_2$ in the marine troposphere, namely, the oxidation of DMS to $SO_2$ by reaction with OH by providing an alternate pathway (BrO + DMS) that reduces the yield of $SO_2$ from DMS oxidation. Thus, the presence of bromine and chlorine in the troposphere lowers gas phase $SO_2$ concentrations and thus the formation of new sulfate particles via the reaction sequence $SO_2 + OH \rightarrow H_2SO_4$.

High mixing ratios of iodine oxide at a coastal site indicate a potentially significant role of iodine for the destruction of $O_3$ and new particle embryo formation (Alicke et al., 1999; O'Dowd et al., 1998). Almost 20 years earlier, Chameides and Davis (1980) suggested that open ocean iodine chemistry would be initiated by the photolysis of $CH_3I$. This was based on the measurements of Lovelock et al. (1973) and Singh et al. (1979), who found volume mixing ratios of $CH_3I$ of 1–5 pmol mol$^{-1}$ over the ocean.

The potentially strong involvement of halogens in tropospheric chemistry was first observed in the Arctic, where strong ozone depletion events were found to coincide with high levels of bromine (Barrie et al., 1988).

The first mid-latitude demonstration of reactive halogen chemistry in the troposphere was made downwind of salt pans in the Dead Sea area, where

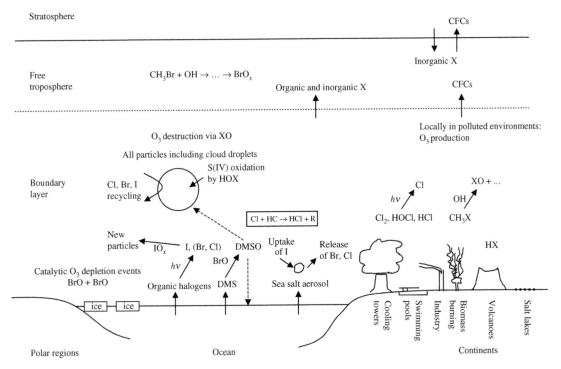

**Figure 1** Schematic depiction of the most important halogen-related processes in the troposphere.

the so far highest atmospheric mixing ratios of BrO were measured (Hebestreit *et al.*, 1999). Volcanoes are sources of halogens as well, mainly in the form of HCl. Biomass burning releases halogens as do industrial processes.

So far we have only mentioned chlorine, bromine, and iodine. This is justified because chemistry of fluorine is of no consequence, as very unreactive HF is efficiently formed in the atmosphere, e.g., via the reaction $F + H_2O \rightarrow HF + OH$. However, several fluorine-containing gases of anthropogenic origin are potentially powerful greenhouse gases, because they absorb strongly in the infrared atmospheric window region near 10 $\mu$m. Fully fluorinated gases—such as $SF_6$, $CF_4$, and $C_2F_6$—have atmospheric lifetimes of the order of thousands of years and thus possess very high global warming potential (GWP). Although their abundance in the atmosphere has not yet grown large enough to be of concern for Earth's climate, their production must ultimately be curtailed in the future. The most abundant fully fluorinated gas, $CF_4$, had an atmospheric volume mixing ratio of $\sim$75 pmol mol$^{-1}$ in 1995 (Warneck, 1999). Because of their higher concentrations in the atmosphere, about 270 pmol mol$^{-1}$ and 530 pmol mol$^{-1}$, respectively, the CFC gases, $CFCl_3$ and $CF_2Cl_2$, already exert a significant radiative greenhouse forcing (Ramanathan, 1975) on Earth's climate. For further discussion about atmospheric fluorine, the reader is referred to a thorough review article by Harnisch (1999).

Several overview articles have been published on tropospheric halogen chemistry since the early 1980s, starting with Cicerone (1981). Wayne *et al.* (1995) list in great detail reaction paths, laboratory data, and atmospheric implications of halogen oxides. A good overview on laboratory measurements was also given by de Haan *et al.* (1999). Reaction cycles involved in tropospheric halogen chemistry and measurements are also thoroughly discussed by Platt (2000) and Platt and Hönninger (2003). Important compilations of laboratory studies that were made to elucidate chemical reaction paths are given by, e.g., DeMore *et al.* (1997), Sander *et al.* (2000), and Atkinson *et al.* (1999, 2000). Emission inventories for chlorine were compiled by Graedel and Keene (1995) and Keene *et al.* (1999).

In Section 4.02.2 of this overview we will first describe the main halogen reaction mechanisms and then discuss, in Section 4.02.3, the springtime surface ozone depletion events in high latitudes that were first observed in the Arctic. Another main part of this chapter is concerned with halogens in the marine boundary layer (Section 4.02.4). In Section 4.02.5 we describe interactions of halogens with some other elements of atmospheric importance. A very recently discovered environment where halogen chemistry plays a large role are salt lakes (Section 4.02.6). There the chemistry bears similarity to that of the high-latitude ozone depletion events. This is followed in Section 4.02.7 by a discussion of halogen chemistry in the free troposphere and in Section 4.02.8 by other sources of halogens such as industry and biomass burning.

## 4.02.2   MAIN REACTION MECHANISMS

To facilitate the discussion in the following sections we start by summarizing the main halogen reaction cycles that are of importance for the troposphere. In the following the species X represents chlorine, bromine, or iodine. If two halogens occur in the same reaction we use X for the first and Y for the second halogen atom.

Halogen radicals are formed in the troposphere by the photolysis of dihalogens ($X_2$ or XY) or other species such as HOX, $XNO_3$, $XNO_2$, or organic halogen precursors. Within a second the resulting halogen atoms react with $O_3$ thereby producing halogen oxides. If the oxides photolyze, both $O_3$ and the halogen radical are reformed and no net reaction occurs:

$$O_3 + X \rightarrow XO + O_2 \qquad (1)$$

$$XO + h\nu \overset{O_2, M}{\rightarrow} X + O_3 \qquad (2)$$

Among the main $O_3$ destruction paths (cycles) involving these radicals are also reactions of the halogen oxides XO with $HO_2$ (cycle I):

$$O_3 + X \rightarrow XO + O_2 \qquad (3)$$

$$XO + HO_2 \rightarrow HOX + O_2 \qquad (4)$$

$$HOX + h\nu \rightarrow X + OH \qquad (5)$$

$$OH + CO \overset{O_2}{\rightarrow} CO_2 + HO_2 \qquad (6)$$

$$\text{I net:}\quad CO + O_3 \rightarrow CO_2 + O_2 \qquad (7)$$

If high levels of BrO are present (e.g., in the polar regions), the two branches of the BrO self-reaction represent a strong catalytic $O_3$ destruction cycle (cycle II):

$$2(Br + O_3 \rightarrow BrO + O_2) \qquad (8)$$

$$BrO + BrO \rightarrow 2Br + O_2 \mid BrO + BrO$$
$$\rightarrow Br_2 + O_2 \qquad (9)$$

$$\mid Br_2 + h\nu \rightarrow 2Br \quad (10)$$

$$\text{II net:}\quad 2O_3 \rightarrow 3O_2 \qquad (11)$$

This cycle was proposed by Barrie *et al.* (1988) for the Arctic. Earlier it had been proposed by Wofsy *et al.* (1975) for the stratosphere, but shown to be unimportant there. Note that in the troposphere only the BrO self-reaction is of importance, whereas in the stratospheric ozone hole the ClO self-reaction (forming the $Cl_2O_2$ dimer) dominates $O_3$ destruction. Halogen oxide cross-reactions are

of importance for stratospheric ozone depletion (e.g., Yung *et al.*, 1980; McElroy *et al.*, 1999). They may also play a role in tropospheric ozone loss (cycle III):

$$X + O_3 \rightarrow XO + O_2 \qquad (12)$$

$$Y + O_3 \rightarrow YO + O_2 \qquad (13)$$

$$XO + YO \rightarrow X + Y + O_2 \qquad (14)$$

$$\text{III net:}\quad 2O_3 \rightarrow 3O_2 \qquad (15)$$

As pointed out by Le Bras and Platt (1995), the reaction BrO + ClO is ~4 times faster than BrO + BrO, making ozone destruction even more efficient if significant amounts of both halogen oxides are present. Other reaction paths for the BrO + ClO reaction yield BrCl and OClO (e.g., Sander *et al.*, 2000), whereas one channel of the BrO + IO reaction and one channel of the self-reaction of IO yield OIO (Gilles *et al.*, 1997; Bedjanian *et al.*, 1998; Misra and Marshall, 1998; Bloss *et al.*, 2001; Rowley *et al.*, 2001). In the self-reaction of IO also $I_2O_2$ can be formed. Formation of OBrO was found to be unimportant (e.g., Rattigan *et al.*, 1995; Rowley *et al.*, 2001).

The formation of halogen nitrates can be important in regions with elevated $NO_x$:

$$XO + NO_2 \rightarrow XNO_3 \overset{\Delta}{\rightarrow} XO + NO_2 \qquad (16)$$

$$\overset{h\nu}{\rightarrow} X + NO_3 \qquad (17)$$

$$\overset{H_2O}{\rightarrow} HOX + HNO_3 \quad (18)$$

Depending on conditions, nitrates either decompose thermally, photolyze, or react with water on surfaces.

The main sink for chlorine radicals is the reaction with hydrocarbons and especially with $CH_4$:

$$Cl + RH \rightarrow HCl + R \qquad (19)$$

Of special interest in the troposphere are reactions that release halogens from salt deposits or sea salt aerosols. Several paths for that have been suggested and confirmed in laboratory studies. Fan and Jacob (1992) proposed an autocatalytic reaction cycle converting $Br^-$ to $Br_2$ (cycle IV):

$$Br_2 + h\nu \rightarrow 2Br, \quad \lambda < 600 \text{ nm} \quad (20)$$

$$2(Br + O_3 \rightarrow BrO + O_2) \qquad (21)$$

$$2(BrO + HO_2 \rightarrow HOBr + O_2) \qquad (22)$$

$$2HOBr \rightarrow 2HOBr_{aq} \qquad (23)$$

$$2(HOBr_{aq} + H^+ + Br^- \rightarrow Br_{2,aq} + H_2O) \quad (24)$$

$$2Br_{2,aq} \rightarrow 2Br_2 \quad (25)$$

$$\text{IV net:} \quad 2O_3 + 2HO_2 + 2Br^- + 2H^+$$
$$\rightarrow Br_2 + 4O_2 + 2H_2O \quad (26)$$

A similar set of reactions was proposed by Vogt *et al.* (1996) (cycle V):

$$BrCl + h\nu \rightarrow Br + Cl,$$
$$\lambda < 560 \text{ nm} \quad (27)$$

$$Br + O_3 \rightarrow BrO + O_2 \quad (28)$$

$$BrO + HO_2 \rightarrow HOBr + O_2 \quad (29)$$

$$HOBr \rightarrow HOBr_{aq} \quad (30)$$

$$HOBr_{aq} + H^+ + Cl^- \rightarrow BrCl_{aq} + H_2O \quad (31)$$

$$BrCl_{aq} \rightarrow BrCl \quad (32)$$

$$\text{V net:} \quad O_3 + HO_2 + Cl^- + H^+$$
$$\rightarrow Cl + 2O_2 + H_2O \quad (33)$$

Here, a number of bromine compounds catalyze the production of chlorine. BrCl can also take part in the following aqueous autocatalytic cycle (when sufficient $Br^-$ is available) leading to the release of $Br_2$ (Vogt *et al.*, 1996) (cycle VI):

$$BrCl + h\nu \rightarrow Br + Cl,$$
$$\lambda < 560 \text{ nm} \quad (34)$$

$$Br + O_3 \rightarrow BrO + O_2 \quad (35)$$

$$BrO + HO_2 \rightarrow HOBr + O_2 \quad (36)$$

$$HOBr \rightarrow HOBr_{aq} \quad (37)$$

$$HOBr_{aq} + H^+ + Cl^- \rightarrow BrCl_{aq} + H_2O \quad (38)$$

$$BrCl_{aq} + Br^- \leftrightarrow Br_2Cl^- \quad (39)$$

$$Br_2Cl^- \leftrightarrow Br_{2,aq} + Cl^- \quad (40)$$

$$Br_{2aq} \leftrightarrow Br_2 \quad (41)$$

$$\text{VI net:} \quad BrCl + O_3 + HO_2 + Br^- + H^+$$
$$\rightarrow Cl + Br_2 + H_2O + 2O_2 \quad (42)$$

In cycles IV and VI, one $Br^-$ ion is released to the gas phase in each cycle upon uptake of HOBr,

which makes these cycles autocatalytic in active bromine. If sufficient $Br^-$ is available and if gas phase bromine loss processes are less than 50%, then these cycles lead to the exponential release of bromine, which is often referred to as "bromine explosion." It is important to note that the reaction cycles IV–VI are acid catalyzed. However, Adams *et al.* (2002) found that, in contrast to the same reaction in the aqueous phase, uptake of HOBr on frozen sea salt leads to the release of dihalogens without the need for acidification. The lifetime of HOBr against photolysis is ~450 s, whereas the lifetime against uptake on sulfate aerosol is ~1,000 s and $10^4$ s for uptake on sea salt aerosol (all data calculated with the model Mistra-MPIC (von Glasow *et al.*, 2002a) for a cloud-free marine boundary layer at 30° latitude at the end of July). Therefore, the autocatalytic cycles are fast enough to compete with photolysis.

Note that in these reactions, for each molecule of active bromine $(Br + BrO + HOBr + Br_2)$ that is created, one $HO_2$ radical is lost to produce $H_2O$ leading to the loss $HO_x$ and a reduction of the $HO_2/OH$ concentration ratio (already discussed for iodine by Jenkin (1993)). Therefore, a significant factor that drives these chains of reactions may be the production of $HO_x$ radicals.

We want to stress that reaction cycles IV and VI are autocatalytic in bromine and that in order for cycles IV–VI to proceed, $O_3$, $HO_x$, and acidity are needed.

Many laboratory studies (e.g., Abbatt and Waschewsky, 1998; Mochida *et al.*, 1998a; Fickert *et al.*, 1999) have investigated the reaction cycles IV–VI with aqueous salt solutions or dry salt crystals and confirmed the rapid uptake of HOBr and the release of $Br_2$ and BrCl depending on the composition of the condensed phase. Behnke *et al.* (1999) examined the uptake of HOBr on sea salt aerosol in a smog chamber with suspended sea salt aerosols of different composition and verified the cycles IV–VI.

Fickert *et al.* (1999) examined the production of $Br_2$ and BrCl from the uptake of HOBr onto aqueous salt solutions in a wetted-wall flow tube reactor. The yield of $Br_2$ and BrCl was found to depend on the $Cl^-$ to $Br^-$ ratio, with more than 90% yield of $Br_2$ when $[Cl^-]/[Br^-]$ (in mol $L^{-1}$) was less than 1,000. With increasing $[Cl^-]/[Br^-]$ BrCl was the main product (see Figure 2). They also found a pH dependence of the outgassing of $Br_2$ and BrCl with greater release rates at lower pH.

This pH dependence was further examined by Keene *et al.* (1998), who used the box model of Sander and Crutzen (1996) and Vogt *et al.* (1996) (see Section 4.02.4.3). They showed that significant sea salt dehalogenation is limited to acidified aerosol but that differences in the results between a pH of 5.5 and 3 are not significant.

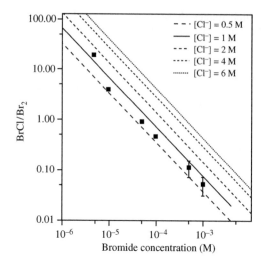

**Figure 2** Laboratory study of the reaction of HOBr with aqueous salt solutions. Shown is the ratio of the gas phase products BrCl and Br$_2$ as a function of Br$^-$ concentration. The data points represented by the symbols were obtained with [Cl$^-$] = 1 M and a pH of 5.5. The lines are simulations of aqueous phase BrCl/Br$_2$ ratios based on the equilibrium constants of Wang and Margerum (1994) (source Fickert *et al.*, 1999).

Several groups (e.g., Hirokawa *et al.*, 1998; Oum *et al.*, 1998b; Mochida *et al.*, 2000) pointed to the potential importance of the reaction:

$$O_3 + Br^- \rightarrow O_2 + BrO^- \qquad (43)$$

for the release of bromine from salt or sea salt aerosol. However, modeling studies (Vogt *et al.*, 1996; von Glasow *et al.*, 2002a) with the reaction rate measured by Haag and Hoigné (1983) (which is similar to other measurements; see Haag and Hoigné (1983) and references therein) could not confirm the importance of this reaction. The laboratory results of Disselkamp *et al.* (1999), however, point to an underestimated rate constant or nonlinear behavior of the O$_3$ + Br$^-$ reaction or other not yet understood processes. Anastasio and Mozurkewich (2002) found that bromide deposits on the glass wall of the reaction chamber was the source of dark production of bromine. They suggested that internally mixed silicate–sea salt particles might be significant in releasing photoactive halogen species in the marine boundary layer. Similarly, Behnke and Zetzsch (1989) found that photolysable chlorine was produced in their reaction chamber on glass walls. Nevertheless, there still seems to be a lack in our understanding of this reaction which urges further investigation.

Relatively early, research started to investigate if nitrogen oxides could be involved in the release of halogens from sea salt aerosols. The reactions of N$_2$O$_5$ and NO$_3$ with NaCl or sea salt have been studied extensively (e.g., Zetzsch *et al.*, 1988; Finlayson-Pitts *et al.*, 1989; Behnke *et al.*, 1991, 1994, 1997; Zetzsch and Behnke, 1992; George *et al.*, 1994; Seisel *et al.*, 1997, 1999; Karlsson and Ljungström, 1998; Schweitzer *et al.*, 1999; Yoshitake, 2000). It was shown that uptake of N$_2$O$_5$ leads to the release of halogen nitrites from sea salt aerosol, competing with the reaction with water:

$$N_2O_5 \xrightarrow{H_2O} 2HNO_{3,aq} \qquad (44)$$

$$\xrightarrow{Br^-} BrNO_2 + NO_3^- \qquad (45)$$

$$\xrightarrow{Cl^-} ClNO_2 + NO_3^- \qquad (46)$$

Photolysis of XNO$_2$ releases halogen radicals to the gas phase. Frenzel *et al.* (1998) showed that the reactions X$_2$ + NO$_2^-$ ↔ XNO$_2$ + X$^-$ (X = Cl, Br) occur on salt solutions. Sander *et al.* (1999) suggested that when sufficient NO$_x$ is available, reactions on the surface of aerosols can also lead to the production of Br$_2$ and BrCl without the need for acid catalysis. The uptake of BrNO$_3$ is followed by conversion to HOBr and the degassing of Br$_2$ and BrCl:

$$BrO + NO_2 \xrightarrow{M} BrNO_3 \qquad (47)$$

$$BrNO_3 \xrightarrow{H_2O} HNO_{3,aq} + HOBr_{aq} \qquad (48)$$

$$BrNO_3 \xrightarrow{Cl^-} NO_3^- + BrCl_{aq} \qquad (49)$$

$$BrNO_3 \xrightarrow{Br^-} NO_3^- + Br_{2,aq} \qquad (50)$$

Similar reactions exist for chlorine and iodine. For the clean marine boundary layer, however, reactions involving N$_2$O$_5$, XNO$_3$, and XNO$_2$ are expected to play a minor role because of their low concentrations.

It is important to note that the condensed phase (i.e., aqueous and dry) halogen chemistry is linked mainly via inter-halogen equilibria like Br$^-$ + BrCl ↔ Br$_2$Cl$^-$ ↔ Br$_2$ + Cl$^-$ (e.g., Fickert *et al.*, 1998; Mochida *et al.*, 1998b). Holmes *et al.* (2001) investigated inter-halogen reactions involving iodine on frozen and dry NaCl and NaBr surfaces and found that reactions involving HOI similar to the reactions discussed above for HOBr occur:

$$HOI + X^- \rightarrow IX + OH^- \qquad (51)$$

with X = Cl, Br. The inter-halogens (IX) are released to the gas phase. They also discussed the release of ICl upon uptake of INO$_3$ and reaction with Cl$^-$.

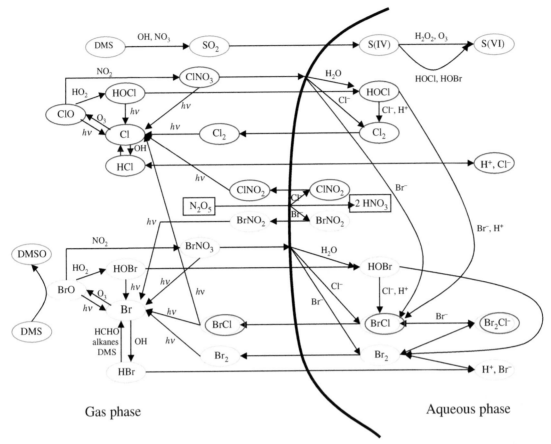

**Figure 3** Schematic diagram of the major bromine, chlorine, and sulfur related reactions in the gas and aqueous phase.

The most important bromine and chlorine reaction cycles are depicted in Figure 3.

### 4.02.3 TROPOSPHERIC OZONE DEPLETION AT POLAR SUNRISE

In the early 1980s, it was discovered that during winter a significant part of the Arctic troposphere becomes highly polluted, largely because of the accumulation of Eurasian pollutants emitted north of the polar front (e.g., Barrie, 1986; Lowenthal and Rahn, 1985; Barrie *et al.*, 1989), creating what became generally known as the "Arctic haze." The haze largely consists of light scattering, submicron sulfate, and soot-containing particles (Herbert, 1989; Bridgman, 1989; Clarke, 1989) which are inefficiently removed from the atmosphere due to low rates of precipitation and a meteorologically very stable boundary layer. Because of its potential climatic impact (Valero *et al.*, 1989), research efforts intensified in the Arctic in the 1980s (Schnell, 1983). A special issue of the *Journal of Atmospheric Chemistry* **9**(1–3), 1989 contains in detail the information that was obtained during the Arctic Gas

and Aerosol Sampling Program (AGASP) in March–April of 1986 over the Canadian and Alaskan Arctic.

Air pollution is often associated with high ozone concentrations and indeed an upward trend in surface ozone by $0.76 \pm 0.60\%$ per year was observed for the period 1973–1984 (Oltmans and Komhyr, 1986). However, remarkably low surface ozone concentrations had also been measured at Barrow, Alaska (71° N, 157° W; Oltmans, 1981; Oltmans and Komhyr, 1986) and at Alert, northern Canada (82.5° N, 62.3° W; Bottenheim *et al.*, 1986) in late winter/early spring, with ozone levels on occasion dropping to below 10 nmol mol$^{-1}$, even reaching zero values (Oltmans, 1981; Barrie *et al.*, 1988). It was soon discovered that the depletion in ozone was anticorrelated with the presence of enhanced amounts of "filterable" bromine (f-Br) compounds that were collected on cellulose filters, most likely HBr (Barrie *et al.*, 1988). To account for the ozone loss in the presence of the high concentrations of bromine, Barrie *et al.* (1988) proposed the bromine radical-catalyzed, BrO–BrO ozone destruction cycle I (see Section 4.02.2).

Barrie *et al.* (1988) hypothesized that the source of the f-Br was the photolysis of bromoform

(CHBr$_3$), which had been observed in the Arctic marine troposphere, maximizing during the winter (Cicerone *et al.*, 1988; Rasmussen and Khalil, 1984). Although the spectrum which they used for the photolysis of bromoform later turned out to be incorrect (Moortgat *et al.*, 1993), leading to an overestimation of active bromine (Br + BrO + Br$_2$ + HOBr + BrCl), further research has clearly shown that complete near-surface ozone loss is indeed largely due to BrO–BrO catalysis and that it is a common feature in the Arctic during early springtime. This remarkable phenomenon has been the motivation for several field campaigns in the Canadian and European Arctic, which have been described in a great number of papers, including special issues of the *Journal of Geophysical Research* **99**(D12) (Barrie *et al.*, 1994), *Tellus* **49B** (Barrie and Platt, 1997), *Journal of Atmospheric Chemistry* **34**(1) (Platt and Moortgat, 1999), and *Atmospheric Environment* **36**(15–16) (Bottenheim *et al.*, 2002).

In the following sections we will discuss the main characteristics of the polar surface ozone depletion events (ODEs).

### 4.02.3.1 Main Features of Polar ODEs

The Arctic (and Antarctic) lower troposphere is highly stratified in the vertical. Since exchange with mid-latitude air masses is also limited, a chemical-reactor-like situation is created. When sunshine returns to the high-latitude regions after the long polar winter, some exceptional and intensive chemistry takes place in this "reactor." Because photochemically active solar radiation is limited under the low-sun conditions, gas phase reactions alone would be of relatively little significance. However, chemical reactions taking place in or on airborne sea salt, the snow pack, or more likely on the Arctic ocean sea ice, produce very reactive, photolabile halogen compounds, especially BrO radicals, which rapidly destroy ozone. This is similar to stratosphere ozone hole chemistry, where mostly chlorine catalysts are involved (see, e.g., Solomon, 1990).

The catalytic ozone destruction chain, with BrO + BrO → 2Br + O$_2$ as the key reaction, was shown to be correct by direct BrO radical measurements, using the long-path differential optical absorption spectrometer (DOAS) technique at Alert (Hausmann and Platt, 1994) and at Spitsbergen (78.9° N, 11.8° E; Tuckermann *et al.*, 1997). At alert, Hausmann and Platt (1994) measured up to 17 pmol mol$^{-1}$ of BrO (see Figure 4). On two occasions, when no ozone could be detected, f-Br reached concentrations as high as 60 ng m$^{-3}$ and 140 ng m$^{-3}$, or 23 pmol mol$^{-1}$ and 53 pmol mol$^{-1}$. This was most likely HBr, formed by reactions of bromine atoms with aldehydes and acetylene, which become most efficient after all ozone is destroyed and no BrO can be formed. Prior to this, BrO radical levels in the above f-Br ranges may well have been present in the boundary layer, causing very rapid ozone depletion.

In ODEs, further O$_3$ destruction can be limited by low availability of HO$_2$, which is participating in O$_3$ destruction and the activation and recycling of bromine (cycles I and IV–VI). The most important source for HO$_2$ radicals is the photolysis of ozone by solar UV radiation, producing O($^1$D) atoms which react with water vapor to produce two OH radicals. The subsequent reaction of OH with O$_3$ produces HO$_2$. If the penetration of photochemically active solar UV radiation, which is needed for the photolysis of ozone and production of HO$_x$, would indeed be a critical factor in BrO$_x$ activation, one may wonder whether the large springtime stratospheric ozone losses during the past decades may have had an influence on the development of the low-altitude springtime ozone depletion events, especially near the Antarctic continent. As of early 2000s this issue remains to be addressed by the modeling community.

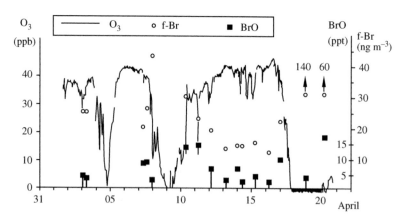

**Figure 4**  O$_3$, BrO, and f-Br as measured at Alert in April 1992 (source Hausmann and Platt, 1994).

Because both water vapor concentrations and solar UV radiation penetrating into the troposphere are small under the cold temperature and low-sun conditions prevailing in late winter/early springtime in the polar regions, the rate of production of OH radicals may be a rate limiting step in the $BrO_x$ autocatalytic reaction cycles. One possible source for $HO_x$ in this situation could also be the photolysis of HONO which is produced within the snowpack by photolysis of $NO_3^-$ as shown by Zhou *et al.* (2001) and Simpson *et al.* (2002). Several authors measured the production of $NO_x$ in snowpack most likely by photolysis of $NO_3^-$ (e.g., Honrath *et al.*, 1999, 2000a,b; Jones *et al.*, 2000, 2001). Furthermore, the photolysis of $CH_2O$, which is also produced in the snowpack (e.g., Sumner and Shepson, 1999; Jacobi *et al.*, 2002), might also supply $HO_x$, especially under the elevated $CH_2O$ levels that were found by, e.g., de Serves (1994). The most important processes occurring in and on top of snowpack are reviewed by Dominé and Shepson (2002). Several groups found that $CH_3Br$ is produced in firn air in the Arctic but not in the Antarctic (e.g., Butler *et al.*, 1999; Sturges *et al.*, 2001; Swanson *et al.*, 2002). This might have consequences for the interpretation of firn and ice core data.

Following the early discoveries of low levels of ozone in the Canadian Arctic during springtime (Oltmans, 1981; Oltmans and Komhyr, 1986; Bottenheim *et al.*, 1986; Barrie *et al.*, 1988; Mickle *et al.*, 1989), ozone depletion episodes have been reported at many other sites in the Arctic, such as the Norwegian Sea (Solberg *et al.*, 1996), Spitsbergen (Tuckermann *et al.*, 1997), and Greenland (Rasmussen *et al.*, 1997; Miller *et al.*, 1997). Ozone depletion events have also been reported off the rim of Antarctica (Wessel, 1996; Frieß, 1997; Krecher *et al.*, 1997; Wagner and Platt, 1998; Wessel *et al.*, 1998; Roscoe *et al.*, 2001). Therefore, as most impressively shown by satellite measurements of vertical column BrO (Wagner *et al.*, 2001; Richter *et al.*, 1998, 2002, see also Section 4.02.3.2), accompanying polar boundary layer springtime ozone depletion events are widespread. They are more pronounced in the lower parts of the boundary layer (Anlauf *et al.*, 1994; Rasmussen *et al.*, 1997).

The vertical extent of BrO layers was investigated by Anlauf *et al.* (1994) at Alert during the Polar Sunrise Experiment (PSE) 1992 with the help of a tethered balloon. They found large vertical extents of the ozone depletions and a strong link between $O_3$ and potential temperature. Anlauf *et al.* (1994) report that ozone depletions at the Alert base camp were associated with thermally neutral boundary layers capped by elevated inversions at several-hundred-meter altitude (as shown by the potential temperature) which was also the upper limit of the ODE. More

recent measurements of the vertical extent of ODEs were made by Strong *et al.* (2002) at Alert. During a nine-day period with basically no surface $O_3$ they found that $O_3$ was depleted up to 1,500 m altitude. It recovered by down-mixing of $O_3$-rich air from above. The rapid onset of the ODE can only be explained by advection of an airmass in which $O_3$ depletion had already occurred. Their analysis of the meteorological circumstances of this event shows that ODEs are strongly influenced by complex interactions between atmospheric transport and chemistry. During the same ODE, Hönninger and Platt (2002) used a passive multi-angle differential optical absorption spectrometer (MAX-DOAS) and concluded that the bulk of the BrO is present in a layer of $1 \pm 0.5$ km depth. If BrO was well mixed within this layer, it would contain $20-30$ pmol mol$^{-1}$ BrO.

During the airborne experiment TOPSE in the Canadian Arctic, very widespread regions with low ozone were found as far south as the Hudson Bay, corresponding to the large areas with elevated BrO that have been found with the GOME satellite (see below). Ozone depletion events with more than 2 km vertical extension were encountered (Ridley *et al.*, 2003).

Meteorological analyses show that ODEs only occurred, when air masses have been in contact with the Arctic Ocean surface (Worthy *et al.*, 1994; Hopper *et al.*, 1994). Hopper *et al.* (1994) report that ODEs occurred during 40% of the time at polar sunrise at their camp site on the ice, 150 km north of Ellesmere island. During a long period of 18 days in April 1992, no ozone could be detected at all at this site. Hopper *et al.* (1994) concluded that ozone destruction takes place over the Arctic Ocean and that the presence of ozone-poor air over the continental stations, such as Alert, is not due to local chemistry, but caused by transport from regions of the Arctic Ocean where earlier bromine activation and related chemical ozone loss had occurred. At Spitsbergen, low ozone mixing ratios, below 10 nmol mol$^{-1}$, occurred most frequently when the air arrived from north to west, i.e., from the Arctic Ocean (Solberg *et al.*, 1996).

During periods when sizable ozone concentrations were nevertheless measured in air over the Arctic Ocean, meteorological analyses showed that turbulence had brought down ozone from the free troposphere. Ozone concentrations were highly variable as well (Hopper *et al.*, 1994). During such periods, transfer of active bromine from the boundary layer to the free troposphere can take place, as observed on the NASA ER-2 high-altitude research aircraft by spectroscopic analysis of backscattered solar UV radiation by McElroy *et al.* (1999). This could conceivably be accompanied by ozone destruction. Very high column BrO densities, as large as

$(7-15) \times 10^{14}$ molecules cm$^{-2}$, were measured on April 26, 1997. On two other flights, May 2 and May 6, however, 3 times smaller, but still high, column densities were measured. A column density of $2 \times 10^{14}$ molecules cm$^{-2}$ corresponds to ~75 pmol mol$^{-1}$ if all BrO were located in the lowest kilometer of the atmosphere, which would cause very rapid ozone depletion. If distributed between 1 km and 5 km altitude, a volume mixing ratio of ~30 pmol mol$^{-1}$ would result, still causing substantial ozone depletion. Transfer of BrO from the Arctic boundary layer to the free troposphere would occur through convection created by evaporation of $H_2O$ in open leads in the Arctic ice (McElroy *et al.*, 1999). Bromine activation may take place on sea-salt-loaded ice surfaces near the open leads, as well as on ice crystals in the atmosphere (McElroy *et al.*, 1999). Sulfate particles may also provide surfaces for such reactions (Wennberg, 1999). With such high atmospheric mixing ratios of BrO, as measured by McElroy *et al.* (1999), substantial ozone destruction will take place also in the free troposphere.

An interesting issue concerns the existence of interannual variability in bromine activation and ODEs. At Spitsbergen, Tuckermann *et al.* (1997) measured up to 30 pmol mol$^{-1}$ of BrO during the spring of 1995, however, much less in 1996 when there was also much less ozone destruction. The reason is probably less transport to the site of air masses that had been in contact with the Arctic Ocean in 1996 than in 1995.

Tarasick and Bottenheim (2002) investigated ozonesonde records from the 1990s to early 2000s and could identify ODEs at several stations in the Arctic. Differences in occurrence seem to be related to differences in average springtime surface temperatures. For Resolute they deduced an increase in the frequency of ODEs over the period 1966–2000 of $0.66 \pm 0.59\%$ per year.

### 4.02.3.2 Satellite Observations

High values in the vertical column densities (VCDs) of BrO in the troposphere have been measured over extended regions around the globe with a DOAS-type instrument (Global Ozone Monitoring Experiment (GOME)), on board of the European Research Satellite (ERS-2) (Richter *et al.*, 1998, 2002; Wagner and Platt, 1998; Wagner *et al.*, 2001; Hegels *et al.*, 1998; Chance, 1998). In the southern hemisphere, after polar sunrise, enhanced VCDs of BrO were located in large patches between 77° S and 59° S, at the edge and equatorwards of the Antarctic continent over the ocean. No major ODEs could be detected over the Antarctic continent proper. In the Arctic, alternatively, highest VCDs of BrO were observed during springtime (February–May) from the North Pole down to 70° N with a patch reaching 60° N in the Hudson Bay region. The events of enhanced tropospheric BrO columns move northwards from winter towards summer. As an important clue to explain these features, Wagner *et al.* (2001) proposed that frozen salt water pools, located on top of the sea ice, are favored locations for BrO formation. A similar proposal had already been made by Tang and McConnell (1996) and McConnell *et al.* (1992). Another region with elevated VCDs of BrO is the frozen part of the Caspian Sea (Wagner *et al.*, 2001). Richter *et al.* (2002) point out that, contrary to the Arctic, in the Antarctic BrO maxima were not restricted to spring but were also present in summer. Nevertheless, the majority of these events occur during spring. It is not yet clear what causes the elevated BrO in summer. GOME data also suggest that BrO is absent over ice that is older than one year (Lindberg *et al.*, 2002).

Figure 5 shows monthly averages of BrO vertical columns in winter/spring for both hemispheres. The very widespread presence of enhanced BrO also over large land distances (e.g., Siberia) is impressive.

### 4.02.3.3 Sources of Active Bromine

Within a few years from the discovery of the ODEs, the BrO + BrO catalytic reaction cycle was confirmed as the main sink for ozone. The main process leading to the release of reactive bromine to the atmosphere was more difficult to identify. It now appears that the main source for the reactive gaseous bromine compounds are chemical reactions in the snowpack and on top of fresh (one-year-old) sea ice.

Upon freezing, the halides get concentrated on the surface of the sea ice, which remains a highly concentrated brine even down to temperatures of 230 K (Koop *et al.*, 2000). Based on laboratory simulations, halides are concentrated in this surface brine layer by a factor of 11 for Cl$^-$ and even 38 for Br$^-$ compared to seawater (Koop *et al.*, 2000). Often so-called frost flowers form on fresh sea ice, which have very large surface areas and likewise consist of highly concentrated brine (e.g., Martin *et al.*, 1995). Further surface segregation effects will be described in Section 4.02.4.5. The importance of fresh sea ice in acting as a halogen source is supported by previously cited GOME observations which indicate that ODEs often occur over fresh (one-year-old) sea ice.

The depletion of halides from the brine layer on top of the sea ice by emission to the atmosphere and the breakup of the sea ice in late spring might be one explanation why ODEs are mainly

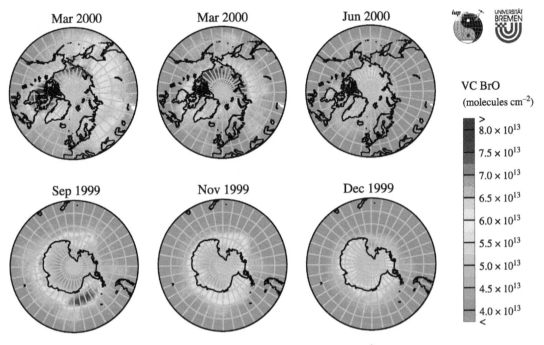

**Figure 5** Monthly averages of BrO vertical columns (in molecules $cm^{-2}$) in winter/spring for the northern hemisphere (upper row) and the southern hemisphere (lower row) (courtesy Andreas Richter).

observed during spring and not during summer. An old observation by "Trapper Steffanson" (1825, cited in Leser (1999)) refers to the loss of salts from the surface of sea ice:

> When sea ice forms it is salty ... in June and July when rains begin and snow melts ... the saltiness disappears ... and the following year this ice is the potential source of the purest possible cooking or drinking water.

An additional physical explanation for salt depletion from the surface of sea ice is that upon freezing of sea ice most of the salt is trapped in the form of concentrated brine droplets that slowly "melt" their way through the ice due to a lowered freezing point and greater density than the surrounding ice (see Open University Course Team, 1999).

Probably less important as sources of reactive bromine gases are reactions of biogenic organo-bromine compounds that have been found in the marine atmosphere by various research groups (Dyrssen and Fogelquist, 1981; Rasmussen and Khalil, 1984; Class and Ballschmiter, 1986; Cicerone *et al.*, 1988; Wever *et al.*, 1991a; Reifenhäuser and Heumann, 1992a; Sturges *et al.*, 1992; Schall and Heumann, 1993; Schall *et al.*, 1994). They are all present in the $<10$ pmol $mol^{-1}$ range, on their own too small to produce BrO concentrations high enough to affect ozone and to explain those measured. Wever (1988) also found emissions of HOBr and $Br_2$ in the water by seaweeds, probably serving as

biocidal defense agents. As is the case with the halocarbon emissions, these too maximize during winter and spring.

Another source for active bromine could be reaction between $N_2O_5$ and sea salt NaBr yielding $BrNO_2$ and $NaNO_3$, followed by rapid photolysis of $BrNO_2$ to produce bromine, as proposed by Finlayson-Pitts *et al.* (1990) and Behnke *et al.* (1994). However, competing reactions of $N_2O_5$ involving much more abundant NaCl and $H_2O$ (Behnke *et al.*, 1992, 1997) and observations showing low $NO_x$ concentrations (Beine *et al.*, 1997) rule out this proposal as a significant direct source of active bromine for the Arctic (and even more so for the Antarctic).

Another possible source would be sea salt aerosol. In general, it seems, however, that the amount of $Br^-$ in sea salt aerosol is too small to account for the measured BrO concentrations. An additional fact to consider is that the aerosol composition measurements of Lehrer *et al.* (1997) showed that during ODEs the aerosol was highly enriched in $Br^-$ compared to seawater, implying the presence of non-sea-salt $Br^-$ in the aerosol. A similar observation was made by Ariya *et al.* (1999) who took surface snow samples on an ice floe on the Arctic Ocean during the April 1994 PSE. They postulated that the observed enrichment of $Br^-$ relative to $Na^+$ by a factor of 75–130 compared to seawater was due to recent scavenging of bromine from the atmosphere. Similarly, Toom-Sauntry and Barrie (2002) found that during Arctic spring uptake of gaseous bromine by snow flakes is important for the removal of bromine from the atmosphere. As of early 2000s, the issue of the sources and sinks of active bromine is, therefore, not quite settled.

When considering the sources for bromine, one has to keep in mind the extent of the BrO clouds as measured by GOME (see Figure 5). Whereas in the Antarctic mainly the coasts show elevated BrO, in the Arctic BrO is found far inland over Canada and Siberia. This could be caused by advection of bromine activated elsewhere or by the combined release of bromine from sea ice and sea salt deposits or other, so far unidentified, processes.

An additional fact that might contribute to the predominance of spring ODEs are differences between spring and summer in photolysis rates caused by different solar elevation. In spring the shortwave radiative spectrum at the ground is shifted to longer wavelengths leading to differences in the photochemistry as explained later in Section 4.02.4.3. Also, as already mentioned, a neutral-to-stable boundary layer is important for the occurrence of ODEs. During spring the occurrence of favorable meteorological situations is more likely than during summer.

### 4.02.3.4 Chlorine Chemistry in ODEs and Br–Cl Interactions

There has been considerable debate as to what degree Cl, like Br, radicals might play a role in Arctic springtime tropospheric chemistry. To obtain information on this, Jobson et al. (1994) collected daily air samples at Alert (82.5° N, 62.3° W) from January 21 to April 19, and on an ice floe 150 km north of Alert during April 2–15, 1992. They derived information on the concentrations of OH, Cl, and Br from the different decay rates of a suite of non-methane hydrocarbons: the so-called "hydrocarbon clock" method. Besides some removal of alkanes by reaction with OH during ODEs, additional alkane losses, consistent with removal by reaction with Cl, were measured.

Furthermore, acetylene concentrations decayed faster than expected from removal by OH and Cl, indicating the presence of bromine atoms with mixing ratios of $\sim$1–2 pmol mol$^{-1}$. The corresponding BrO mixing ratios in the range of several tens of pmol mol$^{-1}$ should cause loss of ozone by $\sim$90% per day by the BrO–BrO reaction cycle. Estimated concentrations of chlorine atoms were $\sim$1,000 times smaller than those of bromine (Jobson et al., 1994). There are some tentative indications of substantial concentrations of ClO radicals, similar to those of BrO, $21 \pm 12$ pmol mol$^{-1}$, at a detection limit of 20 pmol mol$^{-1}$ per individual measurement,

during the spring of 1995. If this were indeed the case, then the ozone destruction cycle via the BrO + ClO reaction cycle III would gain similar importance as the BrO + BrO cycle II.

Considering the relatively large uncertainty in the measurements, this finding of appreciable ClO concentrations is rather speculative. It would imply chlorine atom concentrations (in 1995) 3–7 times larger than the concentrations derived by the "hydrocarbon clock" method (Ramacher et al., 1997), which itself has substantial uncertainties. Röckmann et al. (1999) used the isotopic signature of CO to estimate the integral amount of chlorine radicals that an airmass encountered. For the 1995 campaign the values were similar as those of Ramacher et al. (1997), but for 1996 they exceeded the measurements of Ramacher (1997) by $\sim$30%. Measurements and modeling of tri-chloro- and tetrachloroethane in the Arctic troposphere by Ariya et al. (1997) required the presence of both chlorine and bromine atoms to explain their decline, in agreement with Ramacher et al. (1997).

More recent "hydrocarbon clock" measurements by Ramacher et al. (1999) in Spitsbergen during the ARCTOC 1996 campaign provided new information: at chemically undisturbed background ozone mixing ratios ($>$30 nmol mol$^{-1}$), insignificant chlorine and bromine atom concentrations were measured. However, during ODEs, the time-integrated concentration of chlorine was 1,000 times smaller than that of bromine. This is very similar to earlier findings of Jobson et al. (1994), showing that ozone loss by ClO–BrO catalysis is much smaller than by the BrO–BrO cycle. However, the chemical loss of alkanes is mainly due to reaction with chlorine atoms. Using a novel chemical amplification technique, Perner et al. (1999) and Martinez et al. (1999) derived ClO concentrations below 2 pmol mol$^{-1}$ at Spitsbergen. Total bromine captured on charcoal traps and analyzed by the neutron activation technique gave values of 90 pmol mol$^{-1}$ during a zero surface ozone event. On one occasion, 75 pmol mol$^{-1}$ of bromine was measured together with 8 pmol mol$^{-1}$ of iodine, indicating simultaneous active bromine and iodine production which may involve key reactions of the type (X = Br, I, Cl; Y = Br, I, Cl).

$$HOX + Y^- + H^+ \rightarrow XY + H_2O \quad (52)$$

and various autocatalytic and halogen cross-catalytic reaction chains (Vogt et al., 1996) as discussed in Section 4.02.2.

The situation regarding the importance of chlorine in polar springtime surface ozone chemistry remains rather confusing. Contrary to most information, newer measurements of BrCl volume mixing ratios by up to 35 pmol mol$^{-1}$, comparable to those of Br$_2$ ($\approx$27 pmol mol$^{-1}$), obtained with an atmospheric pressure chemical ionization-mass spectrometer at Alert, Canada during the period February 9 to March 13, 2001, are in favor of a significant role of chlorine in Arctic halogen chemistry (Foster *et al.*, 2001; Spicer *et al.*, 2002). The measured BrCl and $Br_2$ concentrations were anticorrelated with those of ozone, maximizing during ozone depletion events in air masses which came from the northeast crossing the Arctic Ocean, north of Greenland. Because BrCl and $Br_2$ were also measured during complete darkness, the authors proposed a mechanism for BrCl and $Br_2$ formation which does not require solar radiation, but Spicer *et al.* (2002) also state that they cannot rule out a role for radiation for the production of these species during the twilight period. They never measured $Cl_2$ above their detection limit of 2 pmol $mol^{-1}$, however. The set of reactions proposed by Foster *et al.* (2001) to explain their measurements are

$$O_3 + Br^- + H^+ \rightarrow HOBr + O_2 \quad (43')$$

(Hirokawa *et al.*, 1998; Mochida *et al.*, 2000; Oum *et al.*, 1998b) in sea salt containing aerosol and on snowpack/ice surfaces (Ariya *et al.*, 1999), followed by $HOBr + Cl^- + H^+ \rightarrow BrCl + H_2O$ (namely, (31)) (Abbatt, 1994; Abbatt and Nowak, 1997; Sander and Crutzen, 1996; Vogt *et al.*, 1996; Tang and McConnell, 1996; Kirchner *et al.*, 1997; Michalowski *et al.*, 2000; Adams *et al.*, 2002), or by $HOBr + Br^- + H^+ \rightarrow Br_2 + H_2O$ (namely, (24)) (Eigen and Kustin, 1962; Fan and Jacob, 1992; Fickert *et al.*, 1999), or also by halogen ion exchange reactions, such as $BrCl + Br^- \rightarrow Br_2Cl^- \rightarrow Br_2 + Cl^-$ (namely, (39) and (40)).

Note that these reactions are promoted by acidity (see also Langendörfer *et al.*, 1999). However, as we already mentioned earlier, a newer study found that uptake of HOBr on frozen seal salt leads to the release of dihalogens without the need for acidification (Adams *et al.*, 2002). If this process is significant on a larger scale, it would imply that ODEs can occur in pristine environments like the Hudson Bay area without enough gas phase acids to acidify the aerosol particles.

The model study of Sander *et al.* (1997) reproduced measured total inorganic bromine measurements by assuming a source strength of bromine of 17 pmol $mol^{-1} d^{-1}$. The high mixing ratios of $Br_2$ and BrCl observed by Foster *et al.* (2001) were not predicted, suggesting there is some missing understanding of the cycling of halogen species in the polar boundary layer, which might include the snowpack.

Foster *et al.* (2001) analyzed in more detail why $Br_2$ and BrCl are present in roughly equal concentrations (and why BrCl was much more abundant than $Cl_2$), despite the fact that chlorine is

~650 times more abundant than bromine in seawater. Several factors are important for this: a more than 3 times larger enrichment of bromine to chlorine in the ice pack compared to seawater (Koop *et al.*, 2000), stronger surface segregation of bromine (Jungwirth and Tobias, 2001; see also Section 4.02.4.5), and, probably most important, differences in the kinetics and thermodynamics between bromine and chlorine.

Michalowski *et al.* (2000) calculated that the source of bromine from $Br_2$ photolysis is about an order of magnitude larger than bromine release from BrCl after sunrise. Following BrO formation from the reaction of Br with $O_3$ and HOBr formation by $HO_2 + BrO \rightarrow HOBr + O_2$ (namely, (4)), further reactions of HOBr with $Cl^-$ and $Br^-$ would give rise to enhanced levels of $Br_2$ and BrCl. They also found that heterogeneous halogen chemistry within the snowpack was necessary for the $O_3$ depletion events to occur and that the mass transfer of HOBr to the snowpack is a rate-limiting step.

Impey *et al.* (1999) deployed a photoactive halogen detector, which can distinguish between photolyzable chlorine ($Cl_2$ and HOCl) and bromine ($Br_2$ and HOBr), at Alert from March 27 to April 12, 1997. They found substantial amounts of $Br_2$ in the absence of ozone and thus conclude that "$O_3$ is not a necessary requirement for production of $Br_2$" and suggest processes in the snowpack for this. Two measurements showed HOBr volume mixing ratios of as much as 260 pmol $mol^{-1}$, which may indicate an influence from the reaction $Br^- + O_3 + H^+ \rightarrow HOBr + O_2$. Impey *et al.* (1999) also measured $BrO_x$ with a radical amplifier and found that it exceeded the detection limit of 4 pmol $mol^{-1}$ most of the time, including two ODEs. The highest values were 15 pmol $mol^{-1}$ towards the end of an ODE. Occasionally they detected $BrO_x$ during very low UV radiation levels in the night.

Although tremendous progress elucidating the role of halogens in the chemistry during ODEs has been made since the early 1980s, contradictory findings abound and important questions remain open, which must be addressed by further research.

## 4.02.4 MARINE BOUNDARY LAYER

After the involvement of halogen species in $O_3$ destruction was shown for both the stratosphere and the Arctic boundary layer, another potentially very important region where active halogen chemistry can occur received attention: the marine boundary layer (MBL). The MBL is the lowest, 500–1,000 m deep part of the troposphere that is in direct contact with the sea surface. It is separated from the free troposphere by a temperature and humidity inversion and is generally well mixed. Approximately 70% of the Earth's surface is covered by oceans, making processes in the MBL potentially significant for the whole atmosphere.

In the MBL, halogens are very abundant in the form of sea salt aerosols which contain chloride and bromide. If there are ways to release reactive chlorine and bromine (i.e., other than acid displacement which mostly releases HCl) from the sea salt aerosol, then especially bromine could play a very important role in the chemistry of the MBL. Furthermore, organic iodine gases produced in the ocean by microorganisms (see Section 4.02.4.4) are emitted from the sea surface. These alkyl iodides are rapidly photolyzed, releasing iodine atoms, leading to very rapid reaction cycles that can also involve chlorine and bromine.

### 4.02.4.1 Sea Salt Aerosols

Sea salt aerosols are produced at the sea surface by the bursting of air bubbles (e.g., Smith, 1872; Woodcock *et al.*, 1953; Mason, 1954; Blanchard and Woodcock, 1957; Pruppacher and Klett, 1997; see Figure 6). These bubbles are formed by processes such as the breaking of waves, biological processes, melting of icebergs/ice floes, melting of snowflakes on the surface, or rocking of ice floes. In general, the first two processes are the most important ones for bubble generation. A very small amount of background bubbles is always present in the oceans. The areas where bubbles are dense enough to reflect light visibly are referred to as whitecaps. In these regions the number of bubbles is several orders of magnitude larger than in the background ocean. Whitecap coverage depends on wind speed and is often described by

$$W = au_{10}^b \qquad (53)$$

where $W$ is given in percent, the wind speed $u_{10}$ in 10 m altitude in m s$^{-1}$, $a$ is $\sim 10^{-4}$, and $b$ in the range 3.4–4 (e.g., Wu, 1988). This gives a whitecap coverage of $\sim 1\%$ for $u_{10} = 10$ m s$^{-1}$ which is roughly what is estimated globally (e.g., Spillane *et al.*, 1986; Erickson *et al.*, 1986b; Pruppacher and Klett, 1997). The nonlinearity of this function means that $W$ will be highly variable too.

Bubble bursting produces small droplets from the film of the air bubbles as well as large jet droplets (e.g., Mason, 1954; see Figure 6). Even larger spray droplets are produced by strong winds blowing over wind crests (e.g., Pruppacher and Klett, 1997). These very large particles, however, settle quickly back to the water surface. According to Mason (1954) the majority of sea salt particles with dry radii of $\sim 0.15$ $\mu$m (100–200 particles per bursting bubble) are produced by the rupture of the bubble film.

Eriksson (1959a) estimated an initial sea salt production of $\sim 1$ Pg a$^{-1}$ (1 Pg = 1 petagram = $10^{15}$ g), while Blanchard (1963) estimated an order of magnitude higher flux. In a later article Blanchard (1985) states that the correct value is within the range of these values. Based on this number, Duce (1969) proposed that the corresponding flux of 0.6 Pg Cl is a, by far, more important source for tropospheric HCl than emission by volcanoes. According to Jaenicke (1988) and Seinfeld and Pandis (1998), the global flux of sea salt is between 180 Tg a$^{-1}$ and $10^4$ Tg a$^{-1}$, with the latter showing a best estimate of 1,500 Tg a$^{-1}$. Brasseur *et al.* (1999) list a source strength of $10^3$–$10^4$ Tg a$^{-1}$ for particles with diameters $<25$ $\mu$m and 20–100 Tg a$^{-1}$ for particles with diameters $<1$ $\mu$m. Larger particles have very short atmospheric lifetimes of minutes to hours. Kritz and Rancher (1980) estimated that the size-averaged lifetime of sea salt particles during their cruise off

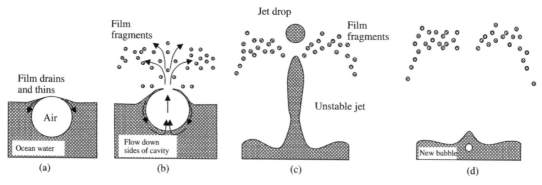

**Figure 6** Four stages in the production of sea salt aerosol by the bubble-burst mechanism. (a) A bubble rises to the ocean surface thereby forming a thin film at the interface which begins to thin. (b) Flow of water down the sides of the cavity further thins the film which eventually ruptures into many small sea spray particles. (c) An unstable jet, produced from water flowing down the sides of the cavity, releases a few large sea spray drops. (d) Tiny salt particles remain airborne as drops evaporate; a new bubble is formed. Note the scale change between (a) to (c) and (d) (after Pruppacher and Klett, 1997).

West Africa was three days, which they found to be similar to the mean residence time of gaseous inorganic chlorine of ~2.25 days. The lifetime of gaseous inorganic bromine, however, was ~7.25 days. Moyers and Duce (1972) estimated that the lifetime of gaseous bromine would be ~7 times that of particulate bromine. These numbers are of course dependent on transport and deposition which vary with time and location. The vertical profile of sea salt aerosol was examined by Woodcock (1953) and Blanchard and Woodcock (1980), who found an exponential decrease with height and a "salt inversion" (i.e., local maximum) near cloud base.

Many studies give a diagnostic expression for the total concentration of sea salt (in $\mu g\ m^{-3}$) (Woodcock, 1953; Exton *et al.*, 1986; Erickson *et al.*, 1986a; O'Dowd and Smith, 1993) as a function of surface wind speed. These formulas are usually in the form

$$c = de^{eu_{10}} \qquad (54)$$

where $c$ is the concentration of (usually dry) sea salt aerosol in $\mu g\ m^{-3}$, and $d$ and $e$ are empirical constants. Only Erickson *et al.* (1986a) use $u_{15}$. The various parametrizations (see, e.g., compilation in Gong *et al.* (1997)) differ by up to an order of magnitude.

Based on field and wind tunnel observations, other groups developed relationships between the wind speed at 10 m height and the flux of sea salt particles to get spectral aerosol source functions in particles per unit area, time, and unit size range (e.g., Monahan *et al.*, 1986; Smith *et al.*, 1993; Andreas, 1998; Smith and Harrison, 1998). As pointed out by Andreas (2002), this is a very difficult task and published spectral sea spray production functions differ by up to six orders of magnitude. Guelle *et al.* (2001) compared the source functions of Monahan *et al.* (1986), Andreas (1998), and Smith and Harrison (1998) within the framework of a global three-dimensional transport model with observations. They concluded that using the Monahan *et al.* (1986), dry particle radii below 4 $\mu m$ and the Smith and Harrison (1998) larger particle values would yield best results. A more recent quantitative investigation on breaking waves and their scales by Melville and Matusov (2002) is laying the foundation for a kinematic characterization of whitecaps that may lead to a better aerosol source function.

Recent work (O'Dowd *et al.*, 1997; Yoon and Brimblecombe, 2002; Mason, 2001, and references therein) suggests that the role of sea salt aerosols as cloud condensation nuclei for the unpolluted MBL has been underestimated in the past. Murphy *et al.* (1998) showed that almost all aerosols larger than 0.13 $\mu m$ dry diameter contained sea salt in the Southern Ocean. Mason (2001) states that the production of sea salt aerosol with radii at 80% relative humidity of less than 0.5 $\mu m$ might be larger than assumed in the past and that these particles, due to their large solubility, provide the majority of cloud condensation nuclei (CCN) in the unpolluted MBL. Combining satellite observations and model calculations, Rosenfeld *et al.* (2002) found that the precipitation suppression effects of small anthropogenic aerosol particles (indirect aerosol effects) are averted by large sea salt aerosols that enable large droplets to grow which efficiently scavenge small aerosols. This would imply that the indirect effects of aerosol on climate were significantly smaller than previously estimated.

It was usually assumed that large sea salt particles with lifetimes of hours could not get acidified (prerequisite for many halogen-release paths) in the atmosphere. Nevertheless, model calculations (Erickson *et al.*, 1999) and measurements in moderately polluted air masses at Bermuda and clean air masses at Hawaii (Keene *et al.*, 2002; Bill Keene, personal communication, 2002) showed that particles with ambient mean geometric diameters of up to 25 $\mu m$ can get acidified. These measurements also showed significant bromide deficits of 60–80% (Bill Keene, personal communication, 2002; see also discussion of bromide deficits in Section 4.02.4.3), so that these particles (given their large mass) could play a significant role for atmospheric chemistry—not only sea salt particles in the micrometer diameter range, as often assumed previously. The very large spume particles (radii up to 500 $\mu m$) affect heat, moisture, and momentum transfer across the sea–air interface (e.g., Andreas, 2002).

Sea salt aerosol initially consists mainly of seawater (see Table 1). Organic carbon is present in sea salt particles as well, typically enriched in smaller sea salt aerosols compared to bulk seawater carbon (e.g., Blanchard, 1964; Hoffman and Duce, 1977; Blanchard and Woodcock, 1980; Middlebrook *et al.*, 1998; Turekian *et al.*, 2003). This organic carbon originates from three

**Table 1** Ionic composition of seawater. The numbers are constant with time due to the long residence times of the ions in the oceans ~$10^8$ years for $Na^+$ and $Cl^-$.

| Ion | $Cl^-$ | $Na^+$ | $Mg^{2+}$ | $SO_4^{2-}$ | $K^+$ | $Ca^{2+}$ | $HCO_3^-$ | $Br^-$ | $I^-$ |
|---|---|---|---|---|---|---|---|---|---|
| Concentration (mmol $L^{-1}$) | 550 | 470 | 53 | 28 | 10 | 10 | 2 | 0.85 | $10^{-3}$ |

Source: Andrews *et al.* (1996), except for $Br^-$ and $I^-$, which are calculated after Jaenicke (1988).

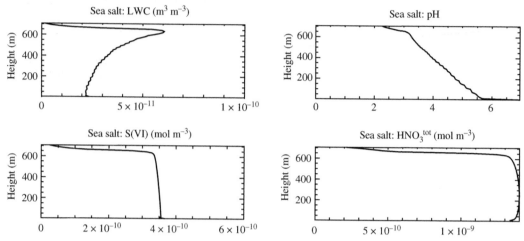

**Figure 7** Vertical profiles of the liquid water content (in $m_{aq}^3\, m_{air}^{-3}$), pH, S(VI) ($H_2SO_{4aq} + HSO_4^- + SO_4^{2-}$), and $HNO_3^{tot}$ ($HNO_{3aq} + NO_3^-$) in the sea salt aerosol (in $mol\, m_{air}^{-3}$) according to the one-dimensional MBL model Mistra-MPIC (source von Glasow and Sander, 2001).

major sources: (i) organic surfactants that got concentrated from bulk seawater on walls of gas bubbles that rise to the surface and scavenge material as they move through the water (e.g., Blanchard and Woodcock, 1980; Tseng *et al.*, 1992); (ii) the surface microlayer of the ocean (e.g., Zhou and Mopper, 1997; Hunter, 1997, and references therein), and (iii) condensation of organic gases on the airborne particles. Coagulation of chemically distinct aerosols (e.g., sea salt and predominantly organic aerosols via cloud processing) can also provide mixed aerosol. Organic compounds can, depending on their nature, both increase (e.g., Facchini *et al.*, 1999) or decrease condensation and evaporation rates of aqueous aerosols (e.g., Hansson *et al.*, 1998; Ming and Russell, 2001; Hegg *et al.*, 2002). However, it still needs to be determined how significant these processes are for cloud microphysics.

The pH of sea salt aerosol is an important property as many important aqueous phase reactions are pH dependent. For example, oxidation of S(IV) ($SO_2 + HSO_3^- + SO_3^{2-}$) by $O_3$ is only important for pH of more than ~6. Sea salt aerosol is buffered with $HCO_3^-$. Uptake of acids from the gas phase leads to acidification of the particles. According to the indirect sea salt aerosol pH determinations by Keene and Savoie (1998, 1999), the pH values for moderately polluted conditions at Bermuda were in the mid-3s to mid-4s. The equilibrium model calculations of Fridlind and Jacobson (2000) estimated marine aerosol pH values of 2–5 for remote conditions during ACE-1. Using a one-dimensional model of the MBL which includes gas phase and aqueous phase chemistry of sulfate and sea salt aerosol particles, von Glasow and Sander (2001) predicted that under the chosen initial conditions the pH of sea salt aerosol decreases from ~6 near

the sea surface with increasing height and correspondingly increasing relative humidity in the MBL to ~2 at the top of the MBL (see Figure 7).

A decrease of sea salt aerosol pH with increasing relative humidity was also described by Chameides and Stelson (1992) and Fridlind and Jacobson (2000) and is also present in the data of Keene and Savoie (1998, 1999). This counterintuitive result was further examined by von Glasow and Sander (2001) with an analytical solution of the reduced $HCl_{gas}-Cl^-$ system. They showed that two prerequisites are: (i) the high concentration of chloride in the particles and (ii) the fact that most undissociated HCl resides in the gas phase and thereby acts as reservoir of acidity for the sea salt aerosol. By increasing the relative humidity, the liquid water content of the particles increases and, therefore, the aqueous fraction of all compounds that are present in both gas and aqueous phase. Caused by the high $Cl^-$ content of the sea salt particles which does not change noticeably due to the uptake of HCl from the gas phase and the subsequent equilibration via $HCl \leftrightarrow Cl^- + H^+$, higher $H^+$ concentrations result in the aqueous phase and, therefore, a lower pH. This effect can only occur as long as the liquid water content of the particle is not too high (~$10^{-4}$–$10^{-2}$ g m$^{-3}$ depending on the salt loading of the air), otherwise dilution occurs and the $H^+$ concentration decreases. Keene *et al.* (2002) measured the pH in size-segregated aerosols under moderately polluted conditions at Bermuda in 1997 and found values for super-micrometer diameter particles in the upper 3s to upper 4s. Using gas phase HCl (see also Keene and Savoie, 1998; Erickson *et al.*, 1999) and the $HSO_4^- \leftrightarrow H^+ + SO_4^{2-}$ equilibrium, they also found evidence for the buffering of sea salt aerosol pH of different sizes to similar values.

It is also important to note that, in addition to other acids, HCl can play an important role in the acidification of fresh sea salt particles ("self-acidification"). HCl is released from previously acidified (i.e., older) sea salt particles by acid displacement or is produced in the gas phase (for this, release of chlorine by surface reactions as described in Section 4.02.4.5 might play a role). This might be important in clean regions with high wind speeds and, therefore, high sea salt alkalinity and low gas phase acidity. In such cases "self-acidification" of sea salt aerosol could start the acid-catalyzed reaction cycles IV–VI which would otherwise probably be rather inefficient.

Upon drying, sea salt particles remain in a metastable highly concentrated solution state below their deliquescence relative humidity of ~75%. Only when they reach their crystallization (or effluescence) point, which is ~45% relative humidity for NaCl, will they assume the crystalline form. This hysteresis effect is well documented by laboratory experiments (e.g., Shaw and Rood, 1990; Tang, 1997; Pruppacher and Klett, 1997; Lee and Hsu, 2000) and implies that, in the MBL, sea salt aerosol will usually be present in an aqueous form. Only in very dry marine regions and in the free troposphere, where the relative humidity is less than 45%, these particles can be expected to be dry. Even then a semiliquid layer can be present on the surface which makes surface reactions easier.

### 4.02.4.2 Reactive Chlorine

Field measurements of sea salt aerosol composition in the MBL have been made for a long time. Many measurements found chlorine deficits (e.g., Duce et al., 1963, 1965; Kritz and Rancher, 1980; Keene et al., 1990; Ayers et al., 1999; Johansen et al., 1999, 2000) which we cannot discuss here in detail. It is well known that the main reason for this chlorine deficit is the release of HCl from sea salt aerosol by acid displacement (Eriksson, 1959a,b):

$$H_2SO_{4,g} + 2NaCl_{aq} \rightarrow 2HCl_g + Na_2SO_{4,aq} \quad (55)$$

Later it was found that other acids—such as $HNO_3$, methanosulfonic acid (MSA), and oxalic acid (e.g., Kerminen et al., 1998)—also play a role in acid displacement.

Volpe et al. (1998) analyzed the isotopic composition of chlorine in marine aerosol and found enrichment of $^{37}Cl$, i.e., the heavier chlorine isotope, in chlorine depleted aerosol, as expected, thereby providing further evidence for acid displacement in mid-size sea salt aerosols (mean aerodynamic diameter up to 2 μm). In larger sea salt aerosol (mean aerodynamic diameter between

3.6 μm and 7.2 μm), however, the more depleted sea salt aerosol particles showed a decrease in $^{37}Cl$, pointing to a mechanism other than acid displacement for chlorine release in this size range.

Based on data from field campaigns in the Arabian Sea and tropical North Atlantic Ocean, Johansen et al. (1999, 2000) also suggest evidence for chlorine release processes other than acid displacement. This conclusion was made based on discrepancies in measured total anion and cation charges. Release of chlorine in the form of a reactive species may not be replaced by a detectable ion (e.g., $CO_3^{2-}$) and this charge difference might, therefore, be used as a measure for nonacid displacement release processes for chlorine.

Several techniques were used to detect reactive chlorine in the MBL. Keene et al. (1993) developed a tandem mist chamber to sample chlorine gases in the MBL. In the inlet both coarse and fine aerosol particles are removed. In the first (acidic mist chamber) the gases HCl, NOCl, ClNO₂, and ClNO₃ (=HCl*) are sampled (HCl goes into solution at a pH greater than 1), whereas in the second (alkaline chamber) Cl₂ and HOCl (=Cl*) are sampled.

This technique was employed at different locations and showed mixing ratios for HCl* of $<38-268$ pmol mol$^{-1}$ in coastal air near Miami, Florida, USA (Keene et al., 1993; Pszenny et al., 1993), 30 to ~1,500 pmol mol$^{-1}$ in Bermuda during spring 1996 (Keene and Savoie, 1998; Keene et al., submitted), and $39-263$ pmol mol$^{-1}$ HCl* in Hawaii in September 1999 under clean conditions (Pszenny et al., 2002). Measured values for Cl* were $<26$ pmol(Cl) mol$^{-1}$ and 254 pmol(Cl) mol$^{-1}$ near Miami, $<30-258$ pmol mol$^{-1}$ in Bermuda, and $<26-48$ pmol mol$^{-1}$ in Hawaii. Near Miami the presence of HOCl could be inferred from the presence of Cl* during daylight. It has to be added that these sometimes high mixing ratios could not be found during all periods, often the values were below detection limit. On Bermuda the detection of Cl* was generally restricted to situations with relative humidity greater than 75%.

The first direct detection of Cl₂ using atmospheric ionization tandem mass spectrometry in the MBL was reported by Spicer et al. (1998). On several consecutive days in June 1996 they detected up to 150 pmol mol$^{-1}$ Cl₂ at a coastal site in Long Island, New York, USA during nighttime. They used backtrajectories to show that the air stayed over the Atlantic Ocean the preceding 1.5 days. These high numbers cannot be reproduced or explained by known chemistry. One suggestion was reaction of O₃ with sea salt in the presence of ferric ions (Sadanaga et al., 2001). A prerequisite for this is the internal mixing of dust, sulfate, and sea salt particles which, according to Sadanaga *et al.* (2001), might explain the sporadic nature of these "high chlorine" events. Similar results were obtained by Behnke *et al.* (2000), who found increased release of chlorine in a smog chamber; however, they used externally mixed metal oxide and sea salt aerosol.

Individual hydrocarbons are removed from the atmosphere by reactions with OH and Cl, with rate coefficients that differ by orders of magnitude. Several groups used this fact to determine the concentration of the Cl radical indirectly in the MBL by measuring changes in hydrocarbon relative abundances, the already mentioned "hydrocarbon clock" method. For estimating Cl concentrations, tetra-chloroethylene ($C_2Cl_4$), which reacts ~100 times faster with Cl than with OH, is often used. Wingenter *et al.* (1996) estimated midnight 11:00 and noontime concentrations of $3.3 \times 10^4$ atoms cm$^{-3}$ and $6.5 \times 10^4$ atoms cm$^{-3}$, respectively, during the ASTEX/MAGE 1992 campaign for an aged continental air mass with high sea salt concentrations in the lower troposphere in the North Atlantic. Singh *et al.* (1996a) used measurements of five different chemicals to estimate early-morning chlorine concentrations for a case in the remote tropical MBL in the Pacific and inferred from these different chemicals a range of $1.8 \times 10^4$ atoms cm$^{-3}$ to $7.6 \times 10^5$ atoms cm$^{-3}$ with a mean of $(2.5 \pm 2.3) \times 10^5$ atoms cm$^{-3}$. Other published data show average chlorine atom concentrations (~$10^3$ atoms cm$^{-3}$) in the MBL (Singh *et al.*, 1996b; Rudolph *et al.*, 1996, 1997; Jobson *et al.*, 1998; Wingenter *et al.*, 1999). Measurements of the isotopic composition of $CH_4$–C by Allan *et al.* (2001b) implied a kinetic isotope fractionation that is inconsistent with OH being the only sink. They inferred summer maximum and yearly mean chlorine concentrations of $6.5 \times 10^3$ atoms cm$^{-3}$ and $3.6 \times 10^3$ atoms cm$^{-3}$, respectively. Chlorine atom concentrations on the order of several $10^4$ atoms cm$^{-3}$ can be reproduced with photochemical models of the MBL without assumption of any additional reaction mechanisms like surface reactions (e.g., von Glasow *et al.*, 2002a). In summary, it is still not clear if Cl radical chemistry plays a role globally or if it is limited to special conditions and/or to the early-morning concentrations. Relatively small global average values do not exclude the existence of regions of increased importance of Cl radical chemistry.

To assess the magnitude of fluxes of reactive chlorine from sea salt aerosol, Erickson *et al.* (1999) used a general circulation-model-based approach to calculate the release of HCl by acid displacement and formation of $ClNO_2$ by surface reaction of $N_2O_5$ from sea salt aerosol in different size classes under different ambient conditions. They found net HCl fluxes from

1 mg(Cl) m$^{-2}$ a$^{-1}$ to 300 mg(Cl) m$^{-2}$ a$^{-1}$ and net $ClNO_2$ fluxes of 1–8 mg(Cl) m$^{-2}$ a$^{-1}$. Strongest HCl release was found to occur in the polluted coastal regions of North America, whereas in the remote oceanic regions the available acidity was insufficient to titrate the sea salt alkalinity, only leading to minor acid displacement. Virtually all $HNO_3$ was scavenged by the sea salt. Seventy percent of the HCl release was predicted to occur in particles between 0.75 μm and 4 μm radius. Once the particle alkalinity has been titrated, the aerosol pH of all size bins equilibrate to the same pH values mainly via exchange of HCl through the gas phase.

### 4.02.4.3 Reactive Bromine

Many field measurements show not only a depletion of Cl$^-$ in aged sea salt but often even more so of Br$^-$, which is best seen by comparing the X$^-$/Na$^+$ ratio of sea salt aerosol with that of seawater. This was found by, e.g., Duce *et al.* (1965, 1983), Duce and Woodcock (1971), Moyers and Duce (1972), Martens *et al.* (1973), Kritz and Rancher (1980), Maenhaut *et al.* (1981), Raemdonck *et al.* (1986), Zhou *et al.* (1990), Ayers *et al.* (1999), Johansen *et al.* (1999, 2000), Keene *et al.* (submitted), Pszenny *et al.* (2002), so that in addition to chlorine (see also compilation by Sander *et al.* (2003)) bromine is released from sea salt aerosols to the gas phase. Early measurements reported Br$^-$/Cl$^-$ ratios to estimate the Br$^-$ loss, which can be misleading as both get released from the aerosol. Many (but not all) studies showed an enrichment of Br$^-$ in submicron (ambient radius) particles and up to complete depletion in larger particles. The degree of depletion tends to decrease with increasing radius. It appears that, on average, at least 50% of the bromide is lost in the sampled aerosols. The effective solubility (i.e., physical solubility and dissociation) for bromide is ~600 times greater than that for chloride (Brimblecombe and Clegg, 1989) so that no acid displacement with HBr occurs. Therefore, as will be discussed below, other mechanisms that involve photochemical processes are the reason for a release of bromine form the aerosol.

The enrichment of bromine in smaller particles was also noted by measurements over the Indian Ocean by Gabriel *et al.* (2002). They proposed that the reason for the accumulation of bromine in small particles close to the Indian coast could have been due to the continued use of leaded gasoline in India in 1999 which contains $C_2H_4Br_2$ as an additive whose oxidation products can end up in particles (see also Duce *et al.* (1965) and Martens *et al.* (1973) and discussion in Lightowlers and Cape (1988)). Nevertheless, other sources for enrichment of bromine in small aerosol particles are probably acting as well, as enrichments were often also found in remote regions. Even though $Br^-$ is relatively depleted a lot more than $Cl^-$, in absolute numbers a lot less Br than Cl is available in the MBL due to the low $Br^-/Cl^-$ ratio of seawater (see Table 1).

In general, however, one has to be careful in interpreting the Br/Na ratio in small particles. Most of the relevant measurements are made with multistage impactors yielding size-segregated aerosol data. This implies that all particles of a certain size with different compositions are lumped together. In general, many non-sea-salt (which do not contain sodium) and only a few sea salt aerosol particles are mixed. Even if the sea salt particles are completely debrominated and only small bromine concentrations are present in non-sea-salt particles (uptake from the gas phase), the mixture might have a higher Br : Na ratio than seawater.

Ayers *et al.* (1999) presented measurements from Cape Grim, Tasmania, that showed bromine deficits of 30–50% on an annual average and maximum monthly mean values of more than 80%. They also showed that $Br^-$ and $Cl^-$ deficits were linked to the availability of strong sulfur acidity in the aerosol, pointing to the importance of acid catalysis in the dehalogenation process as also shown by Fickert *et al.* (1999) in the laboratory (see Section 4.02.2). During the measurements of very high $Cl_2$ mixing ratios near Long Island (Spicer *et al.*, 1998), $Br_2$ was also measured at the detection limit of 6 pmol $mol^{-1}$ but only reported later (Foster *et al.*, 2001; Spicer *et al.*, 2002).

According to the measurements of Rancher and Kritz (1980), the total gas phase bromine concentrations are comparable to the particulate concentrations. In contrast, Moyers and Duce (1972) found 4–10 times higher concentrations of gas phase than particulate bromine. One explanation for the discrepancies might be the differences in dry and wet deposition rates. Rancher and Kritz (1980) could distinguish a diurnal cycle in particulate $Br^-$ with nighttime values that were about twice as high than daytime values.

Using a photochemical box model of the MBL, Sander and Crutzen (1996) showed that the reaction cycles that were proposed for the cycling of halogens on and within aerosols under conditions for the Arctic can also occur under polluted conditions at mid-latitudes. Vogt *et al.* (1996) proposed an autocatalytic mechanism for the release of bromine under clean conditions. They showed that $Br_2$ and BrCl can degass from the particles, photolyze in the gas phase, and destroy $O_3$ in an autocatalytic cycle. HOBr, which is produced during this cycle, can be taken up by the aerosol particles, thereby closing reaction cycles IV and V presented in Section 4.02.2. Predicted BrO mixing ratios were <1 pmol $mol^{-1}$ and total reactive gas phase bromine (dominated by HOBr) was ~10 pmol $mol^{-1}$. When sufficient $Br^-$ is available, BrCl can also take part in the aqueous phase autocatalytic cycle VI leading to the release of $Br_2$. In cycles IV and VI, bromine is released from the sea salt aerosol upon uptake of HOBr which makes these cycles autocatalytic. Note that, in these cycles, $HO_2$ radicals are also destroyed, leading to loss of $HO_x$ and a reduction of the $HO_2/OH$ ratio.

Although $O_3$ is lost in these gas–aqueous reaction cycles, the main halogen-related $O_3$ destruction in the remote MBL occurs via cycle I (see Section 4.02.2). The main difference between the catalytic reaction cycles that are proposed for the MBL and those for the polar sunrise, the ozone hole, and salt pans (see Section 4.02.6) is that self-reactions of the XO radicals—BrO for the Arctic BL and ClO for the ozone hole—are of minor importance in the remote MBL because of the expected relatively small mixing ratios of the XO radicals in the low pmol $mol^{-1}$ range. These reactions do, however, contribute to $O_3$ destruction in the MBL as does the reaction $XO + CH_3OO$, which leads to the production of $CH_2O$ (see also discussion of the impact of halogen chemistry on $CH_2O$ in Wagner *et al.* (2002)).

One should add that the gas phase mixing ratio of $O_3$ is important for these cycles, because the production of XO, and therefore the loss of $O_x$ (via $XO + HO_2$ and $XO + NO_2$), is promoted by $O_3$ (reaction (3)). If $O_3$ mixing ratios are small, halogen activation is slowed down as shown by Wagner *et al.* (2002) and von Glasow *et al.* (2002a).

As we discussed briefly in the previous section, many, especially the longer, alkanes react very rapidly with chlorine atoms, often with rate constants close to the collision-controlled limit (Atkinson *et al.*, 1997). Bromine, however, reacts orders of magnitude slower with most organics than Cl (including $CH_4$, which is the most important sink for chlorine).

Further aspects of the halogen reaction cycles that were discussed in Section 4.02.2 were the subject of numerical modeling studies. Using a box model based on Vogt *et al.* (1996), Dickerson *et al.* (1999) showed for scenarios for the Indian Ocean that $O_3$ diurnal variations in the MBL could partly be explained by bromine reaction cycles. Another process that was found to be important for diurnal cycles of $O_3$ in the MBL is the establishment of an $O_3$ gradient between the continent and the ocean through advection of continental air with elevated $O_3$ and photochemical destruction during the transport (de Laat and Lelieveld, 2000).

A trajectory box model for coastal sea salt aerosol transport over the land including halogen chemistry was developed by Moldanová and Ljungström (2001). They found that in polluted areas the importance of $N_2O_5$ uptake and subsequent release of chlorine is enhanced during winter. This leads to an increase in the importance of chlorine for volatile organic compounds (VOC) oxidation, which might be up to 9% of the total oxidation in wintertime. Using a similar reaction mechanism as Sander *et al.* (1999) they showed that, during winter, reaction of $BrNO_3$ with $Br^-$ yielding $Br_2$ becomes the major pathway for bromine release. The deposition of sea salt aerosols decreases relatively fast with distance from the coast (in accordance with measurements) while the deposition of HCl remains fairly constant, which hints at different lifetimes for particulate and gaseous halogens and/or continued production of HCl as the airmass is advected over land.

Using a box model, Toyota *et al.* (2001) explored the effects of treating the chemistry of sea salt aerosols in separate size bins compared to the bulk model approach, where sea salt aerosols of different sizes are combined. This is the approach used in all other published model studies, except for Erickson *et al.* (1999). Toyota *et al.* (2001) found that significantly less bromine was released in the size-segregated model. Taking the average of all particle sizes, ~45% of the $Br^-$ is depleted in the bulk, compared to 23% in the size-segregated model (numbers from their figure 1). They explain this discrepancy by differences in size-dependent $Br^-$ depletion which is nearly complete in the small particles (leading to a release of BrCl instead of $Br_2$ in accordance with the laboratory results of Fickert *et al.* (1999)). Bromine is thus only recycled in contrast to the additional release of bromine via $Br_2$, that occurs in the bulk approach. A second factor is the lifetime of particulate $Br^-$. For the bulk model they used a mass-weighted-average residence time, whereas in the second approach they used the size-resolved values. The calculated BrO mixing ratios are between $0.5\,\text{pmol mol}^{-1}$ and $2\,\text{pmol mol}^{-1}$ for the size-segregated approach and $2-10\,\text{pmol mol}^{-1}$ for the bulk approach with a very sharp maximum after sunrise (see also next under sunrise ozone destruction (SOD)). Toyota *et al.* (2001), however, do not include sulfate particles in their model calculations that have at least comparable uptake rates for all gas phase species as the sea salt aerosol and that can be important for the recycling of bromine species (e.g., HBr, see also discussion below) due to the low pH of sulfate aerosol particles.

The one-dimensional MBL model of von Glasow *et al.* (2002a) calculated a distinct diurnal variation of BrO for low $NO_x$ regimes, with peaks in the morning and in the evening of $2.5-4\,\text{pmol mol}^{-1}$ and noontime values of $\sim1.5\,\text{pmol mol}^{-1}$ (see Figure 8). This diurnal variation was found to be a consequence of different wavelength dependence of the main BrO production reaction and the main sink reaction. The former is the photolysis of $Br_2$ and BrCl followed by reaction of the Br radical with $O_3$ which is efficient at the longer wavelengths that dominate the atmospheric spectrum during dawn and dusk. The latter is the reaction of BrO with $HO_2$. $HO_2$ is a product of $O_3$ photolysis which occurs only at shorter wavelengths. This might help explain the so-called SOD in the diurnal variation of $O_3$ found by Nagao *et al.* (1999) in data taken over three years on an island in the subtropical western North-Pacific. This SOD could not be explained by $HO_x$ photochemistry alone. They speculated that bromine species might play a role in SOD. A similar feature was found by Galbally *et al.* (2000) in a time series of 13 years of $O_3$ observations at Cape Grim, Tasmania. In the model results of von Glasow *et al.* (2002a) the morning BrO peak is stronger than the evening one, and increased $O_3$ destruction is seen only in the morning.

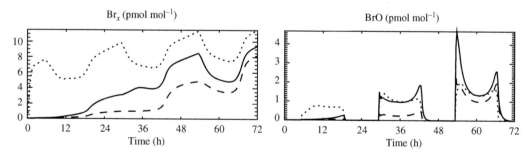

**Figure 8** Results from the one-dimensional model Mistra-MPIC showing the buildup of $Br_x$ (sum of all gas phase bromine species except HBr) and BrO with time. Plotted are runs with conditions for the remote MBL (solid line), a continentally influenced MBL (dotted line), and for reduced initial $O_3$ and $SO_2$ mixing ratios (dashed line), in 50 m height. The time is given in hours from modelstart. For details on the model runs, see von Glasow *et al.* (2002a).

The bromine activation reaction cycles that we explained in Section 4.02.2 require the initial presence of some HOBr or another active bromine compound to start the autocatalytic reaction cycles. As already discussed in Section 4.02.3.3, this is usually the case. Some aqueous phase processes transform bromide into bromine compounds that can degass (e.g., reaction of Br$^-$ with HSO$_5^-$, OH, NO$_3$, or O$_3$). The photochemical breakdown of organobromine compounds, photolysis of HOBr that originates from algal emission, and the reactions of N$_2$O$_5$ with sea salt form the gas phase sources. In theory only one bromine atom in the gas phase could suffice to start the bromine release as the activation cycle is autocatalytic but loss processes are also active, and obviously the "bromine explosion" can only happen when the bromine source from the activation cycles is larger than the sinks. Furthermore, reaction cycles IV–VI require the presence or production of sufficient O$_3$ and HO$_x$. Models of the MBL and the Arctic BL (Vogt *et al.*, 1996; Sander *et al.*, 1997; von Glasow *et al.*, 2002a) show that the bromine activation (=release from sea salt aerosol) takes about two days when no bromine molecules were in the gas phase at model initialization (see Figure 8). This is, of course, dependent on conditions. In polluted regions acidification of the particles is very quick, so that the acidification of sea salt aerosol (and the release of bromine to the gas phase) occurs almost instantaneously (see dotted line in Figure 8).

As a consequence of the previously discussed vertical profile of the sea salt aerosol pH (von Glasow and Sander, 2001), von Glasow *et al.* (2002a) also predicted for the cloud-free MBL that bromide depletion, which is acid catalyzed, is greater at higher altitudes and that the maxima of BrO are not found at ground level but at the levels with the lowest pH (i.e., near the top of the MBL). This might help explain the difficulty in detecting BrO near the surface by the ground-based DOAS technique (see below). Larger BrO concentrations higher up in the MBL could facilitate the transport of gas phase bromine from the MBL to the free troposphere. Although the main source for halogens in the MBL are sea salt aerosol particles, sulfate particles play an important role as well, as bromine recycling (i.e., the transformation of HOBr and HBr to the more easily photolyzable species Br$_2$ and BrCl) is about an order of magnitude faster on sulfate aerosol than on sea salt aerosol (von Glasow *et al.*, 2002a). This is due to the comparable surface areas and low pH values of sulfate aerosol. von Glasow *et al.* (2002a) further showed that during winter, with longer nights and a solar spectrum that is shifted to longer wavelengths due to the lower elevation of the Sun, the activation of bromine from the sea salt aerosol is more efficient than during summer for the same fluxes of gases and sea salt particles.

Shon and Kim (2002) compared data from two Lagrangian flights during ACE-1 with a constrained box model and found a better agreement regarding total O$_3$ loss and the timing (i.e., sunrise ozone destruction) when they included halogen chemistry. The effects were mainly caused by bromine oxide but also by iodine oxide.

von Glasow *et al.* (2002b) investigated the impact of halogen chemistry in a stratocumulus-covered MBL with the same model as von Glasow *et al.* (2002a) and found that the dynamics of stratocumulus clouds has an important impact on the vertical distribution of halogen compounds even below cloud base. During daytime the cloudy part of the MBL is dynamically decoupled from the subcloud layer which confines the active halogen chemistry to the subcloud layers. During nighttime the complete MBL is dynamically coupled leading to rapid uptake of Br$_2$ and BrCl into the very dilute cloud droplets, so that in contrast to the cloud-free MBL very low gas phase halogen mixing ratios are predicted during the night. The diurnal variation of gas and aqueous phase bromine is reversed compared to the cloud-free MBL, which might help resolve discrepancies in the diurnal variations between previous model studies and field experiments.

Direct measurements of reactive bromine are difficult. In the MBL several groups tried to measure BrO with the DOAS technique (Perner and Platt, 1979; Platt, 1999) at Mace Head/Ireland, Hawaii, Tenerife, Cape Grim/Tasmania, and on Crete using an artificial light source. When BrO was measured, it was close to the detection limit of 0.5–10 pmol mol$^{-1}$ (Hönninger (2002) and personal communication with: B. Allan, K. Hebestreit, D. Perner, J. Stutz). Therefore, these measurements only provide upper limits for BrO that are consistent with model results (see above and discussion in Hönninger (2002)). Leser *et al.* (2003) detected BrO for the first time in the MBL using a different experimental setup, where scattered sunlight is used as light source (MAX-DOAS; Hönninger and Platt, 2002; Hönninger, 2002). They measured the vertical column of BrO and calculated that based on an assumed MBL height of 1 km the largest measured individual BrO mixing ratio was 2.4 ± 2.1 pmol mol$^{-1}$ and the statistically averaged value was 0.7 ± 0.2 pmol mol$^{-1}$ at 13° W and between 30° N and 37° N in the North Atlantic Ocean. On the other days of the cruise, the measurements remained below the detection limit of 1–3.6 pmol mol$^{-1}$ for the individual measurements.

#### 4.02.4.4  Reactive Iodine

Like chloride and bromide, iodide is also present in seawater, although in very low concentrations. In sea salt aerosol, however, chlorine and especially bromine are usually depleted whereas iodine is strongly enriched. This was already discussed by Duce *et al.* (1963, 1965, 1967) who found an enrichment of iodine by ~500–1,000 times in rain compared to seawater, indicating a major additional iodine source. The measurements of Seto and Duce (1972) also showed large enrichments of iodine by a factor of 100–1,000 in the aerosol compared to seawater which was thought to partly originate from enhancement of iodine in the oceanic surface film but they also concluded that, at that time unidentified, gaseous precursors could be the cause. From measurements in Puerto Rico, Martens *et al.* (1973) also found I/Na enrichments varying from 100 in the largest to 14,000 in the smaller particles. Therefore, the aerosol appears to be a sink rather than a source for iodine in the MBL. See also the overview in Vogt (1999).

It is not clear, however, in which chemical state the particulate iodine is present. Some studies reported it to be $IO_3^-$ but in others no $IO_3^-$ was found (see references and discussion in McFiggans *et al.* (2000)). Based on measurements of aerosol composition, Baker *et al.* (2000) state that "iodine is present in aerosol in varying proportions as soluble inorganic iodine, soluble organic iodine and insoluble, or unextractable, iodine." Baker *et al.* (2001) measured the deposition of iodine in rainwater and in aerosol at Weybourne, North coast of Norfolk, UK. They found that iodide ($I^-$) constituted 5–100% of total iodine deposition in both rain and aerosol. The rest was found to be iodate ($IO_3^-$) with a very small contribution from $CH_3I$ (<3% of the total iodine deposition).

The main source of iodine in the MBL is emission of biogenic alkyl iodides like $CH_3I$, $C_3H_7I$, $CH_2ClI$, or $CH_2I_2$ by various types of macroalgae and phytoplankton that live in the upper ocean and in tidal areas along the coast. These alkyl iodides are probably produced by these organisms for pest control. Most early studies only considered $CH_3I$ as source for reactive iodine. In the gas phase the iodocarbons are photolyzed to produce iodine atoms that mainly react with $O_3$ to produce IO. The lifetimes for the most important organic iodine compounds for mid-latitudes are roughly: $CH_3I$ 5 days, $C_3H_7I$ 40 h, $CH_2ClI$ 4.5 h, $CH_2I_2$ 5 min. These short lifetimes make estimates of fluxes very difficult, especially for the potentially most important species $CH_2I_2$. For iodopropanes, reaction with OH is also an important loss process (Carl and Crowley, 2002) which in the end might also lead to the release of iodine atoms. Mixing ratios

compiled by Vogt (1999) are typically several pmol $mol^{-1}$ $CH_3I$ (but sometimes up to 43 pmol $mol^{-1}$), between 0.2 pmol $mol^{-1}$ and 2 pmol $mol^{-1}$ $C_3H_7I$, less than 1 pmol $mol^{-1}$ $CH_2ClI$ and 0.5 pmol $mol^{-1}$ $CH_2I_2$. One measurement (Pruvost *et al.*, 2000) during a phytoplankton bloom showed very high mixing ratios of $CH_2I_2$ of up to 57 pmol $mol^{-1}$ in the Bay of Brest, France. This would imply a huge source of iodine which would locally dominate the photochemistry.

Carpenter *et al.* (1999) detected $CH_2IBr$ for the first time in the marine environment at Mace Head, Ireland. They found evidence for common or linked marine sources for the polyhalomethane pairs $CHBr_3$–$CHBr_2Cl$ and $CHBr_3$–$CH_2IBr$ and the monoiodoalkane pair $CH_3I$–$C_2H_5I$. During low tide the concentrations of $CHBr_3$, $CH_2ICl$, and $CH_2IBr$ reached maxima. Comparing field data with results from a two-dimensional model, Carpenter *et al.* (2001) concluded that a contribution from nonlocal/offshore sources of $CH_2I_2$ was needed in order to explain the observed $CH_2I_2$ concentrations. They suggest that sea–air exchange of $CH_2I_2$ occurs in coastal waters but add that this does not necessarily imply an open ocean source of $CH_2I_2$. A strong correlation between the $CH_3I$ flux and sea surface temperature was found by Yokouchi *et al.* (2001) who also suggest that global $CH_3I$ fluxes are larger than previously assumed. Emissions of alkyl iodides are strongly dependent on biological activity, so highest fluxes are expected in coastal regions where kelp is exposed to the air during low tide (e.g., Mace Head, Ireland) or ocean regions with high biological productivity such as upwelling regions. Regions with high biological activity are also found in polar waters, especially around Antarctica. Other sources of iodocarbons are discussed in Section 4.02.8.4.

If the already mentioned measurements of Cauer (1939) of gas phase iodine concentrations that resulted from the industrial burning of kelp of 0.5 $\mu g$ $m^{-3}$ were indeed correct and representative, it may be interesting to find out what impact anthropogenic iodine pollution may have had on the chemistry of the atmosphere during the first quarter of the past century in Europe.

According to the measurements of Rancher and Kritz (1980) off the equatorial African coast, total gaseous inorganic iodine presents a strong diurnal variation which is not coincident with a similar (anticorrelated) variation in particulate iodine measurements. They concluded that during nighttime, gaseous iodine species might exist that could not be efficiently trapped on their LiOH impregnated filters.

One common conclusion of the modeling studies on iodine chemistry in the atmosphere (Zafiriou, 1974; Chameides and Davis, 1980; Jenkin *et al.*, 1985; Chatfield and Crutzen, 1990;

Jenkin, 1993; Solomon *et al.*, 1994; Davis *et al.*, 1996) was that a significant lack of kinetic data and missing information on iodine precursor gas fluxes from the ocean made solid assessments of the importance of iodine chemistry very difficult. Nevertheless, the large potential of iodine for the destruction of $O_3$ was already indicated in these studies. For example, Davis *et al.* (1996) estimated that under conditions typical for the tropical marine environment ~6% of total tropospheric column $O_3$ destruction was due to iodine chemistry and that in regions of high marine biological activity this could be as high as 30%. Solomon *et al.* (1994) and Sander *et al.* (1997) suggested a contribution of iodine chemistry during the ODEs in the Arctic.

More recently iodine in the MBL received considerable attention, mainly due to the detection of IO by DOAS measurements in the coastal MBL. Maximum mixing ratios measured by Alicke *et al.* (1999) were 6 pmol mol$^{-1}$ at Mace Head, Ireland. Allan *et al.* (2000) measured IO at Tenerife and Cape Grim, Tasmania, with a mean mixing ratio of ~1 pmol mol$^{-1}$. The lower mixing ratios are probably due to the fact that no large tidal areas with very high precursor emissions are present at these sites compared to Mace Head. These data may suggest a rather widespread presence of IO although it could not be measured on all days of the measurement campaigns. Hönninger (2002) reports that IO is likely present around the Kerguelen island in the south Indian Ocean at mixing ratios of ~10 pmol mol$^{-1}$. Figure 9 shows how low tide, solar irradiation, and high IO coincide. Low tides are associated with high mixing ratios of, e.g., $CH_2I_2$ and $CH_2IBr$ (not shown).

IO was also detected by spectral analysis of zenith scattered sunlight in polar regions. Wittrock *et al.* (2000) found evidence for the presence of IO both in the troposphere and stratosphere above Spitsbergen. Frieß *et al.* (2001) made measurements at Neumayer Station, Antarctica. Comparison of radiative transfer calculations assuming different vertical profiles of IO with the measurements suggests that the iodine oxide is located in the troposphere. Assuming that all IO is located in the boundary layer, mixing ratios of up to 10 pmol mol$^{-1}$ would have been present. The measurements have been performed continuously and showed higher IO mixing ratios in summer than in winter. Frieß *et al.* (2001) explained the higher summertime values as resulting from shorter distances between the measurement site and the open ocean in summer (due to differences in sea ice coverage), higher biological activity, and greater insolation. Figure 10 shows the main reaction pathways of iodine in the MBL. $O_3$ destruction via iodine is occurring via reaction cycle I (see Section 4.02.2).

OIO was also measured with the DOAS technique in the MBL at Mace Head (Hebestreit *et al.*, 2000) and Tasmania (Allan *et al.*, 2001a) at mixing ratios of $1-3$ pmol mol$^{-1}$. OIO was first detected in the laboratory by Himmelmann *et al.* (1996). It is formed in the reaction of IO with IO or with BrO (Gilles *et al.*, 1997; Bedjanian *et al.*, 1998; Misra and Marshall, 1998; Bloss *et al.*, 2001; Rowley *et al.*, 2001). Further reactions of OIO are highly uncertain. Ingham *et al.* (2000) studied the photolysis of OIO and found that OIO is photochemically very stable with respect to oxygen atom formation in agreement with calculations by Misra and Marshall (1998).

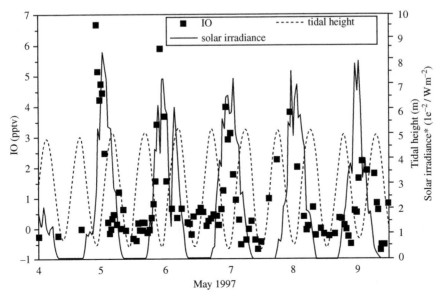

**Figure 9** IO mixing ratio, tidal height, and solar irradiance at Mace Head, Ireland during May 5–9, 1997 (source Carpenter *et al.*, 1999).

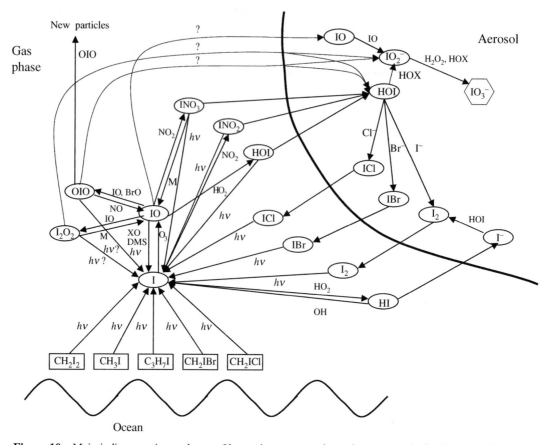

**Figure 10** Main iodine reaction pathways. Uncertain or assumed reactions are marked with a question mark.

Ashworth *et al.* (2002) confirmed that OIO does not photolyze to produce IO and O. They did find evidence, though, for the channel that leads to the production of I and $O_2$. They estimated a very short photolytic lifetime of a few seconds. Photolysis producing $I + O_2$ would enhance the $O_3$ depleting potential of iodine in the MBL.

The field campaigns were accompanied by further modeling studies. Vogt *et al.* (1999) developed a combined gas and aerosol box model using $CH_2I_2$, $CH_2ClI$, $CH_3I$, and $iC_3H_7I$ as precursor gases for iodine. Cycling of HOI, HI, and $I_2O_2$ in the aqueous phase leads to the release of ICl, IBr, and $I_2$ (see Figure 10). Their study suggested that almost all particulate iodine is in the form of $IO_3^-$ because cycling between other aqueous phase species is efficient and creates temporary reservoirs only. As already mentioned, measurements showed that particulate iodine is in the form of $I^-$, $IO_2^-$, or $IO_3^-$, so this points to some missing understanding in the aqueous phase chemistry of iodine. Vogt *et al.* (1999) found a strong impact of gas phase iodine chemistry on the $O_3$ concentration, with destruction rates of up to $1 \ nmol \ mol^{-1} \ d^{-1}$, which under their conditions was faster than $O_3$ destruction by the $O_3 + HO_2$ reaction. They also show strong chemical

coupling between reactive iodine, chlorine, and bromine compounds with the iodine compounds speeding up the release of these compounds from the sea salt aerosol as in the study of Vogt *et al.* (1996). They predicted IO mixing ratios $\sim 1 \ pmol \ mol^{-1}$.

Stutz *et al.* (1999) used a quasi-steady-state box model to explain the measurements of iodine oxides at Mace Head. Their model did not include the aerosol phase. They reproduced the measured IO mixing ratios of $\sim 6 \ pmol \ mol^{-1}$ and found that for these conditions, cycles involving the production of HOI by reaction of IO with $HO_2$ and subsequent photolysis of HOI would lead to an $O_3$ destruction rate of $\sim 12.5 \ nmol \ mol^{-1} \ d^{-1}$, whereas the IO self-reaction would lead to an $O_3$ destruction rate of $\sim 3.8 \ nmol \ mol^{-1} \ d^{-1}$.

McFiggans *et al.* (2000) used a box model that was constrained with data from measurements and showed the importance of the cycling of iodine species on aerosol particles for the Mace Head and Tenerife measurements. They used a modified version of the mechanism suggested by Vogt *et al.* (1999).

von Glasow *et al.* (2002a) employed the mechanisms of Vogt *et al.* (1999) in a one-dimensional model. Some potential reactions of

OIO and interhalogen equilibria between ICl and IBr in the aqueous phase were added to the model based on the laboratory observation of Holmes *et al.* (2001) who found a strong coupling between bromine, chlorine, and iodine chemistry. They tried unsuccessfully to simulate the observed diurnal variation of OIO and concluded that current knowledge of iodine chemistry, especially iodine oxide chemistry, is insufficient to reproduce the field experiments. They did not include new particle formation (see below) or photolysis as a possible sink for OIO in their simulations. According to their model, total inorganic gaseous iodine consists mainly of HOI, IO, and OIO during day and ICl and IBr during night. Maxima of ICl and IBr concentrations were calculated to be located below the inversion that caps the cloud-free MBL (pointing to the importance of acid-catalyzed cycling in the sea salt aerosol), whereas maxima of IO and OIO were predicted to be near the ocean surface, i.e., near the source of alkyl iodides.

During the events with high IO concentrations at Mace Head, Ireland, bursts of aerosol nucleation with number concentrations for particles with radii of 1.5–5 nm of more than $10^5$ part. cm$^{-3}$ were also observed (O'Dowd *et al.*, 1998, 1999; Grenfell *et al.*, 1999; see also special issue of *J. Geophys. Res.*, Sept. 2002), implying nucleation rates of $10^3$–$10^4$ part. cm$^{-3}$ s$^{-1}$. Two conditions appeared to be necessary for the new particle formation: low tide and solar irradiation. The formation of aerosol from iodine oxide radicals produced from the photolysis of CH$_3$I in the presence of O$_3$ was already found by Cox and Coker (1983). Cox *et al.* (1999) suggested that OIO might be involved in the formation of particulate iodine. Hoffmann *et al.* (2001) suggested that the multiple addition reactions of OIO are major steps leading to particle formation. They used CH$_2$I$_2$ as a precursor species in laboratory experiments and found that the newly formed particles were purely composed of iodine oxides.

In follow-up smog chamber studies (O'Dowd *et al.*, 2002; Jimenez *et al.*, 2002), the new particle formation was confirmed and again the three prerequisites—CH$_2$I$_2$ (down to 15 pmol mol$^{-1}$), O$_3$, and UV radiation for the photolysis of CH$_2$I$_2$—were found to be essential. Jimenez *et al.* (2003) also used a photochemical box model to simulate the reactions in the smog chamber and state that the main uncertainties are associated with unknown reactions of I$_2$O$_2$ and OIO. They compare their results to data sampled at Mace Head, Ireland, and find reasonable agreement. One main uncertainty is the concentration of the potential precursor CH$_2$I$_2$ which might have been present in very high concentrations in small "plumes" that could not be resolved by the present analytical instruments (as well as plumes of IO and OIO, Colin O'Dowd, personal communication, 2002). Average mixing ratios of CH$_2$I$_2$, however, are $\sim 0.5$ pmol mol$^{-1}$.

In addition to the just mentioned and well-studied coastal particle formation, O'Dowd *et al.* (2002) also speculated that iodine compounds might play a role in the formation of "sulfate" cloud condensation nuclei over the open ocean in remote marine regions by overcoming thermodynamic growth limitations.

The main conclusions from the early model studies on iodine chemistry remain valid: there is a significant lack of information on the kinetics of reactive iodine (especially IO and OIO reactions; paths to stable particulate iodine) and on fluxes of alkyl iodides from the oceans. Nevertheless, important progress has recently been made and work is currently ongoing in several laboratories worldwide. This is an important area of atmospheric research in need of more attention.

### 4.02.4.5 Surface Segregation Effects

A recent paper by Oum *et al.* (1998a) presented laboratory studies of photolytically induced formation of Cl$_2$ from sea salt particles in an aerosol chamber. They suggested that oxidation of Cl$^-$ by OH, that had been produced from O$_3$ photolysis and subsequent reaction of the product HOCl$^-$ with H$^+$, would lead to formation of chlorine radicals which could lead to the formation of Cl$_2$ via Cl$_2^-$. They also extrapolated their laboratory results to the atmosphere. However, the assumptions that were made to extrapolate the measurements to the atmosphere are questionable, as pointed out by Sander (1998), Behnke *et al.* (1999), and Jacob (2000).

Knipping *et al.* (2000) used a box model to simulate the experiment of Oum *et al.* (1998a) and concluded that "conventional chemical and physical processes do not explain the experimental observations." Molecular dynamics simulations of NaCl solutions showed that Cl$^-$ anions are segregated to the surface, facilitating surface reactions involving chloride. They suggested that reactions on the particle surface would be responsible for the release of Cl$_2$. According to their box model, the contribution of the interfacial mechanism is 40% of the total chlorine release at a pH of 4; its relative importance increases for less acidic particles (see Section 4.02.4.1 for a discussion of sea salt pH).

Knipping and Dabdub (2002) extended the box model study of Knipping *et al.* (2000) and proposed a reaction probability of $\gamma_{OH} = 0.04[Cl^-]$ for the suggested overall surface reactions:

$$OH_g + Cl^-_{aq,surface} \rightarrow 0.5Cl_{2,g} + OH^-_{aq} \quad (56)$$

Experimentally, surface-segregation effects were shown by Ghosal *et al.* (2000) who studied the surfaces of bromine-doped NaCl solids (the Br/Cl ratio was ~33 times higher than that of seawater) by X-ray photoelectron spectroscopy. They found that bromine was predominantly present at the surface. Zangmeister *et al.* (2001) confirmed this result by using a variety of experimental techniques and a range of Br/Cl ratios that covers that of seawater. However, both experiments were performed on solid salts.

In the earlier work of Markovich *et al.* (1994) a combination of photoelectron spectroscopy, molecular dynamics, and *ab initio* calculations was used to investigate the solvation of halides in water clusters. They found evidence that the halides were preferentially present at the surface of the clusters. The molecular dynamics simulations of Knipping *et al.* (2000) and Jungwirth and Tobias (2000) showed that $Cl^-$ was located both at the surface and in the interior of the water slab that they simulated, whereas $Na^+$ was only present inside the slab. They explained that this effect results from the size difference of the ions and the higher polarizability of $Cl^-$. In a follow-up study, Jungwirth and Tobias (2001) showed that this behavior is not unique for $Cl^-$ but is even more pronounced for the larger ions $Br^-$ and $I^-$. Only the small, nonpolarizable $F^-$ did not exhibit this behavior. $Br^-$ and $I^-$ were shown to have higher concentrations in the interfacial region than in the bulk of the slab. These simulations reproduced the experimentally measured increase in surface tension in the presence of halides compared to pure water.

As shown by Knipping *et al.* (2000) and Knipping and Dabdub (2002) for chlorine, surface segregation has the potential of increasing the release of halogens from sea salt aerosol. This should be pursued further, especially for $Br^-$ and $I^-$, where the surface enrichment effects are predicted to be highest. These effects also likely play a role in polar ODEs.

# 4.02.5   IMPACT OF HALOGEN CHEMISTRY ON SPECIES OTHER THAN OZONE

## 4.02.5.1   Alkanes

The presence of halogen-containing radicals can play a substantial role in chemical processes in the atmosphere. Chlorine atoms react much faster with alkanes than OH, the effect of which has been clearly observed in the Arctic troposphere in connection with the sunrise ODEs (Ramacher *et al.*, 1997; Ariya *et al.*, 1997; Song and Carmichael, 1999). The high reactivity of chlorine with hydrocarbons may explain the relatively high $CH_2O$ volume mixing ratios of

$100-700$ pmol mol$^{-1}$ that were measured by de Serves (1994). Production of $CH_2O$ in snowpack (Sumner and Shepson, 1999) is another source that we already discussed in Section 4.02.3.1.

## 4.02.5.2   Mercury

BrO is also involved in oxidation reactions of partially anthropogenic $Hg^0$ vapor to Hg(II), leading to lower gaseous elemental mercury and enhanced particulate mercury concentrations and increased deposition of mercury. These reactions may be particularly important during tropospheric ODEs in the Arctic. It has been proposed that this may have consequences for the health of the aboriginal population and the Arctic ecosystems following flushing of Hg(II) in the melt water during springtime and summer (Schroeder *et al.*, 1998; Lu *et al.*, 2001; Lindberg *et al.*, 2002). Similar processes even may play a role in the Antarctic environment (Ebinghaus *et al.*, 2002). Possible reaction cycles are discussed in Lindberg *et al.* (2002).

The modeling study of Hedgecock and Pirrone (2001) suggests a fundamental role of sea salt aerosol in determining the levels of oxidized mercury species in the gas phase in marine and coastal environments.

## 4.02.5.3   Sulfur

Sulfate aerosols are important for cloud processes and climate because they act, like sea salt aerosol, as CCN. The total number and size distribution of CCN determine the microphysical and, therefore, also radiative properties of clouds. In Section 4.02.4.4 we already discussed the possible involvement of iodine in seeding the formation of sulfate CCN. DMS is the most important precursor for $SO_2$ and sulfate aerosol in the remote MBL. A possible link between iodine and sulfur chemistry was investigated by Chatfield and Crutzen (1990), using a reaction rate coefficient of the IO + DMS reaction available at that time. They concluded that at that rate this reaction could play a significant role for the oxidation of DMS. In an alternative scenario with a slower rate coefficient, they found the iodine and sulfur cycles to be decoupled, in better agreement with field observations. Later the reaction turned out to be even slower than their lower limit (DeMore *et al.*, 1997; Knight and Crowley, 2001).

Another link between DMS and halogen chemistry was suggested by Toumi (1994), based on laboratory studies by Barnes *et al.* (1991), who proposed that the reactions of DMS with BrO could be important for the production of DMSO and loss of DMS:

$$BrO + DMS \rightarrow Br + DMSO$$

$$Br + O_3 \rightarrow BrO + O_2$$

Net:  $DMS + O_3 \rightarrow DMSO + O_2$

Note that this is not a sink for BrO which is reformed in the reaction with $O_3$, but rather a sink for $O_3$ and DMS. Ingham *et al.* (1999) presented updated kinetic data and a model study of the importance of this reaction and proposed to use the DMSO/DMS ratio to detect the presence of BrO. They calculated (taking DMSO loss by scavenging on aerosol and reaction with OH into account) that this ratio should be $\leq 0.01$ in the absence of BrO and $\geq 0.1$ in the presence of $1 \, pmol \, mol^{-1}$ BrO. If DMS is lost by this reaction, less $SO_2$ is being formed with potential implications on sulfate production. von Glasow *et al.* (2002b) studied this question with a one-dimensional MBL model and found that in the remote MBL, where oxidation of DMS by $NO_3$ plays a very minor role, DMS oxidation by BrO can be as high as 44% of the total rate.

Oxidation kinetics of DMSO suggests that the yield of $SO_2$ is very small. The main sinks for DMSO are deposition and uptake by aerosols and clouds droplets. If the production of DMSO by DMS + BrO would be significant, this would imply that model estimates of $SO_2$ and S(IV) yields from DMS would be too high. Boucher *et al.* (2002) investigated in a sensitivity study with a GCM the global importance of DMS oxidation by BrO. As so far no reliable global information on BrO mixing ratios is available, they chose a uniform daytime mixing ratio of $1 \, pmol \, mol^{-1}$ BrO over the oceans and found that 28.8% of the global DMS oxidation is by BrO. Compared to a base model run without the inclusion of bromine chemistry, the DMS burden would decrease by about one-third. During SAGA 3 Huebert *et al.* (1993) found that most DMS was not oxidized to non-sea-salt sulfate which would in agreement with a DMS oxidation pathway like DMS + BrO. This would be of major importance for the cycling of sulfur in the MBL and the precursors of the formation for non-sea-salt CCN.

Another link between halogen and sulfur chemistry is the formation of S(VI) within particles. Aerosol particles grow, among other processes, by uptake of $SO_2$ in cloud droplets where it is oxidized to sulfate. The most important aqueous phase oxidants for S(IV) are often thought to be $H_2O_2$ and $O_3$ (e.g., Seinfeld and Pandis, 1998) with $O_3$ being important only for pH $> 6$. Some authors state that oxidation by $O_3$ is the dominant process for the formation of non-sea-salt sulfate in sea salt particles

(e.g., Chameides and Stelson, 1992; Sievering *et al.*, 1992, 1995). However, as already discussed before (Section 4.02.4.1), the pH of aged sea salt is usually a lot less than 8 which is the pH of seawater, making sulfate production by $O_3$ oxidation only important for fresh, alkaline sea salt aerosol or if additional sources of alkalinity are present.

Vogt *et al.* (1996) suggested that halogens might play a role in the oxidation of S(IV)–S(VI) ($H_2SO_4 + HSO_4^- + SO_4^{2-}$). They examined the potential importance of HOCl and HOBr in the production of sulfate in sea salt aerosol and found that ~20% of the $SO_2$ that is scavenged by the aerosol is oxidized by HOBr and ~40% by HOCl.

The pH dependence of S(IV) oxidation in sea salt aerosols was also investigated by Keene *et al.* (1998) who found that at a pH of 8, oxidation by $O_3$ dominated, whereas at a pH of 5.5, HOCl was most important. This link was further investigated with a one-dimensional model by von Glasow *et al.* (2002b) for the cloud-free and for the stratocumulus-covered MBL. They give values integrated over the whole depth of the MBL and found that in the cloud-free cases between 19% and 57% of the $SO_2$ oxidation within the sea salt aerosol was due to oxidation by HOCl and 0–13% by HOBr. In cloud droplets, however, the relative importance of HOCl and HOBr is reversed compared to that in aerosols in their model. This is caused by the dilution in cloud droplets which makes the backward reaction in the equilibria:

$$HOBr_{aq} + X^- + H^+ \leftrightarrow BrX_{aq} + H_2O \quad (57)$$

where X = Cl or Br (and the same equilibria for HOCl), faster than the forward reaction. This leads to a greater availability of HOX for S(IV) oxidation. In cloud droplets, the faster sulfur oxidation rate of HOBr more than compensates the smaller concentrations of HOBr compared to HOCl. In clouds, most liquid water is associated with droplets that grew on sulfate aerosols, which scavenge halogen compounds very efficiently from the gas phase. Therefore, oxidation by HOX is also important in these droplets. Between 23% and 29% of the total S(IV) oxidation is attributed to HOBr in the sulfate droplets, compared to 5–8% by HOCl and 25–52% by $H_2O_2$. Oxidation by $O_3$ was found to be less important than by HOCl.

The net effect of these two links between sulfur and halogen chemistry is to decrease the gas phase concentration of $SO_2$ via a reduced yield of $SO_2$ from the oxidation of DMS and the stronger aqueous phase sink for S(IV) which results in enhanced uptake of $SO_2$ by droplets and aerosols. A critical prerequisite for new particle formation in the marine troposphere is the reaction chain:

$$SO_2 + OH \rightarrow HSO_3 \quad (58)$$

$$HSO_3 + O_2 \rightarrow HO_2 + SO_3 \qquad (59)$$

$$SO_3 + H_2O \overset{H_2O}{\rightarrow} H_2SO_4 \qquad (60)$$

Reduced availability of $SO_2$ would reduce the formation rate of new particles. Further studies are needed to elucidate the combined effect of these reactions of cloud microphysics and radiative forcing. Regions that cloud be affected most by these changes are the large regions of marine stratocumuli, especially the tradewind systems.

### 4.02.6  SALT LAKES

An environment very different from the afore-mentioned polar regions and the MBL proved to be an active scene of halogen chemistry—salt lakes. In the Dead Sea area, at a site downwind of large salt pans, BrO mixing ratios of up to 90 pmol $mol^{-1}$ (Hebestreit et al., 1999) were measured during spring 1997 using the DOAS technique. These extreme events were also accompanied by low boundary layer $O_3$ mixing ratios. Box model calculations by Stutz et al. (1999) for these conditions upstream of the measurements site show mixing ratios of BrO to be ~800 pmol $mol^{-1}$ and total gas phase bromine of up to 2 nmol $mol^{-1}$, assuming an autocatalytic release process from the salt pans.

In a follow-up study in the summer of 1997 (Matveev et al., 2001) this special situation was studied in more detail. Again the measurement location was ~23–25 km downwind from the largest salt pans at the southern end of the Dead Sea. In addition to *in situ* gas phase measurements, whole air and aerosol samples were taken. The strong anticorrelation of BrO and $O_3$ was confirmed. Sometimes $O_3$ even dropped below the detection limit of 2 nmol $mol^{-1}$. These events were also strongly correlated with the wind direction: only when the wind came from the site of the salt pans, these depletion events were detected. Using a small aircraft, the significant spatial extent of the ODEs was shown. The highest BrO mixing ratio during this campaign was 176 pmol $mol^{-1}$, but measurements in summer 2001 showed mixing ratios of close to 220 pmol $mol^{-1}$ (Jutta Zingler, personal communication, 2002), which are the highest BrO mixing ratios ever measured in the atmosphere. The high $Br^-/Cl^-$ ratio of the salt deposits in the Dead Sea area (0.025 in terms of mass), ~7 times higher than in seawater, might contribute to these high mixing ratios.

Matveev et al. (2001) excluded anthropogenic emissions of bromine as the source for the observed BrO. They concluded that the most likely source is bromine release from the salt deposits by the previously described autocatalytic reaction, where uptake of HOBr is followed by

$$HOBr_{aq} + H^+ + Br^- \rightarrow Br_{2,aq} + H_2O \quad (61)$$

release of $Br_2$, photolysis to Br radicals, and their reaction with $O_3$ to produce BrO. The pH of the Dead Sea water is below 6.5 and strongly buffered, so that enough acidity should be available to make this reaction efficient. Further bromine release mechanisms, like those involving nitrogen oxides as suggested by Finlayson-Pitts et al. (1990) and Behnke et al. (1994), might also contribute.

With the use of a simple photochemical box model, Matveev et al. (2001) suggested that upwind of the DOAS site much higher BrO mixing ratios of the order of $1-2$ nmol $mol^{-1}$ must have been present, which would explain the very rapid $O_3$ destruction.

Stutz et al. (2002) made measurements in October 2000 at the Great Salt Lake in Utah, USA. For the first time they detected ClO in the mid-latitude boundary layer with mixing ratios between 5 pmol $mol^{-1}$ and 15 pmol $mol^{-1}$. BrO was also found at mixing ratios of up to 6 pmol $mol^{-1}$. Again, $O_3$ concentrations were anti-correlated with those of halogen oxides. The mobilization of bromine and chlorine from salt pans seems to be the source of halogen oxides in this region as well. Anthropogenic sources could be excluded with the help of backtrajectory analysis. During the 24 h preceding the measurements light rains were reported which could have affected the release of halogens from the salt flats. According to Stutz et al. (2002), the salinity of the Great Salt Lake in Utah is only a tenth of that of the Dead Sea, furthermore the $Br^-/Cl^-$ is only 0.0007 (in terms of mass), significantly smaller than in ocean water and in the Dead Sea. Other regions like the huge salt lakes in Bolivia (Salar de Uyuni, see Figure 11) or salt deserts (Dasht-e Kevir, or Great Salt Desert, Iran) are likely to be halogen sources of importance at least for local atmospheric chemistry as well. According to data from the Food and Agriculture Organization (FAO) of the United Nations (cited in Stutz et al. (2002)) saline soils cover an area of $\sim 3.6 \times 10^6$ $km^2$ worldwide, or as much as 2.5% of the land surface of the Earth, pointing to a potential global chemical importance of these halogen emissions. Desertification, which might increase in the near future due to global change, might enhance the importance of halogen chemistry.

Although the salt lake and polar environments appear different at first, the source (surface deposits) and gas and aerosol phase cycling are very similar.

**Figure 11** Picture of the Salar de Uyuni in Bolivia, where rains led to the presence of water (5–10 cm deep) on top of the salt. The geometric structure (hexagons) of the salt deposits is clearly visible. Photo by one of the authors (RvG).

## 4.02.7 FREE TROPOSPHERE

### 4.02.7.1 Satellite Observations of Free Tropospheric BrO

Using observations from the satellite-borne instrument, GOME, Pundt *et al.* (2000) and Richter *et al.* (2002) showed the presence of BrO outside the polar regions (see also Richter *et al.* (2002) and references therein). They found global background vertical columns of $4 \times 10^{13}$ molecules cm$^{-2}$, which would correspond to 0.5–2 pmol mol$^{-1}$ BrO if it were uniformly mixed in the troposphere. Comparisons with balloon and ground measurements have been made for mid- and high northern latitudes (between 42°N and 68°N; Harder *et al.*, 1998; Fitzenberger *et al.*, 2000; Van Roozendael *et al.*, 2002) which showed that the tropospheric BrO was mainly located within the free troposphere. Further evidence for free tropospheric BrO in high latitudes comes from high-altitude aircraft observations in the Arctic by McElroy *et al.* (1999). Such observations have not yet been made at lower latitudes.

Although most evidence indicates that the measured BrO is located in the free troposphere, these data do not exclude other contributions, e.g., from the marine boundary layer. Especially over ocean regions the sensitivity of GOME to BrO in lower atmospheric layers in reduced and a contribution of BrO in the MBL as measured by Leser *et al.* (2003) cannot be excluded (see, e.g.,

discussion of air mass factors in Wagner *et al.* (2001)).

The presence of 0.5–2 pmol mol$^{-1}$ BrO in the troposphere would have a considerable effect on atmospheric chemistry as shown by the model studies that we discussed in Section 4.02.4.3. It could help explain events of upper troposphere near-zero $O_3$ mixing ratios which have been found by Kley *et al.* (1996) over the equatorial Pacific. A similar event was discussed by Davies *et al.* (1998) where near-zero $O_3$ concentrations were encountered in the mid-latitude upper troposphere. According to trajectory studies, the airmass with low ozone was advected from the subtropical upper troposphere.

### 4.02.7.2 Possible Source of Free Tropospheric BrO

In the previous section it was shown that significant amounts of BrO might be present in the free troposphere. This could be the result of three processes: up-ward transport of sea-salt-derived bromine from the MBL, *in situ* production from decomposition of organic halogen compounds (especially $CH_3Br$ and $CHBr_3$), or downward transport from the stratosphere, where photolysis releases halogens from very stable anthropogenic organobromine compounds.

As pointed out by Donnell *et al.* (2001), vertical transport processes (especially advection by frontal motions and orographically forced motion) are very efficient in transporting passive tracers from the boundary layer to the free troposphere. Also in regions of deep convection this upward transport can strongly be increased.

CH$_3$Br emanates from several sources including some oceans regions (which, however, in total seem to act as a sink; see, e.g., Lobert *et al.*, 1997; King *et al.*, 2000) and anthropogenic activities (including biomass burning). Because of its low solubility CH$_3$Br is not scavenged by clouds and can, therefore, be transported over relatively long distances. Apart from surface deposition its only significant sink in the troposphere is reaction with OH. According to Singh and Fabian (1999), the mean mixing ratio of CH$_3$Br in surface air is ~10 pmol mol$^{-1}$. Using a rate constant of $1.9 \times 10^{-14}$ cm$^3$ molecules$^{-1}$ s$^{-1}$ (Atkinson *et al.*, 1999; at $T = 270$ K), a 24 h global mean concentration of OH of ~$10^6$ molecules cm$^{-3}$, assuming the same mixing ratio of CH$_3$Br in the free troposphere as in surface air (which is confirmed by, for example, Blake *et al.* (1996)), and a tropopause height of 10 km, we estimate the release rate of active bromine by this process to be ~$10^{11}$ g a$^{-1}$.

To estimate the global release of bromine from sea salt aerosol, we use a sea salt source of 100 Tg a$^{-1}$ as lower and 1,000 Tg a$^{-1}$ as upper limit for particles in size ranges with significant bromine deficits (see Section 4.02.4.1). Assuming a mean Br$^-$ deficit of 50% the annual mass of bromine from sea salt aerosol to the atmosphere is ~$(1–10) \times 10^{11}$ g a$^{-1}$. Many bromine compounds are soluble, so only part of this flux will reach the free troposphere. Based on these—very rough—estimates, sea-salt-derived bromine could, however, be a more important source process than CH$_3$Br-derived bromine. Certainly, both the source and the implications of this tropospheric "background" BrO need further investigation.

The importance of CHBr$_3$ as source for bromine in the free troposphere is difficult to estimate because of the scarce field data and uncertainties regarding the fate of the breakdown products. Assuming an average mixing ratio of 0.5 pmol mol$^{-1}$, a lifetime of four weeks and that two bromine atoms are released, the bromine source would be ~30% of that of CH$_3$Br.

# 4.02.8 ADDITIONAL SOURCES OF REACTIVE HALOGENS

## 4.02.8.1 Volcanoes

Symonds *et al.* (1988) used equilibrium thermodynamic considerations to conclude that the overwhelming fraction of gaseous halogens that is released by volcanic eruptions are in the form of the hydrogen halides—they examined HCl and

HF. They estimated the magnitude of the halogen source of different types of volcanic eruptions and found that up to several hundred terragrams of HCl could be emitted into the atmosphere in the largest eruptions. Only a small fraction (down to ≈0.01%; Tabazadeh and Turco, 1993) of this reaches the stratosphere, most HCl being removed by precipitation in the eruption plume. More passively degassing volcanoes could produce ~$(0.4–4) \times 10^{11}$ g HCl per year which predominantly remains in the troposphere. From stratospheric observations there is no indication of any growth in chlorine or bromine after the eruptions of Pinatubo or El Chichón.

Pinto *et al.* (1989) list an HBr/HCl ratio of $4.3 \times 10^{-4}$ for high temperature, i.e., explosive eruptions. If the Br/Cl ratio is not much higher for more quiescent volcanoes or fumaroles, then volcanoes are a small source for bromine in the troposphere and stratosphere as well. Very recent measurements, however, found up to 1 nmol mol$^{-1}$ BrO in a volcanic plume at Montserrat and the authors concluded that this might be a substantial global source of BrO$_x$ (Bobrowski *et al.* 2003).

## 4.02.8.2 Industry and Fossil Fuel Burning

Many industrial processes involve the use of chlorine. Aucott *et al.* (1999) estimate the sum of industrial CHCl$_3$ emissions to 62 Gg(Cl) a$^{-1}$, with the two biggest contributors being pulp and paper manufacturing and water treatment. The industrial emissions of higher chlorinated hydrocarbons as C$_2$HCl$_3$, C$_2$Cl$_4$, or CH$_2$Cl$_2$ are 195 Gg(Cl) a$^{-1}$, 313 Gg(Cl) a$^{-1}$, and 487 Gg(Cl) a$^{-1}$, respectively (Aucott *et al.*, 1999) compared to 20 Gg(Cl) a$^{-1}$, 20 Gg(Cl) a$^{-1}$, and 160 Gg(Cl) a$^{-1}$, respectively, for oceanic sources (Khalil *et al.*, 1999).

The chlorine content of oil is usually very small, so that it can be neglected as a source, but the chlorine content in coal is significant (Smith, 1872; McCulloch *et al.*, 1999). They estimated the HCl emissions from fossil fuel burning in 1990 to be 4.6 Tg(Cl) a$^{-1}$. Halogens are added as antiknock compounds (ethylene dibromide or ethylene dichloride, C$_2$H$_4$X$_2$) to leaded gasoline (see also Section 4.02.4.3). This leads to the release of chlorine and bromine as lead salts, NaX, or HX (Lightowlers and Cape, 1988), but this is getting less important because the use of leaded gasoline is banned in most countries.

The strong acid, HCl, is also emitted by other processes like incineration of waste, or steel making and might be regionally important for acidification but not on a greater scale. Lightowlers and Cape (1988) estimated a source of 260 Gg a$^{-1}$ for the UK in 1983. On a global scale, release of HCl from sea salt aerosols and volcanoes are much more important than anthropogenic production.

There are many anthropogenic sources of photolyzable chlorine ($Cl_2$, $HOCl$). These include cooling towers (chlorine is used to prevent bacterial growth within the system), water treatment and swimming pools, and industrial processes. Inventories for these processes are only available for limited regions and are urgently needed.

Contrary to the case in unpolluted environments, where halogen chemistry leads to a destruction of $O_3$, emission of chlorine (mainly as photolyzale chlorine like $Cl_2$) can increase catalytic photochemical $O_3$ production in $NO_x$-rich, polluted areas by fast reactions of chlorine with hydrocarbons that produce $RO_2$ radicals. An early study of this was performed by Hov (1985) for southern Norway. Comparing field measurements, smog chamber, and model results, Tanaka *et al.* (2000, 2003, submitted) and Chang *et al.* (2002) found increased ozone formation due to chlorine reactions with hydrocarbons in the highly polluted Houston area. Largest $O_3$ enhancements were shown to occur in the morning (Tanaka *et al.*, submitted). Emissions from cooling towers were found to be primarily responsible for morning $O_3$ increases, while afternoon enhancements were mainly caused by emissions from residential swimming pools (Chang *et al.*, 2002). Based on the emission data for the USA from Chang *et al.* (2001), one can extrapolate (by extrapolating the Houston area emissions to the total USA and assuming that the global emissions are twice as large as the US emissions) the emissions of swimming pools and cooling towers to ~1 Tg(Cl) a$^{-1}$.

Measurements with an off-axis DOAS instrument in France ~100 km North of Marseille (i.e., not in coastal or salt sea environment) showed evidence of free tropospheric BrO but also local emissions in the neighborhood of the station during summer (Van Roozendael *et al.*, 2002; Michel Van Roozendael, personal communication, 2002). Work is ongoing to identify the local source of bromine.

We also note that other important anthropogenic chlorine-containing substances like PCBs are emitted that have severe health effects. In this review, however, we concentrate on reactive halogen compounds that have an impact on tropospheric chemistry.

### 4.02.8.3 Biomass Burning

It has been known for many years that biomass burning is a source of methyl halides (e.g., Lovelock, 1975; Palmer, 1976; Crutzen *et al.*, 1979). In laboratory experiments, Reinhardt and Ward (1996) found that inefficient combustion which is typical of smoldering processes is associated with high rates of $CH_3Cl$ emissions possibly due to the reaction

$$CH_3OH + HCl \rightarrow CH_3Cl + H_2O \quad (62)$$

In flaming combustions most chlorine mass is found in fine particles which might de due to uptake of HCl by the particles.

Emission ratios for methyl halides were published for savanna fires by Andreae *et al.* (1996) who estimated that ~90% of the halogen content was released, mostly as halides, but between 3% and 38% as methyl halides.

Blake *et al.* (1996) estimated methyl halide emissions by biomass burning from measurements in the South Atlantic region and found that roughly 25% and 20% of the global $CH_3Cl$ and $CH_3Br$ emissions are due to biomass burning, whereas the contribution of biomass burning to global $CH_3I$ emissions was negligible.

According to Lobert *et al.* (1999), the sum of all chlorine emissions from biomass burning is ~7 Tg(Cl) a$^{-1}$. About 9% is emitted as $CH_3Cl$, less than 1% as other alkyl chlorides while the largest part is emitted as unspecified volatile inorganic and as particulate chlorine.

In a recent compilation, Andreae and Merlet (2001) list the emission factors (amount of compound released in fire per amount of dry fuel consumed) for different types of biomass burning and the global annual emissions of $CH_3X$. Their estimates for the global production in the late 1990s are: $CH_3Cl$ 450 Gg(Cl) a$^{-1}$, $CH_3Br$ 24 Gg(Br) a$^{-1}$, and $CH_3I$ 12 Gg(I) a$^{-1}$, respectively.

Recent observations in controlled biomass burning experiments showed that less than 50% of volatile inorganic chlorine was in the form of HCl. Most appeared to be $Cl_2$ and HOCl and other alkaline reactive chlorine compounds (Keene *et al.*, 2001).

### 4.02.8.4 Organic Halogen Compounds

Many groups have investigated the emissions, concentrations, and atmospheric fate of organic halogen compounds in the past. Most, however, were concerned with the potential transport of halogens into the stratosphere and the destruction of $O_3$. Some of the organic halogen compounds get photolyzed or react with OH in the troposphere leading to the release of halogen radicals. The main sources for organohalogens are the ocean, terrestrial plants, fungi, biomass burning, and anthropogenic emissions. We will discuss these only briefly here. Gribble (2000) reviews that natural production of organobromine compounds.

It has been known for several decades that oceanic phytoplankton emits $CH_3X$ (e.g., Lovelock *et al.*, 1973; Lovelock, 1975; Reifenhäuser and Heumann, 1992b; Sæmundsdóttir and Matrai, 1998; Giese *et al.*, 1999). Also other halogen methanes and di- and trihalogenated organics can be released in substantial amounts (e.g., Laturnus, 1996; Reifenhäuser and Heumann, 1992a; Schall *et al.*, 1994, 1996, 1997; Schall and Heumann, 1993). The production by macroalgae (e.g., kelp) in coastal regions is expected to be stronger (Goodwin *et al.*, 1997) than in the open ocean. Giese *et al.* (1999) compiled estimates of global organic iodine emissions from the ocean to be $10^{11}–10^{12}$ g a$^{-1}$ with macroalgae contributing $(42–210) \times 10^6$ g a$^{-1}$. The corresponding emissions of bromine are $10^{10}–10^{12}$ g a$^{-1}$ and $10^8–10^9$ g a$^{-1}$, respectively. Wever (1988) and Wever *et al.* (1991a,b) report direct emissions of the reactive HOBr from brown and red seaweed. Bassford *et al.* (1999) showed a very significant degree of covariance of DMS and $CH_3I$ emissions which might imply similar source processes.

Although the maximum of these emissions is likely to occur near coastal or upwelling regions, where maxima of chlorophyll can also be found (Moore and Webb, 1996), it is clear that the open ocean is also a source for organic halogen compounds. A detailed discussion of gas emissions from the ocean is given by Nightingale and Liss (see Chapter 6.03).

Land-based sources of halocarbons also exist. Fungi can convert halides to halomethanes (Harper, 1985). Bacteria might be important in production of $CH_3I$ both in marine and terrestrial ecosystems (Amachi *et al.*, 2001). Soils might also play a role as sources for $CH_3X$. Muramatsu and Yoshida (1995) found that the presence of plants stimulated the emission of iodine. $CH_3I$ was mainly emitted from the plant shoot and not from the soil. Keppler *et al.* (2000) found production of halocarbons during the degradation of organic matter within the soil. A potential, so far neglected, source might also be subsurface ectomycorrhizal fungi which are globally very widespread. Laboratory studies showed significant $CH_3X$ release (Kathleen Treseder, personal communication, 2002). Dimmer *et al.* (2001) measured emissions from Irish peatland and calculated global annual fluxes of about 4.7 Gg a$^{-1}$, 0.9 Gg a$^{-1}$, 5.5 Gg a$^{-1}$, and 1.4 Gg a$^{-1}$ for $CHCl_3$, $CH_3Br$, $CH_3Cl$, and $CH_3I$, respectively. Based on measurement in the northeast of the USA, Varner *et al.* (1999) calculated global fluxes from wetlands of 4.6 Gg a$^{-1}$ and 48 Gg a$^{-1}$ for $CH_3Br$ and $CH_3Cl$, respectively. Salt marshes might account for 10% of global flux of $CH_3Br$ and also emit $CH_3Cl$ (Rhew *et al.*, 2000). Yokouchi *et al.* (2000, 2002) and Lee-Taylor *et al.* (2001)

suggested that emissions of $CH_3Cl$ from tropical lands, especially in coastal regions with higher soil salinity, might help to close the global budget of $CH_3Cl$.

Agriculture can also contribute to halocarbon emissions. Redeker *et al.* (2000) studied the emissions from rice paddies and found that they might contribute 1% of the global $CH_3Br$ and 5% of the global $CH_3I$ emissions. Cattle were also found to emit $CH_3X$ but the global contribution was found to be negligible (Williams *et al.*, 1999).

Quack and Wallace (2003) reviewed the concentration, sources, and fate of $CHBr_3$ in the atmosphere. They list typical mixing ratios of 0.5–1 pmol mol$^{-1}$, with the ocean being the most important source. They estimated that 23% of the sea-to-air flux is from nearshore regions that comprise 0.3% of the ocean surface, whereas the open ocean (88% of ocean surface) only accounts for 29% of the flux. Continental shelfs are responsible for the remaining 40%. Very large mixing ratios were found in localized, high-productivity regions with several hundred pmol mol$^{-1}$. The lifetime of $CHBr_3$ against photolysis is about four weeks (Moortgat *et al.*, 1993) and two to three months against reaction with OH.

### 4.02.8.5 Inventories

Graedel and Keene (1995) compiled an inventory of tropospheric reactive chlorine to derive a global budget for these compounds. They found a typical background mixing ratio of reactive chlorine in the lower troposphere of 1.5 nmol mol$^{-1}$, including HCl, $Cl_2$, HOCl, and some alkyl chlorides. This estimate may appear on the high side, as Vierkorn-Rudolph *et al.* (1984) measured HCl volume mixing ratios over the northeast Atlantic between 0.1 nmol mol$^{-1}$ and 0.5 nmol mol$^{-1}$ below 3 km and 0.05–0.1 nmol mol$^{-1}$ between 3 km and 7 km altitude, Harris *et al.* (1992) found less than 0.25 nmol mol$^{-1}$ at ground level over the Atlantic.

In an international effort, the Reactive Chlorine Emissions Inventory (Keene *et al.*, 1999, and references therein) was compiled which includes chlorine emissions from the following four source-type classes: (i) oceanic and terrestrial biogenic emissions; (ii) sea salt aerosol production and dechlorination; (iii) biomass burning; and (iv) exclusively anthropogenic sources like industry, fossil fuel combustion, and incineration. They provide numbers from atmospheric burdens and fluxes for the individual species and sources.

We will here only repeat a few numbers from this important inventory and refer the reader to the original papers for more details. The total tropospheric chlorine burden is dominated with a rather uncertain 15 Tg(Cl) of particulate chlorine (i.e., sea salt $Cl^-$), followed by $CH_3Cl$ with 2.8 Tg(Cl) and $CH_3CCl_3$ with 2.3 Tg(Cl). Since 1995, the emission of the latter has largely been reduced to very low numbers due to international regulations. Other alkyl chlorides sum up to more than 1 Tg(Cl) and the burden of HCl is ~0.6 Tg(Cl). The flux of chlorine is also dominated by sea salt chloride (1,785 Tg(Cl) $a^{-1}$), part of which is released from the particles as HCl by acid displacement (7.6 Tg(Cl) $a^{-1}$). The next important (in terms of mass) chlorine source is biomass burning (6.3 Tg(Cl) $a^{-1}$ of inorganic chlorine). Volcanoes contribute 2 Tg(Cl) $a^{-1}$ as HCl. The basis year for this inventory was 1990.

## 4.02.9 SUMMARY

Since the 1970s, it has been known that halogens, in particular the photolysis products Cl and ClO of industrially manufactured CFC gases, have a large impact on stratospheric ozone (Molina and Rowland, 1974), culminating in the discovery of the springtime "ozone hole" in the lower stratosphere in the Antarctic (Farman et al., 1985). This issue has been thoroughly reviewed by many authors and international research panels.

Research since the late 1980s has, however, shown that halogen chemistry can also have a large impact on tropospheric chemistry, most clearly shown in the Arctic where, during the springtime, very low, sometimes zero, concentrations of ozone are measured. Based on the observations at Alert, northeast Canada, showing a negative correlation between "filterable bromine" and ozone concentrations, Barrie et al. (1988) proposed a catalytic cycle of reactions, involving a self-reaction of BrO as the rate determining step, which can efficiently destroy ozone. Many subsequent studies which are reviewed in this chapter have confirmed this hypothesis and have furthermore shown that springtime surface ozone depletion is a widespread high-latitude phenomenon in both hemispheres. The source for active bromine is most likely provided by various autocatalytic reactions which take place on salt water deposits on the sea ice. Details about these processes and quantification of them are, however, still in need of further research.

Strong bromine activation and ozone depletion have also been recorded at salt lakes like the Dead Sea and the Great Salt Lake. Besides the regions where the impact of bromine chemistry is clearly noted, there are indications that halogens can also play a significant role in the chemistry of the marine boundary layer and perhaps even the free troposphere. This is supported by measurements showing a general loss of bromine (and to a lesser degree chlorine) from sea salt particles, thus providing a source of bromine in the gas phase. Some measurements have shown the presence of IO in concentrations that are significant for the chemistry in the marine boundary layer near Ireland, Tenerife and Cape Grim (Alicke et al., 1999; Allan et al., 2000), and indicate the presence of an active iodine source from seawater, consisting of photochemical oxidation products of iodocarbons released by kelp or marine algae. The potential role of iodine in particle nucleation is an issue that has attracted attention (O'Dowd et al., 2002).

As explained in this chapter, the release of bromine from sea salt aerosol depends on the pH of the sea salt aerosol particles. One can speculate that due to the lack of anthropogenic emissions of $NO_x$ and $SO_2$ in pre-Industrial times the pH of sea salt aerosols was higher and consequently less bromine was released from sea salt aerosol. This would imply that all potential implications that we have listed are stronger now than in the pre-Industrial era. Furthermore other anthropogenic sources of halogens exist like downmixing of inorganic halogens from the stratosphere where they are released from CFCs and halons.

Changes in polar $O_3$ column densities during springtime might influence the photochemistry during the time of the boundary layer ODEs, especially in the southern hemisphere. Increases in air temperatures will reduce the polar ice coverage (especially in the Arctic). This means the areas of fresh sea ice formation in winter will increase with time, and thereby the most important source for bromine in polar regions.

Altogether, however, even without a major impact by human activity on halogen chemistry of the troposphere, tropospheric halogen chemistry is a highly fascinating scientific issue with many unknowns including possible climate change feedbacks on the source strength of sea salt aerosols or biogenic halogen precursors: it is a highly recommended area for further research with the need for instruments with even lower detection limits, more field campaigns in remote maritime regions, and global model studies.

## ACKNOWLEDGMENTS

We would like to thank the following people for helpful discussions, for making unpublished material available to us and/or for very helpful comments on earlier versions of the manuscript: Elliot Atlas, Wolfgang Behnke, Bob Duce, Gerd Hönninger, Jose Jimenez, Melissa Joseph, Eladio Knipping, Ken Melville, John Plane, Ulrich Platt,

Andreas Richter, Jochen Stutz, Paul Tanaka, Kenjiro Toyota, Kathleen Treseder, Michel Van Roozendael, Thomas Wagner, and Jutta Zingler. Many thanks to Eric Wilcox and Greg Roberts for proofreading.

We want to thank especially John Crowley, Bill Keene, and Rolf Sander for their comments on this manuscript and years of stimulating discussions.

# REFERENCES

Abbatt J. P. D. (1994) Heterogeneous reaction of HOBr with HBr and HCl on ice surfaces at 228 K. *Geophys. Res. Lett.* **21**, 665–668.

Abbatt J. P. D. and Nowak J. B. (1997) Heterogeneous interactions of HBr and HOCl with cold sulfuric acid solutions: implications for Arctic boundary layer bromine chemistry. *J. Phys. Chem. A* **101**, 2131–2137.

Abbatt J. P. D. and Waschewsky G. C. G. (1998) Heterogeneous interactions of HOBr, $HNO_3$, $O_3$ and $NO_2$ with deliquescent NaCl aerosols at room temperature. *J. Phys. Chem. A* **102**, 3719–3725.

Adams J. W., Holmes N. S., and Crowley J. N. (2002) Uptake and reaction of HOBr on frozen and dry salt surfaces. *Atmos. Chem. Phys.* **2**, 79–91.

Alicke B., Hebestreit K., Stutz J., and Platt U. (1999) Iodine oxide in the marine boundary layer. *Nature* **397**, 572–573.

Allan B. J., McFiggans G., Plane J. M. C., and Coe H. (2000) Observation of iodine oxide in the remote marine boundary layer. *J. Geophys. Res.* **105**, 14363–14370.

Allan B. J., Plane J. M. C., and McFiggans G. (2001a) Observations of OIO on the remote marine boundary layer. *Geophys. Res. Lett.* **28**, 1945–1948.

Allan W., Lowe D. C., and Cainey J. M. (2001b) Active chlorine in the remote marine boundary layer: modeling anomalous measurements of $^{13}C$ in methane. *Geophys. Res. Lett.* **28**, 3239–3242.

Amachi S., Kamagata Y., Kanagawa T., and Muramatsu Y. (2001) Bacteria mediate methylation of iodine in marine and terrestrial environments. *Appl. Environ. Microbiol.* **67**, 2718–2722.

Anastasio C. and Mozurkewich M. (2002) Laboratory studies of bromide oxidation in the presence of ozone: evidence for a glass-surface mediated reaction. *J. Atmos. Chem.* **41**, 135–162.

Andreae M. O. and Merlet P. (2001) Emission of trace gases and aerosols from biomass burning. *Global Biogeochem. Cycles* **15**, 955–966.

Andreae M. O., Atlas E., Harris G. W., Helas G., de Kock A., Koppmann R., Maenhaut W., Manø S., Pollock W. H., Rudolph J., Scharffe D., Schebeske G., and Welling M. (1996) Methyl halide emissions from savanna fires in southern Africa. *J. Geophys. Res.* **101**, 23603–23613.

Andreas E. L. (1998) A new sea spray generation function for wind speeds up to $32 \, m \, s^{-1}$. *J. Phys. Oceanogr.* **28**, 2175–2184.

Andreas E. L. (2002) A review of the sea spray generation function for the open ocean. *Atmosphere–Ocean Interactions* (ed. W. Perrie). WIT Press, Southampton, Boston, vol. 1, pp. 1–46.

Andrews J. E., Brimblecombe P., Jickells T. D., and Liss P. S. (1996) *An Introduction to Environmental Chemistry.* Blackwell, Oxford.

Anlauf K. G., Mickle R. E., and Trivett N. B. A. (1994) Measurement of ozone during Polar Sunrise Experiment 1992. *J. Geophys. Res.* **99**, 25345–25353.

Ariya P. A., Catoire V., Sander R., Niki H., and Harris G. W. (1997) Trichloroethene and tetrachloroethene: tropospheric probes for Cl and Br-atom reactions during the polar sunrise. *Tellus* **49B**, 583–591.

Ariya P. A., Hopper J. F., and Harris G. W. (1999) $C_2–C_7$ hydrocarbon concentrations in Arctic snowpack interstitial air: potential presence of active Br within the snowpack. *J. Atmos. Chem.* **34**, 55–64.

Ashworth S. H., Allan B. J., and Plane J. M. C. (2002) High resolution spectroscopy of the OIO radical: implications for the ozone-depletion potential of iodine in the marine boundary layer. *Geophys. Res. Lett.* **29**, doi: 10.1029/2001GL013851.

Atkinson A., Baulch D. L., Cox R. A., Hampson R. F., Jr., Kerr J. A., Rossi M. J., and Troe J. (1997) Evaluated kinetic and photochemical data for atmospheric chemistry: supplement VI. *J. Phys. Chem. Ref. Data* **26**, 1329–1499.

Atkinson R., Baulch D. L., Cox R. A., Hampson R. F., Jr., Kerr J. A., Rossi M. J. and Troe J. (1999) Summary of evaluated kinetic and photochemical data for atmospheric chemistry. *Web Version*, http://www.iupac-kinetic.ch.cam.ac.uk

Atkinson R., Baulch D. L., Cox R. A., Hampson R. F., Jr., Kerr J. A., Rossi M. J., and Troe J. (2000) Evaluated kinetic and photochemical data for atmospheric chemistry: supplement VIII, halogen species. IUPAC subcommittee on gas kinetic data evaluation for atmospheric chemistry. *J. Phys. Chem. Ref. Data* **29**, 167–266.

Aucott M. L., McCulloch A., Graedel T. E., Midgley P., and Li Y.-F. (1999) Anthropogenic emissions of trichloromethane (chloroform, $CHCl_3$ and chlorodifluoromethane (HCFC-22): reactive chlorine emissions inventory. *J. Geophys. Res.* **104**, 8405–8415.

Ayers G. P., Gillett R. W., Cainey J. M., and Dick A. L. (1999) Chloride and bromide loss from sea-salt particles in Southern Ocean air. *J. Atmos. Chem.* **33**, 299–319.

Baker A. R., Thompson D., Campos M. L. A. M., Perry S. J., and Jickells T. D. (2000) Iodine concentration and availability in atmospheric aerosol. *Atmos. Environ.* **34**, 4331–4336.

Baker A. R., Tunnicliffe C., and Jickells T. D. (2001) Iodine speciation and deposition fluxes from the marine atmosphere. *J. Geophys. Res.* **106**, 28743–28749.

Barnes I., Bastian V., Becker K. H., and Overrath R. D. (1991) Kinetic studies of the reactions of IO, BrO and ClO with DMS. *Int. J. Chem. Kinet.* **23**, 579–591.

Barrie L. and Platt U. (1997) Arctic tropospheric chemistry: an overview. *Tellus* **49B**, 450–454.

Barrie L. A. (1986) Arctic air pollution: an overview of current knowledge. *Atmos. Environ.* **20**, 643–663.

Barrie L. A., Bottenheim J. W., and Hart W. R. (1994) Polar Sunrise Experiment 1992 (PSE 1992): Preface. *J. Geophys. Res.* **99**, 25313–25314.

Barrie L. A., Bottenheim J. W., Schnell R. C., Crutzen P. J., and Rasmussen R. A. (1988) Ozone destruction and photochemical reactions at polar sunrise in the lower Arctic atmosphere. *Nature* **334**, 138–141.

Barrie L. A., den Hartog G., Bottenheim J. W., and Landsberger S. (1989) Anthropogenic aerosols and gases in the lower troposphere at Alert, Canada in April 1986. *J. Atmos. Chem.* **9**, 101–127.

Bassford M. R., Nickless G., Simmonds P. G., Lewis A. C., Pilling M. J., and Evans M. J. (1999) The concurrent observation of methyl iodide and dimethyl sulphide in marine air; implications for sources of atmospheric methyl iodide. *Atmos. Environ.* **33**, 2373–2383.

Bedjanian Y., Bras G. L., and Poulet G. (1998) Kinetics and Mechanisms of the IO + BrO Reaction. *J. Phys. Chem. A* **102**, 10501–10511.

Behnke W. and Zetzsch C. (1989) Smog chamber investigations of the influence of NaCl aerosol on the concentrations of $O_3$ in a photosmog system. In *Ozone in the Atmosphere* (eds. R. D. Bojkow and P. Fabian). A. Deepak Publishing, pp. 519–523.

Behnke W., Krüger H.-U., Scheer V., and Zetzsch C. (1991) Formation of atomic Cl from sea spray via photolysis of nitryl chloride: determination of the sticking coefficient of $N_2O_5$ on NaCl aerosol. *J. Aerosol Sci.* **22**, S609–S612.

Behnke W., Krüger H.-U., Scheer V., and Zetzsch C. (1992) Formation of $ClNO_2$ and HONO in the presence of $NO_2$, $O_3$ and wet NaCl aerosol. *J. Aerosol Sci.* **23**, S933–S936.

Behnke W., Scheer V., and Zetzsch C. (1994) Production of $BrNO_2$, $Br_2$ and $ClNO_2$ from the reaction between sea spray aerosol and $N_2O_5$. *J. Aerosol Sci.* **25**(suppl. 1), S277–S278.

Behnke W., George C., Scheer V., and Zetzsch C. (1997) Production and decay of $ClNO_2$ from the reaction of gaseous $N_2O_5$ with NaCl solution: bulk and aerosol experiments. *J. Geophys. Res.* **102**, 3795–3804.

Behnke W., Elend M., Krüger U., and Zetzsch C. (1999) The influence of NaBr/NaCl ratio on the $Br^-$-catalyzed production of halogenated radicals. *J. Atmos. Chem.* **34**, 87–99.

Behnke W., Elend M., Krüger H.-U., and Zetzsch C. (2000) The interaction between sea-salt aerosol and marine gaseous chemistry. In *Annual Report 1999 EUROTRAC2—Chemical Mechanism Development*. International Scientific Secretariat, GSF, München, Germany, pp. 155–160.

Beine H. J., Jaffe D. A., Stordal F., Engardt M., Solberg S., Schmidbauer N., and Holmén K. (1997) $NO_x$ during ozone depletion events in the artic troposphere at Ny-Ålesund, Svalbard. *Tellus* **49B**, 556–565.

Blake N. L., Blake D. R., Sive B. C., Chen T.-Y., Rowland F. S., Collins J. E., Jr., Sachse G. W., and Anderson B. E. (1996) Biomass burning emissions and vertical distribution of atmospheric methyl halides and other reduced carbon gases in the South Atlantic region. *J. Geophys. Res.* **101**, 24151–24164.

Blanchard D. (1963) Electrification of the atmosphere. *Prog. Oceanogr.* **1**, 71–202.

Blanchard D. C. (1964) Sea-to-air transport of surface active material. *Science* **146**, 396–397.

Blanchard D. C. (1985) The oceanic production of atmospheric sea salt. *J. Geophys. Res.* **90**, 961–963.

Blanchard D. C. and Woodcock A. H. (1957) Bubble formation and modification in the sea and its meteorological significance. *Tellus* **9**, 145–158.

Blanchard D. C. and Woodcock A. H. (1980) The production, concentration and vertical distribution of the sea-salt aerosol. *Ann. NY Acad. Sci.* **338**, 330–347.

Bloss W. J., Rowley D. M., Cox R. A., and Jones R. L. (2001) Kinetics and products of the IO self-reaction. *J. Phys. Chem. A* **105**, 7840–7854.

Bobrowski N., Hönninger G., Galle B., and Platt U. (2003) Detection of bromine monoxide in a volcanic plume. *Nature* **423**, 273–276.

Bottenheim J. W., Gallant A. J., and Brice K. A. (1986) Measurements of $NO_y$ species and $O_3$ at 82° N latitude. *Geophys. Res. Lett.* **13**, 113–116.

Bottenheim J. W., Dibb J. E., Honrath R. E., and Shepson P. B. (2002) An introduction to the ALERT 2000 and SUMMIT 2000 Arctic research studies. *Atmos. Environ.* **36**, 2467–2469.

Boucher O., Moulin C., Belviso S., Aumont O., Bopp L., Cosme E., von Kuhlmann R., Lawrence C. G., Pham M., Redyy M. S., Sciare J., and Venkataraman C. (2002) Sensitivity study of dimethylsulphide (DMS) atmospheric concentrations and sulphate aerosol indirect radiative forcing to the DMS source representation and oxidation. *Atmos. Chem. Phys. Discuss.* **2**, 1181–1216.

Brasseur G. P., Orlando J. J., and Tyndall G. S. (eds.) (1999) *Atmospheric Chemistry and Global Change*. Oxford University Press, New York, Oxford.

Bridgman H. A. *et al.* (1989) Meteorology and haze structure during AGASP-II: Part 2. Canadian Arctic flights, 13–16 April 1986. *J. Atmos. Chem.* **9**, 49–70.

Brimblecombe P. and Clegg S. L. (1989) Erratum. *J. Atmos. Chem.* **8**, 95.

Butler J. H., Battle M., Bender M. L., Montzka S. A., Clarke A. D., Saltzman E. S., Sucher C. M., Severinghaus J. P., and Elkins J. W. (1999) A record of atmospheric halocarbons during the twentieth century from polar firn air. *Nature* **399**, 749–755.

Carl S. A. and Crowley J. N. (2002) 298 K rate coefficients for reaction of OH with i-$C_3H_7I$, n-$C_3H_7I$ and $C_3H_8$. *Atmos. Chem. Phys.* **1**, 1–7.

Carpenter L. J., Sturges W. T., Penkett S. A., Liss P. S., Alicke B., Hebestreit K., and Platt U. (1999) Short-lived alkyl iodides and bromides at Mace Head, Ireland: links to biogenic sources and halogen oxide production. *J. Geophys. Res.* **104**, 1679–1689.

Carpenter L. J., Hebestreit K., Platt U., and Liss P. S. (2001) Coastal zone production of IO precursors: a 2-dimensional study. *Atmos. Chem. Phys.* **1**, 9–18.

Cauer H. (1939) Schwankungen der Jodmenge der Luft in Mitteleuropa, deren Ursprung und deren Bedeutung für den Jodgehalt unserer Nahrung. *Angew. Chem.* **52**, 625–628.

Cauer H. (1951) Some problems of atmospheric chemistry. *Compend. Meteorol.* 1126–1138.

Chameides W. L. and Davis D. D. (1980) Iodine: its possible role in tropospheric photochemistry. *J. Geophys. Res.* **85**, 7383–7398.

Chameides W. L. and Stelson A. W. (1992) Aqueous-phase chemical processes in deliquescent sea-salt aerosols: a mechanism that couples the atmospheric cycles of S and sea salt. *J. Geophys. Res.* **97**, 20565–20580.

Chance K. (1998) Analysis of BrO Measurements from the Global Ozone Monitoring Experiment. *Geophys. Res. Lett.* **25**, 3335–3338.

Chang S., Tanaka P., McDonald-Buller E., and Allan D. T. (2001) *Emission Inventory for Atomic Chlorine Precursors in Southeast Texas*. Technical Report, University of Texas.

Chang S., McDonald-Buller E., Kimura Y., Yarwood G., Neece J., Russell M., Tanaka P., and Allen D. (2002) Sensitivity of urban ozone formation to chlorine emission estimates. *Atmos. Environ.* **36**, 4991–5003.

Chatfield R. B. and Crutzen P. J. (1990) Are there interactions of iodine and sulfur species in marine air photochemistry? *J. Geophys. Res.* **95**, 22319–22341.

Cicerone R. (1981) Halogens in the atmosphere. *Rev. Geophys. Space Phys.* **19**, 123–139.

Cicerone R. J., Heidt L. E., and Pollack W. H. (1988) Measurements of atmospheric methyl-bromide and bromoform. *J. Geophys. Res.* **93**, 3745–3750.

Clarke A. D. (1989) *In-situ* measurements of the aerosol size distribution, physiochemistry and light absorption properties of Arctic haze. *J. Atmos. Chem.* **9**, 255–266.

Class T. and Ballschmiter K. (1986) Chemistry of organic traces in air VIII: sources and distribution of bromo- and bromochloromethanes in marine air and surface water of the Atlantic Ocean. *J. Atoms. Chem.* **6**, 35–46.

Cox R. A. and Coker G. B. (1983) Absorption cross section and kinetics of IO in the photolysis of $CH_3I$ in the presence of ozone. *J. Phys. Chem.* **87**, 4478–4484.

Cox R. A., Bloss W. J., Jones R. L., and Rowley D. M. (1999) OIO and the atmospheric cycle of iodine. *Geophys. Res. Lett.* **26**, 1857–1860.

Crutzen P. J., Heidt L. E., Krasnec J. P., Pollock W. H., and Seiler W. (1979) Biomass burning as a source of atmospheric gases: CO, $H_2$, $N_2O$, NO, $CH_3Cl$ and COS. *Nature* **282**, 253–256.

Davies W. E., Vaughan G., and O'Connor F. M. (1998) Observation of near-zero ozone concentrations in the upper troposphere at midlatitudes. *Geophys. Res. Lett.* **25**, 1173–1176.

Davis D., Crawford J., Liu S., McKeen S., Bandy A., Thornton D., Rowland F., and Blake D. (1996) Potential impact of iodine on tropospheric levels of ozone and other critical oxidants. *J. Geophys. Res.* **101**, 2135–2147.

de Haan D. O., Brauers T., Oum K., Stutz J., Nordmeyer T., and Finlayson-Pitts B. J. (1999) Heterogeneous chemistry in the troposphere: experimental approaches and applications to the chemistry of sea salt particles. *Int. Rev. Phys. Chem.* **18**, 343–385.

de Laat A. T. J. and Lelieveld J. (2000) Diurnal ozone cycle in the tropical and subtropical marine boundary layer. *J. Geophys. Res.* **105**, 11547–11559.

DeMore W. B., Sander S. P., Golden D. M., Hampson R. F., Kurylo M. J., Howard C. J., Ravishankara A. R., Kolb C. E., and Molina M. J. (1997) *Chemical Kinetics and Photochemical Data for Use in Stratospheric Modeling*. Technical Report JPL Publications 97-4, Jet Propulsion Laboratory, Pasadena, CA.

de Serves C. (1994) Gas phase formaldehyde and peroxide measurements in the Arctic atmosphere. *J. Geophys. Res.* **99**, 25391–25398.

Dickerson R. R., Rhoads K. P., Carsey T. P., Oltmans S. J., Burrows J. P., and Crutzen P. J. (1999) Ozone in the remote marine boundary layer: a possible role for halogens. *J. Geophys. Res.* **104**, 21385–21395.

Dimmer C. H., Simmonds P. G., Nickless G., and Bassford M. R. (2001) Biogenic fluxes of halomethanes from Irish peatland ecosystems. *Atmos. Environ.* **35**, 321–330.

Disselkamp R. S., Howd C. D., Chapman E. G., Barchet W. R., and Colson S. D. (1999) BrCl production in NaBr/NaCl/HNO$_3$/O$_3$ solutions representative of sea-salt aerosols in the marine boundary layer. *Geophys. Res. Lett.* **26**, 2183–2186.

Dominé F. and Shepson P. B. (2002) Air–snow interactions and atmospheric chemistry. *Science* **297**, 1506–1510.

Donnell E. A., Fish D. J., Dicks E. M., and Thorpe A. J. (2001) Mechanisms for pollutant transport between the boundary layer and the free troposphere. *J. Geophys. Res.* **106**, 7847–7856.

Duce R. A. (1969) On the source of gaseous chlorine in the marine atmosphere. *J. Geophys. Res.* **74**, 4597–4599.

Duce R. A. and Woodcock A. H. (1971) Difference in chemical composition of atmospheric sea salt particles produced in the surf zone and on the open sea in Hawaii. *Tellus* **23**, 427–435.

Duce R. A., Wasson J. T., Winchester J. W., and Burns F. (1963) Atmospheric iodine, bromine, and chlorine. *J. Geophys. Res.* **68**, 3943–3947.

Duce R. A., Winchester J. W., and van Nahl T. W. (1965) Iodine, bromine, and chlorine in the Hawaiian marine atmosphere. *J. Geophys. Res.* **70**, 1775–1799.

Duce R. A., Woodcock A. H., and Moyers J. L. (1967) Variations of ion ratios with size among particles in tropical oceanic air. *Tellus* **19**, 369–379.

Duce R. A., Arimoto R., Ray B. J., Unni C. K., and Harder P. J. (1983) Atmospheric trace elements at Enewetak Atoll: 1. Concentrations, sources, and temporal variability. *J. Geophys. Res.* **88C**, 5321–5342.

Dyrssen D. and Fogelquist E. (1981) Bromoform concentrations of the arctic ocean in the Svalbard area. *Oceanolog. Acta* **4**, 313–317.

Ebinghaus R., Kock H. H., Temme C., Einax J. W., Löwe A. G., Richter A., Burrows J. P., and Schroeder W. H. (2002) Antarctic springtime depletion of atmospheric mercury. *Environ. Sci. Technol.* **36**, 1238–1244.

Eigen M. and Kustin K. (1962) The kinetics of halogen hydrolysis. *J. Am. Chem. Soc.* **84**, 1355–1361.

Erickson D. J., Merrill J. T., and Duce R. A. (1986a) Seasonal estimates of global atmospheric sea-salt distributions. *J. Geophys. Res.* **91**, 1067–1072.

Erickson D. J., Merrill J. T., and Duce R. A. (1986b) Seasonal estimates of global oceanic whitecap coverage. *J. Geophys. Res.* **91**, 12975–12977, 13107.

Erickson D. J., Seuzaret C., Keene W. C., and Gong S. L. (1999) A general circulation model based calculation of HCl and ClNO$_2$ production from sea salt dechlorination: reactive chlorine emissions inventory. *J. Geophys. Res.* **104**, 8347–8372.

Eriksson E. (1959a) The yearly circulation of chloride and sulfur in nature; meteorological, geochemical and pedological implications: Part I. *Tellus* **11**, 375–403.

Eriksson E. (1959b) The yearly circulation of chloride and sulfur in nature; meteorological, geochemical and pedological implications: Part II. *Tellus* **12**, 63–109.

Exton H. J., Latham J., Park P. M., Smith M. H., and Allan R. R. (1986) The production and dispersal of maritime aerosol. In *Oceanic Whitecaps* (eds. E. C. Monahan and G. M. Niocaill). D. Reidel, Norwell, MA, pp. 175–193.

Facchini M. C., Mircea M., Fuzzi S., and Charlson R. J. (1999) Cloud albedo enhancement by surface-active organic solutes in growing droplets. *Nature* **401**, 257–259.

Fan S.-M. and Jacob D. J. (1992) Surface ozone depletion in Arctic spring sustained by bromine reactions on aerosols. *Nature* **359**, 522–524.

Farman J. C., Gardiner B. G., and Shanklin J. D. (1985) Large losses of total ozone in Antarctica reveal seasonal ClO$_x$/NO$_x$ interaction. *Nature* **315**, 207–210.

Fickert S., Helleis F., Adams J. W., Moortgat G. K., and Crowley J. N. (1998) Reactive uptake of ClNO$_2$ on aqueous bromide solutions. *J. Phys. Chem. A* **102**, 10689–10696.

Fickert S., Adams J. W., and Crowley J. N. (1999) Activation of Br$_2$ and BrCl via uptake of HOBr onto aqueous salt solutions. *J. Geophys. Res.* **104**, 23719–23727.

Finlayson-Pitts B. J., Ezell M. J., and Pitts J. N., Jr. (1989) Formation of chemically active chlorine compounds by reactions of atmospheric NaCl particles with gaseous N$_2$O$_5$ and ClONO$_2$. *Nature* **337**, 241–244.

Finlayson-Pitts B. J., Livingston F. E., and Berko H. N. (1990) Ozone destruction and bromine photochemistry at ground level in the Arctic spring. *Nature* **343**, 622–625.

Fitzenberger R., Bösch H., Camy-Peyret C., Chipperfield M. P., Harder H., Platt U., Sinnhuber B.-M., Wagner T., and Pfeilsticker K. (2000) First profile measurements of tropospheric BrO. *Geophys. Res. Lett.* **27**, 2921–2924.

Foster K. L., Plastridge R. A., Bottenheim J. W., Shepson P. B., Finlayson-Pitts B. J., and Spicer C. W. (2001) The Role of Br$_2$ and BrCl in Surface Ozone Destruction at Polar Sunrise. *Science* **291**, 471–474.

Frenzel A., Scheer V., Sikorski R., George C., Behnke W., and Zetzsch C. (1998) Heterogeneous interconversion reactions of BrNO$_2$, ClNO$_2$, Br$_2$, and Cl$_2$. *J. Phys. Chem. A* **102**, 1329–1337.

Fridlind A. M. and Jacobson M. Z. (2000) A study of gas-aerosol equilibrium and aerosol pH in the remote marine boundary layer during the First Aerosol Characterization Experiment (ACE 1). *J. Geophys. Res.* **105**, 17325–17340.

Frieß U. (1997) Spektroskopische Messungen stratosphärischer Spurenstoffe auf der Neumayer-Station (Antarktis) in den Jahren 1994/95. Master's Thesis, Universität Heidelberg, Germany.

Frieß U., Wagner T., Pundt I., Pfeilsticker K., and Platt U. (2001) Spectroscopic measurements of tropospheric iodine oxide at Neumayer Station, Antarctica. *Geophys. Res. Lett.* **28**, 1941–1944.

Gabriel R., von Glasow R., Sander R., Andreae M. O., and Crutzen P. J. (2002) Bromide content of sea-salt aerosol particles collected over the Indian Ocean during IN-DOEX 1999. *J. Geophys. Res.* **107**, 8032, doi: 10.1029/2001JD001133.

Galbally I. E., Bentley S. T., and Meyer C. P. M. (2000) Mid-latitude marine boundary-layer ozone destruction at visible sunrise observed at Cape Grim, Tasmania, 41° S. *Geophys. Res. Lett.* **27**, 3841–3844.

George C., Ponche J. L., Mirabel P., Behnke W., Scheer V., and Zetzsch C. (1994) Study of the uptake of N$_2$O$_5$ by water and NaCl solutions: *J. Phys. Chem.* **98**, 8780–8784.

Ghosal S., Shbeeb A., and Hemminger J. C. (2000) Surface Segregation of Bromine doped NaCl: implications for the seasonal variations in Arctic ozone. *Geophys. Res. Lett.* **27**, 1879–1882.

Giese B., Laturnus F., Adams F. C., and Wiencke C. (1999) Release of volatile iodinated C$_1$–C$_4$ hydrocarbons by marine macroalgae from various climate zones. *Environ. Sci. Technol.* **33**, 2432–2439.

Gilles M. K., Turnipseed A. A., Burkholder J. B., Ravishankara A. R., and Solomon S. (1997) Kinetics of the IO radical: 2. Reaction of IO with BrO. *J. Phys. Chem. A* **101**, 5526–5534.

Gong S. L., Barrie L. A., and Blanchet J.-P. (1997) Modeling sea-salt aerosols in the atmosphere: 1. Model development. *J. Geophys. Res.* **102**, 3805–3818.

Goodwin K. D., North W. J., and Lidstrom M. E. (1997) Production of bromoform and dibromomethane by Giant Kelp: factors affecting release and comparison to anthropogenic bromine sources. *Limnol. Oceanogr.* **42**, 1725–1734.

Graedel T. E. and Keene W. C. (1995) Tropospheric budget of reactive chlorine. *Global Biogeochem. Cycles* **9**, 47–77.

Grenfell J. L., Harrison R. M., Allen A. G., Shi J. P., Penkett S. A., O'Dowd C. D., Smith M. H., Hill M. K., Robertson L., Hewitt C. N., Davison B., Lewis A. C., Creasey D. J., Heard D. E., Hebestreit K., Alicke B., and James J. (1999) An analysis of rapid increases in condensation nuclei concentrations at a remote coastal site in western Ireland. *J. Geophys. Res.* **104**, 13771–13780.

Gribble G. W. (2000) The natural production of organobromine compounds. *Environ. Sci. Pollut. Res.* **7**, 37–49.

Guelle W., Schulz M., Balkanski Y., and Dentener F. (2001) Influence of the source formulation on modeling the atmospheric global distribution of sea salt aerosol. *J. Geophys. Res.* **106**, 27509–27524.

Haag W. R. and Hoigné J. (1983) Ozonation of bromide-containing waters: kinetics of formation of hypobromous acid and bromate. *Environ. Sci. Technol.* **17**, 261–267.

Hansson H.-C., Rood M. J., Koloutsou-Vakakis S., Hämeri K., Orsini D., and Wiedensohler A. (1998) NaCl aerosol particle hygroscopicity dependence on mixing with organic compounds. *J. Atmos. Chem.* **31**, 321–346.

Harder H., Camy-Peyret C., Ferlemann F., Fitzenberger R., Hawat T., Osterkamp H., Schneider M., Perner D., Platt U., Vradelis P., and Pfeilsticker K. (1998) Stratospheric BrO profiles measured at different latitudes and seasons: atmospheric observations. *Geophys. Res. Lett.* **25**, 3843–3846.

Harnisch J. (1999) Reactive fluorine compounds. In *The Handbook of Environmental Chemistry: Volume 4 Part E. Air Pollution* (eds. P. Fabian and O. N. Singh). Springer, Heidelberg, Germany, pp. 81–112.

Harper D. B. (1985) Halomethane from halide ion—a highly efficient fungal conversion of environmental significance. *Nature* **315**, 55–57.

Harris G. W., Klemp D., and Zenker T. (1992) An upper limit on the HCl near-surface mixing ratio over the Atlantic measured using TDLAS. *J. Atmos. Chem.* **15**, 327–332.

Hausmann M. and Platt U. (1994) Spectroscopic measurement of bromine oxide and ozone in the high Arctic during Polar Sunrise Experiment 1992. *J. Geophys. Res.* **99**, 25399–25413.

Hebestreit K., Stutz J., Rosen D., Matveiv V., Peleg M., Luria M., and Platt U. (1999) DOAS measurements of tropospheric bromine oxide in mid-latitudes. *Science* **283**, 55–57.

Hebestreit K., Hönninger G., Stutz J., Alicke B., and Platt U. (2000) Measurements of halogen oxides in the troposphere. *Geophys. Res. Abstr.* **2**, 1061.

Hedgecock I. M. and Pirrone N. (2001) Mercury and photochemistry in the marine boundary layer-modelling studies suggest the *in situ* production of reactive phase mercury. *Atmos. Environ.* **35**, 3055–3062.

Hegels E., Crutzen P. J., Klüpfel T., Perner D., and Burrows J. P. (1998) Global distribution of atmospheric bromine-monoxide from GOME on earth observing satellite ERS-2. *Geophys. Res. Lett.* **25**, 3127–3130.

Hegg D. A., Covert D. S., Crahan K., and Jonssen H. (2002) The dependence of aerosol light-scattering on RH over the Pacific Ocean. *Geophys. Res. Lett.* **29**, doi: 10.1029/2001GL014495.

Herbert G. A. *et al.* (1989) Meteorology and haze structure during AGASP-II: Part I. Alaskan Arctic flights, 2–10 April 1986. *J. Atoms. Chem.* **9**, 17–48.

Himmelmann S., Orphal J., Bovensmann H., Richter A., Ladstätter-Weissenmayer A., and Borrows J. P. (1996) First observation of the OIO molecule by time-resolved flash photolysis absorption spectroscopy. *Chem. Phys. Lett.* **251**, 330–334.

Hirokawa J., Onaka K., Kajii Y., and Akimoto H. (1998) Heterogeneous processes involving sodium halide particles and ozone: molecular bromine release in the marine boundary layer in the absence of nitrogen oxides. *Geophys. Res. Lett.* **25**, 2449–2452.

Hoffman E. J. and Duce R. A. (1977) Organic carbon in marine atmospheric particulate matter: concentration and particle size distribution. *Geophys. Res. Lett.* **4**, 449–452.

Hoffmann T., O'Dowd C. D., and Seinfeld J. H. (2001) Iodine oxide homogeneous nucleation: an explanation for coastal new particle production. *Geophys. Res. Lett.* **28**, 1949–1952.

Holmes N. S., Adams J. W., and Crowley J. N. (2001) Uptake and reaction of HOI and $IONO_2$ on frozen and dry NaCl/NaBr surfaces and $H_2SO_4$. *Phys. Chem. Chem. Phys.* **3**, 1679–1687.

Hönninger G. (2002) Halogen oxide studies in the boundary layer by multi axis differential optical absorption spectroscopy and active longpath-DOAS. PhD Thesis, Universität Heidelberg, Germany, http://www.ub.uni-heidelberg.de/archiv/1940

Hönninger G. and Platt U. (2002) Observations of BrO and its vertical distribution during surface ozone depletion at Alert. *Atoms. Environ.* **36**, 2481–2489.

Honrath R. E., Peterson M. C., Guo S., Dibb J. E., Shepson P. B., and Campbell B. (1999) Evidence of $NO_x$ production within or upon ice particles in the Greenland snowpack. *Geophys. Res. Lett.* **26**, 695–698.

Honrath R. E., Guo S., Peterson M. C., Dziobak M. P., Dibb J. E., and Arsenault M. A. (2000a) Photochemical production of gas phase $NO_x$ from ice crystal $NO_3^-$. *J. Geophys. Res.* **105**, 24183–24190.

Honrath R. E., Peterson M. C., Dziobak M. P., Dibb J. E., Arsenault M. A., and Green S. A. (2000b) Release of $NO_x$ from sunlight-irradiated midlatitude snow. *Geophys. Res. Lett.* **27**, 2237–2240.

Hopper J. F., Peters B., Yokouchi Y., Niki H., Jobson B. T., Shepson P. B., and Muthuramu K. (1994) Chemical and meteorological observations at ice camp SWAN during Polar Sunrise Experiment 1992. *J. Geophys. Res.* **99**, 25489–25498.

Hov Ø. (1985) The effect of chlorine on the formation of photochemical oxidants in southern Telemark, Norway. *Atmos. Environ.* **19**, 471–485.

Huebert B. J., Howell S., Laj P., Johnson J. E., Bates T. S., Quinn P. K., Yegorov V., Clarke A. D., and Porter J. N. (1993) Observations of the atmospheric sulfur cycle on SAGA 3. *J. Geophys. Res.* **98**, 16985–16995.

Hunter K. A. (1997) Chemistry of the sea-surface microlayer. In *The Sea Surface and Global Change* (eds. P. S. Liss and R. A. Duce). Cambridge University Press, Cambridge, UK; New York, pp. 287–319.

Impey G. A., Mihele C. M., Anlauf K. G., Barrie L. A., Hastie D. R., and Shepson P. B. (1999) Measurements of photolyzable halogen compounds and bromine radicals during the Polar Sunrise Experiment 1997. *J. Atmos. Chem.* **34**, 21–37.

Ingham T., Bauer D., Sander R., Crutzen P. J., and Crowley J. N. (1999) Kinetics and products of the reactions BrO + DMS and Br + DMS at 298 K. *J. Phys. Chem. A* **103**, 7199–7209.

Ingham T., Cameron M., and Crowley J. N. (2000) Photodissociation of IO (355 nm) and OIO (532 nm): quantum yields for $O(^3P)$/I production. *J. Phys. Chem. A* **104**, 8001–8010.

Jacob D. J. (2000) Heterogeneous chemistry and tropospheric ozone. *Atmos. Environ.* **34**, 2131–2159.

Jacobi H.-W., Frey M. M., Hutterli M. A., Bales R. C., Schrems O., Cullen N. J., Steffen K., and Koehler C.

(2002) Measurements of hydrogen peroxide and formaldehyde exchange between the atmosphere and surface snow at Summit, Greenland. *Atmos. Environ.* **36**, 2619–2628.

Jaenicke R. (1988) Aerosol physics and chemistry. In *Landolt-Börnstein "Zahlenwerte und Funktionen aus Naturwissenschaften und Technik"*. Springer, vol. 4b, pp. 391–457.

Jenkin M. E. (1993) A comparative assessment of the role of iodine photochemistry in tropospheric ozone depletion. In *The Tropospheric Chemistry of Ozone in the Polar Regions*, NATO ASI Series (eds. H. Niki and K. H. Becker). Springer, Berlin, and New York, vol. 17, pp. 405–416.

Jenkin M. E., Cox R. A., and Candeland D. E. (1985) Photochemical aspects of tropospheric iodine behaviour. *J. Atmos. Chem.* **2**, 359–375.

Jimenez J. L., Cocker D. R., Bahreini R., Zhuang H., Varutbangkul V., Flagan R. C., Seinfeld J. H., O'Dowd C., and Hoffmann T. (2003) New particle formation from photooxidation of diiodomethane ($CH_2I_2$). *J. Geophys. Res.* **108**, 4318, 10.1029/2002JD002452.

Jobson B. T., Niki H., Yokouchi Y., Bottenheim J., Hopper F., and Leaitch R. (1994) Measurements of $C_2$–$C_6$ hydrocarbons during the Polar Sunrise 1992 Experiment: evidence for Cl atom and Br atom chemistry. *J. Geophys. Res.* **99**, 25355–25368.

Jobson B. T., Parrish D. D., Goldan P., Kuster W., Fehsenfeld F. C., Blake D. R., Blake N. J., and Niki H. (1998) Spatial and temporal variability of nonmethane hydrocarbon mixing ratios and their relation to photochemical lifetime. *J. Geophys. Res.* **103**, 13557–13567.

Johansen A. M., Siefert R. L., and Hoffmann M. R. (1999) Chemical characterization of ambient aerosol collected during the southwest monsoon and intermonsoon seasons over the Arabian Sea: anions and cations. *J. Geophys. Res.* **104**, 26325–26347.

Johansen A. M., Siefert R. L., and Hoffmann M. R. (2000) Chemical composition of aerosols collected over the tropical North Atlantic Ocean. *J. Geophys. Res.* **105**, 15277–15312.

Jones A. E., Weller R., Wolff E. W., and Jacobi H.-W. (2000) Speciation and rate of photochemical NO and $NO_2$ production in Antarctic snow. *Geophys. Res. Lett.* **27**, 345–348.

Jones A. E., Weller R., Anderson P. S., Jacobi H.-W., Wolff E. W., Schrems O., and Miller H. (2001) Measurements of $NO_x$ emissions from the Antarctic snowpack. *Geophys. Res. Lett.* **28**, 1499–1502.

Junge C. E. (1963) *Air Chemistry and Radioactivity*. Academic Press, New York and London.

Jungwirth P. and Tobias D. J. (2000) Surface effects on aqueous ionic solvation: a molecular dynamics simulation study of NaCl at the air/water interface from infinite dilution to saturation. *J. Phys. Chem. B* **104**, 7702–7706.

Jungwirth P. and Tobias D. J. (2001) Molecular structure of salt solutions: a new view of the interface with implications for heterogeneous atmospheric chemistry. *J. Phys. Chem. A* **105**, 10468–10472.

Karlsson R. and Ljungström E. (1998) Formation of nitryl chloride from dinitrogen pentoxide in liquid sea salt aerosol. *Atmos. Environ.* **32**, 1711–1717.

Keene W. C. and Savoie D. L. (1998) The pH of deliquesced sea-salt aerosol in polluted marine air. *Geophys. Res. Lett.* **25**, 2181–2184.

Keene W. C. and Savoie D. L. (1999) Correction to Keene and Savoie (1998). *Geophys. Res. Lett.* **26**, 1315–1316.

Keene W. C., Pszenny A. A. P., Jacob D. J., Duce R. A., Galloway J. N., Schultz-Tokos J. J., Sievering H., and Boatman J. F. (1990) The geochemical cycling of reactive chlorine through the marine troposphere. *Global Biogeochem. Cycles* **4**, 407–430.

Keene W. C., Maben J. R., Pszenny A. A. P., and Galloway J. N. (1993) Measurement technique for inorganic chlorine gases in the marine boundary layer. *Environ. Sci. Technol.* **27**, 866–874.

Keene W. C., Sander R., Pszenny A. A. P., Vogt R., Crutzen P. J., and Galloway J. N. (1998) Aerosol pH in the marine boundary layer: a review and model evaluation. *J. Aerosol Sci.* **29**, 339–356.

Keene W. C., Aslam M., Khalil K., Erickson D. J., McCulloch A., Graedel T. E., Lobert J. M., Aucott M. L., Gong S. L., Harper D. B., Kleiman G., Midgley P., Moore R. M., Seuzaret C., Sturges W. T., Benkovitz C. M., Koropalov V., Barrie L. A., and Li Y. F. (1999) Composite global emissions of reactive chlorine from anthropogenic and natural sources: reactive chlorine emissions inventory. *J. Geophys. Res.* **104**, 8429–8440.

Keene W. C., Lobert J. M., Maben J. R., Scharffe D. H., and Crutzen P. J. (2001) Emissions of volatile inorganic halogens, carboxylic acids, $NH_3$, and $SO_2$ from experimental burns of southern African biofuels. *EOS, Trans., AGU* **82**, Abstract a51A-0043.

Keene W. C., Pszenny A. A. P., Maben J. R., and Sander R. (2002) Variation of marine aerosol acidity with particle size. *Geophys. Res. Lett.* **29**, doi: 10.1029/2001GL013881.

Keene W. C., Maben J. R., Savoie D. J., Arimoto R., Merrill J. T., Milne P. J., Pszenny A. A. P., and Galloway J. N. Inorganic halogens in surface marine air at Bermuda during spring. *J. Geophys. Res.* (submitted for publication).

Keppler F., Eiden R., Niedan V., Pracht J., and Schöler H. F. (2000) Halocarbons produced by natural oxidation processes during degradation of organic matter. *Nature* **403**, 298–301.

Kerminen V.-M., Teinilä K., Hillamo R., and Pakkanen T. (1998) Substitution of chloride in sea-salt particles by inorganic and organic anions. *J. Aerosol Sci.* **29**, 929–942.

Khalil M. A. K., Moore R. M., Harper D. B., Lobert J. M., Erickson D. J., Koropalov V., Sturges W. T., and Keene W. C. (1999) Natural emissions of chlorine-containing gases: reactive chlorine emissions inventory. *J. Geophys. Res.* **104**, 8333–8346.

King D. B., Butler J. H., Montzka S. A., Yvon-Lewis S. A., and Elkins J. W. (2000) Implications of methyl bromide supersaturations in the temperature North Atlantic Ocean. *J. Geophys. Res.* **105**, 19763–19769.

Kirchner U., Benter T., and Schindler R. N. (1997) Experimental verification of gas phase bromine enrichment in reactions of HOBr with sea salt doped ice surfaces. *Ber. Bunsenges. Phys. Chem.* **101**, 975–977.

Kley D., Crutzen P. J., Smit H. G. J., Vömel H., Oltmans S. J., Grassl H., and Ramanathan V. (1996) Observations of near-zero ozone concentrations over the convective Pacific: effects on air chemistry. *Science* **274**, 230–233.

Knight G. P. and Crowley J. N. (2001) The reactions of IO with $HO_2$, NO and $CH_3SCH_3$: flow tube studies of kinetics and product formation. *Phys. Chem. Chem. Phys.* **3**, 393–401.

Knipping E. M. and Dabdub D. (2002) Modeling $Cl_2$ formation from aqueous NaCl particles: evidence for interfacial reactions and importance of $Cl_2$ decomposition in alkaline solution. *J. Geophys. Res.* **107**, doi: 10.1029/2001JD000867.

Knipping E. M., Lakin M. J., Foster K. L., Jungwirth P., Tobias D. J., Gerber R. B., Dabdub D., and Finlayson-Pitts B. J. (2000) Experiments and molecular/kinetics simulations of ion-enhanced interfacial chemistry on aqueous NaCl aerosols. *Science* **288**, 301–306.

Koop T., Kapilashrami A., Molina L. T., and Molina M. J. (2000) Phase transitions of sea-salt/water mixtures at low temperatures: implications for ozone chemistry in the polar marine boundary layer. *J. Geophys. Res.* **105**, 26393–26402.

Kreher K., Johnston P. V., Wood S. W., Nardi B., and Platt U. (1997) Ground-based measurements of tropospheric and stratospheric BrO at Arrival Heights, Antarctica. *Geophys. Res. Lett.* **24**, 3021–3024.

Kritz M. and Rancher J. (1980) Circulation of Na, Cl, and Br in the tropical marine atmosphere. *J. Geophys. Res.* **85**, 1633–1639.

Langendörfer U., Lehrer E., Wagenbach D., and Platt U. (1999) Observation of filterable bromine variabilities during Arctic tropospheric ozone depletion events in high (1 hour) time resolution. *J. Atmos. Chem.* **34**, 39–54.

Laturnus F. (1996) Volatile halocarbons released from Arctic macroalgae. *Mar. Chem.* **55**, 359–366.

Le Bras G. and Platt U. (1995) A possible mechanism for combined chlorine and bromine catalyzed destruction of tropospheric ozone in the Arctic. *Geophys. Res. Lett.* **22**, 599–602.

Lee C.-T. and Hsu W.-C. (2000) The measurements of liquid water mass associated with collected hygroscopic particles. *J. Aerosol Sci.* **31**, 189–197.

Lee-Taylor J. M., Brasseur G. P., and Yokouchi Y. (2001) A preliminary three-dimensional global model study of atmospheric methyl chloride distributions. *J. Geophys. Res.* **106**, 34221–34223.

Lehrer E., Wagenbach D., and Platt U. (1997) Aerosol chemical composition during tropospheric ozone depletion at Ny Ålesund/Svalbard. *Tellus* **49B**, 486–495.

Leser E. (1999) Das polare troposphärische Ozonloch. PhD Thesis, Universität Heidelberg, Germany.

Leser H., Hönninger G., and Platt U. (2003) MAX-DOAS measurements of BrO and NO$_2$ in the marine boundary layer. *Geophys. Res. Lett.* **30**, 1537, doi: 10.1029/2002GLO15811.

Lightowlers P. J. and Cape J. N. (1988) Sources and fate of atmospheric HCl in the UK and western Europe. *Atmos. Environ.* **22**, 7–15.

Lindberg S. E., Brooks S., Lin C.-J., Scott K. J., Landis M. S., Stevens R. K., Goodsite M., and Richter A. (2002) Dynamic oxidation of gaseous mercury in the Arctic troposphere at Polar Sunrise. *Environ. Sci. Technol.* **36**, 1245–1256.

Lobert J. M., Yvon-Lewis S. A., Butler J. H., Montzka S. A., and Myers R. C. (1997) Undersaturation of CH$_3$Br in the Southern Ocean. *Geophys. Res. Lett.* **24**, 171–172.

Lobert J. M., Keene W. C., Logan J. A., and Yevich R. (1999) Global chlorine emissions from biomass burning: reactive chlorine emissions inventory. *J. Geophys. Res.* **104**, 8373–8389.

Lovelock J. E. (1975) Natural halocarbons in the air and in the sea. *Nature* **256**, 193–194.

Lovelock J. E., Maggs R. J., and Wade R. J. (1973) Halogenated hydrocarbons in and over the Atlantic. *Nature* **241**, 194–196.

Lowenthal D. H. and Rahn K. A. (1985) Regional sources of pollution aerosol at Barrow, Alaska during winter 1979–1980 as deduced from elemental tracer. *Atmos. Environ.* **19**, 1995–2010.

Lu J. Y., Schroeder W. H., Barrie L. A., Steffen A., Welch H. E., Martin K., Lockhart L., Hunt R. V., Boila G., and Richter A. (2001) Magnification of atmospheric mercury deposition to polar regions in springtime: the link to tropospheric ozone depletion chemistry. *Geophys. Res. Lett.* **28**, 3219–3222.

Maenhaut W., Darzi M., and Winchester J. W. (1981) Seawater and nonseawater aerosol components in the marine atmosphere of Samoa. *J. Geophys. Res.* **86**, 3187–3193.

Marchand E. (1852) Sur la constitution physique et chimique des eaux naturelle. *Comput. Rendu.* **34**, 54–56.

Markovich G., Pollack S., Giniger R., and Cheshnovsky O. (1994) Photoelectron spectroscopy of Cl$^-$, Br$^-$, and I$^-$ solvated in water clusters. *J. Chem. Phys.* **101**, 9344–9353.

Martens C. S., Wesolowski J. J., Harriss R. C., and Kaifer R. (1973) Chlorine loss from Puerto Rican and San Francisco Bay area marine aerosols. *J. Geophys. Res.* **78**, 8778–8792.

Martin S., Drucker R., and Fort M. (1995) A laboratory study of frost flower growth on the surface of young sea-ice. *J. Geophys. Res.* **100**, 7027–7036.

Martinez M., Arnold T., and Perner D. (1999) The role of bromine and chlorine chemistry for arctic ozone depletion events in Ny-Alesund and comparison with model calculations. *Ann. Geophys.* **17**, 941–956.

Mason B. J. (1954) Bursting of air bubbles at the surface of sea water. *Nature* **174**, 470–471.

Mason B. J. (2001) The role of sea-salt particles as cloud condensation nuclei over the remote oceans. *Q. J. R. Meteorol. Soc.* **127**, 2023–2032.

Matveev V., Peleg M., Rosen D., Tov-Alper D. S., Hebestreit K., Stutz J., Platt U., Blake D., and Luria M. (2001) Bromine oxide–Ozone interaction over the Dead Sea. *J. Geophys. Res.* **106**, 10375–10387.

McConnell J. C., Henderson G. S., Barrie L., Bottenheim J., Niki H., Langford C. H., and Templeton E. M. J. (1992) Photochemical bromine production implicated in Arctic boundary-layer ozone depletion. *Nature* **355**, 150–152.

McCulloch A., Aucott M. L., Benkovitz C. M., Graedel T. E., Kleinman G., Midgley P. M., and Li Y.-F. (1999) Global emissions of hydrogen chloride and chloromethane from coal combustion, incineration and industrial activities: reactive chlorine emissions inventory. *J. Geophys. Res.* **104**, 8391–8403.

McElroy C. T., McLinden C. A., and McConnell J. C. (1999) Evidence for bromine monoxide in the free troposphere during the Arctic polar sunrise. *Nature* **397**, 338–341.

McFiggans G., Plane J. M. C., Allan B. J., Carpenter L. J., Coe H., and O'Dowd C. (2000) A modelling study of iodine chemistry in the marine boundary layer. *J. Geophys. Res.* **105**, 14371–14377.

Melville W. K. and Matusov P. (2002) Distribution of breaking waves at the ocean surface. *Nature* **417**, 58–63.

Michalowski B. A., Francisco J. S., Li S.-M., Barrie L. A., Bottenheim J. W., and Shepson P. B. (2000) A computer model study of multiphase chemistry in the Arctic boundary layer during polar sunrise. *J. Geophys. Res.* **105**, 15131–15145.

Mickle R. E., Bottenheim J. W., Leaitch R. W., and Evans W. (1989) Boundary layer ozone depletion during AGASP-II. *Atmos. Environ.* **23**, 2443–2449.

Middlebrook A. M., Murphy D. M., and Thomson D. S. (1998) Observations of organic material in individual marine particles at Cape Grim during the First Aerosol Characterization Experiment (ACE 1). *J. Geophys. Res.* **103**, 16475–16483.

Miller H. L., Weaver A., Sanders R. W., Arpag K., and Solomon S. (1997) Measurements of arctic sunrise surface ozone depletion events at Kangerlussuaq, Greenland (67° N, 51° W). *Tellus* **49B**, 496–509.

Ming Y. and Russell L. M. (2001) Predicted hygroscopic growth of sea salt aerosol. *J. Geophys. Res.* **106**, 28259–28274.

Misra A. and Marshall P. (1998) Computational investigations of iodine oxides. *J. Phys. Chem. A* **102**, 9056–9060.

Mochida M., Akimoto H., van den Bergh H., and Rossi M. J. (1998a) Heterogeneous kinetics of the uptake of HOBr on solid alkali metal halides at ambient temperature. *J. Phys. Chem. A* **102**, 4819–4828.

Mochida M., Hirokawa J., Kajii Y., and Akimoto H. (1998b) Heterogeneous reactions of Cl$_2$ with sea salts at ambient temperature: implications for halogen exchange in the atmosphere. *Geophys. Res. Lett.* **25**, 3927–3930.

Mochida M., Hirokawa J., and Akimoto H. (2000) Unexpected large uptake of O$_3$ on sea salts and the observed Br$_2$ formation. *Geophys. Res. Lett.* **27**, 2629–2632.

Moldanová J. and Ljungström E. (2001) Sea-salt aerosol chemistry in coastal areas: a model study. *J. Geophys. Res.* **106**, 1271–1296.

Molina M. J. and Rowland F. S. (1974) Stratospheric sink for chlorofluoromethanes: chlorine-atom catalysed destruction of ozone. *Nature* **249**, 810–812.

Monahan E. C., Spiel D. E., and Davidson K. L. (1986) A model of marine aerosol generation via whitecaps and wave disruption. In *Oceanic Whitecaps* (eds. E. C. Monahan and G. M. Niocaill). D. Reidel, Norwell, MA, pp. 167–174.

Moore R. M. and Webb M. (1996) The relationship between methyl bromide and chlorophyll $\alpha$ in high latitude ocean waters. *Geophys. Res. Lett.* **23**, 2951–2954.

Moortgat G. K., Meller R., and Schneider W. (1993) Temperature dependence (256–296 K) of the absorption cross sections of bromoform in the wavelength range 285–360 nm. In *Tropospheric Chemistry of Ozone in Polar Regions*, NATO ASI Ser., Subser. 1: Global Environmental Change (eds. H. Niki and K. H. Becker). Springer, New York, pp. 359–370.

Moyers J. L. and Duce R. A. (1972) Gaseous and particulate bromine in the marine atmosphere. *J. Geophys. Res.* **77**, 5330–5338.

Muramatsu Y. and Yoshida S. (1995) Volatilization of methyl iodide from the soil–plant system. *Atmos. Environ.* **29**, 21–25.

Murphy D. M., Anderson J. R., Quinn P. K., McInnes L. M., Brechtel F. J., Kreidenweis S. M., Middlebrook A. M., Pósfai M., Thomson D. S., and Buseck P. R. (1998) Influence of sea-salt on aerosol radiative properties in the Southern Ocean marine boundary layer. *Nature* **392**, 62–65.

Nagao I., Matsumoto K., and Tanaka H. (1999) Sunrise ozone destruction found in the sub-tropical marine boundary layer. *Geophys. Res. Lett.* **26**, 3377–3380.

O'Dowd C. and Smith M. H. (1993) Physicochemical properties of aerosols over the northeast Atlantic: evidence for wind-speed-related submicron sea-salt aerosol production. *J. Geophys. Res.* **98**, 1137–1149.

O'Dowd C., McFiggans G., Creasey D. J., Pirjola L., Hoell C., Smith M. H., Allan B. J., Plane J. M. C., Heard D. E., Lee J. D., Pilling M. J., and Kulmala M. (1999) On the photochemical production of new particles in the coastal boundary layer. *Geophys. Res. Lett.* **26**, 1707–1710.

O'Dowd C. D., Smith M. H., Consterdine I. E., and Lowe J. A. (1997) Marine aerosol, sea-salt, and the marine sulphur cycle: a short review. *Atmos. Environ.* **31**, 73–80.

O'Dowd C. D., Geever M., Hill M. K., Smith M. H., and Jennings S. G. (1998) New particle formation: nucleation rates and spatial scales in the clean marine coastal environment. *Geophys. Res. Lett.* **25**, 1661–1664.

O'Dowd C. D., Jimenez J. L., Bahreini R., Flagan R. C., Seinfeld J. H., Hämeri K., Pirjola L., Kulmala M., Jennings S. G., and Hoffmann T. (2002) Marine particle formation by biogenic iodine emissions. *Nature* **417**, 632–636.

Oltmans S. J. (1981) Surface ozone measurements in clean air. *J. Geophys. Res.* **86**, 1174–1180.

Oltmans S. J. and Komhyr W. (1986) Surface ozone distributions and variations from 1973 to 1984 measurements at the NOAA Geophysical Monitoring for Climate Change Baseline observatories. *J. Geophys. Res.* **91**, 5229–5236.

Open University Course Team (ed.) (1999) *Seawater: Its Composition, Properties and Behaviour*. Butterworth-Heinemann, Oxford.

Oum K., Lakin M. J., Haan D. O. O., Brauers T., and Finlayson-Pitts B. J. (1998a) Formation of molecular chlorine from the photolysis of ozone and aqueous sea-salt particles. *Science* **279**, 74–77.

Oum K. W., Lakin M. J., and Finlayson-Pitts B. J. (1998b) Bromine activation in the troposphere by the dark reaction of $O_3$. *Geophys. Res. Lett.* **25**, 3923–3926.

Palmer T. Y. (1976) Combustion sources of atmospheric chlorine. *Nature* **263**, 44–46.

Perner D. and Platt U. (1979) Detection of nitrous-acid in the atmosphere by differential optical-absorption. *Geophys. Res. Lett.* **6**, 917–920.

Perner D., Arnold T., Crowley J., Klüpfel T., Martinez M., and Seuwen R. (1999) The measurements of active chlorine in the atmosphere by chemical amplification. *J. Atmos. Chem.* **34**, 9–20.

Pinto J. P., Turco R. P., and Toon O. B. (1989) Self-limiting physical and chemical effects in volcanic eruption clouds. *J. Geophys. Res.* **94**, 11165–11174.

Platt U. (1999) Modern methods of the measurement of atmospheric trace gases—invited lecture. *Phys. Chem. Chem. Phys.* **1**, 5409–5415.

Platt U. (2000) Reactive halogen species in the mid-latitude troposphere—recent discoveries. *Water Air Soil Pollut.* **123**, 229–244.

Platt U. and Hönninger G. (2003) The role of halogen species in the troposphere. *Chemosphere* **52**, 325–338.

Platt U. and Moortgat G. K. (1999) Heterogeneous and homogeneous chemistry of reactive halogen compounds in the lower troposphere. *J. Atmos. Chem.* **34**, 1–8.

Pruppacher H. R. and Klett J. D. (1997) *Microphysics of Clouds and Precipitation*. Kluwer Academic, Dordrecht/Boston/London.

Pruvost J., Morin P. and Corre P. L. (2000) Volatile Halogenated Organic Compounds (VHOC) production during a spring phytoplankton bloom in a coastal system (Bay of Brest). Poster presented at the SOLAS Open Science Conference, February 20–24, 2000, Damp, Germany.

Pszenny A. A. P., Keene W. C., Jacob D. J., Fan S., Maben J. R., Zetwo M. P., Springer-Young M., and Galloway J. N. (1993) Evidence of inorganic chlorine gases other hydrogen chloride in marine surface air. *Geophys. Res. Lett.* **20**, 699–702.

Pszenny A. A. P., Keene W. C. and Maben J. R. (2002) Inorganic halogens and aerosol pH in the Hawaiian marine boundary layer. In *7th Scientific Conference of the International Global Atmospheric Chemistry Project (IGAC)*, September 18–25, 2002, Hersonissos, Crete, Greece.

Pundt I., Van Roozendael M., Wagner T., Richter A., Chipperfield M., Burrows J. P., Fayt C., Hendrick F., Pfeilsticker K., Platt U., and Pommereau J.-P. (2000) Simultaneous UV-vis Measurements of BrO form Balloon, Satellite and Ground: implications for tropospheric BrO. In *Proc. 5th European Symp. on Polar Stratospheric Ozone 1999*, Air Poll. Res. Report 73, EUR 19340 (eds. N. R. P. Harris, M. Guirlet, and G. T. Amanatidis). European Commission, Brussels, Belgium, pp. 316–319.

Quack B. and Wallace D. W. R. (2003) The air–sea flux of bromoform: controls, rates and implication. *Global Biogeochem. Cycles* **17**, 1023, doi: 10.1029/2002GB001890.

Raemdonck H., Maenhaut W., and Andreae M. O. (1986) Chemistry of marine aerosol over the tropical and equatorial Pacific. *J. Geophys. Res.* **91**, 8623–8636.

Ramacher B. (1997) Messung organischer Spurengase in der arktischen Troposphäre: Hinweise auf einen regional halogeninduzierten Ozonabbau. PhD Thesis, Forschungszentrum Jülich, Jülich, Germany.

Ramacher B., Rudolph J., and Koppmann R. (1997) Hydrocarbon measurements in the spring arctic troposphere during the ARCTOC 95 campaign. *Tellus* **49B**, 466–485.

Ramacher B., Rudolph J., and Koppmann R. (1999) Hydrocarbon measurements during tropospheric ozone depletion events: evidence for halogen atom chemistry. *J. Geophys. Res.* **104**, 3633–3653.

Ramanathan V. (1975) Greenhouse effect due to chlorofluorocarbons: climatic implications. *Science* **190**, 50–51.

Rancher J. and Kritz M. A. (1980) Diurnal fluctuations of Br and I in the tropical marine atmosphere. *J. Geophys. Res.* **85**, 5581–5587.

Rasmussen A., Kiilisholm S., Sørensen J. H., and Mikkelsen I. S. (1997) Analysis of tropospheric ozone measurements in Greenland. *Tellus* **49B**, 510–521.

Rasmussen R. A. and Khalil M. A. K. (1984) Gaseous bromine in the arctic and arctic haze. *Geophys. Res. Lett.* **11**, 433–436.

Rattigan O. V., Cox R. A., and Jones R. L. (1995) $Br_2$-sensitised decomposition of ozone—kinetics of the reaction $BrO + O_3 \rightarrow$ products. *J. Chem. Soc. Faraday Trans.* **91**, 4189–4197.

Redeker K. R., Wang N.-Y., Lowe J. C., McMillan A., Tyler S. C., and Cicerone R. J. (2000) Emissions of methyl halides and methane from rice paddies. *Science* **290**, 966–969.

Reifenhäuser W. and Heumann K. G. (1992a) Bromo- and Bromochloromethanes in the Antarctic atmosphere and the South Polar Sea. *Chemosphere* **24**, 1293–1300.

Reifenhäuser W. and Heumann K. G. (1992b) Determinations of methyl iodide in the Antarctic atmosphere and the South Polar Sea. *Atmos. Environ.* **26A**, 2905–2912.

Reinhardt T. E. and Ward D. E. (1996) Factors affecting methyl chloride emissions from forest biomass combustion. *Environ. Sci. Technol.* **29**, 825–833.

Rhew R. C., Miller B. R., and Weiss R. F. (2000) Natural methyl bromide and methyl chloride emissions from coastal salt marshes. *Nature* **403**, 292–295.

Richter A., Wittrock F., Eisinger M., and Burrows J. P. (1998) GOME observations of tropospheric BrO in northern hemispheric spring and summer 1997. *Geophys. Res. Lett.* **25**, 2683–2686.

Richter A., Wittrock F., Ladstätter-Weißenmayer A., and Burrows J. P. (2002) GOME measurements of stratospheric and tropospheric BrO. *Adv. Space Res.* **29**, 1667–1672.

Ridley B. A., Atlas E. L., Montzka D. D., Browell E. V., Cantrell C. A., Blake D. R., Blake N. J., Cinquini L., Coffey M. T., Emmons L. K., Cohen R. C., DeYoung R. J., Dibb J. E., Eisele F. L., Flocke F. M., Fried A., Grahek F. E., Grant W. B., Hair J. W., Hannigan J., Heikes B. J., Lefer B. L., Mauldin R. L., Moody J. L., Shetter R. E., Snow J. A., Talbot R. W., Thornton J. A., Walega J. G., Weinheimer A. J., Wert B. P., and Wimmers A. J. (2003) Ozone depletion events observed in the high latitude surface layer during the TOPSE Aircraft Program. *J. Geophys. Res.* **108**, 8356, doi: 10.1029/2001JD001507.

Robbins R. C., Cadle R. D., and Eckhardt D. L. (1959) The conversion of sodium chloride to hydrogen chloride in the atmosphere. *J. Meteor.* **16**, 53–56.

Röckmann T., Brenninkmeijer C. A. M., Crutzen P. J., and Platt U. (1999) Short term variations in the 13C/12C ratio of CO as a measure of Cl activation during tropospheric ozone depletion events in the Arctic. *J. Geophys. Res.* **104**, 1691–1697.

Roscoe H. K., Kreher K., and Friess U. (2001) Ozone loss episodes in the free Antarctic troposphere, suggesting a possible climate feedback. *Geophys. Res. Lett.* **28**, 2911–2914.

Rosenfeld D., Lahav R., Khain A., and Pinsky M. (2002) The role of sea spray in cleansing air pollution over ocean via cloud processes. *Science* **297**, 1667–1670.

Rowley D. M., Bloss W. J., Cox R. A., and Jones R. L. (2001) Kinetics and products of the IO + BrO Reaction. *J. Phys. Chem. A* **105**, 7855–7864.

Rudolph J., Koppmann R., and Plass-Dülmer C. (1996) The budgets of ethane and tetrachloroethene: Is there evidence for an impact of reactions with chlorine atoms in the troposphere? *Atmos. Environ.* **30**, 1887–1894.

Rudolph J., Ramacher B., Plass-Dülmer C., Müller K.-P., and Koppmann R. (1997) The indirect determination of chlorine atom concentration in the troposphere from changes in the patterns of non-methane hydrocarbons. *Tellus* **49B**, 592–601.

Sadanaga Y., Hirokawa J., and Akimoto H. (2001) Formation of molecular chlorine in dark condition: heterogeneous reaction of ozone with sea salt in the presence of ferric ion. *Geophys. Res. Lett.* **28**, 4433–4436.

Sæmundsdóttir S. and Matrai P. A. (1998) Biological production of methyl bromide by cultures of marine phytoplancton. *Limnol. Oceanogr.* **43**, 81–87.

Sander R. (1998) Comment on K. Oum *et al.* (1998a), http://www.mpch-mainz.mpg.de/~sander/res/oumetal.html

Sander R. and Crutzen P. J. (1996) Model study indicating halogen activation and ozone destruction in polluted air masses transported to the sea. *J. Geophys. Res.* **101**, 9121–9138.

Sander R., Vogt R., Harris G. W., and Crutzen P. J. (1997) Modeling the chemistry of ozone, halogen compounds, and hydrocarbons in the arctic troposphere during spring. *Tellus* **49B**, 522–532.

Sander R., Rudich Y., von Glasow R., and Crutzen P. J. (1999) The role of BrNO₃ in marine tropospheric chemistry: a model study. *Geophys. Res. Lett.* **26**, 2857–2860.

Sander S. P., Friedl R. R., DeMore W. B., Golden D. M., Kurylo M. J., Hampson R. F., Huie R. E., Moortgat G. K., Ravishankara A. R., Kolb C. E., and Molina M. J. (2000) *Chemical Kinetics and Photochemical Data for Use in Stratospheric Modeling.* Technical Report JPL Publication 00-3, Jet Propulsion Laboratory, Pasadena, CA.

Sander R., Keene W. C., Pszenny A. A. P., Arimoto R., Ayers G. P., Baboukas E., Cainey J. M., Crutzen P. J., Duce R. A., Hönninger G., Huebert B. J., Maenhaut W., Mihalopoulos N., Turekian V. C., and Van Dingenen R. (2003) Inorganic bromine compounds in the marine boundary layer: a critical review. *Atmos. Chem. Phys. Discuss.* **3**, 2963–3050.

Schall C. and Heumann K. G. (1993) GC determination of volatile organoiodine and organobromine compounds in Arctic seawater and air samples. *Fresenius J. Anal. Chem.* **346**, 717–722.

Schall C., Laturnus F., and Heumann K. G. (1994) Biogenic volatile organoiodine and organobromine compounds released from polar macroalgae. *Chemosphere* **28**, 1315–1324.

Schall C., Heumann K. G., de Mora S., and Lee P. A. (1996) Biogenic brominated and iodinated organic compounds in ponds on the McMurdo Ice Shelf, Antarctica. *Antarctic Sci.* **8**, 45–48.

Schall C., Heumann K. G., and Kirst G. O. (1997) Biogenic volatile organoiodine and organobromine hydrocarbons in the Atlantic Ocean from 42° N to 72° S. *Fresenius J. Anal. Chem.* **359**, 298–305.

Schnell R. C. (1983) Arctic haze and the Arctic Gas and Aerosol Sampling Program (AGASP). *Geophys. Res. Lett.* **11**, 361–364.

Schroeder W. H., Anlauf K. G., Barrie L. A., Lu J. Y., Steffen A., Schneeberger D. R., and Berg T. (1998) Arctic spring-time depletion of mercury. *Nature* **394**, 331–332.

Schweitzer F., Mirabel P., and George C. (1999) Heterogeneous chemistry of nitryl halides in relation to tropospheric halogen activation. *J. Atmos. Chem.* **34**, 101–117.

Seinfeld J. H. and Pandis S. N. (1998) *Atmospheric Chemistry and Physics.* Wiley, New York, Chichester, Weinheim.

Seisel S., Caloz F., Fenter F. F., van den Bergh H., and Rossi M. J. (1997) The heterogeneous reaction of NO₃ with NaCl and KBr: a nonphotolytic source of halogen atoms. *Geophys. Res. Lett.* **24**, 2757–2760.

Seisel S., Flückinger B., Caloz F., and Rossi M. J. (1999) Heterogeneous reactivity of the nitrate radical: reactions on halogen salt at ambient temperature and on ice in the presence of HX (X = Cl, Br, I) at 190 K. *Phys. Chem. Chem. Phys.* **1**, 2257–2266.

Seto F. Y. B. and Duce R. A. (1972) A laboratory study of iodine enrichment on atmospheric sea-salt particles produced by bubbles. *J. Geophys. Res.* **77**, 5339–5349.

Shaw M. A. and Rood M. J. (1990) Measurement of the crystallization humidities of ambient aerosol particles. *Atmos. Environ.* **24A**, 1837–1841.

Shon Z.-H. and Kim N. (2002) A modeling study of halogen chemistry's role in marine boundary layer ozone. *Atmos. Environ.* **36**, 4289–4298.

Sievering H., Boatman J., Gorman E., Kim Y., Anderson L., Ennis G., Luria M., and Pandis S. (1992) Removal of sulphur from the marine boundary layer by ozone oxidation in sea-salt aerosols. *Nature* **360**, 571–573.

Sievering H., Gorman E., Ley T., Pszenny A., Springer-Young M., Boatman J., Kim Y., Nagamoto C., and Wellman D. (1995) Ozone oxidation of sulfur in sea-salt aerosol particles during the Azores Marine Aerosol and Gas Exchange Experiment. *J. Geophys. Res.* **100**, 23075–23082.

Simpson W. R., King M. D., Beine H. J., Honrath R. E., and Zhou X. (2002) Radiation-transfer modeling of snow-pack photochemical processes during ALART 2000. *Atmos. Environ.* **36**, 2663–2670.

Singh H. B., Salas L. J., Shigeishi H., and Scritner E. (1979) Atmospheric halocarbons, hydrocarbons and sulfur hexafluoride: global distribution, source and sinks. *Science* **203**, 899–903.

Singh H. B., Gregory G. L., Anderson B., Browell E., Sachse G. W., Davis D. D., Crawford J., Bradshaw J. D., Talbot R., Blake D. R., Thornton D., Newell R., and Merrill J. (1996a) Low ozone in the marine boundary layer of the tropical Pacific Ocean: photochemical loss, chlorine atoms, and entrainment. *J. Geophys. Res.* **101**, 1907–1917.

Singh H. B., Thakur A. N., Chen Y. E., and Kanakidou M. (1996b) Tetrachloroethylene as an indicator of low Cl atom concentrations in the troposphere. *Geophys. Res. Lett.* **23**, 1529–1532.

Singh O. N. and Fabian P. (1999) Reactive bromine compounds. In *The Handbook of Environmental Chemistry: Volume 4, Part E. Air Pollution* (eds. P. Fabian and O. N. Singh). Springer, Heidelberg, Germany, pp. 1–43.

Smith M. H. and Harrison N. M. (1998) The sea spray generation function. *J. Aerosol Sci.* **29**, S189–S190.

Smith M. H., Park P. M., and Consterdine I. E. (1993) Marine aerosol concentrations and estimated fluxes over the sea. *Q. J. R. Meteorol. Soc.* **119**, 809–824.

Smith R. A. (1872) *Air and Rain: The Beginnings of a Chemical Climatology.* Longmans, Green (London).

Solberg S., Schmidbauer N., Semb A., and Stordal F. (1996) Boundary-layer ozone depletion as seen in the Norwegian arctic in spring. *J. Atmos. Chem.* **23**, 301–332.

Solomon S. (1990) Progress towards a quantitative understanding of Antarctic ozone depletion. *Nature* **347**, 347–354.

Solomon S., Garcia R. R., and Ravishankara A. R. (1994) On the role of iodine in ozone depletion. *J. Geophys. Res.* **99**, 20491–20499.

Solomon S., Garcia R. R., Rowland F. S., and Wuebbles D. J. (1986) On the depletion of Antarctic ozone. *Nature* **321**, 755–758.

Song C. H. and Carmichael G. R. (1999) The aging process of naturally emitted aerosol (sea-salt and mineral aerosol) during long range transport. *Atmos. Environ.* **33**, 2203–2218.

Spicer C. W., Chapman E. G., Finlayson-Pitts B. J., Plastridge R. A., Hubbe J. M., Fast J. D., and Berkowitz C. M. (1998) Unexpectedly high concentrations of molecular chlorine in coastal air. *Nature* **394**, 353–356.

Spicer C. W., Plastridge R. A., Foster K. L., Finlayson-Pitts B. J., Bottenheim J. W., Grannas A. M., and Shepson P. B. (2002) Molecular halogens before and during ozone depletion events in the Arctic at polar sunrise: concentrations and sources. *Atmos. Environ.* **36**, 2721–2731.

Spillane M. C., Monahan E. C., Bowyer P. A., Doyle D. M., and Stabeno P. J. (1986) Whitecaps and global fluxes. In *Oceanic Whitecaps* (eds. E. C. Monahan and G. M. Niocaill). D. Reidel, Norwell, MA, pp. 209–218.

Strong C., Fuentes J. D., Davis R. E., and Bottenheim J. W. (2002) Thermodynamic attributes of Arctic boundary layer ozone depletion. *Atmos. Environ.* **36**, 2641–2652.

Sturges W. T., Cota G. F., and Buckley P. T. (1992) Bromoform emission form Arctic ice algae. *Nature* **358**, 660–662.

Sturges W. T., McIntyre H. P., Penkett S. A., Chappellaz J., Barnola J.-M., Mulvaney R., Atlas E., and Stroud V. (2001) Methyl bromide, other brominated methanes, and methyl iodide in polar firn air. *J. Geophys. Res.* **106**, 1595–1606.

Stutz J., Hebestreit K., Alicke B., and Platt U. (1999) Chemistry of halogen oxides in the troposphere: comparison of model calculations with recent field data. *J. Atmos. Chem.* **34**, 65–85.

Stutz J., Ackermann R., Fast J. D., and Barrie L. (2002) Atmospheric reactive chlorine and bromine at the Great Salt Lake, Utah. *Geophys. Res. Lett.* **29**, doi: 10.1029/2002GL014812.

Sumner A. L. and Shepson P. B. (1999) Snowpack production of formaldehyde and its effect on the Arctic troposphere. *Nature* **398**, 230–233.

Swanson A. L., Blake N. J., Dibb J. E., Albert M. R., Blake D. R., and Rowland F. S. (2002) Photochemically induced production of $CH_3Br$, $CH_3I$, $C_2H_5I$, ethene, and propene within surface snow at Summit, Greenland. *Atmos. Environ.* **36**, 2671–2682.

Symonds R. B., Rose W. I., and Reed M. H. (1988) Contribution of Cl- and F-bearing gases to the atmosphere by volcanoes. *Nature* **334**, 415–418.

Tabazadeh A. and Turco R. P. (1993) Stratospheric chlorine injection by volcanic eruptions: HCl scavenging and implications for ozone. *Science* **260**, 1082–1086.

Tanaka P. L., Oldfield S., Neece J. D., Mullins C. B., and Allen D. T. (2000) Anthropogenic sources of chlorine formation in urban atmospheres. *Environ. Sci. Technol.* **34**, 4470–4473.

Tanaka P. L., Riemer D. D., Chang S., Yarwood G., McDonald-Buller E. C., Apel E. C., Orlando J. J., Silva P. J., Jimenez J. L., Canagaratna M. R., Neece J. D., Mullins C. B., and Allen D. T. (2003) Direct evidence for chlorine-enhanced urban ozone formation in Houston, TX. *Atmos. Environ.* **37**, 1393–1400.

Tanaka P. L., Allen D. T., and Mullins C. B. An environmental chamber investigation of chlorine-enhanced ozone formation in Houston, TX. *J. Geophys. Res.* (submitted for publication).

Tang I. N. (1997) Thermodynamic and optical properties of mixed-salt aerosols of atmospheric interest. *J. Geophys. Res.* **102**, 1883–1893.

Tang T. and McConnell J. C. (1996) Autocatalytic release of bromine from Arctic snow pack during polar sunrise. *Geophys. Res. Lett.* **23**, 2633–2636.

Tarasick D. W. and Bottenheim J. W. (2002) Surface ozone depletion episodes in the Arctic and Antarctic from historical ozonesonde records. *Atmos. Chem. Phys.* **2**, 197–205.

Toom-Sauntry D. and Barrie L. A. (2002) Chemical composition of snowfall in the high Arctic: 1990–1994. *Atmos. Environ.* **36**, 2683–2693.

Toumi R. (1994) BrO as a sink for dimethylsulphide in the marine atmosphere. *Geophys. Res. Lett.* **21**, 117–120.

Toyota K., Takahashi M., and Akimoto H. (2001) Modeling multi-phase halogen chemistry in the marine boundary layer with size-segregated aerosol module: implications for quasi-size-dependent approach. *Geophys. Res. Lett.* **28**, 2899–2902.

Tseng R.-S., Viechnicki J. T., Skop R. A., and Brown J. W. (1992) Sea-to-air transfer of surface-active organic compounds by bursting bubbles. *J. Geophys. Res.* **97**, 5201–5206.

Tuckermann M., Ackermann R., Golz C., Lorenzen-Schmidt H., Senne T., Stutz J., Trost B., Unold W., and Platt U. (1997) DOAS-observation of halogen radical-catalysed arctic boundary layer ozone destruction during the ARCTOC-campaigns 1995 and 1996 in Ny-Alesund, Spitsbergen. *Tellus* **49B**, 533–555.

Turekian V. C., Macko S. A., and Keene W. C. (2003) Concentrations, isotopic compositions, and sources of size-resolved, particulate organic carbon and oxalate in near-surface marine air at Bermuda during spring. *J. Geophys. Res.* **108**, 4157, doi: 10.1029/2002JD002053.

Valero F. P. J., Ackerman T. P., and Gore W. J. (1989) The effects of the Arctic haze as determined from airborne radiometric measurements during AGASP II. *J. Atmos. Chem.* **9**, 225–244.

Van Roozendael M., Wagner T., Richter A., Pundt I., Arlander D. W., Burrows J. P., Chipperfield M., Fayt C., Johnston P. V., Lambert J.-C., Kreher K., Pfeilsticker K., Platt U., Pommereau J.-P., Sinnhuber B.-M., Tørnkvist K. K., and

Wittrock F. (2002) Intercomparison of BrO measurements from ERS-2 GOME, ground-based and balloon platforms. *Adv. Space Res.* **29**, 1661–1666.

Varner R. K., Crill P. M., and Talbot R. W. (1999) Wetlands: a potentially significant source of atmospheric methyl bromide and methyl chloride. *Geophys. Res. Lett.* **26**, 2433–2436.

Vierkorn-Rudolph B., Bachmann K., Schwarz B., and Meixner F. X. (1984) Vertical profiles of hydrogen chloride in the troposphere. *J. Atmos. Chem.* **2**, 47–63.

Vogt R. (1999) Iodine compounds in the atmosphere. In *The handbook of Environmental Chemistry: Volume 4, Part E. Air Pollution* (eds. P. Fabian and O. N. Singh). Springer, Heidelberg, Germany, pp. 113–129.

Vogt R., Crutzen P. J., and Sander R. (1996) A mechanism for halogen release from sea-salt aerosol in the remote marine boundary layer. *Nature* **383**, 327–330.

Vogt R., Sander R., von Glasow R., and Crutzen P. (1999) Iodine chemistry and its role in halogen activation and ozone loss in the marine boundary layer: a model study. *J. Atmos. Chem.* **32**, 375–395.

Volpe C., Wahlen M., Pszenny A. A. P., and Spivack A. J. (1998) Chlorine isotopic composition of marine aerosols: implications for the release of reactive chlorine and HCl cycling rates. *Geophys. Res. Lett.* **25**, 3831–3834.

von Glasow R. and Sander R. (2001) Variation of sea salt aerosol pH with relative humidity. *Geophys. Res. Lett.* **28**, 247–250.

von Glasow R., Sander R., Bott A., and Crutzen P. J. (2002a) Modeling halogen chemistry in the marine boundary layer: 1. Cloud-free MBL. *J. Geophys. Res.* **107**, 4341, doi: 10.1029/2001JD000942.

von Glasow R., Sander R., Bott A., and Crutzen P. J. (2002b) Modeling halogen chemistry in the marine boundary layer: 2. Interactions with sulfur and cloud-covered MBL. *J. Geophys. Res.* **107**, 4323, doi: 10.1029/2001JD000943.

Wagner T. and Platt U. (1998) Satellite mapping of enhanced BrO concentrations in the troposphere. *Nature* **395**, 486–490.

Wagner T., Leue C., Wenig M., Pfeilsticker K., and Platt U. (2001) Spatial and temporal distribution of enhanced boundary layer BrO concentrations measured by the GOME instrument aboard ERS-2. *J. Geophys. Res.* **106**, 24225–24235.

Wagner V., von Glasow R., Fischer H., and Crutzen P. J. (2002) Are CH₂O measurements in the marine boundary layer suitable for testing the current understanding of CH₄ photooxidation? *J. Geophys. Res.* **107**, 4029, doi: 10.1029/2001JD000722.

Wang T. X. and Margerum D. W. (1994) Kinetics of reversible chlorine hydrolysis: temperature dependence and general-acid/base-assisted mechanisms. *Inorg. Chem.* **33**, 1050–1055.

Warneck P. (1999) *Chemistry of the Natural Atmosphere.* Academic Press, San Diego.

Wayne R. P., Poulet G., Biggs P., Burrows J. P., Cox R. A., Crutzen P. J., Hayman G. D., Jenkin M. E., Le Bras G., Moortgat G. K., Platt U., and Schindler R. N. (1995) Halogen oxides: radicals, sources and reservoirs in the laboratory and in the atmosphere. *Atmos. Environ.* **29**, 2677–2884.

Wennberg P. (1999) Bromine explosion. *Nature* **397**, 299–301.

Wessel S. (1996) Troposphärische ozonvariationen in polargebieten. PhD Thesis, Universität Bremen, Germany.

Wessel S., Aoki S., Winkler P., Weller R., Herber A., Gernandt H., and Schrems O. (1998) Tropospheric ozone depletion in polar regions. A comparison of observations in the Arctic and Antarctic. *Tellus* **50B**, 34–50.

Wever R. (1988) Ozone destruction by algae in the Arctic atmosphere. *Nature* **335**, 501.

Wever R., Tromp M. G. M., Kronn B. E., Marjani A., and van Tol M. (1991a) Brominating activity of the seaweed Ascophyllum nodosum: impact on the biosphere. *Environ. Sci. Technol.* **25**, 446–449.

Wever R., Tromp M. G. M., Krenn B. E., Marjani A., and Tol M. V. (1991b) Brominating activity of the seaweed *Ascophyllum nodosum*: impact on the biosphere. *Environ. Sci. Technol.* **25**, 446–449.

Williams J., Wang N.-Y., Cicerone R. J., Yagi K., Kurihara M., and Terada F. (1999) Atmospheric methyl halides and dimethyl sulfide from cattle. *Global Biogeochem. Cycles* **13**, 485–491.

Wingenter O. W., Kubo M. K., Blake N. J., Smith T. W., Jr., Blake D. R., and Rowland F. S. (1996) Hydrocarbon and halocarbon measurements as photochemical and dynamical indicators of atmospheric hydroxyl, atomic chlorine, and vertical mixing obtained during Lagrangian flights. *J. Geophys. Res.* **101**, 4331–4340.

Wingenter O. W., Blake D. R., Blake N. J., Sive B. C., Rowland F. S., Atlas E., and Flocke F. (1999) Tropospheric hydroxyl and atomic chlorine concentrations, and mixing timescales determined from hydrocarbon and halocarbon measurements made over the Southern Ocean. *J. Geophys. Res.* **104**, 21819–21828.

Wittrock F., Müller R., Richter A., Bovensmann H. H., and Burrows J. P. (2000) Measurements of iodine monoxide (IO) above Spitsbergen. *Geophys. Res. Lett.* **27**, 1471–1474.

Wofsy S. C., McElroy M. B., and Yung Y.-L. (1975) The chemistry of atmospheric bromine. *Geophys. Res. Lett.* **2**, 215–218.

Woodcock A. H. (1953) Salt nuclei in marine air as a function of altitude and wind force. *J. Meteor.* **10**, 362–371.

Woodcock A. H., Kientzler C. F., Arons A. B., and Blanchard D. C. (1953) Giant condensation nuclei from bursting bubbles. *Nature* **172**, 1144–1145.

Worthy D. E. J., Trivett N. B. A., Hopper J. F., and Bottenheim J. (1994) Analysis of long-range transport events at Alert, northwest territories, during the Polar Sunrise Experiment. *J. Geophys. Res.* **99**, 25329–25344.

Wu J. (1988) Variations of whitecap coverage with wind stress and water temperature. *J. Phys. Oceanogr.* **18**, 1448–1453.

Yokouchi Y., Noijiri Y., Barrie L. A., Toom-Sauntry D., Machida T., Inuzuka Y., Akimoto H., Li H.-J., Fujinuma Y., and Aoki S. (2000) A strong source of methyl chloride to the atmosphere from tropical coastal land. *Nature* **403**, 295–298.

Yokouchi Y., Nojiri Y., Barrie L. A., Toom-Sauntry D., and Fujinuma Y. (2001) Atmospheric methyl iodide: high correlation with surface seawater temperature and its implications on the sea-to-air flux. *J. Geophys. Res.* **106**, 12661–12668.

Yokouchi Y., Ikeda M., Inuzuka Y., and Yukawa T. (2002) Strong emissions of methyl chloride from tropical plants. *Nature* **416**, 163–165.

Yoon Y. J. and Brimblecombe P. (2002) Modelling the contribution of sea salt and dimethyl sulfide derived aerosol to marine CCN. *Atmos. Chem. Phys.* **2**, 17–30.

Yoshitake H. (2000) Effects of surface water on NO₂–NaCl reaction studied by diffuse reflectance infrared spectroscopy (DRIRS). *Atmos. Environ.* **34**, 2571–2580.

Yung Y. L., Pinto J. P., Watson R. T., and Sander P. (1980) Atmospheric bromine and ozone perturbations in the lower stratosphere. *J. Atmos. Sci.* **37**, 339–353.

Zafiriou O. C. (1974) Photochemistry of halogens in the marine atmosphere. *J. Geophys. Res.* **79**, 2730–2732.

Zangmeister C. D., Turner J. A., and Pemberton J. E. (2001) Segregation of NaBr in NaBr/NaCl crystals grown from aqueous solutions: implications for sea salt surface chemistry. *Geophys. Res. Lett.* **28**, 995–998.

Zetzsch C. and Behnke W. (1992) Heterogeneous photochemical sources of atomic Cl in the troposphere. *Ber. Bunsenges. Phys. Chem.* **96**, 488–493.

Zetzsch C., Pfahler G., and Behnke W. (1988) Heterogeneous formation of chlorine atoms from NaCl in a photosmog system. *J. Aerosol Sci.* **19**, 1203–1206.

Zhou M. Y., Yang S. J., Parungo F. P., and Harris J. M. (1990) Chemistry of marine aerosols over the western Pacific Ocean. *J. Geophys. Res.* **95**, 1779–1787.

Zhou X. and Mopper K. (1997) Photochemical production of low-molecular-weight carbonyl compounds in seawater and surface microlayer and their air–sea exchange. *Mar. Chem.* **56**, 201–213.

Zhou X., Beine H. J., Honrath R. E., Fuentes J. D., Simpson W., Shepson P. B., and Bottenheim J. W. (2001) Snowpack photochemical production of HONO: a major source of OH in the Arctic boundary layer in springtime. *Geophys. Res. Lett.* **28**, 4087–4090.

# 4.03
# Global Methane Biogeochemistry

## W. S. Reeburgh

### University of California Irvine, CA, USA

## 4.03.1 INTRODUCTION

Methane ($CH_4$) has been studied as an atmospheric constituent for over 200 years. A 1776 letter from Alessandro Volta to Father Campi described the first experiments on flammable "air" released by shallow sediments in Lake Maggiore (Wolfe, 1996; King, 1992). The first quantitative measurements of $CH_4$, both involving combustion and gravimetric determination of trapped oxidation products, were reported in French by Boussingault (1834, 1864) and Gautier (1901), who reported $CH_4$ concentrations of 10 ppmv and 0.28 ppmv (seashore) and 95 ppmv (Paris), respectively. The first modern measurements of atmospheric $CH_4$ were the infrared absorption measurements of Migeotte (1948), who estimated an atmospheric concentration of 2.0 ppmv.

Development of gas chromatography and the flame ionization detector in the 1950s led to observations of vertical $CH_4$ distributions in the troposphere and stratosphere, and to establishment of time-series sampling programs in the late 1970s. Results from these sampling programs led to suggestions that the concentration of $CH_4$, as that of $CO_2$, was increasing in the atmosphere. The possible role of $CH_4$ as a greenhouse gas stimulated further research on $CH_4$ sources and sinks. Methane has also been of interest to microbiologists, but findings from microbiology have entered the larger context of the global $CH_4$ budget only recently.

Methane is the most abundant hydrocarbon in the atmosphere. It plays important roles in atmospheric chemistry and the radiative balance of the Earth. Stratospheric oxidation of $CH_4$ provides a means of introducing water vapor above the tropopause. Methane reacts with atomic chlorine in the stratosphere, forming HCl, a reservoir species for chlorine. Some 90% of the $CH_4$ entering the atmosphere is oxidized through reactions initiated by the OH radical. These reactions are discussed in more detail by Wofsy (1976) and Cicerone and Oremland (1988), and are important in controlling the oxidation state of the atmosphere. Methane absorbs infrared radiation in the troposphere, as do $CO_2$ and $H_2O$, and is an important greenhouse gas (Lacis *et al.*, 1981; Ramanathan *et al.*, 1985).

A number of review articles on atmospheric $CH_4$ have appeared during the last 15 years. Cicerone and Oremland (1988) reviewed evidence for the temporal atmospheric increase, updated source estimates in the global $CH_4$ budget, and placed constraints on the global budget, emphasizing that the total is well constrained, but that the constituent sources may be uncertain by a factor of 2 or more. This paper was part of a special section in *Global Biogeochemical Cycles* that resulted from a 1987 American Chemical Society Symposium, "Atmospheric Methane: Formation and Fluxes form the Biosphere and Geosphere." Tyler (1991) and Wahlen (1993) emphasized new information on stable isotopes of $CH_4$ and $^{14}CH_4$, respectively. Several reviews deal with the microbially mediated $CH_4$ oxidation. King (1992) reviewed the ecology of microbial $CH_4$ oxidation, emphasizing the important role of this process in global $CH_4$ dynamics. R. S. Hanson and T. E. Hanson (1996) reviewed the physiology and taxonomy of methylotrophic bacteria, their role in the global carbon cycle, and the ecology of methanotrophic bacteria. Conrad (1996) reviewed the role of soils and soil microbial communities as controllers of $CH_4$ fluxes, as well as those of $H_2$, CO, OCS, $N_2O$, and NO. Two meetings focusing on $CH_4$ biogeochemistry were held in 1991: an NATO Advanced Science Workshop held at

Mt. Hood, OR, and the Tenth International Symposium on Environmental Biogeochemistry (ISEB). A dedicated issue of *Chemosphere* (**26**(1−4), 1993) contains contributions from the NATO workshop; two additional volumes (Khalil, 1993, 2000) contain a report of the workshop and updates of important topics. Contributions to the ISEB meeting are presented in Oremland (1993). Wuebbles and Hayhoe (2002) reviewed the effects of $CH_4$ on atmospheric chemistry and examined the direct and indirect impact of $CH_4$ on climate. The Intergovernmental Panel on Climate Change (IPCC) has published periodic updates (see, e.g., IPCC, 2001).

Substantial advances have resulted from research aimed at understanding the global $CH_4$ mixing ratio increase. Time-series measurements of atmospheric $CH_4$ have continued, new $CH_4$ flux measurements in a range of environments have been reported, and data allowing use of the stable isotope ($^{13}C/^{12}C$, $^2H/^1H$) composition of $CH_4$ as an independent budget constraint have increased. The importance of microbial oxidation of $CH_4$ has been recognized and modeled; the possible role of $CH_4$ clathrate hydrates in the global budget has been clarified with the introduction of new technology. Studies of $CH_4$ trapped in ice cores from the Greenland and Antarctic ice caps have continued, resulting in higher-resolution records and new interpretations of past conditions. There have been few recent changes in our understanding of the atmospheric chemistry of $CH_4$, and since this is covered in Cicerone and Oremland (1988) it will not be covered here. The aim here is not to repeat information contained in the reviews mentioned above, but to present results that have appeared in the literature since their publication, to outline major questions, and to point to promising new approaches.

## 4.03.2 GLOBAL METHANE BUDGET

The first global $CH_4$ budgets were compiled by Ehhalt (1974) and Ehhalt and Schmidt (1978), who used available published information to estimate emissions of $CH_4$ to the atmosphere. They considered paddy fields, freshwater sources (lakes, swamps, and marshes), upland fields and forests, tundra, the ocean, and enteric fermentation by animals as biogenic sources. Anthropogenic sources included industrial natural gas losses and emission from coal mining, and were considered to be $^{14}C$ free. Observations of $^{14}CH_4$ placed an upper limit on anthropogenic sources. Oxidation by the OH radical, as well as loss to the stratosphere by eddy diffusion and Hadley circulation, were presumed to be methane sinks. In spite of lack of data, this work correctly identified the major atmospheric sources and did a remarkable job of estimating the magnitudes of the source terms in the global $CH_4$ budget, but it dismissed oxidation by soils as insignificant.

### 4.03.2.1 Global Methane Increase

The first observations of a possible atmospheric $CH_4$ increase were reported by Graedel and McRae (1980), who analyzed a 10-year continuous record of total hydrocarbon (THC) data collected at three urban/suburban sites by the New Jersey Department of Environmental Protection. An increasing trend in daily and annual minima was advanced somewhat tentatively because of concerns about calibration techniques, the fraction of the THC signal attributable to $CH_4$, and the urban locations. Rasmussen and Khalil (1981) used 22 months of automated GC/FID measurements of atmospheric $CH_4$ at Cape Meares, OR, to show a $CH_4$ concentration increase of ~2% $yr^{-1}$. Rasmussen and Khalil (1981) also compiled independent northern hemisphere GC/FID measurements from 1965 to 1980 to show a similar trend. The Cape Meares station has operated almost continuously since early 1979 and the techniques were used in establishing the NOAA network described below.

Time-series measurements of atmospheric $CH_4$ have been made at fixed stations as well as at globally distributed stations. Sampling of globally distributed stations started in 1978 and has continued since then by two groups using two different sampling approaches. Although a number of time-series sampling programs were initiated and operated for short periods, only two—the University of California Irvine and the NOAA/CMDL—have continued from the late 1970s to early 2000s. The University of California Irvine group started measurements in 1978 (Blake, 1984), conducting quarterly quasi-synoptic sampling at a range of locations in the northern and southern hemispheres followed by laboratory analysis. These measurements were averaged areally and temporally to produce an annual global average $CH_4$ concentration. Figure 1 (Simpson *et al.*, 2002) shows the Irvine $CH_4$ record from 1978 to 2001. Regular sampling conducted since 1983 at a network of clean air stations maintained by the NOAA Global Monitoring for Climate Change Laboratory (GMCC, now Climate Monitoring and Diagnostics Laboratory (CMDL)) in Boulder, CO, gives more detailed temporal and areal coverage (Steele *et al.*, 1987) The CMDL $CH_4$ results through May 1, 2001, are plotted as a "rug" or "flying carpet" diagram in Figure 2. The interhemispheric gradient in $CH_4$ is evident, and results from larger northern hemisphere $CH_4$ emissions. Seasonal

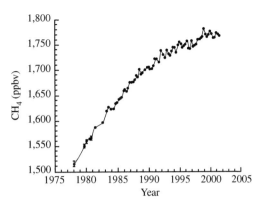

**Figure 1** Seasonally averaged global $CH_4$ mixing ratios (ppbv) from January 1978 to June 2001. The solid line is an interpolated curve fit to the points. Increases and decreases in the global $CH_4$ mixing ratio are evident after 1992 (reproduced by permission of American Geophysical Union from *Geophys. Res. Lett.*, **2002**, *29*, 117-1–117-4).

changes, which are out of phase between hemispheres, are also evident; the $CH_4$ concentration is highest at the end of winter and decreases due to photochemical oxidation in summer. In addition to the interannual changes cited above, measurements of methane trapped in polar ice sheet cores show that the atmospheric concentration of methane has doubled over the past 250 years (Craig and Chou, 1982; Rasmussen and Khalil, 1984).

An increasing trend in atmospheric $CH_4$ was noted by both groups (Blake, 1984; Blake *et al.*, 1982; Blake and Rowland, 1986; Steele *et al.*, 1987). The rate of change has increased (Dlugokencky *et al.*, 1994a, 2001) and decreased (Steele *et al.*, 1992; Dlugokencky *et al.*, 1994b, 1996, 1998). The growth rate increases have been ascribed to changes in natural gas leakage and arctic wildfires, whereas the decreases have been ascribed to volcanic eruptions and decreased northern wetland emission. Khalil and Rasmussen (1985) ascribed the growth rate reductions to decreases in the abundance of OH and increases in sources. Hogan and Harriss (1994) noted the high quality of the time-series data, but noted the difficulty in deducing changes in geographically diverse $CH_4$ sources from the CMDL data. They called for detailed bottom-up analyses to understand changes in source strengths. Dlugokencky *et al.* (1994c) agreed that there are too few data to unequivocally determine causes for variations in the growth rate of $CH_4$ and indicated that their examples were intended to illustrate the magnitude of source strength changes needed to explain the observations. The causes of the $CH_4$ increase as well as interannual fluctuations remain a major question.

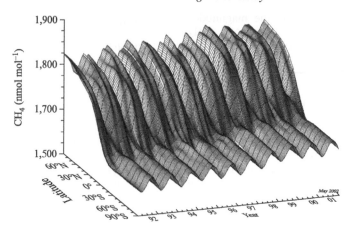

**Figure 2** Global distribution of atmospheric $CH_4$ from 1992 to May 1, 2001. Three-dimensional latitudinal distribution of $CH_4$ in the marine boundary layer is presented. The surface represents data from the NOAA/CMDL cooperative air sampling network smoothed in time and latitude (source National Oceanic and Atmospheric Administration (NOAA), Climate Monitoring and Diagnostics Laboratory (CMDL), Carbon Cycle Greenhouse Gases). Updated versions are available on line at http://www.cmdl.noaa.gov/ccgg/gallery/index?pageType = folder&currDir = ./Data_Figures.

### 4.03.2.2 Methane Budget with Constraints

Cicerone and Oremland (1988) reviewed the information on $CH_4$ sources and sinks that had appeared since Ehhalt's work and emphasized that the atmospheric increase was genuine. They produced a framework of constraints involving $CH_4$ amounts, turnover rates, and isotopes. The quasi-steady-state annual source was estimated to be $500 \pm 90$ Tg yr$^{-1}$ (1 Tg $= 10^{12}$ g). Table 1, which contains entries from Cicerone and Oremland (1988), gives annual emissions from the major sources as well as estimated ranges. Note that the ranges are high for the large rice and wetland terms. Cicerone and Oremland (1988) made estimates of confidence in terms that comprise the global $CH_4$ budget (see their table 3). Confidence in the total amount of $CH_4$, the rate of change, and the residence time is high, resulting in high confidence in the total steady-state source (or sink). Confidence in the fraction of modern biogenic $CH_4$ derived form $^{14}CH_4$ measurements is also high, while confidence in the total steady-state source derived by addition of known sources is low. Summarizing, we know the total budget with high confidence, but how to apportion the individual sources is less well known. Cicerone and Oremland remarked that it was difficult to identify enough sources of radio-carbon-free $CH_4$. Because so little was known at the time about the impact of $CH_4$ hydrates on the budget, they were added to the budget as a "placeholder" term. A budget with constraints like those outlined by Cicerone and Oremland (1988) provides a useful framework for thinking about the magnitudes of terms, and by constraining the magnitude of the total, it served to limit proliferation of source estimates. However, it did not consider uptake by soils, and is much like a snapshot. The budget has no predictive power and provides no information on the causes of the $CH_4$ increase or the interannual variations.

### 4.03.2.3 Gross Methane Budget

All of the above budgets and syntheses consider $CH_4$ actually entering the atmosphere. Field studies have clearly demonstrated that substantial microbially mediated $CH_4$ oxidation occurs within soils and aquatic environments, where it is able to limit and even reverse $CH_4$ fluxes to the atmosphere. Thus, all of the above budgets and syntheses consider *net* $CH_4$ emissions. What role does microbially mediated $CH_4$ oxidation play in these budgets? What is its magnitude in the various source terms? Considering the locations where $CH_4$ oxidation occurs, the only sources where $CH_4$ is introduced directly to the atmosphere with no possibility of microbial oxidation are enteric fermentation, biomass burning, natural gas leaks from production, distribution and flaring, and coal mining. The remaining sources are susceptible to modification by microbial oxidation. Söhngen (1906) recognized that $CH_4$ oxidation occurs in close proximity to $CH_4$ sources. The so-called Söhngen cycle, emission = production − consumption, can be used as a framework to estimate the importance of microbially mediated $CH_4$ oxidation.

Reeburgh *et al.* (1993) estimated the role of microbially mediated $CH_4$ oxidation using limited data on oxidation rates in environments representing the main $CH_4$ budget source terms.

**Table 1** Methane budget source/sink terms (Tg $CH_4$ yr$^{-1}$).

| Source/sink Base year | Cicerone and Oremland (1988) Annual release | Range | Fung et al. (1991)[a] Annual release (1980s) | Hein et al. (1997) Annual release | IPCC (2001) TAR[b] 1998 |
|---|---|---|---|---|---|
| *Natural sources* | | | | | |
| Wetlands | 115 | 100–200 | 115 | 237 | |
| Bogs/Tundra | | | 35 | | |
| Swamps | | | 80 | | |
| Termites | 40 | 10–100 | 20 | | |
| Ocean | 10 | 5–20 | 10[c] | | |
| Freshwaters | 5 | 1–25 | | | |
| Hydrates | 5? | 0–100[d] | 5 | | |
| *Anthropogenic sources* | | | | | |
| Energy | | | 75 | 97 | |
| Mining | 35 | 25–45 | 35 | | |
| Gas drilling, venting | 45 | 25–50 | 40 | | |
| Landfills | 40 | 30–70 | 40 | 35 | |
| Ruminants | 80 | 65–100 | 80 | 90[e] | |
| Waste treatment | | | e | | |
| Rice agriculture | 110 | 60–170 | 100 | 88 | |
| Biomass burning | 55 | 50–70 | 55 | 40 | |
| Other | | | | | |
| Total source | 540 | | 500 | 587 | 598 |
| *Sinks* | | | | | |
| Soils | | | 10 | | 30 |
| Tropospheric OH | | | 450 | 489 | 506 |
| Stratospheric loss | | | | 46 | 40 |
| Total sink | | | 460 | 535 | 576 |

[a] Scenario 7.   [b] TAR budget based on 1.745 ppb, 2.78 Tg ppb$^{-1}$, lifetime of 8.4 yr and an imbalance of $+8$ ppb yr$^{-1}$.   [c] Oceans and freshwaters combined.   [d] Possible future emissions.   [e] Waste treatment included under ruminants.

Reeburgh (1996) considered studies since 1993 and highlighted progress in studies of methane oxidation in wetland, rice, landfills. Table 2 uses the net $CH_4$ budget of Fung *et al.* (1991) (scenario 7), adds oxidation estimated from oxidation rate measurements and mass balances, and produces an estimate of *gross* $CH_4$ production. Although Reeburgh *et al.* (1993) were conservative in their oxidation estimates, neither the oxidation estimates nor the gross production estimates can be constrained by any of the criteria and techniques outlined by Cicerone and Oremland (1988), Fung *et al.* (1991), and Hein *et al.* (1997). Several striking features are evident in Table 2. First, total microbial oxidation is slightly larger than net global emission. More than half of the estimated $CH_4$ production is oxidized by microbes before emission to the atmosphere. Thus, microbial $CH_4$ oxidation is most effective *before* emission, and has remained largely invisible because of our focus on net emissions. The consumption term representing uptake of atmospheric methane is a refinement to the overall budget with a magnitude similar to the annual increment causing the atmospheric increase. Microbial oxidation occurring before emission is clearly more important. This is illustrated by the role of the ocean in the global methane budget. While the ocean and the placeholder hydrate decomposition term have the potential to produce enormous quantities of $CH_4$, a series of very effective microbially mediated oxidation processes, anaerobic oxidation in anoxic sediments and anoxic water columns, followed by aerobic oxidation in the oxic water column, result in the ocean being one of the net global $CH_4$ budget's smallest terms. A similar situation holds for wetlands and rice production. It is clear from Table 2 that future attention must focus not only on net $CH_4$ emission, but also on microbial oxidation.

### 4.03.2.4 Atmospheric Models

Fung *et al.* (1991) attempted to derive a justified global budget for $CH_4$ in the 1980s. Geographic and seasonal distributions of major $CH_4$ sources and sinks were compiled, and oxidation by the OH radical was calculated. A global three-dimensional tracer transport model was used to simulate the atmospheric response to geographic and seasonal changes in $CH_4$ sources and sinks. A number of $CH_4$ budget scenarios were run and tested for their

**Table 2** Global Net $CH_4$ emission, consumption, and gross production ( $Tg\ CH_4\ yr^{-1}$ ).

| Source/sink term | E | + | C | = | P |
|---|---|---|---|---|---|
| | Net emission[a] | | Consumption[b] | | Gross production |
| Animals | 80 | | 0 | | 80 |
| *Wetlands* | 115 | | 27 | | 142 |
| Bogs/tundra (boreal) | 35 | | 15 | | 50 |
| Swamps/alluvial | 80 | | 12 | | 92 |
| Rice production | 100 | | 477 | | 577 |
| Biomass burning | 55 | | 0 | | 55 |
| Termites | 20 | | 24 | | 44 |
| Landfills | 40 | | 22 | | 62 |
| Oceans, freshwaters | 10 | | 75.3 | | 85.3 |
| Hydrates | 5? | | 5 | | 10 |
| Coal production | 35 | | 0 | | 35 |
| Gas production | 40 | | 18 | | 58 |
| Venting, flaring | 10 | | 0 | | 10 |
| Distribution leaks[c] | 30 | | 18 | | 48 |
| Total sources | 500[d] | | | | |
| Chemical destruction | −450 | | | | |
| Soil consumption | −10 | | 40 | | |
| Total sinks | −460[d] | | 688.3 | | 40[e] |
| | | | Total production | | 1,188.3 |

Adapted from Reeburgh *et al.* (1993).
[a] Scenario 7, Fung *et al.* (1991). [b] From table 1, Reeburgh *et al.* (1993). [c] Should be considered *P*. [d] 500−460 = 40 Tg $CH_4\ yr^{-1}$ = annual atmospheric (0.9% $yr^{-1}$) increment. [e] Soil consumption of atmospheric $CH_4$ added to the gross budget as an equivalent production term.

ability to reproduce the meridional gradient as well as the seasonal variations in $CH_4$ concentration observed at GMCC stations. The preferred $CH_4$ emission budget (scenario 7) is included in Table 1. Magnitudes of $CH_4$ emissions from fossil sources, domestic animals, and wetlands and tundra north of 50° N are reasonably constrained geographically. Individual contributions of the landfill, tropical swamp, rice paddy, biomass burning, and termite source terms could not be determined uniquely because of the lack of direct flux and atmospheric variation measurements in the regions where these sources are concentrated.

Hein *et al.* (1997) also used a three-dimensional atmospheric transport model to calculate the global distribution of OH radicals, the main sink for atmospheric $CH_4$. They used an inverse modeling method that allowed selection of the source/sink configuration that gave the best agreement between observed and calculated $CH_4$ concentrations. This allowed objective treatment of the uncertainties in source and sink magnitudes, and reduced the uncertainty of source magnitudes by at least one-third. The Hein *et al.* (1997) results are tabulated for comparison in Table 1. The conclusions of Hein *et al.* (1997) agreed quite well with those of Fung *et al.* (1991) and confirmed that it is not uniquely possible to select only one source–sink combination. The decrease in the atmospheric $CH_4$ growth rate in the early 1990s could not be uniquely associated with changes in particular sources. For comparison, the $CH_4$ budget from the Intergovernmental Panel on

Climate Change Third Assessment Report (IPCC/TAR) (IPCC, 2001) also appears as a column in Table 1.

### 4.03.2.5 Stable Isotopes

Stable isotope ratios of carbon ($^{13}C/^{12}C$) and hydrogen ($^{2}H/^{1}H$) in $CH_4$ were recognized as providing additional constraints for the global $CH_4$ budget (Stevens and Rust, 1982). Limited data suggested that it might be possible to distinguish or discriminate between $CH_4$ sources and that a carbon or hydrogen isotope budget for $CH_4$, parallel to the concentration budget, could be assembled. These measured values are expressed in delta notation (Craig, 1957), which for carbon isotopes is

$$\delta^{13}C = [(R_{sample}/R_{standard}) - 1] \times 1,000$$

where $R$ is the ratio $^{13}C/^{12}C$, sample and standard refer to the carbon isotope ratios in the sample and standard, and the results are expressed as parts per thousand or per mil (‰). Negative values in this notation indicate samples with lower $^{13}C$ content than the standard. Stevens and Rust (1982) proposed that the mass-weighted isotopic composition of all sources should equal the mean isotope composition of the atmosphere ($\delta^{13}C = -47‰$) when corrected for isotope fractionation effects associated with $CH_4$ consumption reactions. Kinetic isotope effects (KIEs) result from the light isotope ($^{12}$C) reacting faster than the heavy isotope ($^{13}$C) and offer a constraint associated with reactions.

Tyler (1991) summarized the $\delta^{13}$C of CH$_4$ ($\delta^{13}$CH$_4$) from a number of sources. The $\delta^{13}$CH$_4$ from a given type of environment covers a fairly large range due to variations in methanogenic substrates and mechanisms, and also to fractionation associated with oxidation, so that many sources of CH$_4$ have overlapping $\delta^{13}$C ranges. In general, CH$_4$ from biogenic sources is isotopically lighter (−40‰ to −80‰) than CH$_4$ from geological or thermogenic sources (−30‰ to −50‰). Plants utilizing the C-3 photosynthetic pathway have carbon isotope composition (−21‰ to −35‰) that differs from plants utilizing the C-4 pathway (−10‰ to −20‰), so that CH$_4$ produced by decomposition of each type of plant may have a distinctive isotopic composition.

Carefully intercalibrated atmospheric time-series measurements of $\delta^{13}$CH$_4$ have been made by Quay *et al.* (1991, 1999) and Tyler *et al.* (1994, 1999) in the northern hemisphere and by Lowe *et al.* (1991, 1994, 1997) in the southern hemisphere. Additional atmospheric $\delta^{13}$CH$_4$ measurements are reported in the modeling study of Gupta *et al.* (1996). This model included KIEs of CH$_4$ sinks, namely, reaction with OH, reaction with Cl, and oxidation by soils, all of which enrich

atmospheric CH$_4$ in $^{13}$C. These models employ a very limited data set and are based on the Oslo two-dimensional (height versus latitude) global tropospheric photochemical model (Gupta *et al.*, 1998). The success of these model treatments of $\delta^{13}$CH$_4$ depends critically on the values of KIEs. Tables 3 and 4 summarize recent KIE determinations for atmospheric and microbially mediated soil and sediment reactions. It should be noted that studies on soil KIEs have only recently reached a point of agreement with a combination of static chamber, soil profile, and landfill emission studies.

Tans (1997) derived relationships showing that the timescale for equilibration of atmospheric $\delta^{13}$CH$_4$ following changes is considerably longer than the timescale for changes in total CH$_4$. Changes in the $\delta^{13}$CH$_4$ result not only from changes in magnitudes of isotopically different sources, but also from isotope dilution by a changing reservoir of total atmospheric CH$_4$. Tans (1997) noted that the dearth of isotope data for atmospheric $\delta^{13}$CH$_4$ is a larger problem than possible misinterpretation of trends. Lassey *et al.* (2000) analyzed archived southern hemisphere samples covering time intervals of 10 years and 17 years, and used models to show that they are compatible with stabilized CH$_4$ sources. An analysis of the sensitivity of the atmospheric

**Table 3** Microbial methane oxidation kinetic isotope fractionation factors.

| Study/(comment) | $\alpha C$ | $\alpha H$ |
|---|---|---|
| *Aerobic* | | |
| Silverman and Oyama (1968) | 1.011 | |
| (calculated by Whiticar and Faber, 1986) | | |
| Coleman *et al.* (1981) | 1.013–1.015 | 1.103–1.325 |
| (enrichment culture, closed system) | | |
| Barker and Fritz (1981) | 1.005–1.103 | |
| (enrichment culture, closed system) | | |
| Whiticar and Faber (1986) | 1.002–1.014 | |
| (model calculations, field data) | | |
| Zyakun *et al.* (1988) | 1.011–1.039 | |
| (*Methylomonas methanica*, flow-through system) | | |
| Happell *et al.* (1994) | 1.003–1.021 | 1.050–1.129 |
| (FL swamp floodwater) | | |
| King *et al.* (1989) (tundra, AK) | 1.016 (4 °C), 1.027 (14 °C) | |
| Tyler *et al.* (1994) (forest soil, NH) | 1.022 ± 0.004 | |
| Bergamaschi and Harris (1995) (landfill cover soil) | 1.008 ± 0.003 | 1.044 ± 0.020 |
| Reeburgh *et al.* (1997) (forest soil, AK) | 1.022 and 1.025 | |
| Bergamaschi *et al.* (1998b) (landfill cover soil) | 1.008 ± 0.004 | 1.039 ± 0.026 |
| Liptay *et al.* (1998) (NE US landfill cover soils) | 1.022 ± 0.008 | 1.046 ± 0.016 |
| Snover and Quay (2000) | | |
| (grassland, WA) | 1.107 ± 0.0010 | 1.009 ± 0.030 |
| (temperate forest, WA) | 1.081 ± 0.0004 | 1.066 ± 0.007 |
| *Anaerobic* | | |
| Alperin *et al.* (1988) (Skan Bay, AK, 4 °C) | 1.0088 ± 0.0013 | 1.157 ± 0.023 |
| maximum AMO rate: 3–4 mM yr$^{-1}$ | | |
| Martens *et al.* (1999) (Ekernförde Bay, FRG, 8 °C) | 1.012 ± 0.001 | 1.120 ± 0.020 |
| maximum AMO rate: 14–16 mM yr$^{-1}$ | | |

**Table 4** KIEs for reactions involving atmospheric $CH_4$.

| Study | Type | $\alpha C$ | $\alpha H$ |
|---|---|---|---|
| **$CH_4$ with Cl** | | | |
| Wallington and Hurley (1992) | Experimental | | $1.47 \pm 0.09$ |
| Saueressig *et al.* (1995) | Experimental | $1.066 \pm 0.002$ | $1.508 \pm 0.041$ |
| Tanaka *et al.* (1996) | Theoretical | $1.026$ | |
| Tanaka *et al.* (1997) | Experimental | $1.013–1.020$ | |
| Gupta *et al.* (1997) | Theoretical | $1.034$ | |
| Roberto-Neto *et al.* (1998) | Theroretical | $1.06$ | $1.45$ |
| Crowley *et al.* (1999) | Experimental | $1.066 \pm 0.002$ | |
| Tyler *et al.* (2000) | Experimental | $1.0621 \pm 0.0004$ | $1.474 \pm 0.026$ |
| **$CH_4$ with OH** | | | |
| Davidson *et al.* (1987) | Experimental | $1.010 \pm 1.007$ | |
| Cantrell *et al.* (1990) | Experimental | $1.0054 \pm 0.0009$ | |
| Gupta *et al.* (1997) | Theoretical | $1.010$ | |
| Saueressig *et al.* (2001) | Experimental | $1.0039 \pm 0.0004$ | $1.294 \pm 0.018$ (D) |

secular response in both concentration and $\delta^{13}CH_4$ to sustained changes in source and sink showed that $\delta^{13}CH_4$ is a potentially powerful indicator of source and sink changes.

The amount of atmospheric $\delta^{13}CH_4$ and $\delta D$-$CH_4$ data available for modeling promises to increase rapidly in the future through application of new techniques. The development of continuous-flow gas chromatography–combustion isotope ratio monitoring mass spectrometry (GC/C/IRMS) (Merritt *et al.*, 1995; Popp *et al.*, 1995) allows measurements of $\delta^{13}CH_4$. Measurements of $\delta D$-$CH_4$ (Rice *et al.*, 2001) are possible with the more recently developed gas chromatography–pyrolysis isotope ratio monitoring mass spectrometry (GC/P/IRMS) technique. Both techniques permit analyses on small ($\sim 10$ cc) atmospheric samples with precision comparable to the classical combustion line techniques using much larger samples. While these techniques promise a large increase in the numbers of samples that can be analyzed, they require great attention to experimental detail and calibration, so careful intercalibrations similar to those conducted by laboratories making the first measurements of atmospheric $\delta^{13}CH_4$ should be carried out before reporting data.

## 4.03.3 TERRESTRIAL STUDIES

Systematic measurements of $CH_4$ fluxes as well as some time series from wetlands, rice agriculture, landfills, and ruminants were in progress but unpublished at the time of the previous reviews. Reviews by Aselmann and Crutzen (1989), Bartlett and Harriss (1993), Harriss *et al.* (1993), and Matthews (2000) integrate and summarize a large number of wetland, tundra, and rice studies and have helped refine our understanding of the

source strength from these systems. This section will focus on studies that emphasize controls on $CH_4$ fluxes and thus provide a basis for development of realistic models.

### 4.03.3.1 Flux Time Series

Many of the available $CH_4$ flux measurements have resulted from short-term campaigns that span only a portion of the growing season; these have been summarized by Bartlett and Harriss (1993) and Harriss *et al.* (1993). Flux time-series measurements that provide information on seasonal changes and interannual changes are much less common. Whalen and Reeburgh (1992) reported results of a four-year time series of weekly measurements conducted at triplicate sites in the University of Alaska Arboretum in Fairbanks, AK. The sites were chosen to represent vegetation and cover types common in tundra systems. Although these sites were not true arctic sites, their growing seasons were only slightly longer than at sites at the Toolik Lake Long-Term Ecological Research (LTER) site, so they were taken as representative. Dise (1993) reported a two-year $CH_4$ flux time series from Minnesota peatland sites, and Shannon and White (1994) reported a three-year $CH_4$ flux time series from a Michigan bog. The longest $CH_4$ flux record (11 years) is from Sallie's Fen in southeastern New Hampshire (Crill, unpublished), where the initial measurements reported in Frolking and Crill (1994) have continued with automated chamber measurements (Goulden and Crill, 1997).

Small positive $CH_4$ fluxes from moss sites were observed from the Alaska sites during winter (Whalen and Reeburgh, 1988). These winter emissions amounted to 41% of the annual emission from the moss sites, which were open and aerated, and had lower summer emissions than other sites studied. Dise (1992) reported winter $CH_4$ fluxes from hummock sites in Minnesota peatlands amounting to 21% of the annual emission, and Melloh and Crill (1996) studied $CH_4$ concentrations at the freezing front and beneath surface ice in Sallie's Fen and estimated that winter emissions ranged between 2% and 9.2% of the annual emission, averaging 4.3%.

### 4.03.3.2 Flux Transects

Flux transects, which provide information on spatial variability and are usually conducted over a month-scale time interval, have been reported by Whalen and Reeburgh (1990a), who made $CH_4$ flux measurements at 10 km intervals along the trans-Alaska pipeline haul road. Christensen *et al.* (1995), using ship and helicopter support, made $CH_4$ flux measurements on a series of short inland transects along the Arctic Ocean coast of Siberia. Crutzen *et al.* (1998); Bergamaschi *et al.* (1998a), and Oberlander *et al.* (2002) report atmospheric concentration and isotope measurements of atmospheric $CH_4$ on a transect across Siberia on the trans-Siberian Railway, which allowed discriminating between wetland, gas production, and biomass burning sources of $CH_4$. Sugawara *et al.* (1996) reported $CH_4$ concentration and isotope measurements on samples collected during aircraft transects in Siberia.

### 4.03.3.3 Process-level Studies

#### *4.03.3.3.1 Vegetation removal experiments*

Vegetation removal experiments were conducted by Torn and Chapin (1993) and Schimel (1995) to understand the role of vascular plants in methane emission and oxidation. King *et al.* (1998) studied sites in the Toolik Lake LTER, where the vegetation was modified (sedge-removal and moss-removal plots) to determine the role of these vegetation types in wetland $CH_4$ emission and to study the gas transport mechanism. The study showed an inverse relationship between root density and pore water $CH_4$ concentration (Figure 3). Parallel experiments involving insertion of gas-permeable silicone rubber tubes show that the tubes are reasonable analogues for the physical process of diffusion through plants.

#### *4.03.3.3.2 Methane flux: net ecosystem exchange relationship*

Much of the above work has been directed toward obtaining reliable data-based estimates of wetland emissions for the global $CH_4$ budget.

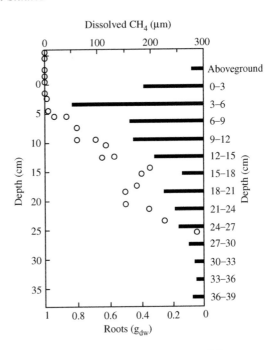

**Figure 3** Comparison of pore water $CH_4$ concentrations (circles) with root density depth distribution (dry weight mass of roots per 3 cm depth interval of a 6.5 cm diameter core) (bars) in a wet meadow site at the Toolik Lake LTER site, 1995 (source King *et al.*, 1998).

It has been very difficult to find a single or simple controlling relationship between $CH_4$ flux and environmental or system variables (Whalen and Reeburgh, 1992). Relationships between the $CH_4$ flux and subsurface properties (water-table depth, thaw depth, soil temperature, $CH_4$ concentration) are site specific and of little value as predictors. Parameters that integrate conditions influencing flux (thaw depth and centimeter degrees, the product of thaw depth and mean soil temperature to permafrost) appear to be the best predictors of $CH_4$ flux.

A relationship between $CH_4$ flux and net ecosystem exchange (NEE = system $CO_2$ uptake − system $CO_2$ release) presented by Whiting and Chanton (1993) showed promise as a basis for estimating $CH_4$ flux using methods other than the reliable but tedious static chamber measurements. Remotely sensed wetland net primary production (NPP) was suggested as a proxy for net ecosystem exchange, raising the possibility of estimating global wetland $CH_4$ flux over large areas with synoptic measurements. The Whiting–Chanton relationship used $CH_4$ flux data from the seasonal maximum in a wide range of wetland types, and implied fairly direct coupling between photosynthesis and $CH_4$ formation and emission. The fraction of modern $^{14}CH_4$ of emitted $CH_4$ is ~0.95 (Aravena *et al.*, 1993; Chanton *et al.*, 1995) and supports the notion of direct coupling by indicating that long-term stored carbon was not the major $CH_4$ source. Subsequent studies show that the Whiting–Chanton relationship applies best to wet systems populated with vascular plants. The relationship between $CH_4$ flux and NEE is less clear in wetland systems where oxidation is more important and no transport system exists, namely those with fluctuating water levels and moss cover.

#### 4.03.3.4 Scaling Up

Scaling up site-based $CH_4$ fluxes to estimate global fluxes is a major problem because of spatial and temporal variability in flux measurements, as well as variability in site types (Matson *et al.*, 1989). The estimate of wetland and tundra emissions by Matthews and Fung (1987) did a very good job of summarizing global areas of likely $CH_4$ sources, but was limited by a very small amount of $CH_4$ flux data. Whalen and Reeburgh (1992) used data from four-year $CH_4$ flux time series and areally weighted fluxes from sites representative of arctic tundra to estimate a high-latitude global $CH_4$ flux that agreed reasonably well ($42 \pm 27$ Tg $CH_4$ yr$^{-1}$) with the Fung *et al.* (1991) estimate (35 Tg $CH_4$ yr$^{-1}$) for high-latitude emissions from tundra. Another study (Reeburgh *et al.*, 1998) attempted to scale up from the site to region level by areally weighting three-year integrated annual $CH_4$ fluxes from sites in the Kuparuk River watershed on Alaska's North Slope. These sites represented specific tundra vegetation and land cover types. The sites were chosen and their integrated fluxes were weighted using a Geographic Information System-based vegetation map (Auerbach *et al.*, 1997). Extending this regional estimate to the circumpolar arctic accounted for only 15% of the Fung *et al.* (1991) high-latitude flux estimate. The Reeburgh *et al.* (1998) work shows that arctic $CH_4$ fluxes are dominated by inundated wetlands, as indicated previously by Matthews and Fung (1987). The poor agreement with the Fung *et al.* (1991) high-latitude $CH_4$ flux estimate obtained by extrapolating this watershed flux estimate suggests that the Hudson's Bay Lowlands (Roulet *et al.*, 1994) and the West Siberian Lowlands, which have received very little study (Harriss *et al.*, 1993), may be the major global contributors. Clearly, better estimates of inundated wetland areas as well as field flux measurements are needed.

Methane flux measurements can be made at scales larger than chambers with aircraft boundary layer measurements (100 km) or micrometeorological measurements using towers (100 m). Chamber, tower, and aircraft measurements were compared in two field campaigns: the Arctic Boundary Layer Experiment (ABLE 3A)

(summarized in the *Journal of Geophysical Research* **97**(D15), 1992) and the Northern Wetlands Study (NOWES/ABLE 3B) (summarized in the *Journal of Geophysical Research* **99**(D1), 1992). During the latter study, $CH_4$ fluxes often agreed to within a factor of 1.2 (Roulet *et al.*, 1994). There is, however, a critical need for a field-portable, rapid-response $CH_4$ detector for a use in aircraft and tower measurements.

#### 4.03.3.5 Wetland Soil Models

A variety of process-based models of methane emission from wetlands have been published, ranging from very complex models requiring a large number of measured input parameters to straightforward special-purpose models involving correlations between measured parameters. These models have been developed to understand controls on $CH_4$ fluxes from a range of environments, as well as to scale fluxes and soil consumption to global scales.

A biologically based approach to the simulation of $CH_4$ transformations was undertaken as part of the Canadian *ecosys* model project. The $CH_4$ studies simulated the interrelated activities of anaerobic fermenters and $H_2$-producing acetogens, acetotrophic and hydrogenic stoichiometries, and energetics of transformations in a series of papers dealing with methanogenesis (Grant, 1998), methanotrophy (Grant, 1999), and finally, $CH_4$ efflux (Grant and Roulet, 2002). The emission model predicted efflux events associated with thawing and warming of soil and pond sediments that would likely be missed by observation programs. However, the *ecosys* model required a knowledge of and assignment of almost 30 parameters, so a great deal of fundamental information about microbial populations, rates, and kinetics must be known.

Segers and Kengen (1998) described a model based on the dynamics of alternative electron acceptors, acetate, and methanogenic biomass that relates $CH_4$ production to anaerobic carbon mineralization. Models incorporating kinetic processes like $CH_4$ production, $CH_4$ oxidation, electron acceptor reduction, electron acceptor reoxidation, and aerobic respiration were employed in a series of subsequent papers describing stepwise scaling up from the single-root scale (Segers and Leffelaar, 2001a), to the soil-layer scale (Segers *et al.*, 2001), and finally to the plot scale (Segers and Leffelaar, 2001b) to estimate methane fluxes.

Frolking and Crill (1994) developed a peat soil climate model driven by daily weather and used correlations of $CH_4$ flux with environmental parameters to investigate how climate and weather control the observed temporal variability in CH$_4$ flux at Sallie's Fen. The model agreed well with field data. Further, the simulated fluxes exhibited three modes of temporal variability that were evident in the field data: seasonal, inter-annual, and event scale (flux suppression by summer rainstorms).

Cao *et al.* (1996) developed a process-based model to estimate seasonal and spatial variations in wetland CH$_4$ emissions at the global scale. The model calculated methanogenesis from primary production and estimated emission by oxidizing a fraction of the CH$_4$ produced as it passed through the oxic zone. The process of oxidation becomes more complicated when the wetland soil is not inundated. Global application of the model yielded total emission of 92 Tg CH$_4$ yr$^{-1}$. Sensitivity analysis showed that the model response of CH$_4$ emission to climate change depended upon the combined effects of carbon storage, rate of decomposition, soil moisture, and methanogen activity.

Walter and Heimann (2000) report application of a one-dimensional process-based climate-sensitive model simulating processes leading to CH$_4$ emission from natural wetlands. The model treats three CH$_4$ transport mechanisms—diffusion, plant transport, and ebullition explicitly—and is forced with daily values of temperature, water table, net primary productivity, and thaw depth at permafrost sites. Their objective was to provide a model that could be applied to simulating CH$_4$ emissions in various regions as a function of the prevailing climate that could also be used on a global scale. The model was tested with time-series data from five different wetland sites. Soil temperature and water-table position explained seasonal variations, but the authors emphasized that the absence of a simple relationship between controlling factors and CH$_4$ emission requires the process-based approach.

Granberg *et al.* (2001) modeled CH$_4$ emission from an oligotrophic lawn community in a boreal mire. Their representation of winter conditions (frost and snow) is the main difference between their model and that of Walter and Heimann (2000). The model was forced with daily mean air temperatures and daily accumulated precipitation. Methane was simulated with a model which separated CH$_4$ emission by ebullition, diffusion, and plant transport. Oxidation was estimated by oxidation potential, which was centered at the water table. This model indicated that the mean level of the water table was the most important predictor of simulated CH$_4$ emission, and that the presence of vascular plants is the most important factor in determining CH$_4$ emission. These authors point out that their model contains tuning parameters that cannot be replaced by measurable parameters in the field. They called for focusing on mechanisms like gas diffusion through plants and ebullition to produce a robust model that works over a broad range of plant communities and climatic conditions.

Ridgwell *et al.* (1999) developed a process-based model for estimating the consumption of atmospheric CH$_4$ by soils, and estimated a global soil sink strength of 20–51 Tg CH$_4$ yr$^{-1}$, with a preferred value of 38 Tg CH$_4$ yr$^{-1}$. Microbial activity, instead of diffusion, limits uptake in tropical ecosystems, which account for almost a third of the total. Del Grosso *et al.* (2000) also modeled methane oxidation as a function of soil water content, temperature, porosity, and field capacity, but emphasized native and cropped grasslands. Using a kinetic model, Cai and Yan (1999) simulated paddy soil methane oxidation activity (assumed to be the product of the number of methanotrophic bacteria and their specific activity) and found it to be dependent on the CH$_4$ concentration.

### 4.03.3.6 Animals, Landfills, and Biomass Burning

Ruminant methane emissions, which account for 95% of animal CH$_4$ emissions, have been reviewed by Johnson *et al.* (2000), who estimated that 81 Tg CH$_4$ yr$^{-1}$ resulted from enteric microbial activity and another 14 Tg CH$_4$ yr$^{-1}$ from manure decomposition. Enteric CH$_4$ production represents a substantial (up to 12%) dietary energy loss, so inhibitors directed at minimizing methanogenesis have been developed and are under study. A common problem has been apparent microbial adaptation, resulting in a return to baseline levels of methane production. Sulfur hexafluoride (SF$_6$) has been used as a conservative internal tracer in individual animals (Johnson *et al.*, 1994) as well as an external tracer in experiments involving groups of animals in barns (Marik and Levin, 1996) or pens (Johnson *et al.*, 1994). In general, diets that are finely ground and pelleted with high grain content result in lower methane emission.

The review of landfill methane emission by Bogner and Matthews (2003) used four scenarios to obtain estimates ranging from 15 Tg CH$_4$ yr$^{-1}$ to 81 Tg CH$_4$ yr$^{-1}$, which is similar to previous estimates. Landfill sites have been a particularly active area for studies of methane oxidation (Whalen *et al.*, 1990; Jones and Nedwell, 1993) and for studies of isotope fractionation of methane during oxidation (Bergamaschi and Harris, 1995; Bergamaschi *et al.* 1998b; Liptay *et al.*, 1998). A laboratory study (Kightley *et al.*, 1995) employing soil microcosms successfully reproduced the "methane: oxygen crossover" zone observed in landfill cover soils and confirmed the high methane-oxidizing capacity of this zone as a control on atmospheric CH$_4$ emissions.

Biomass burning is covered in a separate chapter (see Chapter 4.05). Levine (in Chapter 4.05) estimates that $50 \, \text{Tg yr}^{-1}$ of methane are added by biomass burning.

### 4.03.3.7 Microbial Soil Oxidation

Methane oxidation in soils has received a great deal of attention recently. An example of how rapidly this area has grown is given by two reviews written eight years apart. The first (Conrad, 1984) used existing kinetic information on cultured organisms to make the case that soil organisms could not utilize, grow on, and control the atmospheric concentrations of CO, $CH_4$, and $H_2$. The second (Conrad, 1996) contains some 600 references and clearly shows how soil and atmospheric concentrations of not only $CH_4$, but also $H_2$, CO, $CH_4$, OCS, $N_2O$, and NO are heavily influenced or controlled by microbial activity.

#### 4.03.3.7.1 Moist soils

The first example of soil consumption of $CH_4$ was reported by Harriss *et al.* (1982), who observed seasonal concentration decreases (and increases) in flux chamber headspaces. Keller *et al.* (1990) reported similar observations for tropical soils. This work was followed by measurements in tundra environments by Whalen and Reeburgh (1990b), and by more extensive experimental work using chamber measurements, jar experiments, and $^{14}CH_4$ tracer experiments on boreal forest soils by Whalen *et al.* (1992). Koschorreck and Conrad (1993) devised a multiport device for laboratory studies on soil cores.

A number of experimental studies on controls on soil $CH_4$ oxidation produced relationships that have been incorporated in some of the models discussed earlier. Whalen and Reeburgh (1996) used jar experiments (1 qt (1 qt = 9.463 × $10^{-4} \, \text{m}^3$) canning jars) to determine $CH_4$ uptake kinetics of a number of moist boreal soils. Large samples of soil from the $CH_4$-oxidizing maximum were homogenized and distributed into jars, and first-order $CH_4$ oxidation kinetics were measured at a range at temperature and moisture conditions. Figure 4 shows typical results. The $CH_4$ supply to methane-oxidizing organisms is diffusion controlled and decreased under higher moisture conditions, and oxidation rates are lower.

#### 4.03.3.7.2 Waterlogged soils

Similar rate studies in waterlogged boreal bogs required modified techniques (Whalen and Reeburgh, 2000a). Oxygen microelectrode studies

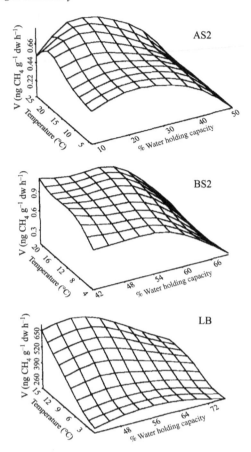

**Figure 4** Methane oxidation rates as a function of moisture–temperature interactions for boreal soils. Methane oxidation rates for AS2 and BS2 soils were calculated for $1.5 \, \mu\text{L} \, CH_4 \, L^{-1}$ from first-order rate constants for each moisture–temperature combination. Methane oxidation rates for LB soil were directly measured in time-course experiments for each moisture–temperature combination (source Whalen and Reeburgh, 1996).

showed that waters become anoxic only millimeters below the level of the water table (Whalen *et al.*, 1995). The water table is a sharp interface (~0.5 cm) between oxic and anoxic conditions and is perhaps the most important location for control of $CH_4$ fluxes by oxidation (Moosavi *et al.*, 1996). The Whalen and Reeburgh (2000a) work focused on concentration and rate distributions adjacent to the water table. Figure 5 gives results from an experiment that attempted to mimic a rapid change in the level of a water table, and shows that the microbial communities can respond within a day to changes from anoxic to oxic conditions. This experiment also shows that methanogens are more sensitive than methanotrophs to adverse oxygen conditions. Methane oxidation is a very effective process in limiting $CH_4$ emission from wetlands inundated with a shallow layer of oxic water (Happell *et al.*, 1994).

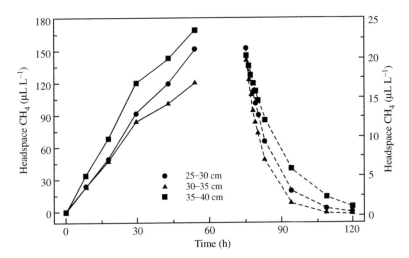

**Figure 5** Time courses for $CH_4$ production (left) and $CH_4$ oxidation (right) in selected 5 cm core sections (site LB2 in Lemeta Bog) that were alternately made anoxic and oxic in an attempt to mimic response to changing water table levels. Methane production by anoxic core sections was measured for 54 h. Core sections were drained for 20 h in an oxic environment and then amended with $\sim20~\mu L~L^{-1}~CH_4$, after which the time course for $CH_4$ oxidation (right) was determined. No $CH_4$ was produced over 48 h by core sections rewetted and made anoxic at 120 h (source Whalen and Reeburgh, 2000a).

### 4.03.3.7.3 Nitrogen fertilization and disturbance

Several early studies on $CH_4$ uptake by soils also noted that $CH_4$ oxidation activity was reduced or inhibited by disturbance or nitrogen additions (Mosier *et al.*, 1991; Steudler *et al.*, 1989; Reay *et al.*, 2001). Enzymes associated with $NH_4^+$ oxidation are similar to those associated with $CH_4$ oxidation, so there was concern that increased nitrogen loading of soils by atmospheric deposition could decrease the soil $CH_4$ sink strength. Nitrogen inhibition of $CH_4$ oxidation is known (Bedard and Knowles, 1989). King (1992) reviewed the responses of soils and concluded that increases in ammonium deposition and water stress should inhibit soil $CH_4$ oxidation. Klüber and Conrad (1998) observed that addition of nitrate to anoxic rice soil slurries caused complete, but reversible inhibition of methanogenesis. No effects on $CH_4$ oxidation were observed in field fertilization experiments on boreal forest soils. These were attributed either to distinct $CH_4$ oxidizer populations (Gulledge *et al.*, 1997) or to the position of the $CH_4$-oxidizing community in the soil column (Whalen and Reeburgh, 2000b).

### 4.03.3.7.4 Effects of drying on paddy soils

Water management has a profound effect on methane emission from rice fields (Sass *et al.*, 1992). Mid-season draining, dressing, and fertilizing Japanese rice fields result in greatly reduced overall $CH_4$ emission (Yagi *et al.*, 1996).

Laboratory studies by Ratering and Conrad (1998) and Klüber and Conrad (1998) demonstrated that short-term drying allows air oxidation of reduced nitrogen, sulfur, and iron species in paddy soils. This oxidized reservoir must be reduced before methanogenesis can resume, resulting in a lag before methane emission is evident.

### 4.03.3.8 New Techniques

Dörr and Munnich (1990) described a technique for estimating gas fluxes from soils using parallel measurements of $CO_2$ and $^{222}Rn$ profiles in soils. This technique does not require chambers and has the advantage of using naturally produced $^{222}Rn$, which is conservative and is affected only by diffusion and radioactive decay. The technique was extended to measurements of $CH_4$ consumption (Born *et al.*, 1990). Dörr *et al.* (1993) used soil texture parametrization to extend local measurements to the global scale, and arrived at an estimate of 28.7 Tg $CH_4$ yr$^{-1}$. Whalen *et al.* (1992) used the $^{222}Rn$ technique in a boreal forest study comparing seven approaches to measuring soil $CH_4$ oxidation.

The above studies are only able to detect net $CH_4$ production or net $CH_4$ consumption, i.e., either production or consumption must exceed the competing process to produce an observable positive or negative concentration change in a flux chamber. To separate the effects of oxidation and production, the effect of $CH_4$ oxidation has been measured by using $CH_3F$, a specific inhibitor of $CH_4$ oxidation (Oremland and Culbertson, 1992; Epp and Chanton, 1993; Moosavi and Crill, 1998) or with tracer techniques using $^{14}CH_4$ (Whalen and Reeburgh, 2000a; Whalen et al., 1992). Two additional isotope techniques have been proposed to determine simultaneous $CH_4$ production and oxidation in soils. Andersen et al. (1998) described laboratory work involving additions of $^{14}CH_4$ to soil columns. They observed that $^{14}CH_4$ was consumed faster than atmospheric $CH_4$, and took this as evidence that although the soil was a net consumer of $CH_4$, there may be another $CH_4$ source. Von Fischer and Hedin (2002) describe a field technique utilizing stable isotope dilution of naturally occurring pools to estimate simultaneous production and consumption. This technique was applied to moist soils; its application to waterlogged soils, especially adjacent to the level of water table, appears to be a natural and needed extension.

One of the questions essential to understanding $CH_4$ biogeochemistry in wetland and rice systems is the source of emitted $CH_4$ and the timescale between photosynthetic production of biomass and methanogenesis. This question has been addressed using pulse-labeling experiments, which involve labeling plants by photosynthetic uptake of $^{14}CO_2$ and quantifying the time course of $^{14}CH_4$ release. Pulse labeling with $^{13}CO_2$ has been used in laboratory experiments on rice plants (Minoda and Kimura, 1994; Minoda et al., 1996) where the product $^{13}CH_4$ was quantified mass spectrometrically. Dannenberg and Conrad (1999) conducted similar laboratory experiments on rice using $^{14}CO_2$. Megonigal et al. (1999) performed a pot-scale $^{14}CO_2$ pulse-labeling experiment on a single *Orontium aquaticam* plant, and King and Reeburgh (2002) performed a field $^{14}CO_2$ pulse-labeling experiment under near *in situ* conditions using arctic plant mesocosms containing communities dominated by *Eriophorum-Carex*. Results from laboratory *Eriophorum-Carex* mesocosm pulse-labeling experiments were reported by King et al. (2002) and Loya et al. (2002). All of these pulse-labeling experiments reported rapid emission of $^{14}CH_4$ (within 24 h). Less than 1% of the $^{14}CO_2$ taken up during photosynthesis was emitted over a two-week period in the nonrice studies. These measurements of net $CH_4$ emission provide little information on pathways and processes involved in methanogenesis as well as $CH_4$ oxidation, but they show clearly that recently fixed photosynthates play a major role in $CH_4$ emission. These findings are consistent with observations of 0.95% modern methane emissions (Aravena et al., 1993; Chanton et al., 1995). Pulse-labeling experiments are labor intensive, so the work above has reported small numbers of samples. King and Reeburgh (2002) point out that the contribution of plant photosynthates to $CH_4$ emission may be dependent on plant phenology, so pulse-labeling studies at intervals over a growing season are needed in the future.

## 4.03.4 MARINE STUDIES

### 4.03.4.1 Ocean Methane Source

The ocean as a $CH_4$ source in the global $CH_4$ budget was re-evaluated by Ehhalt (1978) using a much larger database than his 1974 estimate, but the difference was small and the Ehhalt (1974) estimate was not altered in the Cicerone and Oremland (1988) budget. Two recent reviews re-evaluated open ocean (Lambert and Schmidt, 1993) and coastal and estuarine (Bange et al., 1994) $CH_4$ emissions. Lambert and Schmidt (1993) concluded that the ocean emitted 3.5 Tg $CH_4$ $yr^{-1}$. Bange et al. (1994) summarized all ocean $CH_4$ measurements, including those from shelf and estuarine areas, and applied two air–sea transfer models regionally to estimate that the ocean source lies in the range 11–18 Tg $CH_4$ $yr^{-1}$. Shelf and estuarine areas contribute ~75% of this total. Bange et al. (1994) used a coupled three-layer model to conclude that, even with increasing tropospheric $CH_4$ concentrations, the ocean will maintain close to the present source strength. These estimates lie within the range of values suggested by Cicerone and Oremland (1988).

### 4.03.4.2 Aerobic Methane Oxidation

The ocean has received far less attention than wetlands and soils as a source and sink of $CH_4$. Studies of $CH_4$ distributions in the Eastern Tropical North Pacific showed two $CH_4$ maxima (Burke et al., 1983): a surface maximum presumably associated with methanogensesis in fecal fellets (Karl and Tilbrook, 1994) and a deeper maximum. The mass spectrometric GC/C/IRMS technique discussed earlier allows measurement of $\delta^{13}CH_4$ in small samples of seawater (Holmes et al., 2000). This technique was applied in the Eastern Tropical North Pacific (Sansone et al., 2001) to study methanogenesis and methane oxidation. The Sansone et al. (2001) results show a distinct difference in the $\delta^{13}CH_4$ from the two $CH_4$ maxima, leading to the suggestion that the deeper maximum may result from long distance offshore transport of $CH_4$ with a shelf or seep source similar to that shown in Cynar and Yayanos (1993, figure 3).

Most direct $CH_4$ oxidation rate measurements have focused on anoxic sediments and waters, so there have been very few direct oxidation rate measurements in oxic environments. The study of

Valentine *et al.* (2001), which was conducted near seeps and vents in the Eel River Basin, provided insights and some credence to the notion of long distance offshore $CH_4$ transport. The Valentine *et al.* (2001) study involved tracer measurements of methane oxidation rate using $^3H$-$CH_4$ and showed high oxidation rates adjacent to the vents where $CH_4$ concentrations are high and much lower rates in the remainder of the lower $CH_4$ concentration part of the water column. Figure 6 (Valentine *et al.*, 2001) relates $CH_4$ concentration to turnover time, which for low $CH_4$ concentrations, was over 40 years. This is consistent with $CH_4$ oxidation rates estimated by Scranton and Brewer (1978), who calculated rates using methane analyses and water mass ages determined with $^3H/^3He$ and $^{14}C$ ages to determine "apparent methane utilization" rates. Scranton and Brewer (1978) estimated that the $CH_4$ consumption rate in 150-year-old water masses is $2.2 \times 10^{-4}$ nM yr$^{-1}$. Rehder *et al.* (1999) used CFC-11 distributions and the input function of CFC-11 to deduce a timescale for oxidation of $CH_4$ in North Atlantic Deep Water of ~50 years. All of these studies indicate that $CH_4$ present at low concentrations in the ocean interior is oxidized at very low rates. These studies suggest a threshold $CH_4$ concentration below which $CH_4$ in the ocean cannot be oxidized. It is important to understand what controls this threshold: availability of suitable organisms, the $CH_4$ concentration, or the frequency of oxidizer–$CH_4$ encounters?

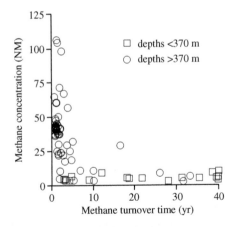

**Figure 6** Plot of methane concentration versus turnover time for samples collected in the Eel River Basin seep area. Turnover time is the $CH_4$ concentration divided by the oxidation rate determined with $C_3H_4$ tracer experiments. The maximum turnover time of 40 years is the lower detection limit for these studies, but it is consistent with similar estimates by Scranton and Brewer (1978) and Rehder *et al.* (1999) (source Valentine *et al.*, 2001).

### 4.03.4.3 Anaerobic Methane Oxidation

Based on $CH_4$ distributions in anoxic marine sediments, anaerobic methane oxidation (AMO) was suggested by Barnes and Goldberg (1976), Reeburgh (1976), and Martens and Berner (1977). The importance of AMO as a $CH_4$ sink was established by measurements of concentration profiles, diagenetic modeling, oxidation rate measurements using radioactive tracers, inhibition experiments, and measurement of stable isotope budget (reviewed by Hoehler *et al.* (1994) and Valentine (2002)). Anaerobic $CH_4$ oxidation occurs in a zone near the point of sulfate depletion in anoxic marine sediments and is a near-quantitative sink for upward fluxes of $CH_4$. Although the process has been shown to be an effective and globally important $CH_4$ sink, it has been controversial because the organisms and mechanism responsible for anaerobic methane oxidation remain unknown.

Hoehler *et al.* (1994) suggested that AMO might be energetically possible through "reverse methanogenesis." According to the "reverse methanogenesis hypothesis," it is energetically possible for existing methanogens to operate in reverse to anaerobically oxidize $CH_4$ when the concentration of $H_2$ is held to concentrations $\leq 0.29$ nM. The reverse methanogenesis hypothesis was particularly attractive because it was consistent with all previous field observations, it offered an energetically feasible means for anaerobic $CH_4$ oxidation, it involved no new organism, and it further explained the close association of the anaerobic $CH_4$ oxidation zone with sulfate reducers. Valentine *et al.* (2000a) devised an apparatus for testing the reverse methanogenesis hypothesis that involved exposing actively growing methanogen cultures to $CH_4$ and controlled and low $H_2$ partial pressures and monitoring the effluent of the reactor for the product $H_2$. Four pure cultures of methanogens were tested for their ability to mediate reverse methanogenesis, but none demonstrated sustained $H_2$ production, so the reverse methanogenesis hypothesis was falsified for these cultures (Valentine *et al.*, 2000b).

Studies reporting anaerobic $CH_4$ oxidation rate and the rate depth distribution studies in anoxic marine sediments subsequent to those of Reeburgh (1980) are summarized by Hoehler *et al.* (1994). Rates of anaerobic $CH_4$ oxidation have been measured directly in the anoxic water columns of the Cariaco Basin (Ward *et al.*, 1987) and the Black Sea (Reeburgh *et al.*, 1991). Two radioactive tracers were used in these studies: $^{14}C$-$CH_4$ for the Cariaco Basin and both $^{14}C$-$CH_4$ and $^3H$-$CH_4$ for the Black Sea. The parallel rate determinations using $^{14}C$-$CH_4$ and $^3H$-$CH_4$ in the Black Sea agreed within a remarkable factor of 2

(Reeburgh *et al.*, 1991). The $^3$H-CH$_4$ has higher specific activity and is most suitable for water column measurements since it can be added in tracer quantities and does not affect the low-concentration ambient CH$_4$ pool size. Samples are incubated in headspace-free vessels, the added tracer is removed by stripping, and the product $^3$H-H$_2$O is counted. Radiocarbon-labeled CH$_4$ has a lower specific activity works best in sediments, where CH$_4$ concentrations are higher than water columns and where overwhelming the ambient pool size is of less concern. These measurements involve injecting tracer into core segments of intact sediment and incubating. Following incubation, a sediment slurry is formed and stripped under basic conditions to remove the unreacted $^{14}$C-CH$_4$ tracer; the slurry is subsequently stripped under acidic conditions and the oxidation product $^{14}$CO$_2$ is trapped and counted.

Curiously, the most striking and extensive examples of anaerobic CH$_4$ oxidation, which has been controversial as a contemporary process, are provided by Deep Sea Drilling Project (DSDP) studies. The depletion of sulfate in DSDP cores (Borowski *et al.*, 1996; Dickens, 2000, 2001b) was ascribed to anaerobic CH$_4$ oxidation. D'Hondt *et al.* (2002) produced global maps based on DSDP cores delineating two provinces of subsurface metabolic activity: a sulfate-rich open ocean province where CH$_4$ is not abundant, and an ocean margin province where sulfate is restricted to shallow depth intervals because of higher sulfate reducing activity, and where CH$_4$ is more abundant. Anaerobic methane oxidation was invoked to explain the observed sulfate and methane distributions and differences in the two provinces.

Anaerobic CH$_4$ oxidation, now referred to as anaerobic oxidation of methane (AOM), has received renewed attention with introduction of new organic geochemical and molecular techniques. Hinrichs *et al.* (1999) applied compound-specific isotope analyses of lipid biomarker molecules associated with specific archea and culture-independent techniques involving 16S rRNA identification studies to samples collected from an Eel River Basin seep. This work showed that the biomarker compounds were so strongly depleted in $^{13}$C that CH$_4$ must be the source rather than the metabolic product for the organisms that produced them. Boetius *et al.* (2000) used culture-independent identification techniques (fluorescent *in situ* hybridization (FISH)) to visualize dense aggregates of ~100 archeal cells surrounded by sulfate reducers. Orphan *et al.* (2001) combined the FISH technique with secondary ion mass spectrometry (SIMS) to measure the stable carbon isotope composition of individual cell aggregates.

This work showed unambiguously that CH$_4$ was the source of cell carbon.

The reviews by Hinrichs and Boetius (2002), Valentine and Reeburgh (2000), and Valentine (2002) point out the uncertainties regarding the nature of the syntrophic association, the biochemical pathway for oxidation of CH$_4$, as well as the interaction of the process with the local chemical and physical environment. Based on observations of higher oxidation rates and larger spatial extent, Hinrichs and Boetius (2002) increased the Reeburgh *et al.* (1993) estimate of the extent of anaerobic CH$_4$ oxidation from 70 Tg CH$_4$ yr$^{-1}$ to 300 Tg CH$_4$ yr$^{-1}$. Valentine (2002) pointed out that anaerobic CH$_4$ oxidation could involve different organisms depending on the supply rate of CH$_4$ and that seeps and vents may differ from diffusive environments. Michaelis *et al.* (2002) described massive microbial mats covering up to 4 m high carbonate buildups at methane seeps in the anoxic waters of the Black Sea shelf. Incubation of the mat material showed anaerobic oxidation of methane coupled to sulfate reduction. Hinrichs and Boetius (2002) indicate that AOM may have influenced the carbon isotope record of the Archean and is thus an important link in the biological cycling of carbon in an anoxic biosphere. Michaelis *et al.* (2002) suggest that these reefs may represent the appearance of large parts of the ancient anoxic ocean. Finally, Nauhaus *et al.* (2002) demonstrated coupling of methane oxidation and sulfate reduction *in vitro*. It appears, then, that we are close to an answer to the question posed over 25 years ago: How does anaerobic CH$_4$ oxidation occur?

### 4.03.4.4 Methane Clathrate Hydrates

Methane clathrate hydrates are nonstoichiometric solid structures composed of cages of water molecules surrounding methane molecules. Estimates of the size of the methane reservoir sequestered as hydrates are enormous. Because these CH$_4$ hydrates are frequently close to the sediment–water interface and because they represent such a large reservoir, hydrate stability has been a major concern. There is extensive evidence of slides and craters on ocean shelves, presumably caused by destabilization and release of gas phase CH$_4$ (Hovland and Judd, 1988), but we have little information on the magnitudes of CH$_4$ release needed to form these features. The global methane budget of Cicerone and Oremland (1988) emphasized how little is known about the role of methane hydrates in the global budget by entering methane hydrates in their budget as a 5 Tg yr$^{-1}$ "placeholder" term.

The physical and thermodynamic properties and stability fields of clathrates, including methane, are presented in Sloane's (1998) text. Methane hydrates have a latent heat of fusion quite similar to that of water (Sloane, 1998), so decomposition of hydrates is a matter of heat flux, not a rise in temperature. Hydrate stability is not a matter of $P–T$ conditions alone, as $CH_4$ concentrations substantially higher than saturation values are required for hydrate formation and stability (Sloane, 1998). Laboratory data on $CH_4$ clathrate stability fields in seawater were presented by Dickens and Quinby-Hunt (1994, 1997).

### 4.03.4.4.1 Methane hydrate reservoir

Reviews by Kvenvolden (1988, 1993) and Kvenvolden and Lorenson (2001) have updated information on the occurrence, stable isotope composition, estimates of total amount, and stability of $CH_4$ clathrate hydrates. Articles in Paull and Dillon (2001) summarize recent work on $CH_4$ hydrates, emphasizing geochemical studies in the Gulf of Mexico and at Hydrate Ridge, Cascadia Margin, and geophysical studies at Blake Ridge and the Peru/Middle American trenches. Gornitz and Fung (1994) calculated the potential size of the ocean $CH_4$ hydrate reservoir using two models for $CH_4$ hydrate formation: *in situ* bacterial production and a fluid expulsion model. Their models resulted in methane hydrate reservoirs of 14 Tt and 75 Tt ($1\ Tt = 10^{18}$ g), respectively, with the most likely value near the lower end of this range. A consensus value of 10 Tt for the amount of $CH_4$ hydrate was suggested by Kvenvolden and Lorenson (2001). Again, this is an enormous number, 2,000-fold larger than the atmospheric $CH_4$ inventory and ~30% of the ocean dissolved inorganic carbon reservoir.

### 4.03.4.4.2 Methane hydrate decomposition rates

Dickens (2003) highlighted the common misconception that present-day $CH_4$ hydrate systems are stable. Hydrates may be in steady state, but they cannot be stable because of differences in the chemical potential of $CH_4$ in the hydrate phase and the adjacent dissolved interstitial methane. Thus, $CH_4$ hydrates must be viewed as dynamic, with large and unknown fluxes to and from the ocean. An understanding of the rate of contemporary hydrate-derived $CH_4$ additions to the ocean is essential to evaluating the role of hydrates in the global $CH_4$ budget, but, as of early 2000s, no measurements permitting this estimate have been made. There are several good reasons for this state of affairs. First, we have no means of

discriminating between hydrate $CH_4$ and diagenetically produced $CH_4$. Second, $CH_4$ released by dissociating hydrates is effectively oxidized by microbially mediated processes in anoxic sediments (Reeburgh, 1980; Alperin and Reeburgh, 1984), as well as in the adjacent oxic water column (Valentine *et al.*, 2001), so elevated $CH_4$ concentrations or "hot spots" are rarely observed. Third, we have been technology-poor in this area; hydrates collected by coring or dredging decompose rapidly, and laboratory studies are difficult if not impossible to conduct at sea.

### 4.03.4.4.3 How can we estimate the rate of $CH_4$ clathrate decomposition?

A USGS-sponsored workshop (Kvenvolden *et al.*, 2001) considered the magnitude of $CH_4$ release from two perspectives: compilation of known seeps, and estimates of the geologic $CH_4$ reservoir's half-life and depletion time. The first approach estimated that $50\ Tg\ CH_4\ yr^{-1}$ was released and that $\sim30\ Tg\ CH_4\ yr^{-1}$ reaches the atmosphere. The reservoir half-life approach resulted in a release estimate of $30\ Tg\ CH_4\ yr^{-1}$ from the seafloor and $10\ CH_4\ Tg\ y^{-1}$ into the atmosphere. Water column oxidation was presumed to cause the difference.

Research on $CH_4$ clathrates has been advanced by using remotely operated vehicles (ROVs), which can be equipped not only to synthesize hydrates *in situ* under appropriate $P–T$ conditions (Brewer *et al.*, 1997, 1998), but to locate seeps for sampling and to place acoustic beacons for longer-term stability experiments as well. Rehder *et al.* (2003) made direct measurements of the decomposition rates of pure $CH_4$ and $CO_2$ hydrates in a seafloor setting nominally within the gas hydrate $P–T$ stability zone. Their rates represent end members because of the flow field and undersaturation. The water column $CH_4$ oxidation rate measurements of Valentine *et al.* (2001) were conducted adjacent to seeps in the Eel River Basin that were previously located with an ROV. These oxidation rate measurements were conducted with tracer additions of $^3$H-$CH_4$, and showed the highest rates near the bottom. These instantaneous rate measurements are difficult to interpret in a dynamic advecting coastal environment, but the fractional turnover rates can be used to infer a $CH_4$ turnover time of ~1.5 years in the deep waters. The depth-integrated rates (5.2 mmol $CH_4\ m^{-2}\ yr^{-1}$) can be used to estimate total oxidation for the 25 $km^2$ vent field. The amount of $CH_4$ emitted is unknown, so it is not possible to estimate the oxidized fraction, but results from a closed system oxidation model applied to $\delta^{13}CH_4$ measurements suggested that ~45% of the added $CH_4$ was oxidized. Grant and Whiticar (2002) concluded from observations of isotopically heavy $\delta^{13}CH_4$ at Hydrate Ridge that substantial aerobic methane oxidation occurred there. These isotopic enrichment measurements indicate cumulative $CH_4$ oxidation, but provide no rate information. Clearly, many more direct methane oxidation rate measurements similar to those of Valentine *et al.* (2001) are needed.

Recent work indicates that it may be possible to discriminate between hydrate methane and diagenetic methane with measurements of $^{14}CH_4$. If hydrates are formed by the fluid expulsion, they will likely involve $CH_4$ from deep sources and will probably contain no radiocarbon; if they are formed by *in situ* bacterial production, the radiocarbon age will depend on the age of the substrate carbon. There have been a few unpublished measurements of the radiocarbon content of hydrate methane (Kastner, personal communication). Winckler *et al.* (2002) reported radiocarbon measurements on $CH_4$ from massive methane hydrates collected at Hydrate Ridge, Cascadia margin, and showed that they, as well as the unpublished results, are devoid of radiocarbon, indicating that there are no recent ($^{14}C$-active) contributions to the hydrate reservoir. Measurements of noble gases in hydrates (Winckler *et al.*, 2002) showed that light noble gases are not incorporated into the hydrate structure, but that significant quantities of the heavier noble gases—argon, krypton, and xenon—are present. The heavy noble gases are potential secondary tracers of hydrate decomposition.

The measurements of Black Sea water column $^{14}CH_4$ (Kessler *et al.*, 2003) suggest another possible approach for estimating hydrate $CH_4$ contributions. This approach is based on exploiting the Black Sea, with its restricted circulation, as a "geochemical bucket." Reeburgh *et al.* (1991) developed a sink-based methane budget for the Black Sea that assumed steady state and was based on the following sink terms: measured water column oxidation rates, measured sediment oxidation rates, calculated rates of evasion to the atmosphere, and export by the Bosporus outflow. Since methane is not produced in the Black Sea water column, the dominant sink (water column oxidation) must be balanced by a sediment source. Shelf sediments were a small methane source, but deep sediments were a methane sink. A rapid turnover time for methane (5–20 years) and stable isotope ($\delta^2H$-$CH_4$, $\delta^{13}C$-$CH_4$) distributions suggested that vents and decomposing hydrates must be the major methane source to the deep Black Sea, and that their rate of addition is equal to the basin-wide water column oxidation rate. Knowledge of the fraction of hydrate methane added to the Black Sea water column and the water column oxidation rate can be used to estimate the hydrate decomposition rate. The measured

Black Sea water column $^{14}CH_4$ profile (Kessler *et al.*, 2003) shows that the fraction of modern carbon is low, ranging from 5% to 15%. The role of vents (Michaelis *et al.*, 2002) is poorly understood, so similar $^{14}CH_4$ measurements must be made in a similar, but hydrate-free, vent-free environment like the Cariaco Basin.

Methane clathrate decomposition has been implicated in the Latest Paleocene Thermal Maximum (~55 Ma ago) by an extraordinary injection of isotopically light carbon into the carbon cycle (Dickens, 2000, 2001a) and in Quaternary interstadials as indicated by observations of isotopically light foraminifera in Santa Barbara Basin sediments (Kennett *et al.*, 2000). Dickens (2001a) compares the functioning of the $CH_4$ hydrate to a bacterially mediated capacitor.

## 4.03.5 ICE CORES

Measurements of $CH_4$ trapped in polar ice cores from Greenland and Antarctica have been reviewed by Raynaud (1993), Raynaud and Chappellaz (1993), Chappellaz *et al.* (2000), and Cuffey and Brook (2000). These studies provide records of atmospheric composition extending to over $4 \times 10^5$ years ago, covering important climate periods like the glacial–interglacial cycles, the Holocene, and the Younger Dryas. The Greenland and Antarctic ice cores agree over $1.1 \times 10^5$ years (Bender *et al.*, 1994; Brook *et al.*, 1996), indicating that the cores are recording global-scale phenomena. The resolution of these records depends on the snow accumulation rate. Since the snow accumulation rate is higher in Greenland than in central Antarctica, the Antarctic cores cover longer time intervals with lower resolution. Previously deposited snow is compacted and sintered under the weight of newly accumulated snow. At the bottom of the porous firn layer, ice forms, and the air is trapped as bubbles. The bubbles disappear with deeper burial. The gases in the firn layer can exchange with the atmosphere until bubble close-off, so the age of the trapped gas is younger than the adjacent ice. The air trapping process in ice sheets, which ranges from 30 years at high ($>100\ cm\ yr^{-1}$) accumulation rates to 2,000 years at low ($<2\ cm\ yr^{-1}$) accumulation rates, can be modeled (Schwander *et al.*, 1997) to estimate the difference between the ice age and the gas age. The close-off process is a progressive phenomenon, resulting in a distribution of gas ages (see Brook *et al.*, 2000, figure 4). This difference between the ice age and the gas age limits our ability to resolve short-term catastrophic events, such as large clathrate-derived $CH_4$ releases, as the gas-trapping process best records sustained global-scale changes. For example, $CH_4$ records in ice core for pre-anthropogenic times have been interpreted as records of changes in wetland $CH_4$ emissions.

Abrupt changes in temperature result in thermal fractionation of gases in the firm layer, which are recorded as anomalies in the stable isotopes of $N_2$ and Ar (Severinghaus *et al.*, 1998). This approach circumvents the problem of differences between the ice age and the gas age by comparing gases with gases in the same sample. This approach was used to show that the $CH_4$ increase lagged the temperature increase at the end of the Younger Dryas by 0–30 years.

A scenario involving instantaneous release of 4,000 Tg of $CH_4$ was presented by Thorpe *et al.* (1996). Brook *et al.* (2000) analyzed the ice core response to such an event and concluded that the high-resolution ice core record is inconsistent with the large and rapid release of methane. Shifts in $\delta^{13}C$ in foraminifera preserved in marine sediments suggest large releases of $CH_4$ that could be important in initiating climate change (Kennett *et al.*, 2000, 2003). There is no question that decomposing methane hydrates contributed to increasing atmospheric methane, but the lag between warming and methane increases (Severinghaus *et al.*, 1998) is difficult to reconcile with the "Clathrate Gun Hypothesis." Relating events recorded in marine sediments with those in ice cores promises to be an area of continuing debate.

## 4.03.6 FUTURE WORK

Using the Cicerone and Oremland (1988) review as modern benchmark, this chapter has attempted to summarize our current knowledge of the global methane budget, our understanding of physical and microbiological controls on $CH_4$ sources and sinks, and how well models are representing these processes. In spite of the remarkable accomplishments since the mid-1980s, we have not been able to ascribe atmospheric $CH_4$ increases and decreases to specific processes. As of early 2000s, we have no predictive capability.

Atmospheric time-series measurements of both $CH_4$ concentration and isotopes should continue as a high priority. As increased $CH_4$ isotope analysis capacity becomes available, we should attempt to produce (with extensive intercalibration) a global $CH_4$ isotope ($\delta^{13}CH_4$ and $\delta D\text{-}CH_4$) time series parallel to the concentration measurements of NOAA/GMCC network.

Wetland and tundra flux time series should also continue and possibly be expanded at carefully selected high-latitude wetland sites, preferably sites in the Hudson's Bay Lowlands and Siberian Lowlands. Sites in areas of discontinuous permafrost as well as other areas susceptible to permafrost melting should be part of the network. Static chamber flux measurements are legendary as labor-intensive, and they should be replaced where possible with automated chambers and supplemented with continuous eddy flux measurements (Hargreaves *et al.*, 2001). Development of a field-portable, rapid-response $CH_4$ sensor suitable for continuous eddy covariance measurements should be a high priority.

More direct measurements of methane oxidation rates, particularly in wetlands and ocean waters, are needed. The use of stable isotope estimates of methane oxidation, which give an indication of total oxidation, should continue, but direct rate measurements using both $^{14}C\text{-}CH_4$ and $^3H\text{-}CH_4$ should be a priority. Pulse-labeling experiments conducted through a growing season are needed to resolve the effect of plant phenology on methane emission. The methane oxidation threshold suggested by a number of open ocean rate measurements should be studied in open ocean samples from areas near and well removed from shelf vent sources.

The role of decomposing methane clathrates in the global methane budget remains a major unsolved problem. Measurements of natural $^{14}C\text{-}CH_4$ will be of great value in identifying and quantifying the source strength of hydrate-derived and seep-derived $CH_4$. The technology (Accelerator Mass Spectrometry) exists to make these measurements on ambient methane in seawater samples as small as 100–200 L of seawater. Stable isotope measurements ($\delta^{13}CH_4$ and $\delta D\text{-}CH_4$) on intact $CH_4$ hydrate samples are needed to better understand the source of $CH_4$ in hydrates. It should be possible to distinguish the ocean and coastal hydrate provinces (D'Hondt *et al.*, 2002) with these measurements. Additional seafloor measurements of hydrate decomposition under realistic flowfield and saturation conditions are needed. Models predicting the natural $CH_4$ hydrate decomposition rate as well as the $CH_4$ hydrate response to temperature increases are needed so that we can confirm measurements of $CH_4$ hydrate decomposition.

The rapidly developing area of culture-independent identification of methanogens and methanotrophs provides a badly needed basis for understanding the microbial ecology of $CH_4$ production as well as consumption processes. This area promises to reveal a large number of archea capable of anaerobically oxidizing $CH_4$ under a range of $CH_4$ supply rates.

The overall result of the measurements and studies suggested above should be an improved understanding of controlling processes and realistic models that represent complex systems and allow predictions of future conditions.

## ACKNOWLEDGMENTS

A draft of this chapter benefited from comments by Patrick Crill, Jennifer King, John Kessler, Stan Tyler, David Valentine, Steve Whalen, and an anonymous reviewer. Don Blake, Tom Graedel, M. A. K. Khalil, and F. S. Rowland provided perspective on obervations documenting the atmospheric methane increase. This work was supported by the National Science Foundation. I thank all of the above for their help.

## REFERENCES

Alperin M. J. and W. S. Reeburgh (1984) Geochemical observations supporting anaerobic methane oxidation. In *Microbial Growth on C-1 Compounds* (eds. R. Crawford and R. Hanson). American Society for Microbiology, Washington, DC, pp. 282–289.

Alperin M. J., Reeburgh W. S., and Whiticar M. J. (1988) Carbon and hydrogen isotope fractionation resulting from anaerobic methane oxidation. *Global Biogeochem. Cycles* 2, 279–288.

Andersen B. L., Bidoglio G., Leip A., and Rembges D. (1998) A new method to study simultaneous methane oxidation and methane production in soils. *Global Biogeochem. Cycles* 12, 587–594.

Aravena R., Warner B. G., Charman D. J., Belyea L. R., Mathur S. P., and Dinel H. (1993) Carbon isotopic composition of deep carbon gases in an ombrogenous peatland, Northwestern Ontario, Canada. *Radiocarbon* 35, 271–276.

Aselmann I. and Crutzen P. J. (1989) Global distribution of natural freshwater wetlands and rice paddies: their primary productivity, seasonality and possible methane emissions. *J. Atmos. Chem.* 8, 307–358.

Auerbach N. A., Walker D. A., and Bockheim J. G. (1997) *Land Cover Map of the Kuparuk River Basin, Alaska.* Inst. Arct. Alp. Res., Univ. of Colo., Boulder.

Bange H. W., Bartell U. H., Rapsomanikis S., and Andreae M. O. (1994) Methane in the Baltic and North Seas and a reassessment of the marine emissions of methane. *Global Biogeochem. Cycles* 8, 465–480.

Barker J. F. and Fritz P. (1981) Carbon isotope fractionation during microbial methane oxidation. *Nature* 293, 289–291.

Barnes R. O. and Goldberg E. D. (1976) Methane production and consumption in anaerobic marine sediments. *Geology* 4, 297–300.

Bartlett K. B. and Harriss R. C. (1993) Review and assessment of methane emissions from wetlands. *Chemosphere* 26, 261–320.

Bedard C. and Knowles R. (1989) Physiology, biochemistry, and specific inhibitors of $CH_4$, $NH_4^+$, and CO oxidation by methanotrophs and nitrifiers. *Microbiol. Rev.* 53, 68–84.

Bender M. L., Sowers T., Dickson M. L., Orchardo J., Grootes P., Mayewski P., and Meese D. A. (1994) Climate correlations between Greenland and Antarctica during the past 100,000 years. *Nature* 372, 663–666.

Bergamaschi P. and Harris G. (1995) Measurements of stable isotope ratios ($^{13}CH_4/^{12}CH_4$): ($^{12}CH_3D/^{12}CH_4$) in landfill methane using a tunable diode laser. *Global Biogeochem. Cycles* 9, 439–447.

Bergamaschi P., Brennickmeijer C. A. M., Hahn M., Röckmann T., Scharffe D. H., Crutzen P. J., Elansky N. F., Belikov I. B., Trivett N. B. A., and Worthy D. E. J. (1998a) Isotope analysis based source identification for atmospheric $CH_4$ and CO sampled across Russia using the Trans-Siberian railroad. *J. Geophys. Res.* 103, 8227–8235.

Bergamaschi P., Lubina C., Königstedt Fischer H., Veltkamp A. C., and Zwaagstra O. (1998b) Stable isotope signatures ($\delta^{13}C$, $\delta D$) of methane from European landfill sites. *J. Geophys. Res.* 103, 8251–8265.

Blake D. R. (1984) Increasing concentrations of atmospheric methane, 1979–1983. PhD Dissertation, Univ. of California, Irvine, 213pp.

Blake D. R. and Rowland F. S. (1986) World-wide increase in tropospheric methane, 1978–1983. *J. Atmos. Chem.* 4, 43–62.

Blake D. R., Mayer E. W., Tyler S. C., Montague D. C., Makide Y., and Rowland F. S. (1982) Global increase in atmospheric methane concentration between 1978 and 1980. *Geophys. Res. Lett.* 9, 477–480.

Boetius A., Ravenschlag K., Schubert C. J., Rickert D., Widdel F., Gieseke A., Amann R., Jørgensen B. B., Witte U., and Pfankuche O. (2000) A marine microbial consortium apparently mediating anaerobic oxidation of methane. *Nature* 407, 623–626.

Bogner J. and Matthews E. (2003) Global methane emissions from landfills: new methodology and annual estimates 1980–1996. *Global Biogeochem. Cycles* (in press).

Born M., Dörr H., and Levin I. (1990) Methane consumption in aerated soils of the temperate zone. *Tellus* 42B, 2–8.

Borowski W., Paull C. K., and Ussler W., III (1996) Marine pore water sulfate profiles indicate methane flux from underlying gas hydrates. *Geology* 24, 655–658.

Boussingault J. B. (1834) *Annales de Chemie et de Physique*, 2e series, t. LVII, p. 171.

Boussingault J. B. (1864) *Annales de Chemie et de Physique*, 7e series, t. LXVI, p. 413.

Brewer P. G., Orr F. M., Jr., Friederich G., Kvenvolden K. A., Orange D. L., McFarlane J., and Kirkwood W. (1997) Deep-ocean field test of methane hydrate formation from a remotely operated vehicle. *Geology* 25, 407–410.

Brewer P. G., Orr F. M., Jr., Friederich G., Kvenvolden K. A., and Orange D. L. (1998) Gas hydrate formation in the deep sea: *in situ* experiments with controlled release of methane, natural gas, and carbon dioxide. *Energy and Fuels* 12, 183–188.

Brook E. J., Sowers T., and Orchardo J. (1996) Rapid variations in atmospheric methane concentration during the past 110,000 years. *Science* 273, 1087–1091.

Brook E. J., Harder S., Severinghaus J., Steig E. J., and Sucher C. M. (2000) On the origin and timing of rapid changes in atmospheric methane during the last glacial period. *Global Biogeochem. Cycles* 14, 559–572.

Burke R. A., Jr., Reid D. F., Books J. M., and Lavoie D. A. (1983) Upper water column methane geochemistry in the eastern tropical North Pacific. *Limnol. Oceanogr.* 28, 19–32.

Cai Z. and Yan A. (1999) Kinetic model for methane oxidation by paddy soil as affected by temperature, moisture and N addition. *Soil Biol. Biochem.* 31, 715–729.

Cantrell A. C., Shetter R. E., McDaniel A. H., Calvert J. G., Davidson J. A., Lowe D. C., Tyler S. C., Cicerone R. J., and Greenberg J. P. (1990) Carbon kinetic isotope effect in the oxidation of methane by the hydroxyl radical. *J. Geophys. Res.* 95, 22455–22462.

Cao M., Marshall S., and Gregson K. (1996) Global carbon exchange and methane emissions from natural wetlands: application of a process-based model. *J. Geophys. Res.* 101, 14399–14414.

Chanton J. P., Bauer J. E., Glaser P. A., Siegel D. I., Kelley C. A., Tyler S. C., Romanowicz E. H., and Lazrus A. (1995) Radiocarbon evidence for the substrates supporting methane formation within Northern Minnesota peatlands. *Geochim. Cosmochim. Acta* 59, 3663–3668.

Chappellaz J., Raynaud D., Blunier T., and Stauffer B. (2000) The ice core record of atmospheric methane. In *Atmospheric Methane* (ed. M. A. K. Khalil). Springer, Berlin, pp. 9–24.

Christensen T. R., Jonasson S., Callaghan T. V., and Havström M. (1995) Spatial variation in a high-latitude methane flux transect across Siberian and European tundra environments. *J. Geophys. Res.* **100**, 21035–21045.

Cicerone R. J. and Oremland R. S. (1988) Biogeochemical aspects of atmospheric methane. *Global Biogeochem. Cycles* **2**, 299–327.

Coleman D. D., Risatti J. B., and Schoell M. (1981) Fractionation of carbon and hydrogen isotopes by methane-oxidizing bacteria. *Geochim. Cosmochim. Acta* **45**, 1033–1037.

Conrad R. (1984) Capacity of aerobic microorganisms to utilize and grow on atmospheric trace gases ($H_2$, CO, $CH_4$). In *Current Perspectives in Microbial Ecology* (eds. M. J. Klug and C. A. Reddy). American Society for Microbiology, Washington, DC, pp. 461–467.

Conrad R. (1996) Soil microorganisms as controllers of atmospheric trace gases ($H_2$, CO, $CH_4$, OCS, $N_2O$, and NO). *Microbiol. Rev.* **60**, 609–640.

Craig H. (1957) Isotopic standards for carbon and oxygen and correction factors for mass-spectrometric analysis of carbon dioxide. *Geochim. Cosmochim. Acta* **12**, 133–149.

Craig H. and Chou C. C. (1982) Methane: the record on polar ice cores. *Geophys. Res. Lett.* **9**, 477–481.

Crowley J. N., Saueressig G., Bergamaschi P., Fischer H., and Harriss G. W. (1999) Carbon kinetic isotope effect in the reaction $CH_4 + Cl$: a relative rate study using FTIR spectroscopy. *Chem. Phys. Lett.* **303**, 268–274.

Crutzen P. J., Elansky N. F., Hahn M., Golitsyn G. S., Brenninkmeijer C. A. M., Scharffe D. H., Belikov I. B., Maiss M., Bergamaschi P., Rockmann T., Grisenko A. M., and Sevostyanov V. M. (1998) Trace gas measurements between Moscow and Vladivostok using the Tran-Siberian Railroad. *J. Atmos. Chem.* **29**, 179–194.

Cuffey K. M. and Brook E. J. (2000) Ice sheets and the ice-core record of climate change. In *Earth System Science from Biogeochemical Cycles to Global Change* (eds. M. C. Jacobson, R. J. Charlson, H. Rodhe, and G. Orians). Academic Press, San Diego, chap. 18, pp. 459–497.

Cynar F. J. and Yayanos A. A. (1993) The oceanic distribution of methane and its flux to the atmosphere over Southern California waters. In *Biogeochemistry of Global Change: Radiatively Active Trace Gases* (ed. R. S. Oremland). Chapman and Hall, New York, pp. 551–573.

Dannenberg S. and Conrad R. (1999) Effect of rice plants on methane production and rhizospheric metabolism in paddy soil. *Biogeochemistry* **45**, 53–71.

Davidson J. A., Cantrell C. A., Tyler S. C., Shetter R. E., Cicerone R. J., and Calvert J. G. (1987) Carbon kinetic isotope effect in the reaction of $CH_4$ with HO. *J. Geophys. Res.* **92**, 2195–2199.

Del Grosso S. J., Parton W. J., Mosier A. R., Potter C. S., Borken W., Brummer R., Butterbach-Bahl K., Crill P. M., Dobbie K., and Smith K. A. (2000) General $CH_4$ oxidation model and comparisons of $CH_4$ oxidation in natural and managed systems. *Global Biogeochem. Cycles* **14**, 999–1019.

D'Hondt S., Rutherford S., and Spivak A. J. (2002) Metabolic activity of subsurface life in deep-sea sediments. *Science* **295**, 2067–2070.

Dickens G. R. (2000) Methane oxidation during the late paleocene thermal maximum. *Bull. Soc. Geol. Fr.* **171**, 37–49.

Dickens G. R. (2001a) Modeling the global carbon cycle with a gas hydrate capacitor: significance for the latest paleocene thermal maximum. In *Geophysical Monograph 124, Natural Gas Hydrates: Occurrence, Distribution, and Detection*, Geophysical Monograph 124 (eds. C. K. Paull and W. D. Dillon). American Geophysical Union, Washington, DC, pp. 19–38.

Dickens G. R. (2001b) Sulfate profiles and barium fronts in sediment on the Blake Ridge: present and past methane fluxes through a large gas hydrate reservoir. *Geochim. Cosmochim Acta* **65**, 529–543.

Dickens G. R. (2003) A methane trigger for climate warming? *Science* **299**, 1017.

Dickens G. R. and Quinby-Hunt M. S. (1994) Methane hydrate stability in seawater. *Geophys. Res. Lett.* **21**, 2115–2118.

Dickens G. R. and Quinby-Hunt M. S. (1997) Methane hydrate stability in pore water: a simple theoretical approach for geophysical applications. *J. Geophys. Res.* **102**, 773–783.

Dise N. (1992) Winter fluxes of methane from Minnesota peatlands. *Biogeochemistry* **16**, 71–83.

Dise N. (1993) Methane emission from Minnesota peatlands: spatial and seasonal variability. *Global Biogeochem. Cycles* **7**, 123–142.

Dlugokencky E. J., Steele L. P., Lang P. M., and Masarie K. A. (1994a) The growth rate and distribution of atmospheric methane. *J. Geophys. Res.* **99**, 17021–17043.

Dlugokencky E. J., Masarie K. A., Lang P. M., Tans P. P., Steele L. P., and Nisbet E. G. (1994b) A dramatic decrease in the growth rate of atmospheric methane in the northern hemisphere during 1992. *Geophys. Res. Lett.* **21**, 45–48.

Dlugokencky E. J., Masarie K. A., Lang P. M., Tans P. P., Steele L. P., and Nisbet E. G. (1994c) Reply to comments by K. B. Hogan and R. C. Harriss on Dlugokencky et al. (1994b). *Geophys. Res. Lett.* **21**, 2447–2448.

Dlugokencky E. J., Dutton E. G., Novelli P. C., Tans P. P., Masarie K. A., Lantz K. O., and Madronich S. (1996) Changes in $CH_4$ and CO growth rates after the eruption of Mt. Pinatubo and their link with changes in tropospheric UV flux. *Geophys. Res. Lett.* **23**, 2761–2764.

Dlugokencky E. J., Masarie K. A., Lang P. M., and Tans P. P. (1998) Continuing decline in the growth rate of atmospheric methane burden. *Nature* **393**, 447–451.

Dlugokencky E. J., Walter B. P., Masarie K. A., Lang P. A., and Kasischke E. S. (2001) Measurements of an anomalous methane increase during 1998. *Geophys. Res. Lett.* **28**, 499–502.

Dörr H. and Munnich K. O. (1990) $^{222}$Rn flux and soil air concentration profiles in West Germany: soil $^{222}$Rn as a tracer for gas transport in the unsaturated zone. *Tellus* **42B**, 20–28.

Dörr H., Katruff L., and Levin I. (1993) Soil texture parameterization of the methane uptake in aerated soils. *Chemosphere* **26**, 697–713.

Ehhalt D. H. (1974) The atmospheric cycle of methane. *Tellus* **26**, 58–70.

Ehhalt D. H. (1978) The $CH_4$ concentration over the ocean and its possible variations with latitude. *Tellus* **30**, 169–176.

Ehhalt D. H. and Schmidt U. (1978) Sources and sinks of atmospheric methane. *Pure Appl. Geophys.* **116**, 452–464.

Epp M. A. and Chanton J. P. (1993) Rhizospheric methane oxidation determined via the methyl fluoride inhibition technique. *J. Geophys. Res.* **98**, 18413–18422.

Frolking S. and Crill P. M. (1994) Climate controls on temporal variability of methane flux from a poor fen in southeastern New-Hampshire—measurement and modeling. *Global Biogeochem Cycles* **8**, 385–397.

Fung I., John J., Lerner J., Matthews E., Prather M., Steele L. P., and Fraser P. J. (1991) Three dimensional model synthesis of the global methane cycle. *J. Geophys. Res.* **96**, 13033–13065.

Gautier A. (1901) *Annales de Chemie et de Physique* **22**, June 1.

Gornitz V. and Fung I. (1994) Potential distribution of methane hydrates in the world's oceans. *Global Biogeochem. Cycles* **8**, 335–347.

Goulden M. L. and Crill P. M. (1997) Automated measurements of $CO_2$ exchange at the moss surface of a black spruce forest. *Tree Physiol.* **17**, 537–542.

Graedel T. E. and McRae J. E. (1980) On the possible increase of the atmospheric methane and carbon monoxide concentrations during the last decade. *Geophys. Res. Lett.* **7**, 977–979.

Granberg G., Ottoson-Löfvenius M., and Grip H. (2001) Effect of climatic variability from 1980 to 1997 on simulated methane emission from a boreal mixed mire in northern Sweden. *Global Biogeochem. Cycles* **15**, 977–991.

Grant N. J. and Whiticar M. J. (2002) Stable carbon isotopic evidence for methane oxidation in plumes above Hydrate Ridge, Cascadia Oregon Margin. *Global Biogeochem. Cycles* **16**, 71-1–71-13, doi: 10.1029/2001GB001851.

Grant R. F. (1998) Simulation of methanogenesis in the mathematical model *ecosys*. *Soil Biol. Biochem.* **30**, 883–896.

Grant R. F. (1999) Simulation of methanotrophy in the mathematical model *ecosys*. *Soil Biol. Biochem.* **31**, 287–297.

Grant R. F. and Roulet N. T. (2002) Methane efflux from boreal wetlands: theory and testing of the ecosystem model *ecosys* with chamber and tower flux measurements. *Global Biogeochem. Cycles* **16**, 2-1–2-16, doi: 10.1029/2001GB 001702.

Gulledge J., Doyle A. P., and Schimel J. P. (1997) Different $NH_4^+$ inhibition patterns of soil $CH_4$ consumption: a result of distinct $CH_4$-oxidizer populations across sites? *Soil Biol. Biochem.* **29**, 13–21.

Gupta M. L., Tyler S., and Cicerone R. (1996) Modeling atmospheric $\delta^{13}CH_4$ and the causes of recent changes in atmospheric $CH_4$ amounts. *J. Geophys. Res.* **101**, 22923–22932.

Gupta M. L., McGrath M. P., Cicerone R. J., Rowland F. S., and Wolfsburg M. (1997) $^{12}C/^{13}C$ kinetic isotope effects in reactions of $CH_4$ with OH and Cl. *Geophys. Res. Lett.* **24**, 2761–2764.

Gupta M. L., Cicerone R. J., Blake D. R., Rowland F. S., and Isaksen I. S. A. (1998) Global atmospheric distributions and source strengths of light hydrocarbons and tetrachloroethylene. *J. Geophys. Res.* **103**, 28219–28235.

Hanson R. S. and Hanson T. E. (1996) Methanotrophic bacteria. *Microbiol. Rev.* **60**, 439–471.

Happell J. D., Chanton J. P., and Showers W. S. (1994) The influence of methane oxidation on the stable isotopic composition of methane emitted from Florida swamp forests. *Geochim. Cosmochim. Acta* **58**, 4377–4388.

Hargreaves K. J., Fowler D., Pircairn C. E. R., and Aurela M. (2001) Annual methane emission from Finnish mires estimated from eddy covariance campaign measurements. *Theor. Appl. Climatol.* **70**, 203–213.

Harriss R. C., Sebacher D. I., and Day F. P., Jr. (1982) Methane flux in the Great Dismal Swamp. *Nature* **297**, 673–674.

Harriss R. C., Bartlett K. S., Frolking S., and Crill P. (1993) Methane emissions from northern high-latitude wetlands. In *Biogeochemistry of Climate Change* (ed. R. S. Oremland). Chapman and Hall, New York, pp. 449–486.

Hein R., Crutzen P. J., and Heimann M. (1997) An inverse modeling approach to investigate the global atmospheric methane cycle. *Global Biogeochem. Cycles* **11**, 43–76.

Hinrichs K.-U. and Boetius A. (2002) The anaerobic oxidation of methane: new insights in microbial ecology and biogeochemistry. In *Ocean Margin Systems* (eds. G. Wefer, D. Billett, B. B. Jorgensen, M. Schluter, and T. van Weering). Springer, Heidelberg, pp. 457–477.

Hinrichs K.-U., Hayes J. M., Sylva S. P., Brewer P. G., and DeLong E. F. (1999) Methane-consuming archaebacteria in marine sediments. *Nature* **398**, 802–805.

Hoehler T. M., Alperin M. J., Alpert D. B., and Martens C. S. (1994) Field and laboratory studies of methane oxidation in an anoxic marine sediment: evidence for a methanogen-sulfate reducer consortium. *Global Biogeochem. Cycles* **8**, 451–463.

Hogan K. B. and Harriss R. C. (1994) Comment on E. J. Dlugokencky *et al.* (1994b). *Geophys. Res. Lett.* **21**, 2445–2446.

Holmes M. E., Sanson F. J., Rust T. M., and Popp B. N. (2000) Methane production, consumption and air–ea exchange in the open ocean: an evaluation based on carbon isotopic ratios. *Global Biogeochem. Cycles* **14**, 1–10.

Hovland M. and Judd A. G. (1988) *Seabed Pockmarks and Seepages: Impact on Geology, Biology and the Marine Environment*. Graham and Trotman, London.

IPCC (2001) *Climate Change 2001: The Scientific Basis*. Contribution of Working Group 1 to the Third Assessment Report of the Intergovernmental Panel on Climate Change, Greenhouse Gases, pp. 241–287.

Johnson D. E., Johnson K. A., Ward G. M., and Branine M. E. (2000) Ruminants and other animals. In *Atmospheric Methane: Its Role in the Global Environment* (ed. M. A. K. Kahlil). Springer, Berlin, pp. 112–133.

Johnson K. A., Huyler M. T., Westberg H. H., Lamb B. K., and Zimmerman P. (1994) Measurement of methane emissions from ruminant livestock using a $SF_6$ tracer technique. *Environ. Sci. Technol.* **28**, 359–362.

Jones H. A. and Nedwell D. B. (1993) Methane emission and methane oxidation in a land-fill cover soil. *FEMS Microbiol. Ecol.* **102**, 185–195.

Karl D. M. and Tilbrook B. D. (1994) Production and transport of methane in oceanic particulate matter. *Nature* **368**, 732–743.

Keller M., Mitre M. E., and Stallard R. F. (1990) Consumption of methane in soils of central Panama. *Global Biogeochem. Cycles* **4**, 21–27.

Kennett J. P., Cannariato K. G., Hendy L. L., and Behl R. J. (2000) Carbon isotope evidence for methane hydrate instability during Quaternary interstadials. *Science* **288**, 129–133.

Kennett J. P., Cannariato K. G., Hendy I. L., and Behl R. J. (2003) *Methane Hydrates in Quaternary Climate Change: The Clathrate Gun Hypothesis*. Special Publication 54. American Geophysical Union, Washington, DC, 216 pp.

Kessler J., Reeburgh W., Southon J., and Tyler S. (2003) Natural radiocarbon in Black Sea water column methane: a tracer of hydrate-derived methane? EGS/AGU/EUG abstracts, 5, 12216, Apr. 6–11, Nice, France.

Khalil M. A. K. (ed.) (1993) Atmospheric Methane: Sources, Sinks and Role in Global Change, In *Proceedings of the NATO Advanced Science Workshop on the Atmospheric Methane Cycle: Sources, Sinks, Distributions, and Role in Global Change*, Held at Mt. Hood near Portland, OR, Oct. 7–11, 1991. Springer, Berlin.

Khalil M. A. K. (ed.) (2000) *Atmospheric Methane: Its Role in the Global Environment*. Springer, Berlin.

Khalil M. A. K. and Rasmussen R. A. (1985) Causes of increasing methane: depletion of hydroxyl radicals and the rise of emissions. *Atmos. Environ.* **19**, 397–407.

Kightley D., Nedwell D. B., and Cooper M. (1995) Capacity for methane oxidation in landfill cover soils measured in laboratory-scale soil microcosms. *Appl. Environ. Microbiol.* **61**, 592–601.

King G. M. (1992) Ecological aspects of methane oxidation, a key determinant of global methane dynamics. *Adv. Microb. Ecol.* **12**, 431–468.

King J. Y. and Reeburgh W. S. (2002) A pulse-labeling experiment to determine the contribution of recent plant photosynthates to net methane emission in arctic wet sedge tundra. *Soil Biol. Biochem.* **34**, 173–180.

King J. Y., Reeburgh W. S., and Regli S. K. (1998) Methane emission and tranport by sedges in Alaska: results of a vegetation removal experiment. *J Geophys. Res.* **102**, 29083–29092.

King J. Y., Reeburgh W. S., Thieler K., Kling G. W., Loya W. M., and Johnson N. K. J. (2002) Pulse-labeling studies of carbon cycling in Arctic tundra ecosystems: the contribution of photosynthesis to methane emission. *Global Biogeochem. Cycles* **16**, 10-1–10-8, doi: 10.1029/2001GB001456.

King S. L., Quay P. D., and Lansdown J. M. (1989) The $^{13}C/^{12}C$ kinetic isotope effect for soil oxidation of methane at ambient atmospheric concentrations. *J. Geophys. Res.* **94**(D15), 18273–18277.

Klüber H. D. and Conrad R. (1998) Effects of nitrate, nitrite, NO and $N_2O$ on methanogenesis and other redox processes in anoxic rice field soil. *FEMS Microbiol. Ecol.* **25**, 301–318.

Koschorreck M. and Conrad R. (1993) Oxidation of atmospheric methane in soil: measurements in the field, in soil cores, and in soil samples. *Global Biogeochem. Cycles* **7**, 109–121.

Kvenvolden K. A. (1993) Gas hydrates: geological perspective and global change. *Rev. Geophys.* **31**, 173–187.

Kvenvolden K. A., Lorenson T. D., and Reeburgh W. S. (2001) Attention turns to naturally occurring methane seepage. *EOS* **82**(40), 457.

Kvenvolden K. K. (1988) Methane hydrates and global change. *Global Biogeochem. Cycles* **3**, 221–229.

Kvenvolden K. K. and Lorenson T. D. (2001) The global occurrence of natural gas hydrate. In *Natural Gas Hydrates:Occurrence, Distribution and Detection*, Geophysical Monograph 124 (eds. C. K. Paull and W. D. Dillon). American Geophysical Union, Washington, DC.

Lacis A., Hansen J., Lee P., Mitchell T., and Lebedeff S. (1981) Greenhouse effect of trace gases 1970–1980. *Geophys. Res. Lett.* **8**, 1035–1038.

Lambert G. and Schmidt S. (1993) Re-evaluation of the oceanic flux of methane: uncertainties and long term variations. *Chemosphere* **26**, 579–589.

Lassey K. R., Lowe D. C., and Manning M. R. (2000) The trend in atmospheric methane $\delta^{13}C$ and implications for isotopic constraints on the global methane budget. *Global Biogeochem. Cycles* **14**, 41–49.

Liptay K., Chanton J., Czepiel P., and Mosher B. (1998) Use of stable isotopes to determine methane oxidation in landfill cover soils. *J. Geophys. Res.* **103**, 8243–8250.

Lowe D. C., Brenninkmeijer C. A. M., Tyler S. C., and Dlugokencky E. J. (1991) Determination of the isotopic composition of atmospheric methane and its application in the Antarctic. *J. Geophys. Res.* **96**, 15455–15467.

Lowe D. C., Brenninkmeijer C. A. M., Brailsford G. W., Lassey K. R., Gomez A. J., and Nisbet E. G. (1994) Concentration and $^{13}C$ records of atmospheric methane in New Zealand and Antarctica: evidence for changes in methane sources. *J. Geophys. Res.* **99**, 16913–16925.

Lowe D. C., Manning M. R., Brailsford G. W., and Bromley A. M. (1997) The 1991–1992 methane anomaly: southern hemispheric $^{13}C$ decrease and growth rate fluctuations. *Geophys. Res. Lett.* **24**, 857–860.

Loya W. M., Johnson L. C., Kling G. W., King J. Y., Reeburgh W. S., and Nadelhoffer K. J. (2002) Pulse-labeling studies of carbon cycling in arctic tundra ecosystems: contribution of photosynthates to soil organic matter. *Global Biogeochem. Cycles* **16**(4), 48-1–48-8, doi: 10.1029/2001GB001464.

Marik T. and Levin I. (1996) A new tracer experiment to estimate methane emissions from a dairy cow shed using sulfur hexafluoride ($SF_6$). *Global Biogeochem. Cycles* **10**, 413–418.

Martens C. S. and Berner R. A. (1977) Interstitial water chemistry of Long Island Sound sediments: I. Dissolved gases. *Limnol. Oceanogr.* **22**, 10–25.

Martens C. S., Albert D. B., and Alperin M. J. (1999) Stable isotope tracing of anaerobic methane oxidation in the gassy sediments of Ekernörde Bay, German Baltic Sea. *Am. J. Sci.* **299**, 589–610.

Matson P. A., Vitousek P. M., and Schimel D. S. (1989) Regional extrapolation of trace gas flux based on soils and ecosystems. In *Exchange of Trace Gases Between Terrestrial Ecosystems and the Atmosphere* (eds. M. O. Andreae and D. S. Schimel). Wiley, New York, pp. 97–108.

Matthews E. (2000) Wetlands. In *Atmospheric Methane* (ed. M. A. K. Khalil). Springer, Berlin, pp. 202–233.

Matthews E. and Fung I. (1987) Methane emission from natural wetlands: global distribution, area, and environmental characteristics of sources. *Global Biogeochem. Cycles* **1**, 61–86.

Megonigal J. P., Whalen S. C., Tissue D. T., Bovard B. D., Albert D. B., and Allen A. S. (1999) A plant-soil-atmosphere microcosm for tracing radiocarbon from photosynthesis through methanogenesis. *Soil Sci. Soc. Am. J.* **63**, 665–671.

Melloh R. A. and Crill P. M. (1996) Winter methane dynamics in a temperate peatland. *Global Biogeochem. Cycles* **10**, 247–254.

Merritt D. A., Hayes J. M., and Des Marais D. M. (1995) Carbon isotopic analysis of atmospheric methane by isotope ratio monitoring gas chromatography-mass spectrometry. *J. Geophys. Res.* **100**, 1317–1326.

Michaelis W., Seifert R., Nauhaus K., Treude T., Thiel V., Blumenberg M., Knittle K., Gieseke A., Peterknecht K., Pape T., Boetius A., Amann R., Jørgensen B. B., Widdel F., Peckman J., Pimenov N., and Gulin M. (2002) Microbial reefs in the Black Sea fueled by anaerobic oxidation of methane. *Science* **297**, 1013–1015.

Migeotte M. V. (1948) Spectroscopic evidence of methane in the Earth's atmosphere. *Phys. Rev.* **73**, 519–520.

Minoda T. and Kimura M. (1994) Contribution of photosynthesized carbon to the methane emitted from paddy fields. *Geophys. Res. Lett.* **21**, 2007–2010.

Minoda T., Kimura M., and Wada E. (1996) Photosynthates as dominant sources of $CH_4$ and $CO_2$ in soil water and $CH_4$ emitted to the atmosphere from paddy fields. *J. Geophys. Res.* **101**, 21091–21097.

Moosavi S. C. and Crill P. M. (1998) $CH_4$ oxidation by tundra wetlands as measured by a selective inhibitor technique. *J. Geophys. Res.* **103**, 29093–29106.

Moosavi S. C., Crill P. M., Pullman E. R., Funk D. W., and Peterson K. M. (1996) Controls on $CH_4$ flux from an Alaskan boreal wetland. *Global Biogeochem. Cycles* **10**, 287–296.

Mosier A., Schimel D., Valentine D., Bronson K., and Parton W. (1991) Methane and nitrogen oxide fluxes in native, fertilized, and cultivated grasslands. *Nature* **350**, 330–332.

Nauhaus K., Boetius A., Kruger M., and Widdel F. (2002) *In vitro* demonstration of anaerobic oxidation of methane coupled to sulfate reduction in sediment from a marine gas hydrate area. *Environ. Microbiol.* **4**, 296–305.

Oberlander E. A., Brenninkmeijer C. A. M., Crutzen P. J., Elansky N. F., Golitsyn G. S., Granberg I. G., Scharffe D. H., Hoffmann R., Belikov I. B., Paretzke H. G., and van Velthoven P. F. J. (2002) Trace gas measurements along the Trans-Siberian railroad: the TROIKA 5 expedition. *J. Geophys. Res. Atmos.* **107**(D14), ACH 13-1–13-15, doi: 10.1029/2001JD000953.

Oremland R. S. (ed.) (1993) *Biogeochemistry of Global Change: Radiatively Active Trace Gases*. Selected papers from the 10th International Symposium on Environmental Biogeochemistry, San Francisco, August 19–24, 1991, Chapman and Hall, New York.

Oremland R. S. and Culbertson C. W. (1992) Importance of methane oxidizing bacteria in the methane budget as revealed by the use of a specific inhibitor. *Nature* **356**, 421–423.

Orphan V. J., House C. H., Hinrichs K.-U., McKeegan K. D., and DeLong E. F. (2001) Methane-consuming Archaea revealed by directly coupled isotopic and phylogenetic analysis. *Science* **293**, 484–487.

Paull C. K. and Dillon W. P. (eds.) (2001) In *Natural Gas Hydrates: Occurrence, Distribution, and Detection*. Geophysical Monograph 124. American Geophysical Union, Washington, DC.

Popp B. N., Sansone F. J., Rust T. M., and Merritt D. A. (1995) Determination of concentration and carbon isotopic composition of dissolved methane in sediments and nearshore waters. *Anal. Chem.* **67**, 405–411.

Quay P. D., King S. L., Stutsman J., Wilbur D. O., Steele L. P., Fung I., Gammon R. H., Brown T. A., Frawell G. W., Grootes P. M., and Schmidt F. H. (1991) Carbon isotopic composition of atmospheric $CH_4$: fossil and biomass burning source strengths. *Global Biogeochem. Cycles* **5**, 25–47.

Quay P. D., Stutsman J., Wilbur D., Snover A., Dlugokencky E., and Brown T. (1999) The isotopic composition of atmospheric methane. *Global Biogeochem. Cycles* **13**, 445–461.

Ramanathan V., Cicerone R. J., Singh H. B., and Kiehl J. T. (1985) Trace gas trends and their potential role in climate change. *J. Geophys. Res.* **90**, 5547–5566.

Rasmussen R. A. and Khalil M. A. K. (1981) Atmospheric methane ($CH_4$): trends and seasonal cycles. *J. Geophys. Res.* **86**, 9286–9832.

Rasmussen R. A. and Khalil M. A. K. (1984) Atmospheric methane in the recent and ancient atmospheres: concentrations, trends, and the interhemispheric gradient. *J. Geophys. Res.* **89**, 11599–11605.

Ratering S. and Conrad R. (1998) Effects of short-term drainage and aeration on the production of methane in submerged rice soil. *Global Change Biol.* **4**, 397–407.

Raynaud D. (1993) Ice core records as a key to understanding the history of atmospheric trace gases. In *Biogeochemistry and Global Change. Radiatively Active Trace Gases* (ed. R. S. Oremland). Chapman and Hall, New York, pp. 29–45.

Raynaud D. and Chappellaz J. (1993) The record of atmospheric methane. In *Atmospheric Methane: Sources, Sinks, and Global Change* (ed. M. A. K. Khalil). Springer, Berlin, pp. 38–61.

Reay D. S., Radajewski S., Murrell J. C., McNamara N., and Nedwell D. B. (2001) Effects of land-use on the activity and diversity of methane oxidizing bacteria in forest soils. *Soil Biol. Biochem.* **33**, 1613–1623.

Reeburgh W. S. (1976) Methane consumption in Cariaco Trench waters and sediments. *Earth Planet. Sci. Lett.* **28**, 337–344.

Reeburgh W. S. (1980) Anaerobic methane oxidation: rate depth distributions in Skan Bay sediments. *Earth Planet. Sci. Lett.* **47**, 345–352.

Reeburgh W. S. (1996) "Soft Spots" in the global methane budget. In *8th International Symposium on Microbial Growth on C-1 Compounds* (eds. M. E. Lidstrom and F. R. Tabita). Kluwer, Dordrecht, pp. 334–342.

Reeburgh W. S., Ward B. B., Whalen S. C., Sandbeck K. A., Kilpatrick K. A., and Kerkhof L. J. (1991) Black Sea methane geochemistry. *Deep-Sea Res.* **38**, S1189–S1210.

Reeburgh W. S., Whalen S. C., and Alperin M. J. (1993) The role of methylotrophy in the global methane budget. In *Microbial Growth on C-1 Compounds* (eds. J. C. Murrell and D. P. Kelly). Intercept Press, Andover, UK, pp. 1–14.

Reeburgh W. S., Hirsch A. I., Sansone F. J., Popp B. N., and Rust T. M. (1997) Carbon kinetic isotope effect accompanying microbial oxidation of methane in boreal forest soils. *Geochim. Cosmochim. Acta* **61**, 4761–4767.

Reeburgh W. S., King J. Y., Regli S. K., Kling G. W., Auerbach N. A., and Walker D. A. (1998) A $CH_4$ emission estimate for the Kuparuk River Basin, Alaska. *J. Geophys. Res.* **103**, 29005–29013.

Rehder G., Keir R. S., Suess E., and Rhein M. (1999) Methane in the northern Atlantic controlled by microbial oxidation and atmospheric history. *Geophys. Res. Lett.* **26**, 587–590.

Rehder G., Kirby S. H., Durham W. B., Stern L. A., Peltzer E. T., Pinkston J., and Brewer P. G. (2003) Dissolution rates of pure methane hydrate and carbon dioxide hydrate in undersaturated seawater at 1000 m depth. *Geochim. Cosmochim. Acta* (in press).

Rice A. L., Gotoh A. A., Ajie H. O., and Tyler S. C. (2001) High-precision continuous flow measurement of $\delta^{13}C$ and $\delta D$ of atmospheric methane. *Anal. Chem* **73**, 4104–4110.

Ridgwell A. J., Marshall S. J., and Gregson K. (1999) Consumption of atmospheric methane by soils: a process-based model. *Global Biogeochem. Cycles* **13**, 59–70.

Roberto-Neto O., Coitino E. L., and Truhlar D. G. (1998) Dual-level direct dynamics calculations of deuterium and carbon-13 kinetic isotope effects for the reaction $Cl + CH_4$. *J. Phys. Chem. A* **102**, 4568–4578.

Roulet N. T., Jano A., Kelly C. A., Moore T. R., Protz R., Ritter J. A., and Rouse W. R. (1994) The role of the Hudson Bay Lowland as a source of atmospheric methane. *J Geophys. Res.* **99**, 1423–1428.

Sansone F. J., Popp B. N., Gasc A., Graham A. W., and Rust T. M. (2001) Highly elevated methane in the eastern tropical North Pacific and associated isotopically enriched fluxes to the atmosphere. *Geophys. Res. Lett.* **28**, 4567–4570.

Sass R. L., Fisher F. M., Wang Y. B., Turner F. T., and Jund M. F. (1992) Methane emission from rice fields: the effect of flood water management. *Global Biogeochem. Cycles* **6**, 249–262.

Saueressig G., Bergamaschi P., Crowley J. N., Fisher H., and Harris G. W. (1995) Carbon kinetic isotope effect in the reaction of $CH_4$ with Cl. *Geophys. Res. Lett.* **22**, 1225–1228.

Saueressig G., Crowley J. N., Bergamaschi P., Bruhl C., Brenninkmeijer C. A. M., and Fischer H. (2001) Carbon 13 and D kinetic isotope effects in the reaction of $CH_4$ with $O(^1D)$ and OH: new laboratory measurements and their implications for the isotopic composition of stratospheric methane. *J. Geophys. Res.* **106**(D19), 23127–23138.

Schimel J. P. (1995) Plant transport and methane production as controls on methane flux from arctic wet meadow tundra. *Biogeochemistry* **110**, 414–422.

Schwander J., Sowers T., Barnola J.-M., Blunier T., Fuchs A., and Malaizé B. (1997) Age scale of air in the summit ice: implication for glacial-interglacial temperature change. *J. Geophys. Res.* **102**, 19483–19493.

Scranton M. I. and Brewer P. G. (1978) Consumption of dissolved methane in the deep ocean. *Limnol. Oceanogr.* **23**, 1207–1213.

Segers R. and Kengen S. W. M. (1998) Methane production as a function of anaerobic carbon mineralization: a process model. *Soil Biol. Biochem.* **30**, 1107–1117.

Segers R. and Leffelaar P. A. (2001a) Modeling methane fluxes in wetlands with gas tranporting plants: 1. Single-root scale. *J. Geophys. Res.* **106**, 3511–3528.

Segers R. and Leffelaar P. A. (2001b) Modeling methane fluxes in wetlands with gas transporting plants: 3. Plot scale. *J. Geophys. Res.* **106**, 3541–3558.

Segers R., Rapplodt C., and Leffelaar P. A. (2001) Modeling methane fluxes in wetlands with gas-transporting plants: 2. Soil layer scale. *J. Geophys. Res.* **106**, 3529–3540.

Severinghaus J. P., Sowers T., Brook E. J., Alley R. B., and Bender M. L. (1998) Timing of abrupt climate change at the end of the Younger Dryas interval from thermally fractionated gases in polar ice. *Science* **391**, 141–146.

Shannon R. D. and White J. R. (1994) A three-year study of controls on methane emissions from two Michigan peatlands. *Biogeochemistry* **27**, 35–60.

Silverman M. P. and Oyama V. I. (1968) Automatic apparatus for sampling and preparing gases for mass spectral studies of carbon isotope fractionation during methane metabolism. *Anal. Chem.* **40**, 1833–1877.

Simpson I. J., Chen T.-Y., Blake D. R., and Rowland F. S. (2002) Implications of the recent fluctuations in the growth rate of tropospheric methane. *Geophys. Res. Lett.* **29**, 117-1–117-4,10.1029/2001GL014521.

Sloane E. D. (1998) *Clathrate Hydrates of Natural Gas*, 2nd edn. Dekker, New York.

Snover A. K. and Quay P. D. (2000) Hydrogen and carbon kinetic isotope effects during soil uptake of atmospheric methane. *Global Biogeochem. Cycles.* **14**, 25–39.

Söhngen N. L. (1906) Über bakterien, welche methan als kohlenstoffnahrung und energiequelle gebrauchen. *Zentralbl. Bakteriol. Z. Abt. Bd.* **15**, 513–517.

Steele L. P., Fraser P. J., Rasmussen R. A., Khalil M. A. K., Conway T. J., Crawford A. J., Gammon R. H., Masarie K. A., and Thoning K. W. (1987) The global distribtuion of methane in the troposphere. *J. Atmos. Chem.* **5**, 125–171.

Steele L. P., Dlugokencky E. J., Lang P. M., Tans P. P., Martin R., and Masarie K. A. (1992) Slowing down of the global accumulation of methane during the 1980s. *Nature* **358**, 313–316.

Steudler P. A., Bowden R. D., Melillo J. M., and Aber J. D. (1989) Influence of nitrogen fertilization on methane uptake in temperate forest soils. *Nature* **341**, 314–316.

Stevens C. M. and Rust F. E. (1982) The carbon isotopic composition of atmospheric methane. *J. Geophys. Res.* **87**, 4879–4882.

Sugawara S., Nakazawa T., Inoue G., Machida T., Mukai H., Vinnichenko N. K., and Khattatov V. U. (1996) Aircraft measurements of the stable carbon isotopic ratio of atmospheric methane over Siberia. *Global Biogeochem. Cycles.* **10**, 223–231.

Tanaka N., Xiao Y., and Lasaga A. C. (1996) *Ab initio* study on carbon kinetic isotope effect (KIE) in the reaction of $CH_4 + Cl$. *J. Atmos. Chem.* **23**, 37–49.

Tanaka N., Xiao Y., Hatakeyama S., and Ueda S. (1997) Determination of carbon KIE of $CH_4 + Cl$ reaction: large reaction chamber experiments and its implication to atmospheric methane chemistry. *EOS Trans. Amer. Geophys. Union* **78** Spring Meeting Supplement S75.

Tans P. P. (1997) A note on isotopic ratios and the global atmospheric methane budget. *Global Biogeochem. Cycles* **11**, 77–81.

Thorpe R. B., Law K. S., Bekki S., Pyle J. A., and Nisbet E. G. (1996) Is methane-driven deglaciation consistent with the ice core record? *J. Geophys. Res.* **101**, 28627–28635.

Torn M. S. and Chapin F. S., III (1993) Environmental and biotic controls over methane flux from arctic tundra. *Chemosphere* **26**, 357–368.

Tyler S. C. (1991) The global methane budget. In *Microbial Production and Consumption of Greenhouse Gases: Methane, Nitrogen Oxides, and Halomethanes* (eds. J. E. Rogers and W. B. Whitman). American Society for Microbiology, Washington, DC, pp. 7–38.

Tyler S. C., Crill P. M., and Brailsford G. W. (1994) $^{13}C/^{12}C$ fractionation of methane during oxidation in a temperate forested soil. *Geochim. Cosmochim. Acta* **58**, 1625–1633.

Tyler S. C., Ajie H. O., Gupta M. L., Cicerone R. J., Blake D. R., and Dlugokencky E. J. (1999) Carbon isotope composition of atmospheric methane: a comparison of surface level and upper tropospheric air. *J. Geophys. Res.* **104**, 13895–13910.

Tyler S. C., Ajie H. O., Rice A. L., Cicerone R. J., and Tuazon E. C. (2000) Experimentally determined kinetic isotope effects in the reaction of $CH_4$ with Cl: implicatons for atmospheric $CH_4$. *Geophys. Res. Lett.* **27**, 1715–1718.

Valentine D. L. (2002) Biogeochemistry and microbial ecology of methane in anoxic environments: a review. *Antonie van Leeuwenhoek* **81**, 271–282.

Valentine D. L. and Reeburgh W. S. (2000) New perspectives on anaerobic methane oxidation. *Environ. Microbiol.* **2**, 477–484.

Valentine D. L., Blanton D. C., and Reeburgh W. S. (2000a) A culture apparatus for maintaining $H_2$ at sub-nanomolar concentrations. *J. Microbial Methods* **39**, 243–251.

Valentine D. L., Blanton D. C., and Reeburgh W. S. (2000b) Hydrogen production by methanogens under low-hydrogen conditions. *Arch. Microbiol.* **174**, 415–421.

Valentine D. L., Blanton D. C., Reeburgh W. S., and Kastner M. (2001) Water column methane oxidation adjacent to an area of active hydrate dissociation, Eel River Basin. *Geochim. Cosmochim. Acta* **65**, 2633–2640.

Von Fischer J. C. and Hedin L. O. (2002) Separating methane production and consumption with a field-based isotope pool dilution technique. *Global Biogeochem. Cycles* **16**(3), 8-1–8-13, doi: 10.1029/2001GB001448.

Wahlen M. (1993) The global methane cycle. *Ann. Rev. Earth Planet. Sci.* **21**, 407–426.

Wallington T. J. and Hurley M. D. (1992) A kinetic study of the chlorine atoms with $CF_3CHCl_2$, $CF_3CH_2F$, $CFCl_2CH_3$, $CF_2ClCH_3$, $CHF_2CH_3$, $CH_3D$, $CH_2D_2$, $CHD_3$, $CD_4$ and $CD_3Cl$ at $295 \pm 2$ K. *Chem. Phys. Lett.* **189**, 437–442.

Walter B. and Heimann M. (2000) A process-based, climate-sensitive model to derive methane emissions from natural wetlands: application to five wetland sites, sensitivity to model parameters, and climate. *Global Biogeochem. Cycles* **14**, 745–765.

Ward B. B., Kilpatrick K. A., Novelli P. C., and Scranton M. I. (1987) Methane oxidation and methane fluxes in the ocean surface layer and deep anoxic waters. *Nature* **327**, 226–229.

Whalen S. C. and Reeburgh W. S. (1988) A methane flux time series for tundra environments. *Global Biogeochem. Cycles* **2**, 399–409.

Whalen S. C. and Reeburgh W. S. (1990a) A methane flux transect along the trans-Alaska pipeline haul road. *Tellus* **42B**, 237–245.

Whalen S. C. and Reeburgh W. S. (1990b) Consumption of atmospheric methane by tundra soils. *Nature* **342**, 160–162.

Whalen S. C. and Reeburgh W. S. (1992) Interannual variations in tundra methane flux: a 4-year time series at fixed sites. *Global Biogeochem. Cycles* **6**, 139–160.

Whalen S. C. and Reeburgh W. S. (1996) Moisture and temperature sensitivity of $CH_4$ oxidation in boreal soils. *Soil Biol. Biochem.* **28**, 1271–1281.

Whalen S. C. and Reeburgh W. S. (2000a) Methane oxidation, production and emission at contrasting sites in a boreal bog. *Geomicrobiol. J.* **17**, 237–251.

Whalen S. C. and Reeburgh W. S. (2000b) The effect of nitrogen fertilization on atmospheric methane oxidation in boreal forest soils. *Chemosphere—Global Change Sci.* **2**, 151–155.

Whalen S. C., Reeburgh W. S., and Sandbeck K. A. (1990) Rapid methane oxidation in a landfill cover soil. *Appl. Environ. Microbiol.* **56**, 3405–3411.

Whalen S. C., Reeburgh W. S., and Barber V. A. (1992) Oxidation of methane in boreal forest soils: a comparison of seven measures. *Biogeochemistry* **16**, 181–211.

Whalen S. C., Reeburgh W. S., and Reimers C. E. (1995) Control of tundra emission by microbial oxidation. In *Landscape Function: Implications for Ecosystem Response to Disturbance*, A Case Study in Arctic Tundra (eds. J. F. Reynolds and J. D. Tenhunen). Springer, New York, pp. 257–274.

Whiticar M. J. and Faber E. (1986) Methane oxidation in sediment and water column environments—isotope evidence. *Org. Geochem.* **10**, 759–768.

Whiting G. J. and Chanton J. P. (1993) Primary production control of methane emission from wetlands. *Nature* **364**, 794–795.

Winckler G., Aeschbach-Hertig W., Holocher J., Kipfer R., Levin I., Poss C., Rehder G., Suess E., and Schlosser P. (2002) Noble gases and radiocarbon in natural gas hydrates. *Geophys. Res. Lett.* **29**, 10, doi: 10.1029/2001GL014013.

Wofsy S. C. (1976) Interactions of $CH_4$ and CO in the Earth's atmosphere. *Ann. Rev. Earth and Planet Sci.* **4**, 441–469.

Wolfe R. S. (1996) 1776–1996: Alessandro Volta's combustible air. *ASM News* **62**(10), 529–534.

Wuebbles D. J. and Hayhoe K. (2002) Atmospheric methane and global change. *Earth Sci. Rev.* **57**, 177–210.

Yagi K., Tsuruta H., Kanda K.-I., and Moinami K. (1996) Effect of water management on methane emission from a Japanese rice paddy field: automated methane monitoring. *Global Biogeochem. Cycles* **10**, 255–267.

Zyakun A. M., Bondar V. A., Mshenskiy Yu. N., Zakharchenko V. N., Gayazov R. R., and Shishkina V. N. (1988) Carbon-isotope fractionation by the methane-oxidizing bacteria *Methylomonas methanica* during its continuous growth. *Geochem. Int.* **25**, 84–90 (Originally published in *Geokhimiya* **7**, 1007–1013).

# 4.04
# Tropospheric Aerosols

P. R. Buseck

*Arizona State University, Tempe, AZ, USA*

and

S. E. Schwartz

*Brookhaven National Laboratory, Upton, NY, USA*

## NOMENCLATURE

| | |
|---|---|
| $A$ | area ($m^2$) |
| $A_c$ | area fraction of clouds |
| $B$ | column burden (mg $m^{-2}$) |
| $D$ | diameter ($\mu m$) |
| $F$ | radiative flux (W $m^{-3}$) |
| $f(RH)$ | humidification growth factor |
| $f_b$ | backscatter fraction |
| $g$ | asymmetry parameter |
| $m$ | index of refraction |
| $M$ | mass concentration ($\mu g^{-3}$) |
| $N$ | number concentration ($cm^{-3}$) |
| $p(r)$ | any particle property |
| $P$ | any integral aerosol property |
| $p(\theta, \phi)$ | phase function (steradian$^{-1}$) |
| $Q$ | Mie scattering kernel |
| $Q_X$ | emission rate of substance X (kg $yr^{-1}$) |
| $R$ | reflectivity |
| $r$ | radius ($\mu m$) |
| $T$ | transmittance |
| $x$ | length variable (m) |
| $Y$ | yield |
| $\alpha$ | Ångström exponent |
| $\alpha_X$ | scattering efficiency of substance X ($m^2\ g^{-1}$) |
| $\beta$ | upscatter fraction |
| $\theta$ | axial angle (rad) |
| $\theta_0$ | solar zenith angle (rad) |
| $\lambda$ | wavelength (nm) |
| $\mu_k$ | $k$th moment of size distribution ($\mu m^k\ cm^{-3}$) |
| $\rho$ | density (g $cm^{-3}$) |
| $\sigma$ | light scattering coefficient ($Mm^{-1}$) |
| $\tau$ | optical thickness |
| $\phi$ | azimuthal angle (rad) |
| $\omega$ | single-scattering albedo |
| $\Theta$ | residence time (s, d) |

*Subscripts*

| | |
|---|---|
| a | absorption |
| e | extinction |
| g | gas |
| p | particle |
| s | scattering |

## 4.04.1  INTRODUCTION

### 4.04.1.1  Overview

It is widely believed that "On a clear day you can see forever," as proclaimed in the 1965 Broadway musical of the same name. While an admittedly beautiful thought, we all know that this concept is only figurative. Aside from Earth's curvature and Rayleigh scattering by air molecules, aerosols—colloidal suspensions of solid or liquid particles in a gas—limit our vision.

Even on the clearest day, there are billions of aerosol particles per cubic meter of air.

Atmospheric aerosols are commonly referred to as smoke, dust, haze, and smog, terms that are loosely reflective of their origin and composition. Aerosol particles have arisen naturally for eons from sea spray, volcanic emissions, wind entrainment of mineral dust, wildfires, and gas-to-particle conversion of hydrocarbons from plants and dimethylsulfide from the oceans. However, over the industrial period, the natural background aerosol has been greatly augmented by anthropogenic contributions, i.e., those produced by human activities. One manifestation of this impact is reduced visibility (Figure 1). Thus, perhaps more than in other realms of geochemistry, when considering the composition of the troposphere one must consider the effects of these activities. The atmosphere has become a reservoir for vast quantities of anthropogenic emissions that exert important perturbations on it and on the planetary ecosystem in general. Consequently, much recent research focuses on the effects of human activities on the atmosphere and, through them, on the environment and Earth's climate. For these reasons consideration of the geochemistry of the atmosphere, and of atmospheric aerosols in particular, must include the effects of human activities.

Although comprising only a small fraction of the mass of Earth's atmosphere, aerosol particles are highly important constituents of the atmosphere. Special interest has focused on aerosols in the troposphere, the lowest part of the atmosphere, extending from the land or ocean surface typically to ~8 km at high latitudes, ~12 km in mid-latitudes, and ~16 km at low latitudes. That interest arises in large part because of the importance of aerosol particles in geophysical processes, human health impairment through inhalation, environmental effects through deposition, visibility degradation, and influences on atmospheric radiation and climate.

Anthropogenic aerosols are thought to exert a substantial influence on Earth's climate, and the need to quantify this influence has sparked much of the current interest in and research on tropospheric aerosols. The principal mechanisms by which aerosols influence the Earth radiation budget are scattering and absorbing solar radiation (the so-called "direct effects") and modifying clouds and precipitation, thereby affecting both radiation and hydrology (the so-called "indirect effects"). Light scattering by aerosols increases the brightness of the planet, producing a cooling influence. Light-absorbing aerosols such as black carbon exert a warming influence. Aerosols increase the reflectivity of clouds, another cooling influence. These radiative influences are quantified as forcings, where a forcing is a perturbation to the energy balance of the atmosphere–Earth

(a)

(b)

**Figure 1**  Impairment of visibility by aerosols. Photographs at Yosemite National Park, California, USA. (a) Low aerosol concentration (particulate matter of aerodynamic diameter less than 2.5 $\mu$m, $PM_{2.5}$ = 0.3 $\mu$g m$^{-3}$; particulate matter of aerodynamic diameter less than 10 $\mu$m, $PM_{10}$ = 1.1 $\mu$g m$^{-3}$; estimated coefficient of light scattering by particulate matter, $\sigma_{ep}$, at 570 nm = 12 Mm$^{-1}$). (b) High aerosol concentration ($PM_{2.5}$ = 43.9 $\mu$g m$^{-3}$; $PM_{10}$ = 83.4 $\mu$g m$^{-3}$; estimated $\sigma_{ep}$ at 570 nm = 245 Mm$^{-1}$) (reproduced by permission of *National Park Service*, **2002**).

system, expressed in units of watts per square meter, W m$^{-2}$. A warming influence is denoted a positive forcing, and a cooling influence, negative. The radiative direct and indirect forcings by anthropogenic aerosols are thought to be of comparable magnitude to the positive forcings resulting from incremental concentrations of greenhouse gases.

The magnitudes and estimated uncertainties of the several forcings over the industrial period are summarized in Figure 2, which was prepared as part of the recent assessment of climate change by

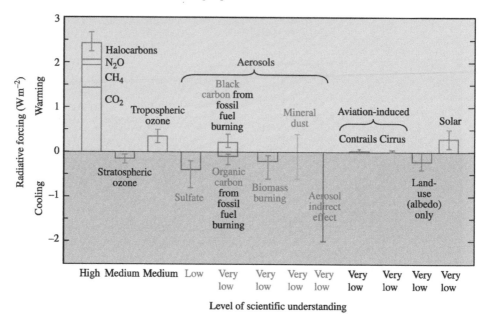

**Figure 2** The effects of various anthropogenic constituents of the atmosphere on the global climate system for the year 2000 relative to 1750 as estimated by the Intergovernmental Panel on Climate Change (IPCC, 2001). The effects are expressed as forcings, which in this case are changes in global mean radiative flux components arising from the indicated perturbing influence. Best estimates are indicated by the bars and uncertainties by "the I-beams". Only an uncertainty range rather than a best estimate is presented for direct aerosol forcing by mineral dust and for indirect aerosol forcing. An assessment of the present level of scientific understanding is indicated at the bottom of the figure (reproduced by permission of Intergovernmental Panel on Climate Change).

the Intergovernmental Panel on Climate Change (IPCC, 2001). This figure shows for each forcing a best estimate of its magnitude and of the associated uncertainty. The uncertainty associated with forcing by the long-lived greenhouse gases is relatively small, reflective of the rather high level of understanding of both the magnitude of the incremental concentrations of these species and of the radiative perturbation per incremental concentration. In marked contrast, the uncertainties associated with the several aerosol forcings are much greater, indicative of a much lesser understanding of the controlling quantities. For direct forcing by dust aerosols, which may be positive or negative, and for indirect radiative forcing by anthropogenic aerosols the IPCC working groups (Penner *et al.*, 2001; Ramaswamy *et al.*, 2001) declined to present best estimates but indicated only possible ranges. This situation is unsatisfying but unavoidable, given the current state of knowledge. Other reviews of aerosol forcings are provided by Ramanathan *et al.* (2001a), Haywood and Boucher (2000), Shine and Forster (1999), Schwartz (1996), and Schwartz and Slingo (1996). Hobbs (1993) provides an introduction to aerosol–cloud interactions.

The importance of atmospheric aerosols to issues of societal concern has motivated much research intended to describe their loading, distribution, and properties and to develop understanding of the controlling processes to address

such issues as air pollution, acid deposition, and climate influences of aerosols. However, description based wholly on measurements will inevitably be limited in its spatial and temporal coverage and in the limited characterization of aerosol properties. These limitations are even more serious for predictions of future emissions and provide motivation for concurrent theoretical studies and development of model-based description of atmospheric aerosols.

An important long-range goal, which has already been partly realized, is to develop quantitative understanding of the processes that control aerosol loading, composition, and microphysical properties as well as the resultant optical and cloud-nucleating properties. An objective is to incorporate these results into chemical transport models that can be used for predictions. Such models are required, for example, to design approaches to achieve air quality standards and to assess and predict aerosol influences on climate change. Much current research is directed toward enhancing this understanding and to evaluating it by comparison of model results and observations. However, compared to gases, models involving particles are far more complex because of the need to specify additional parameters such as particle sizes and size distributions, compositions as a function of size, particle shapes, and temporal and spatial variations, including reactions that occur within the atmosphere. Therefore, the few existing approaches to chemical transformation and aerosol evolution rest heavily on assumptions, for example, that particles are adequately represented as spheres and are homogeneous in composition as a function of particle size, although both assumptions are known to be inaccurate (e.g., Buseck and Pósfai, 1999; Buseck *et al.*, 2002).

This chapter provides an overview of the loading, geographical distribution, and chemical and physical properties of both natural and anthropogenic atmospheric aerosols and of the processes controlling their production, reaction, transport, and ultimate removal—the "life cycle" of tropospheric aerosols. More detailed treatment may be found in texts by Junge (1963), Friedlander (1977), Twomey (1977), Hinds (1982, 1999), Seinfeld and Pandis (1998), and Jacob (1999). We highlight here the effects of aerosols on climate. The effects of aerosols on health, visibility, heterogeneous chemistry, and ozone are examined by Heintzenberg *et al.* (2003), Jacob (2000), Kreidenweis (1995), Anastasio and Martin (2001), Pósfai and Molnár (2000), and Prospero *et al.* (2002). A detailed overview of tropospheric aerosols and their environmental effects is given by EPA (2002). Kaufman *et al.* (2002) provide an overview of satellite measurement of aerosols pertinent to climate change.

### 4.04.1.2 Sources

Aerosols are produced by both natural processes and human activities. Natural aerosol particles include wind-blown mineral dust, smoke from fires, pollens, spores, sea salt arising from spray from the ocean, abraded materials from plants, and certain hazes. Anthropogenic emissions are the dominant contributors to aerosols in and downwind of many urban and industrial locations and are important on local, regional, and sometimes global scales. Examples are smoke from power plants and biomass burning, emissions from industrial activities such as metal smelting, and urban haze resulting in large part from particle formation in the atmosphere. Some major types of aerosols have both natural and anthropogenic sources, for example mineral dust, where the anthropogenic component results in large part from soils disturbed through agricultural activities and animal husbandry.

A rich variety of minerals occurs in and influences the behavior of tropospheric aerosols. These minerals can be reduced to a few major source categories. In addition to salts released from sea spray, volcanic emissions, and wind-blown minerals are major sources of dust to the troposphere. Large volcanic eruptions intermittently spew vast quantities of dust into the atmosphere, in some cases penetrating the tropopause and extending into the stratosphere. Advected dust from arid and semi-arid lands is transported from areas around the globe. The effects of these emissions have been global, both on climate and on additions to the land surface. Many oceanic islands receive important aerosol contributions to their soils (e.g., Hawaii; Parrington *et al.*, 1983; Kurtz *et al.*, 2001).

The effects of volcanic emissions on climate were noted by no less a person than Benjamin Franklin. While living in France, he commented on the persistent haze, dimming of the sun, and unusually cold summer of 1783 and speculated that a volcanic eruption might have been the cause (Franklin, 1789). Although he was not aware of a massive fissure in Iceland at the time, his observation was correct. On Iceland an eruption rained ash that caused death of grass and starvation of livestock and killed 25% of the population from the resulting famine.

Aerosol particles resulting from the 1991 eruption of Mount Pinatubo (Philippines) are widely cited in the recent atmospheric literature because of their global effects on atmospheric radiation and climate (Hansen *et al.*, 1992; McCormick *et al.*, 1995). However, volcanic eruptions occur every year, and some have been far larger than Pinatubo. For comparison, Krakatoa (1883; between Java and Sumatra, Indonesia) and Tambora (1815; Sumbawa, Indonesia), respectively, emitted double and ten times as much pyroclastic debris into the atmosphere as Pinatubo, and significantly reduced sunlight around the globe for months (Sparks *et al.*, 1997). Their effects on the atmosphere have been profound.

A major constituent of volcanic emissions is $SO_2$. Through a series of atmospheric reactions $SO_2$ is oxidized to produce sulfuric acid, $H_2SO_4$. Sulfuric acid and sulfate salts are the dominant aerosol particles in the stratosphere and much of the upper troposphere. They reflect sunlight and produce significant cooling at Earth's surface. Thus, a somewhat surprising effect is that volcanic emissions can produce a net cooling of the surface (Gribbin, 1976; Hansen *et al.*, 1978; Charlock and Sellers, 1980). They are also major components of acid rain. Bryson (1974), after reviewing the global-scale cooling resulting from major volcanic eruptions, likened the enhancement of atmospheric particulate matter arising from human industrial activities to that resulting from moderate global-scale volcanic activity.

Apart from outstanding massive volcanic eruptions, wind-blown mineral dust is by far the largest contributor to the global particulate burden of the lower troposphere (e.g., Andreae, 1995). Prospero *et al.* (2002) provides a good summary of mineral dust generation. Most is derived from a so-called "dust belt" that includes deserts in North Africa, the Middle East, Central and South Asia, and China. Far less dust is entrained from the southern hemisphere, even though areas like Australia have large arid areas. Much dust is derived from topographic lows and areas that have been impacted by human activity such as grazing or farming, although the amounts are in contention (Tegen and Fung, 1995; Prospero et al., 2002).

In addition to the effects on climate and soils, mentioned above, study of deep-sea sediments and ice cores indicate that, during glacial periods, dust deposition rates were 2–20 times the current values (Thompson and Mosley-Thompson, 1981; Hammer et al., 1985; Petit et al., 1990; Rea, 1994; Steffenson, 1997; Reader et al., 1999), possibly because of the effect airborne minerals had on global climate by their interactions with solar and terrestrial radiation (Andreae, 1995, 1996; Duce, 1995; Li et al., 1996; Sokolik and Toon, 1996; Tegen and Lacis, 1996; Mahowald et al., 1999). Possible relations between glacial periods and dust have been the subject of intense speculation.

Mineral dust also has important effects on the amounts of nutrients supplied to the oceans, thereby affecting biological productivity and amounts of $CO_2$ released into the atmosphere (Martin and Gordon, 1988; Watson, 1997). Iron is thought to be a limiting factor for phytoplankton productivity in remote marine areas (Falkowski et al., 1998; Fung et al., 2000; Gao et al., 2001).

### 4.04.1.3 Atmospheric Residence, Transport, and Removal

Aerosol particles can persist in the troposphere for days to weeks, depending on their sizes and compositions. Long-range transport of particles has been documented in a variety of ways. The Chernobyl (Ukraine) accident in 1986 released radionuclides that attached to aerosol particles and were measured in the weeks following the accident at locations throughout the mid-latitudes of the northern hemisphere (Cambray et al., 1987). In situ measurements have established the presence of minerals that can be attributed to specific source regions, for example the transport of Saharan dust to the Caribbean and Florida (Prospero, 1999; Tanré et al., 2001), and dust transported across the Pacific from Asia has been detected across the US. (Husar et al., 2001; Griffin et al., 2002). In recent years, images and other measurements from satellite-borne instruments documenting the presence and transport of aerosols have become routinely available (cf. Figure 3).

Aerosol particles are removed from the atmosphere by wet deposition through incorporation into precipitation either in or below clouds, and by dry deposition through sedimentation and impaction on or diffusion to soil, leaves, or the like.

The dominant removal process is wet deposition for smaller particles (radius less than ~2 μm) and dry deposition for larger particles.

### 4.04.1.4 Primary versus Secondary Particles and Particle Nucleation

Particulate matter may be emitted into the atmosphere as such (primary aerosol particles) or may be formed in the atmosphere by gas-to-particle conversion (secondary aerosol). Primary particles that are produced mechanically, for example, by soil erosion or bursting of bubbles at the ocean surface, tend to be fairly large (1–10 μm radius or greater), whereas those emitted from combustion sources are typically small (down to a few nanometers radius). Although by far the greatest fraction of the mass of primary particulate matter is in particles of diameter greater than 1 μm, sub-micrometer mineral dust and sea salt particles are important constituents of tropospheric aerosols (e.g., Figure 4). Production of secondary aerosol can occur either by new particle formation (nucleation) or by the addition of material to existing aerosol particles. The latter process may take place in clear (non-cloud) air or by reaction within cloud droplets, which upon evaporation return particulate matter to the clear-air aerosol. Secondary particle formation has long been recognized to be responsible for much of the aerosol in regions influenced by anthropogenic emissions.

Nucleation of new particles occurs through the formation of low-vapor-pressure products of gas-phase reactions. The production of this material is generally accompanied by its condensation on existing aerosol particles. In some instances nucleation can be initiated where there is rapid production of new condensable material together with low concentration of aerosol particles and thus low existing aerosol surface area. This combination results in the supersaturation of the condensable vapor, which may build up sufficiently to overcome the free-energy barrier (critical supersaturation) associated with new particle production, thereby initiating nucleation.

### 4.04.1.5 Spatial Homogeneity

Implicit in the above discussion is an assumption that the aerosol is locally homogeneous over some region of space. In fact, aerosol particles that are present at a given time and location are derived from emissions that may have occurred at a variety of times and upwind locations. Spatial inhomogeneities tend to decrease over time because of turbulent mixing. In many studies it has been

**Figure 3** Sea-viewing Wide Field-of-view Sensor (SeaWiFS) satellite image of dust plume extending North across the Mediterranean from the Sahara desert. The Nile River, Gulf of Suez, Gulf of Aqaba, and, south of them, the Red Sea are evident from East to West in the lower part of the image. Note the bright scattering by dust particles relative to the deep blue of the seawater. Image acquired on April 18, 2001 and provided by the SeaWiFS Project, NASA/Goddard Space Flight Center, and ORBIMAGE.

assumed that an aerosol is homogeneous within air masses hundreds of kilometers in extent, and thus that the aerosol that is collected on a filter over some extended period of time, say 24 h, is representative of such a homogeneous aerosol. However, as instruments with improved time resolution are becoming available, it is increasingly being recognized that aerosols may exhibit substantial variation on timescales of tens of minutes and distance scales of a few kilometers in the horizontal and a few tens of meters in the vertical.

The inhomogeneity of aerosol distributions over short spatial and temporal scales is readily demonstrated in lidar (light detection and ranging) retrievals, an example of which is given in Figure 5. This figure shows aerosol backscatter obtained with a down-looking lidar during an aircraft flight over the mountainous terrain of the Lower Fraser Valley near Vancouver, BC, Canada. The intensity of the return signal is a measure of the aerosol concentration. The sharp, black regions are silhouettes of mountains or mountain ranges. The planetary boundary layer

can be discerned below the mountain tops by the thermal inversion below which most tropospheric aerosols are trapped at the time of the observations (afternoon). The several distinct layers of aerosol contain particles typical of haze in the valley and can appear as a result of a stagnant atmospheric condition. They can also result from transport, mountain up-slope flow, or from residual boundary layers. The high aerosol content of some of the valleys results from drainage flow from higher elevations within the valley. Aerosol concentrations decrease in the valleys to the east, further from the population centers of Vancouver and suburbs.

## 4.04.2 AEROSOL PROPERTIES

A notable feature of aerosols is that their collective properties (e.g., light-scattering coefficient) depend on variables that include the bulk properties of the particulate material, its state of dispersion, and size-dependent state of

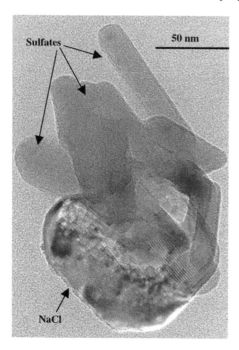

**Figure 4** Transmission electron microscope (TEM) image of an aggregate of NaCl and sulfate salts from a small marine particle. This particle is at the low end of the observed size range of sea salt particles. The sample was collected near Cape Grim, Tasmania, Southern Ocean during the ACE-1 experiment (source Buseck and Pósfai, 1999).

aggregation. Knowledge of the bulk composition of the particulate matter of an aerosol, as might be gained from filter collection and chemical analysis, is necessary to specify the collective properties of the aerosol but is far from sufficient. Also needed is knowledge of the particle size distribution, whether particle compositions and shapes are identical or differ with size, and whether all particles of a given size have the same composition (denoted an *internal mixture*) or whether various particles of a given size have different compositions (*external mixture*).

Numerous aerosol properties are of interest. These may be distinguished as single-particle properties, size-dependent properties, and collective properties. Single-particle properties are those of one particle at a time. They include composition, morphology, homogeneity or lack thereof, and the like. Size-dependent properties are those of the particles within a narrow size range, for example the number concentration, surface area concentration, volume concentration, mass concentration, and composition. These size-dependent properties are commonly expressed as distribution functions. Collective properties represent those of the aerosol as a whole. Mathematically, these collective properties may be thought of as integrals over the size distribution of the aerosol (Section 4.04.2.1). Key among these are particle number concentration,

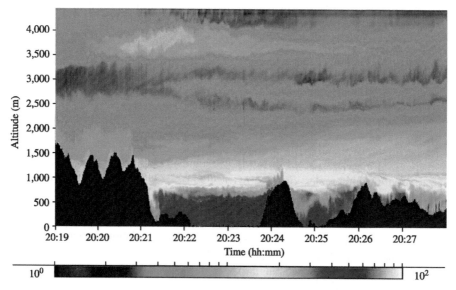

**Figure 5** Down-looking lidar image from aircraft showing high aerosol concentrations in the eastern part of the Lower Fraser Valley near Chilliwack, British Columbia. The mountains (black) are rimmed by down- and up-slope aerosol movement on the left and right sides, respectively, of the largest valley. Tropospheric layering is evident at altitudes above the mountains and valleys. The color bar at the bottom gives the range-normalized lidar return signal on a logarithmic scale covering two orders of magnitude of aerosol concentration; the black areas denote the terrain. The lidar data were obtained at 1,064 nm using a 20-shot average (1 s); vertical resolution —3 m. The horizontal axis is UTC; local time is UTC—07:00 (UTC = coordinated universal time). This axis may be translated into a distance from the aircraft speed, ca. 100 m s$^{-1}$, so 1 min corresponds roughly to 6 km. August 14, 2001 (courtesy of K. Strawbridge, Meteorological Service of Canada).

particle mass concentration, and optical properties such as light-scattering coefficient, light-absorption coefficient, and properties derived from these quantities.

A further useful distinction (Ogren, 1995) is between collective aerosol properties that scale with the abundance of particulate matter (*extensive properties*) and those that are intrinsic properties of the aerosol and do not scale with the local abundance (*intensive properties*). Examples of extensive properties are concentrations of particular substances, number concentration, light scattering, absorption, and extinction coefficients. Examples of intensive properties are material properties such as the composition of each species of the aerosol, index of refraction (real and imaginary components), single-scattering albedo (fraction of extinction due to scattering), Ångström exponent (a measure of the wavelength dependence of scattering coefficient), phase function (angular distribution of light scattering), effective radius (ratio of third to second moment of size distribution; useful in calculating radiative transfer), and mass scattering efficiency (ratio of scattering coefficient to mass concentration). Many intensive properties are ratios of two extensive properties. Table 1 summarizes many of the properties discussed in this and following sections.

### 4.04.2.1 Number and Mass Distributions

Knowledge of the size distribution of the particles comprising an aerosol is required for a comprehensive description. It is implicit in such descriptions that a particle can be characterized by a single size variable, commonly the geometric radius. Although this assumption does not hold for nonspherical particles, it is useful, nonetheless, to employ some characteristic radius. Additionally, because available methods of measuring aerosol particles do not generally yield the geometric radius, often some other characteristic value such as the aerodynamic radius (cf. Section 4.04.3.1 for definition) is used. The measurement community tends to report results in terms of particle diameter, whereas theoretical calculations are typically expressed in terms of radius. We follow these conventions in this chapter.

Commonly the size description is given as the number, $dN$, (or, less frequently, mass, $dM$) of particles per radius interval as a function of radius (Seinfeld and Pandis, 1998, p. 408 ff.). Alternatively, the size description is given as the number, $dN$, (or, less frequently, mass, $dM$) per logarithmic interval of radius as a function of radius. The total number concentration or total mass concentration of aerosol particles is the integral over the size distribution function:

$$N = \int_{r_{min}}^{r_{max}} \left( \frac{dN}{dr} \right) dr = \int_{r_{min}}^{r_{max}} \left( \frac{dN}{d \log r} \right) d \log r$$

$$M = \int_{r_{min}}^{r_{max}} \left( \frac{dM}{dr} \right) dr = \int_{r_{min}}^{r_{max}} \left( \frac{dM}{d \log r} \right) d \log r$$

Several means of displaying the size distribution of particles in an aerosol are shown in Figure 6. A logarithmic abscissa is commonly used to display size-dependent aerosol properties, including concentration, because of the large size range of atmospheric aerosol particles and because aerosol properties vary strongly with size. If a linear scale were used, the distance on the abscissa scale would be the same for 10 nm (0.01 μm) to 1 μm as from 10.01 μm to 11 μm, but the properties of a 10 μm particle are very similar to those of a 11 μm particle, whereas the properties of 10 nm particles are vastly different from those of 1 μm particles. With respect to the vertical scale, a variety of choices are useful. Panel (a) employs a logarithmic ordinate to display the wide range of number concentrations of the three modes that comprise the aerosol. Panels (b), (c), and (d) show number, area, and volume size distributions, respectively, as semi-logarithmic plots, often called "equal area plots" because the area under the curve for a given range of the abscissa is proportional to the quantity plotted. In this example, which is typical of atmospheric aerosols, the smallest particles dominate the number distribution, the intermediate particles dominate the surface-area distribution, and the largest particles dominate the volume distribution. In each case the quantity plotted is a differential quantity, i.e., the fraction of the quantity per unit logarithmic interval of particle radius. As the logarithm is dimensionless, the ordinate is number, area, or volume per volume of air, respectively, where the volume of air is conventionally taken as 1 $cm^3$.

Particles in an aerosol are characterized by a size distribution, and therefore it is evident that any specification of the total mass or number concentration implicitly assumes a size range. In particular, because the size distribution of number concentration generally increases as radius decreases, the lower limit $r_{min}$ is of great importance in specifying total number concentration. Similarly, because the size distribution of mass concentration generally increases as radius increases, the total mass concentration is strongly influenced by the upper limit $r_{max}$. Thus one encounters, for example in air quality regulations, terminology such as $PM_{2.5}$ and $PM_{10}$, referring to particulate matter of aerodynamic diameter less than 2.5 μm and 10 μm, respectively. The mass and number distribution functions are related to

**Table 1** Key aerosol properties, and their symbols, units, and definitions.

| Aerosol property | Symbol | Typical unit | Definition | Comments |
|---|---|---|---|---|
| *Extensive* | | | | |
| Particle number concentration | $N$ | cm$^{-3}$ | $N = \int_{r_{\min}}^{r_{\max}} \left(\dfrac{dN}{dr}\right) dr = \int_{r_{\min}}^{r_{\max}} \left(\dfrac{dN}{d\log r}\right) d\log r$ | Requires size range to be specified. Because the number of small particles is commonly large, $N$ is very sensitive to the lower limit |
| Condensation nucleus (CN) concentration | $N_{CN}$ | cm$^{-3}$ | Number concentration of particles that nucleate droplet formation at a high supersaturation; typically all particles with diameter greater than 10–20 nm | Whether >10 nm or >20 nm is used for the concentration cutoff depends on instrument and settings |
| Cloud condensation nucleus (CCN) concentration | $N_{CCN}$ | cm$^{-3}$ | Number concentration of particles that activate to form droplets at a low supersaturation of water vapor, typically 0.1–1% | Measurements are generally made at a single supersaturation between 0.1% and 1%; more information is gained from measurements at multiple supersaturation values in this range |
| Particle mass concentration | $M$ | µg m$^{-3}$ | $M = \int_{r_{\min}}^{r_{\max}} \left(\dfrac{dM}{dr}\right) dr = \int_{r_{\min}}^{r_{\max}} \left(\dfrac{dM}{d\log r}\right) d\log r$, where $\dfrac{dM}{dr} = \dfrac{4}{3}\pi r^3 \rho(r)\dfrac{dN}{dr}$ | $M$ refers to the particulate component of the aerosol, including associated condensed-phase water. $M$ requires size range to be specified. Because the effects of large particles on total mass is large, $M$ is especially sensitive to the upper size limit. It is sensitive to RH, which must thus be specified |
| Fine particle mass concentration | $PM_{2.5}$ | µg m$^{-3}$ | $PM_{2.5} = \int_0^{2.5\,\mu m} \left(\dfrac{dM}{dD_{aero}}\right) dD_{aero}$, $d_{aero}$ = aerodynamic diameter | Mass concentration of particulate matter having aerodynamic diameter less than 2.5 µm. Used for air quality standards; defined for a standard RH, typically below 45% |
| Coarse plus fine particle mass concentration | $PM_{10}$ | µg m$^{-3}$ | $PM_{10} = \int_0^{10\,\mu m} \left(\dfrac{dM}{dD_{aero}}\right) dD_{aero}$ | Mass concentration of particulate matter having aerodynamic diameter less than 10 µm. Used for air quality standards; defined for a standard RH typically below 45%. Coarse fraction refers to particles of aerodynamic diameter between 2.5 µm and 10 µm |
| Particle light-scattering coefficient | $\sigma_{sp}$ | Mm$^{-1}$ | $\sigma_{sp} = \int \pi r^2 Q_s(r/\lambda, m)\left(\dfrac{dN}{dr}\right) dr$ where $Q_s$ is the scattering kernel[a] | Fraction of light incident upon an aerosol that is removed from the primary beam per unit length by scattering from aerosol particles. Highly sensitive to RH for hygroscopic materials, but often measured and reported for dried aerosol; depends on wavelength $\lambda$ and complex index of refraction |
| Particle light-absorption coefficient | $\sigma_{ap}$ | Mm$^{-1}$ | $\sigma_{ap} = \int \pi r^2 Q_a(r/\lambda, m)\left(\dfrac{dN}{dr}\right) dr$ where $Q_a$ is the absorption kernel[a] | Fraction of light incident upon an aerosol that is removed from the primary beam per unit length by absorption by aerosol particles; depends on RH, wavelength $\lambda$, and complex index of refraction $m$ |

| Property | Symbol | Units | Formula | Comments |
|---|---|---|---|---|
| Particle light-extinction coefficient | $\sigma_{ep}$ | Mm$^{-1}$ | $\sigma_{ep} = \sigma_{sp} + \sigma_{ap}$ | Fraction of light incident upon an aerosol that is removed from the primary beam per unit length by scattering ($\sigma_{sp}$) and absorption ($\sigma_{ap}$) by aerosol particles; depends on wavelength and RH |
| Aerosol particle optical thickness | AOT, $\tau_{ep}$ | Dimensionless | $\tau_{ep} = \int \sigma_{ep}\, dz$ | Total optical thickness (or commonly optical depth) $\tau = -\ln T \cos\theta_0$, where $T$ is transmittance of direct solar beam for solar zenith angle $\theta_0$; optical thickness (or commonly optical depth) due to extinction by aerosol particles $\tau_{ep} = \tau - \tau_g$, where $\tau_g$ is optical thickness arising from scattering and absorption by gases; depends on wavelength |
| Particle mass column burden | $B$ | mg m$^{-2}$ | $B = \int M\, dz$ | Requires specification of range of integration of particle radius under consideration; generally most mass is in lower troposphere. $M$ is sensitive to RH, which is not constant in vertical; column burden commonly refers to mass of dry material |
| *Intensive* | | | | |
| Particle mass scattering efficiency | $\alpha_{sp}$ | m$^2$ g$^{-1}$ | $\alpha_{sp} = \sigma_{sp}/M$ | Both the mass concentration and light-scattering coefficient are sensitive to RH, necessitating specification of RH for both quantities; often reported for dry material, with RH dependence separately specified |
| Single-scattering albedo | $\omega_0$ | Dimensionless | $\omega_0 = \sigma_{sp}/(\sigma_{sp} + \sigma_{ap})$ | Fraction of extinction by particulate matter resulting from scattering. Sensitive to RH mainly because of sensitivity of $\sigma_{sp}$ to RH; depends on wavelength |
| Ångström exponent | $\alpha$ | Dimensionless | $\alpha = -d\log \sigma_{ep}/d\log \lambda$ ($\lambda$ = wavelength of light) or $\alpha = -d\log \tau_p/d\log \lambda$ | A measure of particle size useful in remote sensing; ranges from 0 (cloud droplets) to 4 (Rayleigh scattering by gases) |
| Effective radius | $r_e$ | μm | $r_e = \int r^3 \left(\dfrac{dN}{dr}\right) dr \Big/ \int r^2 \left(\dfrac{dN}{dr}\right) dr$ | Weighted average of the radii of particles contributing to light scattering; useful in radiative transfer calculations |
| Phase function | $p(\theta, \phi)$ | steradian$^{-1}$ | Fraction of scattered radiation by a particle or an aerosol into angle ($\theta, \phi$) ($\theta$, axial; $\phi$, azimuthal) measured from the direction of the incident beam | Often calculated; rarely measured |
| Asymmetry parameter | $g$ | Dimensionless | Average of cosine of scattering angle $\theta$ averaged over the phase function | Useful in radiative transfer calculations |
| Backscatter fraction | $f_b$ | Dimensionless | Fraction of radiation scattered into backward hemisphere | Measured in commercially available nephelometers |
| Upscatter fraction | $\beta(\theta_0)$ | Dimensionless | Fraction of light scattered in upward direction, a function of solar zenith angle $\theta_0$. | Property of aerosol in atmosphere; evaluated by integration of phase function over upward hemisphere |

Measurement of many of these properties is difficult because, among other reasons, of exchange of water and other semivolatile material between particles and gas phase.
[a] Twomey (1977) p. 209.

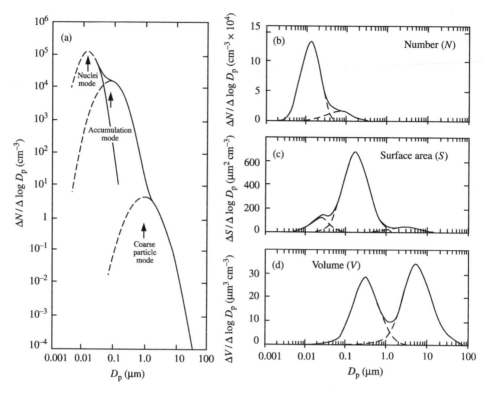

**Figure 6** Several views of the size distribution of a hypothetical aerosol consisting of three modes: denoted nuclei, accumulation, and coarse. Panel (a) is a log–log plot of number distribution of the three modes and their sum as a function of particle diameter, $D_p$. Panels (b)–(d) are semi-log plots of number ($N$), surface area ($S$), and volume ($V$) distributions, respectively (Whitby, 1978) (reproduced by permission of Elsevier from *Atmos. Environ.* **1978**, *12*, 135–159).

each other as

$$\frac{dM}{dr} = \frac{4}{3}\pi r^3 \rho(r)\frac{dN}{dr}$$

where $\rho(r)$ is the density of the particulate matter, explicitly denoted as dependent on particle radius to reflect a possible size dependence of composition. More generally, any extensive aerosol property $P$ may be evaluated as an integral of the individual particle property $p(r)$ over the size distribution:

$$P = \int_{r_{min}}^{r_{max}} p(r)\left(\frac{dN}{dr}\right)dr$$

For example, at a single wavelength $\lambda$ the light scattering coefficient $\sigma_{sp}(\lambda)$ of the particulate matter in an aerosol is

$$\sigma_{sp} = \int \pi r^2 Q_s(r/\lambda, m)\left(\frac{dN}{dr}\right)dr$$

where $m$ is the refractive index and $Q_s(r/\lambda, m)$ is the Mie scattering kernel (see, e.g., Twomey, 1977). In addition to radius, the scattering kernel depends on the index of refraction and hence, implicitly, on the composition. Clearly, the integral must be taken over the entire size

range that contributes appreciably to the scattering.

Both the number concentrations and sizes of aerosol particles directly affect many of their properties and effects. For example, the ability of particles to serve as nuclei for cloud droplet formation depends on their composition as a function of size, although their effectiveness in any given situation depends also on the number of particles present. Knowledge of these aerosol properties is required to evaluate the "indirect" effects (Section 4.04.7.3) of aerosol particles on climate, i.e., the effect of aerosol particles on cloud reflectivity and persistence. Therefore much attention has been and continues to be focused on determining particle number concentrations and size distributions.

**4.04.2.2 Composition**

A full description of an aerosol and its properties requires specification of particle composition and its dependence on particle size. Because the composition of a heterogeneous mixture is difficult to characterize, it has been customary to identify a small number of representative particle types (e.g., d'Almeida *et al.*, 1991) and treat atmospheric aerosols as if they consist of a single type or a mixture of types, depending on location, air-mass back-trajectory, and the like. However, with the advances in ability to characterize aerosol composition in some detail, it is increasingly being recognized that many aerosol particles are heterogeneous in their compositions and size dependence of composition and that categorization by representative types provides at best a fairly crude description. Particle heterogeneity can also be manifested by the clustered aggregation of particles of different types or by radial differences as through surface coatings. An additional complication is that the size distribution of an aerosol and its composition are not constant. Instead, the distribution evolves through new particle formation, exchange of material (especially water) between existing particles and the gas phase, coagulation, and removal of particles from the atmosphere. The rate of particle removal likewise depends on size and composition. Thus, determining the size-dependent composition of an aerosol is complicated and a prime task in characterizing the aerosol. In general, this task is not fully accomplished for any atmospheric aerosol, and the size-dependent composition is thus only approximated to varying degrees of accuracy.

Water plays a key role in influencing aerosol properties (cf. Section 4.04.3.3). Many substances in atmospheric aerosols are hygroscopic: the particles accrete water with increasing relative humidity (RH). They may undergo a phase transition from solid to liquid (deliquescence) if the RH increases, from liquid to solid (efflorescence) if the RH decreases, from one state of hydration to another with changing RH, or freeze or melt with changing temperature (Martin, 2000). Depending on the substance and conditions, the exchange of water between particle and gas phase may or may not be rapid or reversible. Many atmospheric materials commonly occur as super-saturated solutions at RH well below their deliquescence points. This situation gives rise to the possibility that at a given RH particles of a given composition may be either solid or super-saturated solution, depending on their prior history of RH, temperature, and volatile constituents (e.g., $HNO_3$ or $NH_3$).

### 4.04.2.3 Optical Properties

Key optical properties of aerosols are the coefficient of light scattering by particulate matter $\sigma_{sp}$ and its angular dependence, the coefficient of light absorption by particulate matter $\sigma_{ap}$ both functions of wavelength, and quantities derived from them (e.g., Table 1). The light-scattering coefficient is the fraction of incident light that is scattered by the particles per unit length of travel

by the light beam. Similarly, the light-absorption coefficient is the fraction of incident light absorbed by the particles per unit length of travel by the light beam. Both of these quantities are extensive variables. The coefficient of light extinction by particulate matter, $\sigma_{ep}$, is the sum of the scattering and absorption coefficients, $\sigma_{ep} = \sigma_{sp} + \sigma_{ap}$. According to the Beer–Lambert law, the fraction of light transmitted through a path $x$ is $I/I_0 = \exp(-\sigma_e x)$, where $\sigma_e$ is the extinction coefficient, which is the sum of extinction coefficients of the gas (Rayleigh scattering plus absorption) and particulate matter. The dimension of each of these quantities is length$^{-1}$; for atmospheric aerosols the unit $Mm^{-1}$ (per megameter) is convenient.

The single-scattering albedo of an aerosol, $\omega$ is the fraction of particulate extinction resulting from scattering, $\omega = \sigma_{sp}/(\sigma_{sp} + \sigma_{ap})$. As the ratio of two extensive properties, it is an intensive property, independent of the concentration of particulate matter.

The dependence of optical scattering on wavelength $\lambda$ is commonly characterized as an Ångström exponent, $\alpha = -d \log \sigma_{sp}/d \log \lambda$; $\alpha$ ranges from 0 for droplets much larger than the wavelength of light (e.g., cloud droplets) to 4, the Rayleigh limit for particles much smaller than the wavelength of light. The Ångström exponent is thus a measure of particle size and is especially useful in remote sensing.

An important collective aerosol optical property is the scattering efficiency, often reported as the mass scattering efficiency, and evaluated as the light-scattering coefficient of particulate matter divided by the mass concentration of particulate matter. By analogy, it is sometimes reported as the molar scattering efficiency, the scattering coefficient divided by the molar concentration of a particular aerosol species. The scattering efficiency is a further example of an intensive aerosol optical property. The mean mass scattering efficiency of the particulate matter of an aerosol, $\alpha_{sp}$, can be evaluated from the particle size distribution and the mass scattering efficiency as a function of particle radius (Figure 7) as

$$\overline{\alpha_{sp}} = \int \alpha_{sp}(r)\rho(r)r^3 \left(\frac{dN}{dr}\right)dr \Big/ \int \rho(r)r^3 \left(\frac{dN}{dr}\right)dr$$

The symbol $\alpha$ is widely used to represent either the mass scattering efficiency or the Ångström exponent, so care must be exercised to avoid confusion.

Another important aerosol optical property is column aerosol extinction, the vertical integral of the aerosol extinction coefficient, again a function of wavelength. This quantity is often also denoted aerosol optical thickness (AOT) or aerosol optical depth. The wavelength dependence of AOT is also often expressed as an Ångström exponent.

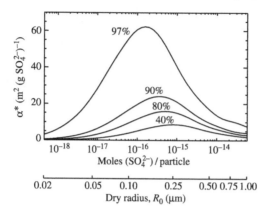

**Figure 7** Dependence of light-scattering efficiency of ammonium sulfate aerosol on the amount of material in the particle and RH. The scattering efficiency is expressed as a scattering coefficient $\sigma_{sp}$, here at wavelength 550 nm, per amount of sulfate. The auxiliary abscissa scale gives the particle radius (Nemesure *et al.*, 1995) (reproduced by permission of American Geophysical Union from *J. Geophys. Res.* **1995**, *100*, 26105–26116).

## 4.04.3 MEASUREMENT OF AEROSOL PROPERTIES

Much current research on atmospheric aerosols consists of determining the size-dependent composition of the particulate matter. The sizes of atmospheric aerosol particles range over a factor of $\sim 10^4$ in radius (and hence $\sim 10^{12}$ in volume or mass), and the physical properties of these particles likewise vary substantially over such a size range. Particles in different size ranges may derive from diverse sources, exhibit different behavior in the atmosphere, and thus exhibit different compositions. Particles of dissimilar size, composition, and shape differ in important effects such as light scattering and health impairment. Historically, most measurements of particulate-matter composition involved collection and subsequent determination, for example filter collection followed by weighing and chemical and physical analysis. However recognition of the important differences in composition and properties as a function of particle size has motivated development of increasingly sensitive methods of study of size-dependent composition.

In view of the large size range spanned by atmospheric particles, the measurement of composition and properties presents many challenges. Common approaches include separation of the aerosol particles into multiple size components, often by taking advantage of differing aerodynamic properties, followed by collection and chemical analysis of the several fractions. Physical means of characterization such as determination of the size distribution by light scattering or mobility are useful but may be ambiguous because

such properties of particles depend on material properties and hence on composition. Thus, most approaches to characterization of aerosol properties can be thought of as successive approximations to the true descriptions. At the limit are various techniques for characterization of individual particles by collection and electron microscopy or by single-particle mass spectrometry. There is thus an increasing tendency to develop and apply methods for the analysis of individual particles and, where possible, to use real-time physical and chemical methods for their characterization. As a consequence, a rich suite of new instruments has become available in the past several years for the characterization of atmospheric aerosol particles. Yet even such approaches are not free of concerns such as changes in particle composition and morphology upon collection and exposure to high vacuum in electron microscopy or fragmentation and selective ionization in single-particle mass spectrometry. Much work therefore remains both in developing techniques for characterization of size-dependent aerosol composition and in application of these techniques to characterize ambient aerosols.

### 4.04.3.1 Particle Sizes, Size Distributions, and Number and Mass Concentrations

Measurements of the total number concentration of aerosol particles and their size distributions are necessarily linked. Such measurements are challenging for several reasons. Counting numbers of particles that span a size range of several orders of magnitude is difficult and requires highly specialized instruments (Table 2). Additionally, atmospheric particles commonly contain water and other volatile or semi-volatile components that can be lost if the particle is heated or placed into a vacuum or irradiated as in an electron microscope. Conversely, hygroscopic particles can gain water if the RH increases, and any such change in water content changes particle sizes and properties.

Particle sizes are commonly reported at low RH to obviate sensitivity to the RH. However, although dry size is a more intrinsic particle property, it does not compare with the actual size in the ambient atmosphere and it is this actual size that is important for light scattering, for example. A further issue is nonsphericity and how it is characterized. Dry particles are commonly irregularly shaped, not compact, or both. Drying of particles on sampling can lead to further problems such as loss during sampling because of lack of adhesion to impactor surfaces, possible fracture of particles, and loss of volatiles. There are also questions of obtaining representative number of measurements to provide reliable values, especially for larger particles that may be sparse as well as difficult to sample.

**Table 2** Key methods for determination of aerosol properties.

| Quantity or property | Method or instrument | Description | On line?[a] |
|---|---|---|---|
| **Size-related properties, size distributions** | | | |
| Aerodynamic diameter | Impactor | Inertial penetration of particle through slip stream as air flow is curved around impactor plate | + |
| Geometric diameter | Transmission electron microscope (TEM) | Use of electron beam to directly measure size, shape, aggregation, composition of dried collected particles | − |
| Diffusion diameter | Diffusion battery | Air is passed through a long, narrow tube as laminar flow; small particles diffuse to and stick to wall; larger particles penetrate further longitudinally along tube; size distribution is obtained by inverting distance distribution | + |
| Mobility diameter | Differential mobility analyzer (DMA) | Separation by differential mobility of charged particles in electric field | + |
| Optical diameter | Optical particle counter (OPC) | Invert angular distribution of light scattering; requires index of refraction | + |
| Condensation nucleus (CN) concentration | CN counter | Expose particles to a high supersaturation, typically of butanol vapor; particles grow to size where they can be optically detected and counted | + |
| Cloud condensation nucleus (CCN) concentration | CCN counter or spectrometer | Measurements commonly made at supersaturations between 0.1% and 1%. Particles activated at the applied supersaturation grow to sizes where they can be optically detected and counted | + |
| **Composition-related properties** | | | |
| Particle mass concentration | Weighing | Filter collection followed by weighing | − |
| | Microbalance | Impacting on oscillator and monitoring frequency change | + |
| Composition of bulk particulate matter | Beta gauge | Filtration followed by attenuation of beta rays | + |
| | Filter collection, Impactor | Chemical analysis of collected sample br (colorimetric, ion-chromatography, mass spectrometry, element-specific analysis, etc.) | − |

(continued)

**Table 2** (continued).

| Quantity or property | Method or instrument | Description | On line?[a] |
|---|---|---|---|
| | Particle into liquid sampler | Exposure to supersaturated vapor; impaction followed by on-line analysis (ion chromatography, total organic carbon by photoxidation followed by conductivity) | + |
| Composition of individual particles | TEM[b] | Energy-dispersive x-ray analysis (EDS); electron energy-loss spectrometry (EELS) | – |
| | Single-particle mass spectrometer | Vaporization, ionization, and time-of-flight mass spectrometry | + |
| Morphology and aggregation state | Electron microscopy (TEM, SEM[c]) | Impaction or filtration followed by analysis | – |
| *Optical properties* | | | |
| Particle light-scattering coefficient | Integrating nephelometer | Scattering intensity from particles within a defined light path as a function of wavelength | + |
| Particle light-absorption coefficient | Absorption photometer | Measurement of light attenuation through a filter | + |
| | Photoacoustic | Measurement of absorbed power as sound intensity | + |
| Particle light-extinction coefficient | Long-path transmittance | Beer's law measurement | + |
| | Ring-down laser | Decrease of laser power due to aerosol extinction in laser-resonant cavity | + |
| Aerosol particle optical thickness (AOT) | Sun photometry | Beer's law using sun as light source | + |
| | Satellite photometry | Inversion of scattered light to infer optical thickness | + |
| 180° backscatter coefficient | Lidar | Intensity as function of time converted to distance by speed of light | + |
| *Hygroscopicity* | | | |
| Hygroscopic growth factor | Increase of any aerosol property or phase change with increasing RH | Light scattering, mobility diameter, etc. | + |

[a] Refers to the ability to obtain a measurement in or near real time.   [b] TEM—transmission electron microscope.   [c] SEM—scanning electron microscope.

Several measures of size are commonly used (McMurry, 2000). To a great extent, the choice depends on the type of instrument utilized to make the measurement:

(i) Geometric diameter (or radius) is perhaps the most intuitive measure of particle size, but it is difficult to determine and, for non-spherical particles, it is not possible to characterize by a single number.

(ii) Aerodynamic diameter characterizes particles by their responses to air flow and is expressed as the diameter of a sphere of density $1 \mathrm{~g} \mathrm{~cm}^{-3}$ having the same aerodynamic property. Aerodynamic size is relevant, for example, to the deposition of particles in lungs.

(iii) Electrical mobility sizing reports the diameter of a sphere having the same diffusion coefficient as the particle.

(iv) Optical sizing implicitly depends on assumed sphericity and index of refraction. Table 2 summarizes selected characteristics of the several types of measurements. In all cases the hygroscopicity and thus RH can strongly affect the measured sizes.

Measurements can be made for particles having diameters ranging from less than 1 nm to greater than 10 μm, and real-time measurements extend over almost that entire range. The instruments are based on widely differing principles and produce data that must be evaluated accordingly.

Depending on the goals, a variety of procedures can be used to facilitate particle detection. In condensation particle counters (also called condensation nucleus (CN) and Aitken nucleus counters), particles as small as 3 nm diameter are detected through size enhancement and optical detection. The particles are introduced into a chamber that contains a supersaturated vapor (water or n-butyl alcohol is commonly used). Condensation on the particles enlarges them typically to several micrometers so that they can be readily detected and counted as they pass through the chamber.

Electrical measurements are widely used (see Flagan, 1998 for a review). Size separation is obtained by charging the particles by exposing them to ions and then flowing them into an electric field between charged coaxial cylinders or parallel plates. Particle concentrations are determined by their current or by counting as a function of size. A size spectrum may be determined in minutes. A single size cut may be sampled for further study, for example, to determine its response to change in RH. Concerns include multiple charging with resultant mis-classification of particles, and uniformity or constancy of charging efficiency as a function of size and composition.

Optical counters depend on measuring light scattered from particles as they traverse a defined path. Such instruments can provide rapid, on-line measurements of the size distribution of particles having diameter larger than ~0.2 μm for visible light (see Gebhart, 1993 for a review). Error can arise from the inversion of the optical signal to a size distribution because the differential light-scattering coefficient is not a monotonic function of size. Additionally, particle properties such as shape and refractive index can complicate the measurements, as can changes in water content in response to heating by the analyzing light beam and RH changes due to mixing.

In plots of particle number concentrations as a function of radius or diameter, the concentration is plotted as the dependent variable against particle size, nominally the independent variable. However, there is commonly as much or more uncertainty in the radius or diameter as in the number concentration within a given size range. Likewise, the separation of particles into size classes is generally not a sharp step function but rather extends over a considerable spread in diameter. Moreover, a rather small error in diameter translates into a much larger error in mass, given that mass varies as the cube of the diameter. For example, a 25% error in diameter translates to almost a factor of two error in mass.

An example of the wealth of information contained in state-of-the-art measurements of aerosol size distributions is shown in Figure 8. Here the distribution of number concentration is shown as a function of particle size versus time of day over a 16 h period. A nucleation event starting at 8 a.m. is indicated by the abrupt increase in concentration of the smallest particles, which are about 3 nm in diameter and at the limit of

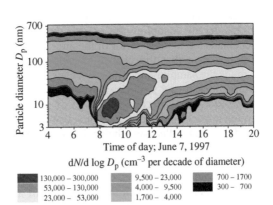

**Figure 8** Contour diagram of aerosol particle concentration as a function of particle diameter (ordinate) and time of day (abscissa). The number concentration, $dN/d \log D_p$, in $\mathrm{cm}^{-3}$ per decade of diameter, is represented by the colors and contour lines. Note the evolution of the particle-size distribution during the day and in particular the nucleation event commencing about 08:00 hours. Measurements were at Melpitz Germany, 50 km NE of Leipzig, Germany (reproduced by permission of American Geophysical Union from *Geophys. Res. Lett.* **2000**, *27*, 3325–3328).

sensitivity of the mobility analyzer used for the measurements. Concentrations of these newly formed particles increased over the next two hours but then decreased, possibly because of growth of particles out of the smallest sizes by condensation. While a detailed interpretation of these results must also consider vertical and horizontal transport, data such as these show the richness of the dynamics of the aerosol over even the course of a single day.

Determining concentrations of cloud condensation nuclei (CCN, particles that activate to form cloud droplets at a given supersaturation; cf. Section 4.04.7.3) at a specified supersaturation and the dependence of CCN concentration on supersaturation is important but difficult (see Hudson, 1993, for a review). The goal in both measurements is to determine the number and size of particles that can contribute to cloud droplet formation. The principle is to maintain a specified supersaturation of water vapor and count the number of particles activated to cloud droplets at that supersaturation. Some instruments establish the desired supersaturation through the use of diffusion chambers that contain a thermal gradient. Others use heated, wetted diffusion tubes or parallel wetted surfaces held at different temperatures through or between which the aerosol is made to pass. In all cases, CCN measurements are challenging, and virtually all instruments are custom-made.

It is often important in atmospheric studies to make measurements aloft to determine vertical or horizontal profiles and aerosol evolution. Thus, it is necessary to collect samples and make measurements from aircraft, which present special challenges because of their rapid movement through the air. Sampling must occur through an inlet connected to the exterior of the aircraft. Care must be taken that the change in velocity that occurs in the sampling inlet does not cause serious perturbations to the size distributions or particle compositions. A key concern is loss of particles through impaction to surfaces, which can result in size-dependent collection efficiencies. Changes in composition can occur through heating as the air is compressed in the inlet, with a potential loss of volatiles, most prominent of which is water. Although measurements on and from aircraft are widely used and are of critical importance, many problems remain (Blomquist *et al.*, 2001; Tyndall *et al.*, 2003).

Certain measurements cannot be made in the field, and for these it is necessary to collect samples for transport to the sorts of instruments that are available only in laboratories. Collection of size-segregated samples can be achieved by the use of a cascade impactor, which consists of a series of apertures of decreasing diameter, each of which is above a matched, stacked collection plate. Particles in the air passing through the apertures, if they have sufficient inertia, strike the collection plates and are collected there for subsequent chemical or physical analysis. Smaller particles flow over and around the plates and pass through a smaller aperture, perhaps to be collected on its collection plate, and so on. The result is a set of size-classified samples, each with its own size cutoff.

Larger particles (typically greater than 1 $\mu$m) can be segregated by use of cyclones or single-stage impactors. In cyclones the air is introduced tangentially into a conical or cylindrical device. Centrifugal force then carries the particles to the walls, where liquid particles adhere and solid particles above a certain size drop onto a collector; smaller particles exit through the top. The size cut of the device can be changed by varying the velocity of the air introduced (Hinds, 1999). Single-stage impactors (Marple and Willeke, 1976) separate particles according to their stopping distance, which depends on aerodynamic diameter. Many other types of particle counters and sizers have been developed for specialized use (see McMurry, 2000, for a review).

### 4.04.3.2 Composition

#### 4.04.3.2.1 Bulk composition

Many aerosol properties depend on composition. Also, attribution of particulate matter to sources and development of strategies to reduce particle concentrations require knowledge of the composition. For these reasons, much effort has been devoted to chemical analysis. Inevitably, analysis of composition involves compromises involving simplicity versus complexity, time resolution, size resolution, chemical speciation, and the like. The composition of particulate matter of aerosols has commonly been measured by collection of large numbers of particles on a filter followed by chemical analysis of the filter extract, yielding the so-called "bulk composition" of the particulate matter. The filter collector is generally preceded by a device to limit the particle size to a range of interest or at least below some upper size limit. The particulate mass and composition of the bulk aerosol sample are then determined by any of a wide variety of analytical techniques. With modern instrumental methods, highly sensitive analyses are possible, and such measurements have been widely used to interpret aerosol properties and effects.

Bulk sampling and analysis has the advantage of simplicity and reproducibility, although it can suffer from artifacts that result from condensation of volatile gases on the filter or from evaporation of particulate material subsequent to deposition on the filter. Further disadvantages are that the measurements are not immediately available, limiting their utility for real-time studies of aerosol processes, and also they provide little information on the size dependence of properties and the variation of properties within size classes.

Determination of mass concentration of particulate matter by weighing of filters is time consuming and labor intensive because of the low masses involved, but such determinations are required in many jurisdictions for the purpose of determining compliance with air quality regulations. They are also useful for providing an overall picture of the substances responsible for the aerosol loading at a given location and time. The results permit comparisons of the time evolution and spatial distribution of the aerosol composition. An example of such results is shown in Figure 9, which displays mass loading and composition of particulate matter having aerodynamic diameter less than 2.5 μm (PM$_{2.5}$) at several sites across the United States. The generally greater concentrations in urban areas than in nearby nonurban areas is evident, as is the prevalence of sulfate in the eastern part of the country compared to carbonaceous materials and nitrate in the western part. Measurements such as these suggest a more-or-less widespread regional aerosol, with incremental contributions from proximate local sources.

An example of the information obtainable from measurements of aerosol particle composition as a function of size is given in Figure 10. In this study of the evolution of aerosol composition in an urban air-shed, a multi-stage impactor was used to separate particles by size for the determination of size-dependent composition. The particle mass concentration increases as air moves inland from Santa Catalina Island to Long Beach, Fullerton, and finally Riverside, as do substances such as nitrate. Differences in particle composition as a function of size at a given site and from site to site are evident.

Detailed information on the molecular composition of the organic component of particulate matter is provided by gas-chromatographic separation followed by mass-spectrometric analysis (GC-MS). Rogge *et al.* (1993) used GC-MS to quantify concentrations of specific organic compounds in aerosol in Los Angeles, CA and vicinity over a one-year period. This approach gives a detailed picture of the organic component of the aerosol, albeit of a fairly small fraction of the total organic matter and an even smaller fraction of the total particulate mass. For example, the annual mean fine particle mass (aerodynamic diameter <2.1 μm) for 1982 at Rubidoux (near Riverside, downwind of the Los Angeles urban area) was 42 μg m$^{-3}$, of which ~7 μg m$^{-3}$ was organic material. Of this organic material, 4 μg m$^{-3}$ was

elutable in gas chromatography, but only 1.1 μg m$^{-3}$ was resolvable into specific peaks. Identified organic species comprised 70% of the resolvable organics or 8% of the total organic matter. Key categories were aliphatic dicarboxylic acids, *n*-alkanoic acids, aromatic polycarboxylic acids, *n*-alkanes, diterpenoid acids, polycyclic aromatic hydrocarbons, and nitrogen-containing compounds such as isoquinoline. Within each of these categories, ~20 compounds were quantified. For example, there were 22 quantified *n*-alkanoic acids, of which the most abundant was the fatty acid hexadecanoic (palmitic) acid followed by octadecanoic (stearic) acid, with quantification of compounds having concentrations as low as 0.5 ng m$^{-3}$. Primary organic aerosol constituents exhibited high winter and low summer concentrations; in contrast, aliphatic dicarboxylic acids of possible secondary origin showed a reverse pattern with high concentrations in late spring and early summer. Obtaining data such as these requires painstaking effort. However, such data provide information that cannot be obtained in any other way on the composition and sources of organic aerosols and thereby serve as tests of understanding of the processes resulting in these materials.

Finally, real-time bulk measurements include an approach in which the frequency of an oscillator such as a quartz crystal is altered as mass is deposited on the oscillating element. Surrogate measurements are also employed, such as inference of mass concentration from the decreasing transmission of electrons from a beta source to a detector as mass is deposited on an intervening substrate (see Gebhart, 1993, for a review).

### 4.04.3.2.2 *Off-line measurements of single-particle compositions*

For many purposes it is important to know the actual chemical species, and determining speciation from bulk compositions is a formidable challenge, especially when the aerosol contains particles of different composition, as invariably occurs in ambient air. At best, bulk analysis yields accurate information on the actual species present only for major particle types having simple compositions. Consequently, there has been an increasing interest in the analysis of individual aerosol particles. However, the attendant problems in such analyses are formidable because of the small sizes and, in many cases, the chemical complexity of individual particles.

A unique and powerful approach to obtaining information regarding individual particles has been the use of focused electron-beam instruments, in large part because great spatial

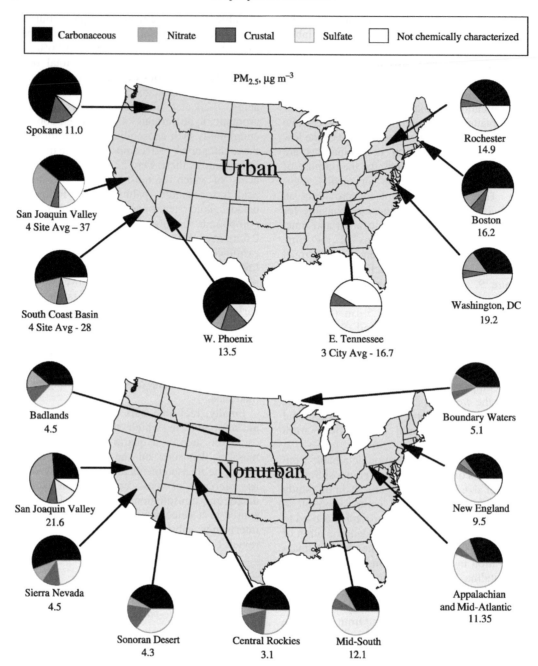

**Figure 9** Annual-average composition and mass concentration ($\mu g\ m^{-3}$) of particulate matter with aerodynamic diameter $<2.5\ \mu m$ at selected urban and nonurban locations in the United States. Concentrations at each location are based on at least one year of monitoring. Urban data are based on one site per city unless otherwise noted; nonurban data represent an average of two or more sites located in the same region except for Sierra Nevada and Badlands. White segments indicate that the sum of the constituents (as determined by separate analyses) was less than the gravimetrically determined mass concentration because certain constituents were not determined (e.g., absence of carbon and nitrate analyses for Tennessee sites) or for other technical reasons (after EPA, 1998).

resolution is possible, so that the beams can be focused into the smallest of particles. Indeed, it is commonly possible to study various parts of individual particles, in effect dissecting them chemically and structurally. The electron microprobe (EMP) was the first instrument that permitted quantitative analysis of individual particles (Armstrong and Buseck, 1975), and it continues to be widely used (Debock *et al.*, 1994; Katrinak *et al.*, 1995; Ebert *et al.*, 2000; Liu *et al.*, 2000). The scanning electron microscope (SEM) provides much the same chemical information while also allowing details of the particle morphology to be examined. A strength of electron-beam

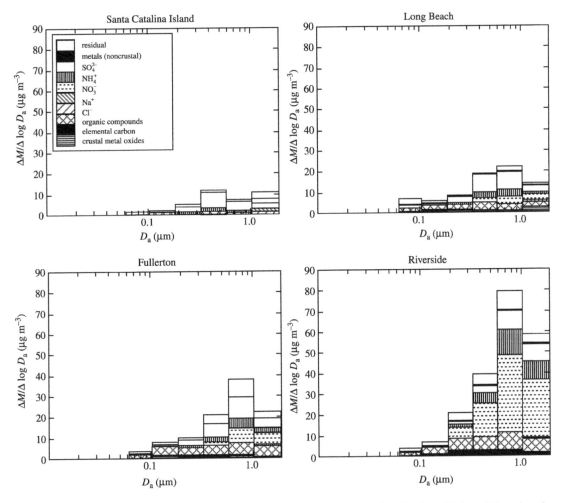

**Figure 10** Size-dependent composition of aerosol particles for several sites in the vicinity of Los Angeles, California over several days in September and October 1996. Particles were sampled with a cascade impactor preceded by a cyclone separator with a cut size of 1.8 μm aerodynamic diameter (Hughes *et al.*, 1999), (reproduced by permission of American Chemical Society from *Environ. Sci. Technol.* **1999**, *33*, 3506–3515).

techniques is their ability to distinguish between particles that consist of a single phase and particles that consist of two or more phases, the latter indicative of an internal mixture of two or more substances.

Both the EMP and SEM have the great benefit that they can be automated for particle analysis (Kim and Hopke, 1988; Germani and Buseck, 1991; Laskin and Cowin, 2001). After establishing certain intensity thresholds, either instrument can be set so that the electron beam scans across a sample grid until it finds an area where the threshold is exceeded. At that point, it goes into analysis mode, scanning back and forth at close intervals across the particle, thereby defining its boundaries. It also can be set to collect emitted X-ray signals that are produced through excitation from the electron beam. These fluorescent X-rays have energies and wavelengths that are characteristic of the elements in the particle. Either

energy-dispersive (EDS) or wavelength-dispersive (WDS) spectrometers may be used for this purpose. The former collect the entire spectrum at one time and so are faster, whereas the latter, which scan across the spectrum, are slower but produce more accurate results. Both types of spectrometers are used, the choice depending on the types of information that is desired.

The transmission electron microscope (TEM) makes possible a far greater degree of sophistication. Not only can it be used to examine smaller particles, but structures and intergrowths can also be imaged directly. For example, TEM images have shown that the great majority of sulfate aerosol particles contain soot inclusions in samples collected above the North Atlantic ocean during a pollution episode, and that even in the clean atmosphere above the Southern Ocean up to half of the sulfate particles contain soot (Buseck and Pósfai, 1999; Pósfai *et al.*, 1999).

Modern TEMs can produce beams with diameter smaller than 1 nm, and thus particles of comparable size can be analyzed for both their composition and structure (Figures 4 and 11). Compositions are determined by either of two methods: EDS, such as is described above, and electron energy-loss spectroscopy (EELS). For the latter, the electrons that pass through the particle are monitored. Some fraction will have interacted inelastically with the particle, transferring energy in the process. The amount of energy that is lost by the electron beam depends on the energy levels and thus types of atoms in the particle.

By monitoring the energy lost by these electrons one can establish the particle composition. Although the signal is harder to quantify than an EDS signal, EELS is applicable to a greater range of elements—all those heavier than lithium, versus sodium, the normal lower limit atomic number for EDS quantification (Garvie *et al.*, 1994; Garvie and Buseck, 1998). Additionally, these results are obtained from areas with dimensions of the cross-section of the electron beam, which, as pointed out above, can be less than 1 nm. EDS signals, in contrast, arise from larger "excited volumes" that can be several times greater than the electron-beam diameter.

The measurements presented so far can provide information on composition and morphology. However, complete characterization of any nonamorphous material also requires knowledge of its crystal structure. Using the TEM, one can obtain electron-diffraction patterns that reveal the crystalline makeup of the particle, thereby providing complete information about the material. In addition, it is possible to image surface coatings, aggregated phases, and, in favorable cases, recognize phases that are frozen in various stages in incomplete reactions. This uniquely powerful method of studying particles also has disadvantages: it is used off-line and is labor intensive because it has, as of early 2000s, not been automated. Because of the requirement for vacuum conditions needed in such analyses (a requirement that also applies to SEM and EMP), particles that are unstable in a vacuum, notably those that occur as aqueous solutions in the atmosphere, may dehydrate or decompose. This technique requires specialized equipment and skilled personnel. Nonetheless, it can provide size, structure, composition, mixing (aggregation) state, and information about coatings on individual particles.

Several types of off-line mass spectrometric techniques have been employed, but they are largely being superseded by the on-line methods described below. Secondary ion mass spectrometry (SIMS) has been used to obtain isotopic data on individual particles (Van Ham *et al.*, 2002). Laser microprobe mass spectrometry (LAMMS) uses a laser beam to excite and desorb the particle or parts of the particle, after which it is passed into a mass spectrometer. By using different power densities it is possible to obtain multiple spectra from given particles and thus get a sense of surface coatings or zonal structures. These methods exhibit the same problem as mass-spectrometric methods (see below), namely, that the fragmentation results in difficulties in inferring the molecular composition of the material. Nonetheless, interesting results have been obtained (Dierck *et al.*, 1992; Hinz *et al.*, 1994; Hara *et al.*, 1996).

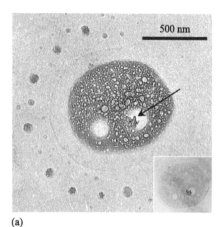

(a)

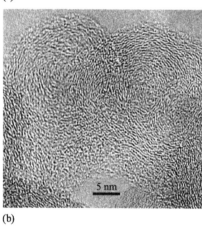

(b)

**Figure 11** (a) TEM images of a single particle consisting mainly of $(NH_4)_2SO_4$ with a soot inclusion, collected from a clean air marine mass near Tasmania. Rings of small $(NH_4)_2SO_4$ crystals and halos formed as the sulfate particle dehydrated within the TEM. The dimensions of such halos can be used to distinguish among particles that had different water contents while airborne. The arrow points to a soot aggregate (source Buseck and Pósfai, 1999). (b) High-resolution TEM image of a similar soot aggregate showing the curved, onion-like disordered graphitic (graphene) layers that are typical of the nanospheres of soot aggregates of smoke from a brush fire in South Africa (Li *et al.*, 2003) (reproduced by permission of American Geophysical Union from *J. Geophys. Res.* **2003**, *108*, 8484).

#### 4.04.3.2.3 On-line measurements

There has been growing emphasis on the development and use of on-line measurements, which provide an abundance of data in real time. Short-term variations in aerosol composition can reveal great detail, as shown in Figure 12, which illustrates marked differences in the composition of the aerosol at an urban site over periods as short as tens of minutes. The mass concentration was measured by an oscillating microbalance whose frequency was altered by deposition of material. Ionic species were measured by subjecting the aerosol to supersaturation, which greatly increased particle sizes and permitted the collection of droplets, followed by analysis by ion chromatography. On some days, the mass was dominated by sulfate, with strong correlation between total mass and sulfate concentrations; on others, sulfate was a much lower fraction of the total mass. Nitrate exhibited short-term excursions on several days, with peak concentrations greatly exceeding the mean. The concentration of acid (hydrogen ion) in the aerosol, which is not measured by the ion-chromatographic technique, is inferred from the difference between the equivalent concentration of sulfate plus nitrate minus ammonium. Clearly, there is a wealth of information in high-frequency measurements that is not revealed

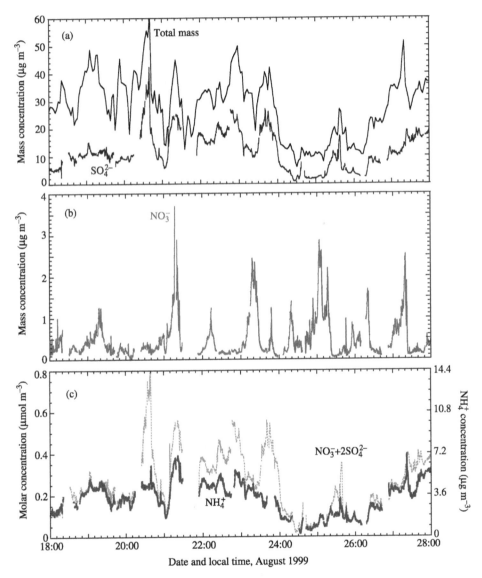

**Figure 12** Concentration of particle mass (panel a) and of major ionic species (panels a and b) in sub-2.5-$\mu$m diameter particles in Atlanta, Georgia during August 18–27, 1999. Panel (c) compares equivalent concentration of anions (sulfate plus nitrate) to that of the ammonium cation. The left axis shows the molar concentrations; the right axis shows mass concentration of ammonium. Time resolution of the measurements is 4 min, with a sample taken every 8 min (after Weber *et al.*, 2001b).

in traditional 24 h sampling. Information such as this can be used to infer mechanisms of aerosol formation. For example, bursts of nitrate generally occurred in late morning, suggesting a photochemical source together with increases in precursors associated with morning traffic. Such information may also be useful in inferring substances responsible for the health impairment produced by aerosols. High-frequency variations in particle composition like those shown in the figure also have implications in consideration of dry deposition of particles to vegetation or terrestrial surfaces. There is a likelihood of strong correlations between concentrations and deposition fluxes, which also exhibit diurnal cycles. Approaches to evaluating dry deposition fluxes based on daily or weekly average aerosol concentrations must therefore be viewed with caution.

Many on-line measurements use one or another type of mass spectrometry, but there are significant differences in their details, depending on the goals. At the time of writing, most current approaches use time-of-flight (TOF) mass spectrometers, in which case the entire mass spectrum for a single particle is obtained almost instantaneously. In contrast, a quadrupole mass filter can determine only the signal corresponding to a single $m/e$ ratio. Particles are introduced into the TOF chamber of the spectrometer, generally with several stages of differential pumping and often with focusing of the aerosol particle beam.

Particles are detected by light scattering and, in some instruments, sized by their velocity. Particles thus detected are subjected to vaporization and ionization under the influence of an intense laser beam, and the masses of the ionized fragments are determined by mass spectrometry (Figure 13) (see Suess and Prather (1999) and Noble and Prather (2000) for reviews). Depending on design, instruments may measure only positively or negatively charged ions from a given particle. With instruments having two mass spectrometers, it is possible to determine both the positively and negatively charged ions from a single particle, thereby gaining a much more detailed picture of particle composition. Measurements are rapid, so that many particles can be analyzed in short times, ten or more per second. Measurements can be made for particles with a wide range of volatilities, and the patterns for simple materials, including organics, produce spectra that can be used as "fingerprints." There are, however, limitations. The fragmentation produced by the laser beam results in complex spectra that can be difficult to interpret, especially for aggregated, coated, or organic particles, so that speciation may be precluded. The mass spectrum can be dominated by minor species, for example, there is extraordinary sensitivity to transition metals. Also, reactions and electron transfer can occur within the plasma produced by the laser, with resultant preferential ionization of certain materials. Inhomogeneities within the laser beam

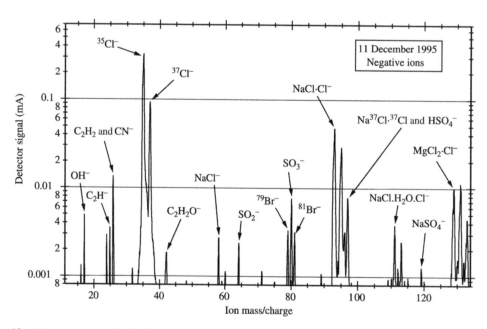

**Figure 13** Mass spectrum of a single marine aerosol particle (collected at Cape Grim, Australia). It shows both the rich chemical detail as well as the complex interpretative back-calculation problem connected with trying to sort out which species were present in such a particle; it presumably contained sea salt, sulfate, and organic material (Murphy *et al.*, 1998; reproduced by permission of Nature Publishing Group from *Nature*, **1998**, *392*, 62–65).

and fluctuations in beam power also contribute to measurement uncertainty.

A newly developed aerosol mass spectrometer (Jayne *et al.*, 2000; Allan *et al.*, 2003) deposits size-selected particles on a heated surface so that volatile and semi-volatile materials are vaporized. These are then analyzed quantitatively on line. The instrument is able to provide quantitative, short-time-resolution measurements of the compositions of particles in selected size ranges. However, the instrument has poor sensitivity for refractory materials such as mineral dust and provides no information on particle-to-particle variability in composition.

Despite their limitations, on-line analyses are becoming widely utilized. The measurements are sensitive and rapid, and can be made using moderately portable instruments. In some instruments the particles are sized by light scattering or by their velocity so that both size and composition are obtained. However, even with the rich variety of instruments and approaches available for the study of individual aerosol particles, major challenges remain. The TEM allows determination of a uniquely wide range of particle characteristics but cannot be operated on-line or in automated mode. Mass spectrometers are unable to determine aggregation, compositional zoning, or the internal structure of a particle. Neither on-line nor off-line techniques are able to quantify water associated with particles.

### 4.04.3.3 Hygroscopicity

Many atmospheric aerosol particles are water soluble and hygroscopic (i.e., absorb water with increasing RH). Interaction with water is important in modifying the compositions and sizes of such particles. Hygroscopic materials may also exhibit phase transitions such as deliquescence and efflorescence (Tang and Munkelwitz, 1977; Martin, 2000).

Hygroscopic behavior has been well characterized in laboratory studies for a variety of materials, for example, ammonium sulfate (Figure 14), an important atmospheric material. When an initially dry particle is exposed to increasing RH it rapidly accretes water at the deliquescence point. If the RH increases further the particle continues to accrete water, consistent with the vapor pressure of water in equilibrium with the solution. The behavior of the solution at RH above the deliquescence point is consistent with the bulk thermodynamic properties of the solution. However, when the RH is lowered below the deliquescence point, rather than crystallize as would a bulk solution, the material in the particle remains as a supersaturated solution to RH well below the deliquescence point. The particle may or may not undergo a phase transition (efflorescence) to give up some or all of the water that has been taken up. For instance, crystalline ammonium sulfate deliquesces at 79.5% RH at 298 K, but it effloresces at a much lower RH, 35% (Tang and Munkelwitz, 1977). This behavior is termed a hysteresis effect, and it can be repeated over many cycles.

There is a rich chemistry of phase transitions of aerosol particles that depends on composition and temperature (Tang *et al.*, 1995; Cziczo *et al.*, 1997; Martin, 2000; Martin *et al.*, 2001). In many cases there is no efflorescence transition at all, but the particle remains a solution essentially to

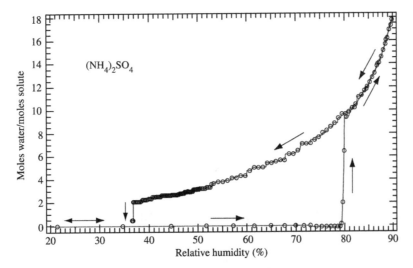

**Figure 14** Effect of RH on the uptake of water by an ammonium sulfate particle. The measurements were made for a single particle suspended in an electrostatic trap. Arrows denote particle response to change in RH in the indicated direction. Temperature, 25 °C (courtesy of T. Onasch, Brookhaven National Laboratory; cf. Tang (1996) and references cited therein).

zero RH. Studies of aerosols in the atmosphere have indicated that particles commonly occur as supersaturated solutions rather than in the thermodynamically more stable dry state (Rood *et al.*, 1989). Although the exact physical state of the particles is unknown, and despite the possibility that they are metastable and aqueous, low-RH particles are often described in the literature as "dry." Even if particles are crystalline they may have aqueous monolayers at the surface (Finlayson-Pitts and Hemminger, 2000).

An important consequence of the hygroscopicity of atmospheric aerosol particles is that the added mass from the uptake of water increases the light-scattering coefficient of the aerosol, thereby decreasing visibility and increasing the influence of the aerosol on the radiation budget of the planet. This enhancement may be substantial. For example, Figure 7 shows the dramatic increase in mass scattering efficiency with increasing RH that arises from the accretion of water to the particles. This dependence of the light-scattering coefficient on RH is responsible for much of the correlation between RH and haziness in industrial regions (Husar *et al.*, 1981). The figure also points out the strong dependence of mass-scattering efficiency on particle radius, with maximum efficiency when the particle diameter is near the wavelength of the light.

Field measurements examining the hygroscopic growth of particles have established that particles of a given dry size, selected with a mobility analyzer (Table 2), commonly exhibit different growth with increasing RH. This effect has been interpreted as indicating different composition, even for particles having the same low-RH radius. Figure 15 provides an example by showing mass spectra of individual particles, both of the same dry radius, which exhibited significantly different RH growth. The particle exhibiting greater growth, i.e., hygroscopicity, shows a much greater proportion of soluble species such as sulfate than the one that exhibited small growth, whose mass spectrum is dominated by elements typical of crustal minerals. Small anhydrous crustal particles inside hygroscopic sulfates and nitrates strongly influence particle efflorescence behavior (Han and Martin, 1999; Martin *et al.*, 2001).

### 4.04.3.4 Optical Properties

As noted in Section 4.04.2.3, key optical properties of aerosols are represented by the light-scattering coefficient, $\sigma_{sp}$, and light-absorption coefficient, $\sigma_{ap}$, both with reference to the particulate matter. The light-scattering coefficient of an aerosol can be measured by using nephelometry (Table 2). The integrating nephelometer (Anderson *et al.*, 1996) exposes an aerosol to a beam of light and collects the light over a large range of scattering angles. In principle, the instrument measures total light scattered by an aerosol, but there are instrumental issues with near-forward and near-backward radiation (Anderson and Ogren, 1998), so that the scattering is typically measured only over the angular range $10-170°$. The scattering by particulate matter is obtained by subtraction of Rayleigh scattering by air, which may be accurately calculated, measured using filtered air, or both. Commercial instruments allow determination of the backscattered fraction, i.e., the fraction of scattering in the back hemisphere, $90-170°$. This quantity is a measure of the angular distribution of the light scattering. Some such instruments also provide scattering coefficients at three different wavelengths. In view of the strong dependence of light scattering by hygroscopic particles on relative humidity, accurate control of RH is required. Typically, measurement is made at low RH to provide an RH-independent, intrinsic scattering coefficient, albeit with possible ambiguity over whether particles have effloresced or remain as metastable supersaturated solutions. Occasionally, measurements are made at a higher RH of interest, and additional information is gained if measurements are made as a function of RH (e.g., Sheridan *et al.*, 2001).

The absorption coefficient, $\sigma_{ap}$, of the particulate matter of an aerosol is typically measured by passing air through a filter while monitoring the diffuse transmittance of light through the filter, with the decrease in transmittance attributed to aerosol absorption (Bond *et al.*, 1999). Concerns over this approach arise because the absorption coefficient may be modified by the filter medium and by the influence of nonabsorbing aerosol on transmittance by the filter. An alternative approach is direct measurement of absorption by the aerosol by measuring the heating of air resulting from the absorption of radiant energy from an intense light beam, typically from a laser. If the laser beam is modulated at an audiofrequency, the absorbed power may be sensitively measured by a microphone as an audio-signal (photoacoustic method, Arnott *et al.* (1999)). Alternatively, the heating has been measured as a change in the local index of refraction (photothermal method, Lin and Campillo (1985)). A possible concern with these approaches is volatilization of aerosol constituents by the probe laser beam. The approach is nonselective; any material (particulate or gaseous) that absorbs at the wavelength of the probe laser will give rise to a signal. Yet another approach is to determine $\sigma_{ap}$ as the difference between $\sigma_{ep}$ and $\sigma_{sp}$. The extinction coefficient may be measured by long path (or folded mirror) transmittance or by change in the rate at which the power of a laser is reduced upon introduction of air into the laser cavity

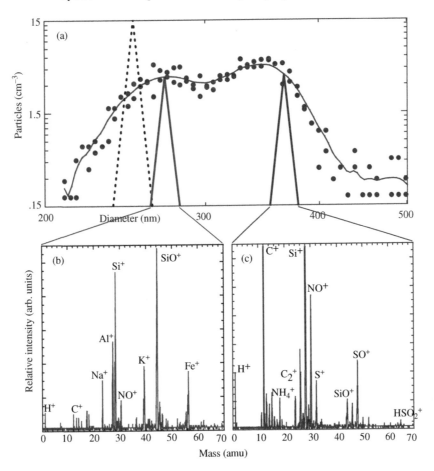

**Figure 15** Relation between aerosol particle size, humidity-dependent growth, and composition. (a) The black triangle shows (on a linear scale) the diameter range of dry particles that were selected for humidification study. The points and smooth fit to the data points show the size distribution of such particles after they were humidified to ca. 90% RH. The red and blue triangles (again, linear scale) show the diameter range, after humidification, of particles selected for mass spectrometric analysis. Mass spectra of typical particles exhibiting (b) low and (c) high humidification growth, respectively, are indicative of crustal material and of primary combustion aerosol with added sulfate or nitrate. Measurements were made at Jeju Island, Republic of South Korea, May, 2001 (Buzorius *et al.*, 2002); (reproduced by permission of American Geophysical Union from *Geophys. Res. Lett.* **2002**, *29*, 1974).

(cavity ring-down method, Strawa *et al.*, 2003). In both cases, the measured extinction must be corrected for scattering and possible absorption by gas molecules. A concern with the latter methods is that absorption is generally a small fraction of extinction, with resultant large uncertainty in the difference.

### 4.04.4 SPATIAL AND TEMPORAL VARIATION OF TROPOSPHERIC AEROSOLS

The nature, intensity, and spatial distribution of sources of tropospheric aerosols are highly variable, the removal processes (mainly in precipitation, but also through dry deposition, especially in arid environments) are intermittent, and the residence times of these aerosols are short, typically days to weeks. Consequently, the distribution of aerosols in the troposphere is heterogeneous in location, altitude, and time. This spatial variability has been documented by *in situ* measurements, often in large-scale field campaigns designed to characterize the amount of the material and determine its composition and microphysical properties. An example is given in Figure 16, which shows number and volume concentrations of aerosol particles measured on a transect from Darwin, Australia to Tokyo. All concentrations show short-scale fluctuations related to air-mass character. Moreover, the several traces exhibit little correlation, demonstrating substantial variation in aerosol microphysical properties. Despite the wealth of information from such *in situ* measurements, this sort of characterization of the distribution and properties of aerosols in three dimensions, and over distances up to the global scale, is impractical. For this reason

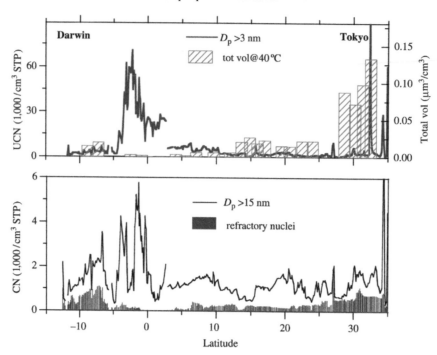

**Figure 16** Concentrations of aerosol particles having diameters ($D_p$) greater than 3 nm and 15 nm (note different scales), concentrations of particles having nonvolatile cores as determined by measurement after heating to 300 °C (refractory nuclei), and concentrations of aerosol particle dry volume (total volume at 40 °C). Measurements were made at an altitude near 8 km on a flight from Darwin to Tokyo on June 1, 1990, as part of the NASA GLOBE mission (Clarke and Kapustin (2002); reproduced by permission of American Meteorological Society from *J. Atmos. Sci.*, **2002**, *59*, 363).

remote sensing is necessary and is increasingly being applied; it is especially useful for indicating spatial and geographical distributions of aerosols. In this section we provide examples of different types of remote measurements.

### 4.04.4.1 Remote Sensing of Aerosol Using Passive Light Sources

A variety of remote-sensing approaches have been used to characterize the spatial distribution, motion, and, in some instances, evolution of aerosols in the atmosphere. Most measurements take advantage of the interaction of the aerosol with light in the visible or near-visible region of the spectrum, so that remote sensing typically yields an optical surrogate of the aerosol loading. It is useful to distinguish passive techniques, which rely on naturally available sources of radiation such as the sun (or moonlight or starlight) from active devices, which may involve intense light sources such as lasers. The geographical distribution of aerosol properties may be obtained from measurements at a network of fixed surface locations or from mobile sensors on aircraft or satellite.

An important remote-sensing approach for determining the local column-integrated

aerosol burden is sun-photometry. Several semi-equivalent terms are in common use. Column aerosol extinction is the vertical integral of aerosol particle extinction coefficient and is often denoted as aerosol optical depth or AOT (cf. Section 4.04.2.3). Total column extinction (gas plus particle) is measured in narrow wavelength bands by atmospheric transmittance of the direct solar beam using the Beer–Lambert law. Calibration of the sensor is achieved by measurements at high-elevation sites or by making the measurements during the course of a day during which the effective path length is changing with the solar zenith angle and extrapolation to zero path-length (Langley method, e.g., Harrison and Michalsky, 1994). The contribution of extinction by aerosol particles to total column extinction is obtained from the measured total column extinction by subtraction of Rayleigh scattering and any absorption by gases, for example, absorption due to water vapor or stratospheric ozone. As Rayleigh scattering can be calculated quite accurately for known surface pressure, such subtraction leads to substantial uncertainty only at low aerosol optical depth (Hansen and Travis, 1974; Michalsky *et al.*, 2001).

An example of the temporal dependence of aerosol loading is shown in Figure 17, which shows the variation of aerosol optical depth determined

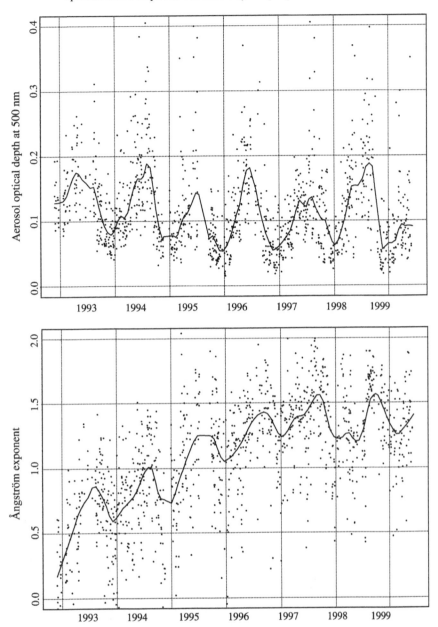

**Figure 17** Time series of AOT (top) and Ångström exponent (bottom) at the Department of Energy Atmospheric Radiation Measurement site in North Central Oklahoma over the period 1993–1999. Measurements are by sunphotometry. The points represent daily averages; the curves are smooth fits of the data to guide the eye (Michalsky *et al.*, 2001) (reproduced by permission of American Geophysical Union from *J. Geophys. Res.* **2001**, *106*, 12099–12107).

by sun-photometry over several years at a mid-continental North American location. A strong annual cycle is prominent. The increased aerosol loading in summer arises, in part, from greater photochemical production. The decrease over the early part of the decade, especially noticeable in winter, arises from the decrease of stratospheric aerosol produced during the eruption of the Mount Pinatubo volcano in the Philippines in June 1991, which injected much $SO_2$ into the stratosphere. The $H_2SO_4$ aerosol formed from this $SO_2$ was removed from the stratosphere over the next

several years by the normal downward circulation of stratospheric air. Superimposed on these longer-term trends is a strong day-to-day variation characteristic of tropospheric aerosols, whose loading is strongly influenced by synoptic-scale meteorology, i.e., transport over continental scales and removal in precipitation. Much of this tropospheric aerosol arises from anthropogenic emissions.

Figure 17 also shows the Ångström exponent (cf. Section 4.04.2) evaluated from the dependence of aerosol optical thickness, $\tau_{ep}$, on wavelength $\lambda$ as $\alpha = -\mathrm{d}\log\tau_{ep}/\mathrm{d}\log\lambda$. A greater Ångström exponent indicates smaller particles. The Ångström exponent increases over the time period as the relatively larger stratospheric aerosol particles are depleted. The higher Ångström exponent in summer than in winter indicates a greater proportion of smaller particles resulting from relatively recent formation by gas-to-particle conversion.

AERONET (Aerosol Robotic Network; Holben, 2003), which is coordinated by NASA (Holben *et al.*, 1998), is an important network of sunphotometers with near global coverage. This network measures time series of the AOT at some 200 locations worldwide, thereby providing the beginning of a global climatology. Additionally, the network serves as ground truth for measurements of AOT by satellite.

Satellite measurements have the advantage of providing global or near-global coverage, albeit with limited time coverage at a given location and with uncertainty arising from assumptions made in translating upwelling radiance to the AOT. Figure 18 provides an example of the global distribution of aerosols determined by satellite measurement of aerosol radiance. The technique is suitable only over homogeneous dark surfaces and is therefore limited to measurements over large water bodies. Although the satellite measurements provide daily global coverage, the requirement of rigorous exclusion of clouds results in rather sparse data recovery, so observations are typically pooled for a period such as one month. The radiance caused by aerosols is

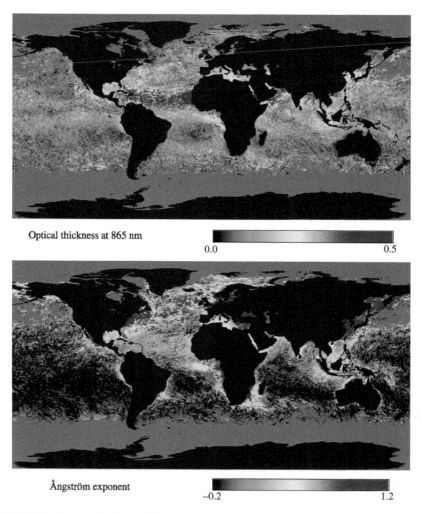

**Figure 18**   Global distribution of AOT $\tau$ at 865 nm (upper panel) and dependence on wavelength $\lambda$ as the Ångström exponent, $\alpha = -d \ln \tau / d \ln \lambda$ (lower panel) for June 1997. The distributions were derived from radiance measurements at 18-km resolution by the POLDER (POLarization and Directionality of the Earth's Reflectance) radiometer aboard the ADEOS (ADvanced Earth Observing Satellite) platform. Retrieval is limited to the atmosphere above water surfaces. Courtesy of Laboratoire d'Optique Atmospherique, Lille, France Laboratoire des Sciences du Climat et de l'Environnement, Gif sur Yvette, France; Centre National d'etudes Spatiales, Toulouse, France; and National Space Development Agency, Japan. For further information see http://earth-sciences.cnes.fr:8060/polder/ Mission. html and Deuzé *et al.* (1999).

converted by a model into the AOT, albeit with considerable uncertainty. Major patterns in the AOT can be readily discerned in the figure: dust plumes off deserts, plumes associated with biomass burning in the tropics, and plumes of industrial aerosol from North America, Europe, and Asia. Note the much greater aerosol loading in the mid-latitudes of the northern than southern hemisphere, which is much less influenced by industrial activities. The lower panel of Figure 18 shows the Ångström exponent. This plot readily distinguishes larger particles in dust plumes (typically greater than 1 μm diameter) from the smaller particles in the plumes resulting from biomass burning, for example westward of equatorial Africa and Indonesia, and those extending eastward from industrial regions.

### 4.04.4.2 Remote Sensing of Aerosol Using an Active Light Source

Lidar is a powerful method for imaging the spatial distribution of aerosols. Also known as laser radar, it allows detection of both aerosol concentration and cloud properties over widely varying scales and in great detail at high spatial and temporal resolutions. It is an active technique that works through the laser emission of high-intensity light along a given path and recording the time it takes backscattered light from aerosol particles to return to the detector. Because it uses light instead of radio waves, it is sensitive to much smaller particles than is radar, and so measures particles whose diameter is comparable to or greater than the wavelength of light. As the technique measures distance from the TOF, resolution can be adjusted by recording the signals at increasingly closely spaced time intervals, consistent with signal-to-noise considerations. Measurements can be made upward from the surface; upward, downward, or horizontally from aircraft; or downward from a satellite. It is possible to obtain signals along paths of arbitrary length, again consistent with signal-to-noise considerations. As the incident and scattered radiation are highly attenuated by optically thick clouds, the lidar images are generally restricted to the region between instrument and cloud. The technique is typically used to measure 180° backscattered signals, but it can be and has been used at other detection angles.

The backscattered signal from aerosol particles is attenuated by traveling (twice) through the air, and this attenuation must be considered in quantitative retrievals. As an optical technique, lidar relies on the optical properties of the aerosol particles, which are generally not known and must be assumed. The difficulty arises because the returned signal is proportional to the backscatter coefficient ($\beta_\pi$), whereas the desired aerosol property is the extinction coefficient, so a conversion coefficient must be obtained. Sometimes it is assumed that the backscatter to extinction ratio is constant. However, this ratio depends on aerosol microphysical properties, most importantly particle size and shape, so it can change along the lidar-viewing path because of changing properties of the aerosol or even variation in RH, which can affect particle size. One approach, which has not yet seen widespread use, is to use Raman scattering from air molecules to obtain path extinction and from its derivative with distance to obtain local extinction (Ferrare *et al.*, 1998a,b).

Remote detection by lidar allows vertical resolution of aerosol concentrations and inference of their transport in dynamic, inhomogeneous regions of the atmosphere. Because aerosol transport is a three-dimensional problem, data obtained from an airborne lidar system provide the opportunity to examine the vertical extent and distribution of aerosols on a regional scale. An example of its use in imaging tropospheric aerosols in a terrestrial environment is given in Figure 5, which shows the effects of mountainous terrain on trapping aerosols in valleys as well as down- and up-slope aerosol movement in response to winds along valley edges.

The remote Arctic troposphere provides another type of example of remote sensing with lidar. Surprisingly, this region is characterized by persistent, relatively dense haze. Lidar images clearly indicate the outlines of haze layers as well as the turbulence of relatively clean air. Strong haze is evident by the yellow and green layer in Figure 19(a). It is presumably produced by emissions from industries in the northern Eurasian landmass (Dreiling and Friederich, 1997; Khattatov *et al.*, 1997; Sirois and Barrie, 1999). The contrast with clearer air is shown in Figure 19(b), which reveals the fine layering in the troposphere plus contrast between the troposphere over ice pack and open water. The marine boundary layer (MBL) above pack ice in the Arctic is shallow and can be seen as the roughly horizontal line marking the change from green to blue at a height of ~200 m in the far left of Figure 19(b). Just to the right (north) of the pack ice is a region of open water, and above it the MBL rises to an altitude of ~400 m. Turbulence over the water produces a jagged character to the top of the MBL. This change in turbulence contrasts with its character over continental masses and oceans farther south, where the greatest atmospheric turbulence occurs over land. The white areas indicate low-level stratus clouds, with tops at an altitude of ~600 m. They are opaque to the lidar signal and thus show no interpretable internal structure. The atmosphere at an altitude of ~2 km is finely stratified and contains multiple layers,

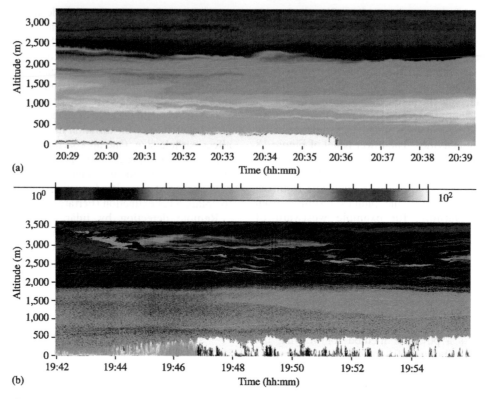

**Figure 19** Down-looking lidar images over pack ice, open water, and clouds in the Arctic. (a) Arctic haze (yellow and green) is evident below the delicately stratified upper troposphere. The white regions are clouds, which are opaque to lidar imaging. (b) The delicate layering of the Arctic troposphere is evident. The ice/water interface occurs at approximately 19:44 UTC, with ice on the left. Open water extends under the cloud cover (white area on right). The MBL is lowest over the pack ice and higher over the open water (cf. image in Curry *et al.*, 2000). Both images were acquired over the Beaufort Sea, northwest of Inuvik, NWT on April 8 and 27, 1998 (top and bottom, respectively) from aircraft at 4,200 m above sea level at an aircraft speed of $\sim$100 m s$^{-1}$. The color bar indicates the lidar backscatter ratio, which is approximately proportional to aerosol concentration; the *x*-axis reflects the passage of the aircraft recording the lidar signal, which was obtained at 1,064 nm using a 20-shot average (1 s); vertical resolution $-3.75$ m. The images were obtained by and are courtesy of K. Strawbridge, Meteorological Service of Canada.

some as thin as a few tens of meters, and each of which has an aerosol concentration different from those in layers immediately above and below. The high stability of the Arctic troposphere, particularly during the winter months, allows these layers to retain their physical integrity for days to weeks. These layers presumably formed as a result of long-range transport of pollutants.

Anthropogenic emissions can also be tracked using lidar. Scanning ground-based lidar systems are capable of fast azimuth- and elevation-scanning profiles of the lower troposphere and can measure such detail as the complicated turbulent activity of a point-source plume and its interaction with the planetary boundary layer (PBL). The largest coal-fired electric power plant in North America is located at Nanticoke, Ontario, Canada. Like many power plants, smelters, and other industrial plants that operate at high temperatures, it uses a tall stack (200 m in this case) for release of emissions into the atmosphere. The intent of this high release is to produce maximum dispersion of the emitted gases and aerosol particles. Figure 20 is a lidar image of its emissions, showing both distance traveled and turbulent intensity, the latter indicated by the irregular margins of the plume. When observed in real time, the plume can be seen to change shape and position continuously. The irregular shape of the plume can also be seen in cross-section in the inserts in Figure 20.

Lidar from space provides another promising technique for global remote sensing of aerosols. Its feasibility was demonstrated by measurements from the Space Shuttle Discovery in September, 1994, with the "Lidar In-space Technology Experiment" (LITE). The satellite instrument on that mission emitted laser pulses at 1,064 nm, 532 nm, and 355 nm at 10 Hz, mapping the vertical distribution of the aerosol from the return scatter. The mission and its results are described by Winker *et al.* (1996) and Strawbridge and Hoff (1996). An extensive interactive web page dealing with this mission (Winker, 1998)

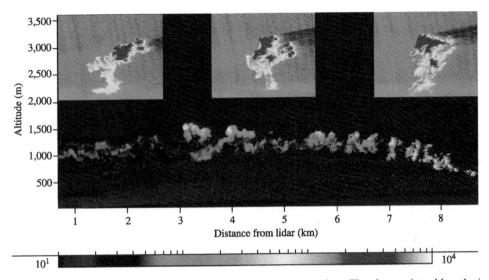

**Figure 20** Lidar images of an aerosol-rich plume emitted from a power plant. The plume, viewed lengthwise from the ground, is above the PBL at an altitude of ~0.3 km. The inserts are enlarged cross-sections of a similar plume, taken at 1 min intervals. The shapes of such sections change continuously with both position along the length of the plume and with time. Nanticoke power plant, Ontario, Canada. The lengthwise and cross-section images were taken on January 22 and 19, 2000, respectively at 1,064 nm, with scan speeds adjusted depending on distance from source and proximity of mobile lab to produce a full-scan image in less than 1 min. The images were obtained by and are courtesy of K. Strawbridge, Meteorological Service of Canada.

permits viewing results and downloading data. Follow-on instruments are being designed for continuous, unmanned operation.

## 4.04.5 AEROSOL PROCESSES

The principal processes that govern the concentration and properties of atmospheric aerosols are emission of aerosol particles and precursor gases, gas-to-particle conversion and other pertinent atmospheric chemical reactions, transport, and processes by which particles are removed from the atmosphere. There is a substantial literature on the characterization of these processes from laboratory studies and field measurements (cf. Section 4.04.1), so only a brief overview is provided here.

### 4.04.5.1 Emissions

Understanding the processes that control atmospheric aerosol concentrations and representing these processes in chemical transport models rests in large part on the accuracy of emissions inventories of aerosols and gaseous precursors. The most widely applied approach to developing such inventories is characterization of emissions per unit of activity (called "emission factors") combined with characterization of the intensity and geographic distribution of these activities. This approach is well developed for some gas-phase species. Emission of $SO_2$ from fossil fuel combustion provides an example. Most sulfur in

fossil fuel is emitted as $SO_2$, with only a small fraction (5–15%) going into bottom ash and fly ash (Benkovitz *et al.*, 1996). Good records of fossil-fuel consumption and sulfur content are maintained in most developed countries on at least an annual basis by location of the source for major combustion facilities. Therefore, with diligence emission inventories can be prepared, as was done by Benkovitz *et al.* (1996) using their gridded global emissions inventories for chemical transport models. The consumption data can be used to obtain typical seasonal and diurnal patterns of emission, but it is difficult to determine departures from these seasonal patterns on specific days, although there is little day-to-day variation in fossil-fuel combustion for electric power production. For regions where emission locations are not known, or for distributed emissions such as nitrogen oxides from vehicles, surrogate distributions of activity level, such as population density, are employed. For nitrogen oxides, additional uncertainty arises because much of the emission results from oxidation of nitrogen in air during high-temperature combustion processes. The extent of this reaction depends sensitively on combustion conditions. For all these reasons the uncertainties of regional and global emissions inventories are substantial.

Ammonia is another important precursor gas whose emission is not well quantified. Atmospheric ammonia results in large part from distributed sources such as animal husbandry, sewage treatment plants, and loss to the atmosphere associated with use of liquid ammonia as fertilizer. In regions of intense animal husbandry, concentrations of ammonia are commonly sufficiently high to be in excess of sulfate, resulting in substantial concentrations of ammonium nitrate (ten Brink *et al.*, 1997).

Characterization and quantification of emissions of particulate matter (primary aerosols) is far more challenging than for gaseous emissions. The difficulty arises because specifying particle emissions requires knowledge not only of the rate of emission of the mass of material but also of size distribution and composition as a function of size, whereas specification of mass emission rates suffices for trace gases. At present, knowledge is generally available at best only for mass emissions, with little information on size distribution or state of mixing of multiple aerosol species. Even mass emissions are much less certain than for gaseous species. For example, only a small fraction of $SO_2$ is oxidized to $H_2SO_4$ in combustion facilities, but this fraction varies with combustion conditions, and its determination is difficult because of the presence of substantial excess $SO_2$ that can be inadvertently oxidized during sampling. Careful measurements suggest that in modern electric generation facilities only about 1% of fuel sulfur is emitted as $H_2SO_4$ (or sulfate, being neutralized by co-emitted metal oxides), but such a figure is probably uncertain to a factor of 2 (Dietz and Wieser, 1983). Many European emission inventories place this figure at 5% (Iversen *et al.*, 1990), suggestive of perhaps a factor of 5 uncertainty.

The situation is still more uncertain for carbonaceous aerosols. Their primary emissions arise from combustion of fossil fuels, bio-fuels, agricultural burning, and uncontrolled wildfires. An important distinction is that between black or elemental carbon and organic carbon, but this is a coarse distinction that is not reproducible by different techniques (Schmid *et al.*, 2001). Mass emission rates per kilogram of fuel combusted are highly variable, depending on fuel type and combustion conditions (Cachier *et al.*, 1989; Cooke *et al.*, 1999; Jacobson *et al.*, 2000). Activity rates are also highly variable in space and time. For example, agricultural combustion is seasonal and sporadic even within the burning season, with decisions to burn influenced by variables such as time of day, wind, weather, and fuel moisture. Yet agricultural burning is a major contributor to submicrometer aerosol in agricultural regions even of developed regions such as Western Europe and the western United states. Biomass burning is a major source of aerosols in underdeveloped areas such as Africa and India and produces large amounts of fine particles (Crutzen and Andreae, 1990; Andreae, 1991; Li *et al.*, 2003; Pósfai *et al.*, 2003).

Characterizing the nature and magnitude of anthropogenic sources of aerosols in urban areas presents great challenges because of the many types of sources and their spatial and temporal variations. A major advance was made by Cass and colleagues in using chemical mass balance receptor modeling based on organic compounds together with detailed characterization of principal source types (Schauer *et al.*, 1996). Key source types were characterized by gas chromatographic separation followed by mass spectrometric analysis, and specific marker compounds were identified for each source type and their emission quantified relative to organic mass emissions. Quantification of these markers at receptor sites permitted attribution of the proportional mass of organic particulate matter at that site to the several source types. Application of this approach at four sites in and downwind of Los Angeles, CA allowed the investigators to ascribe ~85% of the organic fine particle mass (aerodynamic diameter <2 μm) to primary sources on an annual average basis. The largest primary contributions to fine-particle mass concentrations were diesel engine exhaust, paved-roads dust, gasoline-powered vehicle exhaust, emissions from food cooking and wood smoke, with smaller contributions from tire dust, plant fragments, natural-gas combustion aerosol, and cigarette smoke. Key to the approach is that the marker compounds employed are chemically stable in the atmosphere and without appreciable sources or chemical sinks. To be mathematically stable, the receptor modeling must rely on a highly over-determined set of marker compounds that are linearly independent in their emissions from different source types. Secondary organic aerosol represented typically less than 15% of the total fine organic mass concentration, even in the highly photochemically active Los Angeles area. This fine fraction was higher at a downwind site (Rubidoux, near Riverside) and indicated the increasing role of secondary organics with time subsequent to emission and mirroring the increased concentration of other secondary materials such as ammonium nitrate at this site.

## 4.04.5.2 Gas-to-particle Conversions and Other Atmospheric Reactions

Gas-to-particle conversion processes in the atmosphere consist of formation of low-vapor-pressure gases formed by atmospheric chemical reactions followed by new particle formation (nucleation), condensation of the low-vapor-pressure material on existing particles, or both. Chemical reaction producing the low-vapor-pressure product also occurs in cloud droplets, with the product remaining in the condensed phase of clear-air aerosol particles following cloud evaporation. Substances of intermediate vapor pressure may reversibly exchange between the gas and particle phases under conditions of changing temperature or humidity.

Clear-air processes require the formation of new, low-vapor-pressure materials such as $H_2SO_4$ formed by the gas-phase oxidation of $SO_2$ initiated by reaction with hydroxyl radical OH. Modeling the rates of atmospheric chemical reactions responsible for formation of this condensable material requires knowledge of the rate laws of the pertinent reactions, determined mainly from laboratory studies, and specification of the reagent concentrations and pertinent conditions. As the newly formed gaseous $H_2SO_4$ accumulates in the atmosphere, it is removed by diffusion to the surfaces of existing aerosol particles, where it condenses. The concentration of gas-phase $H_2SO_4$ quickly reaches a steady state determined by the rate of production and the time constant for diffusion to and condensation on existing aerosol particles.

Knowledge of the atmospheric concentration of OH (as well as $SO_2$) is necessary to quantify rates of production of gas-phase $H_2SO_4$ and thereby constrain interpretation of nucleation events involving this species. Considerable advances have been made recently in measurement of OH, which occurs in low concentrations in the atmosphere, typically $\sim 10^6$ molecules per cubic centimeter (Mount *et al.*, 1997; Spivakovsky *et al.*, 2000). However, this measurement requires specialized equipment and highly skilled personnel and thus cannot be made routinely.

Developments in the past few years have also resulted in the ability to measure concentrations of gas-phase $H_2SO_4$, again in the range of $10^6$ molecules per cubic centimeter (Weber *et al.*, 2001a). If gas-phase $H_2SO_4$ builds up to sufficiently high concentration, it may nucleate new particles by the process of binary nucleation that occurs in conjunction with condensation of water vapor. The strong affinity of the two substances lowers the free-energy barrier for nucleation, allowing nucleation to occur at lower vapor pressure of $H_2SO_4$ than otherwise. Ammonia is also thought to play a role in nucleation, co-condensing with both $H_2SO_4$ and $H_2O$ (ternary nucleation) with a still lower free-energy barrier (Korhonen *et al.*, 1999). The observation of high concentrations of particles in the 2.7–10 nm diameter range downwind of a penguin rookery on Macquarie Island in the Southern Ocean was interpreted as indicating formation and growth of new sulfate particles enhanced by the presence of ammonia (Weber *et al.*, 1998).

Nucleation of new particles has been demonstrated in laboratory studies from the oxidation of hydrocarbons to form partly oxidized material of lower vapor pressure (e.g., Griffin *et al.*, 1999; Koch *et al.*, 2002). This process also occurs in the atmosphere (O'Dowd *et al.*, 2002), where it is manifested by bursts of particles in the 4–6 nm diameter range. The rate of new particle formation in a forest canopy has been estimated by the eddy correlation method (Buzorius *et al.*, 1998), specifically from the covariance of upward vertical velocity and particle number concentration, which manifests occasional bursts of magnitude several times $10^7$ particles per square meter per second. Nucleation is suppressed where there is substantial pre-existing particulate matter that serves as a sink for condensation of the newly formed low-vapor-pressure gas. New particle formation is important as, together with coagulation and deposition, it governs the number concentration of aerosol particles, influencing the subsequent dynamics of the aerosol and ultimately the cloud droplet formation and precipitation. Key information requirements are the mechanism and rate of new particle formation for substances of interest and the controlling physical–chemical properties of the materials. The latter include the equilibrium vapor pressure of these materials, including effects of curvature of small particles, and kinetic properties such as mass-accommodation coefficients, all as a function of temperature.

Other clear-air, gas-to-particle conversion processes include uptake of acidic gases by basic aerosol (e.g., $SO_2$ uptake by sea salt or carbonate dust) and uptake of ammonia by acidic aerosol. Some such processes are reversible. An example is the co-condensation of $HNO_3$ and $NH_3$ to form ammonium nitrate, for which the equilibrium constant is rather strongly temperature dependent. An example of the release of species from the condensed to the gas phase is the uptake of $HNO_3$ by sea salt, resulting in release of HCl into the gas phase.

Uptake and release of water is an important special case of gas–particle exchange. The importance of water arises because of its ubiquity and abundance, sensitivity of equilibrium liquid water content to RH, and frequent rapid changes of RH in response to changes in temperature associated with vertical motions of air parcels (cf. Section 4.04.3.3 and Figure 14).

In-cloud processes are a second major class of chemical transformations of aerosol particles (cf. Section 4.04.7.3). Clouds are technically aerosols, a special class in which the particles consist mainly of liquid or solid water and the gas phase generally exhibits slight supersaturation with respect to the condensed phase, i.e., an RH slightly greater than 100%. Clouds form in the atmosphere mainly as a consequence of air parcels being cooled below the dew-point of water, the temperature at which water vapor in a given air parcel is saturated with respect to the liquid. Generally as an air parcel rises to lower pressure, thereby expanding, it is cooled as it does work on its surroundings. When the temperature reaches the dew-point, water vapor condenses on existing aerosol particles, which serve as the condensation nuclei for cloud droplet formation. The number of aerosol particles that are activated in this way to form cloud droplets during cloud formation depends on the cooling rate and the number concentration and properties of the aerosol particles. Whether or not a given aerosol particle is activated at a given supersaturation depends on its size and composition. Activation is favored for larger particles and for particles containing soluble materials, a consequence of the interplay of the free energy required to create new surface area and that gained by solute–solvent interaction (Pruppacher and Klett, 1997). During cloud formation, a given air parcel experiences a transient maximum in supersaturation as water condensation becomes favored by the presence of the newly formed cloud droplets. This process controls the number concentration of droplets locally in the cloud (Hobbs, 1993).

Once an aerosol particle is activated, the amount of associated liquid water increases dramatically. For example, a factor of $10^6$ occurs for growth of a 0.1 μm diameter aerosol particle to a 10 μm diameter cloud droplet. This liquid water can serve as a medium for aqueous-phase chemical reaction, importantly oxidation of $SO_2$ by $H_2O_2$ to form $H_2SO_4$. Upon evaporation, which is the fate of most clouds, the droplets again become clear-air aerosol particles. The additional $H_2SO_4$ formed in the cloud droplet remains in the condensed phase. In the absence of reaction, little $SO_2$ would dissolve, and much of that would return to the gas phase upon cloud evaporation. Field measurements suggest that aerosol particles experience numerous cloud condensation and evaporation cycles during their residence in the atmosphere (Pruppacher and Klett, 1997). Much of the growth of aerosol particles into the light-scattering region (diameter $\geq 0.1$ μm) occurs by accretion of material by aqueous reaction during the cloud portion of the cycle.

In addition to particle growth by accretion of mass by chemical reaction, growth can occur as well by coagulation. This process reduces the number concentration of particles, while preserving mass concentration. Coagulation is especially important for small particles shortly after nucleation (Kerminen et al., 2001). Likewise, unactivated particles in clouds can be substantially scavenged by activated cloud droplets because of their high surface areas.

### 4.04.5.3 Long-range Transport

Because of their atmospheric residence times of days to weeks, aerosol particles can be transported

substantial distances, $10^4$ km or more. However, there has not been much unequivocal documentation of this, in part because of the scarcity of markers that can distinguish transported material from material of nearby origin. Nonetheless, instances exist where long-range transport has been reliably demonstrated. Long-range transport of aerosol-borne radionuclides from the Chernobyl accident noted above (Section 4.04.1.3) is one such instance. Much of the aerosol responsible for the so-called Arctic haze (Figure 19) has been transported from sources in Eurasia (Sirois and Barrie, 1999). Aerosol originating from forest fires in northwestern Canada was shown to be responsible for large-scale haze layers above Europe. This aerosol was detected in northern Germany in a remarkably thin (~200 m) aerosol layer with lidar and *in situ* measurements (Forster et al., 2001).

Much attention has been paid in recent years to the importance of mineral dust as a constituent of atmospheric aerosol (Section 4.04.1.2) on local, continental, and subhemispheric scales (Mahowald et al., 1999; Griffin et al., 2002), and in some cases mineral dust aerosol can be ascribed to particular source regions. It can severely influence atmospheric radiation locally, and in the aggregate is thought to exert an appreciable influence on the global scale. Because this material absorbs solar radiation, its influence is greater on surface irradiance than on top-of-the-atmosphere net radiation. The absorption of solar energy results in atmospheric heating, which affects the temperature profile and atmospheric stability. The radiative impacts of mineral dust have been reviewed by Sokolik et al. (2001), and Prospero and colleagues have provided extensive documentation of transport of mineral dust across the Atlantic (Prospero, 1996, 1999).

An extraordinary episode of trans-Pacific transport of mineral aerosol, originating in intense dust storms at the Gobi Desert in April 1998, was documented in satellite imagery, *in situ* measurements, lidar remote sensing, and sun- and sky-photometry at a variety of locations throughout western North America as far east as Minnesota (Husar et al., 2001; Tratt et al., 2001; Vaughan et al., 2001). It was also rather accurately represented in numerical simulations (Tratt et al., 2001; Uno et al., 2001). Visual observers at various non-urban locations on the West Coast of North America reported milky white skies instead of the normal deep blue, as confirmed by a strong increase of diffuse to direct solar irradiance at a number of sites. Direct solar irradiance near Eugene, Oregon was decreased by ~30%. AOT at Reno, NV was 0.3–0.5 for four days, compared to a normal value of less than 0.1. At San Nicolas Island in the Channel Islands off the coast of California, the particle size at the surface was rather narrowly distributed between about 0.6 μm to 4 μm radius, consistent with the low Ångström exponent of the AOT, below 0.5. Lidar backscatter at Salt Lake City, UT showed a distinct 1-km-thick layer at 7.5 km, whereas the vertical extent at Pasadena, CA was much greater, 5–11 km. Compositional and morphological analysis confirmed the origin of the material to the Gobi Desert. Although an episode of this magnitude is extraordinary, allowing the transported aerosol to be readily discerned from material of local origin, which would normally predominate, it is clear that transport of aerosol particles over distance scales of $10^4$ km is not extraordinary. Dust particles can react, e.g., with sulfate and nitrate, and are otherwise aged and changed during their long journeys.

### 4.04.5.4 Removal of Particles from the Atmosphere

Aerosol particles are removed from the atmosphere by dry deposition, below-cloud scavenging, and in-cloud scavenging (Slinn, 1983, 1984). Dry deposition consists of diffusion, impaction, or gravitational settling of particles to soil, vegetation, or the like, followed by attachment to these materials. Dry deposition is slow (weeks) for particles of diameter 0.1–1 μm (a size range for which these processes are inefficient), but somewhat faster for larger and smaller particles. Below-cloud scavenging consists of uptake of aerosol particles by falling hydrometeors (droplets, snow flakes, graupel, drizzle, hail, etc.). It is more efficient for larger particles because they do not as readily move out of the path of interception by falling droplets as do smaller particles. In-cloud scavenging consists of removal of particulate matter that is dissolved or suspended in cloud droplets, as this cloud-water is removed from the atmosphere in precipitation. This scavenging is the most efficient removal process for aerosol particles of diameter 0.1–1 μm, leading to a residence time of about a week in the lower troposphere and somewhat longer at higher altitudes, where precipitation is less frequent. Precipitation scavenging of anthropogenic sulfates and nitrates contributes to acid deposition, which influences terrestrial and freshwater ecosystems, especially in regions with granitic and other poorly buffered soils (Bricker, 1984; Unsworth and Fowler, 1988).

### 4.04.6 REPRESENTATION OF AEROSOL PROCESSES IN CHEMICAL TRANSPORT AND TRANSFORMATION MODELS

Chemical transport models representing the processes governing the concentration and distribution of atmospheric substances are applied to address a wide variety of questions. With suitable inputs, they can be used to describe the properties of atmospheric aerosols and their geographical and temporal distribution at locations and times of interest other than the place and time of measurement. Models are used to determine how aerosol loadings and properties will respond to possible future emissions, information that is needed for informed decisions about emission controls of aerosol particles and their precursors. They are also used to examine changes in the properties and geographical distribution of aerosols over the industrial period, information that is needed to understand radiative forcing of climate change over this period.

The models consist of coupled continuity equations for each of the pertinent atmospheric species and are solved by forward integration. The flow fields can be derived from assimilated meteorological data or from atmospheric general circulation models, and the boundary conditions include emissions and deposition. This approach has been successfully applied to trace atmospheric gases and in principle is well suited also for aerosols, with the requirement that such models must represent not only the mass concentration of the materials but also other pertinent aerosol properties. Representation of aerosol particle concentrations and properties in global-scale models presents many computational challenges. For example, a global model with $1° \times 1°$ horizontal resolution (~100 km at the equator) and 30 levels in the vertical results in nearly two million grid cells, with the requirement that aerosol evolution is represented in each of these grid cells.

Many processes must be represented in chemical transport models for aerosols. Key processes are those which have been discussed above: emissions of aerosol particle and their precursors; vertical and horizontal transport of these materials and associated dilution and mixing; gas-phase reaction forming low-vapor-pressure materials that form new particles or add mass to existing particulate matter; new particle formation; coagulation of particles; exchange of condensable vapors between the gas phase and particles; condensed-phase reaction; particle growth resulting from vapor condensation and condensed phase reaction; exchange of water vapor with the atmosphere and resultant size change of particles; dry deposition of particles and precursor gases; below-cloud scavenging of particles and precursor gases; activation of particles to form cloud droplets during cloud formation; coagulation, uptake of gases, and chemical reactions involving cloud droplets; cloud droplet evaporation during cloud dissipation; and removal of cloud droplets and dissolved and suspended particulate matter during precipitation. Description of these processes needs to track the several chemical substances comprising the aerosol, the distribution of these substances as a function of particle size, and the nature and extent of particle aggregation. New particle formation (nucleation) is profoundly influenced by the concentration and physical properties of the condensed-phase material. Interactions with water vapor such as hygroscopic growth and cloud droplet activation are likewise strongly dependent on composition of the particulate matter and the state of mixing.

Accurate representation of these processes must treat particle size, morphology, and composition. This requirement contrasts with the present situation in which large-scale aerosol models for the most part treat the particle composition as uniform, with properties corresponding to spherical particles having a uniform composition and single effective radius. Studies of individual particles show particles are more complex and that these assumptions are too approximate (e.g., Buseck and Pósfai, 1999; Buseck *et al.*, 2002). In practice, much of the information required to represent aerosol evolution is not known. However, levels of accuracy must be balanced against feasibility and complexity of model appropriate to the problem at hand.

A major complexity in aerosol modeling is representation of the size and composition distributions and their temporal evolution. Continuous, sectional, modal, moment-based, and hybrid approaches have been used.

The *continuous* representation in principle yields a solution of arbitrary accuracy but is computationally intensive. For example, Jacobson (2001) reports a box model that treats nucleation, coagulation, condensation, dissolution, and chemical equilibrium. The model uses 18 size distributions, an average of 30 species per distribution, and 60 size bins (radius ranging from 1 nm to 120 $\mu$m) per species per distribution. The result is some $3.24 \times 10^4$ concentrations in a single spatial grid cell. This approach is, at least for the foreseeable future, impractical in large-scale atmospheric models. It would likewise be impractical to evaluate model performance by comparing so many modeled quantities to observations. In view of the computational demands of the continuous approach, it is necessary, however, to employ alternative approaches that embody a variety of approximations and are hence subject to inaccuracy (Zhang *et al.*, 1999). In this context the continuous approach provides a standard by which such alternative approaches may be evaluated.

The *sectional* approach divides the size distribution into a number of sections, keeping track of the number (or mass or surface area) of all particles in each section and evolving the distribution according to some single particle size within the section. The number of bins required for accurate modeling is large, perhaps thirty per decade of radius (Schulz *et al.*, 1998), with resultant computational burden in atmospheric models. Pirjola *et al.* (2002) presented calculations examining the evolution of aerosol size and two-component composition with a model that treats "27 size bins" × "11 composition bins." Some computational efficiency and accuracy is gained by allowing the radius values characterizing the several sections to evolve as the distribution evolves; however, this approach requires redistribution of material into bins having the same size boundaries to permit representation of transport processes.

The *modal* approach assumes a shape for the particle-size distribution, typically one or more lognormal distributions, and represents evolution of the size distribution as evolution of the parameters characterizing the distribution, i.e., the amplitude, mode radius, and variance for the lognormal distribution (Binkowski and Shankar, 1995; Wilck and Stratmann, 1997). This approach offers the possibility of representing aerosol microphysical properties in models with far fewer variables (modal parameters) than are required in the sectional method (numbers of particles in each bin).

The *moment* approach is based on the evolution of the low-order radial moments of the particle-size distribution, $\mu_k$, defined as

$$\mu_k = \int r^k \left( \frac{\mathrm{d}N}{\mathrm{d}r} \right) \mathrm{d}r$$

A premise of the approach is that aerosol integral properties may be determined from the moments of the size distribution without knowledge of the underlying distribution itself and without *a priori* assumption about the shape of the distribution. This situation contrasts with that of the modal approach, which requires an assumption about the shape of the size distribution. The moment approach retains the advantage of representing the size distribution by a small number of modeled variables, the low-order moments. Six moments seem to suffice to accurately represent aerosol properties and evolution (Wright *et al.*, 2001). A further strength of the approach is that it conserves important aerosol attributes such as number and volume concentration during advection and is free from numerical diffusion in particle size. The moment approach has recently been applied to atmospheric sulfate aerosol on a scale of several hundred kilometers. Particle-size distributions retrieved from the moments agree closely with measurements, especially for radius below 1 $\mu$m, for which sulfate is a dominant component of the aerosol (Wright *et al.*, 2000).

Various *hybrid* approaches are also being investigated. An example is a hybrid sectional–moment approach in which the zeroth and third moments of the size distribution are used in a bin-sectional model to conserve particle number and mass (Adams and Seinfeld, 2002).

At present, all the listed approaches rely on numerous assumptions and approximations, necessitated by practical considerations governing the solution of the large sets of coupled differential equations and by lack of fundamental knowledge regarding emissions and aerosol processes (Seigneur, 2001).

Despite concerns over assumptions and approximations, size-resolved, multicomponent aerosol models are being employed at a variety of scales. An example at the scale of an urban air-shed is the study of Meng *et al.* (1998), which examined aerosol composition and size distribution for the South Coast Air Basin of California. Modeled aerosol substances included $SO_4^{2-}$, $NO_3^-$, $NH_4^+$, $Na^+$, $Cl^-$, and secondary organics. Gas-phase

chemistry included the nitrogen oxide—hydrocarbon—oxidant complex, $SO_2$, $NH_3$, and aromatics, the precursor to secondary organic aerosol.

Figure 21 provides an example of comparison of model results with observations. The model performs credibly in reproducing the observations, here of aerosol ammonium size distributions at three locations. Such favorable comparison lends encouragement to the modeling effort despite the many uncertainties.

An important application of aerosol modeling is at the global scale to determine radiative forcing of climate change by anthropogenic aerosols. Penner *et al.* (2001) reported on an intercomparison involving nine state-of-the-art global-scale aerosol models. An example of the results is shown in Figure 22, in which annual average modeled non-seasalt sulfate (i.e., aerosol sulfate other than that contained in suspended sea-salt aerosol) is shown for several locations in

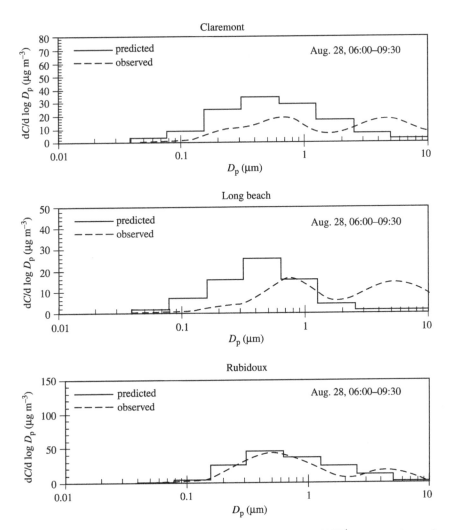

**Figure 21** Comparison of modeled and observed size distributions of aerosol $NH_4^+$ concentration at three locations in the South Coast Air Basin of California on August 28, 1987 (Meng *et al.*, 1998) (reproduced by permission of American Geophysical Union from *J. Geophys. Res.*, **1998**, *103*, 3419–3435).

North and South Atlantic: nss-Sulphate

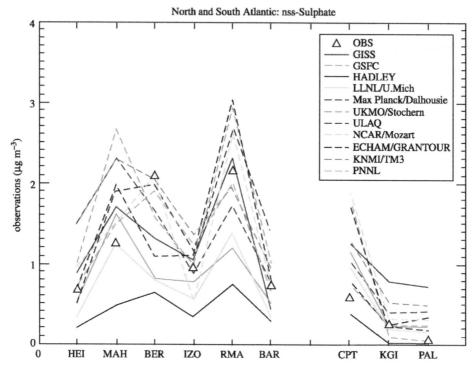

**Figure 22** Observed annual average concentrations of non-seasalt sulfate (cf. text) at stations in the North and South Atlantic (triangles) and predictions from several chemical transport models (line codes and acronyms are shown at upper right). Data were provided by D. Savoie and J. Prospero (University of Miami). Stations along the bottom axis refer to Heimaey, Iceland (HEI); Mace Head, Ireland (MAH); Bermuda (BER); Izania (IZO); Miami, Florida (RMA); Ragged Point, Barbados (BAR); Cape Point, South Africa (CPT); King George Island (KGI); and Palmer Station, Antarctica (PAL). From Penner *et al.* (2001) (reproduced by permission of Intergovernmental Panel on Climate Change).

the North and South Atlantic together with observations. The models all employed the same source emissions, so the spread reflects differences in model representation of aerosol processes. Sulfate is an important test case for model development and comparisons with observation because emissions are rather well characterized (Benkovitz *et al.*, 1996), and the atmospheric processes are thought to be fairly well understood. The results for non-seasalt sulfate (average absolute error in comparison with all observations of 26%) were the best among the several aerosol species examined, compared to black carbon, 179%; organic carbon, 154%; mineral dust, 74%; and sea salt, 46%. Clearly, much improvement is required in understanding the governing processes and in their representation in models.

### 4.04.7 AEROSOL INFLUENCES ON CLIMATE AND CLIMATE CHANGE

#### 4.04.7.1 Background

Increased concentrations of greenhouse gases in the atmosphere and other human-induced changes of Earth's surface and its atmosphere have focused much attention on the issue of climate change. Conventionally this discussion is distinguished

into two major areas: changes in components of the planetary radiation budget, denoted "forcings" (Section 4.04.1.4), and the resultant changes in the climate, denoted "responses." For example, increased concentrations of greenhouse gases increase the amount of infrared radiation emitted by the atmosphere and absorbed at the surface. At present the global annual average radiative forcing from the totality of increases in greenhouse gases over the industrial period is $\sim 2.7$ W m$^{-2}$ (IPCC, 2001). There are many potential climate responses to such radiative forcings, but most interest is focused on the global-mean temperature change. The magnitude of this change that would result from a doubling of $CO_2$ is in the range 1.5–4.5 °C. This climate sensitivity is justifiably of great interest to policymakers (and should be of great interest to all inhabitants of the planet) and is the subject of intense scientific investigation from both observational and modeling approaches.

The attention of the climate change research community was initially focused largely on radiative forcings by increased concentrations of greenhouse gases. However, since about 1990 this community has increasingly become aware that atmospheric aerosols exert a substantial influence on Earth's short-wave radiation budget, i.e., at wavelengths of light comprising the solar spectrum. A result is that any change in atmospheric loadings and properties as a consequence of industrial activity may have exerted a radiative forcing of climate change in parallel with, and additive to, that arising from increased concentrations of greenhouse gases. It is therefore necessary to quantify the totality of radiative forcings over the industrial period to infer the climate sensitivity from the observed record of temperature change.

As noted in Section 4.04.1, atmospheric aerosols influence short-wave radiation in two principal ways: by scattering incident solar radiation and by modifying the reflectivity of clouds (Figure 23, shown for the example of sulfates). Both effects result in an increased fraction of incident solar radiation being scattered upward, out of Earth's atmosphere, rather than being absorbed at the surface. This process, especially when it occurs over a dark surface such as vegetation or the ocean, decreases absorption of solar radiation by Earth and exerts a cooling influence on climate. In contrast, absorption of solar radiation by atmospheric aerosols, especially when it occurs over bright surfaces such as deserts or clouds, exerts a warming influence on the overall radiation budget of Earth.

### 4.04.7.2 Direct Aerosol Shortwave Radiative Forcing

An estimate of the instantaneous radiative forcing $\Delta F_I$ by a nonabsorbing aerosol in the low troposphere, representing the change in solar irradiance absorbed at the surface of Earth (or, equivalently, the change in solar irradiance absorbed by the Earth-atmosphere system), is

$$\Delta F_I = F_T T^2 (1 - R_S)^2 \bar{\beta} \tau$$

where $F_T$ is the solar flux incident at the top of the atmosphere (solar constant); $T$ is the atmospheric transmittance above the aerosol; $R_S$ is the reflectance at Earth's surface; $\bar{\beta}$ is the mean upscatter fraction, the fraction of scattered radiation that is scattered in the upward direction; and $\tau$ is the AOT. The factor $T^2$ accounts for reduction in the incident radiation upon the aerosol and the fraction of radiation scattered by the aerosol that escapes the atmosphere; $(1 - R_S)^2$ accounts for surface reflectance and multiple reflectance between the aerosol and the underlying surface.

This expression is linear in the optical thickness, indicative of an optically thin aerosol for which multiple scattering is unimportant. For a typical AOT of 0.1 (Figure 24), the instantaneous radiative forcing for cloud-free skies is approximately $-17 \text{ W m}^{-2}$ and depends somewhat on particle size and surface reflectance. The negative sign corresponds to a decrease in the amount of radiant energy absorbed by Earth and its atmosphere, indicating that the aerosol exerts a cooling influence on climate. The figure also shows the result of a more accurate calculation, with a radiative transfer model that accounts for multiple scattering, leading to somewhat less forcing per optical thickness as optical thickness increases.

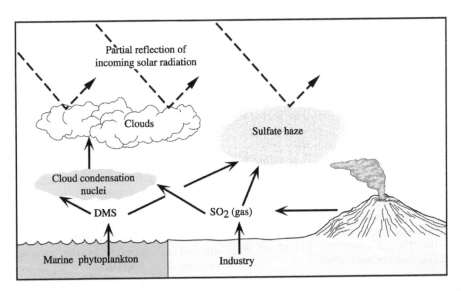

**Figure 23** Schematic diagram of influences of sulfate aerosol on the atmospheric radiation budget. Sulfate from natural and anthropogenic sources scatters light in clear sky and increases the concentration of cloud droplets, thereby increasing cloud reflectivity. Both effects reduce the amount of solar radiation reaching the surface of Earth and thus reduce solar heating of the atmosphere. Increases in sulfate aerosol due to increasing anthropogenic emissions over the industrial period are thought to have exerted a cooling influence on climate. DMS—dimethylsulfide.

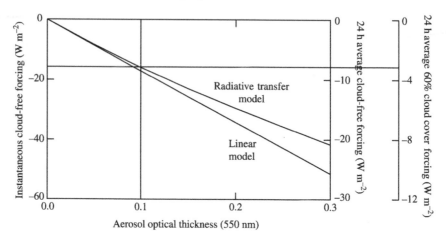

**Figure 24** Dependence of TOA shortwave radiative forcing (W m$^{-2}$) by purely scattering, spherical aerosol particles (single-scatter albedo = 1) on AOT. The negative forcing indicates a cooling influence. The straight line shows the forcing for a linear model (first equation in this section) that omits multiple scattering. The curve indicates the forcing obtained with a radiative transfer model (Hansen and Travis, 1974) for particle radius 85 nm and surface reflectance 0.1. The left axis shows instantaneous, cloud-free forcing for average solar zenith angle, 60°, and the first right axis shows the 24 h average forcing for cloud-free atmosphere. The rightmost axis shows global average forcing for assumed 60% cloud cover. Thin vertical and horizontal lines show forcing for a representative optical thickness of 0.1.

Models of varying sophistication, from simple linear models to complex chemical transport and radiative transfer models, have been used in attempts to estimate the effects of anthropogenic aerosols on global annual mean radiative forcing of climate change. Penner *et al.* (2001) provide an overview of such models as part of the comprehensive examination of radiative forcing of climate change by the Intergovernmental Panel on Climate Change (IPCC, 2001). Nonetheless, the linear model approach, briefly reviewed below, is key to identifying the important processes controlling this forcing and to estimating the contributions of uncertainties in the description of these processes to the overall uncertainty in radiative forcing of climate change by anthropogenic aerosols (Charlson *et al.*, 1992; Penner *et al.*, 1994).

Translating an instantaneous forcing such as that given in Figure 24 into a global average requires accounting for the fact that aerosol scattering and absorption operate only in sunlight and thus only on the illuminated side of the planet. Also, direct scattering is important only in cloud-free regions above dark (absorbing) surfaces. These considerations result in an expression for the average forcing by a nonabsorbing aerosol uniformly distributed over the planet or a hemisphere (northern or southern):

$$\Delta F_{\text{avg}} = -\frac{1}{2} F_T T^2 (1 - A_c)(1 - R_S)^2 \overline{\beta} \tau$$

Here the factor of 1/2 accounts for only half the planet being illuminated by the sun at any given time. The factor $(1 - A_c)$, where $A_c$ is the global average fractional cloudiness, accounts for the forcing occurring only when the aerosol is not under, within, or above a cloud. These factors are taken into account in the two ordinate scales on the right of Figure 24.

An early estimate of the influence of light scattering by anthropogenic sulfate aerosols on the planetary energy budget was given by Charlson *et al.* (1992). Here the change in global-mean short-wave radiation absorbed by the Earth–atmosphere system was given as

$$\overline{\Delta F_R} = -\frac{1}{2} F_T T^2 (1 - A_c)(1 - R_S)^2 \overline{\beta} \alpha_{SO_4^{2-}}$$
$$\times f(\text{RH}) B_{SO_4^{2-}}$$

where, in addition to the symbols previously defined, $\alpha_{SO_4^{2-}}$ is the mass scattering efficiency of sulfate and associated cations at low RH, an intensive property of the aerosol (Section 4.04.3.4; Table 1), and $f(\text{RH})$ is the enhancement in scattering efficiency due to humidity; the product $\alpha_{SO_4^{2-}} f(\text{RH})$ is the scattering efficiency of the particulate matter of the aerosol including the accreted condensed-phase water. $\beta_{SO_4^{2-}}$ is the sulfate mass column burden (Table 1), the pertinent extensive aerosol property indicating the amount of material in the atmosphere. The product $\alpha_{SO_4^{2-}} f(\text{RH}) B_{SO_4^{2-}}$ is equal to the optical thickness of the sulfate aerosol, $\tau_{SO_4^{2-}}$, likewise an extensive aerosol property. Implicit in the treatment is the spatial and temporal averaging necessary to yield a global-mean forcing. Single scattering by the aerosol is assumed, consistent with the linear relation in Figure 24.

The average sulfate mass column burden was related to the rate of emission of the precursor $SO_2$ gas $Q_{SO_2}$ as

$$B_{SO_4^{2-}} = Q_{SO_2} Y_{SO_4^{2-}} \frac{MW_{SO_4^{2-}}}{MW_S} \Theta_{SO_4^{2-}} / A$$

where the yield $Y_{SO_4^{2-}}$ denotes the fraction of emitted $SO_2$ that is converted in the atmosphere to sulfate, $\Theta_{SO_4^{2-}}$ is the mean lifetime of sulfate aerosol in the atmosphere, and $A$ denotes the averaging area, which may be the area of Earth (or of a hemisphere, northern or southern, as the material is not long lived enough in the atmosphere for significant interhemispheric transport). The emission rate of $SO_2$ is typically given in terms of emitted mass of sulfur per unit time. The ratio of molecular weights MW accounts for the fact that emissions are expressed typically as mass of sulfur, whereas mass scattering efficiency of the aerosol is typically expressed in terms of the mass of sulfate ($SO_4^{2-}$).

Expressions such as these are useful in several ways. They permit identification and evaluation of the quantities upon which the forcing depends. They serve as the basis of estimates of forcing of various aerosol species and permit assessment of uncertainties in these estimates from the uncertainties in the component factors. Examination of these uncertainty budgets (Penner *et al.*, 1994) allows identification of the sources of greatest uncertainty and thus guides areas of useful research.

These early estimates yielded a substantial global average radiative forcing by anthropogenic sulfate aerosol ($-1$ W m$^{-2}$) that results from the low average optical thickness, 0.03. More recent estimates with chemical transport models (Penner *et al.*, 2001) tend to average somewhat lower, approximately $-0.5$ W m$^{-2}$. These estimates may be compared to the long-wave radiative forcing of the increment of atmospheric $CO_2$ over the industrial period, 1.6 W m$^{-2}$ (Ramaswamy *et al.*, 2001; Figure 2). Thus, if the estimates of sulfate forcing are accurate, sulfate aerosol alone is negating an appreciable fraction of the greenhouse forcing of $CO_2$. The comparison is even more striking when one considers the mean residence times of the two species, $\sim$100 yr for $CO_2$ and $\sim$1 week for sulfate aerosol. This comparison indicates that the sulfate aerosol from a week's worth of sulfur emissions negates the radiative forcing of a decade or more of anthropogenic $CO_2$ emissions.

It is instructive to compare the average optical thickness for sulfate obtained from such calculations to measured values of AOT such as those in Figure 17. The global AOT for sulfate given by Charlson *et al.* (1992) of 0.03, which gave rise to a forcing of $-1$ W m$^{-2}$, is small relative to the day-to-day fluctuations in AOT that arise from daily changes in loading of tropospheric aerosol at the rural mid-continent location of these measurements. An optical thickness of this magnitude is in fact close to the error limit of such measurements, which with care can approach 0.01, but which is commonly much greater from surface measurements and greater yet from satellite measurements (Wagener *et al.*, 1997). This sensitivity of forcing to slight perturbations in aerosol optical depth makes quantification of this phenomenon extraordinarily challenging.

Examination of the uncertainties in the several factors (Penner *et al.*, 1994), albeit subjective, identifies key areas for further investigation. Most important are issues of the atmospheric chemistry and physics governing the amount of sulfate aerosol in the atmosphere, contributing over a factor-of-2 uncertainty to the estimated forcing. A second area of large uncertainty is the light-scattering efficiency of the aerosol, which depends on the size distribution of the material (Figure 7). Nonetheless, regardless of the uncertainties, calculations such as these establish that the forcing by anthropogenic sulfate aerosol must be recognized as substantial in the context of radiative forcing of climate change over the industrial period.

A major limitation of this approach is that it does not take into account correlations of the several factors. For example, sulfate loadings are greatest in summer, when there is a greater intensity and duration of sunlight (cf. Figure 17). There may also be correlations among the amount of sulfate, RH, cloudiness, and so forth. Additionally, expressions describing these variables do not account for chemical interactions among the several aerosol species in the atmosphere. Accurate treatment of aerosol forcing must also consider multiple scattering, which leads to departures from the approximation that the forcing is linear in column burden and ultimately in emissions (Figure 24). Accounting for such correlations and interactions requires the use of coupled chemical transport and radiative transfer models. Several studies have done this for individual species, such as sulfate (e.g., Feichter *et al.*, 1997; Koch *et al.*, 1999; Ghan *et al.*, 2001) or for multiple species (Penner *et al.*, 2002). However, these studies still embody many simplifications, assumptions, and approximations that result in substantial uncertainties in the calculated radiative forcing. The result is that the calculation of aerosol direct forcing remains far from a solved problem.

Recent estimates of direct radiative forcing by tropospheric aerosols have employed aerosol distributions from global-scale chemical transport models and assumed aerosol properties typical of the several aerosol types. These estimates contain uncertainties both in the modeled aerosol distributions and in the aerosol optical properties. Present understanding and estimated forcing were reviewed and summarized by Penner *et al.* (2001) and Ramaswamy *et al.* (2001) as part of the assessment of radiative forcing of climate change over the industrial period by the IPCC. The direct forcing by industrial sulfate aerosol calculated by 17 models ranged from $-0.26$ W m$^{-2}$ to $-0.82$ W m$^{-2}$ (average $-0.45 \pm 0.18\sigma$). Similar estimates have been made for forcing by other aerosol species (Figure 2).

### 4.04.7.3 Clouds and Indirect Effects

A second class of aerosol influences on climate is through their effect on the concentration and size distribution of droplets in clouds, the so-called indirect effects. As described in Section 4.04.5.2, every cloud droplet forms on an aerosol particle that has acted as a nucleation core. Over polluted continental areas and in areas influenced by continental aerosols, there is a greater abundance of aerosol particles that act as CCN than in pristine marine air masses and, in turn, a greater concentration of cloud droplets (Radke and Hobbs, 1976; Albrecht *et al.*, 1995; Brenguier *et al.*, 2000; Breon *et al.*, 2002; Schwartz *et al.*, 2002). Indirect forcing by aerosols results from this increase in number concentration of cloud droplets associated with an increase in aerosol particle number concentration. The liquid-water concentration is governed by large-scale thermodynamics and precipitation and is, to good approximation, independent of the concentration and properties of the aerosol particles in the air prior to cloud

formation. However, other things being equal, the number concentration of cloud droplets increases with greater concentrations of aerosol particles in the pre-cloud air. An increase in cloud-droplet concentration enhances multiple scattering of incident solar radiation (Twomey, 1974, Twomey, 1977), resulting in more light being scattered upward, out of the cloud, and correspondingly less being transmitted. This cloud-brightening effect of aerosols is shown in Figure 25 for an assumed cloud liquid-water content in which the number concentration of cloud droplets is increased or decreased from a base case by a factor of 8, which corresponds to halving or doubling of the mean cloud-droplet radius, respectively. The effect is greatest for clouds of intermediate reflectance such as marine stratus clouds. This phenomenon is commonly called the Twomey effect or the indirect aerosol influence on climate change and, more recently, the first indirect effect.

An approximate expression for the change in cloud-top reflectivity, $R_{CT}$, is $dR_{CT}/d \ln N_{cd} = 0.075$. This dependence of the change in cloud-top albedo on number concentration served as the basis for an estimate of the global-mean indirect forcing by anthropogenic aerosols (Charlson *et al.*, 1992), Figure 26. The albedo change at cloud top exhibits a logarithmic dependence on the cloud-droplet concentration $N_{cd}$. The lesser albedo change at the top of the atmosphere results from scattering in the atmosphere above the assumed boundary-layer clouds. The global-mean albedo change in Figure 26 was assessed by considering the perturbation to marine stratus clouds under the assumption that 30% of Earth's atmosphere is occupied by marine stratus

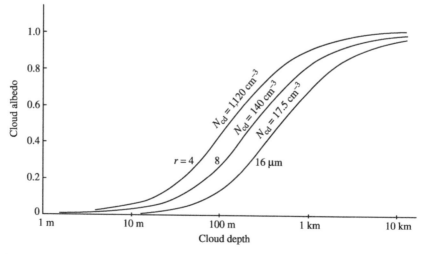

**Figure 25** Dependence of cloud-top albedo on cloud thickness and indicated values of cloud-droplet radius $r$ and number concentration $N_{cd}$. Liquid water content, 0.3 cm$^3$ m$^{-3}$; asymmetry parameter, 0.858 (Twomey, 1977; reproduced by permission of Twomey from *Atmospheric Aerosols*, **1977**).

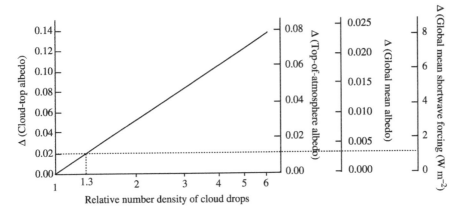

**Figure 26** Calculated relation between albedo and the number concentration of cloud droplets. The sloping line shows the relation between the albedo (at various levels of the atmosphere and globally) as a function of an increase in number concentration of cloud-droplets, $N_{cd}$. The four ordinates show the perturbation in cloud-top albedo (left axis), TOA albedo above marine stratus (first right axis), global-mean albedo (second right axis), and global-mean cloud radiative forcing (far right axis). The fractional atmospheric transmittance of short-wave radiation above the cloud layer was taken as 76%. The dotted line indicates the perturbations resulting from a 30% increase in $N_{cd}$ (Charlson *et al.*, 1992) (reproduced by permission of American Association for the Advancement of Science from *Science*, **1992**, *255*, 423–430).

clouds. Finally, the global- or hemispheric-mean forcing (second axis on right) was evaluated for average solar irradiance at the top of the atmosphere, one-fourth of the solar constant. In that estimate a 30% increase in global-mean $N_{cd}$ was assumed, based on considerations of the difference in sulfate concentrations between the northern and southern hemispheres. The resulting forcing is $-1.1 \text{ W m}^{-2}$.

As with the direct forcing, this simple estimate of the magnitude of indirect forcing by aerosols is comparable to the forcing by greenhouse gases, necessitating more accurate estimates to be taken into consideration of the totality of aerosol forcing over the industrial era. More detailed calculations made using calculations of aerosol loading derived from chemical transport models that consider more aerosol species place this effect broadly in the range $-0.3 \text{ W m}^{-2}$ to $-2 \text{ W m}^{-2}$. Because these calculations had not considered possible cloud darkening effects of black carbon aerosols, the IPCC working group (Ramaswamy *et al.*, 2001) extended the upper limit of the uncertainty range to 0 and declined to present a best estimate of the magnitude (Figure 2).

Processes similar to those in liquid-water clouds govern the formation of ice particles from water vapor, except that only a limited subset of aerosol particles are effective at nucleating ice particles (ice-forming nuclei, IFN; Demott, 2002). For this process to occur, there must be a match between the crystal structure of the particle nucleus and that of ice. Liquid-water clouds commonly form well below 0 °C because of the abundance of CCN and the paucity of IFN, despite the formation of ice particles being thermodynamically favored (Demott, 2002).

### 4.04.7.4 Aerosol Forcing Relative to Other Forcings of Climate Change over the Industrial Period

In addition to the direct and indirect radiative effects of anthropogenic aerosols, there are aerosol influences on the planetary energy budget that are more difficult to quantify or even to express as radiative forcings. Absorption of solar radiation by atmospheric aerosols changes the locus of energy deposition, warming the atmosphere while cooling the Earth surface but exerting much less influence on the top-of-atmosphere (TOA) radiative flux. This phenomenon has been well documented in aerosols advected from India to the Indian Ocean (Ramanathan *et al.*, 2001b). Reduction of solar irradiance at Earth's surface because of absorption and scattering of solar radiation by aerosols alters the surface energy budget, with consequences that include reduction in evapotranspiration and decrease in photosynthetic activity, resulting in complex feedback in the climate system that are, at best, poorly understood.

A further effect of aerosols is the evaporation of clouds as a consequence of atmospheric heating caused by absorption of solar radiation by aerosols. This phenomenon reduces cloudiness, increasing absorption of solar radiation by the Earth–atmosphere system, a so-called "semi-direct" effect (Hansen *et al.*, 1997; Satheesh and Ramanathan, 2000; Lohmann and Feichter, 2001).

The increase in cloud droplet concentration arising from increased number concentrations of aerosol particles tends to inhibit precipitation. This reduction in precipitation is thought to result in longer cloud lifetimes and an increase in cloudiness (Albrecht, 1989; Rosenfeld, 2000).

As low clouds exert a net cooling influence on climate, an increase in concentration of aerosol particles in the lower troposphere leads to a further cooling influence on climate. Possible additional consequences include a shift in the location of precipitation, with resultant influence on the hydrological cycle both locally and globally.

Taken together, the uncertainty range associated with estimates of the several aerosol forcings is the greatest source of uncertainty in estimates of radiative forcing over the industrial period (Figure 2). Perhaps even more important, the uncertainty in the total radiative forcing is so great as to result in a substantial uncertainty in total forcing of climate change over the industrial period, limiting efforts to infer Earth's climate sensitivity from the temperature record over this period (Boucher and Haywood, 2001). For this reason, in recent years much effort has been and continues to be directed to improving estimates of the aerosol forcing.

## 4.04.8  FINAL THOUGHTS

The 1990s have seen a virtual explosion of research on tropospheric aerosols, resulting in large part from recognition of their important but poorly quantified role in climate change. Relative to gas-phase species, specification of emissions of particles requires much more information. Even if emission rates were known by compound, and generally they are not, that would still leave a significant knowledge gap regarding distributions of particle size, composition, and state of mixing as a function of size. Because of the importance of aerosol particles and the number of unsolved problems, it may well be expected that research on aerosol particles will continue at a high level of effort.

A major recent development in aerosol research has been large multi-agency, multinational, multi-platform experiments characterizing aerosol loading, properties, processes, and transport. Key among these experiments have been the several Aerosol Characterization Experiments (ACE 1, Bates *et al.*, 1998; ACE-2, Raes *et al.*, 2000; ACE-Asia, Huebert *et al.*, 2003), Southern African Fire–Atmosphere Research Initiative (SAFARI-2000; Swap *et al.*, 2002), and the Indian Ocean Experiment (INDOEX; Ramanathan *et al.*, 2001b). Such large experiments are essential in view of the complexity of aerosol processes; the great distances, thousands of kilometers, over which they take place; and the resultant need for a large number of highly complex measurement techniques from aircraft, ships, ground-based stations examining aerosol composition, microphysical and optical properties, and radiative influences. Remote sensing by satellite photometry, sun-photometry, and lidar are also necessary to map the geographical and vertical distributions of the aerosols. Key to understanding is chemical transport modeling and integration of the model results with the observations. Such variety of techniques, as well as the need for their deployment over large areas, necessitates large-scale cooperative projects such as these. It is essential that such projects continue if the progress in understanding aerosols required to determine their influences on climate is to be made.

The global abundances and mass emission rates of the various aerosol particle types are important but poorly known. To our thinking, a sense of the contributions of various sources to atmospheric aerosols is best gained by examining the aerosol that is present in the atmosphere at a given location or time, or globally, depending on the application. Maps of AOT obtained from satellite observations, such as Figure 18, provide a global picture of where the aerosol is, how much is there, and, fortuitously, because of the residence times of aerosols of about a week, a sense of the geographical distribution of their sources. Ultimately, at any location of interest one must turn to chemical analysis of the aerosol to identify the materials present and their sources. In such efforts, one is well served by measurements such as those depicted in Figure 9, which identify the important constituents and shows as well the variation of composition with location. From knowledge of composition one can infer sources, but quantitative understanding and descriptive capability requires understanding of the sources, transport, transformation, and deposition of these materials. Developing this understanding is the goal of much current research into tropospheric aerosols.

It is hoped that this chapter has convinced the reader of the importance and complexity of aerosol processes and the interactions of aerosols with other components of the atmosphere and with climate. Many advances may be expected in the study of tropospheric aerosols in the coming years as increasingly focused and precise analytical measurements are made and entered into models used to predict atmospheric changes, including climate, environmental impacts, and health. As procedures are refined, the use of satellite data and global models will increase in utility and impact for understanding the effects of aerosols on the troposphere, the atmosphere in general, and the broad Earth environment. Stay tuned!

## ACKNOWLEDGMENTS

We thank K. Strawbridge for lidar images and S. Martin, E. Lewis, and an anonymous reviewer for comments on the manuscript. Supported by grants NAG5-9838 and NAG5-11552 from the NASA Radiation Science Program (PRB) and Environmental Sciences Division of the US Department of Energy (SES).

# REFERENCES

Adams P. J. and Seinfeld J. H. (2002) Predicting global aerosol size distributions in general circulation models. *J. Geophys. Res.* **107**, 4370, doi: 10.1029/2001JD001010.

Albrecht B. A. (1989) Aerosols, cloud microphysics, and fractional cloudiness. *Science* **245**, 1227–1230.

Albrecht B. A., Bretherton C. S., Johnson D., Schubert W. H., and Frisch A. S. (1995) The Atlantic stratocumulus transition experiment—ASTEX. *Bull. Am. Meteorol. Soc.* **76**, 889–904.

Allan J. D., Jimenez J. L., Williams P. I., Alfarra M. R., Bower K. N., Jayne J. T., Coe H., and Worsnop D. R. (2003) Quantitative sampling using an aerodyne aerosol mass spectrometer: 1. Techniques of data interpretation and error analysis. *J. Geophys. Res.* **108**, 4090, doi:10.1029/2002JD002358.

Anastasio C. and Martin S. T. (2001) Atmospheric nanoparticles. In *Nanoparticles and the Environment* (eds. J. F. Banfield and A. Navrotsky). Mineralogical Society of America, Washington, DC, vol. 44.

Anderson T. L. and Ogren J. A. (1998) Determining aerosol radiative properties using the TSI 3563 integrating nephelometer. *Aerosol Sci. Technol.* **29**, 57–69.

Anderson T. L., Covert D. S., Marshall S. F., Laucks M. L., Charlson R. J., Waggoner A. P., Ogren J. A., Caldow R., Holm R. L., Quant F. R., Sem G. J., Wiedensohler A., Ahlquist N. A., and Bates T. S. (1996) Performance characteristics of a high-sensitivity, three-wavelength, total scatter/backscatter nephelometer. *J. Atmos. Ocean. Technol.* **13**, 967–986.

Andreae M. O. (1991) Biomass burning: its history, use and distribution and its impact on environmental quality and global climate. In *Global Biomass Burning: Atmospheric, Climatic and Biospheric Implications* (ed. J. S. Levine). MIT Press, Cambridge, MA, pp. 3–21.

Andreae M. O. (1995) Climatic effects of changing atmospheric aerosol levels. In *Future Climates of the World: A Modeling Perspective, World Survey of Climatology* (ed. A. Henderson-Sellers). Elsevier, Amsterdam, vol. 16, pp. 341–392.

Andreae M. O. (1996) Raising dust in the greenhouse. *Nature* **380**, 389–390.

Armstrong J. T. and Buseck P. R. (1975) Quantitative chemical analysis of individual microparticles using the electron microprobe: theoretical. *Anal. Chem.* **47**, 2178–2191.

Arnott W. P., Moosmuller H., Rogers C. F., Jin T. F., and Bruch R. (1999) Photoacoustic spectrometer for measuring light absorption by aerosol: instrument description. *Atmos. Environ.* **33**, 2845–2852.

Bates T. S., Huebert B. J., Gras J. L., Griffiths F. B., and Durkee P. A. (1998) International global atmospheric chemistry (IGAC) project's first aerosol characterization experiment (ACE 1): overview. *J. Geophys. Res.* **103**, 16297–16318.

Benkovitz C. M., Scholtz M. T., Pacyna J., Tarrason L., Dignon J., Voldner E. C., Spiro P. A., Logan J. A., and Graedel T. E. (1996) Global gridded inventories of anthropogenic emissions of sulfur and nitrogen. *J. Geophys. Res.* **101**, 29239–29253.

Binkowski F. S. and Shankar U. (1995) The regional particulate matter model:1. Model description and preliminary results. *J. Geophys. Res.* **100**, 26191–26209.

Birmili W. and Wiedensohler A. (2000) New particle formation in the continental boundary layer: meteorological and gas phase parameter influence. *Geophys. Res. Lett.* **27**, 3325–3328.

Blomquist B. W., Huebert B. J., Howell S. G., Litchy M. R., Twohy C. H., Schanot A., Baumgardner D., Lafleur B., Seebauch R., and Laucks M. L. (2001) An evaluation of the community aerosol inlet for the NCAR C-130 research aircraft. *J. Atmos. Ocean. Technol.* **18**, 1387–1397.

Bond T. C., Anderson T. L., and Campbell D. (1999) Calibration and intercomparison of filter-based measurements of visible light absorption by aerosols. *Aerosol Sci. Technol.* **30**, 582–600.

Boucher O. and Haywood J. (2001) On summing the components of radiative forcing of climate change. *Clim. Dyn.* **18**, 297–302.

Brenguier J. L., Chuang P. Y., Fouquart Y., Johnson D. W., Parol F., Pawlowska H., Pelon J., Schuller L., Schroder F., and Snider J. (2000) An overview of the ACE-2 Cloudy-Column closure experiment. *Tellus Ser. B-Chem. Phys. Meteorol.* **52**, 815–827.

Breon F. M., Tanré D., and Generoso S. (2002) Aerosol effect on cloud droplet size monitored from satellite. *Science* **295**, 834–838.

Bricker O. P. (ed.) (1984) *Geological Aspects of Acid Deposition.* Butterworth, Boston.

Bryson R. A. (1974) Perspective on climatic change. *Science* **184**, 753–760.

Buseck P. R. and Pósfai M. (1999) Airborne minerals and related aerosol particles: effects on climate and the environment. *Proc. Natl Acad. Sci. USA* **96**, 3372–3379.

Buseck P. R., Jacob D. J., Pósfai M., Li J., and Anderson J. R. (2002) Minerals in the air: an environmental perspective. In *Frontiers in Geochemistry: Global Inorg. Geochem. (Konrad Krauskopf Vol. 1, Internat'l. Book Series Vol. 5)* (ed. W. G. Ernst). Geol. Soc. Am., Columbia, MD. pp. 106–122.

Buzorius G., Rannik U., Mäkelä J. M., Vesala T., and Kulmala M. (1998) Vertical aerosol particle fluxes measured by eddy covariance technique using condensational particle counter. *J. Aerosol Sci.* **29**, 157–171.

Buzorius G., Zelenyuk A., Brechtel F., and Imre D. (2002) Simultaneous determination of individual ambient particle size, hygroscopicity and composition. *Geophys. Res. Lett.* **29**, 1974, doi:10.1029/2001GL014221.

Cachier H., Bremond M. P., and Buat-Ménard P. (1989) Determination of atmospheric soot carbon with a simple thermal method. *Tellus* **41B**, 379–390.

Cambray R. S., Cawse P. A., Garland J. A., Gibson J. A. B., Johnson P., Lewis G. N. J., Newton D., Salmon L., and Wade B. O. (1987) Observations on radioactivity from the Chernobyl accident. *Nucl. Energy—J. Br. Nucl. Energy Soc.* **26**, 77–101.

Charlock T. P. and Sellers W. D. (1980) Aerosol effects on climate—calculations with time-dependent and steady-state radiative-convective models. *J. Atmos. Sci.* **37**, 1327–1341.

Charlson R. J., Schwartz S. E., Hales J. M., Cess R. D., Coakley J. A., Hansen J. E., and Hofmann D. J. (1992) Climate forcing by anthropogenic aerosols. *Science* **255**, 423–430.

Clarke A. D. and Kapustin V. N. (2002) A Pacific aerosol survey: Part I: a decade of data on particle production, transport, evolution, and mixing in the troposphere. *J. Atmos. Sci.* **59**, 363–382.

Cooke W. F., Liousse C., Cachier H., and Feichter J. (1999) Construction of a $1° \times 1°$ fossil fuel emission data set for carbonaceous aerosol and implementation and radiative impact in the ECHAM4 model. *J. Geophys. Res.* **104**, 22137–22162.

Crutzen P. J. and Andreae M. O. (1990) Biomass burning in the tropics-impact on atmospheric chemistry and biogeochemical cycles. *Science* **250**, 1669–1678.

Curry J. A., Hobbs P. V., King M. D., Randall D. A., and Minnis P. (2000) FIRE arctic clouds experiment. *Bull. Am. Meteorol. Soc.* **81**, 5–29.

Cziczo D. J., Nowak J. B., Hu J. H., and Abbatt J. P. D. (1997) Infrared spectroscopy of model tropospheric aerosols as a function of relative humidity: observation of deliquescence and crystallization. *J. Geophys. Res.* **102**, 18843–18850.

d'Almeida G. A., Koepke P., and Shettle E. P. (1991) *Atmospheric Aerosols: Their Global Climatology and Radiative Characteristics.* Deepak Publishing, Hampton, VA.

Debock L. A., Vanmalderen H., and Vangrieken R. E. (1994) Individual aerosol-particle composition variations in air masses crossing the North Sea. *Environ. Sci. Technol.* **28**, 1513–1520.

Demott P. (2002) Laboratory studies of cirrus cloud processes. In *Cirrus* (eds. D. K. Lynch, K. Sassen, D. O'C. Starr, and G. Stephens). Oxford University Press, Oxford, pp. 102–135.

Deuzé J. L., Herman M., Goloub P., Tanré D., and Marchand A. (1999) Characterization of aerosols over ocean from POLDER/ADEOS-1. *Geophys. Res. Lett.* **26**, 1421–1424.

Dierck I., Michaud D., Wouters L., and Vangrieken R. (1992) Laser microprobe mass analysis of individual North Sea aerosol particles. *Environ. Sci. Technol.* **26**, 802–808.

Dietz. R. N., Wieser, R. F. (1983) Sulfate formation in oil-fired power plant plumes: Vol. 1. Parameters affecting primary sulfate emissions and a model for predicting emissions and plume capacity. Electric Power Research Institute, EPRI EA-3231, Final Report.

Dreiling V. and Friederich B. (1997) Spatial distribution of the arctic haze aerosol size distribution in western and eastern Arctic. *Atmos. Res.* **44**, 133–152.

Duce R. A. (1995) Sources, distributions, and fluxes of mineral aerosols and their relationship to climate. In *Aerosol Forcing of Climate* (eds. R. J. Charlson and J. Heintzenberg). Wiley, New York, pp. 43–72.

Ebert M., Weinbruch S., Hoffmann P., and Ortner H. M. (2000) Chemical characterization of North Sea aerosol particles. *J. Aerosol. Sci.* **31**, 613–632.

EPA (1998). National Air Quality and Emissions Trends Report, 1997. Report 454/R-98-016. US Environmental Protection Agency. Office of Air Quality Planning and Standards. Emissions Monitoring and Analysis Division. Air Quality Trends Analysis Group, Research Triangle Park, NC.

EPA (2002). US Environmental Protection Agency, Air Quality Criteria for Particulate Matter, Third External Review Draft, Report EPA/600/P-99/002aC. National Center for Environmental Assessment, Office of Research and Development, Research Triangle Park, NC.

Falkowski P. G., Barber R. T., and Smetacek V. (1998) Biogeochemical controls and feedbacks on ocean primary production. *Science* **281**, 200–206.

Feichter J., Lohmann U., and Schult I. (1997) The atmospheric sulfur cycle in ECHAM-4 and its impact on the shortwave radiation. *Clim. Dyn.* **13**, 235–246.

Ferrare R. A., Melfi S. H., Whiteman D. N., Evans K. D., and Leifer R. (1998a) Raman lidar measurements of aerosol extinction and backscattering: 1. Methods and comparisons. *J. Geophys. Res.* **103**, 19663–19672.

Ferrare R. A., Melfi S. H., Whiteman D. N., Evans K. D., Poellot M., and Kaufman Y. J. (1998b) Raman lidar measurements of aerosol extinction and backscattering: 2. Derivation of aerosol real refractive index, single-scattering albedo, and humidification factor using Raman lidar and aircraft size distribution measurements. *J. Geophys. Res.* **103**, 19673–19689.

Finlayson-Pitts B. J. and Hemminger J. C. (2000) Physical chemistry of airborne sea salt particles and their components. *J. Phys. Chem. A* **104**, 11463–11477.

Flagan R. C. (1998) History of electrical aerosol measurements. *Aerosol Sci. Technol.* **28**, 301–380.

Forster C., Wandinger U., Wotawa G., James P., Mattis I., Althausen D., Simmonds P., O'Doherty S., Jennings S. G., Kleefeld C., Schneider J., Trickl T., Kreipl S., Jager H., and Stohl A. (2001) Transport of boreal forest fire emissions from Canada to Europe. *J. Geophys. Res.* **106**, 22887–22906.

Franklin B. (1789) Meteorological imaginations and conjectures (Paper read 1784). In *Memoirs of the Literary and Philosophical Society of Manchester* 2nd edn. (ed. T. Cadwell). Strand, London, pp. 373–377.

Friedlander S. (1977) *Smoke, Dust and Haze.* Wiley, New York.

Fung I. Y., Meyn S. K., Tegen I., Doney S. C., John J. G., and Bishop J. K. B. (2000) Iron supply and demand in the upper ocean. *Global Biogeochem. Cycle* **14**, 281–295.

Gao Y., Kaufman Y. J., Tanré D., Kolber D., and Falkowski P. G. (2001) Seasonal distributions of aeolian iron fluxes to the global ocean. *Geophys. Res. Lett.* **28**, 29–32.

Garvie L. A. J. and Buseck P. R. (1998) Ratios of ferrous to ferric iron from nanometre-sized areas in minerals. *Nature* **396**, 667–670.

Garvie L. A. J., Craven A. J., and Brydson R. (1994) Use of electron energy-loss near-edge fine structure in the study of minerals. *Am. Mineral.* **79**, 411–425.

Gebhart J. (1993) Optical direct-reading techniques: light intensity systems. In *Aerosol Measurement: Principles, Techniques, and Applications* (eds. K. Willeke and P. A. Baron). Van Norstrand Reinhold, New York, pp. 313–344.

Germani M. S. and Buseck P. R. (1991) Automated scanning electron-microscopy for atmospheric particle analysis. *Anal. Chem.* **63**, 2232–2237.

Ghan S. J., Easter R. C., Chapman E., Abdul-Razzak H., Zhang Y., Leung R., Laulainen N., Saylor R., and Zaveri R. (2001) A physically-based estimate of radiative forcing by anthropogenic sulfate aerosol. *J. Geophys. Res.* **106**, 5279–5294.

Gribbin J. (1976) Man's influence not yet felt by climate. *Nature* **264**, 608.

Griffin D. W., Kellogg C. A., Garrison V. H., and Shinn E. A. (2002) The global transport of dust—an intercontinental river of dust, microorganisms and toxic chemicals flows through the Earth's atmosphere. *Am. Scientist* **90**, 228–235.

Griffin R. J., Cocker D. R., III, Flagan R. C., and Seinfeld J. H. (1999) Organic aerosol formation from the oxidation of biogenic hydrocarbons. *J. Geophys. Res.* **104**, 3555–3567.

Hammer C. U., Clausen H. B., Dansgaard A., Neftel A., Kristinsdottir P., and Johnson E. (1985) Continuous impurity analysis along the Dye 3 deep core. In *Greenland Ice Core: Geophysics, Geochemistry, and the Environment* (eds. C. C. Langway, Jr. and H. Oeschger, W. Dansgaard). American Geophysical Union, Washington, DC, vol. 33, pp. 90–94.

Han J. H. and Martin S. T. (1999) Heterogeneous nucleation of the efflorescence of $(NH_4)(2)SO_4$ particles internally mixed with $Al_2O_3$, $TiO_2$, and $ZrO_2$. *J. Geophys. Res.* **104**, 3543–3553.

Hansen J., Lacis A., Ruedy R., and Sato M. (1992) Potential climate impact of Mount Pinatubo eruption. *Geophys. Res. Lett.* **19**, 215–218.

Hansen J., Sato M., and Ruedy R. (1997) Radiative forcing and climate response. *J. Geophys. Res.* **102**, 6831–6864.

Hansen J. E. and Travis L. D. (1974) Light-scattering in planetary atmospheres. *Space Sci. Rev.* **16**, 527–610.

Hansen J. E., Wang W. C., and Lacis A. A. (1978) Mount Agung eruption provides test of a global climatic perturbation. *Science* **199**, 1065–1068.

Hara K., Kikuchi T., Furuya K., Hayashi M., and Fujii Y. (1996) Characterization of Antarctic aerosol particles using laser microprobe mass spectrometry. *Environ. Sci. Technol.* **30**, 385–391.

Harrison L. and Michalsky J. (1994) Objective algorithms for the retrieval of optical depths from ground-based measurements. *Appl. Optics* **33**, 5126–5132.

Haywood J. and Boucher O. (2000) Estimates of the direct and indirect radiative forcing due to tropospheric aerosols: a review. *Rev. Geophys.* **38**, 513–543.

Heintzenberg J., Raes F., Schwartz S. E., *et al.* (2003) Tropospheric aerosols. In *Atmospheric Chemistry in a Changing World—an Integration and Synthesis of a Decade of Tropospheric Chemistry Research* (ed. G. P. Brasseur, R. G. Prinn, and A. A. P. Pszenny). Springer, Berlin, pp. 125–156.

Hinds W. C. (1982) *Aerosol Technology: Properties, Behavior, and Measurement of Airborne Particles.* Wiley, New York.

Hinds W. C. (1999) *Aerosol Technology: Properties, Behavior, and Measurement of Airborne Particles*, 2nd edn. Wiley, New York.

Hinz K. P., Kaufmann R., and Spengler B. (1994) Laser-induced mass analysis of single particles in the airborne state. *Anal. Chem.* **66**, 2071–2076.

Hobbs P. V. (1993) Aerosol–cloud interactions. In *Aerosol–Cloud–Climate Interactions* (ed. P. V. Hobbs). Academic Press, San Diego, pp. 33–73.

Holben, B., 2003. AERONET: AErosol RObotic NETwork.http://aeronet.gsfc.nasa.gov:8080/

Holben B. N., Eck T. F., Slutsker I., Tanre D., Buis J. P., Setzer A., Vermote E., Reagan J. A., Kaufman Y. J., Nakajima T., Lavenu F., Jankowiak I., and Smirnov A. (1998) AERONET—a federated instrument network and data archive for aerosol characterization. *Remote Sens. Environ.* **66**, 1–16.

Hudson J. G. (1993) Cloud condensation nuclei. *J. Appl. Meteorol.* **32**, 596–607.

Huebert B., Bates T., Russell P., Shi G., Kim Y. J., Kawamura K., Carmichael G., and Nakajima T. (2003) An overview of ACE-Asia: strategies for quantifying the relationships between Asian aerosols and their climatic impacts. *J. Geophys. Res.* (in press).

Hughes L. S., Allen J. O., Kleeman M. J., Johnson R. J., Cass G. R., Gross D. S., Gard E. E., Galli M. E., Morrical B. D., Fergenson D. P., Dienes T., Noble C. A., Silva P. J., and Prather K. A. (1999) Size and composition distribution of atmospheric particles in southern California. *Environ. Sci. Technol.* **33**, 3506–3515.

Husar R. B., Holloway J. M., Patterson D. E., and Wilson W. E. (1981) Spatial and temporal pattern of eastern-United-States haziness—a summary. *Atmos. Environ.* **15**, 1919–1928.

Husar R. B., Tratt D. M., Schichtel B. A., Falke S. R., Li F., Jaffe D., Gasso S., Gill T., Laulainen N. S., Lu F., Reheis M. C., Chun Y., Westphal D., Holben B. N., Gueymard C., McKendry I., Kuring N., Feldman G. C., McClain C., Frouin R. J., Merrill J., DuBois D., Vignola F., Murayama T., Nickovic S., Wilson W. E., Sassen K., Sugimoto N., and Malm W. C. (2001) Asian dust events of April 1998. *J. Geophys. Res.* **106**, 18317–18330.

IPCC (2001) Climate Change 2001 The Scientific Basis. Contribution of Working Group I to the *Third Assessment Report of the Intergovernmental Panel on Climate Change* (eds. J. T. Houghton, Y. Ding, D. J. Griggs, M. Noguer, P. van der Linden, X. Dai, and K. Maskell). Cambridge University Press, Cambridge.

Iversen T., Halvorsen N. E., Saltbones J., Sandnes H., (1990). Calculated budgets for airborne sulfur and nitrogen in Europe. In *Co-operative Programme for Monitoring and Evaluation of the Long Range Transmission of Air Pollutants in Europe (EMEP)*. The Norwegian Meteorological Institute, Oslo, Norway, EMEP/MSC-W Report 2/90.

Jacob D. J. (1999) *Introduction to Atmospheric Chemistry*. Princeton University Press, Princeton, NY.

Jacob D. J. (2000) Heterogeneous chemistry and tropospheric ozone. *Atmos. Environ.* **34**, 2131–2159.

Jacobson M. C., Hansson H. C., Noone K. J., and Charlson R. J. (2000) Organic atmospheric aerosols: review and state of the science. *Rev. Geophys.* **38**, 267–294.

Jacobson M. Z. (2001) Global direct radiative forcing due to multicomponent anthropogenic and natural aerosols. *J. Geophys. Res.* **106**, 1551–1568.

Jayne J. T., Leard D. C., Zhang X., Davidovits P., Smith K. A., Kolb C. E., and Worsnop D. R. (2000) Development of an aerosol mass spectrometer for size and composition analysis of submicron particles. *Aerosol. Sci. Technol.* **33**, 49–70.

Junge C. E. (1963) *Air Chemistry and Radioactivity*. Academic Press, New York.

Katrinak K. A., Anderson J. R., and Buseck P. R. (1995) Individual particle types in the aerosol of Phoenix, Arizona. *Environ. Sci. Technol.* **29**, 321–329.

Kaufman Y. J., Tanré D., and Boucher O. (2002) A satellite view of aerosols in the climate system. *Nature* **419**, 215–223.

Kerminen V. M., Pirjola L., and Kulmala M. (2001) How significantly does coagulational scavenging limit atmospheric particle production? *J. Geophys. Res.* **106**, 24119–24125.

Khattatov V. U., Tyabotov A. E., Alekseyev A. P., Postnov A. A., and Stulov E. A. (1997) Aircraft lidar studies of the Arctic haze and their meteorological interpretation. *Atmos. Res.* **44**, 99–111.

Kim D. and Hopke P. K. (1988) Classification of individual particles based on computer-controlled scanning electron-microscopy data. *Aerosol Sci. Technol.* **9**, 133–151.

Koch D., Jacob D., Tegen I., Rind D., and Chin M. (1999) Tropospheric sulfur simulation and sulfate direct radiative forcing in the Goddard Institute for Space Studies general circulation model. *J. Geophys. Res.* **104**, 23799–23822.

Koch S., Winterhalter R., Uherek E., Kolloff A., Neeb P., and Moortgat G. K. (2002) Formation of new particles in the gas-phase ozonolysis of monoterpenes. *Atmos. Environ.* **34**, 4031–4042.

Korhonen P., Kulmala M., Laaksonen A., Viisanen Y., McGraw R., and Seinfeld J. H. (1999) Ternary nucleation of $H_2SO_4$, $NH_3$, and $H_2O$ in the atmosphere. *J. Geophys. Res.* **104**, 26349–26353.

Kreidenweis S. M. (1995) Atmospheric aerosol research in the US 1991–1994. *Rev. Geophys.* **33**, 775–784.

Kurtz A. C., Derry L. A., and Chadwick O. A. (2001) Accretion of Asian dust to Hawaiian soils: isotopic, elemental, and mineral mass balances. *Geochim. Cosmochim. Acta* **65**, 1971–1983.

Laskin A. and Cowin J. P. (2001) Automated single particle SEM/EDX analysis of submicrometer particles down to 0.1 μm. *Anal. Chem.* **73**, 1023–1029.

Li J., Pósfai M., Hobbs P. V., and Buseck P. R. (2003) Individual aerosol particles from biomass burning in southern Africa: 2. Composition and aging of inorganic particles. *J. Geophys. Res.* **108**, 8484, doi: 10.1029/2002JD002310.

Li X., Maring H., Savoie D., Voss K., and Prospero J. M. (1996) Dominance of mineral dust in aerosol light-scattering in the North Atlantic trade winds. *Nature* **380**, 416–419.

Lin H. B. and Campillo A. J. (1985) Photothermal aerosol absorption-spectroscopy. *Appl. Optics* **24**, 422–433.

Liu X. D., Van Espen P., Adams F., Cafmeyer J., and Maenhaut W. (2000) Biomass burning in southern Africa: individual particle characterization of atmospheric aerosols and savanna fire samples. *J. Atmos. Chem.* **36**, 135–155.

Lohmann U. and Feichter J. (2001) Can the direct and semi-direct aerosol effect compete with the indirect effect on a global scale? *Geophys. Res. Lett.* **28**, 159–161.

Mahowald N., Kohfeld K., Hansson M., Balkanski Y., Harrison S., Prentice C., Schulz M., and Rodhe H. (1999) Dust sources and deposition during the last glacial maximum and current climate: a comparison of model results with paleodata from ice cores and marine sediments. *J. Geophys. Res.* **104**, 15895–15916.

Marple V. A. and Willeke K. (1976) Impactor design. *Atmos. Environ.* **10**, 891–896.

Martin J. H. and Gordon R. M. (1988) Northeast Pacific iron distributions in relation to phytoplankton productivity. *Deep-Sea Res.* **35**, 177–196.

Martin S. T. (2000) Phase transitions of aqueous atmospheric particles. *Chem. Rev.* **100**, 3403–3453.

Martin S. T., Han J. H., and Hung H. M. (2001) The size effect of hematite and alumina inclusions on the efflorescence relative humidities of aqueous ammonium sulfate particles. *Geophys. Res. Lett.* **28**, 2601–2604.

McCormick M. P., Thomason L. W., and Trepte C. R. (1995) Atmospheric effects of the Mt Pinatubo eruption. *Nature* **373**, 399–404.

McMurry P. H. (2000) A review of atmospheric aerosol measurements. *Atmos. Environ.* **34**, 1959–1999.

Meng Z., Dabdub D., and Seinfeld J. H. (1998) Size-resolved and chemically resolved model of atmospheric aerosol dynamics. *J. Geophys. Res.* **103**, 3419–3435.

Michalsky J. J., Schlemmer F. A., Berkheiser W. E., Berndt J. L., Harrison L. C., Laulainen N. S., Larson N. R., and Barnard J. C. (2001) Multiyear measurements of aerosol optical depth in the atmospheric radiation measurement and quantitative links programs. *J. Geophys. Res.* **106**, 12099–12107.

Mount G. H., Eisele F. L., Tanner D. J., Brault J. W., Johnston P. V., Harder J. W., Williams E. J., Fried A., and Shetter B. R. (1997) An intercomparison of spectroscopic laser long-path and ion-assisted *in situ* measurements of hydroxyl concentrations during the tropospheric OH photochemistry experiment Fall 1993. *J. Geophys. Res.* **102**, 6437–6455.

Murphy D. M., Anderson J. R., Quinn P. K., McInnes L. M., Brechtel F. J., Kreidenweis S. M., Middlebrook A. M., Pósfai M., Thomson D. S., and Buseck P. R. (1998) Influence of sea salt particles on aerosol radiative properties in the Southern Ocean marine boundary layer. *Nature* **392**, 62–65.

National Park Service (2002) http://vista.cira.colostate.edu/DataWarehouse/improve/data/photos/YOSE/html/img0001.htm and /img0005.htm

Nemesure S., Wagener R., and Schwartz S. E. (1995) Direct shortwave forcing of climate by anthropogenic sulfate aerosol: sensitivity to particle size, composition, and relative humidity. *J. Geophys. Res.* **100**, 26105–26116.

Noble C. A. and Prather K. A. (2000) Real-time single particle mass spectrometry: a historical review of a quarter century of the chemical analysis of aerosols. *Mass. Spectrom. Rev.* **19**, 248–274.

O'Dowd C. D., Aalto P., Hameri K., Kulmala M., and Hoffmann T. (2002) Aerosol formation—atmospheric particles from organic vapours. *Nature* **416**, 497–498.

Ogren J. A. (1995) A systematic approach to *in situ* observations of aerosol properties. In *Aerosol Forcing of Climate* (eds. R. J. Charlson and J. Heintzenberg). Wiley, New York, pp. 215–226.

Parrington J. R., Zoller W. H., and Aras N. K. (1983) Asian dust—seasonal transport to the Hawaiian Islands. *Science* **220**, 195–197.

Penner J. E., Charlson R. J., Hales J. M., Laulainen N. S., Leifer R., Novakov T., Ogren J., Radke L. F., Schwartz S. E., and Travis L. (1994) Quantifying and minimizing uncertainty of climate forcing by anthropogenic aerosols. *Bull. Am. Meteorol. Soc.* **75**, 375–400.

Penner J. E., Andreae M., Annegarn H., Barrie L., Feichter J., Hegg D., Jayaraman A., Leaitch R., Murphy D., Nganga J., and Pitari G. (2001) Aerosols, their direct and indirect effects. In *Climate Change 2001: The Scientific Basis. Contribution of Working Group I to the Third Assessment Report of the Intergovernmental Panel on Climate Change* (eds. J. T. Houghton, Y. Ding, D. J. Griggs, M. Noguer, P. van der Linden, X. Dai, and K. Maskell). Cambridge University Press, Cambridge, 2001, pp. 289–348.

Penner J. E., Zhang S. Y., Chin M., Chuang C. C., Feichter J., Feng Y., Geogdzhayev I. V., Ginoux P., Herzog M., Higurashi A., Koch D., Land C., Lohmann U., Mishchenko M., Nakajima T., Pitari G., Soden B., Tegen I., and Stowe L. (2002) A comparison of model- and satellite-derived aerosol optical depth and reflectivity. *J. Atmos. Sci.* **59**, 441–460.

Petit J. R., Mounier L., Jouzel J., Korotkevich Y. S., Kotyakov V. I., and Lorius C. (1990) Paleoclimatological and chronological implications of the Vostok core dust record. *Nature* **343**, 56–58.

Pirjola L., O'Dowd C. D., and Kulmala M. (2002) A model prediction of the yield of cloud condensation nuclei from coastal nucleation events. *J. Geophys. Res.* **107**, article no. 8098.

Pósfai M. and Molnár A. (2000) Aerosol particles in the troposphere: a mineralogical introduction. In *EMU Notes in Mineralogy: Environmental Mineralogy* (eds. D. J. Vaughan and R. A. Wogelius). Eötvös University Press, Budapest, vol. 2, pp. 197–252.

Pósfai M., Anderson J. R., Buseck P. R., and Sievering H. (1999) Soot and sulfate aerosol particles in the remote marine troposphere. *J. Geophys. Res.* **104**, 21685–21693.

Pósfai M., Simonics R., Li J., Hobbs P. V., and Buseck P. R. (2003) Individual aerosol particles from biomass burning in southern Africa: 1. Compositions and size distributions of carbonaceous particles. *J. Geophys. Res.* **108**, 8483 doi:10.1029/2002JD00291.

Prospero J. M. (1996) Saharan dust transport over the North Atlantic Ocean and Mediterranean: an overview. In *The Impact of Desert Dust Across the Mediterranean* (eds. S. Guerzoni and R. Chester). Kluwer, Dordrecht, pp. 133–151.

Prospero J. M. (1999) Long-range transport of mineral dust in the global atmosphere: impact of African dust on the environment of the southeastern United States. *Proc. Natl Acad. Sci. USA* **96**, 3396–3403.

Prospero J. M., Ginoux P., Torres O., Nicholson S. E., and Gill T. E. (2002) Environmental characterization of global sources of atmospheric soil dust identified with the Nimbus 7 total ozone mapping spectrometer (TOMS) absorbing aerosol product. *Rev. Geophys.* **40**, article no. 1002.

Pruppacher H. R. and Klett J. D. (1997) *Microphysics of Clouds and Precipitation*. Kluwer, Dordrecht.

Radke L. F. and Hobbs P. V. (1976) Cloud condensation nuclei on Atlantic Seaboard of United States. *Science* **193**, 999–1002.

Raes F., Bates T. F., McGovern F. M., and van Liedekerke M. (2000) The second aerosol characterization experiment: general overview and main results. *Tellus* **52**, 111–125.

Ramanathan V., Crutzen P. J., Kiehl J. T., and Rosenfeld D. (2001a) Aerosols, climate, and the hydrological cycle. *Science* **294**, 2119–2124.

Ramanathan V., Crutzen P. J., Lelieveld J., Althausen D., Anderson J., Andreae M. O., Cantrell W., Cass G., Chung C. E., Clarke A. D., Collins W. D., Coakley J. A., Dulac F., Heintzenberg J., Heymsfield A. J., Holben B., Hudson J., Jayaraman A., Kiehl J. T., Krishnamurti T. N., Lubin D., Mitra A. P., McFarquhar G., Novakov T., Ogren J. A., Podgorny I. A., Prather K., Prospero J. M., Priestley K., Quinn P. K., Rajeev K., Rasch P. J., Rupert S., Sadourney R., Satheesh S. K., Sheridan P., Shaw G. E., and Valero F. P. J. (2001b) The Indian Ocean experiment: an integrated assessment of the climate forcing and effects of the great Indo-Asian haze. **106**, 28371–28398.

Ramaswamy V., Boucher O., Haigh J., Hauglustaine D., Haywood J. M., Myhre G., Nakajima T., Shi G. Y., and Solomon S. (2001) Radiative forcing of climate change. In *Climate Change 2001: The Scientific Basis. Contribution of Working Group I to the Third Assessment Report of the Intergovernmental Panel on Climate Change* (eds. J. T. Houghton, Y. Ding, D. J. Griggs, M. Noguer, P. van der Linden, X. Dai, and K. Maskell). Cambridge University Press, Cambridge, 2001, pp. 349–416.

Rea D. K. (1994) The paleoclimatic record provided by eolian deposition in the deep sea: the geologic history of wind. *Rev. Geophys.* **32**, 159–195.

Reader M. C., Fung I., and McFarlane N. (1999) The mineral dust aerosol cycle during the Last Glacial Maximum. *J. Geophys. Res.* **104**, 9381–9398.

Rogge W. F., Mazurek M. A., Hildemann L. M., Cass G. R., and Simoneit B. R. T. (1993) Quantification of urban organic aerosols at a molecular level: identification, abundance, and seasonal variation. *Atmos. Environ.* **27A**, 1309–1330.

Rood M. J., Shaw M. A., Larson T. V., and Covert D. S. (1989) Ubiquitous nature of ambient metastable aerosol. *Nature* **337**, 537–539.

Rosenfeld D. (2000) Suppression of rain and snow by urban and industrial air pollution. *Science* **287**, 1793–1796.

Satheesh S. K. and Ramanathan V. (2000) Large differences in tropical aerosol forcing at the top of the atmosphere and Earth's surface. *Nature* **405**, 60–63.

Schauer J. J., Rogge W. F., Hildemann L. M., Mazurek M. A., and Cass G. R. (1996) Source apportionment of airborne particulate matter using organic compounds as tracers. *Atmos. Environ.* **30**, 3837–3855.

Schmid H., Laskus L., Abraham H. J., Baltensperger U., Lavanchy V., Bizjak M., Burba P., Cachier H., Crow D., Chow J., Gnauk T., Even A., ten Brink H. M., Giesen K.-P., Hitzenberger R., Hueglin C., Maenhaut W., Pio C.,

Carvalho A., Putaud J.-P., Toom-Sauntry D., and Puxbaum H. (2001) Results of the "Carbon Conference" international aerosol carbon round robin test stage: I. *Atmos. Environ.* **35**, 2111–2121.

Schulz M., Balkanski Y. J., Guelle W., and Dulac F. (1998) Role of aerosol size distribution and source location in a three-dimensional simulation of a Saharan dust episode test against satellite-derived optical thickness. *J. Geophys. Res.* **103**, 10579–10592.

Schwartz S. E. (1996) The Whitehouse effect—Shortwave radiative forcing of climate by anthropogenic aerosols: an overview. *J. Aerosol Sci.* **27**, 359–382.

Schwartz S. E. and Slingo A. (1996) Enhanced shortwave cloud radiative forcing due to anthropogenic aerosols. In *Clouds, Chemistry, and Climate—Proceedings of NATO Advanced Research Workshop* (eds. P. Crutzen and V. Ramanathan). Springer, Heidelberg, pp. 191–236.

Schwartz S. E., Harshvardhan, and Benkovitz C. (2002) Influence of anthropogenic aerosol on cloud optical depth and albedo shown by satellite measurements and chemical transport modeling. *Proc. Natl Acad. Sci. USA* **99**, 1784–1789.

Seigneur C. (2001) Current status of air quality models for particulate matter. *J. Air Waste Manage. Assoc.* **51**, 1508–1521.

Seinfeld J. H. and Pandis S. N. (1998) *Atmospheric Chemistry and Physics.* Wiley, New York.

Sheridan P. J., Delene D. J., and Ogren J. A. (2001) Four years of continuous surface aerosol measurements from the department of energy's atmospheric radiation measurement program southern great plains cloud and radiation testbed site. *J. Geophys. Res.* **106**, 20735–20747.

Shine K. P. and Forster P. M. D. (1999) The effect of human activity on radiative forcing of climate change: a review of recent developments. *Global Planet. Change* **20**, 205–225.

Sirois A. and Barrie L. A. (1999) Arctic lower tropospheric aerosol trends and composition at Alert, Canada: 1980–1995. *J. Geophys. Res.* **104**, 11599–11618.

Slinn W. G. N. (1983) Air to sea transfer of particles. In *Air-Sea Exchange of Gases and Particles* (eds. P. S. Liss and W. G. N. Slinn). Reidel, Dordrecht, pp. 299–405.

Slinn W. G. N. (1984) Precipitation scavenging. In *Atmospheric Science and Power Production* (ed. D. Randerson). US Department of Energy Report DOE/TIC-27601 (DE84005177), pp. 466–532.

Sokolik I. N. and Toon O. B. (1996) Direct radiative forcing by anthropogenic airborne mineral aerosols. *Nature* **381**, 681–683.

Sokolik I. N., Winker D. M., Bergametti G., Gillette D. A., Carmichael G., Kaufman Y. J., Gomes L., Schuetz L., and Penner J. E. (2001) Outstanding problems in quantifying the radiative impacts of mineral dust. *J. Geophys. Res.* **106**, 18015–18027.

Sparks R. S. J., Bursik M. I., Carey S. N., Gilbert J. S., Glaze L. S., Sigurdsson H., and Woods A. W. (1997) *Volcanic Plumes.* Wiley, Chichester.

Spivakovsky C. M., Logan J. A., Montzka S. A., Balkanski Y. J., Foreman-Fowler M., Jones D. B. A., Horowitz L. W., Fusco A. C., Brenninkmeijer C. A. M., Prather M. J., Wofsy S. C., and McElroy M. B. (2000) Three-dimensional distribution of tropospheric OH: update and evaluation. *J. Geophys. Res.* **105**, 8931–8980.

Steffenson J. P. (1997) The size distribution of microparticles from selected segments of the Greenland Ice Core Project ice core representing different climatic periods. *J. Geophys. Res.* **102**, 26755–26763.

Strawa A. W., Castaneda R., Owano T., Baer D. S., and Paldus B. A. (2003) The measurement of aerosol optical properties using continuous wave cavity ring-down techniques. *J. Atmos. Ocean. Tech.* **20**, 454–465.

Strawbridge K. B. and Hoff R. M. (1996) LITE validation experiment along California's coast: preliminary results. *Geophys. Res. Lett.* **23**, 73–76.

Suess D. T. and Prather K. A. (1999) Mass spectrometry of aerosols. *Chem. Rev.* **99**, 3007ff.

Swap R. J., Annegarn H. J., Suttles J. T., Haywood J., Helmlinger M. C., Hely C., Hobbs P. V., Holben B. N., Ji J., King M. D., Landmann T., Maenhaut W., Otter L., Pak B., Piketh S. J., Platnick S., Privette J., Roy D., Thompson A. M., Ward D., and Yokelson R. (2002) The southern African regional science initiative (SAFARI 2000): overview of the dry season field campaign. *S. Afr. J. Sci.* **98**, 125–130.

Tang I. N. (1996) Chemical and size effects of hygroscopic aerosols on light scattering coefficients. *J. Geophys. Res.* **101**, 19245–19250.

Tang I. N. and Munkelwitz H. R. (1977) Aerosol growth studies: III. Ammonium bisulfate aerosols in a moist atmosphere. *J. Aerosol Sci.* **8**, 321–330.

Tang I. N., Fung K. H., Imre D. G., and Munkelwitz H. R. (1995) Phase transformation and metastability of hygroscopic microparticles. *Aerosol Sci. Technol.* **23**, 443–453.

Tanré D., Bréon F. M., Deuzé J. L., Herman M., Goloub P., Nadal F., and Marchand A. (2001) Global observation of anthropogenic aerosols from satellite. *Geophys. Res. Lett.* **28**, 4555–4558.

Tegen I. and Fung I. (1995) Contribution to the atmospheric mineral aerosol load from land-surface modification. *J. Geophys. Res.* **100**, 18707–18726.

Tegen I. and Lacis A. A. (1996) Modeling of particle size distribution and its influence on the radiative properties of mineral dust aerosol. *J. Geophys. Res.* **101**, 19237–19244.

ten Brink H. M., Kruisz C., Kos G. P. A., and Berner A. (1997) Composition/size of the light-scattering aerosol in The Netherlands. *Atmos. Environ.* **31**, 3955–3962.

Thompson L. G. and Mosley-Thompson E. (1981) Microparticle concentration variations linked with climatic change: evidence from polar ice. *Science* **212**, 812–815.

Tratt D. M., Frouin R. J., and Westphal D. L. (2001) April 1998 Asian dust event: a southern California perspective. *J. Geophys. Res.* **106**, 18371–18379.

Twomey S. (1974) Pollution and planetary albedo. *Atmos. Environ.* **8**, 1251–1256.

Twomey S. (1977) *Atmospheric Aerosols.* Elsevier, New York.

Tyndall G. S., Winker D. M., Anderson T. L., and Eisele F. L. (2003) Advances in laboratory and field measurements. In *Atmospheric Chemistry in a Changing World—an Integration and Synthesis of a Decade of Tropospheric Chemistry Research* (eds. G. P. Brasseur, R. G. Prinn, and A. A. P. Pszenny). Springer, Berlin, pp. 157–184.

Uno I., Amano H., Emori S., Kinoshita K., Matsui I., and Sugimoto N. (2001) Trans-Pacific yellow sand transport observed in April 1998: a numerical simulation. *J. Geophys. Res.* **106**, 18331–18344.

Unsworth M. H. and Fowler D. (eds.) (1988). *Acid Deposition at High Elevation Sites,* NATO ASI Series, C, *Mathematical and Physical Sciences,* Kluwer, Dordrecht, vol. 252.

Van Ham R., Adriaens A., Prati P., Zucchiatti A., Van Vaeck L., and Adams F. (2002) Static secondary ion mass spectrometry as a new analytical tool for measuring atmospheric particles on insulating substrates. *Atmos. Environ.* **36**, 899–909.

Vaughan J. K., Claiborn C., and Finn D. (2001) April 1998 Asian dust event over the Columbia Plateau. *J. Geophys. Res.* **106**, 18381–18402.

Wagener R., Nemesure S., and Schwartz S. E. (1997) Aerosol optical depth over oceans: high space- and time-resolution retrieval and error budget from satellite radiometry. *J. Atmos. Ocean. Technol.* **14**, 577–590.

Watson A. J. (1997) Volcanic iron, $CO_2$, ocean productivity and climate. *Nature* **385**, 587–588.

Weber R. J., McMurry P. H., Mauldin L., Tanner D. J., Eisele F. L., Brechtel F. J., Kreidenweis S. M., Kok G. L., Schillawski R. D., and Baumgardner D. (1998) A study of

new particle formation and growth involving biogenic and
trace gas species measured during ACE 1. *J. Geophys. Res.*
**103**, 16385–16396.

Weber R. J., Chen G., Davis D. D., Mauldin R. L., Tanner D. J.,
Eisele F. L., Clarke A. D., Thornton D. C., and Bandy A. R.
(2001a) Measurements of enhanced $H_2SO_4$ and 3–4 nm
particles near a frontal cloud during the First Aerosol
Characterization Experiment (ACE 1). *J. Geophys. Res.* **106**,
24107–24117.

Weber R. J., Orsini D., Daun Y., Lee Y. N., Klotz P. J., and
Brechtel F. (2001b) A particle-into-liquid collector for rapid
measurement of aerosol bulk chemical composition. *Aerosol
Sci. Technol.* **35**, 718–727.

Whitby K. T. (1978) The physical characteristics of sulfur
aerosols. *Atmos. Environ.* **12**, 135–159.

Wilck M. and Stratmann F. (1997) A 2-D multicomponent
modal aerosol model and its application to laminar flow
reactors. *J. Aerosol. Sci.* **28**, 959–972.

Winker D .M., (1998). LITE—a view from space. Lidar In-
space Technology Experiment. http://www-lite.larc.nasa.
gov/index.html

Winker D. M., Couch R. H., and McCormick M. P. (1996) An
overview of LITE: NASA's Lidar In-space Technology
Experiment. *Proc. IEEE* **84**, 164–180.

Wright D. L., Kasibhatla P. S., McGraw R., and Schwartz S. E.
(2001) Description and evaluation of a six-moment aerosol
microphysical module for use in atmospheric chemical
transport models. *J. Geophys. Res.* **106**, 20275–20291.

Wright D. L., Jr., McGraw R. L., Benkovitz C. M., and Schwartz
S. E. (2000) Six-moment representation of multiple aerosol
populations in a sub-hemispheric chemicaltransformation
model. *Geophys. Res. Lett.* **27**, 967–970.

Zhang Y., Seigneur C., Seinfeld J. H., Jacobson M. Z., and
Binkowski F. S. (1999) Simulation of aerosol dynamics: a
comparative review of algorithms used in air quality models.
*Aerosol Sci. Technol.* **31**, 487–514.

# 4.05
# Biomass Burning: The Cycling of Gases and Particulates from the Biosphere to the Atmosphere

J. S. Levine

*NASA Langley Research Center, Hampton, VA, USA*

## 4.05.1 INTRODUCTION: BIOMASS BURNING, GEOCHEMICAL CYCLING, AND GLOBAL CHANGE

Biomass burning is both a process of geochemical cycling of gases and particulates from the biosphere to the atmosphere and a process of global change. In the preface to the book,

*One Earth, One Future: Our Changing Global Environment* (National Academy of Sciences, 1990), Dr. Frank Press, the President of the National Academy of Sciences, writes:

"Human activities are transforming the global environment, and these global changes have many faces: ozone depletion, tropical deforestation, acid

143

deposition, and increased atmospheric concentrations of gases that trap heat and may warm the global climate."

It is interesting to note that all four global change "faces" identified by Dr. Press have a common thread—they are all caused by biomass burning.

Biomass burning or vegetation burning is the burning of living and dead vegetation and includes human-initiated burning and natural lightning-induced burning. The bulk of the world's biomass burning occurs in the tropics—in the tropical forests of South America and Southeast Asia and in the savannas of Africa and South America. The majority of the biomass burning, primarily in the tropics (perhaps as much as 90%), is believed to be human initiated for land clearing and land-use change. Natural fires triggered by atmospheric lightning only accounts for ~10% of all fires (Andreae, 1991). As will be discussed, a significant amount of biomass burning occurs in the boreal forests of Russia, Canada, and Alaska.

Biomass burning is a significant source of gases and particulates to the regional and global atmosphere (Crutzen *et al.*, 1979; Seiler and Crutzen, 1980; Crutzen and Andreae, 1990; Levine *et al.*, 1995). Its burning is truly a multidiscipline subject, encompassing the following areas: fire ecology, fire measurements, fire modeling, fire combustion, remote sensing, fire combustion gaseous and particulate emissions, the atmospheric transport of these emissions, and the chemical and climatic impacts of these emissions. Recently, a series of dedicated books have documented much of our understanding of biomass burning in different ecosystems. These volumes include: Goldammer (1990), Levine (1991, 1996a,b), Crutzen and Goldammer (1993), Goldammer and Furyaev (1996), van Wilgen *et al.* (1997), Kasischke and Stocks (2000), Innes *et al.* (2000), and Eaton and Radojevic (2001).

## 4.05.2 GLOBAL IMPACTS OF BIOMASS BURNING

On an annual global basis, biomass burning is a significant source of gases and particulates to the atmosphere. The gaseous and particulate emissions produced during biomass burning are dependent on the nature of the biomass matter, which is a function of the ecosystem and the temperature of the fire, which is also ecosystem dependent. In general, biomass is composed mostly of carbon (~45% by weight) and hydrogen and oxygen (~55% by weight), with trace amounts of nitrogen (0.3–3.8% by mass), sulfur (0.1–0.9%), phosphorus (0.01–0.3%), potassium (0.5–3.4%) and still smaller amounts of chlorine, and bromine (Andreae, 1991).

During complete combustion, the burning of biomass matter produces carbon dioxide ($CO_2$) and water vapor as the primary products, according to the reaction

$$CH_2O + O_2 \rightarrow CO_2 + H_2O$$

where $CH_2O$ represents the appropimate average chemical composition of biomass matter. In the more realistic case of incomplete combustion in cooler and/or oxygen-deficient fires, i.e., the smoldering phase of burning, carbon is released in the forms of carbon monoxide (CO), methane ($CH_4$), nonmethane hydrocarbons (NMHCs), and various partially oxidized organic compounds, including aldehydes, alcohols, ketones, and organic acids and particulate black (soot) carbon. Nitrogen is present in biomass mostly as amino groups ($R-NH_2$) in the amino acids of proteins. During combustion the nitrogen is released by pyrolytic decomposition of the organic matter and partially or completely oxidized to various volatile nitrogen compounds, including molecular nitrogen ($N_2$), nitric oxide (NO), nitrous oxide ($N_2O$), ammonia ($NH_3$), hydrogen cyanide (HCN), cyanogen (NCCN), organic nitriles (acetonitrile ($CH_3CN$), acrylonitrile ($CH_2CHCN$), and propionitrile ($CH_3CH_2CN$)), and nitrates. The sulfur in biomass is organically bound in the form of sulfur-containing amino acids in proteins. During burning the sulfur is released mostly in the form of sulfur dioxide ($SO_2$) and smaller amounts of carbonyl sulfide (COS) and nonvolatile sulfate ($SO_4^-$). About one-half of the sulfur in the biomass matter is left in the burn ash, whereas very little of the fuel nitrogen is left in the ash.

Laboratory biomass burning experiments conducted by Lobert *et al.* (1991) have identified the carbon (Table 1) and nitrogen (Table 2) compounds released to the atmosphere by burning. The major gases produced during the biomass burning process include $CO_2$, CO, $CH_4$, oxides of nitrogen ($NO_x = NO + NO_2$), and $NH_3$. Carbon dioxide and $CH_4$ are greenhouse gases, which trap Earth-emitted infrared radiation and lead to global warming. Carbon monoxide, methane, and the oxides of nitrogen lead to the photochemical production of ozone ($O_3$) in the troposphere. In the troposphere, $O_3$ is harmful to both vegetation and humans at concentrations not far above the global background levels. Nitric oxide leads to the chemical production of nitric acid ($HNO_3$) in the troposphere. Nitric acid is the fastest growing component of acidic precipitation. Ammonia is the only basic gaseous species that neutralizes the acidic nature of the troposphere.

It has been reported that the burning of vegetation results in the complete release of mercury contained in the biomass (Friedli *et al.*, 2001). About 95% of the mercury is emitted as

**Table 1** Carbon gases produced during biomass burning.

| Compound | Mean emission factor relative to the fuel C (%) |
|---|---|
| Carbon dioxide ($CO_2$) | 82.58 |
| Carbon monoxide (CO) | 5.73 |
| Methane ($CH_4$) | 0.424 |
| Ethane ($CH_3CH_3$) | 0.061 |
| Ethene ($CH_2=CH_2$) | 0.123 |
| Ethine (CH≡CH) | 0.056 |
| Propane ($C_3H_8$) | 0.019 |
| Propene ($C_3H_6$) | 0.066 |
| *n*-butane ($C_4H_{10}$) | 0.005 |
| 2-butene (*cis*) ($C_4H_8$) | 0.004 |
| 2-butene (*trans*) ($C_4H_8$) | 0.005 |
| *i*-butene, *i*-butene ($C_4H_8 + C_4H_8$) | 0.033 |
| 1,3-butadiene($C_4H_6$) | 0.021 |
| *n*-pentane ($C_3H_{12}$) | 0.007 |
| Isoprene ($C_5H_8$) | 0.008 |
| Benzene ($C_6H_6$) | 0.064 |
| Toluene ($C_7H_8$) | 0.037 |
| *m*-, *p*-xylene ($C_8H_{10}$) | 0.011 |
| *o*-xylene ($C_8H_{10}$) | 0.006 |
| Methyl chloride ($CH_3Cl$) | 0.010 |
| NMHC (As C) ($C_2$–$C_8$) | 1.18 |
| Ash (As C) | 5.00 |
| *Total sum C* | 94.92 (including ash) |

Source: Lobert *et al.* (1991).

**Table 2** Nitrogen gases produced during biomass burning.

| Compound | Mean emission factor relative to the fuel N (%) |
|---|---|
| Nitrogen oxides ($NO_x$) | 13.55 |
| Ammonia ($NH_3$) | 4.15 |
| Hydrogen cyanide (HCN) | 2.64 |
| Acetonitrile ($CH_3CN$) | 1.00 |
| Cyanogen (NCCN) (As N) | 0.023 |
| Acrylonitrile ($CH_2CHCN$) | 0.135 |
| Propionitrile ($CH_3CH_2CN$) | 0.071 |
| Nitrous oxide ($N_2O$) | 0.072 |
| Methylamine ($CH_3NH_2$) | 0.047 |
| Dimethylamine (($CH_3)_2NH$) | 0.030 |
| Ethylamine ($CH_3CH_2NH_2$) | 0.005 |
| Trimethylamine (($CH_3)N$) | 0.02 |
| 2-methyl-1-butylamine ($C_5H_{11}NH_2$) | 0.04 |
| *n*-pentylamine (*n*-$C_5H_{11}NH_2$) | 0.137 |
| Nitrates (70% $HNO_3$) | 1.10 |
| Ash (As N) | 9.94 |
| *Total sum N (As N)* | 33.66 (Including ash) |
| Molecular nitrogen ($N_2$) | 21.60 |
| Higher HC and particles | 20 |

Source: Lobert *et al.* (1991).

elemental mercury and the remainder emitted as particulate mercury. Friedli *et al.* (2001) concluded that the mercury released by burning becomes part of the global mercury reservoir and undergoes chemical transformation in clouds and in the free troposphere and eventually returns to the surface via wet and dry deposition.

Particulates, small (usually ~10 μm or smaller) solid particles, such as smoke or soot particles, are also produced during the burning process and released into the atmosphere. These solid particulates absorb and scatter incoming sunlight and hence impact the local, regional, and global climate. In addition, these particulates (specifically particulates 2.5 μm or smaller) can lead to various human respiratory and general health problems when inhaled. The gases and particulates produced during biomass burning lead to the formation of "smog." The word "smog" was coined as a combination of smoke and fog and is now used to describe any smoky or hazy pollution in the atmosphere.

Gaseous and particulate emissions produced during biomass burning and released into the atmosphere impact the local, regional, and global atmosphere and climate in several different ways as follows.

(i) Biomass burning is a significant global source of $CO_2$ and $CH_4$. Both gases are greenhouse gases that lead to global warming.

(ii) Biomass burning is a significant global source of CO, $CH_4$, nonmethane hydrocarbons, and oxides of nitrogen. These gases lead to the photochemical production of $O_3$ in the troposphere (this chemistry is outlined in a later section). Tropospheric ozone is a pollutant and irritant and has a negative impact on plant, animal, and human life.

(iii) Methyl chloride and methyl bromide, while only released in trace amounts during biomass burning, have a negative impact on stratospheric ozone (this chemistry is outlined in a later section). Methyl chloride and methyl bromide produced during biomass burning will become even more important in the future, as human-produced sources of chlorine and bromine are phased out as a result of the Montreal Protocol banning gases containing chlorine and bromine.

(iv) Particulates produced during biomass burning absorb and scatter incoming solar radiation, which impact climate. In addition, particulates produced by biomass burning lead to reduced atmospheric visibility. Some studies suggest that atmospheric particulates produced during biomass burning may directly enter the stratosphere via strong vertical thermal convective currents produced during the fire.

(v) Particulates produced during biomass burning become cloud condensation nuclei (CCN) and impact the formation and distribution of clouds.

(vi) Particulates produced during biomass burning, particularly particulates of 10 μm or less in diameter, lead to severe respiratory problems when inhaled.

### 4.05.3 ENHANCED BIOGENIC SOIL EMISSIONS OF NITROGEN AND CARBON GASES: A POSTFIRE EFFECT

Measurements have shown that in addition to the instantaneous production of trace gases and particulates resulting from the combustion of biomass matter, burning also enhances the biogenic emissions of NO and $NO_2$ from soil (Anderson *et al.*, 1988; Levine *et al.*, 1988, 1991, 1996a,b) and the biogenic emission of CO from soil (Zepp *et al.*, 1996). It is believed that enhanced biogenic soil emissions of NO and $NO_2$ are related to increased concentrations of ammonium found in soil following burning. Ammonium, a component of the burn ash, is the substrate in nitrification, which is the microbial process believed to be responsible for the production of NO and $NO_2$ (Levine *et al.*, 1988, 1991, 1996a,b). The postfire enhanced biogenic soil emissions of NO and $NO_2$ may be comparable to or even surpass the instantaneous production of these gases during biomass burning (Harris *et al.*, 1996).

### 4.05.4 THE GEOGRAPHICAL DISTRIBUTION OF BIOMASS BURNING

The locations of biomass burning are varied and include tropical savannas (Figure 1), tropical, temperate and boreal forests (Figure 2) and

agricultural lands after the harvest. The burning of fuelwood for domestic use is another source of biomass burning. Global estimates of the annual amounts of biomass burning from these sources are estimated in Table 3 (Andreae, 1991). In Table 3, the unit of biomass burned is Tg dm $yr^{-1}$ (1 Tg $= 10^{12}$ g $= 10^6$ metric tons; dm = dry matter (biomass matter)). As already noted, biomass matter is $\sim$45% by weight composed of carbon. Table 3 also gives estimates of the carbon released (Tg C $yr^{-1}$) by the burning of this biomass (the total biomass burned is multiplied by 45% to determine the amount of carbon released into the atmosphere during burning). Combining estimates of the total amount of biomass matter burned per year (Table 3) with measurements of the gaseous and particulate emissions from biomass burning (Tables 1 and 2) permits estimates of the global production and release into the atmosphere of gases and particulates from burning. Estimates of the global contribution of biomass burning are summarized in Table 4 (Andreae, 1991). The data in Tables 3 and 4 clearly indicate that biomass burning is a truly global process of major importance in the global budgets of atmospheric gases and particulates.

### 4.05.5 BIOMASS BURNING IN THE BOREAL FORESTS

In the past, it was generally assumed that biomass burning was primarily a tropical phenomenon. This is because most of the information that we have on the geographical

**Figure 1** A savanna fire with its characteristic long fire front as it traverses across the savanna in South Africa. The low-intensity fire consumes grass and shrubs and consumes on an annual and global basis more total biomass than any other kind of fire (see Table 3) (photograph by J. S. Levine, NASA).

**Figure 2** A boreal forest fire in Canada. The boreal forest fire is the most intense and energetic fire. The gases and particulates produced during boreal forest fires can be directly transported into the stratosphere by the strong vertical thermal convection currents produced by the fire (photograph by B. J. Stocks, Canadian Forest Service).

**Table 3** Global estimates of annual amounts of biomass burning and of the resulting release of carbon to the atmosphere.

| Source | Biomass burned (Tg dm yr$^{-1}$) | Carbon released (Tg C yr$^{-1}$) |
|---|---|---|
| Savanna | 3,690 | 1,660 |
| Agricultural waste | 2,020 | 910 |
| Fuelwood | 1,430 | 640 |
| Tropical forests | 1,260 | 570 |
| Temperate/boreal forests | *280* | *130* |
| World totals | 8,680 | 3,910 |

Source: Andreae (1991).

and temporal distribution of biomass burning is largely based on tropical burning. Very little information is available on the geographical and temporal distribution on biomass burning in the boreal forests, which cover ~25% of the world's forests. To illustrate how our knowledge of the geographical extent of burning in the world's boreal forests has increased in recent years, consider the following: early estimates based on surface fire records and statistics suggested that 1.5 Mha (1 ha = 2.47 acres) of boreal forests burn annually (Seiler and Crutzen, 1980). Later studies, based on more comprehensive surface fire records and statistics, indicated that earlier values underestimated burning in the world's boreal forests and that an average of 8 Mha burned annually during the 1980s, with great year-to-year fluctuations (Stocks *et al.*, 1991). One of the largest fires ever

measured occurred in the boreal forests of the Heilongjiang province of Northeastern China in May 1987. In less than four weeks, more than 1.3 Mha of boreal forest were burned (Levine *et al.*, 1991; Cahoon *et al.*, 1994). At the same time, extensive fire activity occurred across the Chinese border in Russia, particularly in the area east of Lake Baikal between the Amur and Lena rivers. Estimates based on NOAA AVHRR imagery indicate that 14.4 Mha in China and Siberia were burned in 1987 (Cahoon *et al.*, 1994), dwarfing earlier estimates of boreal forest fire burned area.

While 1987 was an extreme fire year in Eastern Asia, the sparse database may suggest a fire trend. Is burning in the boreal forests increasing with time, or are satellite measurements providing more accurate data? Satellite measurements are certainly providing a more accurate assessment of the extent and frequency of burning in the world's boreal forests. As global warming continues, warmer and dryer conditions in the world's boreal forests are predicted to result in more frequent and larger fires and greater production of $CO_2$ by these fires. The increased burning will have an amplifying effect on global warming!

Calculations using the satellite-derived burn area and measured emission ratios of gases for boreal forest fires indicate that the Chinese and Siberian fires of the 1987 contributed ~20% of the total $CO_2$ produced by savanna burning, 36% of the total CO produced by savanna burning, and 69% of the total $CH_4$ produced by savanna burning

**Table 4** Comparison of global emissions from biomass burning with emissions from all sources (including biomass burning).

| Species | Biomass burning (Tg element[a] yr$^{-1}$) | All sources (Tg element[a] yr$^{-1}$) | % due to biomass burning |
|---|---|---|---|
| $CO_2$ (gross)[b] | 3,500 | 8,700 | 40 |
| $CO_2$ (Net)[c] | 1,800 | 7,000 | 26 |
| CO | 350 | 1,100 | 32 |
| $CH_4$ | 38 | 380 | 10 |
| NMHC[d] | 24 | 100 | 24 |
| $N_2O$ | 0.8 | 13 | 6 |
| $NO_x$ | 8.5 | 40 | 21 |
| $NH_3$ | 5.3 | 44 | 12 |
| Sulfur | 2.8 | 150 | 2 |
| COS | 0.09 | 1.4 | 6 |
| $CH_3Cl$ | 0.51 | 2.3 | 22 |
| $H_2$ | 19 | 75 | 25 |
| Tropospheric $O_3$ | 420 | 1,100 | 38 |
| TPM[e] | 104 | 1,530 | 7 |
| POC[f] | 69 | 180 | 39 |
| EC[g] | 19 | <22 | >86 |

Source: Andreae (1991).

[a] Tg element yr$^{-1}$ where C, N, S, and Cl are the elements. [b] Biomass burning plus fossil fuel burning. [c] Deforestation plus fossil fuel burning. [d] Nonmethane hydrocarbons (excluding isoprene and terpenes). [e] Total particulate matter (Tg yr$^{-1}$). [f] Particulate organic matter (including elemental carbon). [g] (black-soot) carbon.

(Cahoon *et al.*, 1994). Since savanna burning represents the largest component of tropical burning in terms of the vegetation consumed by fire (Table 3), it is apparent that the atmospheric emissions from boreal forest burning must be included in global species budgets.

There are several reasons that burning in the world's boreal forests is very important:

(i) The boreal forests are very susceptible to global warming. Small changes in the surface temperature can significantly influence the ice/snow/albedo feedback. Thus, infrared absorption processes by fire-produced greenhouse gases, as well as fire-induced changes in surface albedo and infrared emissivity in the boreal forest regions are more environmentally significant than in the tropics.

(ii) In the world's boreal forests, global warming will result in warmer and drier conditions. This, in turn, may result in enhanced frequency of fire and the accompanying enhanced production of greenhouse gases that will amplify the greenhouse effect.

(iii) Fires in the boreal forests are the most energetic in nature. The average fuel consumption per unit area in the boreal forest is $\sim 2.5 \times 10^4$ kg ha$^{-1}$, which is about an order of magnitude greater than in the tropics. Large boreal forest fires typically spread very quickly, most often as "crown fires," causing the burning of the entire tree up to and including the crown (Figure 3). Large boreal forest fires release enough energy to generate convective smoke columns that routinely reach well into the upper troposphere, and on

**Figure 3** A crown boreal forest fire in Canada. The boreal forest crown fire consumes the entire tree up to and including the crown (photograph by B. J. Stocks, Canadian Forest Service).

occasion, may directly penetrate across the tropopause into the stratosphere. The tropopause is at a minimum height over the world's boreal forests. As an example, a 1986 forest fire in northwestern Ontario (Red Lake) generated a convective smoke column 12–13 km in height, penetrating across the tropopause into the stratosphere (Stocks and Flannigan, 1987). Fromm *et al.* (2000) have found a strong link between boreal forest fires in Canada and eastern Russia in 1998 and increased stratospheric aerosols during the same period.

(iv) The cold temperature of the troposphere over the world's boreal forests results in low levels of tropospheric water vapor. The deficiency of tropospheric water vapor and the scarcity of incoming solar radiation over most of the year results in very low photochemical production of the hydroxyl (OH) radical over the boreal forests. The OH radical is the overwhelming chemical scavenger in the troposphere and controls the atmospheric lifetime of many tropospheric gases. The very low concentrations of the OH radical over the boreal forests result in enhanced atmospheric lifetimes for most tropospheric gases, including the gases produced by biomass burning. Hence, gases produced by burning, such as CO, $CH_4$, and the oxides of nitrogen, will have enhanced atmospheric lifetimes over the boreal forest.

New information about burning in the world's boreal forests based on satellite measurements was reported by Kasischke *et al.* (1999). Some of the findings reported in this study are summarized here.

(i) Fires in the boreal forest covering at least $10^5$ ha are not uncommon.

(ii) In the boreal forests of North America, most fires (>90%) are crown fires. The remainder are surface fires. Crown fires consume much more fuel (30–40 t of biomass material per hectare burned) than surface fires (8–12 t of biomass material per hectare burned).

(iii) The fire record for North America since the early 1970s clearly shows the episodic nature of fire in the boreal forests. Large fire years occur during extended periods of drought, which allow naturally ignited fires (i.e., lightning ignited fires) to burn large areas. Since 1970, the area burned during six episodic fire years in the North American boreal forest was 6.2 Mha $yr^{-1}$, while 1.5 Mha burned per year in the remaining years. There is evidence that a similar episodic pattern of fire may also exist in the Russia boreal forest.

(iv) The fire data in the North American boreal forest show a significant increase in the annual area burned since the early 1970s, with an average 1.5 Mha $yr^{-1}$ burning during the 1970s and 3.2 Mha $yr^{-1}$ burning during the 1990s. This increase in burning corresponds to rises of 1.0–1.6 °C over the same period (Hansen *et al.*, 1996). The projected 2–4 °C increase in temperature due to projected increases in greenhouse gases should result in high levels of fire activity

throughout the world's foreal forests in the future (Stocks *et al.*, 1991).

(v) During typical years in the boreal forests, the amounts of biomass consumed during fire ranges between 10 t and 20 t $ha^{-1}$. During the drought years with episodic fires, the amounts of biomass consumed during biomass burning may be as high as 50–60 t $ha^{-1}$. Assuming that biomass is ~50% carbon by mass, such amounts would release 450–600 Tg C globally. These amounts are considerably higher than the often-quoted value for total carbon released by biomass burning in the world's boreal and temperate forests of 130 Tg C globally (Andreae, 1991: see Table 3).

## 4.05.6 ESTIMATES OF GLOBAL BURNING AND GLOBAL GASEOUS AND PARTICULATE EMISSIONS

Global estimates of the annual amounts of biomass burning from these sources are estimated in Table 3 (Andreae, 1991). In Table 3, the unit of biomass burned is Tg dm $yr^{-1}$. As already noted, biomass matter is composed of ~45% by weight of carbon. Table 3 also gives estimates of the carbon released (Tg C $yr^{-1}$) by the burning of this biomass. Combining estimates of the total amount of biomass matter burned per year (Table 3) with measurements of the gaseous and particulate emissions from biomass burning (Tables 1 and 2) permits estimates of the global production and release into the atmosphere of gases and particulates from burning. Estimates of the global contribution of biomass burning are summarized in Table 4 (Andreae, 1991). The data in Tables 3 and 4 clearly indicate that biomass burning is a truly global process of major importance in the global budgets of atmospheric gases and particulates.

## 4.05.7 CALCULATION OF GASEOUS AND PARTICULATE EMISSIONS FROM FIRES

To assess both the environmental and health impacts of biomass burning, information is needed on the gaseous and particulate emissions produced during the fire and released into the atmosphere. The calculation of gaseous emissions from vegetation and peat fires can be calculated using a form of an expression from Seiler and Crutzen (1980) for each burning ecosystem/terrain:

$$M = ABE \qquad (1)$$

where $M$ is total mass of vegetation or peat consumed by burning (tons), $A$ the area burned

($km^2$), $B$ the biomass loading ($t\ km^{-2}$), and $E$ the burning efficiency (dimensionless). The total mass of carbon, $M(C)$, released to the atmosphere during burning is related to $M$ by the following expression:

$$M(C) = CM \text{(tons of carbon)} \qquad (2)$$

$C$ is the mass percentage of carbon in the biomass. For tropical vegetation, $C = 0.45$ (Andreae, 1991); for peat, $C = 0.50$ (Yokelson *et al.*, 1996). The mass of $CO_2$, $M(CO_2)$, released during the fire is related to $M(C)$ by the following expression:

$$M(CO_2) = CE\ M(C) \qquad (3)$$

The combustion efficiency (CE) is the fraction of carbon emitted as $CO_2$ relative to the total carbon compounds released during the fire. For tropical vegetation fires, $CE = 0.90$ (Andreae, 1991); for peat fires, $CE = 0.77$ (Yokelson *et al.*, 1997).

Once the mass of $CO_2$ produced by burning is known, the mass of any other species, $X_i$ ($M(X_i)$), produced by burning and released to the atmosphere can be calculated with knowledge of the $CO_2$-normalized species emission ratio ($ER(X_i)$). The emission ratio is the ratio of the production of species $X_i$ to the production of $CO_2$ in the fire. The mass of species, $X_i$, is related to the mass of $CO_2$ by the following expression:

$$M(X_i) = ER(X_i)\ M(CO_2) \text{(tons of } X_i) \qquad (4)$$

where $X_i =$ CO, $CH_4$, $NO_x$, $NH_3$, and $O_3$. It is important to re-emphasize that $O_3$ is not a direct product of biomass burning. However, $O_3$ is produced via photochemical reactions of CO, $CH_4$, and $NO_x$, all of which are produced directly by biomass burning. Hence, the mass of ozone resulting from biomass burning may be calculated by considering the ozone precursor gases produced by biomass burning. Values for emission ratios for tropical forest fires and peat fires are summarized in Table 5.

To calculate the total particulate matter (TPM) released from tropical forest fires and peat fires, we use the following expression (Ward, 1990)

$$TPM = MP \text{(tons of carbon)} \qquad (5)$$

where $P$ is the conversion of biomass matter or peat matter to particulate matter during burning. For the burning of tropical vegetation, $P = 20$ t of TPM per kiloton of biomass consumed by fire; for peat burning, we assume $P = 35$ t of TPM per kiloton of organic soil or peat consumed by fire (Ward, 1990).

Perhaps the major uncertainties in the calculation of gaseous and particulate emissions resulting from fires involve poor or incomplete information about four fire and ecosystem parameters: (i) the area burned (A), (ii) the ecosystem or terrain that burned, i.e., forests, grasslands, agricultural lands, peat lands, etc., and (iii) the biomass loading (B), i.e., the amount of biomass per unit area of the ecosystem prior to burning, and (iv) the fire efficiency (C), i.e., the amount of biomass in the burned ecosystem that was actually consumed by burning.

## 4.05.8  BIOMASS BURNING AND ATMOSPHERIC NITROGEN AND OXYGEN

Biomass burning is both an instantaneous source (combustion of biomass) and a long-term source (enhanced biogenic soil emissions via increased nitrification and denitrification in soil) of gases to the atmosphere. These gases impact the chemistry of the troposphere and stratosphere, as outlined in this chapter. In addition, biomass burning may also impact atmospheric concentrations and the biogeochemical cycling of nitrogen and oxygen, the two major constituents of the atmosphere. Through the process of nitrogen fixation, molecular nitrogen ($N_2$) is transformed to the surface in the form of "fixed" nitrogen, i.e., ammonium

**Table 5**  Emission ratios for tropical forest fires and peat fires.

| Species | Tropical forest fires | References | Peat fires | References |
|---------|----------------------|------------|------------|------------|
| $CO_2$ | 90.00% | Andreae (1991) | 77.05% | Yokelson *et al.* (1997) |
| CO | 8.5% | Andreae *et al.* (1988) | 18.15% | Yokelson *et al.* (1997) |
| $CH_4$ | 0.32% | Blake *et al.* (1996) | 1.04% | Yokelson *et al.* (1997) |
| $NO_x$ | 0.21% | Andreae *et al.* (1988) | 0.46% | Derived from Yokelson *et al.* (1997) (see text) |
| $NH_3$ | 0.09% | Andreae *et al.* (1998) | 1.28% | Yokelson *et al.* (1997) |
| $O_3$ | 0.48% | Andreae *et al.* (1988) | 1.04% | Derived from Yokelson *et al.* (1997) (see text) |
| TPM[a] | $20\ t\ kt^{-1}$ | Ward (1990) | $35\ t\ kt^{-1}$ | Ward (1990) |

[a] Total particulate matter emission ratios are in units of tons/kiloton (tons of total particulate matter/kiloton of biomass or peat material consumed by fire).

($NH_4^+$) and nitrate ($NO_3^-$). Nitrogen fixation results from both natural processes (biological fixation in root modules in certain agricultural crops and atmospheric lightning) and human processes (the production of nitrogen fertilizer and high-temperature combustion). The world's use of industrially fixed nitrogen fertilizer has increased from $\sim 3$ Tg N yr$^{-1}$ in 1940 to $\sim 75$ Tg N yr$^{-1}$ in 1990 (Levine *et al.*, 1996b). The "fixed" nitrogen in the forms of $NH_4^+$ and $NO_3^-$ is returned to the atmosphere mainly in the form of $N_2$, with smaller amounts of $N_2O$, and still smaller amounts of NO by denitrification and in the form of NO by nitrification. Burning or "pyrodenitrification" may also be an important source of nitrogen, mostly in the form of $N_2$, from the biosphere to the atmosphere (Lobert *et al.*, 1991; Levine *et al.*, 1996b). The problem is that it is difficult to quantify the amount of $N_2$ released during burning (Lobert *et al.*, 1991); however, biomass burning or pyrodenitrification may prove to be an important process in the recycling of nitrogen compounds from the biosphere to the atmosphere.

Burning impacts the concentration of atmospheric oxygen in two ways. Carbon released during the burning of biomass combines with atmospheric oxygen to form $CO_2$. Hence, burning is a sink for atmospheric oxygen. In addition, biomass burning destroys the very source of atmospheric oxygen—its production in the biosphere via the process of photosynthesis in the world's forests. In addition, the burial of fire-produced charcoal is a long-term source of oxygen.

# 4.05.9 ATMOSPHERIC CHEMISTRY RESULTING FROM GASEOUS EMISSIONS FROM THE FIRES

## 4.05.9.1 Chemistry of the Hydroxyl Radical (OH) in the Troposphere

The hydroxyl radical (OH) is the major chemical scavenger in the troposphere and it controls the atmospheric lifetime of most gases in the troposphere. The atmospheric lifetime, $t$, of any gas, $x_i$, that reacts with the OH radical is given by the following expression $t = 1/k[OH]$ where $k$ is the kinetic reaction rate for the reaction between OH and $x_i$ and [OH] is the concentration of the OH radical (molecules cm$^{-3}$).

The concentration of the OH radical is controlled by the balance between its chemical production and destruction. The OH radical is formed by the reaction of excited atomic oxygen ($O(^1D)$) with water vapor

$$O(^1D) + H_2O \xrightarrow{k_1} 2OH$$

Tropospheric excited atomic oxygen ($O(^1D)$) is produced by the photolysis of $O_3$. In the troposphere, this photolysis reaction occurs over a very narrow spectral interval, between 290 nm and 310 nm. The production of excited atomic oxygen decreases as the latitude increases, i.e., less incoming solar radiation for photolysis is available. The bulk of the water vapor in the atmosphere resides in the troposphere. The amount of $H_2O$ in the atmosphere is controlled by the saturation vapor pressure, which decreases with decreasing atmospheric temperature, i.e., as altitude or latitude increases.

The OH radical is destroyed via its reactions with CO and $CH_4$, both important products of biomass burning:

$$OH + CO \xrightarrow{k_2} H + CO_2$$

and

$$OH + CH_4 \xrightarrow{k_3} CH_3 + H_2O$$

Assuming that the production of OH is controlled by the reaction of water vapor with excited atomic oxygen and its loss is controlled by its reactions with CO and $CH_4$, the concentration of the OH radical (molecules cm$^{-3}$) is determined by dividing the OH production term by its destruction terms:

$$[OH] = (2k_1[O(^1D)][H_2O])/(k_2[CO] + k_3[CH_4])$$

## 4.05.9.2 Production of $O_3$ in the Troposphere

In addition to controlling the chemical destruction of OH, the oxidation of CO and $CH_4$ by OH outlined above initiates the CO and $CH_4$ oxidation schemes, which lead to the photochemical production of $O_3$ in the troposphere.

### 4.05.9.2.1 CO oxidation chain

$$CO + OH \rightarrow CO_2 + H$$

$$H + O_2 + M \rightarrow HO_2 + M$$

$$HO_2 + NO \rightarrow NO_2 + OH$$

$$NO_2 + hv \rightarrow NO + O$$

$$O + O_2 + M \rightarrow O_3 + M$$

Net reaction: $CO + 2O_2 \rightarrow CO_2 + O_3$

Note that the key gases in the CO oxidation chain leading to the photochemical production of tropospheric ozone, CO and NO are both produced by biomass burning.

### 4.05.9.2.2  *Methane oxidation chain*

$$CH_4 + OH \rightarrow CH_3 + H_2O$$

$$CH_3 + O_2 + M \rightarrow CH_3O_2 + M$$

$$CH_3O_2 + NO \rightarrow CH_3O + NO_2$$

$$CH_3O + O_2 \rightarrow CH_2O + HO_2$$

$$HO_2 + NO \rightarrow NO_2 + OH$$

$$2(NO_2 + h\nu \rightarrow NO + O)$$

$$2(O + O_2 + M \rightarrow O_3 + M)$$

Net reaction : $CH_4 + 4O_2 \rightarrow CH_2O + H_2O + 2O_3$

Note that the key gases in the $CH_4$ oxidation chain leading to the photochemical production of tropospheric ozone, $CH_4$ and NO are both produced by biomass burning.

### 4.05.9.3  Chemistry of Nitrogen Oxides in the Troposphere

In addition to being a key player in the CO and $CH_4$ oxidation chains leading to the chemical production of $O_3$ in the troposphere, NO also leads to the chemical production of $HNO_3$, the fastest growing component of acidic precipitation. NO is chemically transformed to nitrogen dioxide ($NO_2$) and then to NO via the following reactions:

$$NO + O_3 \rightarrow NO_2 + O_2$$

$$NO + HO_2 \rightarrow NO_2 + OH$$

$$NO + CH_3O_2 \rightarrow NO_2 + CH_3O$$

$$NO_2 + OH + M \rightarrow HNO_3 + M$$

### 4.05.9.4  Chemistry of the Stratosphere

Approximately 90% of the ozone in the atmosphere is found in the stratosphere (15–50 km), with only ~10% in the troposphere (0–15 km). Stratospheric ozone is very important because it absorbs ultraviolet radiation (200–300 nm) from the Sun and shields the surface from this biologically lethal radiation. Stratospheric ozone is destroyed via a series of chemical reactions involving NO, OH, Cl, and Br. These species destroy stratospheric ozone through the following catalytic cycle where $X$ may be any of the following: NO, OH, Cl, or Br (Wayne, 1991)

$$X + O_3 \rightarrow XO + O_2$$

$$XO + O \rightarrow X + O_2$$

Net reaction: $O + O_3 \rightarrow 2O_2$

Another competing $O_3$ stratospheric catalytic cycle where $X$ = OH, Cl, or Br and $Y$ = OH, Cl, or Br is represented by (Wayne, 1991)

$$X + O_3 \rightarrow XO + O_2$$

$$Y + O_3 \rightarrow YO + O_2$$

$$XO + YO \rightarrow XY + O_2$$

$$XY + h\nu \rightarrow X + Y$$

Net reaction: $O + O_3 \rightarrow 2O_2$

### 4.05.10  A CASE STUDY OF BIOMASS BURNING: THE 1997 WILDFIRES IN SOUTHEAST ASIA

Extensive and widespread tropical forest and peat fires swept throughout Kalimantan and Sumatra, Indonesia, between August and December 1997 (Brauer and Hisham-Hishman, 1988; Hamilton *et al.*, 2000). The fires resulted from burning for land clearing and land-use change. However, the severe drought conditions resulting from El Nino caused small land-clearing fires to become large uncontrolled wildfires. Based on satellite imagery, it has been estimated that a total of $4.56 \times 10^4$ km$^2$ burned on Kalimantan and Sumatra between August and December 1997 (Liew *et al.*, 1998). The gaseous and particulate emissions produced in these fires and released into the atmosphere reduced the atmospheric visibility, impacted the composition and chemistry of the atmosphere, and affected human health. Some of the consequences of the fires in Southeast Asia were: (i) more than 200 million people were exposed to high levels of air pollution and particulates produced during the fires, (ii) more than 20 million smoke-related health problems were recorded, (iii) fire-related damage cost in excess of $4 billion, (iv) on September 26, 1997, a commercial airliner (Garuda Airlines Airbus 300-B4) crashed in Sumatra owing to very poor visibility due to smoke from the fires on landing with 234 passengers killed, and (v) on September 27, 1997, two ships collided at sea due to poor visibility in the Strait of Malacca, off the coast of Malaysia, with 29 crew members killed. International concern about the environmental and health impacts of these fires was great. Three different agencies of the United Nations organized workshops and reports on the environmental and health impacts of these fires: The World Meteorological Organization (WMO) Workshop on Regional Transboundary Smoke and Haze in Southeast Asia, Singapore, June 2–5, 1998, The World Health Organization (WHO) Health Guidelines for Forest Fires Episodic Events, Lima, Peru, October 6–9, 1998, and the United Nations

Environmental Program (UNEP) *Report on Wild-land Fires and the Environment: A Global Synthesis*, published in February, 1999 (Levine *et al.*, 1999). The Indonesian fires formed the basis of an article in *National Geographic* magazine, entitled, *Indonesia's Plague of Fire* (Simons, 1998).

Indonesia ranks third, after Brazil and the Democratic Republic of the Congo (formerly Zaire), in its area of tropical forest. Of Indonesia's total land area of 1.9 Mkm$^2$, current forest cover estimates range from 0.9 Mkm$^2$ to 1.2 Mkm$^2$, or 48% to 69% of the total. Forests dominate the landscape of Indonesia (Makarim *et al.*, 1998). Large areas of Indonesian forests burned in 1982 and 1983. In Kalimantan alone, the fires burned from 2.4 Mha to 3.6 Mha of forests (Makarim *et al.*, 1998). It is interesting to note that there is an uncertainty of 1.2 Mha or an uncertainty of 50% in our knowledge of the burned area of fires that occurred 16 yr ago!

Liew *et al.* (1998) analyzed 766 SPOT "quick-look" images with almost complete coverage of Kalimantan and Sumatra from August to December 1997. Liew *et al.* (1998) estimate the burned area in Kalimantan to be $3.06 \times 10^4$ km$^2$ and the burned area in Sumatra to be $1.5 \times 10^4$ km$^2$, for a total burned area of $4.56 \times 10^4$ km$^2$. (This is equivalent to the combined areas of the states of Rhode Island, Delaware, Connecticut, and New Jersey, in the United States.) The estimate of Liew *et al.* (1998) represents only a lower limit estimate of the area burned in Southeast Asia in 1997, since the SPOT data only covered Kalimantan and Sumatra and did not include fires on the other Indonesian islands of Irian Jaya, Sulawesi, Java, Sumbawa, Komodo, Flores, Sumba, Timor, and Wetar or the fires in the neighboring countries of Malaysia and Brunei.

What is the nature of the ecosystem/terrain that burned in Kalimantan and Sumatra? In October 1997, NOAA satellite monitoring produced the following distribution of fire hot spots in Indonesia (UNDAC, 1998): agricultural and plantation areas—45.95%; bush and peat soil areas—24.27%; productive forests—15.49%; timber estate areas—8.51%; protected areas—4.58%; and transmigration sites—1.20% (the three forest/timber areas add up to a total of 28.58% of the area burned). While the distribution of fire hot spots is not an actual index for area burned, the NOAA satellite-derived hot spot distribution is quite similar to the ecosystem/terrain distribution of burned area deduced by Liew *et al.* (1998) based on SPOT images of the actual burned areas: agricultural and plantation areas—50%; forests and bushes—30%; and peat swamp forests—20%. Since the estimates of burned ecosystem/terrain of Liew *et al.* (1998) are based on actual SPOT images of the burned area, their estimates were adopted in our calculations.

What is the biomass loading for the three terrain classifications identified by Liew *et al.* (1998)? Values for biomass loading or fuel load for various tropical ecosystems are summarized in Table 6. The biomass loading for tropical forests in Southeast Asia ranges from 5,000 t km$^{-2}$ to $5.5 \times 10^4$ t km$^{-2}$, with a mean value of $2.3 \times 10^4$ t km$^{-2}$ (Brown and Gaston, 1996). However, in our calculations we have used a value of $10^4$ t km$^2$ to be conservative. The biomass loading for agricultural and plantation areas (mainly rubber trees and oil palms) of 5,000 t km$^{-2}$ is also a conservative value (Liew *et al.*, 1998). Nichol (1997) has investigated the peat deposits of Kalimantan and Sumatra and used a biomass loading value of $9.75 \times 10^4$ t km$^{-2}$ (Supardi and Subekty, 1993) for the dry peat deposits 1.5 m thick as representative of the Indonesian peat in her study. Brunig (1997)

**Table 6** Biomass load range and burning efficiency in tropical ecosystems.

| Vegetation type | Biomass load range (t km$^{-2}$) | Burning efficiency |
|---|---|---|
| Peat[a] | 97,500 | 0.50 |
| Tropical rainforests[b] | 5,000–55,000 | 0.20 |
| Evergreen forests | 5,000–10,000 | 0.30 |
| Plantations | 500–10,000 | 0.40 |
| Dry forests | 3,000–7,000 | 0.40 |
| Fynbos | 2,000–4,500 | 0.50 |
| Wetlands | 340–1,000 | 0.70 |
| Fertile grasslands | 150–550 | 0.96 |
| Forest/savanna mosaic | 150–500 | 0.45 |
| Infertile savannas | 150–500 | 0.95 |
| Fertile savannas | 150–500 | 0.95 |
| Infertile grasslands | 150–350 | 0.96 |
| Shrublands | 50–200 | 0.95 |

Source: Scholes *et al.* (1996) (except as noted).
[a]Brunig (1997) and Supardi *et al.* (1993).    [b]Brown and Gaston (1996).

**Table 7**  Parameters used in calculations.

(i) Total area burned in Kalimantan and Sumatra,
    Indonesia in 1997: 45,600 km$^2$
(ii) Distribution of burned areas, biomass loading,
     and combustion efficiency
     (a) Agricultural and plantation areas 50%,
         5,000 t km$^{-2}$, 0.20
     (b) Forests and bushes 30%, 10,000 t km$^{-2}$, 0.20
     (c) Peat swamp forests 20%, 97,500 t km$^{-2}$, 0.50

gives a similar value for peat biomass loading.
The combustion efficiency for forests is estimated
at 0.20 and for peat is estimated at 0.50 (Levine
and Cofer, 2000). Based on the discussions
presented in this section, the values for burned
area, biomass loading, and combustion efficiency
used in the calculations are summarized in Table 7.

### 4.05.11  RESULTS OF CALCULATIONS: GASEOUS AND PARTICULATE EMISSIONS FROM THE FIRES IN KALIMANTAN AND SUMATRA, INDONESIA, AUGUST TO DECEMBER 1997

The calculated gaseous and particulate emissions
from the fires in Kalimantan and Sumatra, from
August to December 1997, are summarized in
Table 8 (Levine, 1999). (It is important to keep in
mind that wildfires continued throughout Southeast
Asia from January to April 1998 and that the fires
covered much more of the region than Kalimantan
and Sumatra.) For each of the seven species listed,
the emissions due to agricultural/plantation burning
(A), forest burning (F), and peat burning (P) are
given. The total (T) of all three components
(A + F + P) is also given. The "best estimate"
total emissions are: $CO_2$—191.485 million metric
tons of C (Mt C); CO—32.794 Mt C; $CH_4$—
1.845 Mt C; $NO_x$—5.898 Mt N; $NH_3$—2.585 Mt
N; $O_3$—7.100 Mt $O_3$; and total particulate matter—
16.154 Mt C. The $CO_2$ emissions from these fires
is ~2.2% of the global annual production of $CO_2$
from all sources (see Table 4 for global annual
production of $CO_2$, which is 8,700 Mt C and
191.485 Mt C / 8,700 Mt C = 2.2%). The percentage for other gases produced by these fires
compared to the global annual production from
all sources is: CO—2.98%; $CH_4$—0.48%; oxides
of nitrogen—2.43%; $NH_3$—5.87%; and total
particulate matter—1.08%.

Scholes *et al.* (1996) calculated the biomass
consumed by burning in 11 different ecosystems
in another tropical ecosystem, southern Africa,
and performed a detailed statistical analysis of the
errors associated with the calculated values of
biomass consumed by fire using a statistical
procedure of Nelson (1992), which assumes that

all error terms ($e$) are independent. In the error
analysis of Scholes *et al.* (1996), the total error
($E_{total}$), which corresponds to the three-sigma
(99%) confidence level, is estimated using the
following expression:

$$E_{total} = (e^2_{burned\ area} + e^2_{fuel\ load} + e^2_{combustion\ completeness})^{1/2} \qquad (6)$$

Scholes *et al.* (1996) assumed the following
uncertainties for each calculation parameter:
$e_{burned\ area} = 30\%$, $e_{fuel\ load} = 30\%$, and $e_{combustion\ completeness} = 25\%$. In the error analysis of the
calculations presented in this chapter, the error
associated with uncertainties in the emission ratio
($e_{emission\ ratio}$) (30%) has also been included. The
uncertainties in the calculation parameters in the
detailed error of analysis of Scholes *et al.* (1996)
were adopted in this study, with the exception of
the uncertainty in the burned area of 30%. The
Scholes *et al.* (1996) determination of burned area
was based on satellite measurements of active
fires that were converted to such area, which is
not a simple one-to-one transformation and
introduces errors. The burned area determination
used in this chapter was more straightforward
since it was based on direct satellite photography
of burned areas using SPOT images (Liew *et al.*,
1998). Hence, there is less uncertainty in this
burned area determination and an uncertainty of
10% was assumed (Liew *et al.* (1998) did not give
a burned area uncertainty in their paper). Using
Equation (6), the calculated uncertainty in the
emission calculations presented in this chapter is
50.2%. The uncertainty range for each species
emission is shown in parentheses under the
"best estimate" value in Table 8.

However, it is important to re-emphasize that
these emission calculations represent lower limit
values since the calculations are only based on
burning in Kalimantan and Sumatra in 1997. The
calculations do not include burning in Java,
Sulawesi, Irian Jaya, Sumbawa, Komodo, Flores,
Sumba, Timor, and Wetar in Indonesia or in
neighboring Malaysia and Brunei.

It is interesting to compare the gaseous and
particulate emissions from the 1997 fires in
Kalimantan and Sumatra with those from the
Kuwait oil fires of 1991, described as a major
environmental catastrophe. Laursen *et al.* (1992)
have calculated the emissions of $CO_2$, CO, $CH_4$,
$NO_x$, and particulates from the Kuwait oil fires in
units of metric tons per day. The Laursen *et al.*
(1992) calculations are summarized in Table 9. To
compare these calculations with the calculations
for Kalimantan and Sumatra (Table 8), we
have normalized our calculations by the total
number of days of burning. The SPOT images
(Liew *et al.*, 1998) covered a period of 5
months (August–December 1997) or ~150 d.

**Table 8** Gaseous and particulate emissions from the fires in Kalimantan and Sumatra in 1997 (for total burned area $= 4.56 \times 10^4$ km$^2$) (Levine, 1999).

|  | *Agricultural/plantation fire emissions* | *Forest fire emissions* | *Peat fire emissions* | *Total fire emissions* |
|---|---|---|---|---|
| $CO_2$ | 9.234 (4.617–13.851) | 11.080 (5.54–16.62) | 171.170 (85.585–256.755) | 191.485 (95.742–287.226) |
| CO | 0.785 (0.392–1.177) | 0.942 (0.471–1.413) | 31.067 (15.533–46.600) | 32.794 (16.397–49.191) |
| $CH_4$ | 0.030 (0.015–0.045) | 0.035 (0.017–0.052) | 1.780 (0.89–2.67) | 1.845 (0.922–2.767) |
| $NO_x$ | 0.023 (0.011–0.034) | 0.027 (0.013–0.040) | 0.921 (0.460–1.381) | 0.971 (0.485–1.456) |
| $NH_3$ | 0.010 (0.005–0.015) | 0.012 (0.006–0.018) | 2.563 (1.281–3.844) | 2.585 (1.292–3.877) |
| $O_3$ | 0.177 (0.088–0.265) | 0.213 (0.106 –0.319) | 6.710 (3.35–10.06) | 7.100 (3.55–10.65) |
| TPM | 0.460 (0.23–0.69) | 0.547 (0.273–0.820) | 15.561 (7.780–23.341) | 16.568 (8.284–24.852) |

For each species, the best estimate emission value is on the first line and the range of emission values in parentheses to the right (see text for discussion of emission estimate range and uncertainty calculations). Units of emissions: Million metric tons of C (Mt C) for $CO_2$, CO, and $CH_4$; Mt N for $NO_x$ and $NH_3$; Mt $O_3$ for $O_3$; and Mt C for particulates (1 Mt $= 10^{12}$ g $= 1$ Tg).

**Table 9** Comparison of gaseous and particulate emissions: the Indonesian fires and the Kuwait oil fires.

| *Species* | *Indonesian fires* | *Kuwait oil fires*[a] |
|---|---|---|
| $CO_2$ | $1.28 \times 10^6$ | $5.0 \times 10^5$ |
| CO | $2.19 \times 10^5$ | $4.4 \times 10^3$ |
| $CH_4$ | $1.23 \times 10^4$ | $1.5 \times 10^3$ |
| $NO_x$ | $6.19 \times 10^3$ | $2.0 \times 10^2$ |
| Particulates | $1.08 \times 10^5$ | $1.2 \times 10^4$ |

Units of emissions: Metric tons per day of C for $CO_2$, CO, and $CH_4$; metric tons per day of N for $NO_x$; metric tons per day for particulates (1 Mt $= 10^{12}$ g $= 1$ Tg).
[a] Laursen *et al.* (1992).

For comparison with the Kuwait fire emissions, we divided our calculated emissions by 150 d. These values are summarized in Table 9. The gaseous and particulate emissions from the fires in Kalimantan and Sumatra significantly exceeded the emissions from the Kuwait oil fires. The 1997 fires in Kalimantan and Sumatra were a significant source of gaseous and particulate emissions to the local, regional, and global atmosphere.

## 4.05.12 THE IMPACT OF THE SOUTHEASTERN ASIA FIRES ON THE COMPOSITION AND CHEMISTRY OF THE ATMOSPHERE

A series of papers have discussed the impact of the Southeastern Asia fires on the composition and chemistry of the atmosphere (Fujiwara *et al.*, 1999; Hauglustaine *et al.*, 1999; Matsueda and Inoue, 1999; Nakajima *et al.*, 1999; Rinsland *et al.*, 1999; Sawa *et al.*, 1999; Tsutsumi *et al.*, 1999). The results of these studies are summarized below.

### 4.05.12.1 Modeling $O_3$ and CO over Indonesia

Hauglustaine *et al.* (1999) used MOZART, a global chemical transport model to investigate the impact of fire emissions calculated by Levine (1999) (presented in Table 8) on the photochemical production of $O_3$ over Indonesia. The calculations indicate that the tropospheric $O_3$ column density increased by 20–25 Dobson Units (DU) and the $O_3$ mixing ratio reached 50 ppbv in the mid-troposphere in November. (These model calculations are consistent with *in situ* $O_3$ measurements obtained by Fujiwara *et al.* (1999) (described in the next section), and satellite measurements of $O_3$ obtained with the Earth Probe/TOMS (Chandra *et al.*, 1998) and the ERS-2/GOME (Burrows *et al.*, 1999)). South of the source region, low $O_3$ mixing ratios of 20–25 ppbv were calculated in the boundary layer due to marine air influence and reduced photochemical activity in the presence of biomass burning aerosols.

Normal, nonfire surface concentrations of CO over Indonesia are usually less than $\sim$100 ppbv. The calculations of Hauglustaine *et al.* (1999) indicate increases of CO by up 2,900 ppbv at the surface over Kalimantan and Suimatra. A perturbation of 50–1,000 ppbv was calculated over the entire region of Indonesia. In the free troposphere, CO increased by 1,000 ppbv over Indonesia due to the upward transport prevailing in this region. The calculated high levels of CO are consistent with actual measurements of CO obtained over Indonesia during the fires by Sawa *et al.* (1999).

### 4.05.12.2 Measurements over Indonesia

Aircraft measurements over Kalimantan, Indonesia, on October 13, 1997, indicated high concentrations of $O_3$, $NO_x$, CO, and aerosols (Tsutsumi *et al.*, 1999). The maximum concentration of $O_3$ (80.5 ppbv) was found in the middle layer of the smoke haze and very low concentrations ($\sim$20 ppbv) were measured in the lower smoke layer. The authors concluded that the low $O_3$ concentrations near the surface may be caused by the reduction in solar flux due to aerosols high in the haze layer and the loss of $O_3$ due to large aerosol surface area in the lower haze layer.

Pronounced enhancements of total and tropospheric $O_3$ were observed with a Brewer spectrophotometer and ozonesondes at Watukosek, Indonesia in October 1997 (Fujiwara *et al.*, 1999). The integrated tropospheric $O_3$ increased from 20 DU to 55 DU in October 1997. On October 22, 1997, the $O_3$ concentrations were more than 50 ppbv throughout the troposphere and exceeded 100 ppbv at several altitudes. The authors conclude that the enhanced levels of $O_3$ measured in October 1997, resulted from $O_3$ precursors, produce during the fires and released into the atmosphere.

Sawa *et al.* (1999) measured CO and hydrogen over Kalimantan during the fire. They reported CO concentrations in the range of 3,000–9,000 ppbv below an altitude of 2.6 km. Above 4.4 km altitude, the CO concentration was 500–1,200 ppbv.

### 4.05.12.3 Measurements between Singapore and Japan

Aircraft measurements of $CO_2$, CO, and $CH_4$ were obtained between Japan and Singapore in October 1993, 1996, and 1997 (Matsueda and Inoue, 1999). The mixing ratios of all three gases at 9–12 km were enhanced over the South China Sea in 1997 compared with measurements in 1993 and 1996.

### 4.05.12.4 Measurements over Hawaii

Infrared solar spectral measurements of CO, ethane ($C_2H_6$), and hydrogen cyanide (HCN) were obtained for over 250 days between August 1995 and February 1998 above Mauna Loa, Hawaii (Rinsland *et al.*, 1999). Correlated variations of CO, $C_2H_6$, and HCN and unusual seasonal cycles observed above Mauna Loa, Hawaii, during the second half of 1997 suggest a common origin for these emissions. Back-trajectory model calculations indicate that the source of these anomalous concentrations over Hawaii was the fires in Southeast Asia.

From August 1997 through early 1998, the entire region of Southeast Asia, but particularly Indonesia and Malaysia, experienced extensive and widespread wildfires. These fires were initiated for land clearing and land-use change, as is the usual custom in this region of the world. However, in 1997, the severe drought conditions resulting from a very severe El Nino Southern Oscillation resulted in very widespread, uncontrolled wildfires that lasted for months, well into 1998. These emissions significantly reduced atmospheric visibility and produced record high levels of atmospheric pollution. These gaseous and particulate emissions also impacted human health. Three different agencies of the United Nations (UNEP, WMO, and WHO) conducted studies and workshops and issued reports on the environmental and health impacts of these fires. Many lessons were learned by much work needs to be done to avoid such catastrophes in the future.

## REFERENCES

Anderson I. C., Levine J. S., Poth M. A., and Riggan P. J. (1988) Enhanced biogenic emissions of nitric oxide and nitrous oxide following surface biomass burning. *J. Geophys. Res.* **93**, 3893–3898.

Andreae M. O. (1991) Biomass burning: its history, use, and distribution and its impact on environmental quality and global climate. In *Global Biomass Burning: Atmospheric, Climatic, and Biospheric Implications* (ed. J. S. Levine). MIT Press, Cambridge, MA, pp. 3–21.

Andreae M. O., Browell E. V., Garstang M., Gregory G. L., Harriss R. C., Hill G. F., Jacob D. J., Pereira M. C., Sachse G. W., Setzer A. W., Silva Dias P. L., Talbot R. W., Torres A. L., and Wofsy S. C. (1998) Biomass burning emission and associated haze layers over Amazonia. *J. Geophys. Res.* **93**, 1509–1527.

Blake N. J., Blake D. R., Sive B. C., Chen T.-Y., Rowland F. S., Collins J. E., Sachse G. W., and Anderson B. E. (1996) Biomass burning emissions and vertical distribution of atmospheric methyl halides and other reduced carbon gases in the south Atlantic region. *J. Geophys. Res.* **101**, 24151–24164.

Brauer M. and Hisham-Hishman J. (1988) Fires in Indonesia: crises and reaction. *Environ. Sci. Technol.* **22**, 404A–407A.

Brown S. and Gaston G. (1996) Estimates of biomass density for tropical forests. In *Biomass Burning and Global Change* (ed. J. S. Levine). MIT Press, Cambridge, MA, vol. 1, pp. 133–139.

Brunig E. F. (1997) The tropical rainforest—a wasted asset or an essential biospheric resource? *Ambio* **6**, 187–191.

Burrows J. P., Weber M., Buchwitz M., Rozanov V., Ladstatter-Weibenmayer A., Richter A., DeBeek R., Hoogen R., Bramstedt K., Eichmann K.-U., and Eisinger M. (1999) The global ozone monitoring experiment (GOME): mission concept and first scientific results. *J. Atmos. Sci.* **56**, 151–175.

Cahoon D. R., Stocks B. J., Levine J. S., Cofer W. R., and Pierson J. M. (1994) Satellite analysis of the severe 1987 forest fires in northern China and southeastern Siberia. *J. Geophys. Res.* **99**, 18627–18638.

Chandra S., Ziemke J. R., Min W., and Read W. G. (1998) Effects of 1997–1998 El Nino on tropospheric ozone and water vapor. *Geophys. Res. Lett.* **25**, 3867–3870.

Crutzen P. J. and Andreae M. O. (1990) Biomass burning in the tropics: impact on atmospheric chemistry and biogeochemical cycles. *Science* **250**, 1678–1679.

Crutzen P. J. and Goldammer J. G. (eds.) (1993) *Fire in the Environment: The Ecological, Atmospheric, and Climatic Importance of Vegetation Fires*. Wiley, Chichester, England, 400pp.

Crutzen P. J., Heidt L. E., Krasnec J. P., Pollock W. H., and Seiler W. (1979) Biomass burning as a source of atmospheric gases CO, $H_2$, $N_2O$, NO, $CH_3Cl$, COS. *Nature* **282**, 253–256.

Eaton P. and Radojevic M. (eds.) (2001) *Forest Fires and Regional Haze in Southeast Asia*. Nova Science, Huntington, NY, 280pp.

Friedli H. R., Radke L. F., and Lu J. Y. (2001) Mercury in smoke from biomass burning. *Geophys. Res. Lett.* **28**, 3223–3226.

Fromm M., Alfred J., Hoppel K., Hornstein J., Bevilacqua R., Shettle E., Servranckx R., Li Z., and Stocks B. (2000) Observations of boreal forest fire smoke in the stratosphere by POAM III, SAGE II, and lidar in 1998. *Geophys. Res. Lett.* **27**, 27–30.

Fujiwara M., Kita K., Kawakami S., Ogawa T., Komala N., Saraspriya S., and Suripto A. (1999) Tropospheric ozone enhancements during the Indonesian forest fire events in 1994 and in 1997 as revealed by ground-based observations. *Geophys. Res. Lett.* **26**, 2417–2420.

Goldammer J. G. (ed.) (1990) *Fire in the Tropical Biota: Ecosystem Processes and Global Challenges.* Springer, Berlin, Germany, 497pp.

Goldammer J. G. and Furyaev V. V. (eds.) (1996) *Fire in Ecosystems of Boreal Eurasia.* Kluwer Academic, Dordrecht, The Netherlands, 528pp.

Hamilton M. S., Miller R. O., and Whitehouse A. (2000) Continuing fire threat in Southeast Asia. *Environ. Sci. Technol.* **34**, 82A–85A.

Hansen J., Ruedy R., Sato M., and Reynolds R. (1996) Global surface air temperature in 1995: return to Pre-Pinatubo level. *Geophys. Res. Lett.* **23**, 1665–1668.

Harris G. W., Wienhold F. G., and Zenker T. (1996) Airborne observations of strong biogenic $NO_x$ emissions from the Namibian Savanna at the end of the dry season. *J. Geophys. Res.* **101**, 23707–23711.

Hauglustaine D. A., Brasseur G. P., and Levine J. S. (1999) A sensitivity simulation of tropospheric ozone changes due to the 1997 Indonesian fire emissions. *Geophys. Res. Lett.* **26**, 3305–3308.

Innes J. L., Beniston M., and Verstraet M. M. (eds.) (2000) *Biomass Burning and its Inter-relationships with the Climate System.* Kluwer Academic, Dordrecht, 358pp.

Kasischke E. S. and Stocks B. J. (eds.) (2000) *Fire, Climate Change, and Carbon Cycling in the Boreal Forest.* Springer Ecological Studies, New York, vol. 138, 461pp.

Kasischke E. S., Bergen K., Fennimore R., Sotelo F., Stephens G., Janetos A., and Shugart H. H. (1999) Satellite imagery gives clear picture of Russia's boreal forest fires. *EOS, Trans., AGU* **80**, 141–147.

Laursen K. K., Ferek R. J., and Hobbs P. V. (1992) Emission factors for particulates, elemental carbon, and trace gases from the Kuwait oil fires. *J. Geophys. Res.* **97**, 14491–14497.

Levine J. S. (ed.) (1991) *Global Biomass Burning: Atmospheric, Climatic, and Biospheric Implications.* MIT Press, Cambridge, MA, 569pp.

Levine J. S. (ed.) (1996a) *Biomass Burning and Global Change: Remote Sensing, Modeling and Inventory Development, and Biomass Burning in Africa.* MIT Press, Cambridge, MA, 581pp.

Levine J. S. (ed.) (1996b) *Biomass Burning and Global Change: Biomass Burning in South America, Southeast Asia, and Temperate and Boreal Ecosystems, and the Oil Fires of Kuwait.* MIT Press, Cambridge, MA, 377pp.

Levine J. S. (1999) The 1997 fires in Kalimantan and Sumatra, Indonesia: gaseous and particulate emissions. *Geophys. Res. Lett.* **26**, 815–818.

Levine J. S. and Cofer W. R. (2000) Boreal forest fire emissions and the chemistry of the atmosphere. In *Fire, Climate Change and Carbon Cycling in the North American Boreal Forests*, Ecological Studies Series (eds. E. S. Kasischke and B. J. Stocks). Springer, New York.

Levine J. S., Cofer W. R., Sebacher D. I., Winstead E. L., Sebacher S., and Boston P. J. (1988) The effects of fire on biogenic soil emissions of nitric oxide and nitrous oxide. *Global Biogeochem. Cycles* **2**, 445–449.

Levine J. S., Cofer W. R., Winstead E. L., Rhinehart R. P., Cahoon D. R., Sebacher D. I., Sebacher S., and Stocks B. J. (1991) Biomass burning: combustion emissions, satellite imagery, and biogenic emissions. In *Global Biomass Burning: Atmospheric, Climatic, and Biospheric Implications* (ed. J. S. Levine). MIT Press, Cambridge, MA, pp. 264–272.

Levine J. S., Cofer W. R., Cahoon D. R., and Winstead E. L. (1995) Biomass burning: a driver for global change. *Environ. Sci. Technol.* **29**, 120A–125A.

Levine J. S., Winstead E. L., Parsons D. A. B., Scholes M. C., Scholes R. J., Cofer W. R., Cahoon D. R., and Sebacher D. I. (1996a) Biogenic soil emissions of nitric oxide (NO) and nitrous oxide ($N_2O$) from savannas in South Africa: the impact of wetting and burning. *J. Geophys. Res.* **101**, 23689–23697.

Levine J. S., Cofer W. R., Cahoon D. R., Winstead E. L., Sebacher D. I., Scholes M. C., Parsons D. A. B., and Scholes R. J. (1996b) Biomass burning, biogenic soil emissions, and the global nitrogen budget. In *Biomass Burning and Global Change* (ed. J. S. Levine). MIT Press, Cambridge, MA, vol. 1, pp. 370–380.

Levine J. S., Bobbe T., Ray N., Witt R. G., and Singh A. (1999) *Wildland Fires and the Environment: A Global Synthesis.* Environment Information and Assessment Technical Report 99-1, The United Nations Environmental Program, Nairobi, Kenya, 46pp.

Liew S. C., Lim O. K., Kwoh L. K., and Lim H. (1998) A study of the 1997 fires in SouthEast Asia using SPOT quicklook mosaics. Paper presented at the 1998 International Geosience and Remote Sensing Symposium, July 6–10, Seattle, Washington, 3pp.

Lobert J. M., Scharffe D. H., Hao W.-M., Kuhlbusch T. A., Seuwen R., Warneck P., and Crutzen P. J. (1991) Experimental evaluation of biomass burning emissions: nitrogen and carbon containing compounds. In *Global Biomass Burning: Atmospheric, Climatic, and Biospheric Implications* (ed. J. S. Levine). MIT Press, Cambridge, MA, pp. 289–304.

Makarim N., Arbai Y. A., Deddy A., and Brady M. (1998) Assessment of the 1997 Land and forest fires in Indonesia: national coordination. In *International Forest Fire News*, United Nations Economic Commission for Europe and the Food and Agriculture Organization of the United Nations, Geneva, Switzerland, No. 18, pp. 4–12.

Matsueda H. and Inoue H. Y. (1999) Aircraft measurements of trace gases between Japan and Singapore in October of 1993, 1996, and 1997. *Geophys. Res. Lett.* **26**, 2413–2416.

Nakajima T., Higurashi A., Takeuchi N., and Herman J. R. (1999) Satellite and ground-based study of optical properties of 1997 Indonesian forest fire aerosols. *Geophys. Res. Lett.* **26**, 2421–2424.

Nelson L. S. (1992) Technical aids: propagation of error. *J. Qualit. Control* **24**, 232–235.

Nichol J. (1997) Bioclimatic impacts of the 1994 smoke haze event in southeast Asia. *Atmos. Environ.* **44**, 1209–1219.

Rinsland C. P., Goldman A., Murcray F. J., Stephen T. M., Pougatchev N. S., Fishman J., David S. J., Blatherwick R. D., Novelli P. C., Jones N. B., and Connor B. J. (1999) Infrared solar spectroscopic measurements of free tropospheric CO, $C_2H_6$, and HCN above Mauna Loa, Hawaii: seasonal variations and evidence for enhanced emissions from the Southeast Asian tropical fires of 1997–1998. *J. Geophys. Res.* **104**, 18667–18680.

Sawa Y., Matsueda H., Tsutsumi Y., Jensen J. B., Inoue H. Y., and Makino Y. (1999) Tropospheric carbon monoxide and hydrogen measurements over Kalimantan in Indonesia and northern Australia during October, 1997. *Geophys. Res. Lett.* **26**, 1389–1392.

Scholes R. G., Kendall J., and Justice C. O. (1996) The quantity of biomass burned in southern Africa. *J. Geophys. Res.* **101**, 23667–23676.

Seiler W. and Crutzen P. J. (1980) Estimates of gross and net fluxes of carbon between the biosphere and the atmosphere from biomass burning. *Climat. Change* **2**, 207–247.

Silver C. S. and DeFries R. S. (1990) *One Earth, One Future: Our Changing Global Environment.* Academic press, Washington, DC, pp. i–ix.

Simons L. M. (1998) Indonesia's plague of fire. *Natl. Geogr.* **194**(2), 100–119.

Stocks B. J. and Flannigan M. D. (1987) Ninth Conference on Fire and Forest Meteorology. Society of American Foresters, Bethesda, MD, pp. 23–30.

Stocks B. J., Fosberg M. A., Lyman T. J., Means L., Wotton, B. M., Yang Q., In J.-Z., Lawrence K., Hartley G. J., Mason J. G., and McKenney D. T. (1991) Climate change and forest fire potential in Russian and Canadian boreal forests. *Climat. Change* **38** 1–13.

Supardi A. D. and Subekty S. S. (1993) General geology and peat resources of the Siak Kanan and Bengkalis Island peat deposits, Sumatra, Indonesia. In *Modern and Ancient Coal Forming Environments*. (eds. J. C. Cobb and C. B. Cecil). Geological Society of America, Washington, DC, vol. 86, pp. 45–61.

Tsutsumi Y., Sawa Y., Makino Y., Jensen J. B., Gras J. L., Ryan B. F., Diharto S., and Harjanto H. (1999) Aircraft measurements of ozone, $NO_x$, CO, and aerosol concentrations in biomass burning smoke over Indonesia and Australia in October 1997: depleted ozone layer at low altitude over Indonesia. *Geophys. Res. Lett.* **26**, 595–598.

UNDAC (United Nations Disaster Assessment and Coordination Team) (1998) Mission on Forest Fires, Indonesia, September–November 1997. *International Forest Fire News*, United Nations Economic Commission for Europe and the Food and Agriculture Organization of the United Nations, Geneva, Switzerland, No. 18, January 1998, pp. 13–26.

van Wilgen B. W., Andreae M. O.,Goldammer J. G., and Lindesay J. A. (eds.) (1997) *Fire in Southern African Savannas: Ecological and Atmospheric Perspectives.* Witwatersrand University Press, Johannesburg, SA, 256pp.

Ward D. E. (1990) Factors influencing the emissions of gases and particulate matter from biomass burning. In *Fire in the Tropical Biota: Ecosystem Processes and Global Challenges*, Ecological Studies (ed. J. G. Goldammer). Springer, Berlin, vol. 84, pp. 418–436.

Wayne R. P. (1991) *Chemistry of Atmospheres,* 2nd edn. Oxford University Press, Oxford, England, pp. 160–164.

Yokelson R. J., Griffith D. W. T., and Ward D. E. (1996) Open-path Fourier transform infrared studies of large-scale laboratory biomass fires. *J. Geophys. Res.* **101**, 21067–21080.

Yokelson R. J., Susott R., Ward D. E., Reardon J., and Griffith D. W. T. (1997) Emissions from smouldering combustion from biomass measured by open-path Fourier transform infrared spectroscopy. *J. Geophys. Res.* **102**, 18865–18877.

Zepp R. G., Miller W. L., Burke R. A., Parsons D. A. B., and Scholes M. C. (1996) Effects of moisture and burning on soil–atmosphere exchange of trace carbon gases in a southern African savanna. *J. Geophys. Res.* **101**, 23699–23706.

# 4.06
# Non-mass-dependent Isotopic Fractionation Processes: Mechanisms and Recent Observations in Terrestrial and Extra-terrestrial Environments

M. H. Thiemens

*University of California, San Diego, CA, USA*

## 4.06.1 GENERAL INTRODUCTION

There were several key observations, which were made almost around the same time, that led to studies in the fields of stable isotope geochemistry and cosmochemistry. First, it was the physicochemical formalism for isotope effects, particularly the determination of the position of equilibrium in isotopic exchange reactions. Urey (1947) and Bigeleisen and Mayer (1947) demonstrated that the position of isotope exchange in a chemical reaction may be calculated with high precision. The difference in chemical behavior for

isotopically substituted molecules in this specific instance arises from quantum mechanical effects. The vibrating molecule is energetically represented as a harmonic oscillator. Thus, in the case of isotopic substituted molecules, the quantized energy levels vary with the heavier species possessing lower vibrational frequencies. As a result, the isotopically substituted molecules possess slightly stronger bond strengths, which can be calculated with high precision. These can also be measured spectroscopically. Furthermore, since vibrational frequencies are temperature dependent, the partitioning of isotopes between two molecules is temperature dependent and the observed isotopic differences may be used as a temperature index for the equilibrium process of interest.

This phenomenon forms the basis for the formulations of Urey (1947) and Bigeleisen and Mayer (1947) for the temperature dependence of isotopic exchange between two molecules. With the nearly simultaneous development of the isotope-ratio mass spectrometer by Nier *et al.* (1947), the potential for application of stable isotopes was created. Other isotopic fractionation processes are observed in kinetics, diffusion, evaporation–condensation, crystallization, and biology (e.g., photosynthesis, respiration, nitrogen fixation, sulfate reduction, and transpiration). The concomitant isotopic fractionations can also be used to provide details of the relevant process.

Although, the conventional isotope effects vary widely in the physicochemical basis for their origin, they all are dependent upon relative isotopic mass differences. The first quantitative discussion of the mass dependence of isotope effects was given by Hulston and Thode (1965), who showed that conventional isotope effects alter the isotope ratios in a manner strictly dependent upon relative mass differences. Specifically, the changes in $\delta^{33}S$ and $\delta^{34}S$ are shown to be highly correlated such that the change in $\delta^{33}S$ is half that of $\delta^{34}S$. The mass range in the $\delta^{33}S$ is 1 amu (33 − 32 amu) and for $\delta^{34}S$ is 2 amu (34 − 32 amu). Thus,

$$\delta^{33}S \approx 0.5\delta^{34}S$$

is observed and arises due to the magnitude of the relative isotopic mass differences. The focus of the paper of Hulston and Thode was to utilize meteoritic stable-isotope-ratio measurements as a mechanism to resolve nuclear (nucleosynthetic or spallogenic) processes from non-nuclear processes. The fundamental assumption in this work was, based upon the assumption that all conventional physicochemical processes produce correlated mass-dependent arrays, that any deviation from this relation must reflect a nuclear process. The core assumption for this assertion was that any physical or chemical process may not lead to a variance in stable isotope ratio that does not depend upon mass differences.

Clayton *et al.* (1973) observed that the high-temperature calcium–aluminum inclusions (CAIs) present in the carbonaceous chondrite Allende possess an oxygen isotope composition of $\delta^{17}O \approx \delta^{18}O$, rather than the expected mass-dependent $\delta^{17}O \approx 0.5\delta^{18}O$. It was suggested that this anomalous isotopic composition must derive from a nuclear, rather than chemical process. In general, an equality $\delta^{17}O = \delta^{18}O$ may arise in two ways, either by alteration of $^{17}O$ and $^{18}O$ by equal amounts or, by addition/subtraction of pure

$^{16}O$. Models for supernova processes had shown that for certain conditions, essentially pure $^{16}O$ is produced. Clayton *et al.* argued that it is unlikely that $^{17}O$ and $^{18}O$ would be equally altered, thus the supernova event was deemed the most plausible mechanism to account for the Allende isotopic observations.

It is now known that these basic assumptions are incorrect. Thiemens and Heidenreich (1983) demonstrated that, in the process of ozone formation from molecular oxygen, ozone is equally enriched in $^{17}O$ and $^{18}O$, or $\delta^{17}O = \delta^{18}O$. This has several consequences. The assumption that only a nuclear process may produce a mass-independent isotopic composition is invalid. Most significantly, the relation $\delta^{17}O = \delta^{18}O$ is identical to that for the Allende CAIs. These observations immediately raise the vital issue as to the source of the anomalous oxygen-isotopic composition observed in the CAI of Allende. In fact, because the bulk oxygen-isotopic compositions of nearly all meteorites do not define arrays associated with a common oxygen reservoir and associated mass-dependent fractionations. As discussed in Thiemens and Heidenreich (1983), the meteoritic oxygen-isotopic compositions more likely result from chemical processes. A more detailed discussion of this possibility was given by Thiemens and Jackson (1988). The literature on meteoritic oxygen-isotopic observations is extensive, and most importantly, includes high spatial resolution analyses. This has derived from the development of secondary ion-microprobe (SIMs), nano-SiMS, and continuous-flow isotope-ratio mass spectrometry. These new techniques allow highly precise multi-oxygen isotopic analysis at micrometer (and smaller) spatial scales. An outcome of these observations has been that there is no observational support for a nuclear source of the meteorite oxygen-isotopic compositions, which would have included identification of $^{16}O$ carrier grains or, a direct correlation with other supernova-produced isotopes. A chemical process is now indicated. Contributions from isotopic self-shielding in the nebula by an early active Sun (Thiemens and Heidenreich, 1983), or from chemical reactions (e.g., Thiemens, 1999) are all under consideration at present.

There are other facets of chemical mass-independent processes and their meteoritic applications, and these will be a focus of this chapter. The general observations are:

(i) They are highly ubiquitous on Earth and offer a mechanism by which an extraordinary range of natural processes may be investigated.

(ii) The physicochemical fractionation process itself has opened a new avenue of study of physical–chemical processes, particularly of gas-phase reactions and molecular transition state theory.

(iii) Isotopic observations of minerals from martian meteorites has opened a new mechanism to understand atmospheric–regolithic processes on Mars.

There are several review papers on mass-independent chemical processes and their applications. Thiemens and Weston reviewed the progress in understanding the physical chemistry of gas-phase mass-independent processes and their observation on Earth and meteorites. Thiemens *et al.* (2001) reviewed the observations of mass-independent isotopic composition in various solid reservoirs of Earth and Mars, including both oxygen and sulfur isotopes. A more recent review (Thiemens, 2002) has summarized the theoretical and laboratory studies of the physical chemistry of mass-independent isotope effects and their observation on Earth and Mars, subsequent to the review of Thiemens *et al.* (2001).

This chapter will not recount in detail material already covered elsewhere. There have been a number of recent advances in mass-independent chemistry, which we predominantly address here.

### 4.06.1.1 The Physical Chemistry of Mass-independent Isotope Effects

When the mass-independent isotopic fractionation chemical process was first discovered by Thiemens and Heidenreich (1983), there existed no physical–chemical mechanism that accounted for the ozone observations. In this paper, a mechanism based upon optical self-shielding was proposed. Although this mechanism may not account for the experimental results, there are potential cosmochemical environments where self-shielding may be operative, as discussed in this paper. These potential applications will be discussed in detail in a later section of this chapter.

Heidenreich and Thiemens developed a theory for the mass-independent theory based upon molecular symmetry considerations. In particular, the differential chemistry for asymmetric $^{16}O^{16}O$ $^{17}O$, $^{16}O^{16}O^{18}O$ versus symmetric $^{16}O^{16}O^{16}O$ was proposed as the basis for the observed mass-independent effect. In this consideration, an equal $\delta^{17}O$, $\delta^{18}O$ fractionation occurs as a result of the identical chemistry for the asymmetric isotopomeric species with respect to symmetric $^{16}O^{16}O^{16}O$. It was in fact a rotational symmetry effect that originally led to the discovery of oxygen isotopes. In the observations of $O_2$ absorption through the atmosphere, with the sun as the background irradiance source, it was observed that there are extra rotational absorption lines for $^{16}O^{18}O$ as compared to $^{16}O^{16}O$. This arises because of the well-known line doubling in

rotational spectra for asymmetric species. This quantum mechanical phenomenon also led to the discovery of deuterium by Urey. Heidenreich and Thiemens (1986) suggested that the rotational symmetry produces a longer lifetime for the asymmetric isotopomer and, in turn, a greater probability of stabilization. This hypothesis was based upon the following assumptions:

(i) The probability of an excited $O_3^*$ intermediate energetically quenching via collisions is determined by the product of its lifetime ($\tau$) and it's collisional frequency.

(ii) The collisional frequency is well known from classical kinetic theory to derive from mass-dependent effects, such as velocity, therefore the source of the mass independence does not arise from the collisional terms for stabilization.

(iii) The lifetime of the $^{17}O$ and $^{18}O$ species ($^{16}O^{16}O^{17}O$, $^{16}O^{16}O^{18}O$) are equal and longer than symmetric $^{16}O^{16}O^{16}O$. This leads to a product equally enriched in $^{17}O$ and $^{18}O$, i.e., $\delta^{17}O = \delta^{18}O$.

The exact mechanism arises in the process of inverse pre-dissociation, as discussed in detail by Herzberg (1966). During an atom–molecule collision, the reactants interact with one another subject to the relevant potential energy surface. The lifetime of this excited intermediate is on the order of molecular vibrational periods, or $\sim 10^{-13}$ s. The lifetime is a complex function of the chemical reaction dynamics, which in turn depends on the number of available states. In this specific instance, there is a state dependence for the isotopically substituted species. Ozone of pure $^{16}O$ has a $C_{2v}$ symmetry and has half the rotational complement of the asymmetric isotopomers. As a result, it was suggested that the extended lifetime for the asymmetric species leads to a greater probability of stabilization. While these assumptions are valid for a gas phase molecular reaction, they do not sufficiently account for the totality of the experimental ozone isotopic observations. Reviews by Weston (1999) and Thiemens (1999) have detailed the physical–chemical reasons.

A comprehensive quantum mechanical model for the effect has been developed by Marcus and his colleagues at the California Institute of Technology. The Gao and Marcus (2001, 2002) model accounts for many of the experimental observations and utilizes classical quantum mechanical RRKM theory in its development. Statistical RRKM theory quantitatively describes the energetics of gas phase atom–molecule encounters and the relevant parameters which lead to either stabilization and product formation or re-dissociation to atomic and molecular species. This is a well-developed theory and will not be described in detail here. An important application of this theory is that it determines individual rate constants for isotopically substituted species. In a conventional RRKM approach, all parameters are mass dependent, such as collisional frequency, bond strength, and vibrational and rotational energies. Gao and Marcus (2001, 2002) provide an expression that captures an apparent symmetry dependence. During the stabilization process of the excited intermediate (e.g., $O_3^*$), there is a partial dependence upon the rate at which excess energy is dispersed. If this does not happen sufficiently rapidly, stabilization does not occur and the excited $O_3^*$ will re-dissociate to the original O and $O_2$ reactants. Part of this energy dispersal process depends upon symmetry factors. The greatest contribution is the standard mass-dependent energy process. In the new isotopic model, there is an inclusion of an "$\eta$ effect," which is a modest deviation from the statistical density of states for symmetric versus asymmetric species (Gao and Marcus, 2002). There exists a partitioning of energy that derives from very slight differences in zero-point energies for the exit channels for the dissociation of the asymmetric ozone isotopomers. These exit channels do not appear in isotopically enriched experiments (Hathorn and Marcus, 1999; Gao and Marcus, 2002). For the source of the mass-independent isotope effect, this then leaves only the "$\eta$ effect." As stated by Gao and Marcus (2002), this "can be regarded as symmetry driven." In a more recent paper, Gao and Marcus (2002) have examined the effect of pressure in the context of the experimental data. There exist extensive experimental data (reviewed by Thiemens (1999, 2002), Thiemens *et al.* (2001), and Weston (1999)) that describe these details, in particular the effects of pressure, temperature, and oxygen source dependence. An important advancement has been the utilization of isotopic enrichment to determine the explicit rate constants of individual isotopomeric reactions. One of the first experiments was reported by Anderson *et al.* (1997), a striking feature of which was the unexpected observation that the relative rates for $^{16}O + {}^{16}O{}^{16}O$ vary by a factor 50% (greater) from the reaction $^{16}O + {}^{18}O{}^{18}O$. This is consistent with the models of Gao and Marcus (2001, 2002), which predict a greater rate for the asymmetric species. In addition, including the "$\eta$ effect" quantitatively accounts for the difference in magnitude for the different measured rate constants. Other recent observations by Mauersberger *et al.* (1999) and Janssen and co-workers (1999, 2001) report the rate differences for a number of the isotopically substituted reactions. These measurements have been of great utility in developing a theory for the mass-independent isotopic effect in ozone. Guenther *et al.* (1999, 2000) have further resolved the effect

of pressure and third-body composition on the isotopic fractionation process.

The most recent theoretical advances in resolving the quantum basis of the mass-independent effect as described by Marcus and collaborators have been of fundamental importance. This is true for both understanding the fundamental chemical physics along with the applications in nature. There remain, however, significant challenges to understanding the physical–chemical mechanism. As Janssen (2001) concluded, there are both theoretical and experimental obstacles: for example, the temperature dependence for the individual isotopomeric reactions needs to be determined. There also exist other reactions for which rate constants are presently unknown, such as those for the $^{17}O$ variants.

This chapter focuses upon some recent observations of mass-independent isotopic processes in nature. As discussed by Thiemens *et al.* (2001) and Thiemens (2002), there exist other mass-independent isotope effects in nature that derive from non-ozone reactions. For example, $CO_2$ photolysis produces a large mass-independent isotope effect that, in part, may account for observations in the SNC (martian) meteorites and the synthesis of their secondary minerals. UV photolysis of $SO_2$ produces new isotopic fractional effect. An accompanying mass-independent isotopic composition determines the evolution of oxygen in the Earth's earliest atmosphere.

## 4.06.2 OBSERVATIONS IN NATURE

There now exist numerous observations of mass-independent isotopic compositions in nature. Most of these have recently been reviewed and will not be repeated here. When the first laboratory measurements of the mass-independent isotope effect were reported by Thiemens and Heidenreich (1983), their occurrence in nature was not expected, except possibly for the early solar system to produce the observed meteoritic CAI data. It is significant to note that, at present, all oxygen-bearing molecules in the atmosphere (except water) possess mass-independent isotopic compositions. These molecules include $O_2$, $O_3$, $CO_2$, $CO$, $N_2O$, $H_2O_2$, and aerosol nitrate and sulfate. Mass-independent sulfur isotopic compositions are also observed in aerosol (solid) sulfates and nitrates and sulfide and sulfate minerals from the Precambrian, Miocene volcanic sulfates, Antarctica dry valley sulfates, Namibian Gypretes, and Chilean nitrates. In addition, martian (SNC meteorites) carbonates and sulfates possess both mass-independent sulfur and oxygen isotopic compositions. These studies have been reviewed recently (Thiemens *et al.*, 2001; Thiemens, 1999).

In all of the examples cited above, insights into terrestrial atmospheric or martian cycles were obtained that could not have otherwise been obtained. In the remainder of this chapter, recent progress in individual natural processes and mass-independent chemistry are reported.

### 4.06.2.1 Atmospheric Ozone

The importance of ozone in the Earth's atmosphere is well established. Its presence in the stratosphere serves as a shield of UV light, which is vital for the sustenance of life, particularly land-based, on Earth. In the troposphere, ozone serves as a source of electronically excited atomic oxygen $O(^1D)$, which subsequently, via the reaction

$$O(^1D) + H_2O \rightarrow 2OH$$

creates tropospheric hydroxyl radicals. The hydroxyl free radical (OH) serves as the dominant mediator of the lifetime of most reduced molecular species in the troposphere. Thus, the OH radical serves as a control agent for the oxidative capacity of the atmosphere.

Mauersberger (1981), utilizing *in situ* mass spectrometric measurements, demonstrated that ozone possesses a large $^{18}O$ enrichment. The $^{17}O$ isotopic composition was not determined and the mass-independent isotopic composition could not be detected. As reviewed by Thiemens (1999), Weston (1999), and Johnston and Thiemens (2003), there now exists an extensive literature on stratospheric ozone isotopic measurements obtained by different techniques. Measurements by Mauersberger (1987) confirmed that stratospheric ozone was mass-independently fractionated as displayed in the 1983 laboratory experiments of Thiemens and Heidenreich. Return ozone isotopic analysis by Schueler *et al.* (1990) demonstrated that stratospheric ozone possessed an isotopic composition entirely consistent with laboratory observations. Tropospheric ozone has also been studied for its $\delta^{17}O$, $\delta^{18}O$ isotopic composition. Krankowsky *et al.* (1995) developed a collection and isotopic analytical procedure that revealed that tropospheric ozone possesses a $\delta^{17}O$, $\delta^{18}O$ enrichment of 70–90‰ (with respect to air molecular oxygen). Their observations are completely consistent with laboratory observations, including the dependence on temperature and pressure. Johnston and Thiemens reported ozone isotopic measurements from several locations in the southwestern United States. The tropospheric measurements have shown that these measurements may provide a mechanism to develop further resolution of the rather complex tropospheric ozone cycle and particularly the complex $NO_x$ interactions. Recent measurements

of tropospheric sulfate and nitrate aerosols by the La Jolla group demonstrate that the complexities of the $NO_x$–$SO_x$–$O_3$ cycle (chemical transformation and transport) could be significantly better understood from $\delta^{17}O$, $\delta^{18}O$ measurements. This has opened up the possibility that the isotopic composition of polar ice nitrate and sulfate could provide a unique means by which paleo-ozone and oxidative levels maybe qualified.

There have been a new series of papers on stratospheric ozone. Krankowsky *et al.* (2000) and Johnson *et al.* (2001) have discussed in detail a range of stratospheric ozone isotopes. Johnston and Thiemens (2003) have reviewed the literature on atmospheric ozone isotopic measurements. Krankowsky *et al.* (2000) reported the isotopic composition of ozone collected from four balloon flights in the 22–33 km range. The heavy-isotope enrichment extends between 70‰ and 110‰, somewhat greater than the reported tropospheric range. The actual collection process of stratospheric ozone for return analysis is extraordinarily challenging. Details of the procedure for collection have been reported by Stehr *et al.* (1996). The collection apparatus consists of a cryogenic device, with air passed through a condenser at 80 K and a second one maintained at 63 K. The first trap removes the water (ppm level in the stratosphere) and $CO_2$, and the second trap collects ozone. At the relevant flow and pressure regime, only ozone is collected in the second trap, all other stratospheric gases are noncondensable. The data reported by Krankowsky *et al.* (2000) utilizing this sampling device at Aire-sur-l'Adour France (43.7° N, 0.3° W) and Kiruna, Sweden (67.9° N, 21.1° E) reveal a range in heavy-isotope enrichment. There are several significant conclusions drawn by these measurements.

- The large $^{18}O$ ozone enrichments reported earlier (Mauersberger, 1981) were not confirmed by the more precise measurements.
- The new data are well within laboratory observations, particularly in consideration of the temperature and pressure dependence.
- The data suggest that the early *in situ* measurements may have had a problem associated with collection and/or analysis procedures.
- The data further confirm that ozone isotopic measurements ($\delta^{17}O$, $\delta^{18}O$) provide a means by which stratospheric photochemistry and transport may be elucidated.
- Johnson *et al.* (2001) employed the Smithsonian astrophysical far-infrared spectrometer on seven balloon flights to measure stratospheric ozone isotopic compositions. There are two important aspects to these observations. First, they confirm and are consistent with the mass spectrometric and laboratory observations. Second,

the measurements provide structural information, namely, the asymmetric to symmetric ratio: $^{50}O_3$ (or $^{18}O$) to $^{49}O_3$ ($^{17}O$).

The Smithsonian instrument (FIRS-2) is a remote sensing Fourier transform spectrometer that detects molecular thermal emission in the atmosphere. Johnson *et al.* (2001) report data taken from seven balloon flights from 1980 to 1997. The last flight is from 68° N; the others were between 30° N and 35° N. An important feature of the spectrometer is its ability to resolve spectra at 0.004 cm$^{-1}$, which allows for resolution of $^{16}O^{16}O^{16}O$, $^{17}O^{16}O^{16}O$, $^{16}O^{17}O^{16}O$, $^{18}O^{16}O^{16}O$, and $^{16}O^{18}O^{16}O$ species. It is observed, over the altitudinal range 25–35 km, that the average enhancements for symmetric, asymmetric, and total $^{50}O_3$ are $61 \pm 18$‰, $122 \pm 10$‰, and $102 \pm 9$‰, respectively (Johnson *et al.*, 2001). The average enhancement of $^{17}O$ enrichment ($^{49}O_3$) are $16 \pm 76$‰, $80 \pm 52$‰, and $73 \pm 43$‰, respectively. These values are in agreement with previous measurements by other techniques. Most importantly, the data provide information on isotopomers. Irion *et al.* (1996) utilized a solar occultation spectrometer to provide vertical profiles for both $^{16}O^{18}O^{16}O$ and $^{16}O^{16}O^{18}O$. Their work demonstrated that there is a significant difference between the symmetric and asymmetric species. The enhancements, on average, revealed an enhancement in the asymmetric over symmetric species by a factor of 1.7. The recent measurements of Johnson *et al.* (2001) also indicate that asymmetric ozone is significantly more enriched than the symmetric one. These enrichment factors are in agreement with laboratory measurements (Mauersberger *et al.*, 1999; Anderson *et al.*, 1989, 1997; Janssen *et al.*, 1999).

In conclusion, with respect to atmospheric ozone isotopes, there have been significant advancements in their measurements. As in, e.g., nuclear astrophysics, observations in nature provide important insight into fundamental physics or chemistry, which often is not easily possible under experimental conditions. However, significant details and observations remain pending, including ozone vertical and seasonal isotopic variability. The ozone enrichment process is also known to cascade through other atmospheric molecular species (see, e.g., Lyons, 2001). There are two new and important applications to earth science of the stratospheric ozone phenomena. First is the interaction with carbon dioxide in the stratosphere. Second, a heavy-isotope enrichment in stratospheric ozone creates a compensating depletion in the precursor molecule oxygen. Although by material balance consideration, the effect is small, it creates a unique tracer of primary biological productivity

in the ocean, freshwater, and from ice core measurements of paleo-global productivity. Recent advances in atmospheric $CO_2$ and $O_2$ applications will be reviewed.

### 4.06.2.2  Stratospheric Carbon Dioxide

Prior to 1991, numerous isotopic measurements of stratospheric carbon dioxide were made, though only for $\delta^{18}O$. A mass-independent isotopic composition was later observed in samples from 26 km to 35.5 km, with a deviation from the expected isotopic composition (Thiemens *et al.*, 1991). It has been established that the mass independence arises from isotopic exchange between electronically excited atomic oxygen, $O(^1D)$, and $CO_2$:

$$O(^1D) + CO_2 = CO_2 + O(^3P)$$

In this process, the $^{17}O$, $^{18}O$ enrichment of atomic oxygen, inherited from stratospheric ozone photolysis, is passed on to $CO_2$ via exchange, as proposed by Yung *et al.* (1991, 1997). A fundamentally important but unresolved issue associated with the exchange process is that it occurs via a short-lived transition state, $CO_3^*$. From the laboratory experiments of Wen and Thiemens (1993), it is apparent that the isotopic composition of $CO_2$ at steady state is not simply derived from statistical transfer. This has subsequently been addressed by Barth and Zahn (1997) and Johnston *et al.* (2000). On the basis of existing observations and models, it is not possible to account for some of the important aspects of the exchange process.

Utilizing a rocket-borne cryogenic whole-air collection system, Thiemens *et al.* (1995) demonstrated that stratospheric and mesospheric $CO_2$ possess a mass-independent isotopic composition that derives from $O(^1D)$. This is confirmed by the observed inverse correlation of the magnitude of the $\delta^{17}O$, $\delta^{18}O$ anomaly with $CH_4$ and $N_2O$ concentrations. Methane and nitrous oxide are both (in part) removed from the stratosphere by reaction with $O(^1D)$, and the enhancement in $\delta^{17}O$, $\delta^{18}O$ from interaction with the same species. This has the significance of providing a new, quantitative measure of atomic oxygen, in large part a driving factor in upper atmospheric chemistry.

Alexander *et al.* (2001) advanced the applications of stratospheric $CO_2$ $\delta^{17}O$, $\delta^{18}O$ measurements. The observations were reported for samples collected from within the Arctic polar vortex. The samples were obtained utilizing a balloon-borne cryogenic air sampler on February 11, 1997 (Strung *et al.*, 2000). The samples were obtained in an altitude range extending from

12 km to 21.1 km. Aside from these representing the first vortex $\delta^{17}O$, $\delta^{18}O$ measurements, concentrations of $SF_6$, $CCl_3F$ (CFC-11), $CCl_2F_2$ (CFC-12) and $Cl_2FCClF_2$ (CFC-113) were measured. An important aspect of the work of Alexander *et al.* (2001) is that the $CO_2$ oxygen isotopic relation with $SF_6$ is clearly observed. Sulfur hexafluoride has a lifetime in excess of a thousand years and is only removed in the mesosphere and above by highly energetic processes such as photodecomposition and/or electron attachment. This long lifetime renders $SF_6$ as a valuable tool as a conservative tracer of stratospheric air mass movement, much in the same way as chloride is used to track oceanic currents and water masses. The observation of an inverse relation between the magnitude of mass independence and $SF_6$ concentration now combines an age factor to the chemical information obtained from the isotopic measurements which provides a measure of the age of the air mass and the integrated odd oxygen chemical activity. Future measurements will be instrumental in providing new details of upper atmospheric chemistry and dynamics. The same is true for the combined measurement of fluorocarbons and $CO_2$ isotopes.

More recent measurements have been reported by Lämmerzahl *et al.* (2002). This work reports simultaneous $CO_2$ and $O_3$ isotope ratios from eight balloon flights from Kiruna, Sweden and Aire-sur-l'Adour, France. There is a correlation observed between $\delta^{17}O$ and $\delta^{18}O$ of stratospheric $CO_2$. The observed $\delta^{17}O/\delta^{18}O$ ratio is 1.71 ± 0.03, independent of the altitude range sampled. The data are reported with respect to its enrichment above tropospheric $CO_2$ ($\delta^{18}O = 41$, $\delta^{17}O = 21$ per mil). This assumes that tropospheric $CO_2$ is single valued, which however is not the case. $\delta^{18}O$ is well known to vary due to its temperature-dependent isotopic exchange with $H_2O$. The observations are significant in that they include both $CO_2$ and $O_3$ isotopes. These data are of importance in providing a quantitative model for establishing the physical–chemical isotopic relation between stratospheric $CO_2$ and $O_3$.

Another important application of the $CO_2$ isotopic measurements is their use as a measure of stratosphere–troposphere mixing. Stratospheric $CO_2$ is mass-independently fractionated due to its coupling with ozone, while tropospheric is mass dependent because of its equilibrium with water. As originally described by Urey (1947), this is a purely mass-dependent process and has been confirmed by laboratory measurements (Thiemens *et al.*, 1995). This feature provides an ideal marker for the two individual atmospheric reservoirs.

An important aspect of the $\delta^{17}O$, $\delta^{18}O$ isotopic observations is the actual isotopic measurement analytical technique. In the first measurements, Thiemens *et al.* (1991) reported a technique that quantitatively converts $CO_2$ to $CF_4$ and $O_2$. In this technique $CO_2$ is reacted with $BrF_5$ at 800 °C for 48 h. At the end of the reaction, $O_2$ is separated from $CF_4$ by using a molecular sieve 13 X powder at the melting point of pure ethyl alcohol. This step is vital as trace amounts of $CF_4$ may cause error in the $\delta^{17}O$ measurements, presumably due to ion molecule reactions in the source of the mass spectrometer. Conversion to $O_2$ is required because measurement of $CO_2$ has isobaric and uncorrectable interference from the contribution at mass 45 from $^{13}C$. A second technique has recently been developed that converts $CO_2$ to methane and water, with subsequent chemical conversion of water to HF and $O_2$ by reaction with hydrogen (Brenninkmeijer and Röckmann, 1998). Subsequently, Assknov and Brenninkmeijer (2001) have reported another new technique. In this technique, $CO_2$ is measured at high precision for the $\delta^{13}C$, $\delta^{18}O$ isotopic composition. Following isotopic measurement, $CO_2$ is exchanged with solid $CeO_2$, which causes exchange of the oxygen isotopes. In the second isotopic measurement, only the 45/44 mass ratio is altered due to the exchange of oxygen isotopes. Since the $^{13}C/^{12}C$ ratio does not alter, the $\delta^{17}O$ may be determined by calculation. This technique has several advantages: (i) the mass 44, 45, 46 ratios are commonly done on isotope ratio mass spectrometers; (ii) it is fast and safe, not requiring $BrF_5$; and (iii) the technique also results in higher-quality $\delta^{13}C$ measurements.

### 4.06.3 AEROSOL SULFATE: PRESENT EARTH ATMOSPHERE

Thiemens *et al.* (2001) and co-workers have reviewed the progress in measurement of the mass-independent oxygen isotopic composition of aerosol sulfate. The importance of sulfate in the Earth's atmosphere and environment is well established. Aerosol sulfate alters radiative processes due to its role in increasing the Earth's albedo and as a cloud condensation nucleus (CCN). While these important roles are recognized, there remain significant gaps in understanding the role of aerosols in these processes. Sulfate and nitrate aerosols are also known to be agents in increasing the incidence of cardiovascular disease. In addition, nitrate is known to be an agent in alteration of terrestrial biodiversity. As a consequence, quantification of sources, transformation mechanisms, and chemical interactions are quite important. Stable isotope ratio measurements have been of limited utility in addressing these issues in the past due to the nonspecificity of single isotope ratio such as $\delta^{34}S$ or $\delta^{18}O$.

Another environmental issue with respect to sulfur is that 70–80% of the atmospheric sulfur species present in the northern hemisphere is anthropogenic (Rasch *et al.*, 2000). A gap in understanding the atmospheric sulfur cycle arises from the inability to adequately define the oxidation process of $SO_2$ to sulfate (Kasibhatla *et al.*, 1997). In particular, model calculations underestimate the observed sulfate concentrations in northern latitudes, and it appears that enhanced heterogeneous chemical activity could be important (Lelieveld *et al.*, 1997; Kasibhatla *et al.*, 1997) in bringing together model calculations and observations. Measurement of oxygen and sulfur mass-independent isotopic compositions of aerosols has provided new insights into these cycles.

It is now well known that mass-independent isotopic compositions are observed in atmospheric aerosol sulfates (Lee and Thiemens, 1997, 2001; Lee *et al.*, 1998, 2001a,b, 2002; Savarino *et al.*, 2000; Johnson *et al.*, 2001). The details of these papers are reviewed in Thiemens *et al.* (2001) and Thiemens (2002). The measurements of the $\delta^{17}O$, $\delta^{18}O$ isotopic composition of aerosol sulfate have revealed the following:

(i) Coupled with isotopic measurements of atmospheric $H_2O_2$ (Savarino and Thiemens, 1999a,b) and knowledge of reaction rate constants, pH reaction rate dependence, and isotopic fractionation factors (Savarino *et al.*, 2000), the homogenous and heterogeneous reaction pathways may be quantified. This has been a goal for decades and provides a new quantitative means to supplement modeling efforts.

(ii) Source processes of aerosol sulfates may be identified. For example, during the INDOEX (Indian Ocean Experiment), the La Jolla group demonstrated that the Inter-tropical Convergence Zone is a source of new aerosol particles. This was previously unrecognized and a definition of a new aerosol generation process is of major significance in global climate models.

(iii) Mass-independent sulfate laboratory measurements have further elucidated the free-radical oxidation mechanism of liquid-phase $SO_2$ to sulfate.

(iv) The mass-independent isotopic composition of sulfate may be used to quantify the relative contribution of $O_3$ and $H_2O_2$ in sulfur oxidative processes. This is a significant parameter in determining transport phenomena, particularly over long (hemispheric) ranges.

(v) In specific locations, the isotopic measurements provide unique process information. For example, Lee and Thiemens (1997) used combined radiogenic [35]S (87.2 d half-life) and $\delta^{17}O$, $\delta^{18}O$ isotopic measurements to recognize that in winter months at the White Mountain Research Station (12,500 ft) there is a significant upper atmospheric source of sulfate. Johnson *et al.* (2001) have used the $\delta^{17}O$, $\delta^{18}O$ measurements to provide new insight into surface water interactions and sources in a high alpine region of the Rocky Mountains.

The most recent studies of sulfate have furthered our understanding of the natural sulfur cycle. Lee *et al.* (2002) performed sulfur and oxygen isotopic measurements of sulfates produced in controlled combustion processes. These experimental observations were conducted at the Centre de Recherches Atmospheriques in Lanne-mezan, France in a dark combustion chamber. This system was specifically constructed to study aerosol and gas emissions derived from fuel combustion and emission. The chamber has a volume of $\sim 160\ m^3$, which effectively minimizes wall effects and maximizes mixing. The collections are done in the dark to minimize any potential photochemical effects. Fossil fuel and vegetation burns were performed, including Savanna grasses, Lamto grass, rice grass, hay, diesel fuel, and charcoal. All of these processes produce sulfur and oxygen isotopic compositions that are strictly mass dependent. This experimentally eliminates these processes as the source of the observed mass-independent isotopic composition of aerosol sulfate.

### 4.06.3.1 Other Mass-independent Sulfate Isotopic Compositions

There are several other natural repositories of mass-independent sulfates, all of which have recently been reviewed (Thiemens *et al.*, 2001; Thiemens, 2002). Namibian desert gypsum ($CaSO_4 \cdot 2H_2O$) is a representative example. Bao *et al.* (2000a,b) reported mass-independent oxygen isotopic compositions of the massive ($3 \times 10^4\ km^2$) gypsum deposit of Namibia. In these measurements, isotopic trends, as a function of distance from the ocean, were observed. In a careful analysis of the various parameters, Bao *et al.* (2000a,b) determined that the source of the oxygen isotopic anomalies derives from the atmospheric photo-oxidation of dimethyl sulfide from the nearby upwelling oceanic region. These measurements therefore have provided a new mechanism to examine paleo-oceanic variations of biologic activity on tens of millions of year timescales.

Bao *et al.* (2000b) extracted sulfate from vertical profiles of Antarctic dry valley samples and utilized the oxygen isotopic composition to investigate sources. The variance in the mass-independent oxygen isotopic composition was defined as the source of the anomalous sulfate composition and source. In particular, the source, particle size dependence, and relative amounts of sulfate from sea salt aerosols and biogenic sources were resolved. Observations of a mass-independent isotopic composition in sulfate from Miocene volcanogenic samples also displayed large mass-independent isotopic compositions. These results are particularly intriguing. The anomaly is quite large and associated with the Miocene event; however, other volcanic activities do not possess any mass-independent isotopic composition. Clearly, in this there remain some outstanding questions, and further measurements are needed to advance our understanding of the origin of these anomalies and how they may be exploited to further understand the natural processes associated with these events. The details of these measurements are reviewed in Thiemens *et al.* (2001) and Thiemens (2002). For applications of sulfate mass-independent isotopic compositions as a means for understanding the process of desert varnish foundation, Bao *et al.* (2001a,b) report details of the measurements and applications.

## 4.06.4 ATMOSPHERIC MOLECULAR OXYGEN

It has been known for decades that photosynthesis and respiration predominantly establish the steady-state levels of $O_2$ and $CO_2$ in the Earth's atmosphere. From this standpoint, it might be argued that these processes are among the most important of all global biogeochemical cycles. For this reason, careful and precise establishment of the rates of photosynthetic and primary productivity in the world's oceans is essential. This rate, as reviewed by Bender (2000), is intimately linked to the primary productivity of the oceans, carbon cycling, and $CO_2$ levels in the ocean and atmosphere. Traditionally, in spite of this enormous importance, quantification has been difficult. In part, this stems from the magnitude of the oxygen reservoir and the hysteresis for production of a measurable change. Luz *et al.* (1999) reported an entirely new mechanism to evaluate biospheric productivity. This new technique is based upon isotopic material balance constraints on the creation of an anomalous mass-independent oxygen isotopic reservoirs in ozone, $CO_2$, and $O_2$. As discussed, this anomaly is transferred from $O_3$ to stratospheric $CO_2$. This occurs in two ways: first as a statistical nuclidic exchange and, second, due to the isotopic effort associated with creation of the short-lived $CO_3^*$ transition state. The magnitude of the anomaly is measurable in $CO_2$ (~360 ppm) but for $O_2$, (~20%), the steady-state effect on the $\delta^{17}O$, $\delta^{18}O$ of atmospheric $O_2$ is not measurable. However, there is a significantly longer lifetime for $O_2$ than $CO_2$ (factor of

hundreds), which is eventually lost by photosynthesis in the upper ocean. As a result, the positive $^{17}O$, $^{18}O$ mass-independent anomaly in $CO_2$ produces a small but measurable isotopic anomaly in $O_2$, which accrues over time because of the differential lifetimes. The consequence of the field and laboratory measurements is that measurement of the mass-independent $O_2$ isotopic composition in oceanic vertical profiles provides a new, quantitative means by which primary productivity may be evaluated. The $O_2$ anomaly is only removed by respiration in the oceans. Thus, its rate of disappearance is directly linked to primary productivity. Measurement of $\delta^{17}O$, $\delta^{18}O$ of $O_2$ trapped in polar ice may also be used as a means of determining global biospheric primary productivity. Luz and Barkan (2000) furthered these studies by developing a new methodology for determining oceanic production rates on week timescales.

Subsequent to the review by Thiemens (2002), there have been further advances in the utilization of the molecular $O_2$ isotopic composition (Angert *et al.*, 2001). Though they did not measure $\delta^{17}O$, they have determined the oxygen isotopic fractionation by respiration and diffusion in soils, measurements critical in quantifying the atmospheric $O_2$ cycle. The well-known $^{18}O$ enhancement of atmospheric $O_2$ is utilized to determine paleo-variations in the relative $O_2$ contributing proportions of marine and terrestrial sources. Quantification of the oxygen cycle is mandatory, in order to resolve the tightly coupled carbon cycle. The experiments of Angert *et al.* (2001) provide high-precision isotopic measurements requisite for a sound modeling of both the carbon and oxygen biogeochemical cycles. More recently, Angert *et al.* (2002) have reported one of the most extensive evaluations of the $\delta^{17}O$, $\delta^{18}O$ for the individual processes associated with photosynthesis and respiration. This is a key component in extending the utilization of mass-independent $O_2$ isotopic compositions as a means to estimate changes in global productivity. In particular, Angert *et al.* (2002) have evaluated the values of $\theta$ for the most important processes that determine the isotopic composition of molecular oxygen:

$$\theta \equiv \frac{\ln(^{17}\alpha)}{\ln(^{18}\alpha)}$$

which follows the relation described by Mook (2000). In this instance, $^{17}\alpha$ and $^{18}\alpha$ are the isotopic fractionation factors. The value of $\theta$ has been determined for the dark respiration factor, i.e., the cytochrome pathway (0.516 ± 0.001) and its mechanistic alternative (0.514 ± 0.001). A modestly higher value for diffusion in air (0.521 ± 0.001) was determined. The precise determination of these $\theta$-factors leads to the determination of the steady-state value for diffusion and respiration in the atmosphere. The $\theta$ value of photorespiration ($0.506 \pm 0.003$) is less than for dark respiration. These evaluations represent a major step forward in the development of $\delta^{17}O$, $\delta^{18}O$ measurements of atmospheric $O_2$ as a means to study global rates of photorespiration, both at present and on glacial timescales. In addition, as Angert *et al.* (2002) discuss in detail, the individual processes associated with photo-respiration have significantly been resolved. A final important aspect is that these results will ultimately lead to a closure of the difference between the modeled and measured Dole effect. This has been an important issue since the first report of the Dole effect.

In summary, the $O_2$ $\delta^{17}O$, $\delta^{18}O$ measurements have significantly advanced our understanding of the oxygen and carbon global biogeochemical cycles on present and past timescales. These are enhancements in understanding made possible by the utilization of, albeit small, mass-independent isotopic composition of $O_2$. Finally, it is seen that the exact value of $\theta$, or mass fractionation law, may be determined at extraordinarily high precision. The very slight variances in mass fractionation laws have now led to new applications, which are reviewed, in a subsequent section.

### 4.06.5 THE ATMOSPHERIC AEROSOL NITRATE AND THE NITROGEN CYCLE

It is well established that atmospheric nitrate expresses its importance in a number of vital biogeochemical cycles. These include:

(i) its role as a major sink for nearly all $NO_x$ species;

(ii) its increasingly important role as an acid agent in the environment;

(iii) participation as a heterogeneous reactant in polar stratosphere clouds (PSCs) and destruction of ozone in the Antarctic polar vortex;

(iv) as a source of fixed nitrogen to key ecosystems (Paerl, 1993); and

(v) as a possible agent in radiative transfer in the Earth's atmosphere.

In spite of these highly significant roles, there remain serious limitations in the quantification of the nitrogen cycle. This includes, e.g.,

- depositional rates (Kendall and McDonnell, 1998; Cress *et al.*, 1995);
- chemical transformation mechanisms in the troposphere (lower and upper);
- source variability and identification; and
- long-range transport importance.

It has been shown that atmospheric aerosol nitrate possesses one of the largest mass-independent oxygen isotopic compositions observed in nature

(Michalski and Thiemens, 2000). An important development in atmospheric nitrate studies has been the development of the ability to perform the $\delta^{17}O$, $\delta^{18}O$ measurements in aerosol nitrate. In the process of conversion of nitrate to $O_2$ for mass spectrometer analysis, high purity is essential. Ion chromatography is utilized to simultaneously isolate and concentrate nitrate. The nitrate is subsequently reacted under a proper pH regime to convert to $HNO_3$. Reaction with $Ag_2O$ quantitatively converts nitric acid to $AgNO_3$, which is filtered and dried. $AgNO_3$ is thermally decomposed to $O_2$, $NO_2$, and Ag at a constant chemical and isotopic branching ratio. Following a final purification process, the $O_2$ is measured for the $\delta^{17}O$, $\delta^{18}O$ composition. A variety of standards have been analyzed, including USGS-35, a sample of $NaNO_3$ from the Atacama desert of Chile and IAEA-N3, an internationally distributed standard sample $KNO_3$, with a range of reported $\delta^{18}O$ values (Revesz *et al.*, 1997; Kornexyl *et al.*, 1999).

As Michalski and Thiemens (2000) have shown, aerosol nitrate possesses an extraordinarily large mass-independent isotopic composition. As for aerosol sulfate, this isotopic signature has already been shown to provide a new means to elucidate source and chemical transformation processes. This has proved to be an important technique by which the nitrate biogeochemical cycle may be understood further. For example, the massive mass-independent isotopic signature observed in Chilean desert nitrates uniquely reveals that these nitrates must be atmospherically derived since all other sources (by measurement) possess mass-dependent isotopic compositions. In addition, these measurements, coupled with contemporary aerosol nitrate measurements reveal that the oxygen isotopic signatures are stable on million year timescales. This is particularly valuable, as this permits measurement of nitrate in polar ice samples to examine paleo-variations in nitrate and in general, $NO_x$ chemistry. As discussed in the next section, linked with sulfate oxygen isotopic measurements, an entirely new mechanism to study paleo-atmospheric oxidative capacity variation.

Finally, in the case of present-day aerosol nitrate isotopic measurements, samples collected as a function of particle size provide another level of detection in the resolution of sources and atmospheric transformation mechanisms. Large particles (1–10 $\mu$m) typically are crustal or oceanic sea spray, depending upon where the particular samples are collected. Small particles (less than 0.1 $\mu$m) generally are gas-to-particle conversion process products. Using combined multi-isotope ratio measurements and size-fractionated collection processes, it is possible to provide sophisticated details of atmospheric aerosol fates.

## 4.06.6 PALEO-OXYGEN VARIATIONS AND POLAR ICE MEASUREMENTS

As discussed earlier, the mass-independent isotopic compositions of sulfate and nitrate aerosols have enhanced our understanding of formation processes. For example, as discussed by Savarino *et al.* (2000), and Lee *et al.* (2001a,b), the sulfate mass-independent oxygen isotopic composition quantitatively describes the proportional contributions of gas (OH radical) and liquid phase ($O_3$, $H_2O_2$) oxidative processes. These proportions are vital parameters in evaluating paleo-climate variations. In addition, evaluation of the mechanism of aerosol formation for both nitrate and sulfate will influence the indirect aerosol component of climate models. For these reasons, $\delta^{17}O$, $\delta^{18}O$ measurements of ice core sulfate and nitrate represent a new horizon in climate research. Alexander *et al.* (1999, 2000, 2002a,b,c) and Savarino *et al.* (2002) have reported mass-independent isotopic compositions of sulfate in polar ice samples. In Alexander *et al.* (2002b), the oxygen isotopic composition of sulfate in eight samples from the Vostok Antarctic ice core sample were reported. These samples spanned sufficient temporal range to include a full climate cycle. This data clearly demonstrated that during the glacial period the relative proportion of gas versus aqueous phase oxidation was significantly greater than during interglacial periods. The work confirmed that the oxidative capacity of the atmosphere might be identified from such measurements. This work also highlights the relative importance of cloud processing, an important parameter that has been historically difficult to assess. In addition, such measurements coupled with $CH_4$, $H_2O_2$, and $MSA/SO_4$ ratio measurements of polar ice species, provide a powerful tool to evaluate the variation of paleo-climates and define the quantitative coupling between atmospheric chemistry and climate.

Alexander *et al.* (2002a,b,c) have measured the $\delta^{17}O$ and $\delta^{18}O$ of both sulfate and nitrate from samples at Greenland ice core site A (70° 45' N, 35° 57.5', elevation 3,145 m). These samples were particularly interesting because, temporally, they surround the Industrial Revolution. The years spanned by these measurements are from around 1692 to 1976. The details of ice decontamination and chemistry are presented in Alexander *et al.* (2002a,b,c). These measurements have distinctly shown that biomass-burning events prior to the Industrial Revolution altered the atmospheric oxidative mechanisms in the northern hemisphere. This work is significant in that: (i) it further demonstrates how $\delta^{17}O$, $\delta^{18}O$ measurements of sulfate and nitrate provide a new means by which the oxidative capacity of the atmosphere on extended timescales may be determined; (ii) the biomass burning emissions were highly significant, even in comparison with fossil fuel combustion. The results underscore the need to include precise values for the magnitude of biomass burning in paleo-climatological models. Finally, the newest results depict the importance of mass-independent isotopic compositional measurements in ice core studies as a complement to concentration and other stable isotope ratio measurements.

## 4.06.7 MASS-INDEPENDENT ISOTOPIC COMPOSITIONS IN SOLIDS

There are few examples of mass-independent isotopic fractionations, as yet, associated with the formation of solids. The first example was by Colman *et al.* (1996), who demonstrated that in the irradiation (313 nm) of gas phase $CS_2$, a solid aerosol $(CS_2)_x$ was generated. It was found that this polymeric/solid is highly mass-independently fractionated, with a $^{33}S$ excess of 5–10‰, and a $^{36}S$ deficit of 61–84‰. An interesting aspect of these observations was that the optical properties of this solid were quite similar to those observed for solids produced by comet Shumaker–Levy 9(SL9) during the collision with the outer atmosphere of Jupiter. Colman *et al.* (1996) suggested that the mechanism responsible for the observed sulfur isotopic fractionation process involves the differing rates of nonradiative transfer from the initial absorbing rate to the final reactive state. In processes such as this, the rate is predominantly dependent upon Franck–Condon factors and, hence, is not directly dependant upon mass. In a follow-up investigation, Zmolek *et al.* (1999) reported the results of experiments that further detailed the mechanism responsible for the production of the photopolymeric $(CS_2)_x$ solid from $CS_2$ photolysis. In these experiments, it was demonstrated that the sulfur isotopic composition in products from photolysis of $^{12}CS_2$ and $^{13}CS_2$ differ significantly. This observation immediately rules out any involvement of a symmetry-dependent fractionation effect. Zmolek *et al.* (1999) suggest Franck–Condon and vibronic coupling effects as the source of the observed effect. It is specifically the nonradiative decay and intersystem crossing rates for the lowest excited states which give rise to the highly anomalous sulfur isotopic composition observed in the solid product.

A similar fractionation process was reported by Bhattacharya *et al.* (2000). Photodissociation of $CO_2$ by UV light at 185 nm produces CO and $O_2$ enriched in $^{17}O$ by more than 100‰. As reported, this dissociation arises via a spin-forbidden process during the singlet to triplet transition. The mass independence derives from the reliance of the reaction rate on essentially resonant spin–orbit coupling of the low-energy vibrational states of the $^{16}O^{12}C^{17}O$ molecule of the singlet with the triplet state. This hypothesis was confirmed by demonstration that the alteration of the oxygen isotopic fractionation results by simple substitution of $^{13}C$ for $^{12}C$. These experiments also demonstrate that the mass-independent isotopic fractionation process is restricted neither to symmetry nor to ozone.

A completely new variety of mass-independent isotopic fractionation has been reported by Miller *et al.* (2002). It was observed that in the thermal decomposition of naturally occurring carbonates (calcium and magnesium), the product $CO_2$ and the residual solid oxides possess mass-independent isotopic compositions. The product $CO_2$ is depleted in $^{17}O$ by ~0.2‰ and the residual CaO or MgO enriched by a comparable factor. Numerous control experiments were performed in two different laboratories to demonstrate that the effect is associated with the thermolysis of the carbonate and not a result of secondary gas-phase reaction or experimental artifact. The effect is also observed for an extended suite of carbonate samples of widely varying $\delta^{18}O$ isotopic composition. This is the first report of a mass-independent effect associated with a solid reactant. As discussed by Miller *et al.* (2002), the physical–chemical mechanism responsible for the observed effect is unknown, though of considerable interest. That paper discusses the limited likelihood of detecting this effect in nature and therefore, at present the effect is restricted to a significant result in the fundamental physical chemistry of mass-independent isotope effects. Future experiments are needed to determine the physical–chemical basis for this effect in solids. Measurements on other solids are needed, particularly for molecules possessing oxygen bound in an equilateral triangular arrangement. Theoretical considerations of this effect are also needed if the potential occurrence in nature is to be critically explored. As Miller *et al.* discuss, detection of calcite and dolomite in dust shells surrounding evolved stars (Kemper *et al.*, 2002) may provide a cosmochemical environment where an anomalous isotopic fractionation process could occur. As discussed in the next section, it is well established that preserved interstellar grains occur in meteoritic material.

## 4.06.8 MASS-INDEPENDENT ISOTOPIC FRACTIONATION PROCESSES IN THE EARLY SOLAR SYSTEM

The suggestion by Clayton *et al.* (1973) that the $\delta^{17}O = \delta^{18}O$ isotopic composition observed in the meteorite Allende CAIs was due to a nucleosynthetic process was based upon the inability of a chemical process to produce a mass-independent isotopic composition. With the discovery by Thiemens and Heidenreich (1983) of a chemical process capable of producing the same $\delta^{17}O = \delta^{18}O$ isotopic composition, the fundamental assumption was shown to be incorrect. As discussed in Thiemens (1999), one of the fundamental requirements for the nucleosynthetic mechanism was that it requires that the $^{16}O$ carrier be added as a dust grain. Clayton *et al.* (1973) hypothesized that this should produce a correlation with another element, such as magnesium, silicon, or aluminum. After a quarter of century of measurements, no such correlation exists and oxygen remains the only element with meteoritic anomalies present at the bulk level. Thiemens (1999) has reviewed potential chemical mechanisms by which the meteoritic anomaly may be produced. Of particular significance is the position of oxygen in the periodic chart. It is the only element capable of producing a symmetry isotopic fractionation, namely, it coordinates most rock-forming elements. Any other element, except noble gases, is coordinated by oxygen, such as silicates $(SiO_4)$ the major rock-forming minerals. Elements such as hydrogen and carbon could ostensibly produce mass-independent symmetry effects; however, they only possess two stable isotopes and observation of such effects is impossible. Sulfur has the possibility, but its multiple valence states and inherent volatility render it difficult to maintain its original isotopic signature. The only meteoritic class that has combined low sulfate content and is isotopically anomalous at the bulk level are the Ureitites. These are achondritic meteorites, made of mostly olivines and pyroxenes. They also possess graphite, diamond, sulfides, metal, and minor silicates (Mittlefehldt *et al.*, 1998). The source of these meteorites has remained enigmatic, and based upon their bulk isotopic composition, possess highly unusual oxygen (Clayton and Mayeda, 1988). An excellent review of this highly unusual class of meteorites is provided by Goodrich *et al.* (1987). Thiemens *et al.* (1994) and Farquhar *et al.* (2000) have reported that the sulfur isotopic composition of ureilities is mass independent, with a slight $^{33}S$ excess. As discussed by Farquhar *et al.* (2000), it is possible that this $^{33}S$ enrichment may be derived from a chemical process. These particular meteorites have exceedingly low sulfur content, and perhaps these features may allow for preservation of a nebular signature. At present, the source of these anomalies remains unidentified. Future measurements, particularly those that search for correlations of the sulfur isotopes with other isotopic signatures, are essential.

It is now well known that there exist preserved interstellar grains in meteoritic material (see, e.g., Nittler *et al.*, 1999). There is extensive literature on this subject and will not be reviewed here. However, with respect to the oxygen nucleosynthetic model there are two key points. First, although there are literally thousands of isotopic measurements of the oxygen isotopic composition of interstellar grains, there is no evidence for the presence of a $^{16}O$ carrier. Second, the concentration of these grains is quite low, typically parts per million or less. Therefore, there is a material balance issue as well, given that the oxygen isotopic anomalies exist at the bulk level in the major meteoritic element.

It is clear that nucleosynthetic models may not account for the observed meteoritic oxygen isotopic anomalies, and chemical production processes are needed. Clayton (2002) has abandoned the nucleosynthetic model as an explanation for the observed meteoritic oxygen isotopic anomalies and proposed that isotopic self-shielding in the nebula accounts for the observed isotopic anomaly. The concept of self-shielding in the nebula as a mechanism to account for the oxygen isotopic anomalies was first discussed in Thiemens and Heidenreich (1983). In addition, the possibility that the $^{15}N$ anomalies could arise from this process was also discussed by Thiemens and Heidenrich (1983). The restriction on isotopic self-shielding for oxygen is that, subsequent to production of an anomalous oxygen in the dissociative event, its isotopic character is removed by isotopic exchange before it may be preserved in a stable product. This issue was discussed in great detail by Navon and Wasserburg (1985). It is implausible for this reason, as well as opacity constraints, that self-shielding accounts for the meteoritic oxygen isotopic anomalies. A further issue is that the $\delta^{17}O = \delta^{18}O$ composition that would be produced by self-shielding only occurs over a restricted carbon monoxide column density. At greater capacities, the different $^{18}O/^{17}O$ abundance ratio (5.5) becomes significant and varying isotopic compositions are produced. In these regions, the $\delta^{17}O = \delta^{18}O$ fractionation is not produced. Given that the anomalous oxygen isotopic effect is known to occur in the actual chemical reaction step, the issue of exchange is moot, as it is the reaction leading to product itself where the effect occurs (Thiemens, 1999). This removes restrictions on the preparation made of reactants. The recent successful models of Marcus and colleagues are of particular importance, as they also demonstrate that the source of the $\delta^{17}O = \delta^{18}O$ fractionation is a general feature of a gas phase reaction for oxygen and not restricted to ozone fromation. It is also this aspect that accounts for the issue as to why, of all the elements on the periodic chart, oxygen is unique. It is now clear that chemical production mechanisms are responsible for the production of meteoritic oxygen isotopic anomalies. This is a major solar-system process, since oxygen is the major element and the isotope effect large, existing at the bulk level. What is needed are relevant experiments examining in detail the isotopic fractionations associated with the formation of solid oxides. Such experiments have proved to be formidably difficult due to the required reaction conditions; however, the need persists.

## 4.06.9 CONCLUDING COMMENTS

Since the discovery of a chemically produced mass-independent isotope effect by Thiemens and Heidenreich (1983), it has been demonstrated that there exists an extensive range of applications and observations in nature. Resolution of the actual physical–chemical mechanism has significantly advanced our understanding of gas-phase transition states and chemical reactions. Mass-independent oxygen isotopic compositions have been observed in essentially all atmospheric molecules, except water. This includes: $O_2$, $O_3$, $CO_2$, $N_2O$, $H_2O_2$, and aerosol sulfate and nitrate. In each specific instance, our understanding of its biogeochemical cycle has been enhanced as a result of the particular and specific isotopic signature. Measurements of sulfate and nitrate in ice core samples have provided new details of the past oxidative capacity of the Earth's atmosphere. Oxygen isotopic compositions of ancient sulfates (e.g. Miocene) have provided new paleo-atmospheric information on the desert and nearby ocean in Namibia, volcanogenic processes, Antarctic dry valley sources, and desert varnish formation processes. Sulfur mass-independent isotopic anomalies in the Earth's oldest rocks ($3.8 \times 10^9$ yr) have provided a new means by which the evolution of oxygen-ozone, and consequently life has evolved until $\sim 2.2 \times 10^9$ yr before present. Measurement of sulfur and oxygen isotopic anomalies in secondary minerals from martian meteorites have provided new insights into crust–atmosphere interactions on Mars. Finally, the chemical mass-independent process appears to be responsible for the production of the anomalous oxygen isotopic compositions observed in meteorites and thus was a major process in the formation of the solar system.

There remain many new horizons in mass-independent chemistry and its applications. This includes theory, laboratory experiments, and new applications in nature, extending from Earth to the solar system, and in time, from the present to 4.55 Ga.

## ACKNOWLEDGMENTS

P.I. gratefully acknowledges the generous support of the NSF and NASA. Without this support, most of this work would not have been possible. He also thanks several of his colleagues who have worked with him over the years to develop new and exciting avenues of application in nature.

## REFERENCES

Alexander B., Savarino J., Farquhar J., Thiemens M. H., Jourdain B., and Legrand M. (1999) *EOS, Trans., AGU* **80**, F134.

Alexander B., Savarino J., Thiemens M. H., and Delmas R. (2000) *EOS, Trans., AGU* **81**, F38.

Alexander B., Vollmer M. K., Jackson T., Weiss R. F., and Thiemens M. H. (2001) *Geophys. Res. Lett.* **28**, 4103.

Alexander B., Savarino J., Sneed S., Kreutz K., and Thiemens M. H. (2002a) *EOS, Trans., AGU.*

Alexander B., Savarino J., Barkov N. I., Delmas R. J., and Thiemens M. H. (2002b) *Geophys. Res. Lett.*

Alexander B., Savarino J., Kreutz K., and Thiemens M. H. (2002c) *Science* (submitted).

Anderson S. M., Morton J., and Mauersberger K. (1989) *Chem. Phys. Lett.* **156**, 175.

Anderson S. M., Hüsebusch D., and Mauersberger K. (1997) *J. Chem. Phys.* **107**, 5385.

Angert A., Luz B., and Yakir D. (2001) *Global Biogeochem. Cycles* **15**, 871.

Angert A., Rachmileuitch S., Barcan E., and Luz B. (2002) *Global Biogeochem. Cycles* (submitted).

Assknov S. S. and Brenninkmeijer C. A. M. (2001) Rapid Communication. *Mass Spectrom.* **15**, 2426.

Bao H., Thiemens M. H., Farquhar J., Campbell D. A., Lee C. C.-W., Heine K., and Loope D. B. (2000a) *Nature* **406**, 176.

Bao H., Campbell D. A., Böckheim J. G., and Thiemens M. H. (2000b) *Nature* **407**, 499.

Bao H., Thiemens M. H., and Heine K. (2001a) *Earth Planet Sci. Lett.* **192**, 125.

Bao H., Michalski G. M., and Thiemens M. H. (2001b) *Geochim. Cosmochim. Acta* **65**, 2029.

Barth V. and Zahn A. (1997) *J. Chem. Phys.* **102**(12), 995.

Bender M. L. (2000) *Science* **288**, 5473.

Bhattacharya S., Savarino J., and Thiemens M. H. (2000) *Geophys. Res. Lett.* **27**, 1459.

Bigeleisen J. and Mayer M. (1947) *J. Chem. Phys.* **15**, 261.

Brenninkmeijer C. A. M. and Röckmann T. (1998) Rapid Communication. *Mass Spectrom.* **12**, 479.

Clayton R. N. (2002) *Lunar Planet. Sci. Conf.*, Houston, Texas (abstr.) **XXXIII**, 1326.

Clayton R. N. and Mayeda T. K. (1988) *Geochim. Cosmochim. Acta* **52**.

Clayton R. N., Grossman L., and Mayeda T. K. (1973) *Science* **182**, 485.

Colman J. J., Xu J. P., Thiemens M. H., and Trogler W. C. (1996) *Science* **273**, 774.

Cress R. G., Williams M. W., and Sieuering H. (1995) *IAHS Publ.* **228**, 33.

Farquhar J., Jackson T. L., and Thiemens M. H. (2000) *Geochim. Cosmochim. Acta* **64**, 1819.

Gao Y. Q. and Marcus R. A. (2001) *Science* **293**, 259.

Gao Y. Q. and Marcus R. A. (2002) *J. Chem. Phys.* **116**, 137.

Goodrich C. A., Jones J. H., and Berkley S. L. (1987) *Geochim. Cosmochim. Acta* **51**, 2255.

Guenther J., Erbacher B., Krankowsky D., and Mauersberger K. (1999) **306**, 209.

Guenther J., Krankowsky D., and Mauersberger K. (2000) *Chem. Phys. Lett.* **324**, 31.

Hathorn R. C. and Marcus R. A. (1999) *J. Chem. Phys.* **111**, 4087.

Herzberg G. (1966) Molecular spectra and molecular structure: III. Electronic spectra and electronic structures of polyatomic molecules. Van Nostrand Reinhold, New York.

Hulston J. R. and Thode H. G. (1965) *J. Geophys. Res.* **70**, 3475.

Irion F. W., Gunson M. R., Rinsland C. P., Yung Y. L., Abrams M. C., Change A. Y., and Goldman A. (1996) *Geophys. Res. Lett.* **23**, 2377.

Janssen C. (2001) *Science* **294**, 951.

Janssen C., Günther J., Krankowsky K., and Mauersberger K. (1999) *J. Chem. Phys.* **111**, 7179.

Johnson C. A., Mast M. A., and Kester C. L. (2001) *Geophys. Res. Lett.* **28**, 4483.

Johnston J. C. and Thiemens M. H. (2003) In *Handbook of Stable Isotope Techniques* (ed. P. de Grout). Elsevier, Holland.

Johnston J. C., Röckmann T., and Brenninkmeijer C. A. M. (2000) *J. Geophys. Res.* **105**(15), 213.

Kasibhatla P., Chameiles W. L., and St. John J. (1997) *J. Geophys. Res.* **102**(D3), 3737.

Kemper F., Jager C., Waters L., Henning T., Molster T. J., Banlow M. D., Lim T., and de Koter A. (2002) *Nature* **415**, 295.

Kendall C. and McDonnell J. J. (eds.) (1998) In *Isotope Tracks in Catchment Hydrology.*

Kornexyl B. E., Gehre M., Hofling R., and Werner R. A. (1999) Rapid Communication. *Mass Spectrom.* **13**, 1685.

Krankowsky P., Bartecki F., Klees G. G., Mauersberger K., and Stehr J. (1995) *Geophys. Res. Lett.* **27**, 1713.

Krankowsky P., Lämmerzahl P., and Mauersberger K. (2000) *Geophys. Res. Lett.* **27**, 2593.

Lämmerzahl P., Rockmann T., Brenninkmeijer C. M. M., Krankowsky D., and Mauersberger K. (2002) *Geophys. Res. Lett.* **29**, 1–23.

Lee C. C.-W. and Thiemens M. H. (1997) *EOS, Trans., AGU* **78**, F111.

Lee C. C.-W., Savarino J. P., and Thiemens M. H. (1998) *EOS, Trans., AGU* **79**, F91.

Lee C.-W., Savarino J., and Thiemens M. H. (2001a) *Geophys. Res. Lett.* **28**, 1783.

Lee C.-W., Savarino J., and Thiemens M. H. (2001b) *Geophys. Res. Lett.* **106**, 17359.

Lee C.-W., Savarino J., Cachier H., and Thiemens M. H. (2002) *Tellus* **54B**, 193.

Lelieveld J., Roelofs G. J., Ganzeweld L., Feichter J., and Rodhe H. (1997) *Phil. Trans. Roy. Soc. London, Ser. B* **352**, 149.

Luz B. and Barkan E. (2000) *Science* **288**, 2028.

Luz B., Barkan E., Bender M. L., Thiemens M. H., and Boering K. A. (1999) *Nature* **400**, 547.

Lyons J. R. (2001) *Geophys. Res. Lett.* **28**, 3231.

Mauersberger K. (1981) *Geophys. Res. Lett.* **8**, 935.

Mauersberger K. (1987) *Geophys. Res. Lett.* **14**, 80.

Mauersberger K., Erbacher B., Krankowsky D., Günther J., and Nickel R. (1999) *Science* **283**, 370.

Michalski G. and Thiemens M. H. (2000) *EOS, Trans., AGU* **81**, 120.

Miller M. F., Franchi I. A., Thiemens M. H., Jackson T. L., Brass A., Kurat G., and Dilliger C. T. (2002) *Proc. Natl. Acad. Sci.* (accepted).

Mittlefehldt D. W., McCoy T. J., Goodrich C. A., and Kracher E. A. (1998) *Planet. Mater.* **36**, 4-1–4-195.

Mook W. O. (2000) Environmental isotopes in the hydrological cycle: principles and applications: Vol. I. Introduction—theory, methods review, UUESCO/IAEA, Paris.

Navon O. and Wasserburg G. J. (1985) *Earth Planet. Sci. Lett.* **73**, 1.

Nier A. O., Ney F. P., and Ingraham M. G. (1947) *Rev. Sci. Instr.* **18**, 294.

Nittler L., O'D. Alexander C., Gao X., Walker R. M., and Zinner E. K. (1999) *Nature* **370**, 443.

Paerl H. W. (1993) *Can. J. Fish. Aquatic. Sci.* **50**, 2254.

Rasch P. J., Barth M. C., Kiehl J. T., Schwartz S. E., and Benkovitz C. M. (2000) *J. Geophys. Res.* **105**, 1367.

Revesz K., Boehlice J. K., and Yoshinavi T. (1997) *Anal. Chem.* **69**, 4375.

Savarino J. and Thiemens M. H. (1999a) *Atmos. Environ.* **33**, 3683.

Savarino J. and Thiemens M. H. (1999b) *J. Phys. Chem.* **103**, 9221.

Savarino J., Lee C. C.-W., and Thiemens M. H. (2000) *J. Geophys. Res.* **105**, 29079.

Savarino J., Alexander B., Sneed S., Kreutz K., and Thiemens M. H. (2002) *EOS, Trans., AGU* **83**(19), 545.

Schueler B., Morton J., and Mauersberger K. (1990) *Geophys. Res. Lett.* **17**, 1295.

Stehr J., Krankowsky D., and Mauersberger K. (1996) *J. Atmos. Chem.* **24**, 317.

Strunk M., Engel A., Schmidt U., Volk C. M., Wetter T., Levin I., and Glatzel-Mattheiz H. (2000) *Geophys. Res. Lett.* **27**, 341.

Thiemens M. H. (1999) *Science* **283**, 341–345.

Thiemens M. H. (2002) *Environ. Chem. Spec. Issue, Israel J.l Chem.* (accepted).

Thiemens M. H. and Heidenreich J. E., III (1983) *Science* **219**, 1073.

Thiemens M. H. and Jackson T. L. (1988) *Geophys. Res. Lett.* **15**, 639.

Thiemens M. H., Jackson T. L., Mauersberger K., Schuler B., and Morton (1991) *J. Geophys. Res. Lett.* **18**, 669.

Thiemens M. H., Brearly A., Jackson T., and Bobias G. (1994) *Meteoritics* **29**, 540.

Thiemens M. H., Jackson T., Zipf E. C., Erdman P. N., and Vansgmond C. (1995) *Science* **270**, 969.

Thiemens M. H., Savarino J., Farquhar J., and Bao H. (2001) *Acc. Chem. Res.* **34**, 645–652.

Urey H. C. (1947) *J. Chem. Soc.* **1947**, 562.

Wen J. S. and Thiemens M. H. (1993) *J. Geophys. Res.* **98**, 12801.

Yung Y. L., DeMore W. B., and Pinto J. P. (1991) *Geophys. Rev. Lett.* **18**, 13.

Yung Y. L., Lee A. Y. T., Irion W. B., DeMore W. B., and Wen J. (1997) *J. Geophys. Res.* **102**, 10857.

Zmolek P., Xu X., Jackson T., Thiemens M. H., and Trogler W. C. (1999) *J. Phys. Chem.* **103**, 2477.

# 4.07

# The Stable Isotopic Composition of Atmospheric CO₂

D. Yakir

*Weizmann Institute of Science, Rehovot, Israel*

## 4.07.1 INTRODUCTION

When a bean leaf was sealed in a closed chamber under a lamp (Rooney, 1988), in two hours the atmospheric $CO_2$ in the microcosm reached an isotopic steady state with a $^{13}C$ abundance astonishingly similar to the global mean value of atmospheric $CO_2$ at that time ($-7.5‰$ in the $\delta^{13}C$ notation introduced below). Almost concurrently, another research group sealed a suspension of asparagus cells in a different type of microcosm in which within about two hours the atmospheric $O_2$ reached an isotopic steady state with $^{18}O$ enrichment relative to water in the microcosm that was, too, remarkably similar to the global-scale offset between atmospheric $O_2$ and mean ocean water ($21‰$ versus $23.5‰$ in the $\delta^{18}O$ notation

introduced below; Guy *et al.*, 1987). These classic experiments capture some of the foundations underlying the isotopic composition of atmospheric $CO_2$ and $O_2$. First, in both cases the biological system rapidly imposed a unique isotopic value on the microcosms' atmosphere via their massive photosynthetic and respiratory exchange of $CO_2$ and $O_2$. Second, in both cases the biological system acted on materials with isotopic signals previously formed by the global carbon and hydrological cycles. That is, the bean leaf introduced its previously formed organic matter (the source of the $CO_2$ respired into microcosm's atmosphere), and the asparagus cells were introduced complete with local tap water (from which photosynthesis released molecular oxygen). Therefore, while the isotopic composition of the biological system used was slave to long-term processes, intense metabolic processes centered on few specific enzymes (Yakir, 2002) dictated the short-term atmospheric composition.

In a similar vein, on geological timescales of millions of years, the atmosphere and its isotopic composition are integral parts of essentially a single dynamic ocean–atmosphere–biosphere system. This dynamic system exchanges material, such as carbon and oxygen, with the sediments and the lithosphere via slow processes that roughly follow the cycle of: weathering of rock and carbon uptake from the atmosphere, transport to the ocean, sedimentation, plate tectonics, metamorphism, and volcanism—leading to carbon release back to the atmosphere. But on a shorter timescale of years to millennia, the very slow geological processes retreat to the background, against which other massive fluxes control the rapid exchange of carbon and oxygen within the ocean–atmosphere–biosphere system. It is this timescale that is relevant to the well-being of our human society, and is a major focus in much of the research on the carbon cycle.

Isotopes were discovered in 1911 (Urey, 1948) and the implications of isotopic substitution in chemical reactions were realized sometime later (Bigeleisen, 1965). In practice, the use of stable isotopes in geochemistry and biogeochemistry (e.g., Craig, 1953, 1954) awaited the development of the isotope ratio mass spectrometer (McKinney et al., 1950; Nier, 1947) that provided the necessary precision. Over the 50 years following this breakthrough, the application of stable isotopes has made tremendous progress in the scope of applications, as well as in the resolution and precision of the measurements. The carbon isotopic composition of rocks and sediments was measured intensively since the early 1950s (Hoefs, 1987). Isotope hydrology caught up quickly (Clark and Fritz, 1997), followed by the application of stable isotopes in biology and ecology (Rundel et al., 1988; Griffiths, 1998; Ehleringer et al., 1992). Today, stable isotope measurements have become an indispensable and integral part of atmospheric measurement programs (e.g., Francey et al., 2001; Masarie et al., 2001; Trolier et al., 1996). Efforts to develop analytical and numerical models that incorporate the cycling of stable isotopes in $CO_2$ expanded in parallel (e.g., Bolin, 1981; Ciais et al., 1997a,b; Enting et al., 1995). Recently, the consideration of mass-independent isotope phenomena in nature (Thiemens, 1999; see Chapter 4.06, and of triple stable isotopes in geochemistry (e.g., Blunier et al., 2002; Luz et al., 1999; Luz and Barkan, 2000) greatly extended the potential of stable isotope applications.

The chemical and isotopic composition of the atmosphere has drawn particular attention in climate-related research both because it is the most accessible component in the tightly coupled land–ocean–atmosphere system, and because the chemical composition of the atmosphere influences climate, particularly via the concentrations of the radiatively active greenhouse gases, such as $CO_2$, $O_3$, $CH_4$, $N_2O$, and water vapor. Information obtained by measurements of the atmospheric concentration of these gases alone is limited; the additional measurements of the stable isotopic composition provide information that cannot be obtained otherwise. Isotopic fractionations during chemical, physical, and biological process in the ocean, land, and the atmosphere result in unique natural labels. Tracing these labels in time and space allows us both to identify specific fluxes of these gases, and to gain insights into the processes influencing the observed fluxes. Quantitative use of $^{18}O$ and $^{13}C$ in $CO_2$ must rely on precise observations, on experimentation addressing the isotope effects underlying these observations, and on modeling that tests basic assumptions and extends applications beyond our measuring capabilities. Progress is still needed on all of these fronts. But the importance of this still developing science of stable isotopes in environmental research is indisputable.

## 4.07.2 METHODOLOGY AND TERMINOLOGY

Elements in nature come in forms called isotopes that differ only in the number of their neutrons. Most isotopes are stable and can be distinguished from their counterparts simply by their masses. Remarkably, isotopes are associated with a few simple and mass-dependent traits that result in a wide range of useful isotopic signals in natural processes. Coupled with the invention of the isotope ratio mass spectrometer in 1940s (McKinney et al., 1950; Nier, 1947) stable isotope signals provide the basis for application of stable isotopes to environmental sciences. Stable isotopes are denoted by their atomic mass such as $^{13}C$ and $^{12}C$ for the two stable isotopes of carbon, and $^{18}O$, $^{17}O$, and $^{16}O$ for the stable isotopes of oxygen. Because the heavy isotope is normally rare (e.g., ~1.1% for $^{13}C$, 0.2% for $^{18}O$, and 0.04% for $^{17}O$), routine measurements of the absolute isotopic concentrations is difficult and not reliable. Alternatively, the ratio, $R$, of the rare to the abundant isotopes is measured, such as

$$^{13}R = {}^{13}C/{}^{12}C \qquad (1)$$

for carbon or $^{18}R = {}^{18}O/{}^{16}O$ for oxygen, etc. Note that isotope ratios are distinct from isotope concentrations and, for example, $^{13}C$

concentration $[^{13}C]$ can be defined as $^{13}R'$,

$$^{13}R' = {}^{13}C/[{}^{12}C + {}^{13}C + {}^{14}C] \approx {}^{13}R/(1 + {}^{13}R) \quad (2)$$

The value of $^{13}R'$ can be used to estimate, for example, $^{13}C$ content in the atmosphere, which is currently about 4 ppm.

Isotope ratios are normally not measured on the pure element, but rather on a gas containing the element that is convenient for the mass-spectrometric analysis. Carbon and oxygen isotope ratios are normally measured in $CO_2$, which means $^{13}R = {}^{13}CO_2/{}^{12}CO_2$ and $^{18}R = C^{18}O^{16}O/C^{16}O^{16}O$ based on measurements of the mass ratios 45/44 and 46/44 (the mass to charge ratio, $m/z$, of the ionized form of the molecules is determined in practice). This is significant because there can be contributions to these mass units other than the measured molecules. For $CO_2$, corrections must be made, for example, for the occurrence of $C^{17}O^{16}O$ (mass 45; the "Craig correction"; Craig (1957), Mook and Grootes (1973), and Santrock *et al.* (1985)), and contaminations such as by $N_2O$ (mass 44; Craig (1963), Friedli and Siegenthaler (1988), Mook *et al.* (1974), and Mook and Van der Hoek (1983)). To reduce the effects of contaminants, gas samples are carefully purified, by bulk separation and/or by gas chromatography, prior to the mass spectrometric analysis. Furthermore, producing the convenient gas for isotopic analysis, such as $CO_2$, often involves a chemical reaction that modifies the isotopic ratio in question. In this case, the modifications must be carefully characterized and corrected for as part of the isotopic analysis. Such is the case, for example, in producing $CO_2$ from carbonates by acid treatment (Hoefs, 1987). For the determination of $^{18}O$ in water, the conventional practice is to first equilibrate a subsample with pure $CO_2$, determine the isotopic ratio of the $CO_2$ and reconstruct from it the isotopic ratio of the water (Gat, 1996; Gat and Gonfiantini, 1981; Mook, 2000).

To reduce variations due to instrumental drifts and instabilities, the common procedure in stable isotope analyses is to compare the ratio in a sample to that of a standard that is assumed to have a constant and known isotopic ratio. Isotope ratios are therefore reported in their normalized difference form as "delta values," where

$$\delta = \frac{R_{sample} - R_{standard}}{R_{standard}} = \frac{R_{sample}}{R_{standard}} - 1 \quad (3)$$

Note that $\delta$ values are practical, dimensionless parameters. Since values of $\delta$ in nature are small, they are generally given in ‰ (per mil)—which is a $10^3$ factor, not a unit. Further, only the arbitrary choice of standard determines whether a $\delta$ value of a sample is positive or negative (Mook, 2000).

To allow efficient comparisons of results among different labs around the world, a series of common international standards has been established and all samples are normalized to such standards. For carbon, the common standard was originally a carbonate of the Pee Dee belemnite carbonate formation in South Carolina. Naturally, as the original standards are exhausted, appropriate substitutes have been introduced. The two main sources for international standards today are the International Atomic Energy Association (IAEA) in Vienna, and the US National Institute of Standards and Technology (NIST). Two main international standards will be used in this chapter; the first, Vienna Standard Mean Ocean Water (V-SMOW), is used primarily for water isotopes (oxygen and hydrogen). The second, the Vienna Pee-Dee Belemnite (V-PDB), is used primarily for carbon isotopes (Allison *et al.*, 1995; Coplen, 1995; Coplen *et al.*, 1983; Craig, 1957). As mentioned above, $CO_2$ is produced from carbonates by a chemical reaction with associated isotopic modification. Accordingly, the V-PDB-$CO_2$ standard is defined based on the isotopic ratio of $CO_2$ produced from PDB, rather than the original ratio of the carbonate. It turned out that the $^{18}O/^{16}O$ ratio of V-PDB-$CO_2$ is nearly identical (0.3‰ difference) to the $^{18}O/^{16}O$ ratio of $CO_2$ in equilibrium with seawater at 25 °C (i.e., V-SMOW-$CO_2$). Expressing $\delta^{18}O$ values of water on the SMOW scale and that of $CO_2$ in equilibrium with the same water on the V-PDB-$CO_2$ scale is convenient, producing nearly the same $\delta^{18}O$ values (Figure 1).

Modifications in isotopic ratios considered in this chapter occur mainly due to "kinetic" and "equilibrium" isotope fractionations. A kinetic isotope fractionation, in a one-way process, occurs because a molecule containing the lighter isotope has faster translational velocities, allowing it to move preferentially, at higher rates, through the molecules of the medium, leaving behind residue that becomes progressively enriched in the heavy isotope. A good example of this effect is molecular diffusion, in which case the isotopic discrimination can be accurately predicted based on the reduced masses of the gas species of interest and that of the media through which it diffuses (Craig, 1953). Kinetic isotope effects also occur because chemical reactions are sensitive to differences in dissociation energies of molecules. It is usually easier to form, or break, the bonds of molecules containing the lighter isotope, because the vibrational frequency of such bonds tends to be higher. This leads to a difference in the rate of reaction for the isotopic species, and to isotopic fractionation in which molecules containing lighter isotopes will be preferentially incorporated in the product of incomplete reactions, while the heavy isotopes will become

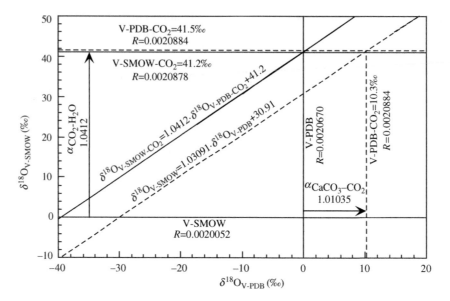

**Figure 1** Relationships between stable isotope scales. Measured isotope ratios are normalized to international standards, such as V-PDB or V-SMOW. Oxygen isotope ratios may be related to the V-SMOW, V-PDB, or the V-PDB-CO$_2$ scale (see text for detail). It is particularly convenient to compare oxygen isotope ratios of CO$_2$ on the V-PDB scale with the oxygen isotope ratio of water on the V-SMOW scale. The difference between the scales almost exactly compensates for the oxygen isotopic equilibrium fractionation between CO$_2$ and water at 25 °C.

enriched in the unreacted residue. Of course, if reactions are complete and if all reactants are converted to product, no fractionation will be observed.

An equilibrium (thermodynamic) isotope effect is observed when a chemical or physical equilibrium between two compounds or phases is permitted. In this case, the equilibrium fractionation is the net effect of the two unidirectional isotope fractionations. Equilibrium fractionation usually results in the heavy isotope accumulating preferentially where it is bound more strongly. Vapor–liquid equilibrium fractionation reflects the fact that molecules containing the light isotopes are more volatile. As a result, the heavy isotopes are preferentially concentrated in the liquid phase. Both kinetic and equilibrium fractionations can be predicted from knowledge of the binding energies of atoms and molecules (Bigeleisen and Mayer, 1947; Bigeleisen and Wolfsberg, 1958; Urey, 1947), but often must be determined empirically (e.g., Friedman and O'Neil, 1977).

The isotope *fractionation factor*, $\alpha$, is defined as the ratio of two isotopic ratios:

$$\alpha_{a-b} = \frac{R_b}{R_a} \qquad (4)$$

where a and b are often, but not always, defined as source (reactant) and product, respectively. Fractionation factors greater or smaller than 1 indicate isotopic enrichment or depletion, respectively,

in going from a to b. In general, isotope fractionation factors are small and $\alpha$ is ~1. Therefore, *fractionation*, $\varepsilon$, the deviation of $\alpha$ from 1, is often used:

$$\varepsilon = \alpha - 1 \qquad (5)$$

As $\varepsilon$ values are small, they are generally given as $\varepsilon$‰ (i.e., $\varepsilon 10^3$). Note first that fractionation factors are multiplicative ($\alpha_{total} = \alpha_1 \alpha_2 \cdots \alpha_n$), while the fractionation $\varepsilon$ is approximately additive ($\varepsilon_{total} \approx \varepsilon_1 + \varepsilon_2 + \cdots + \varepsilon_i$). Second, $\varepsilon$ is often negative, reflecting isotopic depletion in a process, and $\varepsilon_{a-b} = -\varepsilon_{b-a}/(1 + \varepsilon_{b-a}) \approx -\varepsilon_{b-a}$.

Some variations in the isotopic nomenclature should be noted. For example, the widely adopted model of Farquhar *et al.* (1982) for $^{13}$C fractionation uses the definition $\alpha = R_{source}/R_{product}$ and the deviations from 1 as *discrimination*, $\Delta$. While having the same numerical value, $\Delta$ is normally positive (i.e., $\Delta = -\varepsilon$). In the literature, $\Delta$ is often used for the overall discrimination of a system, while specific discrimination steps are assigned symbols such as "a" for steps involving diffusion, "b" for irreversible biochemical reactions and "e" for steps which show an equilibrium fractionation. Note that fractionations or discriminations are independent of reference materials. They are easily related to the measured delta values according to

$$\Delta = \frac{\delta_{source} - \delta_{product}}{1 + \delta_{product}} \qquad (6)$$

## 4.07.3 CARBON-13 IN ATMOSPHERIC CO$_2$

The isotopic composition of carbon has been an important research tool in the study of the carbon cycle on all timescales. In crude terms, we can observe that crustal carbon, derived from the Earth's mantle (inherited, in turn, from the parent solar nebula) has a bulk $\delta^{13}C$ value of around $-5‰$, resulting in carbonate sediments having $\delta^{13}C$ value around $0‰$, and following the evolution of photosynthetic organisms discriminating against $^{13}C$ in organic matter with $\delta^{13}C$ values of around $-25‰$, leaving a small atmospheric CO$_2$ pool with a $\delta^{13}C$ of around $-7‰$ (Schidlowski, 1988). Global isotopic budgets require, therefore, that the partitioning of carbon between the organic and inorganic components of the dominant sedimentary carbon pools be of the order of 20% and 80%, respectively (i.e., $0.2(-25‰) + 0.8(0‰) = -5‰$). The isotopic partitioning among the major carbon cycle components thus provides a powerful indicator for carbon isotopic composition of the atmosphere, as well as of

changes in carbon fluxes and underlying processes, over the past 3–4 billion years (Beerling and Woodward, 1998; Hayes *et al.*, 1999; Holland, 1984; Raven, 1998; Schidlowski, 1988). Focusing on the more recent geological past, carbon isotopic compositions in marine and terrestrial sediments have also been an important paleoclimatic indicator in investigating the details of the glacial cycles (Leuenberger *et al.*, 1992) and the Holocene (Feng, 1999; Francey *et al.*, 1999; Friedli *et al.*, 1986; Hemming *et al.*, 1998; Keeling *et al.*, 1979; Marino *et al.*, 1992). The use of carbon isotopes in studying the current carbon cycle is greatly enhanced by taking advantage of direct access to atmospheric CO$_2$ (Keeling, 1961; Keeling *et al.*, 1980, 1979; Mook *et al.*, 1983; Figure 2).

A pressing question in any attempt to predict future variations in the atmospheric CO$_2$ concentrations is determining what fraction of the CO$_2$ input to the atmosphere remains in the atmosphere (Broecker *et al.*, 1979), is absorbed by the ocean, or is taken up the land biosphere (Ciais *et al.*, 1995a; Enting *et al.*, 1995; Francey *et al.*, 1995;

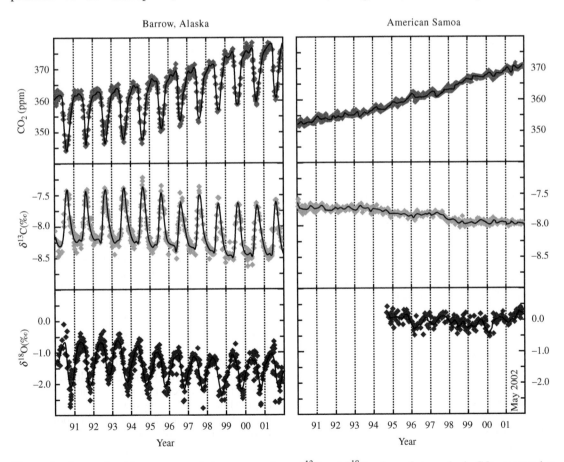

**Figure 2** Seasonal and long-term trends in concentrations, $\delta^{13}C$ and $\delta^{18}O$ values of atmospheric CO$_2$ measured at high latitudes in the northern (Barrow, Alaska) and southern (American Samoa) hemispheres by the NOAA/CMDL-CU flask network (www.cmdl.noaa.gov/ccgg/index.html). The terrestrial biosphere, concentrated in the northern hemisphere, dominants the seasonal cycle. Fossil fuel emission dominates the long-term trends in CO$_2$ and in $\delta^{13}C$ and their interhemispheric gradient.

Heimann and Maier-Reimer, 1996; Morimoto et al., 2000; Tans et al., 1990, 1993; Battle et al., 2000; Still et al., 2003). As is the case in the geological timescale studies, carbon isotopes offer a simple and powerful approach that is based, to a great extent, on the existence of the minute but easily detected biological discrimination against $^{13}C$ in photosynthesis. This isotopic signal is useful in two ways:

(i) There is a major difference in the expression of the photosynthetic discrimination against $^{13}C$ during $CO_2$ exchange in the atmosphere–land, or in the atmosphere–ocean systems. While exchange of $CO_2$ with land plants strongly reflect the photosynthetic discrimination against $^{13}C$, $CO_2$ exchange with the ocean is based predominantly on physical equilibrium, as most of the biological processes occur with respect to the dissolved inorganic carbon in the ocean water. As a result, the two exchange fluxes are isotopically distinct and their relative contribution to changes in the atmospheric $\delta^{13}C$ can be elucidated (Ciais et al., 1995b; Enting et al., 1995; Francey et al., 1995; Fung et al., 1997; Tans et al., 1993).

(ii) The $\delta^{13}C$ values of fossil fuel, and $CO_2$ from land use changes are imprinted with photosynthetically derived, $^{13}C$-depleted carbon. The rapid addition of anthropogenic, $^{13}C$-depleted $CO_2$ to the atmosphere, has resulted in a decreasing trend of atmospheric $\delta^{13}C$ of $\sim 1.5\%$ over the past $\sim 100$ years (e.g., Francey et al., 1999). This trend slowly infiltrates into the other carbon pools on land and in the ocean, providing a powerful tracer for the fate of this carbon.

Tans (1980) laid out a useful approach to assess the isotopic mass balance of atmospheric $CO_2$ as a function of the fluxes, isotopic compositions and isotopic fractionations involved in the transfer of $CO_2$ between the atmosphere, the ocean and the biosphere. The simple formulation shown below involves some approximation that is justified in light of the uncertainties in the system (Tans, 1980). Accordingly, the mean temporal change in atmospheric $CO_2$ content ($C_a$) can be described by

$$\frac{d}{dt}C_a = F_{net-o} + F_{net-1} \qquad (7)$$

indicating that temporal changes in $C_a$, are influenced by the net exchange of all fluxes, in the atmosphere–ocean, $F_{net-o}$, and in the atmosphere–land, $F_{net-o}$, systems. Temporal changes in the $^{13}C$ content in atmospheric $CO_2$ are, accordingly,

$$\frac{d}{dt}{}^{13}C = F_{ne-ot}R'_{net-o} + F_{net-1}R'_{net-1} \qquad (8)$$

where $R'$ indicates the isotopic concentrations associated with a net flux. It is immediately apparent that a significant difference in the

isotopic ratios for the ocean and the land respectively, should permit the use of the two equation system (Equations (7) and (8)), together with observed variations in atmospheric $CO_2$ and $^{13}C$, to partition the exchange of $CO_2$ between the atmosphere and the Earth into its oceanic and land components. It turns out that the terrestrial discrimination is $\sim 10$ times larger than that in the ocean discrimination. However, unperturbed (such as by anthropogenic effects), the biosphere is normally near steady state and the net fluxes depicted above must be near zero. The associated isotopic impact on the atmosphere, averaged over time (e.g., annually) is faint (often a few 1/100s of a per mil in $\delta^{13}C$ values), requiring extremely high sampling density and analytical precision. These capabilities, continuously improving over the fourth quarter of the twentieth century, have become routine only in recent years (Francey et al., 1999; Masarie et al., 2001; Mook, 1986; Thoning et al., 1989; Trolier et al., 1996). Technological progress notwithstanding, the introduction of stable isotope measurements combined with more traditional $CO_2$ concentration measurement methods offered a major advance in our ability to probe the modern carbon cycle (e.g., Battle et al., 2000; Ciais et al., 1995b; Enting et al., 1995; Ito, 2003; Still et al., 2003).

In reality, we recognize that the net exchange fluxes in Equations (7) and (8) are more complex, such that Equation (7) is more commonly broken down according to

$$\frac{d}{dt}C_a = \overset{\text{land}}{[F_{ER} - F_A]} + \overset{\text{ocean}}{[F_{oa} - F_{ao}]}$$
$$+ \overset{\text{man}}{[F_{ff} + F_{bb}]} + \overset{\text{geology}}{[F_V - F_W]} \qquad (9)$$

where $C_a$ is the atmospheric pool of carbon in $CO_2$ (e.g., moles, or Pg of carbon in $CO_2$, where $Pg = 10^{15}$ g) and $F$ denotes a flux (usually measured in Pg C, on annual basis; and negative fluxes, by convention, are out of the atmosphere). The atmospheric $CO_2$ mass balance depicted above reflects the major components of the global carbon cycle:

(i) The long-term control on atmospheric $CO_2$ is achieved by the uptake and dissolution of atmospheric $CO_2$ in surface waters, and its participation in the weathering/sedimentation of carbonate rocks ($F_W$) and $CO_2$ released to the atmosphere via tectonic processes and volcanism ($F_V$). Although providing communication between the atmosphere and the vast carbon pools in the Earth system, it involves less than 1% of the fluxes into or out of the atmosphere on an annual scale, and are either treated separately or neglected in this context.

(ii) The dominant components of the short-term $CO_2$ balance of the atmosphere are gross primary production (GPP) and respiration (R) by the land biosphere, and the air–sea gas exchange ($F_{oa}$ and $F_{ao}$ for the ocean–atmosphere and atmosphere–ocean fluxes, respectively). While by far the largest components of the budget, these fluxes are also near equilibrium, with only small effects on the atmospheric budget when integrated over the annual cycle and the entire globe.

(iii) The main components of the human perturbation of the contemporary carbon cycle are fossil fuel burning ($F_{ff}$), and biomass burning and land-use change (pooled under $F_{bb}$), which release CO$_2$ to the atmosphere. Only part of this CO$_2$ stays in the atmosphere (about one half); the rest is taken up by the land (plant and soil) or by the ocean. The main anthropogenic flux of fossil fuels constitutes essentially an acceleration of the biological carbon cycle, "respiring" over a few centuries vast quantities of carbon that accumulated in the Earth sediments over million of years. The absorption of anthropogenically derived CO$_2$ on land and in the ocean "carry over" some of the imbalances into the natural two-way fluxes between the atmosphere and Earth.

Such processes can induce feedback effects on atmospheric CO$_2$ concentrations, and consequently on climate, over the years to millennia timescales (e.g., Cox *et al.*, 2000), which are motivating much of the research in this field.

It follows from the above that the mean temporal change in the $^{13}$C content of atmospheric CO$_2$ can be described by

$$\frac{d}{dt}\,^{13}C = [F_{ER}R'_{ER} - F_A R'_A] + [F_{oa}R'_{oa}$$
$$- F_{ao}R'_{ao}] + [F_{ff}R'_{ff} + F_{bb}R'_{bb}]$$
$$+ [F_V R'_V - F_W R'_W] \quad (10)$$

where $^{13}$C is the average atmospheric $^{13}$CO$_2$ concentrations ($\sim$1.1% of CO$_2$) and $R' = {}^{13}C/({}^{12}C + {}^{13}C)$ for each specific flux component. The terms in four square brackets on the right-hand side correspond to the land, ocean, anthropogenic, and the geologic contributions. The change in atmospheric $^{13}$C can be separated into the temporal change of two components, isotopic ratio and CO$_2$ concentrations:

$$\frac{d}{dt}\,^{13}C = \frac{d}{dt}(C_a R'_a) = C_a \frac{d}{dt}R'_a + R'_a \frac{d}{dt}C_a \quad (11)$$

We focus here on the change in isotopic composition, $R'$, which is normally based on measured values, where $R = {}^{13}C/{}^{12}C$, introducing a small approximation (usually less than 0.2‰; see Section 4.07.2 for $R$ to $R'$ conversion). Further, $R$ is conventionally reported in the delta notation (see Section 4.07.2). By introducing the delta notation, rearranging Equations (7)–(11), ignoring the negligible geological fluxes, and omitting terms not very different from 1, a more practical description of the isotopic mass balance

of atmospheric CO$_2$ is obtained, which is often used for considering the contemporary $^{13}$C balance of the atmosphere (Tans, 1980; Tans *et al.*, 1993; Fung *et al.*, 1997):

$$\frac{d\delta_a}{dt} = \frac{1}{C_a}[F_{ER}(\delta_{ER} - \delta_a) - F_{Ph}(\delta_{Ph} - \delta_a)$$
$$+ F_{oa}(\delta_{oa} - \delta_a) - F_{ao}(\delta_{ao} - \delta_a)$$
$$+ F_{ff}(\delta_{ff} - \delta_a) + F_{bb}(\delta_{bb} - \delta_a)] \quad (12)$$

This isotopic mass balance approach indicates some of the potential advantages in adding the isotopic analysis to atmospheric measurements. Isotopically constrained, measurements of changes in $\delta_a$ and $C_a$ may allow us to solve for more than one unknown flux. Alternatively, when flux estimates are sufficiently constrained, solving for isotopic signals can provide valuable information on the processes underlying observed changes. Equation (12) also indicates that temporal changes in $\delta_a$ values are influenced by the isotopic "forcing" of the various CO$_2$ sinks and sources, i.e., the relative size of the flux times the deviations of its isotopic signature from that of the mean atmosphere. For example (using IPCC flux and pool size values), a typical anthropogenic flux of 7.1 Pg C yr$^{-1}$ having a mean $\delta^{13}$C value of $-28$‰ into an atmospheric pool of $\sim$730 Pg C with a mean $\delta^{13}$C value of $-8$‰ will have a "forcing" of $(F_{ff}/C_a)(\delta_{ff} - \delta_a) = (7.1/730)$ $(-28 + 8) = -0.2$‰ yr$^{-1}$ (Figure 3). Similarly, if 120 Pg C yr$^{-1}$ flux having a mean $\delta^{13}$C value of $-26$‰ is taken up (negative flux) in land photosynthesis, its "forcing" would be $\sim$+3‰ yr$^{-1}$. Such forcing is nearly balanced by a respiration and biomass burning of $\sim$119 Pg C yr$^{-1}$ with $\delta^{13}$C value of $\sim-25.7$‰. A net terrestrial sink of 2 Pg C yr$^{-1}$ (as estimated) would create a net forcing of $+0.05$‰. Such isotopic signals are small but within the current precision of the isotopic analyses of atmospheric CO$_2$ (e.g., Trolier *et al.*, 1996), provided appropriate long-term calibrations are maintained (e.g., Masarie *et al.*, 2001). Comparison of such estimated forcing with observation provides critical information on the balancing act in the atmosphere CO$_2$ budget.

In Equation (12), the isotopic signals of the individual fluxes are required, but are not normally measured directly. Instead, the isotopic composition of the source CO$_2$ (often atmospheric CO$_2$) is measured with precision, and is combined with extensive knowledge of the isotopic fractionations in the transfer process involved. Much effort must be invested therefore to develop this process-based knowledge of isotopic fractionation for the quantitative use of the isotope approach. For example, the isotopic composition of the CO$_2$ flux from the atmosphere to the ocean, $R_{ao}$, will reflect the isotopic composition of atmospheric

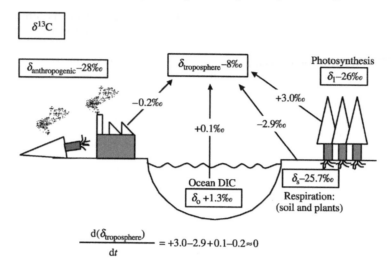

$$\frac{d(\delta_{\text{troposphere}})}{dt} = +3.0-2.9+0.1-0.2\approx0$$

**Figure 3** Annual mean "forcings" on the $\delta^{13}C$ value of atmospheric $CO_2$. Each specific forcing component is the product of the flux involved, $F$, diluted by the atmospheric carbon pool, $C_a$, and the difference in $\delta^{13}C$ values between that of the flux and the mean atmosphere (i.e., forcing $= (F/C_a)(\delta_x - \delta_a)$). Values are order of magnitude estimates, as discussed in the text, and involve significant uncertainties. Typical values for the carbon reservoirs involved are given in boxes. The land biosphere has the strongest forcing, with contrasting effects of photosynthesis and respiration. Net photosynthetic carbon sink tends to enrich the atmosphere in $^{13}C$. The ocean forcing is almost an order of magnitude smaller than the land effect, providing a basis for distinguishing between land and ocean sink and sources. Fossil fuel emission tends to dilute the atmospheric $^{13}C$. Although the different forcing nearly balance, in reality, the fossil fuel effect dominates the atmospheric forcing resulting with a net trend of $\sim -0.02\%_o$ yr$^{-1}$ (see Figure 4).

$CO_2$, $R_a$, and the isotopic fractionation of $CO_2$ associated with air–sea transfer, $\alpha_{ao}$:

$$R_{ao} = \alpha_{ao}R_a \qquad (13)$$

In introducing the delta notation, we also adopt the term $\varepsilon$, (where $\varepsilon = \alpha - 1$), and the close approximation of the $\delta^{13}C$ of the $CO_2$ flux into the ocean:

$$\delta_{ao} = \delta_a + \varepsilon_{ao} \qquad (14)$$

Similar terms can be derived for other flux components based on knowledge of the isotopic composition of the source material (e.g., atmospheric $CO_2$, ocean-surface inorganic carbon, soil organic carbon) and knowledge of the associated, process-based, isotopic fractionation (e.g., $\varepsilon_P$ for photosynthesis, $\varepsilon_R$ for respiration). Note also that terms such as "$\delta_a + \varepsilon_{ao}$" in Equation (14) are also approximations of "$\Delta$."

### 4.07.3.1 Fossil Fuel Input

The term fossil fuel usually refers to coal and lignite, petroleum and natural gas, and usually includes also the production of cement. Cement is produced from limestone with a $\delta^{13}C$ value close to zero, because the V-PDB standard itself is derived from limestone. The other fuel components are all derived from organic matter

depleted in $^{13}C$. Adopted values for coal and lignite are $-24\%_o$ and $-27\%_o$, while natural gas is $-41\%_o$ with relatively large uncertainty. In particular, values of methane range from $-20\%_o$ to $-75\%_o$ with an estimated mean value of $-43\%_o$. Other less abundant gases such as ethane, propane, and butane ($\sim16\%$ of total carbon from natural gas) are less depleted in $^{13}C$. Historical trends in the production and consumption of the various components of fossil fuel, result in a temporal change in the $\delta^{13}C$ value of the combined, global-scale, fossil fuel flux from $\sim -24.0\%_o$ in 1850 to $\sim -28.5\%_o$ for 1991, partly reflecting the shift in consumption pattern from coal through petroleum to natural gas (Andres *et al.*, 2000; Tans, 1981). It is generally assumed that the fossil fuel combustion does not involve additional fractionation, and mean $\delta^{13}C$ values of $-28\%_o$ to $-29\%_o$ are often assigned to Equation (12) as $\delta_{ff}$.

### 4.07.3.2 Exchange with the Ocean

The existence and direction of a $CO_2$ flux between the atmosphere and the ocean can be determined by the difference in partial pressure of $CO_2$ ($p_{CO_2}$) in seawater and in the overlying atmosphere. $p_{CO_2}$ in seawater depends on its solubility and the concentration of dissolved $CO_2$ (with solubility inversely related to temperature and salinity). $CO_2$ gas diffusing into water enters the so-called carbonate system and is partitioned among several species that, combined, are termed "dissolved inorganic carbon" (DIC). It first goes into a dissolved or aqueous form ($CO_2(aq)$) and is rapidly hydrated to form carbonic acid ($H_2CO_3$). Carbonic acid is dissociated first to bicarbonate ($HCO_3^-$) and further to carbonate ($CO_3^{2-}$):

$$CO_2(atm) \overset{H_2O}{\longleftrightarrow} CO_2(aq) + H_2CO_3 \overset{H^+}{\longleftrightarrow}$$
$$HCO_3^- \overset{H^+}{\longleftrightarrow} CO_3^{2-} \qquad (15)$$

An alternative pathway to bicarbonate is possible due to reaction of $CO_2$ and $OH^-$ in water but is normally less important (Skirrow, 1975). The proton concentration is also influenced by total alkalinity and water dissociation, which, in turn, will influence the details of the simplified chemical process depicted above. The relative concentrations of the DIC species in Equation (15) are mainly a function of pH, temperature, and salinity. In seawater, bicarbonate is the dominant species (Skirrow, 1975), and bicarbonate and carbonate are the main components of alkalinity (Broecker and Peng, 1974). The chemical equilibrium for each of the steps in the DIC system in Equation (15) is determined from the specific temperature sensitive reaction constants ($K_T$). Note that in the complex system of seawater, empirical $K_T$ values, rather than thermodynamic theoretical values, are usually adopted. Such reaction constants are of course sensitive to the effects of molecular mass, with molecules containing heavy isotopes favoring slower reactions rates, and are therefore associated with temperature-sensitive isotopic fractionations. Typical equilibrium fractionation values for the carbonate system in dilute solution at 25 °C are (Deuser and Degens, 1967; Mook, 1986; Mook *et al.*, 1974):

$$CO_2(atm) \overset{-1.1\%_0}{\longleftrightarrow} H_2CO_3 \overset{+9.0\%_0}{\longleftrightarrow} HCO_3^- \overset{-0.4\%_0}{\longleftrightarrow} CO_3^{2-}$$
$$(16)$$

Equilibrium isotopic fractions reflect the combined, unidirectional kinetic isotopic fractionations. In considering the one-way fluxes as in Equation (12), estimates of the one-way kinetic fractionations are needed. Knowledge of kinetic effects is obtained from controlled experiment with pure $CO_2$ and water or with salt solutions, but for the chemically complex system that is ocean-water, empirical values are adopted. For example, laboratory experiments show that the hydration of aqueous $CO_2$ to bicarbonate involves fractionation of ~13‰ and the dehydration reaction fractionate by ~22‰ (O'Leary *et al.*, 1992). The difference between these two kinetic fractionations, 9‰, corresponds to the equilibrium fractionation depicted above (Marlier and O'Leary, 1984; O'Leary *et al.*, 1992). In practice, it was estimated

that for $CO_2$ taken up by the sea surface, and for $CO_2$ released from seawater DIC, the kinetic fractionations are smaller (Inoue and Sugimura, 1985; Mook, 1986; Siegenthaler and Münich, 1981) and are (for 20 °C, pH = 8.2 and S = 35):

$$^{13}\varepsilon_k(sea \longrightarrow air) = -10.3 \pm 0.3\%_0$$

$$^{13}\varepsilon_k(air \longrightarrow sea) = -2.0 \pm 0.2\%_0 \qquad (17)$$

The sea to air fractionation, $^{13}\varepsilon_{oa}$, starting from DIC, includes both an equilibrium and a kinetic one, while the air to sea, $^{13}\varepsilon_{ao}$, is purely a kinetic fractionation. These isotopic fractionations are applied in Equation (12) to estimate the $\delta^{13}C$ value of the $CO_2$ flux from the atmosphere to the ocean, $\delta_{ao}$, ($\delta_{ao} = \delta_a + {}^{13}\varepsilon_{ao}$). Isotopic fractionation due to diffusion in water may apply too, but are expected to be small (O'Leary, 1984).

Note that due to the kinetic effects discussed here, the flux into the ocean tends to increase the atmospheric $\delta^{13}C$ and the flux to the atmosphere tends to deplete the atmospheric $\delta^{13}C$ (Broecker and Maier-Reimer, 1992; Kroopnic, 1985; Kroopnick, 1980). At equilibrium, the net effect should result with a depletion of atmospheric $\delta^{13}C$ by ~8‰ relative to the ocean. As an example, ocean surface water (mean temperature 20 °C) having DIC with $\delta^{13}C$ values of, say, +1.3‰, will be in equilibrium with atmospheric $CO_2$ having a $\delta^{13}C$ value of −7.0‰, with $\delta^{13}C$ values for the one-way fluxes of: $\delta_{ao} \approx -7.0 - 2.0 = -9.0\%_0$, $\delta_{oa} \approx +1.3 - 10.3 = -9.0\%_0$ (Mook, 1986). In reality, the ocean–atmosphere system is not at full equilibrium (Broecker and Peng (1974) and see below). The most obvious reason is that the residence time of surface water is approximately an order of magnitude smaller than the time needed for isotopic equilibrium with the atmosphere (~1 yr versus 10 yr, respectively; Lynch-Stieglitz *et al.*, 1995; Quay *et al.*, 2003). The picture is further complicated as the $\delta^{13}C$ values of ocean surface DIC reflect a balance between several factors, including the thermodynamic effects, ocean circulation, and biological processes (Gruber *et al.*, 1999).

The mean fractionation values discussed above, and consequently the ocean–atmosphere isotopic fluxes, vary significantly both on spatial and temporal scales. Isotopic fractionations are temperature sensitive (~0.1‰ °C⁻¹; Mook, 1986) and surface ocean temperature varies in the range of 30 °C (Kalnay *et al.*, 1996), resulting with ~3‰ range in the thermodynamic effect. Further, $CO_2$ exchange between the ocean and the atmosphere is not uniformly distributed. Ocean circulation patterns, biological activity, and large spatial variations in atmospheric turbulence and consequently in the air–sea transfer coefficient, result in a unique pattern of ocean regions with net

$CO_2$ degassing and others with net $CO_2$ absorption (Tans *et al.*, 1990).

### 4.07.3.3 Ocean Biology

Biological activities also increase the carbon content of the deep ocean at the expense of the surface ocean and atmosphere (Gildor and Follows, 2002; Sarmiento and Bender, 1994). This, in turn, is the net effect of two contrasting processes: first the production of organic matter that consumes DIC and second, production of calcium carbonate skeletal material that reduces alkalinity (e.g., consuming $Ca^{2+}$) and increases $p_{CO_2}$ in surface water. By influencing the exchange of $CO_2$ between the ocean and the atmosphere, biology can of course affect atmospheric $^{13}C$ through the kinetic isotope effects discussed above. But probably the main effect of ocean biology on the ocean–atmosphere isotopic exchange is through its effect on the $\delta^{13}C$ value of surface ocean DIC. While, as noted above, atmospheric $CO_2$ tends to deplete surface ocean DIC in $^{13}C$, biological processes have the opposite effect.

Photosynthesis in the ocean discriminates against $^{13}C$ in a similar way to land photosynthesis, with mean discrimination estimated at 22.0‰ (e.g., Tans *et al.*, 1993), with typical values of $\delta^{13}C$ of marine organic matter in the range of $-20$‰ in low- and mid-latitudes to $-30$‰ in the southern ocean (Goericke and Fry, 1994; Rau *et al.*, 1989). This can change, however, under conditions where $CO_2$ availability dramatically decreases (e.g., algal bloom). Under such conditions, some marine photosynthetic organisms can induce a $CO_2$ pump that, similar to $C_4$ plants on land, concentrate $CO_2$ to high levels in a specific compartment where it is almost quantitatively consumed, greatly reducing isotopic discrimination (Falkowski, 1991; Raven, 1992, 1998; Sharkey and Berry, 1985). As marine photosynthesis increases in summer and decreases in winter, it causes some seasonality in surface ocean $\delta^{13}C$ values (but not nearly as much as the seasonality on land). Consequently, the $\delta^{13}C$ values of DIC in surface water will tend to be higher in summer than in winter, as marine organisms take up $^{12}C$ preferentially, enriching $^{13}C$ in the DIC left behind. This may be important when estimates of ocean $\delta^{13}C$ values are based on samplings done during a particular period. For example, Ciais *et al.* (1995b) estimated that global mean summer versus winter $^{13}C$ enrichment of surface ocean DIC of 0.5‰, and particularly large effect are observed in high latitudes (Gruber *et al.*, 1999). Isotopic fractionation is also associated with carbonate shell formation in marine organisms, but this effect is very small (Emrich and Vogel, 1970; Zhang *et al.*, 1995).

The relatively slow air–sea gas exchange diminishes the seasonality effect on the atmosphere.

Biologically produced, $^{13}C$-depleted, organic matter can also influence the $\delta^{13}C$ of surface ocean DIC via large-scale ocean circulation. Cold water in the deep ocean accumulates large quantities of carbon derived from respiration and mineralization of organic matter. Upwelling of carbon-rich deep water in some regions will tend to reduce the $\delta^{13}C$ of surface water DIC, and consequently also of atmospheric $CO_2$ (Lynch-Stieglitz *et al.*, 1995). As discussed in the next section, the large-scale and slow turnover rates of organic matter in the global ocean introduces another important aspect in the isotopic interactions between the ocean and the atmosphere. The penetration of any long-term trend in the $\delta^{13}C$ values of atmospheric $CO_2$, such as observed today, critically depends on ocean biology that provides the major mechanism to "pump" carbon across the ocean depths.

### 4.07.3.4 Atmosphere–Ocean Disequilibrium

Isotopic methods such as based on Equation (12), interpret changes in atmospheric $\delta^{13}C$ mainly as an indicator for terrestrial $CO_2$ exchange because of its dominant fractionation (Figure 3). Any $^{13}C$ enrichment of the atmosphere is ascribed to net biotic uptake, and $^{13}C$ depletion to biotic $CO_2$ release. To implement such an approach, changes in atmospheric $^{13}C$ that are not due to the changes in biotic sink or source, must be accounted for. Such is the case with respect to the $\sim 1.5$‰ decrease in atmospheric $\delta^{13}C$ since the beginning of the industrial period ($\sim -0.02$‰$yr^{-1}$ on average; Feng, 1999; Francey *et al.*, 1995) due to the emission of $^{13}C$-depleted carbon from fossil fuel burning and land use changes. This atmospheric change is redistributed in the atmosphere–ocean–land system, but over a significantly slower timescale. In the ocean, the delay is due to the slow rate of ocean–atmosphere isotopic equilibrium, and because the $\delta^{13}C$ value of surface ocean DIC is influenced by the remineraliztion of the ocean total organic matter with slow turnover rates. Therefore, new carbon entering the ocean remains slightly more depleted in $^{13}C$ relative to the older carbon released from the ocean to the atmosphere and $\delta_{ao} > \delta_{oa}$. This signal, termed the isotopic ocean disequilibriuim, $D_o$ (Figure 4), is difficult to measure directly on a large scale, but must be accurately estimated in atmospheric isotopic budgets (Equation (12); Gruber *et al.*, 1999; Heimann and Maier-Reimer, 1996; Quay *et al.*, 1992, 2003; Tans *et al.*, 1993; Körtzinger and Quay, 2003). A 0.1‰ error in estimating this disequilibrium could result in an error in the order of 0.5 Pg C $yr^{-1}$

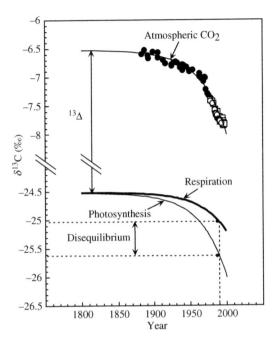

**Figure 4** The global "disequilibrium" effect. $\delta^{13}C$ value of CO$_2$ currently fixed into plants (associated with photosynthetic discrimination, $^{13}\Delta$) is lower than that of older CO$_2$ respired back to the atmospheric CO$_2$ (no fractionation is assumed). This is due to the rapid decrease in atmospheric $\delta^{13}C$ associated with fossil fuel emissions, on the one hand, and to the slow turnover of carbon in the biosphere, on the other hand. A similar disequilibrium occurs in the ocean where the atmospheric trend influences the $\delta^{13}C$ values of newly formed DIC, while the ocean mean DIC pool lags behind this equilibrium values due to slow turnover rates (not shown). The atmospheric trend shown is based on the best fit line to the data of Francey *et al.* (1999); the land organic matter trend is obtained by applying global mean $^{13}\Delta = 18\%_0$, and moving it back in time by 27 yr, the first order estimate of global mean soil carbon turnover time. The resulting $\sim 0.6\%_0$ disequilibrium for the 1990s is within the range of current estimates for both land and ocean.

in estimating carbon sinks and sources in the biosphere.

The disequilibrium term can be defined (Fung *et al.*, 1997; Tans *et al.*, 1993) by comparing the atmospheric $\delta^{13}C$ value, $\delta_a$, with that expected at thermodynamic equilibrium with actual ocean surface DIC, $\delta_a^{eq}$, according to

$$D_o = (\delta_a^{eq} - \delta_a) \qquad (18)$$

where $\delta_a^{eq} = \delta_{DIC} + {}^{13}\varepsilon_{eq}$ (with $\varepsilon_{eq} = \varepsilon_{oa} + \varepsilon_{ao}$). The atmosphere–ocean isotopic exchange depicted in Equation (12) can be substituted by terms that distinguish changes in the atmospheric $\delta^{13}C$ value brought about by the net fluxes and those caused by the "isoflux" that reflect the isotopic disequilibrium (Ciais *et al.*, 1995b;

Tans *et al.*, 1993):

$$F_{oa}(\delta_{oa} - \delta_a) - F_{ao}(\delta_{ao} - \delta_a)$$
$$\approx (F_{ao} - F_{oa})\varepsilon_{ao} + F_{oa}D_o \qquad (19)$$

where the disequilibrium effect on the atmosphere is estimated from the isoflux $F_{oa}D_o$ Pg C‰ yr$^{-1}$ ("isoflux" combines flux and its isotopic composition, but since the $\delta$ notation is often employed, it is equivalent to, not the actual, isotopic flux). Note that $D_o$ can be defined either relative to the atmosphere or relative to a surface reservoir (i.e., $D_o^* = (\delta_{DIC}^{eq} - \delta_{DIC})$; Fung *et al.*, 1997), with the choice of the definition selected for convenience and data availability. There have been a few approaches to estimating the ocean disequilibrium. Focusing on the air–sea isotopic disequilibrium yielded lower estimates (Tans *et al.*, 1993) than attempting to estimate the change over time of the $\delta^{13}C$ of total ocean column DIC (Quay *et al.*, 1992, 2003). A combination approach has also been recently proposed (Heimann and Maier-Reimer, 1996). Estimates of $D_o$ values reported in the literature range between 0.43‰ and 0.79‰ (Gruber *et al.*, 1999; Heimann and Maier-Reimer, 1996; Quay *et al.*, 1992, 2003; Tans *et al.*, 1993; Wittenberg and Esser, 1997). Considerable uncertainties are still associated with these estimates. In all cases, it is also associated with uncertainties regarding the values and constancy of the fractionation in the sea–air exchange and $F_{oa}$.

### 4.07.3.5 Photosynthetic CO$_2$ Uptake on Land

The largest annual flux of CO$_2$ from the atmosphere is the supply of substrate for land plants photosynthesis. The entire land biosphere is estimated to take up $\sim 10^{17}$gC annually (the adopted value is 120 Pg C yr$^{-1}$; IPCC, 2001). Typically, a single illuminated green leaf takes up only $\sim 10^{-7}$ g C s$^{-1}$. But, notably, it is the processes that take place at the leaf scale that are ultimately responsible for the effect of land photosynthesis on the isotopic composition of atmospheric CO$_2$ (Park and Epstein, 1960). The key isotope effect is the preferential uptake of $^{12}$CO$_2$ by the primary photosynthetic enzymes, Rubisco (Ribulose bi-Phosphate Carboxylase/Oxygenase) and PEP-c (Phospho-Enol-Pyruvate-carboxylase). Because such large quantities of CO$_2$ are involved, this enzyme discrimination has a measurable impact on the CO$_2$ remaining in the atmosphere. It is clearly evident in seasonal timescale as photosynthetic removal of $^{13}$C-depleted carbon peaks in the northern hemisphere summer months (Figure 2). Similar effects are also observed on geological timescales, when large quantities of $^{13}$C-depleted photosynthates are deposited and stored in the soil and sediments

(Hayes *et al.*, 1999; Schidlowski, 1988). But, notably, the extent of discrimination against $^{13}C$ in the two carboxylation enzymes is very different (29‰ versus 7‰ for Rubisco and PEP-c) respectively (Guy *et al.*, 1993; O'Leary, 1981; Roeske and O'Leary, 1984). Most plants, including all trees, use Rubisco as the primary photosynthetic enzyme, producing 3-carbon sugars as the primary product and are therefore called C$_3$ plants. But ~23% of global plant productivity (Still *et al.*, 2003), mostly in grassland, savannas, and agriculture, is due to PEP-c as the primary enzyme, acting on bicarbonate as a substrate and producing 4-carbon organic acids, and are called C$_4$ plants. The C$_4$ pathway is a relatively late achievement in the evolution of flowering plants (probably associated with declining atmospheric CO$_2$ over geological times) and is based on a specific, compartmentalized, leaf anatomy designed primarily to counteract the low affinity of Rubisco for its substrate, CO$_2$, and the unfavorable effects of photorespiration in C$_3$ plants, where part of the newly fixed carbon is lost to the atmosphere as CO$_2$. Succulent plants with crassulacean acid metabolism (CAM) form yet another, minor, photosynthetic group that can fix carbon by both the RuBP and PEP carboxylase reactions, but at different times. Photosynthetic bacteria utilize additional forms of photosynthesis with low fractionations, approaching that of C$_4$ plants. This can be due to a different type of carboxylation enzymes, such as in photosynthetic sulfur bacteria, or because of inducible CO$_2$ concentration mechanism (CCM, such as in cyanobacteria) that "pump" CO$_2$ into an internal compartment where it is consumed almost quantitatively (Kaplan and Reinhold, 1999). Note that in C$_4$ plants and CCM containing microorganisms, Rubisco ultimately fixes the CO$_2$ concentrated in the specialized compartments, but since CO$_2$ is almost quantitatively consumed in this case, little fractionation is expressed. The combined, net discrimination of land plants ($\Delta_{Ph}$ in Equation (12)) is sensitive to the proportional contributions of the various photosynthetic pathways, largely to the C$_3$/C$_4$ distribution (Lloyd and Farquhar, 1994; Fung *et al.*, 1997; Still *et al.*, 2003). As discussed below, within the major photosynthetic groups, the extent to which the enzyme discrimination is expressed also depends on physiological and environmental parameters.

Our current and rather extensive knowledge of $^{13}C$ discrimination during photosynthesis is based primarily on the pioneering works of O'Leary (1981), Vogel (1980, 1993), and Farquhar *et al.* (1982), and with more recent updates such as Farquhar and Lloyd (1993), Evans *et al.* (1986), and Lloyd and Farquhar (1994). These works provide similar analytical treatment to isotopic fractionation during photosynthesis but are expressed in different ways (often reflecting variations among chemical, geological, and physiological nomenclatures; see Section 4.07.2). In photosynthesis, CO$_2$ diffuses into active green leaves where a carboxylation reaction incorporates the carbon into an organic substance. The simplified two-step diffusion–carboxylation sequence in photosynthesis is useful to lay down the principles underlying estimates of the $^{13}C$ discrimination involved (Figure 5). The first step is a reversible process, followed by the irreversible carboxylation of an acceptor (cf. Berry, 1988; Farquhar *et al.*, 1982; Vogel, 1993):

$$[CO_2]_a \underset{F_3}{\overset{F_1}{\longleftrightarrow}} [CO_2]_c \overset{F_2}{\longrightarrow} R - COOH \quad (20)$$

and considering the isotope ratios:

$$R_a \underset{F_3}{\overset{F_1}{\longleftrightarrow}} R_c \overset{F_2}{\longrightarrow} R_P \quad (21)$$

where, $F_1 - F_3 = F_2$, $F$ represent flux and $R$ isotopic ratio and subscripts a, c, and p represent atmosphere, leaf chloroplasts, and photosynthetic products, respectively. The two-step *fractionations* are $\alpha_a = R_a/R_c$ for the diffusion step and $\alpha_b = R_c/R_A$ for the enzymatic assimilation (A) step, and the overall photosynthetic fractionation is then $\alpha_P = R_a/R_A$. Using these definitions in Equations (20) and (21) lead to

$$F_1 \frac{R_S}{\alpha_a} - F_3 \frac{R_c}{\alpha_a} = F_2 \frac{R_c}{\alpha_b} \quad (22)$$

Substituting $R_s = R_P\alpha_P$, and $R_c = R_a\alpha_b$ rearranges to

$$\alpha_P = \frac{F_2}{F_1}\alpha_a + \frac{F_3}{F_1}\alpha_b \quad (23)$$

The unidirectional fluxes involved in the simplified sequence above are proportional to the gross CO$_2$ concentration gradient times the conductance, $g$, or the reaction rate constant, $k$, for each step, such that we can define $F_1 = gc_a$, $F_3 = gc_c$ and $F_2 = k_bc_c \sim g(c_a - c_c)$, where $c_a$ and $c_c$ are respectively the ambient and chloroplast concentrations of CO$_2$. As noted above it is common to use *discriminations*, $\alpha - 1$, represented by $a$, $b$, and $\Delta$ for the diffusion step, the enzymatic step, and the overall sequence, respectively. These terms can be substituted into Equation (23) that rearranges to

$$\Delta_P = \frac{c_a - c_c}{c_a}a + \frac{c_c}{c_a}b = a + (b - a)\frac{c_c}{c_a} \quad (24)$$

This equation is now widely used to estimate isotopic discrimination in photosynthesis, and shows the two-step diffusion/reaction nature of the process and the dependence of each step on the local CO$_2$ concentration gradient. Typical values used in this model for C$_3$ plants are $a = 4.4$‰

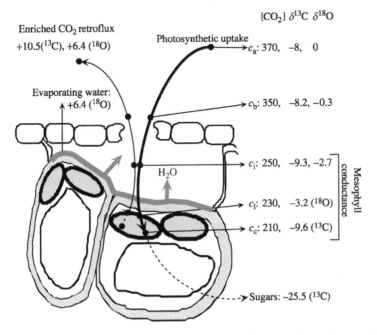

**Figure 5** A simplified cross-section in a leaf indicating the $CO_2$ diffusion pathway between the atmosphere and the chloroplast. Changes in $CO_2$ concentrations and the $\delta^{13}C$ and $\delta^{18}O$ values along this pathway are indicated for the atmosphere, $c_a$, the leaf boundary layer, $c_b$, substomatal internal air spaces, $c_i$, the chloroplast surface, $c_l$ and the center of the chloroplast, $c_c$, are indicated (pathway between $c_i$ and $c_c$ constitutes the internal, mesophyll, resistance to $CO_2$ diffusion, inadequately constrained at present). Values for $\delta^{18}O$ are based on $\delta^{18}O$ of liquid water at the evaporating surfaces (near the chloroplast membranes) of $+6.4\permil$. The arrows indicate the one-way $CO_2$ fluxes into the leaf (controlled by $c_a$), and retro-diffusion out of the leaf (controlled by $c_c$ and $c_l$ in the case of $^{13}C$ and $^{18}O$, respectively). The difference between the two one-way fluxes constitutes $CO_2$ assimilation into sugars (values for $[CO_2]$ in ppm, for isotopes in $\delta\permil$).

and $b = 29\permil$, $c_c/c_a \sim 0.6$ yielding $\Delta_p$ of $\sim 18\permil$. For C₄ plants, $c_c/c_a \sim 0.4$ and $b = 7\permil$, leading to $\Delta_p$ of $\sim 5\permil$.

A more rigorous treatment of $\Delta_p$ is summarized by Farquhar and Lloyd (1993) and Lloyd and Farquhar (1994). In this case, it is recognized that $CO_2$ diffusion into the leaf is going through a complex diffusion and dissolution pathway. The biochemical part considers the fact that while Rubisco discrimination in its isolated form is $\sim 29\permil$ (Guy *et al.*, 1993; Roeske and O'Leary, 1984), $\sim 5\%$ of the flux goes via PEPc even in C₃ plants, potentially reducing the apparent discrimination by $1-2\permil$. In plants that belong to the C₄ type, all the $CO_2$ assimilation is first by PEPc. The primary product is then transported to a specialized, $CO_2$ impermeable, bundle-sheath compartment, where it is de-carboxylated and the concentrated $CO_2$ re-assimilated by Rubisco. If all the $CO_2$ in the bundle sheath is re-assimilated, Rubisco's discrimination will not be expressed. But a certain degree of "leakiness" associated with possible discrimination must be considered (Farquhar, 1983; Henderson *et al.*, 1992). The overall discrimination by leaves of C₄ plants can be estimated according to Farquhar (1983):

$$\Delta = a + [b_4 + \phi(b_3 - s) - a]c_c/c_a \quad (25)$$

where $b_4$, the effective biochemical fractionation in the C₄ process is $\sim 12\permil$, $\phi$ is leakiness factor ($\sim 0.2$) and the associated fractionation, $s(\sim 1.8\permil)$, and $c_i$ is the internal $CO_2$ concentration.

Concurrently to the assimilation in leaves there is always $CO_2$ production in respiration and photorespiration (the first occurs all the time and involves the mitochondria, the latter occurs only in the light and involves also the oxygenase activity of Rubisco in the chloroplasts). Any discrimination in the respiration process will be reflected in the isotopic composition of the $CO_2$ produced and mixed into the leaf internal $CO_2$ pool and its effect must, therefore, be subtracted from $\Delta_P$. The combined effect of the two respiratory components can be subtracted from Equation (24) as (Farquhar and Lloyd, 1993; Farquhar *et al.*, 1982)

$$-d = e\frac{R_d/k}{c_a} + f\frac{\Gamma^*}{c_a} \quad (26)$$

where $R_d$ is the rate of dark respiration in presence of light, $k$ is the carboxylation efficiency, $\Gamma^*$ is the compensation point (the $CO_2$ concentration at which respiration exactly matches photosynthesis) and $e$ and $f$ are the discriminations associated with respiration and photorespiration. Typical values used are

$e = 0\%_0$ (Lin and Ehleringer, 1997), $f = 7\%_0$ (Rooney, 1988), and $d \approx -0.5\%_0$ for C$_3$ plants and $d \approx 0\%_0$ for C$_4$ plants. Recent studies indicate, however, that the discrimination associated with dark respiration, $e$, may be different from zero and as large as 6‰ (Tcherkez *et al.*, 2003). It may reflect significant $\delta^{13}$C heterogeneity among organic compounds (Duranceau *et al.*, 1999; Ghashghaie *et al.*, 2001), as well as intra-molecular $^{13}$C heterogeneity of specific substrates (Gleixner and Schmidt, 1997; Melzer and Schmidt, 1987; Rinaldi *et al.*, 1974). The complex diffusion-reaction process responsible for the overall discrimination by C$_3$ plants (Figure 5) can therefore be summarized according to Farquhar and Lloyd (1993):

$$-------\text{gas phase}--------/----$$

$$\underset{\substack{370 \text{ ppm}}}{\text{CO}_2\text{-atm.}} \overset{a_b}{\longleftrightarrow} \underset{\substack{350 \text{ ppm}}}{\text{leaf-surface}} \overset{a}{\longleftrightarrow} \underset{\substack{230 \text{ ppm}}}{\text{cell-walls}}$$

$$----\text{liquid phase}--------------/$$

$$\overset{e_s + a_l}{\longleftrightarrow} \underset{\substack{210 \text{ ppm}}}{\text{chloroplasts}} \overset{b}{\longrightarrow} \text{A-CO}_2 \overset{-d}{\longrightarrow} \text{CO}_2(\text{R})$$

and estimated according to

$$\Delta_p = \xi_b a_b + \xi_a a + \xi_l(e_s + a_l) + \frac{c_c}{c_a} b_3 - d \quad (27)$$

where $\xi$ is the weighting factor for each discrimination step, based on the associated CO$_2$ drawdown (i.e., $\xi_n = (c_n - c_{n-1})/c_a$ and $c_n$ is the CO$_2$ concentration at a step $n(1-4)$ along the diffusion pathway from the mixed atmosphere across the leaf laminar boundary layer; the leaf internal stagnant internal air spaces; and then across the liquid phase to the site of carboxylation (Figure 5). Accordingly, the diffusion discriminations are $a_b(\sim 2.9\%_0)$, $a(4.4\%_0)$, and $a_l[0.7\%_0$; (O'Leary, 1984)], for the laminar boundary layer, air spaces and liquid phase respectively. In going from air into liquid there is also an equilibrium fractionation $e_s$ [1.1‰; (Mook *et al.*, 1974)]. For C$_4$ plants the term $c_c$ must be replaced with $c_m$, for the CO$_2$ concentrations in the mesophyll cytoplasm where PEP-c is active (Farquhar, 1983) (with typical values $c_m \approx 125$ ppm) and $b_3$ is replaced with the appropriate expression from Equation (25). This will reflect the large differences in $\Delta_P$ for C$_3$ and C$_4$ plants, with typical values, based on the above examples, of $\Delta_P$ of $\sim 18\%_0$ and $\sim 5\%_0$ for C$_3$ and C$_4$ plants respectively. Inspection of Equation (27) clearly indicates the sensitivity of $\Delta_P$ to estimates of $c_c$, which can vary with environmental conditions, and among plant species, but cannot be measured directly. Best estimates are obtained by leaf-scale gas exchange measurements (Evans *et al.*, 1986).

In the framework of Equation (12), we are interested in global-scale estimates of $\Delta_p$ (or $\delta_a + \varepsilon_p$). Such estimates are often obtained either by forward modeling that scale up processes discussed above (e.g., Buchmann and Kaplan, 2001; Fung *et al.*, 1997; Lloyd and Farquhar, 1994; Kaplan *et al.*, 2002; Ito, 2003; Still *et al.*, 2003), as well as by inverse modeling constrained by observations (e.g., Bousquet *et al.*, 1999; Keeling *et al.*, 1989; Quay *et al.*, 1992; Tans *et al.*, 1993; Francey *et al.*, 1995; Enting *et al.*, 1995; Ciais *et al.*, 1995a,b; Battle *et al.*, 2000; Randerson *et al.*, 2002; Still *et al.*, 2003). Global-scale estimates of plant discrimination are still associated with large uncertainties and range between $\sim 15\%_0$ and $\sim 20\%_0$ in the studies mentioned above, with the lowest values estimated by the process-based models and the highest by the inverse studies.

Considering the large differences in discrimination between C$_3$ and C$_4$ plants it is clearly important to the consideration of the relative contributions of C$_3$ and C$_4$ vegetations to global GPP and global $\Delta_P$. This is not a trivial task and introduces significant uncertainty to the isotopic budget. The C$_4$ photosynthetic pathway is advantageous under low ambient CO$_2$ concentrations and in hot and dry conditions because it concentrates CO$_2$, and allows efficient CO$_2$ uptake when leaf stomatal conductance is low, thereby reducing water loss. It was therefore shown that the occurrence of C$_4$ vegetation can be predicted based on climatic conditions (Collatz *et al.*, 1998; Ehleringer *et al.*, 1997) and can continuously change (e.g., Schwartz *et al.*, 1996). Detailed vegetation maps are also often used as the basis for estimating C$_4$ plant distribution, but for CO$_2$ flux studies, estimates must also be based on the proportional productivity, and not on species distribution (Still *et al.*, 2003). Estimates indicate that C$_4$ vegetation contributes 20–25% to the global terrestrial productivity (Lloyd and Farquhar, 1994; Fung *et al.*, 1997; Still *et al.*, 2003; Ito, 2003). This proportion may have been larger during glacial times, and may decrease in future, high CO$_2$ world (Collatz *et al.*, 1998).

### 4.07.3.6 CO$_2$ Release in Respiration

Atmospheric CO$_2$ enters the land biosphere through leaves, and an equivalent, but not necessarily identical, flux leaves the biosphere as respiration (Raich and Potter, 1995; Raich and Schlesinger, 1992). Since practically all carbon is eventually respired, it can be expected that the $\delta^{13}$C value of respired CO$_2$ by an ecosystem, $\delta_{ER}$, will be near that of the photosynthesis flux, $\delta_{ER} = \delta_a - \varepsilon_p$, where $\delta_a$ is the $\delta^{13}$C of atmospheric CO$_2$ and $\varepsilon_p$ is the fractionations associated with photosynthesis. On annual or shorter timescales, however, this is not the case, because respiration acts on carbon that is undergoing chemical and isotopic modifications and is distributed among several distinct pools with widely different turnover rates, ranging from less than a year to many hundreds years (Trumbore, 2000; Trumbore *et al.*, 1995), and reality tends to be better described by

$$\delta_{ER} = \delta_a - \varepsilon_p - \varepsilon_R \qquad (28)$$

where $\varepsilon_R$ is the apparent fractionation associated with respiration. Much uncertainty is also associated with the possible isotopic fractionation involved in respiration, $\varepsilon_R$. Conventionally, it is assumed to be zero (Lin and Ehleringer, 1997), but recent studies in leaves indicate possible fractionations (Tcherkez *et al.*, 2003).

A basic distinction regarding respiration is made between autotrophic and heterotrophic respiration. Autotrophic respiration is carried out by the living plants and can be subdivided into aboveground (stem, branches, leaves) and belowground (roots, often considered together with their closely associated bacteria in the rhizosphere (Hanson *et al.*, 2000; Hogberg *et al.*, 2001; Rochette and Flanagan, 1997)). Autotrophic respiration consumes newly assimilated carbon with short turnover rates from few days (Ekblad and Högberg, 2001; Bowling *et al.*, 2002; Pataki *et al.*, 2003) to ~50 yr, mainly covering the range for herbaceous and woody tissues in different geographical locations (Ciais *et al.*, 1999). The $\delta^{13}C$ value of autotrophic respiration is generally assumed to be similar to that of recent photosynthates. The $\delta^{13}C$ values of plant organic matter vary significantly due to environmental and physiological parameters (e.g., Broadmeadow and Griffiths, 1993; Francey and Farquhar, 1982). In addition, there are fractionations during plant metabolism that result with changes in the $\delta^{13}C$ value of the living biomass, both on a spatial (higher values in top versus bottom of canopies, and in roots versus leaves etc. (Brugnoli *et al.*, 1998)) and on temporal (early and late season carbon; Hemming *et al.*, 2001) scales. The living biomass eventually dies, and the organic matter is translocated, transformed and decomposed in a complex chain of reactions (Boutton and Yamasaki, 1996).

Heterotrophic respiration processes dead plant material by microorganisms, which themselves are made of the carbon they consume (Gleixner *et al.*, 1993). The dead plant materials begin as litter, both above- and belowground, but are transformed into various forms of soil organic matter. Such transformation involves the loss of the more labile material as $CO_2$ and accumulation of residual material with longer turnover time, a process that is generally associated with increasing $\delta^{13}C$ values of soil organic matter (Balesdent and Mariotti, 1996; Ehleringer *et al.*, 2000; Ladyman and Harkness, 1980). Distinction among the specific compartments of soil organic carbon is based primarily on turnover rates

(Schimel *et al.*, 1996; Trumbore, 2000), and is the basis for various schemes that are based on modeling soil processes and used to describe soil respiration and its the $\delta^{13}C$ values (e.g., Ciais *et al.*, 1999; Fung *et al.*, 1997; Potter *et al.*, 1993; Randerson *et al.*, 1996; Schimel *et al.*, 1994), with similar underlying principles. For example, soil organic matter can be broadly separated into detrital, microbial, slow, and passive carbon. Detrital and microbial pools consist of metabolic and structural carbon compounds rapidly decomposed and recycled to the atmosphere in less than 10 years. The slow carbon pool consists of more resistant compounds, with high lignin content, that turn over in 10–100 years (Bird *et al.*, 1996). The passive carbon pool turns over in more than 100 years. The turnover time of each pool can vary and is influenced by factors such as temperature, precipitation and soil moisture, soil texture and microbial community. Note that although soil organic matter shows wide variations in age, up to thousands of years, the respiration flux is strongly dominated by short-lived soil organic matter components. The mean age of soil respired $CO_2$, $\tau$, can be approximated from independent estimates of the global soil carbon pool, SC, and the heterotrophic respiration flux, $F_h$ (1,500 Pg C yr$^{-1}$ and 55 Pg C yr$^{-1}$ respectively; IPCC): $\tau = SC/F_h = 27$ yr, which involves significant uncertainty (e.g., Trumbore, 2000; Ito, 2003).

For the atmospheric $^{13}C$ budget (Equation (12)) an estimate that integrates the $\delta^{13}C$ value of ecosystem respiration is required. Autotrophic respiration $\delta^{13}C$ values may be reasonably estimated from Equation (28), although there may be short time gaps (ca. 5 d) between corresponding photosynthesis and respiration signals (Bowling *et al.*, 2002; Pataki *et al.*, 2003). The $\delta^{13}C$ value of heterotrophic respiration, however, cannot be estimated with Equation (28), and must integrate the relative contributions of the multiple carbon pools in the soil and their unique $\delta^{13}C$ value (Ciais *et al.*, 1995b, 1999; Fung *et al.*, 1997).

It is important to note that the variations in the $\delta^{13}C$ values of soil organic matter, while considered as complications above, can serve as useful tracers in partitioning the contributions of specific components to the respiration flux or in estimating the turnover rates of such components (Balesdent and Mariotti, 1996; Cerling and Wang, 1996; Ehleringer *et al.*, 2000). Such studies are often based on measurements of soil $CO_2$, which is enriched in $^{13}C$, by 4.4‰ relative to the mean $\delta^{13}C$ value of soil organic matter or that of the soil respired $CO_2$, due to diffusion fractionations (Cerling *et al.*, 1991; Davidson, 1995). Note that this diffusion fractionation, while important for $^{18}O$ in $CO_2$ studies as discussed below, does not enter into the atmospheric $\delta^{13}C$ budget because

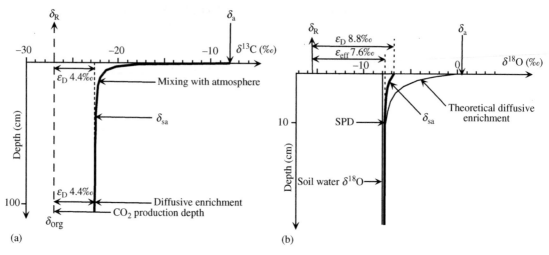

**Figure 6** A schematic representation of the evolution of (a) $\delta^{13}C$ and (b) $\delta^{18}O$ values of soil respired CO$_2$. Decomposition of soil organic matter produces CO$_2$ with similar isotopic composition. Diffusion in the soil, with fractionation $\varepsilon_D$, results in $^{13}$C-depleted respired CO$_2$ preferentially escaping to the atmosphere ($\delta_R$), leaving behind $^{13}$C-enriched CO$_2$ in the soil air ($\delta_{sa}$). This process reaches steady state when the respiration flux has the same $\delta^{13}C$ value as the source material. Mixing with $^{13}$C-depleted atmospheric CO$_2$ ($\delta_a$), influence the shape of the $\delta_{sa}$ profile near the soil surface. The $\delta^{18}O$ of CO$_2$ reflect the $\delta^{18}O$ of soil water up to a setting point depth (SPD) that represents the minimum effective depth at which CO$_2$ equilibrate with soil water before escaping to the atmosphere. Near the surface, $^{18}$O-depleted CO$_2$ escapes by diffusion to the atmosphere with a fractionation, $\varepsilon_D$, leaving behind $^{18}$O-enriched CO$_2$ (with a profile that could follow the thin solid line leading to a steady-state enrichment of soil CO$_2$ of $\varepsilon_D = 8.8‰$). However, if the soil near the surface is not completely dry, some isotopic exchange can occur leading to a profile such as depicted by the thick solid line. If isotopic equilibrium is catalyzed and occurs up to the surface itself, the profile would follow the waterline. While a diffusion fractionation of $\varepsilon_D = 8.8$ takes place in all scenarios, a smaller "effective fractionation" is observed when comparing the $\delta^{18}O$ of soil respired CO$_2$ with that of measured soil water at depth, or of precipitation water. (This simplified diagram ignores additional complications due to soil water enrichment near the surface).

soils are normally near isotopic steady state such that the $^{13}$C value of soil respired CO$_2$ is similar to that of the source carbon (Figure 6(a)).

### 4.07.3.7 The Land Disequilibrium

Based on the discussion in the previous section it is apparent that plant photosynthesis produces organic matter from today's atmosphere, but respiration releases carbon that has been stored for different periods in plant tissues and soil organic matter. It was also noted already that over the industrial era, the $\delta^{13}C$ of the atmosphere has been declining at a mean rate of ~0.02‰ yr$^{-1}$. The carbon respired today is, therefore, enriched in $^{13}$C, compared to current photosynthate, to an extent that depends on the residence time of the carbon in the biosphere (Figure 4). For example, adopting the estimated global mean $\tau$ of 27 yr mentioned above, soil respired CO$_2$ could be ~27 × 0.02 = 0.5‰ more enriched in $^{13}$C with respect to current atmospheric CO$_2$. Under such circumstances, an isotopic flux between the land biosphere and the atmosphere will exist even if there is no net CO$_2$ flux. Similar to the case for the ocean, this represents the land disequilibrium, $D_L$. Note that $D_L$ will tend to enrich the atmosphere in

$^{13}$C, in the same way that photosynthetic CO$_2$ uptake would (due to discrimination against $^{13}$C). It is obvious, therefore, that ignoring $D_L$ would result with over-estimating $^{13}$C-derived estimate of terrestrial carbon sink (or under-estimate a source). The error can be significant, in the order of 0.6 Pg C yr$^{-1}$ globally, and proportionally larger in regions such as the tropics (Ciais et al., 1999).

In a similar manner to the ocean, the land disequilibrium can be defined as

$$D_L = (\delta_{al}^{eq} - \delta_a) \qquad (29)$$

where $\delta_{al}^{eq}$ is the $\delta^{13}C$ value of the atmosphere that would be in isotopic equilibrium with the biosphere (i.e., $\delta_{al}^{eq} = (\delta_b - \varepsilon_p)$, where $\delta_b$ is the mean $\delta^{13}C$ of the biosphere and $\varepsilon_p$ is the photosynthetic fractionation) (Enting et al., 1995; Ciais, 1995; Fung et al., 1997; Tans et al., 1993). Global-scale modeling of this effect suggests that it is actually in the range of 0.20–0.33‰ (Fung et al., 1997; Quay et al., 1992; Tans et al., 1993), with large latitudinal variations between ~0.1‰ and ~0.6‰ (Ciais et al., 1999). Interestingly, large-scale land use changes in the tropics release large quantities of relatively old (~50 yr) carbon to the atmosphere, enhancing the land disequilibrium. The significant discrepancy between the first-order estimate of $D_L = 0.5‰$ made above and the recent model estimates of $0.20-0.33‰$ may reflect the difficulties in accurately estimating turnover rates of the various soil carbon pools and its global distribution, as well as in estimating the total amount of carbon in soils. Modifying the way carbon turnover is treated in models, e.g., from a "pipe" approach to a "well mixed pools" approach, reduced global $D_L$ values, but only by $\sim 0.1‰$ (Ciais *et al.*, 1999). Further reduction in the terrestrial disequilibrium may arise from the replacement of $C_3$ forest subjected to de-forestation by $C_4$ grassland, with a rather large effect on the local scale (Ciais *et al.*, 1999). Replacing $C_3$ with $C_4$ vegetation would result, initially, with photosynthetic uptake with $^{13}C$ of $\sim -12.4‰$ and respiration $^{13}C$ of $\sim -26‰$, in contrast to the fossil fuel effect in $C_3$ stands where uptake is at $\sim -26.0‰$ and release at $\sim -25.7‰$.

### 4.07.3.8 Ecosystem Discrimination

The apparent complexity in the respiratory $^{13}C$ signal, discussed in the previous section, provides an exciting aspect for ecosystem-scale studies. The differences in $\delta^{13}C$ values between the photosynthetic (A) and respiratory (R) $CO_2$ fluxes allow the partitioning of net ecosystem exchange (NEE, where $NEE = A + R$) into its two components, which is critical to understanding the land ecosystem response to change (Bowling *et al.*, 2001; Raupach, 2001; Yakir and Wang, 1996; Lloyd *et al.*, 2001; Styles *et al.*, 2002). Furthermore, ecosystem-scale measurements can provide experimentally based constrain on plant and soil discrimination, parameters that contribute the main uncertainties to the isotopic approach depicted in Equation (12).

To estimate isotopic discrimination at the ecosystem, local or regional scales, an estimate of the $^{13}C$ exchange flux, or the $\delta^{13}C$ value of NEE, at the relevant scale is required. Such estimates are obtained by sampling air in and above plant canopies, or in the and above the atmospheric boundary layer (ABL). A simple and powerful approach employs a two-member mixing model as first proposed by Keeling (1958, 1961). The equation used in the Keeling approach is derived from the basic assumption that the atmospheric concentration of a substance in an ecosystem reflects the combination of some background amount of the substance that is already present in the atmosphere and some amount of substance that is added or removed by a source or sink in the ecosystem:

$$C_E = C_a + C_s \qquad (30)$$

where $C_E$, $C_a$, and $C_s$ are the concentrations of the substance in the ecosystem, in the atmosphere and that contributed by ecosystem sources respectively. Isotope ratios of these different components can be expressed by a simple mass balance equation:

$$C_E \delta_E = C_a \delta_a + C_s \delta_s \qquad (31)$$

where $\delta_E$, $\delta_a$, and $\delta_s$ represent the isotopic composition of the substance in the ecosystem, in the atmosphere and of the sources respectively. Combining Equations (30) and (31):

$$\delta_E = C_a \cdot (\delta_a - \delta_s) \cdot (1/C_E) + \delta_s \qquad (32)$$

Equation (32) describes a linear relationship, with an intercept $\delta_s$, which represents the $\delta^{13}C$ signature of the net sources/sinks in the ecosystem. Note that even if the ecosystem source/sink is composed of several different sub-sources/sinks the Keeling type plot can still be used as long as the relative contribution of each of these sub-components remain fixed.

This relationship was first used by Keeling (1958, 1961) to interpret carbon isotope ratios of ambient $CO_2$ and to identify the sources that contribute to increases in atmospheric $CO_2$ concentrations at a regional basis. It is now widely applied on a wide range of studies and on a wide range of scales (Figure 7; and see Pataki *et al.*, 2003). Miller and Tans (2003) proposed some modifications to the conventional approach by re-arranging Equation (31) to allow consideration of changes in $\delta_a C_a$ (e.g., over the seasonal cycle) and derive $\delta_s$ from the slope of the best fit line. Perhaps the most common application of the Keeling expression is to identify the isotopic composition of respired $CO_2$ in forest ecosystems (Bowling *et al.*, 2002; Buchmann *et al.*, 1997a,b; Harwood *et al.*, 1999; Pataki *et al.*, 2003; Quay *et al.*, 1989; Sternberg, 1989; 1997). The derivation of the isotopic composition of respired $CO_2$ has been used to determine ecosystem carbon isotope fractionation ($\Delta_e$) by the following equation:

$$\Delta_e = (\delta_{\text{trop}} - \delta_{\text{resp}})/(1 + \delta_{\text{resp}}) \qquad (33)$$

where $\delta_{\text{trop}}$, $\delta_{\text{resp}}$ represent the $\delta^{13}C$ values of tropospheric and respired $CO_2$ (Buchmann *et al.*, 1998; Buchmann and Kaplan, 2001; Flanagan *et al.*, 1996). Note that in the equations, the traditional use of $\delta_{\text{resp}}$ (Keeling, 1961) is applied for both nighttime and daytime measurements. This term would be better represented by $\delta_s$ (Equation (16)), which includes effects of both respiration and photosynthesis and will be equal to $\delta_{\text{resp}}$ only for nighttime measurements. Note also that recycling of respired $CO_2$ within canopies can influence estimates of the $\delta$ values of respired $CO_2$ (Lloyd *et al.*, 1996; Sternberg, 1989). A modified equation ($\Delta_e = \delta_{\text{atm}} - \delta_{\text{resp}}$) was also

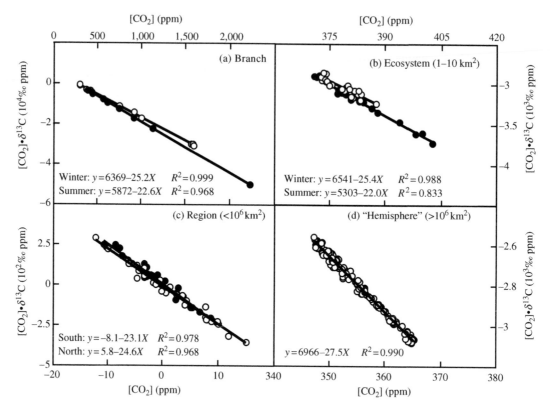

**Figure 7** Modified Keeling plots identifying the $^{13}$C signatures in CO$_2$ exchanged with the atmosphere on different spatial and timescales. A series of air samples is analyzed for changes in CO$_2$ concentration and its $\delta^{13}$C, and plotted in such a way that the slope of the best-fit lines provides estimate of the $^{13}$C signature of the surface flux (see text for detail). Single branches are sealed for a few minutes in a chamber and allowed to respire in the dark while air samples are withdrawn. The results show seasonal change in the $^{13}$C signature of respired CO$_2$ between the wet winter and the dry summer in the semi-arid Yatir forest, Israel. Nocturnal canopy air samples taken at about the same time in the Yatir forest show that ecosystem-scale respiration reflected the same $^{13}$C signature as a single branch (D. Hemming and K. Maseyk, unpublished). Tall, ~400 m, TV towers are used to collect air samples that represent regional-scale footprints. Data from a tall tower in Wisconsin (Bakwin, unpublished results; see Bakwin *et al.*, 1998) indicate the possible influence of the extensive corn cultivation (C$_4$ plants with low $^{13}$C discrimination) to the south of the tower on the predominantly C$_3$ forest area around the tower site. Background troposphere air samples are taken in permanent network stations around the world on a weekly basis by flushing air through glass flasks, which are then sent for analysis in a central laboratory. Data for the summer months (June through August) from Chemya, Alaska (NOAA/CMDL-CU network) collected during 1993–2002 show a typical mid to high latitude biospheric $^{13}$C signature. The range of values shown here ($-22.6$‰ to $-27.5$‰) covers most of the observed range in the biological $^{13}$C signatures. Data for the tall tower were de-trended to remove long-term trends and seasonal cycle in CO$_2$ and $^{13}$C (see Figure 2) and only the "residual" variations in CO$_2$ and $^{13}$C were considered, enhancing the influence of the local biology on the results; the ten years long data set from Chemya were "de-trended" only for the long-term mean trends in CO$_2$ and $\delta^{13}$C enhancing large-scale effects (see Bakwin *et al.*, 1998a,b; Miller and Tans, 2003).

used by Bakwin *et al.* (1998) to estimate regional-scale biological discrimination (Bakwin *et al.*, 1995, 1998a,b; Miller and Tans, 2003; Miller *et al.*, 2003). A more general approach to canopy-scale inverse methods, and in particular the potential in the inverse Langrangian method is discussed by Raupach (2001).

Whole ecosystem and regional-scale discrimination is also assessed by evaluating the isotopic composition of CO$_2$ in the atmospheric boundary layer (ABL; also called planetary boundary layer (PBL), as well as convective boundary layer (CBL), when considered during times of convective enhancement by surface heating during the day).

The major advantage of the ABL budget is in its integration of surface fluxes and discrimination over large areas ($10^4$–$10^6$ km$^2$ compared with $1$–$10$ km$^2$ for measurements associated with canopy flux towers). The ABL approach relies on the turbulence and efficient mixing throughout the ABL height (~$1$–$2$ km), and on the ABL very distinctive upper limit, below the free troposphere (Figure 8). Few measurements over time within the homogeneous ABL (usually relying on small aircrafts), and knowledge of air transport through its top, provide an efficient way to report large-scale changes in the storage of constituents, such as CO$_2$ and $^{13}$CO$_2$, in the ABL that, in turn must

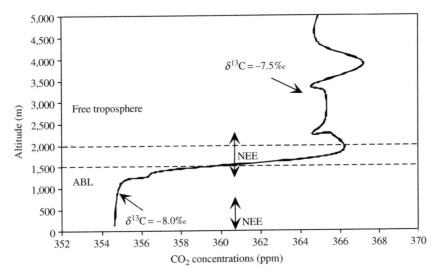

**Figure 8** A representative atmospheric boundary layer (ABL) profile of CO$_2$ (obtained by airplane) during a summer mid-afternoon at a forest study site in Wisconsin (Helliker and Berry, unpublished). Representative $\delta^{13}$C values for the mixed ABL and the free troposphere are indicated. The distinct ABL and its homogeneity provide a basis to estimate changes in CO$_2$ storage and its isotopic composition in the entire ABL column above large surface areas. Alternatively, the large concentration and isotopic gradients at the top of the ABL provide a robust basis for estimating the flux across the top of the ABL, which is assumed to be similar to that at the bottom of the ABL (i.e., the land surface NEE; see text).

reflect surface fluxes and isotopic discrimination (Lloyd *et al.*, 1996, 2001; Nakazawa *et al.*, 1997; Styles *et al.*, 2002a; Ramonet *et al.*, 2002). Whole ecosystem and regional-scale discrimination greatly increase the window of observation and provide greater integration of the mosaic of surface activities. This approach can help, for example, in quantifying the contribution of C$_3$ or C$_4$ vegetation to productivity such as in tropical areas (Lloyd *et al.*, 1996; Miranda *et al.*, 1997). It is also valuable, in constraining large-scale estimates of discrimination associated with photosynthesis and respiration (e.g., Fung *et al.*, 1997). But note that canopy and CBL-scale net discrimination must still be partitioned to photosynthetic and respiration components for mass balance analysis such as based on Equation (12).

An elegant simplification of the ABL approach to estimate regional-scale fluxes and isotopic discrimination has been proposed by Helliker and Berry unpublished. Instead of following changes in storage within the ABL, this approach takes advantage of the sharp concentration gradients at top of the distinctive ABL boundary (Figure 8). Such easily measured gradients in, say, CO$_2$ and $^{13}$CO$_2$, when time averaged over weeks to months, permit precise estimates of the net flux across the top of the ABL and, based on conservation of mass, must reflect the integrated fluxes from/to the surface (e.g., NEE for CO$_2$). As done in small-scale studies (cf. Equations (20)–(23)), the net CO$_2$ exchange flux across the top of the ABL can be simply estimated from:

$$\text{NEE} = \{[\text{CO}_2]_{\text{ABL}} - [\text{CO}_2]_{\text{troposphere}}\}/r_{\text{CO}_2-\text{ABL}},$$

where $r_{\text{CO}_2-\text{ABL}}$ is the resistance to CO$_2$ transport within the ABL that must be independently estimated (for example from radon or meteorological re-analysis products). Similarly, the $\delta^{13}$C of NEE, and consequently regional-scale discrimination (see above), can also be obtained from

$$\delta^{13}C_{\text{NEE}} = \{[\text{CO}_2]_{\text{ABL}}\delta^{13}C_{\text{ABL}}$$
$$- [\text{CO}_2]_{\text{troposphere}}\delta^{13}C_{\text{troposphere}}\} \Big/$$
$$([\text{CO}_2]_{\text{ABL}} - [\text{CO}_2]_{\text{troposphere}}) \quad (34)$$

As discussed below, knowledge of NEE and $\delta^{13}C_{\text{NEE}}$ are prerequisites for the partitioning of NEE into its photosynthetic and respiratory components. Preliminary results of using this approach gave close agreement with surface-based, eddy covariance and isotopic measurements.

### 4.07.3.9 Incorporating Isotopes in Flux Measurements

The dynamics of canopy-scale net fluxes of water and CO$_2$ exchanged between vegetation and the atmosphere are routinely measured today with micrometeorological methods (e.g., with eddy covariance; www.daac.ornl.gov/FLUXNET/ fluxnet.html). Combining these methods with isotopic measurements allows to partition a net flux into its gross flux components. The approach here is similar to that used on the global scale to asses, for example, ocean versus land sink/source relationships (Equation (12)), but focuses instead on the local, ecosystem, scale and considers only ecosystem respiration and photosynthesis. In simplified terms and expression, this was first employed by Yakir and Wang (1996) over crop fields, and can be developed by assuming $F_N$ as a net flux of $CO_2$ (say, photosynthetic uptake and soil respiration combined) such that $F_N = F_1 + F_2$. Each flux carries a unique isotopic identity, $\delta_1$, $\delta_2$, and $\delta_N$ (e.g., the $\delta$ values of $CO_2$ taken up in photosynthesis, released in soil respiration and of the combined net flux, respectively) and isotopic mass balance takes the form

$$F_N \cdot \delta_N = F_1 \cdot \delta_1 + F_2 \cdot \delta_2 \qquad (35)$$

By rearrangement we obtain estimates of the component gross fluxes:

$$F_1 = F_N \cdot \phi_1; \qquad F_2 = F_N \cdot \phi_2 \qquad (36)$$

where $\phi_1 = (\delta_N - \delta_2)/(\delta_1 - \delta_2)$ and $\phi_2 = (\delta_1 - \delta_N)/(\delta_1 - \delta_2)$. Thus, knowledge of the net flux, $F_N$, and the isotopic signatures $\delta_N$, $\delta_1$, and $\delta_2$ allows us to derive an estimate of the component gross fluxes. $F_N$ and $\delta_N$ may be directly estimated by micrometeorological techniques, values of $\delta_N$ can also be estimated from Keeling type plots. But the specific isotopic signatures, $\delta_1$ and $\delta_2$, must be independently estimated from plant and soil samples, or from leaf-scale gas exchange and soil chamber measurements. Sampling and estimates of $\delta_1$ and $\delta_2$ must also consider heterogeneity in the ecosystem and be representative of the relevant scale (determined by the type of micrometeorological measurement).

Micrometeorological measurements of trace gas exchange are commonly made using the flux-gradient or eddy correlation techniques. The flux-gradient technique combines the eddy diffusivity ($K_c$) and concentration gradient with height ($dx/dz$) to calculate the flux according to

$$F_N = K_c \cdot (dx/dz) \qquad (37)$$

Eddy diffusivity is estimated, for example, by wind profile measurements as described by Baldocchi *et al.* (1988). The gradient $dx/dz$ is directly measured by the appropriate instruments at different heights above the canopy and averaged over an appropriate time interval (e.g., 30 min). The main advantages of the gradient approach are that large gradients can be observed over several meters between sampling heights, and that it does not require fast responding instruments. Measurements inside plant canopies, however, are complicated by possible counter-gradient transport and spatial heterogeneity (Baldocchi *et al.*, 1988). This flux-gradient technique can be readily adapted to include isotopic measurements. Air samples, or air moisture samples, are collected over the same time intervals and shipped

to a stable isotope laboratory. The water and/or $CO_2$ samples are analyzed for isotopic composition to provide time integrated mean values. The concentration and isotopic data can then be used to produce a Keeling-plot, which may give an estimate of the isotopic identity of the net flux, $\delta_N$ (cf. $\delta_s$ in Equation (32)). Sampling of soil, stem and leaf water, and soil organics representative of the underlying vegetation can provide the necessary information to estimate the relevant isotopic signatures of the soil and leaf exchange flux ($\delta_1$ and $\delta_2$ for either water or $CO_2$). More directly, some of these signatures can be obtained from leaf-scale gas-exchange and/or isotopic measurements carried out concomitantly in the field (e.g., Harwood *et al.*, 1998; Wang *et al.*, 1998), or by Keeling plots for data obtained within the canopies to determine isotopic identities of soil fluxes. Yakir and Wang (1996) have successfully used the gradient approach to partition net ecosystem exchange of $CO_2$ into photosynthetic assimilation and respiration for several crop species. Such studies provide confidence in the feasibility of the flux-isotope approach, which is supported also by other canopy-scale isotopic studies (Buchmann and Ehleringer, 1998; Flanagan *et al.*, 1997; Moreira, 1998).

The flux-profile relationships utilized above crop fields and grasslands (cf. Cellier and Brunet, 1992) may not be valid, and are therefore not generally accepted, above forests. In forests, the eddy correlation method offers a more direct approach to flux measurements. In this case, a fast responding sonic anemometer is used to indicate the velocity and direction of the vertical air movement component (up or down eddies). A fast responding analyzer records the concentration of the trace gas of interest in conjunction with anemometer measurements. Once data are collected over an appropriate time interval (e.g., 30 min), the concentration measurements can be separated to those related to eddies moving up from the canopy or down into the canopy. The net eddy flux is then calculated according to

$$F_N = \overline{\rho w' c'} \qquad (38)$$

where $\rho$ is density of dry air, $w$ is the vertical wind velocity component (positive and negative values correspond to up/down) and $c$ is the mixing ratio or mole fraction of the chemical constituent with respect to dry air. An overbar indicates a time-averaged quantity and a prime indicates deviation from the mean. Note that using this equation requires several assumptions about the nature of the turbulence and the nature of the landscape (Baldocchi *et al.*, 1988). Given that the exchange between the canopy and the atmosphere can be measured with this technique, then it follows that NEE, which considers also the canopy storage flux can be described as (Baldocchi, 1997; Wofsy *et al.*, 1993):

$$NEE = A + R = \overline{\rho w' c'} + dc_i/dt \quad (39)$$

where $A$ and $R$ are the photosynthetic and ecosystem respiration fluxes and $dc_i/dt$ is the change in canopy storage. Numerous studies have used the eddy correlation approach to provide estimates of $F_N$ or NEE (but note that these measurements in some cases are difficult to validate with actual changes in biomass, e.g., Rochette *et al.*, 1999).

Fast responding isotope analyzers with the required precision are not available at present and the eddy correlation technique cannot be directly combined with *in situ* stable isotope measurements to estimate $\delta_N$ (Equations (34)–(36)). Alternatively, independent methods, such as the Keeling plot, with certain caveats, can be used to estimate $\delta_N$. This requires that the data used to construct the Keeling plot represent the same time interval and footprint as the eddy correlation system. Isotopic data for Keeling plot estimates can be obtained from slow flask sampling and laboratory analysis. Bowling *et al.* (1999a) have recently argued that in the range and precision of data obtained for Keeling plots, the relationships between $CO_2$ concentration and $\delta^{18}O$ or $\delta^{13}C$ are practically linear such that the equation $\delta = mc + b$ (termed reciprocal Keeling equation, where $\delta$ and $c$ are the measured isotopic and concentration values) can be used to predict the isotopic composition based on the component concentration. If the data are representative, such a relationship can be used to produce the isotopic flux of $^{13}CO_2$, $F_{13}$:

$$F_{13} = \overline{\rho \omega' [c(mc + b)]'} \quad (40)$$

where $c$ is measured by the eddy correlation system and the constants $m$ and $b$ are derived from a reciprocal Keeling plot based on concurrent flask sampling. Assuming the $^{13}C$ signature of the biological system, $\delta_N$, is a constant at the measurement scale, $F_{13}$ correspond to $F_N \delta_N$.

There are difficulties in precise estimates of the components of Equation (40). Recent studies have addressed the differences between the Keeling relationship as determined from slow, whole-air flask sampling, and that of eddy covariance sampling (Bowling *et al.*, 1999b). Additionally, isotopic analyses are incorporated into studies using conditional sampling techniques (relaxed eddy accumulation, REA), where fast responding analytical instruments are replaced with a fast responding valve system and an accumulator that collect separate air samples from "updraft" and "downdraft" eddies over a time interval; Bowling *et al.*, 1999b, 1998; Businger and Oncley, 1990).

## 4.07.4 OXYGEN-18 IN CO₂

As stable isotope measurements of atmospheric $CO_2$ were added to the arsenal of environmental measurements, modern, triple-collector isotope ratio mass spectrometers have been used. Notably, these instruments routinely recorded the contributions of molecular masses 44, 45, and 46 in $CO_2$ (representing $^{12}C^{16}O_2$, $^{13}C^{16}O_2$, and $^{12}C^{16}O^{18}O$ and ignoring or as correcting for minor contributions from $^{17}O$ and doubly labeled molecules). Initially, however, attention was focused mainly on the ratio of 45/44, providing information on $^{13}C$ content in $CO_2$ (e.g., Keeling, 1958, 1961; Mook *et al.*, 1983), while the 46/44 ratios, and the $^{18}O$ content in $CO_2$, were hardly addressed (Bottinga and Craig, 1969; Mook *et al.*, 1983). This has changed noticeably in 1987 when two studies noted that the neglected ratios report a strong, large-scale and unique signal (Francey and Tans, 1987; Friedli *et al.*, 1987). These studies showed that the observed latitudinal and altitudinal gradients in atmospheric $C^{18}O^{16}O$ are quantitatively consistent with the expected effects of $C^{18}O^{16}O$ exchange between the atmosphere and the land biosphere, and that there are, in fact, no other fluxes of this magnitude that offer alternative explanations.

It is now recognized that the $^{18}O$ signal in $CO_2$ reflects a unique coupling between the global hydrological and carbon cycles, and can provide important tracer of sink and sources of $CO_2$ particularly in the terrestrial biosphere. The use of $^{18}O$ relies on the fact that $CO_2$ readily dissolves in water, allowing $CO_2$–$H_2O$ oxygen exchange to occur. Water in leaves is highly enriched in $^{18}O$, relative to soil water, due to evaporative fractionation (Dongmann, 1974; Förstel, 1978). Consequently, $CO_2$–$H_2O$ exchange in leaves (associated with photosynthesis) or in soil and trunks (associated with soil, roots and plant respiration) produces contrasting $^{18}O$ signals in the $CO_2$ which is released to the atmosphere (Ciais *et al.*, 1997a,b; Farquhar *et al.*, 1993; Yakir and Wang, 1996). Because the enzyme carbonic anhydrase (CA) is present in all plant leaves and rapidly catalyzes $CO_2$ hydration and isotopic exchange, in spite of the short residence time of $CO_2$ in leaves, $CO_2$ involved in photosynthesis is extensively re-labeled by $^{18}O$-enriched leaf water. Such differences should allow identification of $CO_2$ sources and sinks in the ecosystem and that may allow estimates of the individual photosynthetic and respiratory fluxes. Indeed the limited ability to partition net ecosystem productivity (NEP) into gross primary productivity (GPP; equivalent to net leaf assimilation, A) and ecosystem respiration (R; including aboveground plant respiration and belowground root and microbial respiration) has been a critical limitation in obtaining insights into the causes of the observed inter-annual variations of global NEP, as well as in explaining past variations and predict future levels of terrestrial carbon storage (Tans and White, 1998). The potential in using $^{18}O$ in $CO_2$ as a tracer of gross fluxes in the terrestrial biosphere has been experimentally demonstrated in field-scale measurements (Flanagan *et al.*, 1997; Flanagan and Varney, 1995; Yakir and Wang, 1996), and in global-scale modeling (Ciais *et al.*, 1997a,b; Farquhar *et al.*, 1993; Ishizawa *et al.*, 2002; Peylin *et al.*, 1996, 1999), constrained by the accumulating atmospheric $^{18}O$ data (Conway *et al.*, 1994; Masarie *et al.*, 2001; Trolier *et al.*, 1996).

The primary control on the $\delta^{18}O$ of $CO_2$ is the $\delta^{18}O$ of the liquid water with which it was last in contact. $CO_2$ isotopically equilibrates with water according to the following reaction:

$$H_2^{18}O_{(l)} + CO_{2(g)} \longleftrightarrow H^+ + [HCO_2^{18}O]^-_{(aq)}$$
$$\longleftrightarrow H_2O_{(l)} + CO^{18}O_{(g)} \quad (41)$$

The temperature-dependent value for the equilibrium fractionation $\varepsilon_{eq}$ (where $\varepsilon_{eq} = (\alpha_{eq} - 1) \times 1,000$) between the oxygen in the $CO_2$ and water is

$$\varepsilon_{eq}(T) = 17,604/T - 17.93 \quad (42)$$

where $d\varepsilon_{eq}/dT = -0.20\%_0 \,^{\circ}C^{-1}$, so that at $25\,^{\circ}C$, $\varepsilon_{eq} = 41.11\%_0$ (Brenninkmeijer *et al.*, 1983). Water must be in the liquid phase for the hydration reaction to occur (Gemery *et al.*, 1996). Equilibration does not occur in the atmosphere because the liquid water content is generally too small, given the slow rate for the uncatalyzed forward reaction ($0.012\, s^{-1}$, at $25\,^{\circ}C$). The rate constant for the isotopic reaction is one-third the rate of the chemical reaction because of the three oxygen atoms present in the intermediate bicarbonate species (Mills and Urey, 1940). With the presence of carbonic anhydrase, ubiquitous in leaves, equilibrium (Equation (41)) can be reached nearly instantaneously (with turnover rate of up to $10^6\, s^{-1}$, (Silverman, 1982), and typical rates of $100-1,400\, \mu mol\, CO_2\, m^{-2}\, s^{-1}$ on leaf area basis; Gillon and Yakir, 2001). The quantity of water usually involved in $CO_2$–water interaction is many orders of magnitude greater than that of the $CO_2$ present, so isotopically equilibrated $CO_2$ will take on the oxygen isotopic ratio of the water in which it is dissolved (plus the temperature-dependent equilibrium fractionation), regardless of its initial $\delta^{18}O$ value.

$CO_2$ hydration is associated with all fluxes between the atmosphere and the Earth surface, and the $^{18}O$–$CO_2$ budget may be expressed, as done for $^{13}C$, as a mass balance with respect to the annual atmospheric trend (Ciais *et al.*, 1997a; Farquhar *et al.*, 1993; Miller *et al.*, 1999):

$$\frac{d\delta_a}{dt} = \frac{1}{c_a}[F_L(\delta_l - \delta_a + \varepsilon_l) + F_R(\delta_a - \delta_s)$$
$$+ \varepsilon_{eff}) + F_I(\delta_a - \delta_s) + F_{oa}(\delta_o - \delta_a)$$
$$+ \varepsilon_w(F_{oa} - F_{ao}) + F_{ff}(\delta_{ff} - \delta_a)$$
$$+ F_{bb}(\delta_{bb} - \delta_a) + F_{str}(\delta_a - \delta_{str})] \quad (43)$$

where

$$F_L = \frac{c_l}{c_a - c_l}A, \qquad A = F_{la} - F_{al} \text{ and}$$
$$F_I = vc'_a$$

and $c$ refers to concentrations, $F$ to one-way flux to or from the atmosphere, $\delta$ to isotopic ratios, $A$ to net photosynthetic assimilation (equivalent to GPP; and $F_L$ the "retro-diffusion" flux from the leaf associated with GPP), and $v$ to "piston velocity" of $CO_2$ diffusion into the soil and equilibrating with soil water (Tans, 1998). Subscripts used are: R for respiration (autotrophic and heterotrophic), A for assimilation, a for atmosphere ($c'_a$ refers to concentrations near the soil surface, o for oceans, ff for fossil fuels, bb for biomass burning, l for leaves, I for atmospheric invasion into soils (allowing for oxygen exchange with soil water, but with no net flux), s for soil and str for stratosphere; $\varepsilon$ refers to the kinetic fractionations associated with fluxes between reservoirs ($\varepsilon_W$ for water to air, and $\varepsilon_{eff}$ represents an effective diffusion fractionation in the soil–atmosphere interface).

Note that in Equation (43) $\delta_a$ and $c_a$ are known accurately by measurement (Conway *et al.*, 1994; Trolier *et al.*, 1996). The fossil fuel flux is well known and its oxygen isotopic composition is assumed to be that of atmospheric $O_2$ used in its combustion ($+23.5\%_0$ versus SMOW; Kroopnic and Craig, 1972). The biomass burning flux is relatively small and its isotopic signature is also assumed to be that of atmospheric $O_2$. Although the oceanic exchange flux ($F_{oa}$) is large, the net oceanic exchange ($F_{oa} - F_{ao}$) is near zero, and the isotopic disequilibrium between the ocean and atmosphere ($\delta_o - \delta_a$) is relatively small, and so is the overall oceanic influence on the atmospheric $^{18}O$ (i.e., $F_{oa}(\delta_o - \delta_a)/c_a$). The same is true for the troposphere–stratosphere exchange, where $F_{str}$ is large ($\sim 95$ Pg C $yr^{-1}$, but with no net flux) and the fractionation is uncertain at present but likely small (Blunier *et al.*, 2002). It is the leaf and soil (respiration and invasion) components that dominate the uncertainty and the magnitude of the $^{18}O$ mass balance. Soil and leaf isotopic fluxes each contribute roughly five times more to the atmospheric signal of $\delta^{18}O$ than either the oceanic or fossil fuel components, which are the next most significant terms. Combining estimates of NEP (from $^{13}C$ or $O_2$) with accurate measurements and modeling of the $^{18}O$ signals of soils and leaves should allow the use of Equation (43) to better estimate $F_R$ and $F_A$ and consequently GPP and R.

### 4.07.4.1 The Soil Component

The $\delta^{18}O$ of $CO_2$ from both root respiration and decomposition of soil organic matter is strongly influenced by the oxygen isotopic composition of the water with which it is in contact. The $\delta^{18}O$ value of precipitation (Jouzel *et al.*, 1987 and this volume; Mathieu and Bariac, 1996b; Mathieu *et al.*, 2002) is the single-most important environmental control on the oxygen isotopic composition of $CO_2$. It transforms into soil water and feeds plants (Gat, 1996; Leroux *et al.*, 1995). The isotopic composition of precipitation is influenced by fractionation in the hydrological cycle (Craig, 1961; Dansgaard, 1964; Epstein and Mayeda, 1953; Gat, 1996), which is simulated in detail by global GCMs. Jouzel *et al.* (1987) and Yurtsever and Gat (1981), employed a simpler empirical approach using data from the global precipitation survey conducted by the IAEA to develop a multiple regression analysis that predicts the $\delta^{18}O$ of precipitation, $\delta_p$, based on the two main influencing parameters, temperature, $T$ (°C), and the amount of precipitation, $P$ (mm month$^{-1}$) approximated by

$$\delta_p = -11.78 + (0.418 \pm 0.033)T - 0.0084$$
$$\pm 0.0048)P \quad (44)$$

Generally, $\delta_p$ decreases in going from the equator (around 0‰ on the SMOW scale) toward the poles ($\sim -25‰$), and from the coast into continental regions (Craig, 1961; Gat, 1996; Rozanski *et al.*, 1993, 1992).

In the simplest case, the $\delta^{18}O$ value of rainwater directly translates to the $\delta^{18}O$ value of soil water and, after equilibrium and kinetic fractionations are taken into account, into $\delta^{18}O$ of the $CO_2$ flux from the soil to the atmosphere ($\delta_{sa}$). Even in such simple cases, seasonal variations in soil water occur also due to input from snowmelt water or floodwater, introducing seasonal, indirect precipitation signals (Hesterberg and Siegenthaler, 1991; Miller *et al.*, 1999). A quantitative treatment of the conversion from the $\delta^{18}O$ of soil water to $\delta_s$ was first proposed by Hesterberg and Siegenthaler (1991), which was further tested by Stern *et al.* (1999, 2001); (see also Amundson *et al.*, 1998). Tans (1998) expanded this treatment by considering additional factors influencing $\delta_{sa}$ and Miller *et al.* (1999) experimentally tested and quantified such parameters. The effect of soils to atmosphere

flux, $F_{sa}$, on the $\delta^{18}O$ value of atmospheric $CO_2$, $F_{sa}(\delta_a - \delta_{sa})$ can be summarized as:

$$\frac{1}{c_a}F_{sa}(\delta_a - \delta_{sa}) = \frac{1}{c_a}F_{sa}\Delta_s$$
$$= \frac{1}{c_a}[F_R(\delta_a - \delta_s + \varepsilon_{eff}) + F_I(\delta_a - \delta_s)] \quad (45)$$

Representing the two components of the soil–atmosphere flux, respiration and atmospheric invasion, the associated $CO_2$ in equilibrium with soil water, $\delta_s$, and the effective diffusion fractionation in going from the soil into the atmosphere, $\varepsilon_{eff}$. These components, and what influence them is briefly discussed below.

Water in drying soils can become highly enriched in $^{18}O$ near the surface relative to its initial water source, producing steep isotopic gradients in soil water (enrichment of 10–15‰ are common; Allison and Barnes, 1983; Allison *et al.*, 1983; Barnes and Allison, 1983; Mathieu and Bariac, 1996a,b). Since it is difficult to estimate this gradient, it poses a potential complication in estimating $\delta_s$. Miller *et al.* (1999), however, showed that although soil drying can have an effect on $\delta_s$ values, the $^{18}O$ enrichment in the top 5–10 cm of the soil cannot impart all of its isotopic signature to the $CO_2$ leaving the soil. This is because diffusion of $CO_2$ out of the soil occurs more quickly than uncatalyzed hydration. The half time for isotopic equilibration of $CO_2$ in water in soil pores (with typical values of air porosity and water content) is $\sim 110$ s, while the mean time expected for $CO_2$ molecules to diffuse through 1 cm of soil is only $\sim 14$ s. In the top 5–10 cm, where the $\delta^{18}O$ of soil water changes by at least 1‰ cm$^{-1}$, the diffusing $CO_2$ cannot fully equilibrate before escaping to the atmosphere. The lower water content of the isotopically enriched near-surface soil further inhibits the ability of the $CO_2$ to attain isotopic equilibrium with the soil water. In Equation (45), $\delta_s$ is defined as a weighted average of $CO_2$ isotopically equilibrated with soil water along a depth range. In practice, it can be regarded as the isotopic value of the $CO_2$ in equilibrium with soil water at a specific "setting point depth" (SPD) such that $(\delta_s + \varepsilon_D f)$ is consistent with the $CO_2$ respired out of the soil. Miller *et al.* (1999) found that SPD ranged between 5 cm and 15 cm. Above 5 cm, $CO_2$ escapes from the soil faster than it can be isotopically hydrated; below 15 cm, the isotopic composition of $CO_2$ is reset as it diffuses toward the surface. Therefore, the steepest gradient in the $\delta^{18}O$ of soil water, often observed near the surface, does not play a large role in the isotopic composition of respired $CO_2$. While these results simplify estimates of $\delta_s$, Kesselmeier and Hubert (2002) and Kesselmeier *et al.* (1999) reported some evidence for carbonic anhydrase catalytic effects in soils. In cases where catalysis is present, the value of $\delta_s$ would still be influenced by the gradients near the soil surface, and accelerate the effect of atmospheric invasion into the soil (see below). Carbonic anhydrase, can increase the hydration rate of $CO_2$ by several orders of magnitude (Reed and Graham, 1981) allowing $CO_2$ to equilibrate more fully prior to leaving the soil. In practice, the $\delta_s$ was parameterized in different models to reflect either surface soil water (Ciais et al., 1997a) soil water at depth (Flanagan et al., 1997; Flanagan and Varney, 1995), or mean precipitation (Farquhar et al., 1993).

The type of soil in which $CO_2$ hydration occurs may also be important. Implicit in the hydration reaction is the assumption that all the water in soils is chemically and physically unbound, except for hydrogen bonding between water molecules. However, this assumption may be quite limited. Clays and organic matter have large surface areas to which some fraction of the soil water may be strongly adsorbed (Hsieh et al., 1998; Sposito et al., 1999) altering the equilibrium fractionation, $\varepsilon_{eq}$. Water adsorbed to a surface probably contains $^{18}O$ bound more strongly than when in solution, reducing the value of $\varepsilon_{eq}$ (Miller et al., 1999). Indications for such effects can be noticed in several studies using low moisture soils (below ~10%; Hsieh et al., 1998; McConville et al., 1999; Miller et al., 1999; Scrimgeour, 1995), and was recently addressed by Isaac (2001). The hydration rate, $k_h$, when water is adsorbed to the soil may also be influenced. First, the large surface to volume ratio imposed on water in the soil can significantly enhance the hydration rates (Isaac, 2001). But this effect can be reversed when the additional dissociation step from the surface required prior to the hydration reaction become a significant component in the exchange process. Different components of the same soil might have widely varying adsorptive energies for water resulting in a range of $k_h$ and $\varepsilon_{eq}$ values.

Respired $CO_2$ exit the soil largely via molecular diffusion, resulting in the more rapid escape of the lighter $CO_2$ molecule with an expected kinetic fractionation, $\varepsilon_D$, of 8.8 ‰ between soil $CO_2$ at the soil–air interface and the flux of $CO_2$ into the atmosphere. This theoretical value of 8.8‰ is based on simple kinetic theory and the reduced mass of $CO_2$ and air (Craig, 1953) but has not been measured, and it does not consider any effect of pressure gradients between the soil and the atmosphere. $\varepsilon_D$ can only be fully expressed if $CO_2$ and water are in complete isotopic equilibrium until the soil surface. Since $CO_2$ diffuses out of the soil faster than it can equilibrate with soil water, $CO_2$ remaining in the soil will be enriched in $^{18}O$, relative to its equilibrium value with water, as also predicted for $^{13}C$ in soils (Cerling,

1984; Cerling et al., 1991; Davidson, 1995). This near-surface enrichment diminishes the observed kinetic fractionation, measured relative to $CO_2$ in equilibrium with soil water at lower depths (Figure 6(b)). Global syntheses of $\delta^{18}O$ of $CO_2$ have made simple assumptions regarding the controls on $\delta^{18}O$ of respired $CO_2$. Both Farquhar et al. (1993) and Ciais et al. (1997a) solved versions of Equation (43) for the kinetic fractionation of $CO_2$ diffusing out of soil, $\varepsilon_D$, assuming that all other terms were well known. These studies arrived at different results: 5.0‰ (Ciais et al., 1997a) and 7.6‰ (Farquhar et al., 1993). Miller et al. (1999), using an experimental approach, estimated the effective diffusion discrimination $\varepsilon_{eff} = \varepsilon_D f$, such that $f$ will change as a function of the rates of $CO_2$ hydration and diffusion which are, in turn, functions of the temperature, tortuosity, and the air and water content of the soil. Catalysis of the hydration reaction, e.g., by carbonic anhydrase activity, near the soil surface would render $f$ close to one. As expected, estimates of $\varepsilon_{eff}$ correlated strongly with soil moisture content in the top 5 cm, but never reach $f = 1$ in realistic soil moisture levels. A mean value of $\varepsilon_{eff} = 7.2‰$ was proposed as a robust estimate that represents a wide range of environmental conditions. While the discussion above focuses on $CO_2$ diffusion, $CO_2$ can also leave the soil by advection due to pressure gradient. This aspect is rarely considered, and its isotopic effect was estimated to be small (Stern et al., 1999).

Notably, the $\delta^{18}O$ of soil water affects more than just the isotopic composition of respired $CO_2$. It also affects the isotopic composition of ambient $CO_2$, which makes contact with soil water by diffusing into and out of the soil. Depending upon its residence time in the soil, the diffusing ambient $CO_2$ can partially or fully equilibrate with the soil water. This diffusion–equilibration-retro-diffusion process is termed the "invasion effect" (Tans, 1998) and was demonstrated experimentally by Miller et al. (1999). The impact of this effect on atmospheric $\delta^{18}O$ is defined in Equation (43) as $F_I(\delta_a - \delta_s)/c_a$. This non-biological invasion flux, $F_I$, is a function of the $CO_2$ concentration near the soil surface, $c_a'$ and the speed with which $CO_2$ in the air above the soil diffuses into the soil and equilibrates with soil water, $\nu$ (the "piston velocity"; with values for "typical" soil of ~0.012 cm s$^{-1}$ and effective penetration of invading $CO_2$ of ~3 cm):

$$F_I = c_a' \nu \qquad (46)$$

The piston velocity was defined by Tans (1998) as $\nu = \sqrt{\theta_a B \theta_w \kappa D_{air} k_{h\text{-iso}}}$, where, $\theta_a$ and $\theta_w$ are the volume fractions of air and water in the soil, $B$ is the Bunsen solubility coefficient, $\kappa$ is the soil tortuosity, $D_{air}$ is diffusivity of $CO_2$ in air, and

$k_{h\text{-iso}}$ is the rate constant for the oxygen isotopic equilibration of $CO_2$ and water. Invasion does not distinguish the source of the invading air, original invading background or re-invading soil-respired $CO_2$, but the more invasion there is, the more the $CO_2$ above the soil approaches equilibrium with the $\delta^{18}O$ of the soil water. While in Equation (45) the invasion effect is estimated as a discrete variable, Stern *et al.* (2001) suggested to include it as a correction to $\varepsilon_{eff}$ that would then vary between ~1‰ and ~10‰ among biomes and environmental conditions.

Gillon and Yakir (2001) included the invasion effect in their global $^{18}O$ budget of atmospheric $CO_2$, with relatively small overall effect (<1‰). Stern *et al.* (2001), using simulation models, estimate that while the invasion flux is usually significant, its effect on the $\delta^{18}O$ of atmospheric $CO_2$ is highly variable among different biomes. Generally, if the top several cm of the soil are bone-dry, or if the soil surface is covered by several cm of dry litter, dry sand, or even snow, $\nu$ and $F_I$ will be small because of their dependency on $\theta_w$. The invasion term scales with the difference between $\delta_a$ and $\delta_s$, so soil water with low or high near-surface $\delta^{18}O$ values may magnify the effect of $CO_2$ diffusion into the soil. However, in deserts or other biomes where large differences between $\delta_a$ and $\delta_s$ are accompanied by low water content, the invasion effect likely will be small. In order for a significant amount of ambient $CO_2$ to invade from a given parcel of air, the $CO_2$ residence time in the well-mixed surface layer should be large, and the height of the layer should be small. Usually, if the residence time is not greater than tens of seconds, or if $h > 1$ m, the contribution of invasion to the $\delta^{18}O$ of atmospheric $CO_2$ will be small. In most settings, we would expect this to be the case. One exception may be at night, near the forest floor, where temperature inversions allow respired $CO_2$ to accumulate. Additionally, catalysis of $CO_2$ hydration, such as by the presence of carbonic anhydrase near the soil surface, would greatly accelerate the invasion process. Isaac (2001) reported a new and little explored aspect of the invasion effect in dry soils. Dry soils, especially with high clay content, have 1–7% (w/w) residual water that is not removed even by oven drying at 100 °C, but that still show hydration of $CO_2$ in rates faster than similar quantities in the pure water form. Further, as expected for bound water effect, the observed equilibrium fractionation was smaller than expected for pure water, with the discrepancy increasing inversely to the residual soil water content.

Order of magnitude estimates of soils impact on the $\delta^{18}O$ of atmospheric $CO_2$ can be obtained using Equation (43) by assuming global respiration flux of ~100 Pg, kinetic fractionation of −7.2‰, mean δs values of −7.9 and a global invasion flux of, say, 7 Pg, to yield −2.5‰ yr$^{-1}$ (Figure 9). As discussed below, on global scale, this value is similar in magnitude but opposite in sign to that of the vegetation.

### 4.07.4.2 The Leaf Component

$CO_2$ diffusing into leaves during photosynthesis dissolves and rapidly exchanges its oxygen with water in the chloroplast, a process that depends on catalysis by the enzyme carbonic anhydrase (Reed and Graham, 1981; Silverman, 1982), because of the short residence time of $CO_2$ in leaves (seconds or less). About one-third of the $CO_2$ diffusing into the leaves is fixed in $C_3$ photosynthesis. The remaining two-third diffuse back to the atmosphere after isotopic exchange with chloroplast water. The retro-diffusion $CO_2$ flux is enriched in $^{18}O$ relative to the soil water that feeds the plant because of the evaporative enrichment of leaf water (Gonfiantini *et al.*, 1965). During evaporation, water containing the lighter isotope are preferentially lost and the water remaining in the leaves is enriched in the heavy isotopes (both $^{18}O$ and deuterium; Craig and Gordon, 1965). Because of the normally large ratio of the transpiration flux through the leaf to the leaf water volume, it is usually assumed that leaves are always near isotopic steady state with respect to environmental conditions (but see Harwood *et al.*, 1998, 1999; Wang and Yakir, 1995; Wang *et al.*, 1998). At isotopic steady state, the $^{18}O$ enrichment of leaf water is usually estimated with the widely used evaporative enrichment model of Craig and Gordon (1965), adapted to leaves (e.g., Dongmann *et al.*, 1974; Farquhar and Lloyd, 1993; Flanagan *et al.*, 1991, 1993; Yakir, 1992)

$$\delta_{lw} = \delta_i \varepsilon_{LV} + \varepsilon_K + h(\delta_{vap} - \delta_I - \varepsilon_K) \quad (47)$$

where $h$ is relative humidity at the leaf surface (equivalent to $e_a/e_l$, the vapor pressure ratio in the air and in the leaf), $\varepsilon_{LV}$ the fractionation for liquid to vapor phase transition of water, $\varepsilon_K$ the kinetic fractionation in the diffusion of water vapor across the stomatal cavity and the leaf boundary layer, and subscripts lw, I, and vap representing leaf water, input water from the soil, and atmospheric water vapor, respectively. Note that Equation (47) does not estimate the isotopic composition of bulk leaf water, which is a complex system that involves marked isotopic heterogeneity (Helliker and Ehleringer, 2000; Roden and Ehleringer, 1999; White, 1988; Yakir, 1998). Rather, it is assumed that $\delta_{lw}$ represents the isotopic enrichment of water at the evaporating surfaces inside the leaves (Figure 5), and that the chloroplasts are always near these surfaces to facilitate gas exchange.

Carbonic anhydrase, in turn, is located predominantly in the chloroplasts, and $\delta_{lw}$ provides therefore a reasonable estimate of the specific $\delta^{18}O$ of water fraction with which $CO_2$ equilibrate in the leaves, and $\delta_l$ used in Equation (43) will simply be $\delta_l = \delta_{lw} + \varepsilon_{eq}$, where $\varepsilon_{eq}$ is the water–$CO_2$ equilibrium fractionation. In diffusing out of the leaf, a diffusional fractionation, $\varepsilon_l$ must also be considered, such that the final $\delta^{18}O$ value of the $CO_2$ reaching the atmosphere is $\delta_l + \varepsilon_l$. Note that considerable uncertainty is still associated with estimating the $\delta^{18}O$ value of leaf water on large scale and current estimates ranged between 4.4‰ and ~8.8‰ (Bender et al., 1985, 1994; Dongmann, 1974; Förstel, 1978).

Consider that the net effect of the leaf flux on atmospheric $^{18}O$, i.e., the term $(\delta_l - \delta_a + \varepsilon_l)$ in Equation (43), represents discrimination against $^{18}O$. As $CO_2$ is removed from the atmosphere by the photosynthesis flux, $F_A$, the remaining $CO_2$ in the atmosphere becomes enriched in $^{18}O$ due to the leaf water effect. Unlike the case for $^{13}C$, the apparent discrimination, $^{18}\Delta = (\delta_l - \delta_a + \varepsilon_l)$, is not reflected in the $\delta^{18}O$ of plant organic matter (rather, the access $^{16}O$ is washed away by water), but it can be treated mathematical in the same way (Farquhar and Lloyd, 1993) according to

$$^{18}\Delta = \frac{R_a}{R_l} - 1 = \frac{\bar{a} + \xi' \cdot (\delta_l - \delta_a)}{1 - \xi' \cdot (\delta_l - \delta_a)/1,000}$$
$$\approx \bar{a} + \xi' \cdot (\delta_l - \delta_a) \qquad (48)$$

where $R_a$ and $R_A$ are the oxygen isotope ratios of $CO_2$ in the air and in the $CO_2$ flux taken up by assimilation, $\bar{a}$ is the weighted average fractionation during diffusion of $CO_2$ from the atmosphere to the chloroplast (similar to $\varepsilon_l$ above with values of 8.8‰ in stagnant air and 0.8‰ in solution), $\xi' = c_l/(c_a - c_l)$ and $c_a$ and $c_l$ are the $CO_2$ concentrations in the atmosphere and in the site of oxygen exchange between $CO_2$ and water in the leaves, respectively ($\xi'$ represents the retro-diffusion flux back to the atmosphere after $^{18}O$ exchange with leaf water), and $\delta_a$ and $\delta_l$ are the $\delta^{18}O$ values of $CO_2$ in the atmosphere and in equilibrium with chloroplast water, respectively.

Estimates of $^{18}\Delta$ using Equation (48) rely on the assumption of complete isotopic equilibrium between $CO_2$ and water in the leaf chloroplasts, due to the action of carbonic anhydrase. This assumption has recently been tested, and revised (Gillon and Yakir, 2001, 2000a,b). The extent of $CO_2$–$H_2O$ equilibrium may be derived from Mills and Urey (1940) as

$$\theta_{eq} = 1 - e^{-k\tau/3} \qquad (49)$$

which describes the fractional approach to full equilibrium (where $\theta_{eq} = 1$), as a function of the number of hydration reactions achieved per $CO_2$

molecule, $k\tau$ (divided by 3 for the isotopic exchange). This "coefficient of $CO_2$ hydration" is estimated for a leaf by calculating the rate constant $k$ from biochemical measurements of carbonic anhydrase activity, and the residence time of $CO_2$ in the leaf, $\tau$, from leaf-scale photosynthetic flux measurements of $CO_2$ (Gillon and Yakir, 2000a,b). A survey of $\theta_{eq}$ in a wide range of plant species showed large variations in the activity of carbonic anhydrase among major plant groups that cause variations in $\theta_{eq}$ ($\theta_{eq} = 1$ reflect full equilibrium). On average, $\theta_{eq}$ was 0.93 for C3 trees, 0.70 for C3 grasses and 0.38 for C4 plants, with a global weighted mean of $\theta_{eq} = 0.78$. Put simply, this indicates that, in contrast to earlier assumptions, only ~80% of the diffusional $CO_2$ backflux from plants to the atmosphere reflects the leaf water signal relevant to atmospheric $C^{18}OO$ budgets. Ignoring such incomplete equilibrium could result in ~20% underestimation of the gross $CO_2$ exchange with plants derived from atmospheric measurements of $\delta^{18}O$ values (Ciais et al., 1997b; Farquhar et al., 1993; Francey and Tans, 1987; Yakir and Wang, 1996).

Accordingly, the model depicted in Equation (48) was extended to allow for these variations:

$$^{18}\Delta = \frac{\bar{a} + \varepsilon\Big(\theta_{eq}(\delta_l - \delta_a) - (1 - \theta_{eq})\bar{a}/(\xi' + 1)\Big)}{1 - \varepsilon\Big(\theta_{eq}(\delta_l - \delta_a) - (1 - \theta_{eq})\bar{a}/(\xi' + 1)\Big)/1,000}$$
$$\approx \bar{a} + \xi'[\theta_{eq}(\delta_l - \delta_a) - (1 - \theta_{eq})\bar{a}/(\xi' + 1)] \qquad (50)$$

where $\theta_{eq}$ represents the extent of $^{18}O$ equilibrium between $CO_2$ and water in leaves. The extent of equilibrium, $\theta_{eq}$, can be estimated from direct measurements of carbonic anhydrase activity in leaves, or using leaf-scale measurements of $^{18}\Delta$ and solving Equation (50) for $\theta_{eq}$ (Gillon and Yakir, 2000b). Farquhar and Lloyd (1993) also proposed an indirect estimate of the extent of equilibrium related to the ratio, $\rho$, of Rubisco to carbonic anhydrase activities.

Note that Equations (48) and (50) can also be solved for $\delta_l$ and the $\delta^{18}O$ of leaf water. In this case, Equation (50) will yield a higher $\delta_l$ value than Equation (48). On a global scale, Gillon and Yakir (2001) estimated $\theta_{eq}$ to be 0.8, which would increase global estimates of global mean $\delta^{18}O$ value of leaf water of ~+4.4‰ (Bender et al., 1994; Farquhar et al., 1993) by ~2‰, providing better consistency with estimates based on studies of the $\delta^{18}O$ of atmospheric $O_2$ (see also Beerling, 1999).

Equation (50) indicates that $^{18}\Delta$ is regulated in leaves by three main factors; the isotopic composition of water (as reflected in $\delta_l$), the $CO_2$ concentration at the $CO_2$–$H_2O$ exchange sites ($c_l$), as well as by the extent of isotopic equilibrium between $CO_2$ and $H_2O$ ($\theta_{eq}$).

The recent estimate of global mean $\theta_{eq}$ of $\theta_{eq} \approx 0.8$ (Gillon and Yakir, 2001) reflected low $\theta_{eq}$ in grasses and C4 plants ($\theta_{eq} \approx 0.4$) and high $\theta_{eq}$ in non-grasses C3 plants ($\theta_{eq} > 0.9$). An important source of uncertainty is the $CO_2$ concentration inside leaves ($c_l$), which cannot be directly measured. Recent studies of $c_l$ also provide another demonstration of the implications of processes at cellular and sub-cellular scales to large-scale studies using stable isotopes. Leaf-scale measurements indicated that the $CO_2$ draw-down from the sub-stomata cavities ($c_i$, readily estimated in leaf-scale measurements and often shows $c_i/c_a$ ratios of 0.6–0.7) to the chloroplast, $c_c$, generally yields $(c_i - c_c)/c_a \approx 0.2$ and $c_c/c_a \approx 0.55$, (Epron *et al.*, 1995; Lloyd *et al.*, 1992; Loreto *et al.*, 1992; Von Caemmerer and Evans, 1991). In C4 plants internal $CO_2$ concentrations are much lower ($c_i/c_a$ is typically ~0.3–0.4 with further drawdown to the mesophyll cells where $CO_2$ is fixed by the enzyme PEP-carboxylase). Such estimates of $c_c$ rely mostly on $^{13}C$ discrimination by leaves and represent the $CO_2$ concentrations at the site of Rubisco, the primary photosynthetic enzyme, in the chloroplast. But $CO_2$ hydration is controlled by carbonic anhydrase, and the relevant site of $CO_2$–$H_2O$ equilibration is at the limit of its activity, the chloroplast or cellular membrane. The $CO_2$ concentrations at this point, $c_l$, are distinct from those at the Rubisco site, $c_c$, by 0–40 ppm, (depending on plant species and rates of activities) due to large internal resistances to $CO_2$ diffusion. It was generally observed that the $CO_2$ concentration at the site of $CO_2$–$H_2O$ equilibrium in leaves ($c_l$) was at about mid-way between $c_i$ and $c_c$, $(c_i - c_l)/c_a \approx 0.1$, significantly influencing estimates of $^{18}\Delta$ using Equation (50).

Values of $^{18}\Delta$ are highly variable among plant species and environmental conditions. Modeled $^{18}\Delta$ values for different biomes can range from $-20‰$ to $+32‰$ (Farquhar *et al.*, 1993). Only few comparisons of such modeled $^{18}\Delta$ values were compared with actual, measured ecosystem-scale $^{18}\Delta$. Making such comparison in the Amazon basin (cf. Sternberg *et al.*, 1998) indicated that further comparisons with other ecosystems are necessary. Nevertheless, $^{18}O$ in $CO_2$ is potentially powerful tracer of $CO_2$ fluxes, distinct from $^{13}C$ and influenced by different process. The $^{18}O$ signal in $CO_2$ is comparable in magnitude to that of $^{13}C$ on a global scale (Figures 3 and 9), and often larger on a local scale. Using global mean estimates of $^{18}\Delta = 13.7‰$ (Farquhar *et al.*, 1993), GPP of 120 Pg C y$^{-1}$ and atmospheric carbon pool of ~750 Pg (IPCC), an order of magnitude estimate of the annual scale land vegetation impact on the $\delta^{18}O$ of atmospheric $CO_2$ can be obtained from, $F_A{}^{18}\Delta/c_a = (120 \times 13.7)/730 = +2.2‰$ yr$^{-1}$ (Figure 9).

### 4.07.4.3 The Minor Components

#### 4.07.4.3.1 The ocean component

The Ocean has an enrichment effect on the $\delta^{18}O$ of atmospheric $CO_2$. This is the result of two opposing effects: first, the variations in the $\delta^{18}O$ of ocean water ($\delta_w$), and second, temperature effects on the $CO_2$–$H_2O$ equilibrium fractionation.

Mean ocean water $\delta^{18}O$, $\delta_w$, between 60° N and 60° S, is near 0‰, but there are spatial variations due to salinity/fresh water effects. $\delta_w$ decreases where large amounts of freshwater are delivered to the ocean, because of the isotopic fractionation in the hydrological cycle (global mean precipitation is ~$-7.9‰$, polar ice ~$-25‰$). In particular, $\delta_w$ is depleted by ~1‰ at high latitudes around Antarctica and Greenland because of the massive discharge of icebergs, and near estuaries of large rivers.

In contrast, the isotopic equilibrium fractionation between $CO_2$ and water is inversely correlated with temperature (see above) and has the effect of increasing $\delta^{18}O$ of $CO_2$ in equilibrium with ocean water at high latitudes. The overall results of the opposing salinity and temperature effects is an increase in $\delta^{18}O$ of $CO_2$ in equilibrium with ocean water ($\delta_o$) as a function of latitude from around 0‰ in the tropics (PDB-$CO_2$ scale, see Introduction), to a maximum value around 5‰ near sea ice margins (Ciais *et al.*, 1997a; Ishizawa *et al.*, 2002). Note that the fractionation associated with diffusion and hydration across the sea–atmosphere interface is small, $\varepsilon_w = 0.8‰$. Global mean $\delta^{18}O$ of $CO_2$ in equilibrium with ocean water is ~$+2‰$, the ocean isotopic forcing on $\delta_a$, using Equation (43), and assuming ~90 Pg C yr$^{-1}$ and 88 Pg C yr$^{-1}$ gross ocean uptake and release fluxes (IPCC, 2001), the ocean impact on $\delta_a$ is ~$+0.2‰$ yr$^{-1}$ (Figure 9).

#### 4.07.4.3.2 Anthropogenic emissions

Both fossil fuels and burned biomass are combusted with atmospheric oxygen, and it is generally assumed that the resulting $CO_2$ acquire its $\delta^{18}O$ value of $-17.3‰$ PDB-$CO_2$ ($+23.5$ SMOW; Blunier *et al.*, 2002). Note that in both cases the one-way surface to atmosphere fluxes are considered for their unique $^{18}O$ effect on the atmosphere. Net biomass burning, and deforestation, fluxes include $CO_2$ uptake through re-growth that largely balance the emission. So while only the net loss of carbon contribute to changes in atmospheric $\delta^{13}C$ and $c_a$, the gross burning rate with its unique $^{18}O$ signal constitute the main effect on atmospheric $\delta^{18}O$ (for $^{18}O$, re-growth can be considered as part of the land vegetation GPP; Gillon and Yakir (2001)). Assuming an

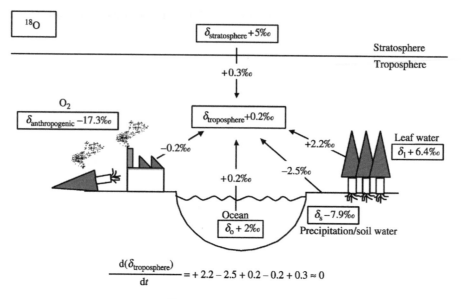

**Figure 9** Annual mean "forcings" on the $\delta^{18}O$ value of atmospheric $CO_2$. Each specific forcing component is the product of the flux involved, $F$, diluted by the atmospheric carbon pool, $C_a$, and the difference in $\delta^{18}O$ values between that of the flux and the mean atmosphere (i.e., forcing $= [(F/C_a)(\delta_x - \delta_a)]$. Values are order of magnitude estimates, as discussed in the text, and involve significant uncertainties. Typical values for the reservoirs involved are given in boxes and reflect the $\delta^{18}O$ value of the water in that reservoir, accept for the anthropogenic reservoir that is represented by the $\delta^{18}O$ value of atmospheric oxygen. The land biosphere has the strongest forcing, with contrasting effects of photosynthesis and respiration. Land photosynthesis tends to enrich the atmosphere in $^{18}O$ and soil respiration has the opposite effect, providing a basis for partitioning photosynthesis and respiration fluxes. The ocean forcing is an order of magnitude smaller. Fossil fuel emission tends to decrease $^{18}O$, but the effect is "washed away" by the massive exchange of oxygen in $CO_2$ with water and long-term trend in atmospheric $\delta^{18}O$-$CO_2$ is not usually observed.

anthropogenic flux (fossil fuels plus combusted biomass) of $\sim 9.4$ Pg C yr$^{-1}$ (IPCC, 2001), the order of magnitude of the fossil fuel forcing on the $\delta^{18}O$ of atmospheric $CO_2$, is $\sim -0.2\permil$ yr$^{-1}$ (Figure 9).

### 4.07.4.3.3 Troposphere–Stratosphere exchange

Stratospheric $CO_2$ is known to posses an unusual isotopic composition, enriched in both $^{18}O$ and $^{17}O$ (Gamo *et al.*, 1989; Thiemens, 1999; Thiemens *et al.*, 1991; Wen and Thiemens, 1993; Yung *et al.*, 1991). $^{18}O$ enrichment of $\sim 2\permil$ is observed in $CO_2$ at altitudes of 20–25 km, and of more than $10\permil$ at altitudes of 30–35 km (Gamo *et al.*, 1989; Thiemens and Jackson, 1990). This enrichment is likely to pass on from ozone $^{18}O$ enrichment. It is based on the formation of a short-lived $CO_3$ transition state, which is created by the interaction of $CO_2$ with electronically exited $O(^1D)$, derived from photolysis of $O_3$ in the stratosphere. The $CO_3$ intermediate breaks apart to $CO_2$ randomly losing oxygen, which is processed back to $O_2$. This mechanism therefore effectively transfers the $^{18}O$ (and $^{17}O$) enrichment from ozone to $CO_2$.

With respect to the troposphere, the stratospheric $^{18}O$ enrichments can be regarded as "invasion effect" (similar to that considered for soils). While there is no net flux involved, it depends on the gross exchange flux of $CO_2$ in and out of the stratosphere and the $^{18}O$ fractionation that $CO_2$ undergoes in the stratosphere, associated with the $O_3$–$CO_2$ oxygen exchange. The troposphere–stratosphere $CO_2$ exchange flux can be estimated from Appenzeller *et al.* (1996) to be $\sim 100$ Pg C yr$^{-1}$. The $^{18}O$ fractionation is not known at present but from recent estimates of model runs can be assumed to be small (Blunier *et al.*, 2002). Alternatively, Peylin *et al.* (1996) used the observed altitudinal gradient in the $\delta^{18}O$ of $CO_2$ from near $0\permil$ (PDB-$CO_2$ scale), to $\sim 2\permil$ at 19 km (Gamo *et al.*, 1989) and $\sim 10\permil$ at 35 km (Thiemens and Jackson, 1990), or stratospheric average of, say, $+5\permil$). Fitting this gradient, the overall stratospheric effect on the annual troposphere $\delta^{18}O$ budget was $\sim +0.3\permil$.

### 4.07.4.4 Spatial and Temporal Patterns

Although some measurements were available as early as the 1970s and before (see Francey *et al.*, 1999 for a recent summary of stable isotope air sampling efforts), large-scale, high precision measurements of the $\delta^{18}O$ of atmospheric $CO_2$ picked up in the early 1990s, carried out by centers such as the NOAA-CU, Scripps-SIO, Groningen University-CIO, and CSRIO. These measurements show several prominent features. First, the long-term global mean $\delta^{18}O$ of atmospheric $CO_2$ is near steady state (i.e., $d(\delta^{18}O)/dt = 0$). Notably, however, there were periods with a significant temporal trends, such as the $>0.5‰$ decreasing trend during the 1990s. Second, a large, nearly $2‰$, Arctic–Antarctic gradient is observed in annual mean values of the $\delta^{18}O$ of atmospheric $CO_2$ (Ciais et al., 1997b). Third, a similarly large (ca. $1.5‰$) seasonal cycle is observed at high northern latitudes, with a maximum in early summer and a minimum in early winter. Fourth, the seasonal $^{18}O$ cycle is clearly off phase (by about a month or more) with respect to the seasonal cycle in $CO_2$ concentrations, and in $^{13}C$ (Figure 2). Most of these features have now been modeled, initially with simple models (Farquhar et al., 1993; Francey and Tans, 1987; Ishizawa et al., 2002) and then with GCMs and 3D transport models coupled to biosphere models such as Sib2 (Ciais et al., 1997a,b; Ishizawa et al., 2002; Peylin et al., 1999; 1996). Although significant uncertainties remain in many of the models parameters, the reasonable agreement of model output and observations indicates that some of the major processes underlying the observations are captured.

The strong global $^{18}O$ gradient, dominated by $\sim{-}1.2‰$ gradient in the northern hemisphere, between Mauna Loa and Pt. Barrow, was noted by Francey and Tans (1987). They pointed out that a large northern hemisphere isotope exchange flux, in the order of $200\,Pg\,C\,yr^{-1}$, is required to maintain such gradient against the vigorous mixing of the atmosphere. Such flux is consistent with the retro-diffusion flux from leaves during photosynthesis (Figure 5), complemented by $CO_2$ exchange with soil water. Combined, these fluxes largely reflect the isotopic gradient in precipitation water on land. This early notion has remained a valid basis for the observed meridional gradient. Later refinements indicated that it is, in fact, the respiration of the biosphere, mostly above $30°$ N, that is the dominant control over the mean latitudinal gradient (Ciais et al., 1997b; Farquhar et al., 1993; Peylin et al., 1999).

Seasonality in the land biosphere is a well known phenomenon and it is clearly reflected in the $CO_2$ and $^{13}C$ records, with a minimum in $CO_2$ concentration and maximum in $^{13}C$, nearly coinciding with peak net ecosystem $CO_2$ uptake associated with peak photosynthetic activity (Figures 2(a) and (b); Trolier et al., 1996). Early studies, such as that of Francey and Tans, were not able to clearly detect a seasonal

cycle in atmospheric $^{18}O$, although it would clearly be consistent with $^{18}O$ exchange with the terrestrial biosphere. With increasing precision and frequency of measurements, a strong seasonal cycle in $^{18}O$ became apparent (Figure 2(c); Cuntz et al., 2002; Flanagan et al., 1997; Peylin et al., 1999), which also showed inter-hemispheric gradient in its amplitude. While in the southern hemisphere seasonality in $^{18}O$ was barely detected ($\sim0.2‰$), it reaches amplitude of ca. $1.5‰$, and in some cases more than $2‰$, in mid- and high-latitudes in the northern hemisphere, consistent with the distribution of land ecosystem activities around the globe (Peylin et al., 1999; Cuntz et al., 2002). Peylin et al. (1999), examining the factors influencing the phase and amplitude of the $\delta^{18}O$ seasonal cycle, observed that the large seasonal cycle at high latitudes is mainly due to the respiratory fluxes (with its negative $\delta^{18}O$ values) of all extra-tropical ecosystems.

In contrast with seasonality in $c_a$, which is clearly dominated by net ecosystem $CO_2$ exchange (the relative contributions of photosynthesis and respiration), seasonality in the $\delta^{18}O$ of $CO_2$ is more complex. It couples variations in $CO_2$ fluxes with variations in the $\delta^{18}O$ of precipitation, in leaf discrimination and in soil discrimination. This coupled effect can be best discussed in terms of isofluxes or isotopic "forcing" (forcing = isoflux/atm. pool) on the atmospheric $^{18}O$ of the gross $CO_2$ flux and the $^{18}O$ discrimination associated with it (i.e., Pg C‰; cf. Figure 7). Local $\delta_a$ is influenced by the opposing effects of the generally $^{18}O$ enriched photosynthesis isoflux (reflecting $^{18}O$ enrichment of leaf water), and the invariably depleted soil isoflux (reflecting $^{18}O$ depleted soil water) ($F_A\,^{18}\Delta_A + F_{sa}(\delta_{sa} - \delta_a)$). The simulation results of Peylin et al. (1999) indicated in simple terms (Figure 10) that the soil forcing is generally greater than that of photosynthesis, and therefore dominates the seasonal cycle. The possible reason for the relatively small photosynthetic isoflux in many locations is that while soil water is strongly depleted with increasing latitude, concurrent decrease in relative humidity and temperature result also in strong leaf water enrichments, diminishing the difference between the $\delta^{18}O$ of $CO_2$ equilibrated with leaf water and that of the atmosphere, reducing leaf discrimination ($^{18}\Delta_A$, Equations (48) and (50)). The simulation results demonstrate two additional interesting features associated with the seasonal cycle. First, ecosystems with similar net ecosystem exchange $CO_2$ flux can have very different effect on the $\delta^{18}O$ of atmospheric $CO_2$. For example, more continental Siberian ecosystems receive more depleted precipitation, as compared, for example, with Canadian ecosystems (3–7‰ difference). With similar net ecosystem $CO_2$ exchange and leaf water enrichments during the growing season, the Siberian ecosystems will have a much larger soil isoflux during the growing season. Second, due to efficient lower troposphere mixing, associated with weak vertical transport at extra-tropical latitudes, local $\delta_a$ is often dominated by distant ecosystems with large isoflux (but not necessarily large $CO_2$ flux). Such is the case of the disproportionate effects of the Northern Siberian taiga on regional and the global seasonal $^{18}O$ cycle (but not on that in $CO_2$). While most northern hemisphere ecosystems have seasonal amplitude in isoflux of ca. 10Pg C‰, the Northern Siberian taiga has ~20 Pg C‰. Note, therefore, that there is unique relationship between the isoflux and the $CO_2$ flux at any given location (Figure 9 and 10).

The phase of the $^{18}O$ seasonal cycle also reflects the interplay between the photosynthetic and the soil isoflux, and not the seasonality in net ecosystem $CO_2$ exchange. Leaf discrimination generally peaks in the spring due to high leaf water enrichment and high leaf internal $CO_2$ concentration (i.e., high retrodiffusion flux from the leaves to the atmosphere). During the summer, leaf discrimination decreases, often because

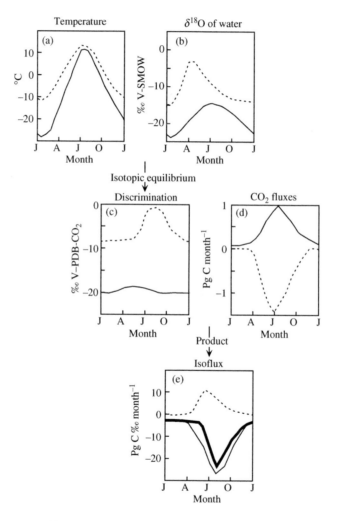

**Figure 10** Development of the $^{18}O$ seasonal cycle in a high latitude forest biome. Water is enriched in $^{18}O$ due to evaporation under decreasing relative humidity in the spring and summer (top right panel; broken and solid lines refer to leaves and soil, respectively). This is transuded to $CO_2$ via isotopic exchange the magnitude of which is inversely correlated with temperature (top left panel). Combined, water $^{18}O$ content and temperature-sensitive isotopic exchange, are the major controls over the apparent $^{18}O$ discrimination associated with soil and leaves (middle left panel). The effect on atmospheric $CO_2$ is weighted by the $CO_2$ fluxes involved (middle right panel; Thick solid line is the combined, net, effect). The flux carrying the $^{18}O$ signature is termed isoflux; using isotopic ratios would yield actual $^{18}O$ flux, using, as often done, $\delta^{18}O$ values provide good indicators of the $^{18}O$ flux. Note that isoflux taken up by leaf photosynthesis (negative flux in middle right panel) actually enrich the atmosphere in $^{18}O$ (bottom panel). The interacting effects of the seasonal patterns in evaporating enrichment, temperature and flux result in a negative $^{18}O$ signature of the biome on the atmosphere that reaches its seasonal peak about one months later that the peak in the actual flux (after Peylin *et al.*, 1999).

reduced leaf water enrichment associated with increasing relative humidity, and because of lower leaf internal $CO_2$ concentrations. As a result, leaf isofluxes show a weak dominance early in the growing season, bringing about a seasonal $^{18}O$ peak in early summer. As the leaf isoflux slowly diminishes, the respiration isoflux, that has lagged photosynthesis due to the slow increase in soil temperature, takes over bringing about a seasonal minimum in late summer/early fall. Notably, this minimum, dominating the seasonal cycle, arrives about one month after the minima in $CO_2$ concentrations associated with the seasonal peak photosynthesis. Here too, dominant regions, such as Siberia with their large soil isoflux, have disproportionate effect in controlling the phase of the northern hemisphere seasonal cycle and the timing of its minimum (Peylin *et al.*, 1999). Detailed understanding of the processes underlying the $^{18}O$ pattern discussed above, should allow using "inverse techniques," and the $\delta^{18}O$ and $CO_2$ data from global sampling networks to derive separately values for assimilation and respiration fluxes of large ecosystems.

## ACKNOWLEDGMENTS

This work would not have been completed without the help of Hagit Affek in preparing the figures and the literature list. John Miller, Leo Sternberg, and Ralph Keeling provided helpful comments. Unpublished data were kindly provided by Peter Bakwin and NOAA/CMDL, and Joe Berry.

## REFERENCES

Allison C. E., Francey R. J., and Meijer H. A. (1995) Recommendations for the reporting of stable isotope measurements of carbon and oxygen in $CO_2$ gas. In *References and Intercomparison Materials for Stable Isotopes of Light Elements, Proceeding of a Consultation Meeting Held in Vienna, December 1–3, 1993.* IAEA, pp. 155–162.

Allison G. B. and Barnes C. J. (1983) Estimation of Evaporation from non-vegetated surfaces using natural deuterium. *Nature* **301**(5896), 143–145.

Allison G. B., Barnes C. J., and Hughes M. W. (1983) The distribution of deuterium and $^{18}O$ in dry soils: 2. Experimental. *J. Hydrol.* **64**(1–4), 377–397.

Amundson R., Stern L., Baisden T., and Wang Y. (1998) The isotopic composition of soil and soil-respired $CO_2$. *Geoderma* **82**(1–3), 83–114.

Andres R. J., Marland G., Boden T., and Bischof S. (2000) Carbon dioxide emissions from fossil fuel consumption and cement manufacture, 1751–1991, and an estimate of their isotopic composition and latitudinal distribution. In *The Carbon Cycle* (eds. T. M. L. Wigley and D. S. Schimel). Cambridge University Press, pp. 53–62.

Appenzeller C., Holton J. R., and Rosenlof K. H. (1996) Seasonal variation of mass transport across the tropopause. *J. Geophys. Res. Atmos.* **101**(D10), 15071–15078.

Bakwin P., Zhao C., Ussler W., III, Tans P. P., and Quesnell E. (1995) Measurements of carbon dioxide on a very tall tower. *Tellus* **47B**, 535–549.

Bakwin P. S., Tans P. P., White J. W. C., and Andres R. J. (1998a) Determination of the isotopic ($^{13}C/^{12}C$) discrimination by terrestrial biology from a global network of observations. *Global Biogeochem. Cycles* **12**(3), 555–562.

Bakwin P. S., Tans P. P., Hurst D. F., and Zhao C. (1998b) Measurements of carbon dioxide on very tall towers: results of the NOAA/CMDL program. *Tellus* **50B**, 401–415.

Baldocchi D. (1997) Measuring and modelling carbon dioxide and water vapor exchange over a temperate broad-leaved forest during the 1995 summer drought. *Plant Cell Environ.* **20**(9), 1108–1122.

Baldocchi D. D., Hicks B. B., and Meyers T. P. (1988) Measuring biosphere-atmosphere exchanges of biologically related gases with micrometeorological methods. *Ecology* **69**(5), 1331–1340.

Balesdent J. and Mariotti A. (1996) Measurement of soil organic matter turnover using $^{13}C$ natural abundance. In *Mass Spectrometry of Soils* (eds. T. W. Boutton and S. I. Yamasaki). Dekker.

Barnes C. J. and Allison G. B. (1983) The distribution of deuterium and $^{18}O$ in dry soils: 1. Theory. *J. Hydrol.* **60**(1–4), 141–156.

Battle M., Bender M. L., Tans P. P., White J. W. C., Ellis J. T., Conway T., and Francey R. J. (2000) Global carbon sinks and their variability inferred from atmospheric $O_2$ and $\delta^{13}C$. *Science* **287**(5462), 2467–2470.

Beerling D. J. (1999) The influence of vegetation activity on the dole effect and its implications for changes in biospheric productivity in the mid-holocene. *Proc. Roy. Soc. of London Ser. B-biol. Sci.* **266**(1419), 627–632.

Beerling D. J. and Woodward F. I. (1998) Modelling changes in the plant function over the phanerozoic. In *Stable Isotopes: Integration of Biological, Ecological and Geochemical Processes* (ed. H. Griffiths). BIOS Scientific Publishers Limited, pp. 347–361.

Bender M., Labeyrie L. D., Raynaud D., and Lorius C. (1985) Isotopic composition of atmospheric $O_2$ in ice linked with deglaciation and global primary productivity. *Nature* **318**(6044), 349–352.

Bender M., Sowers T., and Labeyrie L. (1994) The dole effect and its variations during the last 130,000 years as measured in the Vostok ice core. *Global Biogeochem. Cycles* **8**(3), 363–376.

Berry J. A. (1988) Studied of mechanisms affecting the fractionation of carbon isotopes in photosynthesis. In *Stable Isotopes in Ecological Research*, vol. 68 (eds. P. W. Rundel, J. R. Ehleringer, and K. A. Nagy). Springer, pp. 82–94.

Bigeleisen J. (1965) Chemistry of isotopes. *Science* **147**(3657), 463–471.

Bigeleisen J. and Mayer M. (1947) Calculation of equilibrium constants for isotopic exchange reactions. *J. Chem. Phys.* **15**, 261–267.

Bigeleisen J. and Wolfsberg M. (1958) Theoretical and experimental aspects of isotope effects in chemical kinetics. *Adv. Chem. Phys.* **1**, 15–76.

Bird M. I., Chivas A. R., and Head J. (1996) A latitudinal gradient in carbon turnover times in forest soils. *Nature* **381**(6578), 143–146.

Blunier T., Barnett B., Bender M. L., and Hendricks M. B. (2002) Biological oxygen productivity during the last 60,000 years from triple oxygen isotope measurements. *Global Biogeochem. Cycles* **16**(3), article no. 1029.

Bolin B. (1981) *Carbon Cycle Modelling*. SCOPE 16. Wiley.

Bottinga Y. and Craig H. (1969) Oxygen isotope fractionation between $CO_2$ and water and isotopic composition of marine atmospheric $CO_2$. *Earth Planet. Sci. Lett.* **5**(5) 285-and.

Bousquet P., Peylin P., Ciais P., Ramonet M., and Monfray P. (1999) Inverse modelling of annual atmospheric $CO_2$ sources and sinks: 2. Sensitivity study. *J. Geophys. Res. Atmos.* **104**(D21), 26179–26193.

Boutton T. W. and Yamasaki S. (eds.) (1996) *Mass Spectrometry* of soils Dekker, New York.

Bowling D. R., Turnipseed A. A., Delany A. C., Baldocchi D. D., Greenberg J. P., and Monson R. K. (1998) The use of relaxed eddy accumulation to measure biosphere-atmosphere exchange of isoprene and of other biological trace gasses. *Oecologia* **116**(3), 306–315.

Bowling D. R., Baldocchi D. D., and Monson R. K. (1999a) Dynamics of isotopic exchange of carbon dioxide in a Tennessee deciduous forest. *Global Biogeochem. Cycles* **13**(4), 903–922.

Bowling D. R., Delany A. C., Turnipseed A. A., Baldocchi D. D., and Monson R. K. (1999b) Modification of the relaxed eddy accumulation technique to maximize measured scalar mixing ratio differences in updrafts and downdrafts. *J. Geophys. Res. Atmos.* **104**(D8), 9121–9133.

Bowling D. R., Tans P. P., and Monson R. K. (2001) Partitioning net ecosystem carbon exchange with isotopic fluxes of CO$_2$. *Global Change Biol.* **7**(2), 127–145.

Bowling D. R., McDowell N. G., Bond B. J., Law B. E., and Ehleringer J. R. (2002) $^{13}$C content of ecosystem respiration is linked to precipitation and vapor pressure deficit. *Oecologia* **131**(1), 113–124.

Brenninkmeijer C. A. M., Kraft P., and Mook W. G. (1983) Oxygen isotope fractionation between CO$_2$ and H$_2$O. *Isotope Geosci.* **1**(2), 181–190.

Broadmeadow M. S. J. and Griffiths H. (1993) Carbon isotope discrimination and the coupling of CO$_2$ fluxes within canopies. In *Stable Isotopes and Plant Carbon-water Relations* (eds. J. R. Ehleringer, A. E. Hall, and G. D. Farquhar). Academic Press, pp. 109–130.

Broecker W. S. and Maier-Reimer E. (1992) The influence of air and sea exchange on the carbon isotope distribution in the sea. *Global Biogeochem. Cycles* **6**(3), 315–320.

Broecker W. S. and Peng T. H. (1974) Tracers in the Sea.

Broecker W. S., Takahashi T., Simpson H. J., and Peng T. H. (1979) Fate of fossil-fuel carbon-dioxide and the global carbon budget. *Science* **206**(4417), 409–418.

Brugnoli E., Scartazza A., Lauteri M., Monteverdi M. C., and Máguas C. (1998) Carbon isotope discrimination in structural and non-structural carbohydrates in relation to productivity and adaptation to unfavorable conditions. In *Stable Isotopes: Integration of Biological, Ecological and Geochemical Processes* (ed. H. Griffiths). BIOS Scientific Publishers, pp. 133–146.

Buchmann N. and Ehleringer J. R. (1998) CO$_2$ concentration profiles, and carbon and oxygen isotopes in C$_3$, and C$_4$ crop canopies. *Agri. Forest Meteorol.* **89**(1), 45–58.

Buchmann N. and Kaplan J. O. (2001) Carbon isotope discrimination of terrestrial ecosystems—how well do observed and modeled results match? In *Global Biogeochemical Cycles in the Climate System* (eds. E. D. Schulze, M. Heimann, S. Harrison, E. Holland, J. Lloyd, I. C. Prentice, and D. Schimel). Academic Press, pp. 253–266.

Buchmann N., Guehl J. M., Barigah T. S., and Ehleringer J. R. (1997a) Interseasonal comparison of CO$_2$ concentrations, isotopic composition, and carbon dynamics in an Amazonian Rainforest (French Guiana). *Oecologia* **110**(1), 120–131.

Buchmann N., Kao W. Y., and Ehleringer J. (1997b) Influence of stand structure on Ccarbon-13 of vegetation, soils, and canopy air within deciduous and Evergreen Forests in Utah, US. *Oecologia* **110**(1), 109–119.

Buchmann N., Brooks J. R., Flanagan L. B., and Ehleringer J. (1998) Carbon isotope discrimination of terrestrial ecosystems. In *Stable Isotopes: Integration of Biological, Ecological, and Geochemical Processes* (ed. H. Griffiths). BIOS Scientific Publishers, pp. 203–222.

Businger J. A. and Oncley S. P. (1990) Flux measurement with conditional sampling. *J. Atmos. Ocean. Technol.* **7**(2), 349–352.

Cellier P. and Brunet Y. (1992) Flux gradient relationships above tall plant canopies. *Agri. Forest Meteorol.* **58**(1–2), 93–117.

Cerling T. E. (1984) The stable isotopic composition of modern soil carbonate and its relationship to climate. *Earth Planet. Sci. Lett.* **71**(2), 229–240.

Cerling T. E. and Wang Y. (1996) Stable carbon and oxygen isotopes in soil CO$_2$ and soil carbonate: theory, practice, and application to some prairie soils of upper Midwestern North America. In *Mass Spectrometry of Soils* (eds. T. W. Boutton and S. I. Yamasaki). Dekker, pp. 113–131.

Cerling T. E., Solomon D. K., Quade J., and Bowman J. R. (1991) On the isotopic composition of carbon in soil carbon-dioxides. *Geochim. Cosmochim. Acta* **55**(11), 3403–3405.

Ciais P., Tans P. P., Trolier M., White J. W. C., and Francey R. J. (1995a) A large northern-hemisphere terrestrial CO$_2$ sink indicated by the $^{13}$C/$^{12}$C ratio of atmospheric CO$_2$. *Science* **269**(5227), 1098–1102.

Ciais P., Tans P. P., White J. W. C., Trolier M., Francey R. J., Berry J. A., Randall D. R., Sellers P. J., Collatz J. G., and Schimel D. S. (1995b) Partitioning of ocean and land uptake of CO$_2$ as inferred by $\delta^{13}$C measurements from the NOAA climate monitoring and diagnostics laboratory global air sampling network. *J. Geophys. Res. Atmos.* **100**(D3), 5051–5070.

Ciais P., Denning A. S., Tans P. P., Berry J. A., Randall D. A., Collatz G. J., Sellers P. J., White J. W. C., Trolier M., Meijer H. A. J., Francey R. J., Monfray P., and Heimann M. (1997a) A three-dimensional synthesis study of $\delta^{18}$O in atmospheric CO$_2$: 1. Surface fluxes. *J. Geophys. Res. Atmos.* **102**(D5), 5857–5872.

Ciais P., Tans P. P., Denning A. S., Francey R. J., Trolier M., Meijer H. A. J., White J. W. C., Berry J. A., Randall D. A., Collatz G. J., Sellers P. J., Monfray P., and Heimann M. (1997b) A three-dimensional synthesis study of $\delta^{18}$O in atmospheric CO$_2$: 2. Simulations with the TM2 transport model. *J. Geophys. Res. Atmos.* **102**(D5), 5873–5883.

Ciais P., Friedlingstein P., Schimel D. S., and Tans P. P. (1999) A global calculation of the $\delta^{13}$C of soil respired carbon: implications for the biospheric uptake of anthropogenic CO$_2$. *Global Biogeochem. Cycles* **13**(2), 519–530.

Clark I. and Fritz P. (1997) *Environmental Isotopes in Hydrology*. Lewis Publishers.

Collatz G. J., Berry J. A., and Clark J. S. (1998) Effects of climate and atmospheric CO$_2$ partial pressure on the global distribution of C$_4$ grasses: present, past, and future. *Oecologia* **114**(4), 441–454.

Conway T. J., Tans P. P., Waterman L. S., and Thoning K. W. (1994) Evidence for interannual variability of the carbon cycle from the national oceanic and atmospheric administration climate monitoring and diagnostics laboratory global air sampling network. *J. Geophys. Res. Atmos.* **99**(D11), 22831–22855.

Coplen T. B. (1995) Discontinuance of SMOW and PDB. *Nature* **375**(6529), 285–285.

Coplen T. B., Kendall C., and Hopple J. (1983) Comparison of stable isotope reference samples. *Nature* **302**, 236–238.

Cox P. M., Betts R. A., Jones C. D., Spall S. A., and Totterdel I. J. (2000) Acceleration of global warming due to carbon-cycle feedbacks in a coupled climate model. *Nature* **408**, 184–187.

Craig H. (1953) The geochemistry of the stable carbon isotopes. *Geochim. Cosmochim. Acta* **3**(53–92).

Craig H. (1954) Carbon-13 variations in sequoia rings and the atmosphere. *Science* **119**, 141–143.

Craig H. (1957) Isotopic standards for carbon and oxygen and correction factors for mass-spectrometric analysis of carbon dioxide. *Geochim. Cosmochim. Acta* **12**, 133–149.

Craig H. (1961) Isotopic variations in meteoric waters. *Science* **133**, 1702–1703.

Craig H. (1963) The effects of atmospheric NO$_2$ on the measured isotopic composition of atmospheric CO$_2$. *Geochim. Cosmochim. Acta* **27**(549–551).

Craig H. and Gordon L. I. (1965) Deuterium and oxygen-18 variations in the ocean and marine atmosphere. In *Stable*

*Isotopes in Oceano-graphic Studies and Paleo-Temperatures*, 9–130.

Cuntz M., Ciais P., and Hoffmann G. (2002) Modelling the continental effect of oxygen isotopes over Eurasia. *Tellus Ser. B: Chem. Phys. Meteorol.* **54**(5), 895–909.

Dansgaard W. (1964) Stable isotopes in precipitation. *Tellus* **16**(4), 436–468.

Davidson G. R. (1995) The stable isotopic composition and measurement of carbon in soil $CO_2$. *Geochim. Cosmochim. Acta* **59**(12), 2485–2489.

Deuser W. G. and Degens E. T. (1967) Carbon isotope fractionation in system $CO_2(Gas)$–$CO_2(Aqueous)$–$HCO_3^-$(Aqueous). *Nature* **215**(5105), 1033.

Dongmann G. (1974) Contribution of land photosynthesis to stationary enrichment of $^{18}O$ in atmosphere. *Radiat. Environ. Biophys.* **11**(3), 219–225.

Dongmann G., Nurnberg H. W., Forstel H., and Wagener K. (1974) Enrichment of $H_2^{18}O$ in leaves of transpiring plants. *Radiat. Environ. Biophys.* **11**(1), 41–52.

Duranceau M., Ghashghaie J., Badeck F., Deleens E., and Cornic G. (1999) $\delta^{13}C$ of $CO_2$ Respired in the dark in relation to $\delta^{13}C$ of leaf carbohydrates in *Phaseolus vulgaris* L—under progressive drought. *Plant Cell Environ.* **22**(5), 515–523.

Ehleringer J. R., Hall A. E., and Farquhar G. D. (1992) Stable isotope and plant water relations. Academic Press, San Diego.

Ehleringer J. R., Cerling T. E., and Helliker B. R. (1997) $C_4$ photosynthesis, atmospheric $CO_2$ and climate. *Oecologia* **112**(3), 285–299.

Ehleringer J. R., Buchmann N., and Flanagan L. B. (2000) Carbon isotope ratios in belowground carbon cycle processes. *Ecol. Appl.* **10**(2), 412–422.

Ekblad A. and Högberg P. (2001) Natural abundance C-13 in $CO_2$ respired from forest soils reveals speed of link between tree photosynthesis and root respiration. *Oecologia* **127**, 305–308.

Emrich K. and Vogel J. C. (1970) Carbon isotope fractionation during precipitation of calcium carbonate. *Earth Planet. Sci. Lett.* **8**(5), 363–371.

Enting I. G., Trudinger C. M., and Francey R. J. (1995) A synthesis inversion of the concentration and $\delta^{13}C$ of atmospheric $CO_2$. *Tellus Ser. B: Chem. Phys. Meteorol.* **47**(1–2), 35–52.

Epron D., Godard D., Cornic G., and Genty B. (1995) Limitation of net $CO_2$ assimilation rate by internal resistances to $CO_2$ transfer in the leaves of 2 tree species (*Fagus Sylvatica* L and *Castanea-sativa* Mill). *Plant Cell Environ.* **18**(1), 43–51.

Epstein S. and Mayeda T. (1953) Variations of $^{18}O$ content of waters from natural sources. *Geochim. Cosmochim. Acta* **4**, 213–224.

Evans J. R., Sharkey T. D., Berry J. A., and Farquhar G. D. (1986) Carbon isotope discrimination measured concurrently with gas-exchange to investigate $CO_2$ diffusion in leaves of higher-plants. *Austral. J. Plant Physiol.* **13**(2), 281–292.

Falkowski P. G. (1991) Species variability in the fractionation of $^{13}C$ and $^{12}C$ by marine phytoplankton. *J. Plankton Res.* **13**, 21–28.

Farquhar G. D. (1983) On the nature of carbon isotope discrimination in $C_4$ species. *Austral. J. Plant Physiol.* **10**(2), 205–226.

Farquhar G. D. and Lloyd J. (1993) Carbon and oxygen isotope effects in the exchange of carbon dioxide between terrestrial plants and the atmosphere. In *Stable Isotopes and Plant Carbon—Water Relations* (eds. J. R. Ehleringer, A. E. Hall, and G. D. Farquhar). Academic Press, pp. 47–70.

Farquhar G. D., Olseary M. H., and Berry J. A. (1982) On the relationship between carbon isotope discrimination and the intercellular carbon dioxide concentration in leaves. *Austral. J. Plant Physiol.* **9**(2), 121–137.

Farquhar G. D., Lloyd J., Taylor J. A., Flanagan L. B., Syvertsen J. P., Hubick K. T., Wong S. C., and Ehleringer J. R. (1993) Vegetation effects on the isotope composition of oxygen in atmospheric $CO_2$. *Nature* **363**(6428), 439–443.

Feng X. H. (1999) Trends in intrinsic water-use efficiency of natural trees for the past 100–200 years: a response to atmospheric $CO_2$ concentration. *Geochim. Cosmochim. Acta* **63**(13/14), 1891–1903.

Flanagan L. B. and Varney G. T. (1995) Influence of vegetation and soil $CO_2$ exchange on the concentration and stable oxygen-isotope ratio of atmospheric $CO_2$ within a Pinus-resinosa canopy. *Oecologia* **101**(1), 37–44.

Flanagan L. B., Bain J. F., and Ehleringer J. R. (1991) Stable oxygen and hydrogen isotope composition of leaf water in $C_3$ and $C_4$ plant-species under field conditions. *Oecologia* **88**(3), 394–400.

Flanagan L. B., Marshall J. D., and Ehleringer J. R. (1993) Photosynthetic gas-exchange and the stable-isotope composition of leaf water—comparison of a xylem-tapping mistletoe and its host. *Plant Cell Environ.* **16**(6), 623–631.

Flanagan L. B., Brooks J. R., Varney G. T., Berry S. C., and Ehleringer J. R. (1996) Carbon isotope discrimination during photosynthesis and the isotope ratio of respired $CO_2$ in boreal forest ecosystems. *Global Biogeochem. Cycles* **10**(4), 629–640.

Flanagan L. B., Brooks J. R., Varney G. T., and Ehleringer J. R. (1997) Discrimination against $C^{18}O^{16}O$ during photosynthesis and the oxygen isotope ratio of respired $CO_2$ in Bboreal forest ecosystems. *Global Biogeochem. Cycles* **11**(1), 83–98.

Förstel H. (1978) The enrichment of $^{18}O$ in leaf water under natural conditions. *Radiat. Environ. Biophys.* **15**, 323–344.

Francey R. J. and Farquhar G. D. (1982) An explanation of $^{13}C/^{12}C$ variations in tree rings. *Nature* **297**(5861), 28–31.

Francey R. J. and Tans P. P. (1987) Latitudinal variation in oxygen-18 of atmospheric $CO_2$. *Nature* **327**(6122), 495–497.

Francey R. J., Tans P. P., Allison C. E., Enting I. G., White J. W. C., and Trolier M. (1995) Changes in oceanic and terrestrial carbon uptake since 1982. *Nature* **373**(6512), 326–330.

Francey R. J., Allison C. E., Etheridge D. M., Trudinger C. M., Enting I. G., Leuenberger M., Langenfelds R. L., Michel E., and Steele L. P. (1999) A 1000-year high precision record of $\delta^{13}C$ in atmospheric $CO_2$. *Tellus Ser. B: Chem. Phys. Meteorol.* **51**(2), 170–193.

Francey R. J., Rayner P. J., and Allison C. E. (2001) Constraining the global carbon budget from global to regional scales—the measurement challenge. In *Global Biogeochemical Cycles in the Climate System* (eds. E. D. Schulze, M. Heimann, S. Harrison, E. Holland, J. Lloyd, I. C. Prentice, and D. Schimel). Academic Press, pp. 245–252.

Friedli H. and Siegenthaler U. (1988) Influence of $N_2O$ on isotope analyses in $CO_2$ and mass-spectrometric determination of $N_2O$ in air samples. *Tellus* **40B**, 129–133.

Friedli H., Lötscher H., Oeschger H., Siegenthaler U., and Stauffer B. (1986) Ice core record of the $^{13}C/^{12}C$ ratio of atmospheric $CO_2$ in the past two centuries. *Nature* **324**, 237–238.

Friedli H., Siegenthaler U., Rauber D., and Oeschger H. (1987) Measurements of concentration, $^{13}C/^{12}C$ and $^{18}O/^{16}O$ ratios of tropospheric carbon dioxide over Switzerland. *Tellus* **39B**, 80–88.

Friedman I. and O'Neil J. R. (1977) Compilation of stable isotope fractionation factors of geological interest. In *Data of Geochemistry*, 6th. edn. Geological Survey Professional Paper 44-KK.

Fung I., Field C. B., Berry J. A., Thompson M. V., Randerson J. T., Malmstrom C. M., Vitousek P. M., Collatz G. J., Sellers P. J., Randall D. A., Denning A. S., Badeck F., and John J. (1997) Carbon-13 exchanges between the atmosphere and biosphere. *Global Biogeochem. Cycles* **11**(4), 507–533.

Gamo T., Tsutsumi M., Sakai H., Nakazawa T., Tanaka M., Honda H., and Kubo H. (1989) Carbon and oxygen isotopic ratios of carbon dioxide of a stratospheric profile over Japan. *Tellus* **41B**, 127–133.

Gat J. R. (1996) Oxygen and hydrogen isotopes in the hydrologic cycle. *Ann. Rev. Earth Planet. Sci.* **24**, 225–262.

Gat J. R. and Gonfiantini R. (1981) *Stable Isotope Hydrology: Deuterium and Oxygen-18 in the Water Cycle.* Technical Reports Series No. 210, IAEA..

Gemery P. A., Trolier M., and White J. W. C. (1996) Oxygen isotope exchange between carbon dioxide and water following atmospheric sampling using glass flasks. *J. Geophys. Res. Atmos.* **101**(D9), 14415–14420.

Ghashghaie J., Duranceau M., Badeck F. W., Cornic G., Adeline M. T., and Deleens E. (2001) $\delta^{13}C$ of CO$_2$ Respired in the dark in relation to $\delta^{13}C$ of leaf metabolites: comparison between *Nicotiana sylvestris* and *Helianthus annuus* under drought. *Plant Cell Environ.* **24**(5), 505–515.

Gildor H. and Follows M. J. (2002) Two-way interactions between ocean biota and climate mediated by biogeochemical cycles. *Israel J. Chem.* **42**, 15–27.

Gillon J. and Yakir D. (2001) Influence of carbonic anhydrase activity in terrestrial vegetation on the $^{18}O$ content of atmospheric CO$_2$. *Science* **291**(5513), 2584–2587.

Gillon J. S. and Yakir D. (2000a) Internal conductance to CO$_2$ diffusion and C$^{18}$OO discrimination in C$_3$ leaves. *Plant Physiol.* **123**(1), 201–213.

Gillon J. S. and Yakir D. (2000b) Naturally low carbonic anhydrase activity in C$_4$ and C$_3$ plants limits discrimination against C$^{18}$OO during photosynthesis. *Plant Cell Environ.* **23**(9), 903–915.

Gleixner G. and Schmidt H. L. (1997) Carbon isotope effects on the fructose-1, 6-bisphosphate aldolase reaction, origin for non-statistical $^{13}C$ distribution in carbohydrates. *J. Biol. Chem.* **272**(9), 5382–5387.

Gleixner G., Danier H. J., Werner R. A., and Schmidt H. L. (1993) Correlations between the $^{13}C$ content of primary and secondary plant-products in different cell compartments and that in decomposing basidiomycetes. *Plant Physiol.* **102**(4), 1287–1290.

Goericke R. and Fry B. (1994) Variations of marine plankton $\delta^{13}C$ with latitude, temperature, and dissolved CO$_2$ in the world ocean. *Global Biogeochem. Cycles* **8**(1), 85–90.

Gonfiantini R., Gratziu S., and Tongiorgi E. (1965) Oxygen isotope composition of water in leaves. In *Isotopes and Radiation in Soil-plant Nutrition Studies.* IAEA pp. 405–410.

Gruber N., Keeling C. D., Bacastow R. B., Guenther P. R., Lueker T. J., Wahlen M., Meijer H. A. J., Mook W. G., and Stocker T. F. (1999) Spatiotemporal patterns of carbon-13 in the global surface oceans and the oceanic Seuss effect. *Global Biogeochem. Cycles* **13**(2), 307–335.

Guy R. D., Fogel M. F., Berry J. A., and Hoering T. C. (1987) Isotope fractionation during oxygen production and consumption by plants. In *Progress in Photosynthetic Research III* (ed. J. Biggins), pp. 597–600.

Guy R. D., Fogel M. L., and Berry J. A. (1993) Photosynthetic fractionation of the stable isotopes of oxygen and carbon. *Plant Physiol.* **101**(1), 37–47.

Griffiths H. (1998) Stable isotopes. In *Integration of Biological, Ecological and Geochemical Processes.* Bios Scientific publishers, Oxford.

Hanson P. J., Edwards N. T., Garten C. T., and Andrews J. A. (2000) Separating root and soil microbial contributions to soil respiration: a review of methods and observations. *Biogeochemistry* **48**(1), 115–146.

Harwood K. G., Gillon J. S., Griffiths H., and Broadmeadow M. S. J. (1998) Diurnal variation of $\Delta^{13}CO_2$, $\Delta C^{18}O^{16}O$ and evaporative site enrichment of $\delta H_2^{18}O$ in *Piper aduncum* under field conditions in Trinidad. *Plant Cell Environ.* **21**(3), 269–283.

Harwood K. G., Gillon J. S., Roberts A., and Griffiths H. (1999) Determinants of isotopic coupling of CO$_2$ and water vapor within a *Quercus petraea* forest canopy. *Oecologia* **119**(1), 109–119.

Hayes J. M., Strauss H., and Kaufman A. J. (1999) The abundance of $^{13}C$ in marine organic matter and isotopic fractionation in the global biogeochemical cycle of carbon during the past 800 Ma. *Chem. Geol.* **161**(1–3), 103–125.

Heimann M. and Maier-Reimer E. (1996) On the relations between the oceanic uptake of CO$_2$ and its carbon isotopes. *Global Biogeochem. Cycles* **10**, 89–110.

Helliker B. R. and Ehleringer J. R. (2000) Establishing a grassland signature in veins: $^{18}O$ in the leaf water of C$_3$ and C$_4$ grasses. In *Proceedings of the National Academy of Sciences of the United States of America* **97**(14), 7894–7898.

Hemming D. L., Switsur V. R., Waterhouse J. S., and Heaton T. H. E. (1998) Climate variations and the stable carbon isotope composition of tree ring cellulose: an intercomparison of *quercus robur, fagus sylvatica* and *pinus silvestris*. *Tellus* **50B**, 25–33.

Hemming D. L., Fritts H., Leavitt S. W., Wright W., Long A., and Shashkin A. (2001) Modelling tree-ring $\delta^{13}C$. *Dendrochronologia* **19**(1), 23–38.

Henderson S. A., Voncaemmerer S., and Farquhar G. D. (1992) Short-term measurements of carbon isotope discrimination in several C$_4$ species. *Austral. J. Plant Physiol.* **19**(3), 263–285.

Hesterberg R. and Siegenthaler U. (1991) Production and stable isotopic composition of CO$_2$ in a soil near Bern, Switzerland. *Tellus Ser. B: Chem. Phys. Meteorol.* **43**(2), 197–205.

Hoefs J. (1987) *Stable Isotope Ceochemistry*. Springer.

Hogberg P., Nordgren A., Buchmann N., Taylor A. F. S., Ekblad A., Hogberg M. N., Nyberg G., Ottosson-Lofvenius M., and Read D. J. (2001) Large-scale forest girdling shows that current photosynthesis drives soil respiration. *Nature* **411**(6839), 789–792.

Holland E. (1984) *The Chemical Evolution of the Atmosphere and Oceans*. Princeton University Press.

Hsieh J. C. C., Savin S. M., Kelly E. F., and Chadwick O. A. (1998) Measurement of soil–water $\delta^{18}O$ values by direct equilibration with CO$_2$. *Geoderma* **82**(1–3), 255–268.

Inoue H. and Sugimura Y. (1985) Carbon isotopic fractionation during the CO$_2$ exchange process between air and seawater under equilibrium and kinetic conditions. *Geochim. Cosmochim. Acta* **49**(11), 2453–2460.

IPCC (2001) Climate change 2001. In *The Scientific Basis* (eds. J. T. Houghton, Y. Ding, D. J. Griggs, M. Noguer, P. J. van der Linden, X. Dai, K. Maskell, and C. A. Johnson). Cambridge University Press.

Isaac M. (2001) *The Oxygen Isotopic Signal of CO$_2$ in Low Moisture Soils*. MSc, Weizmann Institute of Science.

Ishizawa M., Nakazawa T., and Higuchi K. (2002) A multi-box model study of the role of the biospheric metabolism in the recent decline of $\delta^8O$ in atmospheric CO$_2$. *Tellus Ser. B: Chem. Phys. Meteorol.* **54**(4), 307–324.

Ito A. (2003) A global-scale simulation of the CO$_2$ exchange between the atmosphere and the terrestrial biosphere with a mechanistic model including stable carbon isotopes, 1953–1999. *Tellus* **55B**, 596–612.

Jouzel J., Russell G. L., Suozzo R. J., Koster R. D., White J. W. C., and Broecker W. S. (1987) Simulations of the HDO and H$_2^{18}$O atmospheric cycles using the Nasa Giss general-circulation model—the seasonal cycle for present-day conditions. *J. Geophys. Res. Atmos.* **92**(D12), 14739–14760.

Kalnay E., Kanamitsu M., Kistler R., Collins W., Deaven D., Gandin L., Iredell M., Saha S., White G., Woollen J., Zhu Y., Chelliah M., Ebisuzaki W., Higgins W., Janowiak J., Mo K. C., Ropelewski C., Wang J., Leetmaa A., Reynolds R., Jenne R., and Joseph D. (1996) The NCEP/NCAR 40-year reanalysis project. *Bull. Am. Meteorol. Soc.* **77**(3), 437–471.

Kaplan A. and Reinhold L. (1999) CO$_2$ concentrating mechanisms in photosynthetic microorganisms. *Ann. Rev. Plant Physiol. Plant Mol. Biol.* **50**, 539–570.

Kaplan J. O., Prentice I. C., and Buchmann N. (2002) The stable carbon isotope composition of the terrestrial biosphere: modelling at scales from the leaf to the globe. *Global Biogeochem. Cycles* **16**(4), 1060.

Keeling C. D. (1958) The concentration and isotopic abundance of carbon dioxide in rural areas. *Geochim. Cosmochim. Acta* **13**, 299–313.

Keeling C. D. (1961) The concentration and isotopic abundance of atmospheric carbon dioxide in rural and marine air. *Geochim. Cosmochim. Acta* **24**, 277–298.

Keeling C. D., Mook W. G., and Tans P. P. (1979) Recent trends in the $^{13}$C–$^{12}$C ratio of atmospheric carbon-dioxide. *Nature* **277**(5692), 121–123.

Keeling C. D., Bacastow R. B., and Tans P. P. (1980) Predicted shift in the $^{13}$C–$^{12}$C ratio of atmospheric carbon-dioxide. *Geophys. Res. Lett.* **7**(7), 505–508.

Kesselmeier J. and Hubert A. (2002) Exchange of reduced volatile sulfur compounds between leaf litter and the atmosphere. *Atmos. Environ.* **36**(29), 4679–4686.

Kesselmeier J., Teusch N., and Kuhn U. (1999) Controlling variables for the uptake of atmospheric carbonyl sulfide by soil. *J. Geophys. Res. Atmos.* **104**(D9), 11577–11584.

Körtzinger A. and Quay P. (2003) Relationship between anthropogenic CO$_2$ and the $^{13}$C Seuss effect in the North Atlantic Ocean. *Global Biogeochem. Cycles* **17**(1), 1005.

Kroopnic P. (1985) The distribution of $^{13}$C on TCO$_2$ in the world ocean. *Deep Sea Res.* **32**, 57–84.

Kroopnic P. and Craig H. (1972) Atmospheric oxygen—isotopic composition and solubility fractionation. *Science* **175**(4017), 54–55.

Kroopnick P. (1980) The distribution of $^{13}$C in the atlantic-ocean. *Earth Planet Sci. Lett.* **49**(2), 469–484.

Ladyman S. J. and Harkness D. D. (1980) Carbon isotope measurements as an index of soil development. *Radiocarbon* **22**, 885.

Leroux X., Bariac T., and Mariotti A. (1995) Spatial partitioning of the soil–water resource between grass and shrub components in a West-African humid Savanna. *Oecologia* **104**(2), 147–155.

Leuenberger M., Siegenthaler U., and Langway C. C. (1992) Carbon isotope composition of atmospheric CO$_2$ during the last ice-age from an Antarctic ice core. *Nature* **357**(6378), 488–490.

Lin G. H. and Ehleringer J. R. (1997) Carbon isotopic fractionation does not occur during dark respiration in C$_3$ and C$_4$. *Plant Physiol.* **114**(1), 391–394.

Lloyd J. and Farquhar G. D. (1994) $^{13}$C Discrimination during CO$_2$ assimilation by the terrestrial biosphere. *Oecologia* **99**(3–4), 201–215.

Lloyd J., Syvertsen J. P., Kriedemann P. E., and Farquhar G. D. (1992) Low conductance's for CO$_2$ diffusion from stomata to the sites of carboxylation in leaves of woody species. *Plant Cell Environ.* **15**(8), 873–899.

Lloyd J., Kruijt B., Hollinger D. Y., Grace J., Francey R. J., Wong S. C., Kelliher F. M., Miranda A. C., Farquhar G. D., Gash J. H. C., Vygodskaya N. N., Wright I. R., Miranda H. S., and Schulze E. D. (1996) Vegetation effects on the isotopic composition of atmospheric CO$_2$ at local and regional scales: theoretical aspects and a comparison between rain forest in Amazonia and a Boreal forest in Siberia. *Austral. J. Plant Physiol.* **23**(3), 371–399.

Lloyd J., Francey R. J., Sogachev A., Mollicone D., Raupach M. R., Sogachev A., Arneth A., Byers J. N., Kelliher F. M., Rebmann C., Valentini R., Wong S. C., Bauer G., and Schulze E. D. (2001) Vertical profiles, boundary layer budgets, and regional flux estimates for CO$_2$ and its $^{13}$C/$^{12}$C ratio and for water vapour above a forest/bog mosaic in central Siberia. *Global Biogeochem. Cycles* **15**, 267–284.

Loreto F., Harley P. C., Dimarco G., and Sharkey T. D. (1992) Estimation of mesophyll conductance to CO$_2$ flux by 3 different methods. *Plant Physiol.* **98**(4), 1437–1443.

Lynch-Stieglitz J., Stocker T. F., Broecker W. S., and Fairbanks R. G. (1995) The influence of air-sea exchange on the isotopic composition of oceanic carbon—observations and modelling. *Global Biogeochem. Cycles* **9**(4), 653–665.

Luz B. and Barkan E. (2000) Assessment of oceanic productivity with the triple-isotope composition of dissolved oxygen. *Science* **288**, 2028–2031.

Luz B., Barkan E., Bender M. L., Thiemens M. H., and Boering K. A. (1999) Triple-isotope composition of atmospheric oxygen as a tracer of biosphere productivity. *Nature* **400**, 547–550.

Marino B. D., McElroy M. B., Salawitch R. J., and Spaulding W. G. (1992) Glacial-to-interglacial variations in the carbon isotopic composition of atmospheric CO$_2$. *Nature* **357**(6378), 461–466.

Marlier J. F. and O'Leary M. H. (1984) Carbon kinetic isotope effects on the hydration of carbon-dioxide and the dehydration of bicarbonate ion. *J. Am. Chem. Soc.* **106**(18), 5054–5057.

Masarie K. A., Langenfelds R. L., Allison C. E., Conway T. J., Dlugokencky E. J., Francey R. J., Novelli P. C., Steele L. P., Tans P. P., Vaughn B., and White J. W. C. (2001) NOAA/CSIRO flask air intercomparison experiment: a strategy for directly assessing consistency among atmospheric measurements made by independent laboratories. *J. Geophys. Res. Atmos.* **106**(D17), 20445–20464.

Mathieu R. and Bariac T. (1996a) An isotopic study ($^2$H and $^{18}$O) of water movements in clayey soils under a semi-arid climate. *Water Resour. Res.* **32**(4), 779–789.

Mathieu R. and Bariac T. (1996b) A numerical model for the simulation of stable isotope profiles in drying soils. *J. Geophys. Res. Atmos.* **101**(D7), 12685–12696.

Mathieu R., Pollard D., Cole J. E., White J. W. C., Webb R. S., and Thompson S. L. (2002) Simulation of stable water isotope variations by the GENESIS GCM for modern conditions. *J. Geophys. Res. Atmos.* **107**(D4) article no. 4037.

McConville C., Kalin R. M., and Flood D. (1999) Direct equilibration of soil water for $\delta^{18}$O analysis and its application to tracer studies. *Rapid Commun. Mass Spectrom.* **13**(13), 1339–1345.

McKinney C. R., McCrea J. M., Epstein S., Allen H. A., and Urey H. C. (1950) Improvements in mass spectrometers for the measurement of small differences in isotope abundance ratios. *Rev. Sci. Instr.* **21**(8), 724–730.

Melzer E. and Schmidt H. L. (1987) Carbon isotope effects on the pyruvate dehydrogenase reaction and their importance for relative carbon-13 depletion in lipids. *J. Biol. Chem.* **262**(17), 8159–8164.

Miller J. B. and Tans P. P. (2003) Calculating isotopic discrimination from atmospheric measurements at various scales. *Tellus* **55B**, 207–214.

Miller J. B., Yakir D., White J. W. C., and Tans P. P. (1999) Measurement of $^{18}$O/$^{16}$O in the soil-atmosphere CO$_2$ flux. *Global Biogeochem. Cycles* **13**(3), 761–774.

Miller J. B., Tans P. P., White J. W. C., Conway T. J., and Vaughn B. W. (2003) The atmospheric signal of terrestrial isotopic discrimination and its implication for carbon fluxes. *Tellus* **55B**, 197–206.

Mills G. A. and Urey H. C. (1940) The kinetics of isotopic exchange between carbon dioxide, bicarbonate ion, carbonate ion and water. *J. Am. Chem. Soc.* **62**, 1019–1028.

Miranda A. C., Miranda H. S., Lloyd J., Grace J., Francey R. J., McIntyre J. A., Meir P., Riggan P., Lockwood R., and Brass J. (1997) Fluxes of carbon, water, and energy over Brazilian Cerrado: an analysis using eddy covariance and stable isotopes. *Plant Cell Environ.* **20**(3), 315–328.

Mook W. G. (1986) $^{13}$C in atmospheric CO$_2$. *Neth. J. Sea Res.* **20**(2–3), 211–223.

Mook W. G. (2000) *Environmental Isotopes in the Hydrological Cycle: Principles and Applications*. IAEA.

Mook W. G. and Grootes P. M. (1973) The measuring procedure and corrections for the high-precision mass-spectrometric analysis of isotopic abundance ratios, especially referring to carbon, oxygen and nitrogen. *Int. J. Mass Spectrom. Ion Phys.* **12**, 198–273.

Mook W. G. and Van der Hoek S. (1983) The $N_2O$ correction in the carbon and oxygen isotopic analysis of atmospheric $CO_2$. *Isotopic Geosci.* **1**, 169–176.

Mook W. G., Bommerso Jc., and Staverma Wh. (1974) Carbon isotope fractionation between dissolved bicarbonate and gaseous carbon-dioxide. *Earth Planet. Sci. Lett.* **22**(2), 169–176.

Mook W. G., Koopmans M., Carter A. F., and Keeling C. D. (1983) Seasonal, latitudinal, and secular variations in the abundance and isotopic-ratios of atmospheric carbon-dioxide: 1. Results from land stations. *J. Geophys. Res.: Ocean. Atmos.* **88**(C15), 915–933.

Moreira M. Z. (1998) Contribution of vegetation to the water cycle of the Amazon basin: an isotopic study of plant transpiration and its water source. PhD Thesis, University of Miami.

Morimoto S., Nakazawa T., Higuchi K., and Aoki S. (2000) Latitudinal distribution of atmospheric $CO_2$ sources and sinks inferred by $\delta^{13}C$ measurements from 1985 to 1991. *J. Geophys. Res. Atmos.* **105**(D19), 24315–24326.

Nakazawa T., Sugawara S., Inoue G., Machida T., Makshyutov S., and Mukai H. (1997) Aircraft measurements of the concentrations of $CO_2$, $CH_4$, $N_2O$, and CO and the carbon and oxygen isotopic ratios of $CO_2$ in the troposphere over Russia. *J. Geophys. Res. Atmos.* **102**(D3), 3843–3859.

Nier A. O. (1947) A mass spectrometer for isotope and gas analysis. *Rev. Sci. Instr.* **19**(6), 398–411.

O'Leary M. H. (1981) Carbon isotope fractionation in plants. *Phytochemistry* **20**(4), 553–567.

O'Leary M. H. (1984) Measurement of the isotope fractionation associated with diffusion of carbon-dioxide in aqueous-solution. *J. Phys. Chem.* **88**(4), 823–825.

O'Leary M. H., Madhavan S., and Paneth P. (1992) Physical and chemical basis of carbon isotope fractionation in plants. *Plant Cell Environ.* **15**(9), 1099–1104.

Park R. and Epstein S. (1960) Carbon isotope fractionation during photosynthesis. *Geochim. Cosmochim. Acta* **44**, 5–15.

Pataki D. E., Ehleringer J. R., Flanagan L. B., Yakir D., Bowling D. R., Still C. J., Buchmann N., Kaplan J. O., and Berry J. A. (2003) The application and interpretation of keeling plots in terrestrial carbon cycle research. *Global Biogeochem. Cycles* **17**(1), 1022.

Peylin P., Ciais P., Tans P. P., Six K., Berry J. A., and Denning A. S. (1996) $^{18}O$ in atmospheric $CO_2$ simulated by a 3-d transport model: a sensitivity study to vegetation and soil fractionation factors. *Phys. Chem. Earth* **21**(5–6), 463–469.

Peylin P., Ciais P., Denning A. S., Tans P. P., Berry J. A., and White J. W. C. (1999) A 3-dimensional study of $\delta^{18}O$ in atmospheric $CO_2$: contribution of different land ecosystems. *Tellus Ser. B: Chem. Phys. Meteorol.* **51**(3), 642–667.

Potter C. S., Randerson J. T., Field C. B., Matson P. A., Vitousek P. M., Mooney H. A., and Klooster S. A. (1993) Terrestrial ecosystem production—a process model-based on global satellite and surface data. *Global Biogeochem. Cycles* **7**(4), 811–841.

Quay P., King S., Wilbur D., Wofsy S., and Richey J. (1989) $^{13}C/^{12}C$ of atmospheric $CO_2$ in the Amazon basin—forest and river sources. *J. Geophys. Res. Atmos.* **94**(D15), 18327–18336.

Quay P. D., Tilbrook B., and Wong C. S. (1992) Oceanic uptake of fossil-fuel $CO_2$—$^{13}C$ evidence. *Science* **256**(5053), 74–79.

Quay P., Sonnerup R., Westby T., Stutsman J., and McNichol A. (2003) Changes in the $^{13}C/^{12}C$ of dissolved inorganic carbon in the ocean as a tracer of anthropogenic $CO_2$ uptake. *Global Biogeochem. Cycles* **17**(1), 1004.

Raich J. W. and Potter C. S. (1995) Global patterns of carbon-dioxide emissions from soils. *Global Biogeochem. Cycles* **9**(1), 23–36.

Raich J. W. and Schlesinger W. H. (1992) The global carbon-dioxide flux in soil respiration and its relationship to vegetation and climate. *Tellus Ser. B: Chem. Phys. Meteorol.* **44**(2), 81–99.

Ramonet M., Ciais P., Nepomniachii I., Sidrov K., Neubert R. E. M., Lanendöfer U., Picard D., Kazan V., Biaud S., Gust M., Kolle O., and Schulze E.-D. (2002) Three years of aircraft-based trace gas measurements over the Fyodorovskye southern taiga forest, 300 km north-west of Moscow. *Tellus* **54B**, 713–734.

Randerson J. T., Thompson M. V., Malmstrom C. M., Field C. B., and Fung I. Y. (1996) Substrate limitations for heterotrophs: implications for models that estimate the seasonal cycle of atmospheric $CO_2$. *Global Biogeochem. Cycles* **10**(4), 585–602.

Randerson J. T., Still C. J., Balle J. J., Fung I. Y., Doney S. C., Tans P. P., Conway T. J., White J. W. C., Vaughn B., Suits N., and Denning A. S. (2002) Carbon isotope discrimination of arctic and boreal biomes inferred from remote atmospheric measurements and a biosphere-atmosphere model. *Global Biogeochem. Cycles* **16**(3), 1028.

Rau G. H., Takahashi T., and Marais D. J. D. (1989) Latitudinal variations in plankton $\delta^{13}C$—implications for $CO_2$ and productivity in past oceans. *Nature* **341**(6242), 516–518.

Raupach M. (2001) Inferring biogeochemical sources and sinks from atmospheric concentrations: general consideration and applications in vegetation canopies. In *Global Biogeochemical Cycles in the Climate System* (eds. E. D. Schulze, M. Heimann, S. Harrison, E. Holland, J. Lloyd, I. C. Prentice, and D. Schimel). Academic Press, pp. 41–60.

Raven J. A. (1992) Present and potential uses of the natural abundance of stable isotopes in plant science, with illustrations from the marine environment: commissioned review. *Plant Cell Environ.* **15**(9), 1083–1090.

Raven J. A. (1998) Phylogeny, palaeoatmospheres and the evolution of phototrophy. In *Stable Isotopes: Integration of Biological, Ecological, and Geochemical Processes* (ed. H. Griffiths). BIOS Scientific Publishers, pp. 323–346.

Reed M. L. and Graham D. (1981) Carbonic anhydrase in plants: distribution and possible physiological roles. In *Progress in Phytochemistry* (eds. L. Reinhold, J. Harborn, and T. Swain). Pergamon, pp. 47–94.

Rinaldi G., Meinschein W. G., and Hayes J. M. (1974) Intramolecular carbon isotopic distribution in biologically produced acetoin. *Biomed. Mass Spectrom.* **1**, 415–417.

Rochette P. and Flanagan L. B. (1997) Quantifying rhizosphere respiration in a corn crop under field conditions. *Soil Sci. Soc. Am. J.* **61**(2), 466–474.

Rochette P., Flanagan L. B., and Gregorich E. G. (1999) Separating soil respiration into plant and soil components using analyses of the natural abundance of carbon-13. *Soil Sci. Soc. Am. J.* **63**(5), 1207–1213.

Roden J. S. and Ehleringer J. R. (1999) Observations of hydrogen and oxygen isotopes in leaf water confirm the Craig-Gordon model under wide ranging environmental conditions. *Plant Physiol.* **120**(4), 1165–1173.

Roeske C. A. and O'Leary M. H. (1984) Carbon isotope effects on the enzyme-catalyzed carboxylation of ribulose bisphosphate. *Biochemistry* **23**, 6275–6284.

Rooney M. A. (1988) Short-term carbon isotope fractionation by plants. PhD Thesis, University of Wisconsin, Madison.

Rozanski K., Araguasaraguas L., and Gonfiantini R. (1992) Relation between long-term trends of $^{18}O$ isotope composition of precipitation and climate. *Science* **258**(5084), 981–985.

Rozanski K., Araguás-Araguás L., and Gonfiantini R. (1993) Isotopic patterns in modern global precipitation. In *Climate Change in Continental Isotopic Records* (eds. P. K. Swart,

K. C. Lohmann, J. McKenzie, and S. Savin). American Geophysical Union, pp. 1–36.

Rundel P. W., Ehleringer J. R., and Nagy K. A. (1988) *Stable Isotopes in Ecological Research.* Ecological Studies 68, Springer, NY.

Santrock J., Studley S. A., and Hayes J. M. (1985) Isotopic analyses based on the mass-spectrum of carbon-dioxide. *Analyt. Chem.* **57**(7), 1444–1448.

Sarmiento J. L. and Bender M. (1994) Carbon biogeochemistry and climate change. *Photosyn. Res.* **39**(3), 209–234.

Schidlowski M. (1988) A 3,800-million-year isotopic record of life from carbon in sedimentary rocks. *Nature* **333**(6171), 313–318.

Schimel D. S., Braswell B. H., Holland E. A., McKeown R., Ojima D. S., Painter T. H., Parton W. J., and Townsend A. R. (1994) Climatic, edaphic, and biotic controls over storage and turnover of carbon in soils. *Global Biogeochem. Cycles* **8**(3), 279–293.

Schimel D. S., Braswell B. H., McKeown R., Ojima D. S., Parton W. J., and Pulliam W. (1996) Climate and nitrogen controls on the geography and timescales of terrestrial biogeochemical cycling. *Global Biogeochem. Cycles* **10**(4), 677–692.

Schwartz D., deForesta H., Mariotti A., Balesdent J., Massimba J. P., and Girardin C. (1996) Present dynamics of the Savanna-forest boundary in the Congolese mayombe: a pedological, botanical and isotopic ($^{13}$C and $^{14}$C) study. *Oecologia* **106**(4), 516–524.

Scrimgeour C. M. (1995) Measurement of plant and soil–water isotope composition by direct equilibration methods. *J. Hydrol.* **172**(1–4), 261–274.

Sharkey T. D. and Berry J. A. (1985) Carbon isotope fractionation of algal as influenced by an inducible $CO_2$ concentrating mechanism. In *Inorganic Carbon Uptake by Aquatic Photosynthetic Organisms* (eds. W. J. Lucas and J. A. Berry). American Society of Plant Physiologists, pp. 389–401.

Siegenthaler U. and Münich K. O. (1981) $^{13}C/^{12}C$ fractionation during $CO_2$ transfer from air to sea. In *Carbon Cycle Modelling*, SCOPE 16 (ed. B. Bolin). Wiley, pp. 249–257.

Silverman D. N. (1982) Carbonic-anhydrase—$^{18}O$ exchange catalyzed by an enzyme with rate contributing proton transfer steps. *Meth. Enzymol.* **87**, 732–752.

Skirrow G. (1975) The dissolved gases—carbon dioxide. In *Chemical Oceanography* (eds. J. P. Riley and G. Skirrow). Academic Press, vol. 2, pp. 1–192.

Sposito G., Skipper N. T., Sutton R., Park S. H., Soper A. K., and Greathouse J. A. (1999) Surface geochemistry of the clay minerals. *Proceedings of the National Academy of Sciences of the United States of America* **96**(7), 3358–3364.

Stern L., Baisden W. T., and Amundson R. (1999) Processes controlling the oxygen isotope ratio of soil $CO_2$: analytic and numerical modelling. *Geochim. Cosmochim. Acta* **63**(6), 799–814.

Stern L. A., Amundson R., and Baisden W. T. (2001) Influence of soils on oxygen isotope ratio of atmospheric $CO_2$. *Global Biogeochem. Cycles* **15**(3), 753–759.

Sternberg L. (1989) A model to estimate carbon-dioxide recycling in forests using $^{13}C/^{12}C$ ratios and concentrations of ambient carbon-dioxide. *Agri. Forest Meteorol.* **48**(1–2), 163–173.

Sternberg L. D. L. (1997) Interpretation of recycling indexes—comment. *Austral. J. Plant Physiol.* **24**(3), 395–398.

Sternberg L. D., Moreira M. Z., Martinelli L. A., Victoria R. L., Barbosa E. M., Bonates L. C. M., and Nepstad D. (1998) The relationship between $^{18}O/^{16}O$ and $^{13}C/^{12}C$ ratios of ambient $CO_2$ in two Amazonian tropical forests. *Tellus Ser. B: Chem. Phys. Meteorol.* **50**(4), 366–376.

Still C. J., Berry J. A., Collatz G. J., and DeFries R. S. (2003) Global distribution of $C_3$ and $C_4$ vegetation: carbon cycle implications. *Global Biogeochem. Cycles* **17**(1) 10.1029/2001GB001807.

Styles J. M., Lloyd J., Zolotoukhine D., Lanbon K. A., Tchebakova N., Francey R. J., Arneth A., Salamakho D., Kolle O., and Schulze E. D. (2002a) Estimates of regional surface carbon dioxide exchange and carbon and oxygen isotope discrimination during photosynthesis from concentration profiles in the atmospheric boundary layer. *Tellus* **54B**, 768–783.

Tans P. P. (1980) On calculating the transfer of $^{13}C$ in reservoir models of the carbon-cycle. *Tellus* **32**(5), 464–469.

Tans P. P. (1981) $^{13}C/^{12}C$ of industrial $CO_2$. In *Carbon Cycle Modelling*, SCOPE 16 (ed. B. Bolin). Wiley, pp. 127–129.

Tans P. P. (1998) Oxygen isotopic equilibrium between carbon dioxide and water in soils. *Tellus Ser. B: Chem. Phys. Meteorol.* **50**(2), 163–178.

Tans P. P. and White J. W. C. (1998) The global carbon cycle—in balance, with a little help from the plants. *Science* **281**(5374), 183–184.

Tans P. P., Fung I. Y., and Takahashi T. (1990) Observational constraints on the global atmospheric $CO_2$ budget. *Science* **247**(4949), 1431–1438.

Tans P. P., Berry J. A., and Keeling R. F. (1993) Oceanic $^{13}C/^{12}C$ observations: a new window on ocean $CO_2$ uptake. *Global Biogeochem. Cycles* **7**(2), 353–368.

Tcherkez G., Nogues S., Bleton J., Cornic G., Badeck F., and Ghashghaie J. (2003) Metabolic origin of carbon isotope composition of leaf dark-respired $CO_2$ in French bean. *Plant Physiol.* **131**(1), 237–244.

Thiemens M. H. (1999) Atmosphere science—mass-independent isotope effects in planetary atmospheres and the early solar system. *Science* **283**(5400), 341–345.

Thiemens M. H. and Jackson T. (1990) Pressure dependency for heavy isotope enhancement in ozone formation. *Geophys. Res. Lett.* **17**(6), 717–719.

Thiemens M. H., Jackson T., Mauersberger K., Schueler B., and Morton J. (1991) Oxygen isotope fractionation in stratospheric $CO_2$. *Geophys. Res. Lett.* **18**(4), 669–672.

Thoning K. W., Tans P. P., and Komyr W. D. (1989) Atmospheric carbon dioxide at Mauna Loa observatory: 2. Analysis of the NOAA GMCC data, 1974, 1985. *J. Geophys. Res.* **94**(8549–8565).

Trolier M., White J. W. C., Tans P. P., Masarie K. A., and Gemery P. A. (1996) Monitoring the isotopic composition of atmospheric $CO_2$: measurements from the NOAA global air sampling network. *J. Geophys. Res. Atmos.* **101**(D20), 25897–25916.

Trumbore S. (2000) Age of soil organic matter and soil respiration: radiocarbon constraints on belowground C dynamics. *Ecol. Appl.* **10**(2), 399–411.

Trumbore S. E., Davidson E. A., Decamargo P. B., Nepstad D. C., and Martinelli L. A. (1995) Belowground cycling of carbon in forests and pastures of eastern Amazonia. *Global Biogeochem. Cycles* **9**(4), 515–528.

Urey H. C. (1947) The Thermodynamic properties of isotopic substances. *J. Chem. Soc.* 562.

Urey H. C. (1948) Oxygen isotopes in nature and in the laboratory. *Science* **108**, 489–496.

Vogel J. C. (1980) Fractionation of the carbon isotopes during photosynthesis. In *Sitzungsberichte der Heidelberger Akademie der Wissenschaften, Mathematisch-Naturwissenschaftliche Klasse Jahrgang 1980. 3. Abhandlung* Springer, pp. 111–135.

Vogel J. C. (1993) Variability of carbon isotope fractionation during photosynthesis. In *Stable Isotopes and Plant Carbon: Water Relations* (eds. J. R. Ehleringer, A. E. Hall, and G. D. Farquhar). Academic Press, pp. 29–46.

Von Caemmerer S. and Evans J. R. (1991) Determination of the average partial–pressure of $CO_2$ in chloroplasts from leaves of several $C_3$ plants. *Austral. J. Plant Physiol.* **18**(3), 287–305.

Wang X. F. and Yakir D. (1995) Temporal and spatial variations in the oxygen-18 content of leaf water in different plant species. *Plant Cell Environ.* **18**(12), 1377–1385.

Wang X. F., Yakir D., and Avishai M. (1998) Non-climatic variations in the oxygen isotopic compositions of plants. *Global Change Biol.* **4**(8), 835–849.

Wen J. and Thiemens M. H. (1993) Multi-isotope study of the O($^1$D) + CO$_2$ exchange and stratospheric consequences. *J. Geophys. Res. Atmos.* **98**(D7), 12801–12808.

White J. W. C. (1988) Stable hydrogen isotope ratio in plants: a review of current theory and some potential applications. In *Stable Isotopes in Ecological Research* (eds. P. W. Rundel, J. R. Ehleringer, and K. A. Nagi). Springer, pp. 142–162.

Wittenberg U. and Esser G. (1997) Evaluation of the isotopic disequilibrium in the terrestrial biosphere by a global carbon isotope model. *Tellus Ser. B: Chem. Phys. Meteorol.* **49**(3), 263–269.

Wofsy S. C., Goulden M. L., Munger J. W., Fan S. M., Bakwin P. S., Daube B. C., Bassow S. L., and Bazzaz F. A. (1993) Net exchange of CO$_2$ in a midlatitude forest. *Science* **260**(5112), 1314–1317.

Yakir D. (1992) Variations in the natural abundance of oxygen-18 and deuterium in plant carbohydrates: commissioned review. *Plant Cell Environ.* **15**(9), 1005–1020.

Yakir D. (1998) Oxygen-18 of leaf water: a crossroad for plant-associated isotopic signals. In *Stable Isotopes: Integration of Biological, Ecological and Geochemical Processes* (ed. H. Griffiths). BIOS Scientific Publishers Limited, pp. 147–183.

Yakir D. (2002) Global enzymes: sphere of influence. *Nature* **416**, 795.

Yakir D. and Wang X. F. (1996) Fluxes of CO$_2$ and water between terrestrial vegetation and the atmosphere estimated from isotope measurements. *Nature* **380**(6574), 515–517.

Yung Y. L., Demore W. B., and Pinto J. P. (1991) Isotopic exchange between carbon-dioxide and ozone via O($^1$D) in the stratosphere. *Geophys. Res. Lett.* **18**(1), 13–16.

Yurtsever Y. and Gat J. R. (1981) Atmospheric waters. In *Stable Isotope Hydrology*, Tech. Rep. Ser. 210 (eds. J. R. Gat and R. Gonfiantini). IAEA, pp. 103–42.

Zhang J., Quay P. D., and Wilbur D. O. (1995) Carbon-isotope fractionation during gas-water exchange and dissolution of CO$_2$. *Geochim. Cosmochim. Acta* **59**(1), 107–114.

# 4.08

# Water Stable Isotopes: Atmospheric Composition and Applications in Polar Ice Core Studies

J. Jouzel

*Institut Pierre Simon Laplace, Saclay, France*

## NOMENCLATURE

| | |
|---|---|
| $Corr_{ocean}$ | correction due to change in oceanic $\delta$ |
| $d$ | deuterium excess defined as $d = \delta D - 8\delta^{18}O$ |
| D | deuterium (heavy hydrogen, $^2H$) when used in the formula of water: HDO |
| $D$ | diffusion coefficient of water in air (formulas (3) and (4)) |
| $D_i$ | diffusion coefficient of water isotopes in air |
| $F$ | ratio of the remaining to the initial amount of water vapor in an air mass |
| $h$ | relative humidity |
| H | hydrogen |
| $k$ | kinetic fractionation parameter |
| $n_L$ | number of moles of liquid |
| $n'_L$ | number of moles of liquid for isotopic species |
| $n_p$ | number of moles of precipitation |
| $n'_p$ | number of moles of precipitation for isotopic species |
| $n_v$ | number of moles of water vapor |
| $n'_v$ | number of moles of water vapor for isotopic species |
| O | oxygen |

$R$          atomic ratios (D/H and $^{18}O/^{16}O$, respectively)

$R_{slopes}$   $S_{spat}/S_{temp}$

$S_i$         saturation ratio with respect to ice

$S_{spat}$     spatial $\delta/T$ slope

$S_{temp}$    temporal $\delta/T$ slope

$t$          time

$t_a$         isotopic relaxation time (drop or droplet)

$T_c$         condensation temperature

$T_s$         surface temperature site

$T_{s(inv)}$    surface temperature site determined with the inversion of simple isotopic model

$T_{s(spat)}$   surface temperature site determined with $S_{spat}$

$T_w$        moisture source temperature

V-SMOW   Vienna Standard Mean Ocean Water with D/H and $^{18}O/^{16}O$ atomic ratios of $T_s$

$\alpha$         isotopic fractionation coefficient

$\alpha_c$        isotopic fractionation coefficient when condensation occurs

$\alpha_k$        kinetic fractionation coefficient

$\alpha_L$        isotopic fractionation coefficient between liquid and vapor

$\alpha_s$        isotopic fractionation coefficient between solid and vapor

$\delta$         unit of isotopic ratio defined as $(R_{sample} - R_{SMOW})/R_{SMOW}$ and expressed in per mil (defined for deuterium, $\delta D$, and oxygen-18, $\delta^{18}O$)

$\delta D, \delta^{18}O$   isotopic content in per mil with respect to the Standard Mean Ocean Water

$\delta_L$        isotopic content of a liquid cloud (stands either for $\delta D$ or for $\delta^{18}O$)

$\delta_p$        isotopic content of a precipitation(stands either for $\delta D$ or for $\delta^{18}O$)

$\delta_v$        isotopic content of water vapor (stands either for $\delta D$ or for $\delta^{18}O$)

$\Delta$         difference between two time periods

$\Delta_{age}$     age difference between the ice and entrapped air bubbles

## 4.08.1  INTRODUCTION

Natural waters formed of ~99.7% of $H_2^{16}O$ are also constituted of other stable isotopic molecules, mainly $H_2^{18}O$ (~2‰), $H_2^{17}O$ (~0.5‰), and $HD^{16}O$ (~0.3‰), where H and D (deuterium) correspond to $^1H$ and $^2H$, respectively. Owing to slight differences in physical properties of these molecules, essentially their saturation vapor pressure, and their molecular diffusivity in air, fractionation processes occur at each phase change of the water except sublimation and melting of compact ice. As a result, the distribution of these water isotopes varies both spatially and temporally in the atmosphere, in the precipitation, and, in turn, in the various reservoirs of the hydrosphere and of the cryosphere. These isotopic variations have applications in such fields as climatology and cloud physics. More importantly, they are at the origin of two now well-established disciplines: isotope hydrology and isotope paleoclimatology. The various aspects dealing with isotope hydrology are reviewed by Kendall (see Chapter 5.11). In this chapter, we focus on this field known as "isotope paleoclimatology." As the behavior of $H_2^{17}O$ in the atmospheric water is very similar to that of $H_2^{18}O$ (more abundant and easier to precisely determine), isotope paleoclimatology is only based on the changes in concentrations of HDO and $H_2^{18}O$. These concentrations are given with respect to a standard as $\delta = (R_{sample} - R_{SMOW})/R_{SMOW}$ and expressed in per mil $\delta$ units ($\delta D$ and $\delta^{18}O$, respectively). In this definition, $R_{sample}$ and $R_{SMOW}$ are the isotopic ratios of the sample and of the Vienna Standard Mean Ocean Water (V-SMOW) with D/H and $^{18}O/^{16}O$ atomic ratios of $155.76 \times 10^{-6}$ and $2005.2 \times 10^{-6}$, respectively (Hageman *et al.*, 1970; Baerstchi, 1976; Gonfiantini, 1978).

The use of water stable isotopes in paleoclimatology is based on the fact that their present-day distribution in precipitation is strongly related to climatological parameters. Of primary interest is the linear relationship between annual values of $\delta D$ and $\delta^{18}O$ and mean annual temperature at the precipitation site, $T_s$, that is observed at middle and high latitudes (Figure 1). This relationship, which, as discussed in Section 4.08.3, is well explained by both simple and complex isotopic models, has given rise to the notion of "isotopic paleothermometer." In a conventional approach, the present-day spatial relationship between the isotopic concentration of the precipitation $\delta_p$ (where $\delta_p$ stands either for $\delta D$ or for $\delta^{18}O$ of the precipitation, which can indifferently be used as paleothermometers) and $T_s$, defined over a certain region, is assumed to hold in time throughout this region. In this approach, it is assumed that the temporal slope, which applies to the isotope–temperature relationship through different climates over time at a single geographic location and should be used to interpret isotopic variations, observed at this site in terms of temperature changes, and the spatial slope ($S_{spat} = d\delta_p/dT_s$) are similar. A so-called "modern analogue method" is thus used, similar to that adopted in most other methods for reconstructing paleoclimates. Of course, the fact that present-day isotope concentrations and local temperatures are correlated is not sufficient to validate this critical assumption. Such factors as the evaporative origin and the seasonality of precipitation can also affect $\delta D$ and $\delta^{18}O$. If these factors change markedly under different climates, the spatial slope can no

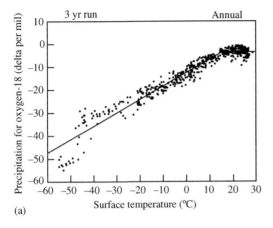

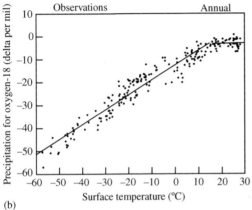

**Figure 1** Annual $\delta^{18}O$ in precipitation versus annual surface temperature for: (a) 3 yr run and (b) observations as simulated by the NASA/GISS isotopic GCM (after Jouzel *et al.*, 1987a).

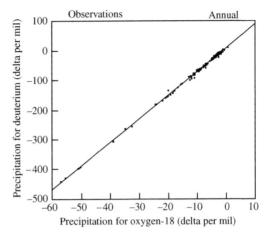

**Figure 2** Annual $\delta D$ in precipitation versus annual $\delta^{18}O$ in precipitation (after Jouzel *et al.*, 1987a).

longer be taken as a reliable surrogate of the temporal slope for interpreting the isotopic signal. For example, there is now ample evidence that temporal slopes are considerably lower (by up to a factor of 2) than the observed present-day spatial slope, for Greenland sites.

Present-day $\delta_p$ distributions are characterized by two other interesting large-scale properties. First, there is no clear relationship between $\delta_p$ and the temperature of the site in tropical and equatorial regions. There, $\delta_p$ is more significantly influenced by precipitation amount. Second (see Figure 2), $\delta D$ and $\delta^{18}O$ are linearly related to each other throughout the world with a slope of ~8 and a deuterium excess ($d = \delta D - 8\delta^{18}O$) of ~10‰ (Craig, 1961; Dansgaard, 1964). Although there has been a lot of potential interest in characterizing modifications in precipitation patterns such as those linked with changes in moonsonal activity or in the El Niño Southern Oscillation, the first property has, up to now, only been exploited in a very limited number of studies. In contrast, the fact that the deuterium excess of a precipitation is influenced by the conditions prevailing in the oceanic moisture source region

(temperature, relative humidity, and, to a lesser degree, wind speed) is now widely used to reconstruct the changes in the temperature of the evaporative source, $T_w$.

A large variety of isotope paleodata is available. Isotope signatures are measured directly in ice cores, groundwaters, and fluid inclusions in speleothems, and indirectly in precipitated calcite, tree ring cellulose, and other organic materials, particularly those in lake sediments. Polar ice cores are particularly suited for paleoclimate reconstructions. First, $\delta_p$ and $T_s$ are strongly correlated in these regions, as illustrated in Figure 3 for Greenland and Antarctica. Second, they give direct access to the precipitation with little postdepositional change at least when the signal can be averaged over a certain number of years. Third, they provide continuous and potentially very detailed sequences with the longest record covering the last 420 ka (thousands of years) at the Vostok site in Antarctica (and possibly even older periods in the recently drilled EPICA core at the Dome C site also in Antarctica). Fourth, they allow measurements of both $\delta D$ and $\delta^{18}O$ and thus are ideal to exploit the deuterium excess as an additional paleoindicator. In turn, although information from other archives and from tropical ice cores is now available, the focus of our chapter will be on the paleoclimatic interpretation of deep ice core isotopic profiles from Greenland and Antarctica.

This chapter is organized as follows. We first summarize what is known about present-day distribution of $\delta D$ and $\delta^{18}O$ in atmospheric water vapor and in precipitation and how this distribution relates with climatic parameters (Section 4.08.2). The next section deals with the physics of isotopes, i.e., how differences in physical properties affect the isotopic concentration in the various water phases. In Section 4.08.4, we review the various

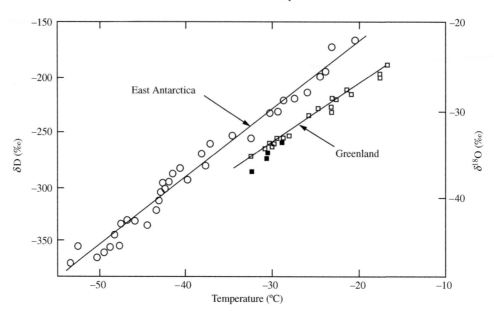

**Figure 3** Isotope content of snow versus local temperature (annual average). Antarctic data ($\delta$D, left scale) are from Lorius and Merlivat (1977) and Greenland data ($\delta^{18}$O, right scale) are from Johnsen *et al.* (1989).

models that allow us to describe isotopic distributions in water vapor and precipitation, and their ability to account for present-day observations. The next three sections are dedicated to isotopic paleodata and their paleoclimatic interpretation. First (Section 4.08.5), we review available isotope ice core data and then (Sections 4.08.6 and 4.08.7) the status of their interpretation in terms of paleotemperature change. In these two sections, we focus on Greenland and Antarctic deep ice cores, dedicating Section 4.08.6 to the conventional approach and a large part of Section 4.08.7 to the calibration of the "isotopic paleothermometer" through the comparison with other methods allowing one to estimate temperature changes in polar regions. The conventional approach underestimates temperature changes in Greenland, whereas available results tend to support its use for Antarctica. Our understanding of why the situation differs between both ice sheets is discussed in the light of model results and of the additional information that can be derived from the deuterium-excess parameter. Finally, we highlight some aspects of isotope modeling and the prospects of isotope climatology for the coming years.

### 4.08.2 PRESENT-DAY OBSERVATIONS

Precise measurements of the natural abundance of deuterium and $^{18}$O in meteoric waters started shortly after World War II and the interest of such measurements for studying various aspects

of the hydrological cycle was soon realized (Dansgaard, 1953; Friedman, 1953; Epstein and Mayeda, 1953). In 1961, a first attempt to summarize data was published by Craig (1961), who defined the Meteoric Water Line (MWL: $\delta$D = 8$\delta^{18}$O + 10) on which falls nonevaporated continental precipitation (Figure 2). The same year, IAEA and WMO initiated a worldwide survey of the isotope composition of monthly precipitation that has been in operation since then (Rozanski *et al.*, 1993). This survey has been the basis of several comprehensive studies in which the isotopic composition of global precipitation was discussed (Dansgaard, 1964; Friedman *et al.*, 1964; Craig and Gordon, 1965; Merlivat and Jouzel, 1979; Gat, 1980; Yurtserver and Gat, 1981; Rozanski *et al.*, 1992, 1993). The reader interested in a detailed description of available data should refer to the very comprehensive review of Rozanski *et al.* (1993). For our purpose, oriented towards a model/data comparison, we limit ourselves to describe the main characteristics of observed distributions. With this in mind, we successively examine the temporal and spatial patterns and then the relationship with meteorological parameters.

This presentation of data based on $\delta^{18}$O concentrations would also apply for $\delta$D as these two parameters are very strongly correlated. Analysis of the long-term annual mean $\delta$D and $\delta^{18}$O values of all IAEA/WMO network confirms that the MWL line defined by Craig (1961) is a good approximation of the locus of points representing average isotopic composition of freshwaters worldwide (Rozanski *et al.*, 1993). Beyond this general relationship, the analysis of both δD and $\delta^{18}O$ in a given precipitation, however, brings additional information about the processes leading to its formation (Section 4.08.8). The deuterium excess, defined by Dansgaard (1964) as $d = \delta D - 8\delta^{18}O$, is a very useful parameter to examine this information.

Studies of individual precipitation events revealed that stable isotope composition collected during single events may vary dramatically (see Rozanski *et al.*, 1993, and references therein). However, examination of IAEA/WMO data shows that a clear temporal pattern (also seen in Greenland and Antarctic snow) emerges after averaging on a monthly basis. As illustrated in Figure 4 for a few selected sites, precipitation at continental mid- and high-latitude stations exhibits a seasonal cycle with that isotopically depleted in winter and enriched in summer, whereas there is a general absence of a defined seasonal cycle for island stations. These seasonal differences are due to several factors (Rozanski *et al.*, 1993): (i) seasonally changing temperature at mid- and high latitudes, with only minor fluctuations in the tropics (Figure 7 shows that a simple Rayleigh model predicts a decrease of the isotopic content of a precipitation when its temperature of formation decreases); (ii) seasonally modulated evapotranspiration flux over the continents induces seasonal differences in the atmospheric water balance; and (iii) seasonally changing source areas of the vapor and/or different storm trajectories. Over continental areas, there is a gradual enhancement of the seasonal variations with increasing distance from the coast with, for example (Rozanski *et al.*, 1993), an amplitude of the $\delta^{18}O$ signal of 2.5‰ at Valentia station in Ireland and of ~10‰ at Moskow, 3,200 km inland. This is largely due to an increase of the seasonal temperature range when going inland. Note also that the deuterium excess may vary seasonally in a given site which, in turn, indicates that the local $\delta D/\delta^{18}O$ slope differs from the value of 8 that applies to a global scale. For example, in Vienna *d* is higher in winter than in summer, which results in a slope lower than 8 (Rozanski *et al.*, 1993).

Figure 5 displays the annual average $\delta^{18}O$ of precipitation obtained from IAEA/WMO network observations and complementary data (Jouzel *et al.*, 1987a). There is a clear latitudinal pattern, with $\delta^{18}O$ decreasing as one approaches the poles. This latitudinal pattern is modulated by a continental effect, i.e., at a given latitude, $\delta^{18}O$ decreases when moving inland. Another aspect of these spatial variations, which is of practical significance for hydrological applications, is the altitude effect (in a given region $\delta^{18}O$ at higher altitudes generally will be more negative).

The magnitude of the altitude effect depends on local climate and topography, with gradients in $\delta^{18}O$ of 0.15–0.50‰ per 100 m (Yurtserver and Gat, 1981).

There is a strong correlation between $\delta^{18}O$ and surface temperature fields between mid- and high latitudes (Figure 1), which does not hold true for tropical and equatorial regions. The $\delta^{18}O$ surface temperature plot of Figure 3 illustrates the strength of this correlation for Greenland and Antarctic sites, whereas Figure 1 was established using data available at IAEA/WMO stations and complementary data from polar sites (Jouzel *et al.*, 1987a). For temperatures below 15 °C, the $\delta^{18}O/T_s$ gradient is of 0.64‰/°C and the correlation coefficient is of 0.96. Another way to relate $\delta^{18}O$ to temperature is to obtain the $\delta^{18}O/T_s$ from monthly values at a single site. This seasonal slope is well defined over continental areas, but its value is generally lower from that calculated from spatial $\delta^{18}O/T_s$ gradient (Rozanski *et al.*, 1992). The relationship between mean annual values of temperature and precipitation $\delta^{18}O$ essentially vanishes above ~15 °C. In contrast, for tropical and equatorial regions, there is, at least for oceanic areas, some relationship between the annual amount of precipitation and its $\delta^{18}O$, the so-called "amount effect" (Dansgaard, 1964), with rainy regions having low $\delta^{18}O$ and dry regions having high $\delta^{18}O$.

## 4.08.3 PHYSICS OF WATER ISOTOPES

### 4.08.3.1 Fractionation Processes

The saturation vapor pressures of HDO and $H_2^{18}O$ are lower than those of $H_2^{16}O$, both over liquid and solid phases. These differences play an important role in the course of the atmospheric water cycle as they cause fractionation effects at vapor/liquid and vapor/solid phase changes, with the condensed phase in equilibrium with vapor being enriched in heavy isotopes. The fractionation coefficient $\alpha$ is defined as the ratio of D/H or $^{18}O/^{16}O$ in the condensed phase to the value of the same parameter in the vapor phase. As reviewed in Jouzel (1986), many determinations of $\alpha$ have been made for the liquid/vapor fractionation at temperature greater than 0 °C. In contrast, there are only a few determinations for low temperatures, which are important in cloud physics, because supercooled drops and droplets commonly exist in natural clouds down to at least −15 °C (Rogers, 1979) and even below, and for the vapor/solid equilibrium. Fractionation coefficients measured by Merlivat and Nief (1967) and by Majoube (1971a,b), generally adopted in isotopic studies dealing with natural processes, are given in Table 1 for

*Water Stable Isotopes*

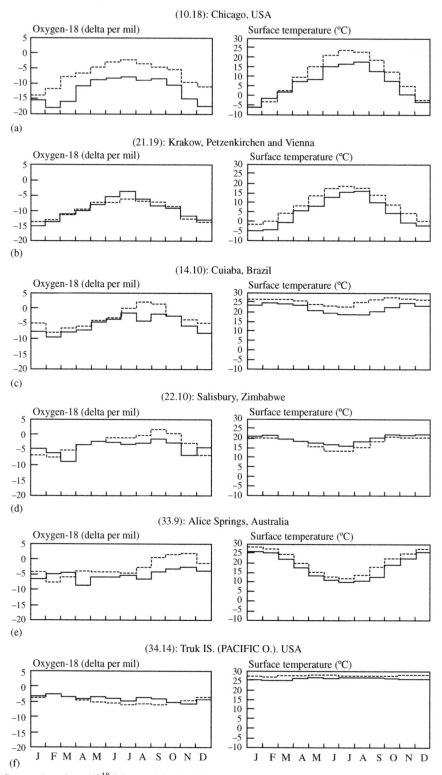

**Figure 4** Seasonal cycles of $\delta^{18}O$ in precipitation for selected sites along with the surface temperature seasonal cycle. The continuous lines correspond to observations and the dotted lines to the seasonal cycles as simulated by the NASA/GISS isotopic GCM in the corresponding grid box (after Jouzel *et al.*, 1987a).

temperatures of 20 °C, 0 °C, and −20 °C. The $\alpha - 1$ ratios reported in this table illustrate that the equilibrium isotopic effect is 8−10 times higher for HDO than for $H_2^{18}O$.

Differences in molecular diffusivities of water molecules in air, $D_{H_2^{16}O}$, $D_{HDO}$, and $D_{H_2^{18}O}$ give rise to what is known as the kinetic isotopic effect. Theoretical and experimental

$\delta^{18}O$ in precipitation (per mil)                   Observations                      Annual

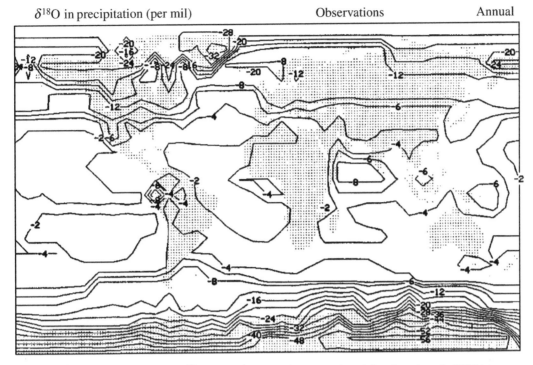

**Figure 5** Map of annual $\delta^{18}O$ in precipitation for observations (after Jouzel *et al.*, 1987a).

**Table 1** HDO/$H_2{}^{16}O$ and $H_2{}^{18}O$/$H_2{}^{16}O$ fractionation coefficients as a function of temperature and of phase change along with the $\alpha-1$ ratio.

| Temperature | Liquid/vapor equilibrium | | | Solid/vapor equilibrium | | |
|---|---|---|---|---|---|---|
| | $\alpha_D$ | $\alpha^{18}O$ | $(\alpha_D - 1)/(\alpha^{18}O - 1)$ | $\alpha_D$ | $\alpha^{18}O$ | $(\alpha_D - 1)/(\alpha^{18}O - 1)$ |
| +20 | 1.0850 | 1.0098 | 8.7 | | | |
| 0 | 1.1123 | 1.0117 | 9.6 | 1.1330 | 1.0152 | 8.8 |
| −20 | 1.1492 | 1.0141 | 10.6 | 1.1744 | 1.0187 | 9.2 |

After Jouzel (1986).

determinations have been obtained by several authors (Craig and Gordon, 1965; Ehhalt and Knott, 1965; Merlivat, 1978). Molecular diffusivity ratios $D_i$ ($D_{HDO}/D_{H_2{}^{16}O}$ and $D_{HDO}/D_{H_2{}^{18}O}$) are practically independent of temperature. They equal 0.9755 and 0.9723, respectively (Merlivat, 1978). The isotopic kinetic effect thus has almost the same value for HDO as for $H_2{}^{18}O$ (the ratio of $1-D_i$ is equal to 0.88). The difference between the relative importance of equilibrium and kinetic effects for HDO and $H_2{}^{18}O$ (8–10 compared to 0.88) is the basic reason for the different behavior between the two isotopes, when nonequilibrium processes take place.

The fractionation effects occurring between liquid and solid phases are linked to variations of other physical properties such as heat capacities and latent heat of fusion. The values generally adopted in studies where these processes are involved are 1.0192 and 1.003 (O'Neil, 1968) for HDO and $H_2{}^{18}O$, respectively.

### 4.08.3.2 Growth of Individual Elements

We now summarize how these slightly different physical properties affect the isotopic composition of water during the growth of individual elements (droplets, drops, ice crystals, hailstones, etc.). In a first stage, the liquid phase of clouds is composed of water droplets formed by condensation of water vapor on a nucleus by heterogeneous nucleation. After this initial stage, growth of droplets and drops is mainly due to collision and coalescence. Larger elements fall faster than smaller ones and capture those lying in their paths (Rogers, 1979). The equations describing how the isotopic concentration of droplets and drops evolves are easy to establish, as they are the same for three water molecules, but for the values of the saturation vapor pressure or of the molecular diffusivity in air (Friedman *et al.*, 1962; Jouzel *et al.*, 1975; Stewart, 1975). One interesting aspect is that the ratio of molecular diffusivities no longer plays a role in the absence of supersaturation ($S = 1$), when the droplet or drop is neither losing nor gaining weight. The isotopic content of the liquid phase $\delta_L$ is

$$\delta_L = \alpha(1 + \delta_v) - 1 + \{(1 + \delta_0) - \alpha(1 + \delta_v)\} \\ \times \exp(-t/t_a) \tag{1}$$

where $\delta_0$ is the isotopic concentration in the liquid at time $t = 0$ and $\delta_v$ the isotopic concentration in the vapor. In this equation, $t_a$ is the adjustment or relaxation time, which increases as the square of the radius and decreases with decreasing temperatures. Typical values of $t_a$ are of a few seconds for droplets and of up to half an hour for large supercooled drops. Jouzel et al. (1975) have demonstrated that droplets with a radius of less than 30 μm can be considered at any time in isotopic equilibrium with the vapor phase:

$$\delta_L = \alpha(1 + \delta_v) - 1 \tag{2}$$

The complete equation applicable when growth is due to vapor transfer, and collection of droplets (coaslescence) has been derived by Jouzel et al. (1975). It allows one to derive an equivalent isotopic relaxation time, $t_a'$ similar to $t_a$, which accounts both for condensation and coaslescence. For a drop of 1 mm radius, $t_a'$ can be up to 15 times lower than $t_a$. This indicates that the collection of droplets plays an important role in maintaining a drop relatively close to isotopic equilibrium during its evolution in a cloud. Under these conditions, the deuterium content of a single drop deviates by less than 10% from isotopic equilibrium during its ascent in typical hailcloud (Jouzel, 1986). As shown in Figure 6, this deviation can be considerably higher when a population of drops is

considered (Federer et al., 1982), owing to the fact that in that case a drop grows from smaller drops not in isotopic equilibrium. In summary, the isotopic equilibrium between the liquid and the vapor phase is only reached for a cloud composed of small droplets. The fact that saturation is very close 1 from means that kinetic effects are not involved.

Kinetic effects have to be taken into account when drops fall below cloud base in a subsaturated environment and they become important for precipitation falling through a dry atmosphere. This process has been the subject of both experimental and theoretical studies (Ehhalt et al., 1963; Dansgaard, 1964; Ehhalt and Knott, 1965; Stewart, 1975). Evaporation results in an isotopic enrichment of the drop both for $\delta D$ and $\delta^{18}O$. However, as the kinetic effect has relatively minor influence with respect to the equilibrium effect for $\delta D$, whereas the two effects have comparable sizes for $\delta^{18}O$, evaporation proceeds on a $\delta D/\delta^{18}O$ line with a slope $<8$, characteristic of the MWL. From isotopic growth equations (Stewart, 1975), it follows that a drop evaporating without collecting droplets evolves along on a slope close to 4. There is one interesting case where kinetic effects should be considered for a liquid phase inside a cloud. It concerns the growth of hail and hailstones in which it is possible to distinguish three types of deposits depending on thermodynamical conditions (Mason, 1971): (i) porous ice formed when the captured droplets freeze rapidly; (ii) compact ice formed when the droplets have time to spread over the surface and form a continuous film before freezing; and (iii) spongy ice produced when the heat exchange is not sufficiently rapid to allow all of the deposited water to freeze. In this latter case, only a fraction of the collected water freezes immediately and produces the skeletal framework of ice which retains the unfrozen water, whereby the whole mixture is maintained at 0 °C. This liquid phase, then much warmer than its environment, is subject to evaporation, and its $\delta D$ and $\delta^{18}O$ concentrations are modified accordingly (again along a $\delta D/\delta^{18}O$ line close to 4 as first studied by Bailey et al. (1969) from analyses of samples of accreted ice formed in an icing tunnel). This effect can also be observed for natural hailstones (Jouzel and Merlivat, 1980; Jouzel et al., 1985). Indeed, hailstones are natural samplers of isotopic conditions prevailing in large convective hailclouds. In turn, the isotopic distribution in hail gives unique insight in mechanisms of formation involved. Such hailstone studies, very interesting from the viewpoint of isotopes and isotopic models, were most active in the 1960s and 1970s (see Jouzel (1986) for a review).

The fact that kinetic effects have to be taken into account when ice crystals formed from water

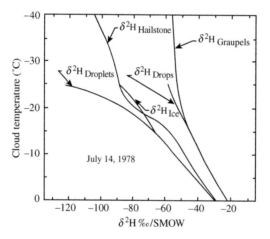

**Figure 6** Isotopic model including liquid and solid phases (Jouzel et al., 1980; Federer et al., 1982) with deuterium content variations of the droplets, drops, ice, and graupels as a function of cloud temperature. $\delta^2 H_{Hailstone}$ is the deuterium content of the water collected by hailstones.

vapor deposition has long been overlooked. It was assumed that snow is formed at isotopic equilibrium (Dansgaard, 1964). However, considering snow is formed in a supersaturated environment with respect to the saturation vapor pressure over ice, and kinetic effects have to be considered in addition to the equilibrium fractionation with respect to ice (Jouzel and Merlivat, 1984; Fisher, 1991). Jouzel and Merlivat (1984) showed that the isotopic concentration of the solid phase, $\delta_s$, can be simply expressed as

$$\delta_s = \alpha_s \alpha_k (1 + \delta_v) - 1 \qquad (3)$$

In this equation, $\alpha_s$ is the equilibrium fractionation between solid and vapor, and

$$\alpha_k = \frac{S_i D_i \alpha_s}{D \alpha_s (S_i - 1) + D_i} \qquad (4)$$

is the kinetic fractionation coefficient, with $S_i$ being the supersaturation of water vapor with respect to ice. The validity of this formulation can be easily demonstrated from laboratory experiments consisting in measuring the condensate formed by water vapor deposited on a cold surface (Jouzel and Merlivat, 1984). In natural conditions $\alpha_k$ depends first on the supersaturation over ice ($\alpha_k$ is equal to 1 when the environment is just saturated with respect to ice, and decreases when supersaturation over ice increases). It depends, to a lesser degree, on pressure, temperature, and on the shape of the growing ice crystal.

Finally, some processes do not give rise to any isotopic fractionation. This is the case when supercooled drops or droplets freeze, because this freezing is instantaneous and there is no time for vapor exchange during the process. It is also the case during melting of compact ice because of the low diffusivity of water which ensures that there is no isotopic homogenization in ice. In contrast, an isotopic fractionation is associated with the freezing of water when the process is sufficiently slow for isotopes to homogeneize, at least partially, in the liquid phase, as it is the case in many natural processes (Jouzel and Souchez, 1982; Jouzel *et al.*, 1999).

### 4.08.3.3 Isotopic Processes in Clouds

The focus in this review is on water isotopes in present-day precipitation and in paleoprecipitation. From an isotopic point of view, understanding and modeling isotopic processes as they occur in clouds, where precipitation is formed, is key to link the growth of individual elements to global-scale distributions. The complexity of cloud isotopic processes depends on the type of cloud considered.

The mechanism of cloud formation (Mason, 1971) is the cooling of moist air below its dewpoint, with the additional requirement for atmospheric clouds, that the air also contains aerosol particles which can serve as condensation nuclei. There are three types of clouds. The stratiform and cumuliform clouds formed, respectively, in stable and in convectively unstable atmospheres, whereas the cirriform clouds are the ice clouds which are, in general, higher and more tenuous, and less clearly reveal the kind of air motion which leads to their formation. As a result of the different rates of cooling and in aerosol content, liquid water contents range from less than $0.01 \text{ g m}^{-3}$ to $10 \text{ g m}^{-3}$, with low values in stratiform clouds (within the narrow range $0.05-0.25 \text{ g m}^{-3}$) and higher values in cumulus clouds (up to $10 \text{ g m}^{-3}$ in localized regions of large cumulus congestus). The ice content in cirriform clouds can be very low in cold regions such as central Antarctica.

Cumulus cloud development is conveniently described in terms of a parcel of air undergoing expansion while being lifted vertically. Its ultimate stage of development is the cumulonimbus, in which hail and hailstones are generated. Very schematically, the condensed phase of a cumulonimbus cloud can be divided into four types of elements: cloud droplets, large drops, ice crystals, and large ice particles (including hail and hailstones). There is mixing of the ascending parcel with the outside air, but in large cumulus clouds, there probably exists a central zone referred to as the updraft core which is unaffected by entrainment (Chilshom, 1973). It is more difficult to discuss stratiform clouds, as the point of origin and the subsequent path of an air parcel are less distinct than for cumulus. However, conceptually at least, stratus can be treated in the same way as cumulus. The same can be said for isotopic modeling: the microphysical processes which lead to fractionation effects, both in stratiform and cirriform clouds, are completely described in the more complete cumulus. Before briefly mentioning how the complexity of isotopic processes is taken into account in cumulus, we first describe the simplest case of a cloud where water vapor and water droplets coexist.

One often assumes that the condensed phase is immediately removed after its formation and that it leaves the air parcel in isotopic equilibrium with the vapor phase (Dansgaard, 1964; Friedman *et al.*, 1964; Taylor, 1972). An opposite approach is to consider the cloud as a closed system. Between these two extreme cases, a variable proportion of the condensed phase, assumed to be in isotopic equilibrium with the vapor, can be kept in the cloud (Craig and Gordon, 1965). To derive the equations for calculating the isotopic concentration in the vapor, $\delta_v$, and in the precipitation $\delta_p$, we use the mass conservation equation for water and water-isotopes (Equations (5) and (6)), and the assumption that the isotopic composition of the liquid kept in the cloud (Equation (7)) and of the precipitation leaving the cloud (Equation (8)) are in isotopic equilibrium with the vapor. Denoting by $n_v$, $n_L$, and $n_p$, the number of moles of vapor, water kept in the cloud, and precipitation, respectively, and $n_v'$, $n_L'$, and $n_p'$, the corresponding number of moles of water isotopes, we can write

$$dn_v + dn_L + dn_p = 0 \qquad (5)$$

$$dn_v' + dn_L' + dn_p' = 0 \qquad (6)$$

$$n_L'/n_L = \alpha_L n_v'/n_v \qquad (7)$$

$$dn_p'/dn_p = \alpha_L n_v'/n_v \qquad (8)$$

where $\alpha_L$ is the fractionation coefficient between liquid and vapor. After logarithmic differentiation of Equation (7), the combination of these equations and of the one defining the $\delta$ values allows us to derive $\delta_v$ (see, e.g., Legrand *et al.* (1994)), and then $\delta_p$ applying Equation (2):

$$\frac{d\delta_v}{1 + \delta_v} = \frac{(\alpha - 1)dn_v + n_L d\alpha_L}{n_v + \alpha n_L} \qquad (9)$$

If no liquid is kept within the cloud ($n_L = 0$), these equations correspond to the classical Rayleigh model, and to a closed system if all liquid is kept. The isotopic composition of the precipitation can be calculated all along the condensation process, if the amount of water leaving the cloud at each step is known. The largest range of $\delta$-variation corresponds to the immediate removal of the condensate (Rayleigh process), and there is a decrease of this range by about a factor of 2, when all liquid formed is kept in the cloud (Jouzel, 1986).

A further complexity arises in cumulus clouds owing to the presence of drops and ice particles which are not in isotopic equilibrium due to the relative fall-speed of these elements with respect to water vapor and cloud droplets. Furthermore, at a given cloud level, the distance to isotopic equilibrium depends on the size and cloud history of these large elements (drops and ice particles). To treat this complexity, Jouzel *et al.* (1980) and Federer *et al.* (1982) used a one-dimensional steady-state cloud model that takes into account mixing with outside air, precipitation fallout, and interactions between water vapor, cloud droplets, large drops, and ice particles. It accounts both for equilibrium and kinetic effects that govern the composition of the droplets and ice crystals, respectively. It includes additional terms corresponding to the isotopic transfer between vapor and drops, calculated for a population of drops assuming a Marshall–Palmer drop distribution (Kessler, 1969), and to the mixing with outside air. This equation is very general and applicable to any

cloud model provided that steady-state conditions exist. As an illustration, the variation of the deuterium content versus temperature of a hailstorm is given in Figure 6 with the curves relative to droplets, drops, cloud ice, graupels (large ice particles), and to the water collected by hailstones. As expected, large drops which are not in isotopic equilibrium with water vapor are enriched in heavy isotopes with respect to cloud droplets.

Gedzelman and Arnold (1994) built on this isotopic approach, but with a more realistic two-dimensional, non-steady-state, cloud model. The model was run for several idealized, classical stratiform and convective storm situations and the resulting isotope ratios of precipitation and water vapor estimated and compared to observations. The model reproduces many of the salient features of isotope meteorology when applied to snowstorms, stratiform rain, and convective precipitation. Also noticeable is the fact that isotope ratios are particularly low when the rain derives from a recirculation process in which air previously charged by vapor from falling rain subsequently rises. This provides a reasonable explanation for extraordinary low isotope ratios observed in some hurricanes and organized thunderstorms.

## 4.08.4 MODELING THE WATER ISOTOPE ATMOSPHERIC CYCLE

Water isotopes have been incorporated into a hierarchy of models, including dynamically simple Rayleigh-type models, and more complex models using atmospheric general circulation models (GCMs) or two-dimensional models. Through a short description of simple distillation models, we first examine the main climate parameters or processes that influence the global distribution of isotopes in precipitation. We then review the current state of development of isotopic GCMs and their performance in simulating present-day distribution of water stable isotopes in precipitation.

### 4.08.4.1 Rayleigh-type Models

A Rayleigh model (Dansgaard, 1964) considers the isotopic fractionation occurring in an isolated air parcel traveling from an oceanic source towards a polar region. The condensed phase is assumed to form in isotopic equilibrium with the surrounding vapor and to be removed immediately from the parcel ($n_L = 0$ in Equation (9)):

$$\frac{d\delta_v}{1 + \delta_v} = (\alpha - 1)\frac{dn_v}{n_v} \qquad (10)$$

This equation can be integrated under isothermal conditions ($\alpha$ which stands here for $\alpha_L$ being then constant). In this case, the isotope content of the precipitation is a unique function of the initial isotopic composition of the water, $\delta_0$, and of the ratio, $F$, of the initial vapor mass within the air parcel and of the water-vapor mass remaining when the precipitation forms:

$$\delta_p = \alpha((1 + \delta_0)F^{(\alpha-1)}) - 1 \qquad (11)$$

However, in nature, condensation processes are not isothermal and mainly occur by cooling of air masses. An exact integration of Equation (10) is no longer possible but $\delta_p$ is very well approximated by

$$\delta_p = \alpha_c((1 + \delta_0)F^{(\alpha_m-1)}) - 1 \qquad (12)$$

in which $\alpha_m$ is the mean $\alpha_L$ value, and $\alpha_c$ its value when condensation occurs. The parcel's water vapor content is proportional to the saturation vapor pressure, a function of temperature and phase change, and is inversely proportional to the air pressure. Thus, in this simple model, the isotope content of precipitation can be expressed as a function of the initial water vapor isotopic concentration and of the initial and final condensation temperatures and air pressures.

Merlivat and Jouzel (1979) showed that the initial isotope concentration in an air parcel above the ocean, $\delta_{v0}$, may, under certain simplifying assumptions, be expressed as a function of the fractionation coefficient $\alpha$ at sea surface temperature, $T_w$, of relative humidity, $h$, of a kinetic fractionation parameter $k$ depending on wind speed (Merlivat, 1978), and of the isotopic composition of the sea surface, $\delta_{ocean}$, as

$$\delta_{v0} = \frac{(1 + \delta_{ocean})(1 - k)}{\alpha(1 - kh)} - 1 \qquad (13)$$

They extended the Rayleigh model by allowing some of the liquid condensate to remain in the parcel (Craig and Gordon, 1965) and by accounting for the kinetic fractionation associated with snow formation by inverse sublimation in a supersaturated environment (Jouzel and Merlivat, 1984). In the extended model, then, the isotope content of precipitation essentially depends on the evaporative source conditions, $h$ and $T_w$, on the proportion of liquid kept in the cloud, and on the condensation temperature, $T_c$. Figure 7 illustrates the observed global-scale decrease of $\delta_p$ with $T_c$ and thus with the surface temperature at the precipitation site, $T_s$, which varies with $T_c$ over most of the globe (polar regions, particularly Antarctica, feature a strong temperature inversion and thus are an exception (not an exception to the assertion that $T_c$ varies with $T_s$, just $T_c > T_s$). The straight line in Figure 7 representing Antarctic data (Dumont d'Urville–Dôme c-axis) is reported with respect to both $T_s$ and the temperature at the

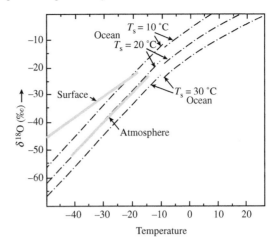

**Figure 7** $\delta^{18}O$ in precipitation as calculated with a Rayleigh model. The three sets of curves are for different initial sea surface temperatures ($T_w$). The approach of Merlivat and Jouzel (1979) is used for the liquid phase, and that of Jouzel and Merlivat (1984) is used for snow formation. The solid lines correspond to East Antarctic data plotted with respect to either $T_s$ (surface temperature) or $T_i$ (temperature in the atmosphere at the inversion level).

inversion level, $T_i$, which is very close to $T_c$ (Robin, 1977). The observed slope obtained in the latter case (1.12‰/°C) is very close to that predicted by the Rayleigh-type model assuming a source temperature of 20 °C (1.1–1.2‰/°C). Thus, given adequate values for its parameters, the Rayleigh-type model can reproduce the data observed over East Antarctica and, more generally, can reproduce the relationship between $\delta_p$ and local temperature for mid- and high latitudes.

Figure 7 also shows the influence of sea surface temperature, $T_w$, on $\delta_p$. Merlivat and Jouzel (1979), Johnsen *et al.* (1989), and Petit *et al.* (1991) show that source conditions (temperature and humidity) also influence the relative amounts of HDO and $H_2^{18}O$ in the parcel and thus the deuterium excess in precipitation. The model reproduces deuterium-excess values observed in Greenland (Johnsen *et al.*, 1989) and in Antarctica, where $d$ becomes higher than 15‰ in central regions (Petit *et al.*, 1991; Dahe *et al.*, 1994). These results as well as those concerning the isotope–temperature relationship were further confirmed by Ciais and Jouzel (1994), who introduced mixed clouds into the Rayleigh-type model, thereby allowing supercooled liquid droplets and ice crystals to coexist between $\sim -15$ °C and $-40$ °C. Such coexistence could be important because the differing saturation conditions over water and ice allows liquid droplets to evaporate while water vapor condenses on ice crystals. Accounting for the associated isotopic fractionation processes in the mixed clouds, however, did not significantly modify the simulated $\delta D$ and $\delta^{18}O$ of the condensed phase.

Jouzel and Koster (1996) later pointed out to a limitation of the formulation proposed by Merlivat and Jouzel (1979) to estimate the isotopic composition of water vapor above the ocean (Equation (13)). This equation is derived by assuming that the mean annual $\delta$ value of the vapor over a given oceanic region can be equated by the mean value of the water which evaporates over this region. This closure equation, which was made by Merlivat and Jouzel (1979) because they were addressing a global-scale problem, is however not valid on a regional scale, for which it has later been often used. The water vapor over a given region is indeed also influenced by distant sources of water vapor transported by the atmospheric circulation. To be correctly treated the problem thus needs to fully account for the transport of water isotopes in the atmosphere as described in the Section 4.08.4.2. Using such an isotopic GCM, Jouzel and Koster (1996) showed that Equation (13) is a correct approximation for global-scale problems such as the simulation of the relationship between $\delta D$ and $\delta^{18}O$. They, however, recommended that the initial conditions for simple models be preferably taken from the GCM estimates of $\delta_{v0}$ over the ocean.

The same physical principles are utilized to develop isotopic models which better account for the transport of air masses at a regional scale, such as done by Fisher (1992) using a regional stable isotope model coupled to a zonally averaged global model. Other authors such as Eriksson (1965) and more recently Hendricks *et al.* (2000) considered the transport of water both by advective and eddy diffusive processes, the latter inducing less fractionation.

As discussed in Section 4.08.3.3, Rayleigh-type fractionation models cannot account for the complexity of large convective systems, such as those occurring in the tropics, for which $\delta_p$ depends on precipitation amount rather than temperature. Despite such limitations, they are able to reproduce the basic behavior of $\delta D$ and $\delta^{18}O$ in precipitation, at least in mid- and high latitudes, where large convective systems do not dominate precipitation production.

### 4.08.4.2 Isotope Modeling with GCMs

Atmospheric GCMs simulate the time evolution of various atmospheric fields (wind speed, temperature, surface pressure, and specific humidity), discretized over the globe, through the integration of the basic physical equations: the hydrostatic equation of motion, the thermodynamic equation of state, the mass continuity equation, and a water vapor transport equation. To reproduce the observed regime of atmospheric circulation, these equations are supplemented with parametrizations for radiative transfer, surface fluxes of momentum, latent heat and sensible heat, latent heat release through condensation, and various internal processes that operate at scales not resolved by the relatively coarse mesh size of the model. These latter processes include turbulence in the boundary layer and cumulus convection, which drives convective precipitation and which redistributes momentum, heat, and water vapor over an atmospheric column. Parametrizations for nonconvective precipitation are also included, as are treatments of heat and water storage in land and ice reservoirs. A full discussion of general circulation modeling, of course, is beyond the scope of this chapter.

The incorporation of the HDO and $H_2^{18}O$ cycles into a GCM involves following the two isotopes through every stage of the GCMs water cycle. Simply put, the model transports the water isotopes between the atmospheric grid boxes and among the surface reservoirs with the same processes used to transport regular water. Isotopic fractionation, including both equilibrium and kinetic effects, is accounted for at every change of phase, i.e., during surface evaporation, atmospheric condensation, and re-evaporation of falling precipitation. The formulations implemented for isotopic fractionation distinguish between convective and nonconvective systems and are largely based on what is used in, or has been learnt from, the simple Rayleigh-type models described above. Although other parts of the hydrological cycle, such as surface hydrological processes and water vapor transport, do not involve fractionation, they still must be extended to include water isotopes. Indeed, a realistic transport scheme for advecting water vapor and isotopes between grid boxes is absolutely critical to the reproduction of observed isotope fields (Joussaume *et al.*, 1984; Jouzel *et al.*, 1991). In particular, the occurrence of negative water mass, which is not a serious problem for some GCMs, is catastrophic for isotope modeling.

Joussaume *et al.* (1984) pioneered GCM isotope modeling, producing global fields of $\delta D$ and $\delta^{18}O$ for present-day January climate using a low-resolution version (32 points in longitude, 24 points in latitude) of the LMD GCM (Laboratoire de Météorologie Dynamique, Paris). Jouzel *et al.* (1987a) generated a full annual cycle of isotope fields with the $8° \times 10°$ (36 points in longitude, 24 points in latitude) GISS GCM (NASA Goddard Institute for Space Studies, New York) and examined the robustness of the approach through an extensive sensitivity study (Jouzel *et al.*, 1991). Simulations using finer spatial resolutions have also been performed for February and August with the LMD model (Joussaume and Jouzel, 1993) and for the full annual cycle with the GISS model (Charles *et al.*, 1994). Water isotopes have then been incorporated into a third model, the ECHAM GCM (Hoffmann and Heimann, 1993), which is the Hamburg version of the European Centre for Medium-Range Weather Forecast GCM. Both coarse grid and fine grid versions of the model have been used. Simulations for present-day climate have been produced using the ECHAM 3 version at two resolutions (T42, T21) corresponding to an LMD, ECHAM, and GISS models, i.e., LMD 4 (Andersen, 1997), ECHAM 4 (Werner *et al.*, 2000; Werner and Heimann, 2002), and the $4° \times 5°$ version of the GISS model (Cole *et al.*, 1999; Charles *et al.*, 2001). Two other isotopic GCMs, based, respectively, on the GENESIS NCAR GCM (Mathieu *et al.*, 2002) and on the CSIRO GCM (Noone and Simmonds, 2002a,b), have been implemented, whereas new isotopic versions of the GISS and LMD GCMs are under development.

The first aim of these simulations is to determine how well isotope GCMs can reproduce observed present-day isotope distributions. The comparison with present-day data is indeed quite satisfying as illustrated from the detailed model/data comparison discussed in Hoffmann *et al.* (2000) for the GISS and ECHAM isotopic models. Figure 8 shows the annual mean $\delta^{18}O$ for the two models. The isotopic composition in the mid- and high latitudes is dominated by the "temperature effect," generating precipitation isotopically more and more depleted with lower temperatures. Both models calculate the most depleted precipitation over the ice sheets of Greenland (about $-30‰$) and Antarctica (about $-55‰$) which is in good agreement with observations, i.e., $-55.5‰$ at Vostok, East Antarctica (Lorius *et al.*, 1985) and $-35‰$ at Summit, central Greenland (Dansgaard *et al.*, 1993; Grootes *et al.*, 1993). The relatively low values over the interior of North America and Siberia demonstrate the strong influence that continentality has on the rainout of air masses and the corresponding isotopic depletion of rain. Owing to its finer orographic resolution the ECHAM simulates a strong uplift and, therefore, a more complete rainout of air masses at large mountain chains as the Andes and the Rocky mountains. Globally, the ECHAM calculated an isotope-latitude gradient of $-0.13‰$ per 100 m, which is a little too low compared with observations. Considering the large differences between the two models in the physical parametrization of sub-grid-scale processes, the resemblance of the simulated water isotope is quite astonishing. Obviously the principal processes controlling the isotopic composition of rain, such as the transport of air masses and the rainout of vapor as a result of cooling, are simulated in a similar and, as the comparison with observations shows, satisfying way.

The general resemblance between results of the GISS and the ECHAM model and their excellent correspondence with observations is also confirmed by the consideration of the zonal mean $\delta^{18}O$ values in precipitation (Hoffmann *et al.*, 2000). The difference between the two models and the corresponding zonal means of $\delta^{18}O$ of precipitation from the IAEA network hardly exceeds 1‰ over nearly all latitudes.

Jouzel *et al.* (2000) have compared the simulated relationships between annual mean isotope concentration and annual mean temperature for the ECHAM, GISS, and LMD isotopic models. The linear regressions between these parameters are performed over two temperature ranges, namely, temperatures above and temperatures below 15 °C (0 °C for ECHAM). The upper range encompasses tropical sites for which the isotope content of precipitation is controlled mostly by the amount of precipitation and not by temperature. Isotopic GCMs are successful in simulating this effect. However, predicted slopes between $\delta^{18}O$ and the amount of precipitation may be lower than the observed ones as seen for tropical islands with the ECHAM 3 model (Hoffmann and Heimann, 1997; Hoffmann *et al.*, 1998) and for tropical and equatorial regions taken as a whole with the GISS model (Jouzel *et al.*, 1987a). In the lower range, the models correctly simulate the observed linear relationship between $\delta^{18}O$ and temperature, as illustrated in Figure 1 for the GISS model. The observed and predicted gradients are within ~10%, except for the ECHAM model, which predicts a slightly lower value (Jouzel *et al.*, 2000). At the regional level, recent simulations performed by Werner *et al.* (1998) and Werner and Heimann (2002) using ECHAM 4 have clearly shown that the quality of an isotopic GCM in terms of its ability to simulate correctly the observed isotopic distribution can be excellent when using a high-resolution model.

A successful isotope GCM must also reproduce the observed seasonal cycles of $\delta^{18}O$. Seasonal cycles generated with the GISS $8° \times 10°$ model, the first to simulate a full seasonal cycle (Jouzel *et al.*, 1987a), are generally realistic, as shown in Figure 4. The seasonal amplitude of $\delta^{18}O$ in precipitation is generally larger at continental stations than at island stations (IAEA, 1981, 1992), and these larger amplitudes, though slightly underestimated in Canada and Siberia, are well simulated in central North America. Simulated amplitudes for Western Europe, Northeast America, and Southeast Asia are also quite reasonable. The GCM reproduces the general absence of a defined seasonal cycle over the island stations, the northern hemisphere character of the cycles in South America and southern Africa, and the early spring maximum in Australia. The higher-resolution LMD simulation also produces

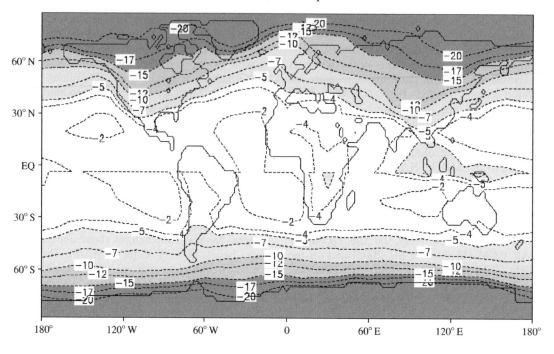

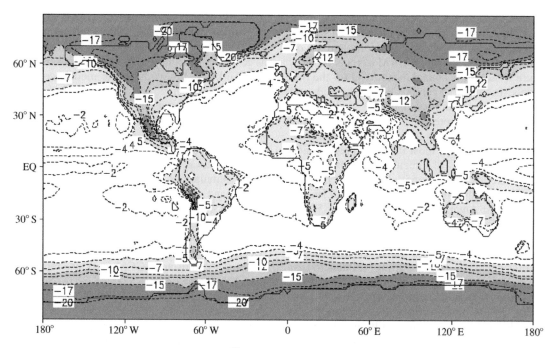

**Figure 8**  Plot of present-day annual average $\delta^{18}O$ in precipitation as simulated by the GISS and ECHAM isotopic models (after Hoffmann *et al.*, 2000).

a realistic seasonal contrast (Joussaume and Jouzel, 1993). The ECHAM model (Hoffmann and Heimann, 1993) produced mixed results, with realistic cycles simulated over central North America and poorer cycles simulated over the western Pacific and central Europe. Some of the coarse grid ECHAM model's deficiencies in this regard (e.g., over the western Pacific) can easily be explained by deficiencies in the simulated climate

itself, whereas others require a different explanation, such as low model resolution. As summarized in Hoffmann *et al.* (2000), a comparison with observations reveals that for all regions the seasonal gradient is lower than the spatial gradient for both the observations and the model results. This fact is usually interpreted as the influence of seasonally changing conditions in the main source regions, and model results of the seasonal gradients are generally in fair agreement with observations (deviations less than 25%).

The observed linear relationship between $\delta D$ and $\delta^{18}O$ (Figure 2) is also well reproduced, both the $\delta D / \delta^{18}O$ slope and the intercept being predicted correctly; for example, the relationship obtained in the GISS model $\delta D = 8.06 \delta^{18}O + 10.4$ is very close to the MWL. This model also captures some of the regional characteristics of the $\delta D / \delta^{18}O$ relationship as the lower slope observed for tropical islands (Jouzel *et al.*, 1987a). The agreement between observed and predicted deuterium-excess values is also relatively good (the difference does not exceed a few per mil) when one considers that *d* is a second-order parameter and that the observations are less complete than those of $\delta^{18}O$ alone. Globally, the deuterium excess simulated with the ECHAM model agrees also fairly well with observations (Hoffmann *et al.*, 1998).

Finally, the GCM approach is particularly well suited for examining the link between the evaporative origin of a precipitation mass and its isotope content. Water evaporating from a well-defined source region on the Earth's surface can be "tagged" in the GCM and followed through the atmosphere until it precipitates. Through this approach, the relative contributions of many different evaporative regions to a given region's precipitation can be quantified exactly. Joussaume *et al.* (1986) determined the evaporative contribution of 10 global divisions to local continental precipitation in the LMD model. The GISS model was used to determine the sources of local precipitation in the northern hemisphere (Koster *et al.*, 1986) and the differences in the sources of Sahelian precipitation during wet and dry years (Druyan and Koster, 1989). Using the GISS model, Koster *et al.* (1993) found that a region's $\delta^{18}O$ in precipitation is significantly related to the extent of continental water recycling in the region. Koster *et al.* (1992) showed that the deuterium content of Antarctic precipitation decreases as the temperature of the evaporative source for the water increases, by about the amount predicted by simple Rayleigh-type models. The situation appears to be more complex over Greenland. Charles *et al.* (1994) performed a similar experiment with the $4° \times 5°$ version of the GISS model, focusing on Greenland. As for Antarctica, several evaporative source regions contribute to the precipitation and the isotope contents of the different contributions vary significantly. For example, moisture from the North Pacific source arrives at the Greenland coast with a $\delta^{18}O$ value ~15‰ lower than its North Atlantic counterpart, a difference attributed to the fact that North Pacific moisture is advected along a much colder path before reaching Greenland (Charles *et al.*, 1994). The tagging of moisture sources has been further

exploited both with the GISS (Armengaud *et al.*, 1998; Delaygue *et al.*, 2000a,b) and the ECHAM (Werner *et al.*, 2001) models (see also Section 4.08.8.2).

These modeling efforts have a further common objective, the reconstruction of paleoclimatic isotope fields to help in the interpretation of paleodata. This important application of water isotope models will be examined in Section 4.08.8.3 after reviewing existing ice core isotopic records (Section 4.08.5), their conventional temperature interpretation (Section 4.08.6), and how this interpretation compares with other estimates of temperature changes (Section 4.08.7).

## 4.08.5 ICE CORE ISOTOPIC RECORDS

Since the early 1950s, ice cores have provided a wealth of information about past climatic and environmental changes. They give access to paleoclimate series that include local temperature and precipitation rate, moisture sources conditions, wind strength and aerosol fluxes of marine, volcanic, terrestrial, cosmogenic, and anthropogenic origin. Elemental composition of the air bubbles gives access to changes in greenhouse gases and in other gaseous compounds, whereas their isotopic composition also contains climate related information. One such example concerns the isotopic composition of the $O_2$ in air which is related to the hydrological cycle in the tropics and to sea-level change.

Ice cores from Greenland, Antarctica, and other glaciers encompass a variety of timescales. The longer timescales, in which we are interested in here, are covered by deep ice cores, the number of which is still very limited. In Greenland (Figure 9), there are only six cores reaching back the last glacial maximum (LGM) 20 kyr ago: Camp Century (Dansgaard *et al.*, 1969), Dye 3 (Dansgaard *et al.*, 1982), GRIP (Johnsen *et al.*, 1992a, 1997; Dansgaard *et al.*, 1993), GISP 2 (Grootes *et al.*, 1993), Renland (Johnsen *et al.*, 1992b), and the ongoing North GRIP project (Johnsen *et al.*, 2001). Deep drilling in Antarctica (Figure 10) started with Byrd (Dansgaard *et al.*, 1969; Epstein and Sharp, 1970) and was then followed by Dome C (Lorius *et al.*, 1979) and the series of cores drilled at the Vostok station, the last one being stopped at a depth record of 3,623 m (Petit *et al.*, 1999). Three other East Antarctic inland cores reached back the LGM at Dome B (Jouzel *et al.*, 1995), Dome F (Watanabe *et al.*, 2003), and Dome C (Jouzel *et al.*, 2001), in the frame of the ongoing European Project for Ice Coring in Antarctica (EPICA). The LGM period is also covered by three other cores drilled in more coastal sites at Taylor Dome (Steig *et al.*, 1998),

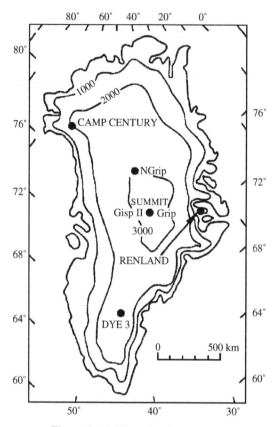

**Figure 9** Drilling sites in Greenland.

based on a multiparametric approach is extensively used for Greenland cores and for high accumulation Antarctic sites. It is not feasible in low accumulation areas such as central Antarctica where other approaches must be employed. For example, the Vostok core has been dated combining an ice flow and an accumulation model (Petit *et al.*, 1999), but other methods (orbital tuning, comparison with other series, and inverse modeling) have been applied (see Parrenin *et al.*, 2001, and references therein). Finally, the age of the gas is younger than that of the ice due to the fact that air bubbles are trapped when firn closes off at depth. The ice age–gas age difference, $\Delta_{age}$, is currently estimated as a function of temperature and accumulation through a firnification model (Barnola *et al.*, 1991; Schwander *et al.*, 1997).

The longest published isotopic records are those of Vostok which cover the last four climatic cycles extending back to ~420 kyr BP (Petit *et al.*, 1999). The next longest is currently Dome F which extends back to ~320 kyr BP (Watanabe *et al.*, 2003), but the timescale covered in Antarctica should be soon extended to at least five climatic cycles, thanks to the new EPICA Dome C drilling. This core reached a depth of 2,871 m in January 2002 with an estimated age of 520 kyr BP (unpublished). Central Greenland records from GRIP (Dansgaard *et al.*, 1993) and GISP 2 (Grootes *et al.*, 1993) provide continuous and reliable isotope climate records back to ~100 kyr BP, but not beyond because of ice flow disturbances due to the proximity of the bedrock (Grootes *et al.*, 1993; Taylor *et al.*, 1993). Other cores cover the last glacial period or, at least, most of it (Byrd, Taylor Dome, Law Dome, Siple Dome, Camp Century, Dye 3, Renland, and North GRIP), and some cores only extend over the LGM or slightly beyond (old Dome C, Dome B, and tropical records). Taken together these ice core isotopic profiles have the following main characteristics.

First, the glacial–interglacial isotopic change, taken between the LGM and the more recent period (i.e., the last thousand years or so), is relatively similar for all these records. The range of variations is remarkably small, if we limit the comparison to inland sites. One exception is Camp Century, with the likely explanation that part of the signal results from a glacial–interglacial increase of the altitude of the Greenland ice sheet at this site (Raynaud and Lorius, 1973). That tropical ice cores show shifts similar to the ones observed in polar areas might be due to the fact that these cores are formed from snow that condenses at the end of the Rayleigh process, as for polar snow. Although these cores are in tropical regions where isotopic concentrations are linked with the amount of precipitation

Law Dome (Morgan *et al.*, 2002), and Siple Dome (Severinghaus *et al.*, 2003). Recently, several tropical ice cores have been obtained that reached back this period either in the Himalayas or in the Andes (Thompson *et al.*, 1995, 1998; Ramirez *et al.*, 2003).

Either $\delta D$ or $\delta^{18}O$ profiles can be indifferently used as a climatic record. Different choices have been made by various teams. The climate reconstruction is based on the interpretation of the $\delta D$ profile for Vostok, Dome B, old Dome C and EPICA Dome C, and on $\delta^{18}O$ for all the other cores. Interestingly, measuring both isotopes on the same core brings additional information about the changes affecting the oceanic sources of Antarctic precipitation through the deuterium-excess parameter. Most of the ice core projects now include measurements of both isotopes. Based on the Vostok core results, we illustrate in Section 4.08.8 how this co-isotopic approach can be used.

Various methods are used to date ice cores each of them having advantages and drawbacks. They fall into four categories: (i) layer counting, (ii) glaciological modeling, (iii) use of time markers and correlation with other dated time series, and (iv) comparison with insolation changes (i.e., orbital tuning). Layer counting

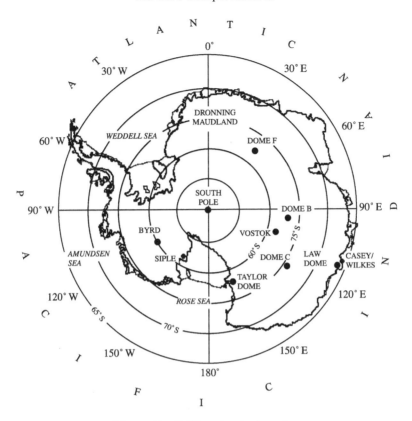

**Figure 10** Drilling sites in Antactica.

(Section 4.08.2), the $\delta_p$ values of high elevation precipitation are most probably controlled by temperature changes.

Second, the minimum isotopic values reached during the coldest parts of the glacial periods (glacial maxima and cold interstadials) are remarkably similar all along the longest Vostok and Dome F records, whereas highest values can be more variable from an interglacial to the next. For example, the last interglacial $\delta^{18}O$ was ~2‰ higher than the recent Holocene which was also true for the interglacial corresponding to marine stage 9 dated ~320 kyr BP (Petit *et al.*, 1999). The two long Antarctic cores Vostok and Dome F show a high degree of similarity both concerning those large changes and interglacials and smaller events, which indeed is remarkable given the distance (~1,500 km) between the two sites (Watanabe *et al.*, 2003). In the more recent period, i.e., since the LGM, similarities are remarkable for East Antarctic inland cores (Jouzel *et al.*, 2001), but there may be some differences between these records and those obtained at sites such as Taylor Dome (Steig *et al.*, 1998) and Siple Dome (Severinghaus *et al.*, 2003).

Third, the most important characteristic of the Greenland records deals with the existence of rapid climatic changes during the last glacial period and the last transition. These "Dansgaard–Oeschger" events were discovered in the Camp

Century and Dye 3 Greenland cores (Dansgaard *et al.*, 1984). Rapid isotopic changes, often more than half of those corresponding to the glacial–interglacial difference and taking place in a few decades, are followed by a slower cooling and a generally rapid return to glacial conditions. The existence and characteristics of these events were fully confirmed at GRIP (Dansgaard *et al.*, 1993) and GISP 2 (Grootes *et al.*, 1993): 24 of such Dansgaard–Oeschger (D/O) events lasting between 500 yr and 2,000 yr occurred during the last glacial period and the last transition. These events (Jouzel *et al.*, 1994; Blunier *et al.*, 1998), and probably all of them (Bender *et al.*, 1994, 1999; Blunier and Brook, 2001), have smooth counterparts in the Antarctic record with, in general, the onset of isotopic changes preceding the onset in Greenland by 1,500–3,000 yr, whereas their maxima are apparently coincident. Raisbeck *et al.* (2002) have, however, pointed out that the situation can differ from at least one of these D/O events, i.e., D/O event 10 that appears to be in phase with its Antarctic counterpart.

In this review, dealing with the temperature interpretation of isotopic ice core records, we will focus mainly on the Vostok core in central East Antarctica and on the GRIP and GISP 2 cores from central Greenland. The reason is that the effort undertaken to calibrate the isotopic paleothermometer through other estimates of temperature changes has so far concentrated on these three cores. We first discuss the conventional approach (Section 4.08.6) and then examine it in the light of other available temperature change estimates (Section 4.08.7).

## 4.08.6 THE CONVENTIONAL APPROACH FOR INTERPRETING WATER ISOTOPES IN ICE CORES

Before examining the "conventional approach" of interpreting ice core $\delta$ profiles, based on the assumption that the spatial and temporal slopes are similar (see Introduction), we note here that the notion of temporal slope covers different timescales. The seasonal slope is deduced from the comparison of the isotope and temperature yearly cycles at a given site (e.g., van Ommen and Morgan, 1997). The short-term (interannual) slope is based on the comparison of mean annual isotope and temperature values at sites where temperature records are available (see, e.g., Jouzel *et al.* (1983)), whereas the long-term slope mainly refers to glacial–interglacial changes. As discussed in Jouzel *et al.* (1997), these three types of slopes generally differ. It is also important to note that some uncertainty is associated with the estimation of the spatial slope for a given region and that this spatial slope may vary from one region to another. For example, slopes of 6.04‰/°C for $\delta D$ and 0.75‰/°C for $\delta^{18}O$ (defined with respect to surface temperature, $T_s$) are used for interpreting Dome C and Vostok isotopic profiles. Recent estimates (Delmotte, 1997) suggest that these values are not defined to be better than ±10%; they clearly may be higher in other regions of Antarctica, e.g., up to 1‰/°C for $\delta^{18}O$ in some areas of Antarctica.

In this review, we will deal mainly with the comparison between the spatial slope (defined in the region where the site is located) and the long-term temporal slope. We will thus focus on the temperature interpretation of large isotopic changes associated with glacial–interglacial changes, or occurring during glacial periods. Hereafter, the ratio of the spatial to the temporal slopes is denoted as $R_{slopes} = S_{spat}/S_{temp}$, e.g., when $R_{slopes} > 1$, the true temperature change $\Delta T_{s(spat)}$ is larger by a factor $R_{slopes}$ than the one estimated by the conventional approach and vice versa ($\Delta$ denoting the difference between two different periods). An illustration of the conventional approach is provided by the Vostok temperature profile derived from the deuterium record (Figure 11) with

$$\Delta T_{s(spat)} = (\Delta\delta D - Corr_{ocean})/6.04 \qquad (14)$$

In this equation, $Corr_{ocean}$ is a correction applied for the change, $\Delta\delta D_{ocean}$, of the deuterium composition of the ocean resulting from the waxing and the waning of the continental ice sheets during the last four climatic cycles. It is calculated as $Corr_{ocean} = 8\Delta\delta^{18}O_{ocean}$ using the $\delta^{18}O$ oceanic record from Bassinot *et al.* (1994) after appropriate scaling. Note that this oceanic correction was incorrectly applied in Petit *et al.* (1999) and in previous estimates of the Vostok temperature record (Jouzel *et al.*, 1987b, 1993, 1996). It did not accounted for the fact that the influence of any isotopic change at the ocean surface weakens as an air mass becomes isotopically depleted. In a Rayleigh model describing the isotopic behavior of an air mass from its oceanic origin to the precipitation site, the isotopic content of a precipitation (here $\delta D_{ice}$) can be written as $1 + \delta D_{ice} = f(1 + \delta D_{ocean})$, where $f$ is a function of climatological parameters and fractionation

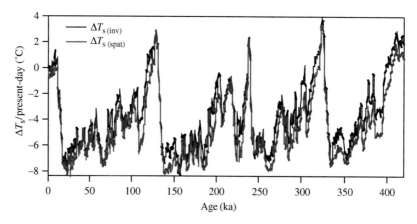

**Figure 11** Vostok temperature changes from present-day values back to 420 kyr BP, estimated either $\Delta T_{s(spat)}$ (in red) by the conventional approach based on the $\delta D$ profile alone (Petit *et al.*, 1999) accounting correctly for the oceanic correction (see text), or $\Delta T_{s(inv)}$ (in green) from the inverse method based on the use of deuterium excess to account for moisture source changes (source Vimeux *et al.*, 2002).

coefficients only. Applying this equation for present-day $\delta D_{ocean(0)}$ and for a certain period in the past, $\delta D_{ocean(t)} = \delta D_{ocean(0)} + \Delta\delta D_{ocean}$, shows that $Corr_{ocean}$ equals $\Delta\delta D_{ocean}(1 + \delta D_{ice})/(1 + \Delta\delta D_{ocean})$ and not $\Delta\delta D_{ocean}$ as previously done for interpreting Vostok records. As illustrated in Figure 12, this would have only a minor impact if the temperature record had been inferred from the $\delta^{18}O$ ice record ($1 + \delta^{18}O_{ice}$ is very close to 1). This is no longer true for the deuterium correction as $1 + \delta D_{ice}$ is between 0.5 and 0.6 in central East Antarctica. Consequently, the glacial–interglacial deuterium oceanic correction which in Petit *et al.* (1999) is slightly below 10‰ for the Vostok site decreases, when correctly applied changes to $\sim 5.5‰$. This reasoning based on a simple model is fully confirmed using an isotopic GCM as checked by Delaygue (2000), who performed two experiments which differ only by a change in the deuterium (9.6‰) and oxygen-18 (1.2‰) of surface oceanic waters, with the isotopic version of the NASA/GISS GCM (Jouzel *et al.*, 1987a). Values are quite similar to those calculated from the above formula (Figure 12) which can thus effectively be used to estimate $Corr_{ocean}$. This is expected because all atmospheric fractionation processes are independent of the isotopic concentrations themselves (small departures with respect to this line are largely due to the contribution of water vapor evaporating over continents).

Of course, the fact that present-day isotope concentrations and local temperature are strongly correlated, as illustrated in Figure 3 for Greenland and Antarctica, does not validate the critical assumption that the present-day spatial slope can be used as a surrogate for the temporal slope. Glaciologists are well aware of, and have systematically pointed out to, the limitations of this conventional approach. As noted by many authors (see Jouzel *et al.*, 1997, and references therein), application of Rayleigh models first points to the combined influence of the temperature of the oceanic source ($T_w$) and of the temperature of condensation ($T_c$) on the isotopic content of a precipitation. Strictly speaking, applying the conventional approach ($R_{slopes} = 1$) requires that there is no change in the temperature of the oceanic source (Figure 13). In contrast, a parallel change in $T_c$ and $T_w$ will, e.g., result in an increase of $R_{slopes}$ (Aristarain *et al.*, 1986; Boyle, 1997). To illustrate this point, we use the formulation derived by Stenni *et al.* (2001) for the EPICA Dome C site in central East Antarctica. We assume (Jouzel and Merlivat, 1984) that the snow is formed just above the inversion layer which allows us to relate $\Delta T_c$ and $\Delta T_s$ through $\Delta T_c = 0.67\Delta T_s$ (Jouzel and Merlivat, 1984). Using the mixed cloud isotopic model (Ciais and Jouzel, 1994) allows us to express the deuterium change at the site, $\Delta\delta D$, as

$$\Delta\delta D = 7.6\Delta T_s - 3.6\Delta T_w + Corr_{ocean} \quad (15)$$

A deuterium change of 60‰ (including the oceanic correction), typical of the

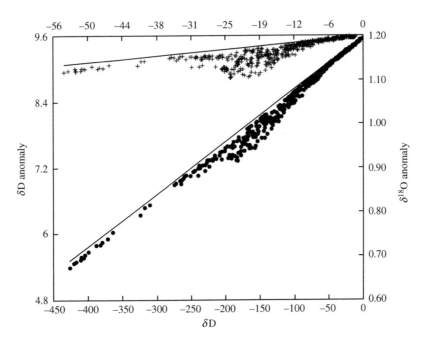

**Figure 12** This plot illustrates how the oceanic correction can be estimated from a simple linear relationship directly derived from a Rayleigh-type model. The points represent the oceanic correction as calculated from the experiments (see legend of Figure 1) performed by Delaygue (2000), all model grid points being represented both for $\delta D$ and $\delta^{18}O$. The two lines are directly derived from a Rayleigh-type model (see text).

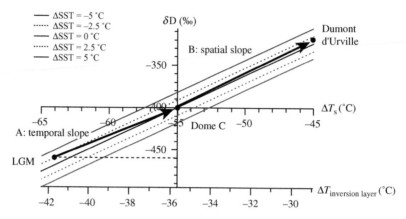

**Figure 13** The influence of the source temperature on the isotopic content of the Antarctic precipitation. Line B corresponds to the observed present-day spatial slope between Dumont d'Urville and Dome C expressed with respect to the temperature of snow formation (i.e., above the inversion layer). Line A represents the temporal slope assuming that the temperature change at the oceanic source is half of that at the Dome C site.

glacial–interglacial transition in central Antarctica, will be interpreted as a $\Delta T_{s(spat)}$ of 7.9 °C if the dependence on the source temperature is neglected as done in the conventional interpretation. In contrast, if we assume that the change in $\Delta T_s$ is accompanied by a concurrent change in $\Delta T_w$ of half its size (assuming a polar amplification of the oceanic change by a factor of 2), and account for it, the resulting estimate of $\Delta T_s$ increases to 10.3 °C ($R_{slopes} = 1.31$; see Figure 13).

Many other factors can obviously influence $R_{slopes}$. They can be linked to other source characteristics controlling the evaporation kinetics such as relative humidity and wind speed (Merlivat and Jouzel, 1979) or to the microphysical processes prevailing in clouds such as the saturation value at snow formation (Fisher, 1991). They can result from changes in the seasonality and intermittency of precipitation fallout, which both bias the temperature sampling by this precipitation, or the wind erosion (Gallée *et al.*, 2001) which can affect the isotopic signal in a different way for an interglacial than for a glacial. Also, as the spatial slope is defined with respect to $T_s$, any change in the strength of the vertical inversion, between, e.g., a glacial and an interglacial, will influence $R_{slopes}$. Changes in cyclonic activity (Holdsworth, 2001) and in the ratio between advection and eddy transport (Hendricks *et al.*, 2000; Noone and Simmonds, 2002b) can also play a role.

### 4.08.7   ESTIMATES OF TEMPERATURE CHANGES IN GREENLAND AND ANTARCTICA

Despite the many factors potentially influencing the temporal slope, the conventional approach has

been used both for Antarctic and Greenland cores until independent estimates of temperature changes became available. However, this was somewhat a surprise in our community when the interpretation of the borehole-temperature profile clearly showed that the present-day spatial slope underestimates glacial–interglacial temperature changes in central Greenland by a factor 2 or so (Cuffey *et al.*, 1995; Johnsen *et al.*, 1995). A second method, based on the detection of anomalies in the isotopic composition of nitrogen and argon in the entrapped air bubbles, has later been exploited (Severinghaus *et al.*, 1998). The use of these two approaches has also been explored for Antarctica where other methods, such as those aiming to derive temperature information from dating constraints, bring useful information. We now summarize these various estimates of temperature changes for Greenland and Antarctica.

#### 4.08.7.1   Greenland

Borehole paleothermometry was the first method allowing precise estimates of glacial–interglacial changes. Because of heat diffusion, conversion of the depth–temperature record of a borehole-to a surface-temperature history involves some difficulties. The borehole-temperature record of old, high-frequency events is lost entirely. More than one possible climate history would produce a given measured borehole profile, although a range of plausible surface-temperature histories can be excluded. A very powerful technique was suggested by Paterson and Clarke (1978). This involves calibrating the isotopic paleothermometer against the borehole profile using an inverse procedure to adjust the constants in the calibration.

This exercise has been independently conducted for the GISP 2 (Cuffey *et al.*, 1995) and GRIP (Johnsen *et al.*, 1995) fluid-filled deep boreholes. The two teams assumed different forms for the calibration curves: Cuffey *et al.* (1995) assumed a linear relation between isotope and temperature but allowed some time variations, while Johnsen *et al.* (1995) kept constant this linear coefficient but allowed for an additional term proportional to δ2. They, however, ended up with similar leading results: (i) the isotopic ratios are an excellent paleothermometer over all time intervals studied and (ii) the calibration is different for recent or short times, i.e., for the last few centuries to millenia (~0.5‰/°C and 0.6‰/°C at GISP 2 and GRIP, respectively, for $\delta^{18}O$) than for older or longer times (~0.33‰/°C and 0.23‰/°C, respectively). The isotopic GRIP and GISP 2 profiles indicate a temperature increase from glacial maximum to Holocene higher than 20 °C and up to 25 °C at Summit, taking into account the lapse-rate effect of thickening associated with accumulation increase at the end of the ice age and the change in seawater $\delta^{18}O$. These large temperature changes were essentially confirmed by Dahl-Jensen *et al.* (1998), who have calculated an independent inverse solution for the GRIP borehole.

Borehole paleothermometry does not allow direct calibration of the isotopic changes at such rapid climatic reorganizations as the termination of the Younger Dryas cold event (Firestone, 1995) or the onset and termination of the stadial phases of the D/O oscillations. This limitation did not hold true for a new method developed by Severinghaus *et al.* (1996, 1998) and Severinghaus and Brook (1999) using a temperature change indicator in the gas phase. Whereas heat diffusion prevents borehole paleothermometry to retain information about the numerous abrupt climatic changes documented in the Greenland isotopic record, this rapidity (a few decades or less) allows the isotopic analysis of air bubbles to provide estimates of abrupt temperature changes.

Due to compaction, the density of snow increases with depth. The entrapped air is thus younger than the ice matrix with the age difference, $\Delta_{age}$, depending both on accumulation and temperature (in central Greenland $\Delta_{age}$ varies between 200 yr and 900 yr). Air composition is very slightly modified by physical processes in them the gravitational and thermal fractionation. The latter depends on the thermal diffusion sensitivity and on the temperature difference between the surface and the close-off depth. Gases diffuse ~10 times faster than heat and in the case of a rapid change, this temperature difference is temporarily modified which causes a detectable anomaly in the isotopic composition of nitrogen and argon. Both the depth of this anomaly, which allows estimate of $\Delta_{age}$ by comparison with the ice $\delta^{18}O$ record, and its strength provide estimates of rapid changes. Focusing first on the rapid warming at the end of the Younger Dryas, 11.5 kyr BP (Preboreal transition), Severinghaus *et al.* (1996, 1998) demonstrated that thermally driven isotopic anomalies are detectable in ice core air bubbles. The inferred $\Delta_{age}$ value indicated that the Younger Dryas was 15 ± 3 °C colder than today, a value about twice larger than $\Delta T_{spat}$. Severinghaus and Brook (1999) then studied the abrupt warming that terminated the glacial period 14.6 kyr BP and led to the Bølling. The warming, directly estimated from the size of the isotopic anomalies, amounts to ~10 °C again about twice larger than $\Delta T_{spat}$.

The abrupt warmings that marked the start of the numerous D/O events during the last glacial period are also larger than initially thought, as shown by Lang *et al.* (1999) from a detailed study of the $^{15}N/^{14}N$ anomaly associated with D/O events 19 (~71 kyr BP). These authors elegantly combine a depth/age model which assumes that accumulation depends on temperature, and a firnification model, to simulate this $^{15}N/^{14}N$ anomaly. The best fit with observations is obtained assuming an abrupt warming of 16 °C, 60% higher than $\Delta T_{spat}$. This model estimate agrees with preliminary estimates proposed for D/O events between 40 kyr BP and 20 kyr BP based on the assumption that $\Delta_{age}$ can be evaluated assuming that methane and temperature changes associated with D/O events are coeval (Schwander *et al.*, 1997).

All the results derived either from borehole paleothermometry or from isotopic anomalies show that the use of the spatial slope underestimate temperature changes in Central Greenland, thus challenging the conventional approach. This is illustrated in Figure 14 (Jouzel, 1999) which includes an additional result concerning the rapid cooling that occurred 8.2 kyr BP (Leuenberger *et al.*, 1999). Temperature changes are consistently higher than $\Delta T_{spat}$, by a factor of ~2 for the LGM, for the Bølling and Preboreal transitions, and for the 8.2 kyr BP cooling, and by ~50% for the investigated D/O events. Note, however, that a recent study of D/O events 12 which combines argon and nitrogen measurements, thus allowing to separate the gravitational and thermal anomalies, indicates warmings only 20–30% larger than derived from the conventional approach (Caillon, 2001).

### 4.08.7.2 Antarctica

In Antarctica, both paleothermometry and the use of nitrogen and argon isotopes pose some problems. First, the low accumulation that prevails at East Antarctic inland sites such as Vostok

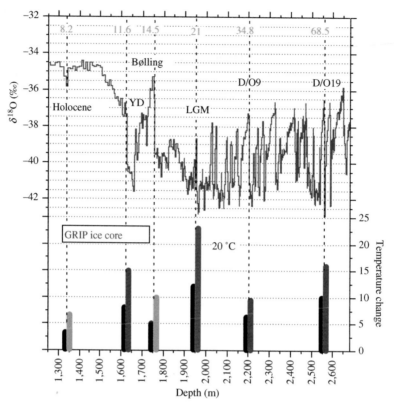

**Figure 14** The bars compare, for the time periods discussed in the text, the absolute temperature changes, $\Delta T^{18}O$, derived from the Greenland $\delta^{18}O$ record using the present-day spatial slope of 0.67‰/°C (black bars) with independent estimates from paleothermometry (blue bar), from estimates of $\Delta_{age}$ (red bars), and from the thermal isotopic anomaly (green bars). Absolute values are reported for the 8,200 BP event which corresponds to a cooling. We have accounted for a 1‰ $\delta^{18}O$ oceanic change when calculating the present-day/LGM temperature difference. The upper continuous curve corresponds to the GRIP $\delta^{18}O$ record (average 100 yr values reported on a linear depth scale; the GRIP and GISP2 $\delta^{18}O$ records are similar once accounted for slight depth adjustments).

erases the glacial–interglacial surface tempera-ture signal at the depth of the LGM. Second, unlike central Greenland, central Antarctica did not experience abrupt temperature changes, and gas isotopic anomalies due to thermal diffusion are, in principle, difficult to detect. Thus, there is no perfect alternative to calibrate the isotopic paleothermometer there. Still there are useful arguments coming from the isotopic composition of the air bubbles (Caillon *et al.*, 2001) and from constraints with respect to ice core chronologies (Parrenin *et al.*, 2001). As recently reviewed by Jouzel *et al.* (2003), they converge towards the idea that the observed present-day spatial slope can be used to interpret Antarctic isotopic profiles.

A straightforward application of an inverse method clearly shows that no useful information can be directly retrieved from borehole paleother-mometry at low accumulation sites like Vostok (Rommelaëre, 1997). To overcome this problem, Salamatin *et al.* (1998 and references therein) developed an inverse procedure based on the assumption that the inferable components of the surface temperature at Vostok can be expressed as

a sum of harmonics of Milankovitch periods. Doing so, the conventional approach of using the present-day spatial slope is confirmed in general (Salamatin *et al.*, 1998), with, however, a significant mismatch between modeled and bore-hole temperatures. This mismatch decreases noticeably if surface temperature is assumed to undergo more intensive precession oscillations than temperature at the inversion level. With this additional assumption, it is inferred that surface temperatures changes were larger by ~30% ($R_{slopes} = 1.3$) and even by up to 50%, depending on assumptions made for the inversion procedure. The fact that Vostok paleothermometry provides such high estimates of changes in glacial–interglacial temperature is now often cited. However, there is currently no clear argument going into support of this additional assumption on which this higher estimate is based.

Although the Antarctic climate is not charac-terized by such abrupt changes as observed in Greenland, Caillon *et al.* (2001) have undertaken a detailed study of the most rapid isotopic warming event that occurred between 107 kyr BP and

108 kyr BP. They successfully measured a small but detectable anomaly in both nitrogen and argon isotopic compositions, resulting possibly from a gravitational signal due to a change in the firn thickness. The position of this anomaly gives a direct estimate of the close-off depth, from which it is inferred that the use of the spatial slope slightly underestimates temperature changes but by no more than $20 \pm 15\%$ (i.e., $R_{slopes} = 1.20$).

In a different study, Blunier *et al.* (in press) proposed to compare the isotopic and methane profiles between Byrd and Vostok in order to constrain the $\Delta_{age}$ at Vostok. They first assumed a large glacial–interglacial temperature (15 °C) and estimated $\Delta_{age}$ for two scenarios with different formulations of the accumulation. The experimentally deduced $\Delta_{age}$ is not compatible with either of those two scenarios. Instead, calculations made assuming the conventional temperature interpretation, and accumulation as in Petit *et al.* (1999), provide $\Delta_{age}$ estimates in agreement with experimentally deduced values (within the error bars). These authors (Blunier *et al.*, in press) conclude that the conventional interpretation is the more probable one.

One way to derive ice core chronologies (Ritz, 1992; Petit *et al.*, 1999; Schwander *et al.*, 2001) is to combine an accumulation history and an ice flow model. As a first approximation, the accumulation rate can be estimated as being proportional to the derivative of the water vapor saturation pressure above the precipitation site (Jouzel *et al.*, 1987b). This reasoning is quite certainly too simplistic. Accumulation and temperature are clearly linked over Antarctica, but the proportionality with the derivative of the saturation vapor pressure is not warranted and other parameters than temperature such as the intensity of the atmospheric circulation may influence accumulation rates. There is no guarantee either on the use of the spatial slope as a surrogate of the temporal slope for interpreting isotopic profiles. With this in mind, Parrenin *et al.* (2001) recently developed an inverse method they first applied to the Vostok ice core. Rather than assuming a linear relationship between the ice deuterium content and temperature with a prescribed slope, they simply use a second-order relationship with two free parameters, thus making no assumption on the amplitude of the glacial–interglacial temperature change. They also assumed that the present-day accumulation upstream of Vostok is a second-order function of the distance to Vostok and discussed the possible influence of atmospheric circulation changes. In turn (see Parrenin *et al.* (2001) for a detailed description), the application of the inverse method provides information both on temperature and accumulation changes. Simply assuming that the number of precessional cycles can be correctly counted in the Vostok profiles, they showed that the

use of the spatial slope slightly underestimates temperature changes but probably by no more than 10–20% over the full range of observed $\delta D$ changes. In this same line, we note the result derived by Schwander *et al.* (2001) from their dating of the EPICA Dome C. These authors inferred that using the spatial slope also underestimates the glacial–interglacial surface temperature change at this site by ~20% (depending on the reference taken for the $S_{spat}$).

## 4.08.8 DISCUSSION

Taken together, this ensemble of results clearly challenges the conventional approach of interpreting ice core isotopic profiles in Greenland with probably a larger bias for the glacial–interglacial change than for the warmings associated with D/O events. However, the results tend to support it in Antarctica, with the exception of borehole paleothermometry. We will now try to understand why the situation differs between both polar ice sheets. To do so, we separately examine the influence of two key parameters, the origin and the seasonality of the precipitation, and then discuss estimates of the temporal slope derived from experiments performed with isotopic GCMs.

### 4.08.8.1 Influence of the Seasonality of the Precipitation

There is practically no information on the seasonality of Antarctic and Greenland precipitation for a period such as the LGM and, even for present-day, only limited data for inland Antarctica (Ekaykin *et al.*, 2002). This information can only be derived from GCM experiments as closely examined by Krinner *et al.* (1997). For this purpose, these authors implemented the LMDs stretched-grid GCM adapted to have a high resolution in polar regions with various diagnoses providing the precipitation weighted temperature of the model layers where the precipitation forms. The difference between the glacial–interglacial change in the condensation temperature and $\Delta T_s$ gives thus an indication of the bias introduced by seasonality and inversion when interpreting the isotopic signal of polar precipitation. The bias due to seasonality is large for Greenland, where the model does not simulate a clear seasonality of present-day precipitation but shows a clear summer precipitation maximum for the LGM. These seasonality changes can largely explain the fact that using the present-day spatial slope for interpreting GRIP and GISP 2 isotopic profiles underestimates glacial–interglacial temperature changes by a factor of 2 (Krinner *et al.*, 1997). This conclusion is supported by experiments performed using the GENESIS/NCAR model (Fawcett *et al.*, 1997). Unlike Krinner *et al.* (1997) and Fawcett *et al.* (1997), Werner *et al.* (2000, 2001) used a GCM implemented with water isotopes (ECHAM) to examine this problem. They clearly showed that the conventional approach is biased due to a substantially increased seasonality of the precipitation during the LGM. During the glacial winter a much more zonal circulation prevents the effective transport of moisture to the Greenland ice sheet, and therefore reduces the contribution of isotopically strongly depleted winter snow to the annual mean isotope signal (Werner *et al.*, 2000). Significant changes are seen in the seasonal cycle of water from polar seas and the North Atlantic which under LGM climate is no longer transported to Greenland during autumn and winter, but during summer season (Werner *et al.*, 2001).

In contrast to Greenland, the experiments of Krinner *et al.* (1997) indicate that the condensation temperature seasonal cycle remains close to the modern level on the East Antarctic Plateau, and that there is only a weak bias due to seasonality (and little influence of other local parameters such as the intermittency of the precipitation and the strength of the inversion). Using the same diagnoses, Delaygue *et al.* (2000b) arrived at a similar conclusion for the GISS model: glacial conditions decrease the winter contribution to annual precipitation inducing a limited 15% decrease of $R_{slopes}$. Werner *et al.* (2001) also noted that in the ECHAM model, glacial–interglacial changes in the seasonal distribution of precipitation are much smaller for Antarctica than for Greenland, and have thus less influence on $R_{slopes}$. In turn, whereas seasonality changes are a plausible explanation of the underestimation of Greenland temperature changes, all available GCM experiments point to a limited impact of these processes for central Antarctica.

### 4.08.8.2 Influence of the Origin of the Precipitation

There are two complementary ways to assess the influence of the origin of a precipitation on its isotopic content. First, the combined measurement of both $\delta D$ and $\delta^{18}O$ enables the calculation of a second-order isotopic parameter, the deuterium excess ($d = \delta D - 8\delta^{18}O$) which (see Section 4.08.4.1) depends on the temperature and relative humidity of the evaporative source (and, to a lesser degree, on the wind speed). In turn, this parameter contains information about conditions prevailing in these source regions and it has been applied, as of early 2000s, only for Antarctic sites (Cuffey and Vimeux, 2001; Stenni *et al.*, 2001; Vimeux *et al.*, 2002), to correct the conventional approach for source temperature

changes and to provide an estimate of those changes. Second, it is possible to perform GCM experiments in which the origin of the precipitation, and in most experiments its isotopic composition, is tagged and then followed from its source to the precipitation site. This approach has now been applied in several experiments addressing the relationship between the origin and the isotopic content of the Antarctic or Greenland precipitation for present-day (Koster *et al.*, 1992; Armengaud *et al.*, 1998; Noone and Simmonds, 2002b) and glacial climates (Charles *et al.*, 1994; Delaygue *et al.*, 2000a,b; Werner *et al.*, 2001).

To illustrate the first approach (Figure 11), Jouzel *et al.* (2003) compared the Vostok temperature records as reconstructed using the conventional approach (see above) and applying a source correction derived from the deuterium-excess profile (Vimeux *et al.*, 2002). Basically, a linear inversion procedure is applied. It allows for the extraction of both $\Delta T_s$, denoted hereafter $\Delta T_{s(inv)}$, and $\Delta T_w$ from these two sets of parameters (Cuffey and Vimeux, 2001; Stenni *et al.*, 2001). Unlike the conventional approach, which relies on present-day observations, this inversion is thus a purely model-based approach. The inversion procedure in which there is an attempt to account for the moisture source changes (Vimeux *et al.*, 2002) and the conventional approach in which the influence of those changes is ignored, provide results very close to each other. For example, the estimates of the successive glacial–interglacial changes are slightly larger when estimated by the conventional approach but by no more than ~10% on the average (with some differences from one cycle to another). As previously noted by Cuffey and Vimeux (2001), the most noticeable difference occurs during glacial inceptions with the consequence that applying a deuterium-excess correction improves the degree of covariation between carbon dioxide and temperature.

Experiments performed with the isotopic version of the NASA/GISS model lead to a similar conclusion (Delaygue *et al.*, 2000b). Unlike simple models that generally consider a unique source, the GCM enables the explicit tagging of the moisture providing from multiple sources (19 in each hemisphere in this particular study). This approach confirms the simple model result that moisture originating from warmer sources provides lower $\delta_p$ but shows additional impact of atmospheric circulation on $\delta_p$ which also depends on the distance between the source and the precipitation site. However, due to changes in the contributions from those various sources, the Antarctic mean source temperature does not significantly change between modern and glacial climates (Delaygue *et al.*, 2000a). This results in a relatively limited increase of $R_{slopes}$ (10–30%).

While changes in seasonality now appear to provide an explanation for the underestimation of Greenland temperatures, the focus there was initially on the origin of the Greenland precipitation. This is based largely on Boyle's (1997) article, where a brief description popularized the idea that a parallel change in source and site temperatures explains that the temporal slope is lower than the spatial slope. The earlier article published by Charles *et al.* (1994), showing that moisture from the North Pacific source arrives at the Greenland coast with a $\delta^{18}O$ value $\sim 15\%o$ lower than its North Atlantic counterpart (see Section 4.08.4.2), should also retain attention. Charles *et al.* (1994) attributed the lower $\delta^{18}O$ to the fact that North Pacific moisture is advected along a much colder path before reaching Greenland. As an extreme example they pointed out that a $\delta^{18}O$ anomaly of $7\%o$ would be generated at a Greenland site if no climate change (including temperature) other than a shift from a pure North Atlantic contribution to an even mixture of North Atlantic and Pacific moisture occurred. This example opened a debate on the relative extents to which local temperature changes and changes in the evaporative sources of precipitation define the isotope shifts recorded in Greenland ice cores. The example, however, must not be misinterpreted, as the GCM results do not suggest such extreme moisture source changes for central Greenland actually occurred. Indeed, the present-day/LGM change in the contribution of North Pacific moisture is only 2%, not 50% as in the above example (see also Charles *et al.* (1995)). A simple calculation that accounts for the simulated changes in moisture sources shows that in the GCM, even a $30\%o$ difference between the isotope contents of Pacific-derived and Atlantic-derived water translates into only a $1\%o$ or $2\%o$ net present-day/LGM change, which is still relatively small with respect to the $5-12\%o$ shifts recorded in Greenland cores.

To explain Greenland observations, Boyle (1997) pointed out the key role of cooler tropics which, as opposed to the classical CLIMAP (1981) reconstruction, has gained large support. This simple explanation is very attractive but, as noted above, Greenland results can be understood only by calling upon changes in the seasonality of the precipitation in central Greenland. In this respect, the results obtained by Delaygue *et al.* (2000b) should be considered as preliminary over Greenland, because the performances of the coarse GISS model are not satisfying (Jouzel *et al.*, 1987a). Still they reinforce the role of local parameters (i.e., seasonal cycle) at the expense of the tropical cooling proposed by Boyle (1997). This is confirmed by Werner *et al.* (2000), who showed, as predicted by Boyle (1997), that

cooler sea surface temperature (SST) shifts the temperature–isotope relationship over Greenland, but the effect is small owing to a parallel change in the geographic origin of the vapor.

### 4.08.8.3 Estimating the Temporal Slope from Isotopic GCMs

Experiments and probes designed to identify separately the influence of specific factors on $R_{slopes}$, such as the origin and seasonality of the precipitation have been performed only recently. One initial objective of isotopic GCMs was to provide a direct comparison between spatial and temporal slopes by simulating different climatic periods and this was done as soon as the ability of isotopic GCMs to reproduce the main characteristics of $\delta D$ and $\delta^{18}O$ in precipitation has been judged to be sufficient. This approach was pioneered by Joussaume and Jouzel (1993) from present-day and LGM simulations using the LMD isotopic model. However, the simulation was limited to a perpetual January and July which did not enable a reliable estimate of $R_{slopes}$ at the yearly scale. Various present-day and LGM experiments covering several years were then performed both with the NASA/GISS (Jouzel *et al.*, 1994; Delaygue *et al.*, 2000b; Charles *et al.*, 2001) and the ECHAM Hamburg (Hoffmann *et al.*, 1997; Werner *et al.*, 2000, 2001) isotopic models.

The first comparison between present-day and LGM simulations showed indeed a low temporal slope ($0.43\%o/°C$ in $\delta^{18}O$) over Greenland (Jouzel *et al.*, 1994), but these authors did not infer from this result that glacial–interglacial change was underestimated applying the conventional approach. One reason was the fact that the two simulations then available (LMD model and coarse grid version of the GISS model) showed differences on a regional basis (Joussaume and Jouzel, 1993). Clearly not too much weight could be given to a conclusion which is model dependent. Rather, Jouzel *et al.* (1994) emphasized one result that is common to both models, i.e., that over polar ice sheets, the predicted temporal and spatial slopes are within 30%, for a given model. Indeed, such a cautious approach finds justification in the fact that simulations performed with versions of the GISS model differing only by their resolution showed that the fine grid version predicts higher temporal slopes over Greenland than the coarse grid version. The results obtained later using the ECHAM model were of better quality in general, over Greenland in particular. Indeed using this model, Werner *et al.* (2000) convincingly showed that the temporal slope is 60% smaller than the model spatial slope in the grid box enclosing Summit ($R_{slopes} = 2.5$), illustrating that the observed discrepancy between borehole and isotope temperatures is clearly reproduced in those simulations.

As far as Antarctica is concerned, the NASA/GISS model predicts spatial slopes lower than temporal slopes over most Antarctic grid points; however, whereas the difference is limited over East Antarctica ($R_{slopes} = 0.80$), it is much larger for West Antarctica ($R_{slopes} = 0.6$). However, in this latter region there are large changes in the prescribed topography of the ice cap which are probably not realistic and bias the estimate of $R_{slopes}$ (which does not account for local changes such as the altitude of the site). Note also that the present-day spatial slope predicted with the NASA/GISS model is higher than the one observed over East Antarctica quite probably because of the very weak simulated inversion strength. In turn, the comparison with data is good when the temperature of condensation, $T_c$, is considered but deteriorates when $T_s$ is taken into account (Jouzel *et al.*, 1994). Although this explanation is satisfying from an isotopic point of view, as it is $T_c$ and not $T_s$ that governs the isotopic content of the precipitation, this discrepancy between observed and predicted inversion strength should be kept in mind. In contrast, ECHAM results, obtained with a more realistic topography and with higher spatial resolution, show an excellent agreement between observed and predicted $S_{spat}$, both for East and West Antarctica (Hoffmann and Heimann, 1997). Comparison with the LGM run (Hoffmann *et al.*, 2000) provides estimates of $R_{slopes} \sim 0.9$. One can thus be confident that temporal and spatial slopes are relatively close to each other in Antarctica when the glacial–modern change is considered. These present-day and LGM ECHAM experiments have now been completed by a series of isotopic simulations with various boundary conditions corresponding to climates intermediate between present-day and the LGM. These new experiments (Hoffmann, personal communication) generally confirm the results presented in Hoffmann *et al.* (2000).

Finally, although not based on a GCM approach, the recent modeling performed by Hendricks *et al.* (2000) is worth mentioning in this section. These authors have developed a one-dimensional model of meridional water vapor transport to evaluate the factors that control spatial and temporal variations of $\delta D$ and $\delta^{18}O$ in global precipitation. They found a good agreement between $S_{spat}$ and $S_{temp}$ for inland sites such as Vostok and South Pole but significantly lower $S_{temp}$ for more coastal regions. Interestingly, this feature of $S_{temp}$ increasing inland is also seen, but much less pronounced, in GCM experiments (Hoffmann *et al.*, 2000; Jouzel *et al.*, 2000).

## 4.08.9 CONCLUSION

The main objective of this chapter was to critically review the paleoclimatic interpretation of large isotopic changes recorded in deep ice cores from Greenland and Antarctica. For this purpose, we have combined various sources of information derived from water isotope models and from independent estimates of temperature changes in polar regions. Before doing so, we have shown how simple isotopic models can be useful tools to understand the distribution of water isotopes ($\delta D$ and $\delta^{18}O$) in precipitation and assessed the ability of isotopic GCMs to correctly simulate present-day isotopic fields both spatially and seasonally. This success justifies the use of these complex isotopic models for simulating isotopic distributions under different climates. These simulations provide model estimates of the temporal isotope/temperature slope to be used for interpreting isotopic paleodata. They offer the possibility of examining the influence of such factors as the seasonality and, provided the different water sources are tagged, the origin of the precipitation. Through this modeling approach we have now a reasonable explanation of why, as shown by other estimates of temperature change, the situation differs between Antarctica and Greenland with respect to the validity of the conventional approach of interpreting ice core isotopic profiles.

In Antarctica, both the origin and the seasonality of the precipitation have a low influence on the temporal slope (Sections 4.08.8.1 and 4.08.8.2). Estimates based on the comparison of LGM and present-day isotopic GCM experiments give temperature changes slightly lower than the conventional approach ($R_{slopes} = \sim 0.8$ and 0.88 with the NASA/GISS and ECHAM models, respectively). Instead, slightly higher temperature changes are suggested by the information inferred for Vostok both from ice core chronologies ($R_{slopes} = \sim 1.2$) and gas-age–ice-age differences ($R_{slopes} = \sim 1.15$–1.2). The paleothermometer calibration based on the borehole-temperature profile provides estimates outside this range ($R_{slopes} \sim 1.3$ and up to 1.5). We suggest that this discrepancy results from assumptions made to invert this temperature profile not fully satisfied, most probably the existence of more intensive precession oscillations at the surface than in the atmosphere during glacial. We, however, note that other empirical estimates (dating and gas age–ice age constraints) point to slightly stronger temperature changes. Also we keep in mind that the lowest estimate of $R_{slopes}$ (0.80) is obtained with the low-resolution NASA/GISS GCM, which does not provide a fully satisfying picture of isotopic distribution for present day (Section 4.08.8.3). In turn, we propose a value of $1.1 \pm 0.2$ as our best current estimate of $R_{slopes}$. We thus conclude that the present-day spatial slope can be taken, within $10 \pm 20\%$, as a surrogate of the temporal slope to interpret isotopic profiles from the East Antarctic Plateau (Jouzel *et al.*, 2003).

In Greenland, there is no doubt that the conventional approach underestimates temperature changes (Figure 14). At first sight, a good candidate to explain this underestimation would be that source and site temperatures have varied simultaneously (Boyle, 1997). Rather, all GCM experiments point to a key role of seasonality and it is this explanation which now appears more likely. In Greenland, seasonality has indeed a strong impact due to the large asymmetrical sea-ice and atmospheric circulation changes associated with a glacial–interglacial transition, whereas East Antarctica is more circularly symmetric with no such large associated changes.

We have focused our model/data comparison on the three cores for which we have independent estimates of temperature changes, Vostok, GRIP, and GISP 2. Our conclusion about the validity of the conventional approach can be easily extended to other cores from the East Antarctic Plateau (EPICA and old Dome C, Dome B, and Dome F), which are remarkably similar for their glacial–interglacial changes (Jouzel *et al.*, 2001; Watanabe *et al.*, 2003). We will be more cautious for other cores from near coastal sites of East Antarctica (Taylor Dome, Law Dome) or from West Antarctica (Byrd and Siple Dome), as isotopic profiles may either be influenced by changes in moisture sources conditions or/and be affected by regional changes in the elevation of the ice cap. In the same way, GRIP and GISP 2 results cannot be extended to other Greenland cores. We are also not inclined to extend our conclusions to shorter timescales as the Holocene. Both model (Cole *et al.*, 1999) and empirical estimates (Jouzel *et al.*, 1997) show that such "short-term" isotopic variability differs from that associated with the cooling during the last ice age.

## ACKNOWLEDGMENTS

The author would like to thank many colleagues for very stimulating discussions on the topics reviewed in this chapter, in particular, Wally Broecker, Nicolas Caillon, Gilles Delaygue, Georg Hoffmann, Sigfus Johnsen, Sylvie Joussaume, Randy Koster, Claude Lorius, Valérie Masson, Liliane Merlivat, Frédéric Parennin, Jean-Robert Petit, Gary Russell, Barbara Stenni, Bob Suozzo, Françoise Vimeux, and Jim White. This work is supported in France by the Programme National d'Études de la Dynamique du Climat (PNEDC) and by the European program Pole-Ocean-Pole (POP EVK2-2000-00089).

## REFERENCES

Andersen U. (1997) Modeling the stable water isotopes in precipitation using the LMD.5.3 atmospheric general circulation model. PhD Thesis, NBIAPG, University of Copenhagen, 100pp.

Aristarain A. J., Jouzel J., and Pourchet M. (1986) Past Antarctic peninsula climate (1850–1980) deduced from an ice core isotope record. *Clim. Change* **8**, 69–89.

Armengaud A., Koster R., Jouzel J., and Ciais P. (1998) Deuterium excess in Greenland snow: analysis with simple and complex models. *J. Geophys. Res.* **103**, 8653–8947.

Baerstchi P. (1976) Absolute $^{18}O$ content of Standard Mean Ocean Water. *Earth Planet. Sci. Lett.* **31**, 341–344.

Bailey I. H., Hulston J. R., Macklin W. C., and Stewart J. R. (1969) On the isotopic composition of hailstones. *J. Atmos. Sci.* **26**, 689–694.

Barnola J. M., Pimienta P., Raynaud D., and Korotkevich Y. S. (1991) $CO_2$ climate relationship as deduced from the Vostok ice core: a re-examination based on new measurements and on a re-evaluation of the air dating. *Tellus* **43B**, 83–91.

Bassinot F. C., Labeyrie L. D., Vincent E., Quidelleur X., Shackleton N. J., and Lancelot Y. (1994) The astronomical theory of climate and the age of the Brunhes-Matuyama magnetic reversal. *Earth Planet. Sci. Lett.* **126**, 91–108.

Bender M., Sowers T., Dickson M. L., Orchado J., Grootes P., Mayewski P. A., and Meese D. A. (1994) Climate connection between Greenland and Antarctica during the last 100,000 years. *Nature* **372**, 663–666.

Bender M., Malaize B., Orchado J., Sowers T., and Jouzel J. (1999) High precision correlations of Greenland and Antarctic ice core records over the last 100 kyr. In *Geophysical Monograph, 112, Mechanisms of Global Climate Change at Millenial Timescales* (eds. P. U. Clark, R. S. Webb, and L. D. Keigwin). American Geophysical Union. pp. 149–164.

Blunier T. and Brook E. J. (2001) Timing of millenial-scale climate change in Antarctica and Greenland during the last glacial period. *Science* **291**, 109–112.

Blunier T., Chappellaz J., Schwander J., Dallenbäch A., Stauffer B., Stocker T., Raynaud D., Jouzel J., Clausen H. B., Hammer C. U., and Johnsen S. J. (1998) Asynchrony of Antarctic and Greenland climate change during the last glacial period. *Nature* **394**, 739–743.

Blunier T., Schwander J., Chappellaz J., and Parrenin F. Antarctic last glacial temperature deduced from Δage. *Earth Planet. Sci. Lett.* (in press).

Boyle E. A. (1997) Cool tropical temperatures shift the global $\delta^{18}O$-T relationship: an explanation for the ice core $\delta^{18}O$ borehole thermometry conflict? *Geophys. Res. Lett.* **24**, 273–276.

Caillon N. (2001) Composition isotopique de l'air piégé dans les glaces polaires: outil de paléothermométrie. Thèse de Doctorat de l'Université Paris VI, University Pierre et Marie Curie, Paris, 269pp.

Caillon N., severinghaus J. P., Barnola J. M., Chappellaz J. C., Jouzel J., and Parrenin F. (2001) Estimation of temperature change and of gas age–ice age difference, 108 kyr BP, at Vostok, Antarctica. *J. Geophys. Res.* **106**, 31893–31901.

Charles C., Rind D., Jouzel J., Koster R., and Fairbanks R. (1994) Glacial interglacial changes in moisture sources for Greenland: influences on the ice core record of climate. *Science* **261**, 508–511.

Charles C., Rind D., Jouzel J., Koster R., and Fairbanks R. (1995) Seasonal precipitation timing and ice core records. *Science* **269**, 247–248.

Charles C., Rind D., Healy R., and Webb R. (2001) Tropical cooling and the isotopic content of precipitation in general circulation model simulations of the ice age climate. *Clim. Dyn.* **17**, 489–502.

Chilshom A. J. (1973) Alberta hailstorms: 1. Radar case studies models. *Meteorol. Monogr.* **14**, 37–95.

Ciais P. and Jouzel J. (1994) Deuterium and Oxygen 18 in precipitation: an isotopic model including mixed cloud processes. *J. Geophys. Res.* **99**, 16793–16803.

CLIMAP (1981) *Seasonal Reconstructions of the Earth's Surface at the Last Glacial Maximum.* Geol. Soc. Am.

Cole J., Rind D., Jouzel J., Webb R. S., and Healy R. (1999) Global controls on interannual variability of precipitation $\delta^{18}O$: the relative roles of temperature, precipitation amount, and vapor source region. *J. Geophys. Res.* **104**, 14223–14235.

Craig H. (1961) Isotopic variations in meteoric waters. *Science* **133**, 1702–1703.

Craig H. and Gordon A. (1965) Deuterium and oxygen 18 variations in the ocean and the marine atmosphere. In *Stable Isotopes in Oceanic Studies and Paleotemperatures.* Consiglio nazionale delle Ricerche, Laboratorio di Geologia Nucleare, Pisa, Italy, pp. 9–130.

Cuffey K. M. and Vimeux F. (2001) Covariation of carbon dioxide and temperature from the Vostok ice core after deuterium-excess correction. *Nature* **421**, 523–527.

Cuffey K. M., Clow G. D., Alley R. B., Stuiver M., Waddington E. D., and Saltus R. W. (1995) Large Arctic temperature change at the Winconsin-Holocene glacial transition. *Science* **270**, 455–458.

Dahe Q., Petit J. R., Jouzel J., and Stievenard M. (1994) Distribution of stable isotopes in surface snow along the route of the 1990 International Trans-Antarctica Expedition. *J. Glaciol.* **40**, 107–118.

Dahl-Jensen D., Mosegaard K., Gundestrup N., Clow G. D., Johnsen S. J., Hansen A. W., and Balling N. (1998) Past temperatures directly from the Greenland ice sheet. *Science* **282**, 268–271.

Dansgaard W. (1953) The abundance of $^{18}O$ in atmospheric water and water vapour. *Tellus* **5**, 461–469.

Dansgaard W. (1964) Stable isotopes in precipitation. *Tellus* **16**, 436–468.

Dansgaard W., Johnsen S. J., Moller J., and Langway C. C. J. (1969) One thousand centuries of climatic record from Camp Century on the Greenland ice sheet. *Science* **166**, 377–381.

Dansgaard W., Clausen H. B., Gundestrup N., Hammer C. U., Johnsen S. J., Krinstindottir P., and Reeh N. (1982) A new Greenland deep ice core. *Science* **218**, 1273–1277.

Dansgaard W., Johnsen S., Clausen H. B., Dahl-Jensen D., Gundestrup N., Hammer C. U., and Oeschger H. (1984) North Atlantic climatic oscillations revealed by deep Greenland ice cores. In *Climate Processes and Climate Sensitivity* (eds. J. E. Hansen and T. Takahashi). American Geophysical Union, Washington, DC, pp. 288–298.

Dansgaard W., Johnsen S. J., Clausen H. B., Dahl-Jensen D., Gunderstrup N. S., Hammer C. U., Steffensen J. P., Sveinbjörnsdottir A., Jouzel J., and Bond G. (1993) Evidence for general instability of past climate from a 250-kyr ice-core record. *Nature* **364**, 218–220.

Delaygue G. (2000) Relationship between the oceanic surface and the isotopic content of Antarctic precipitation: simulations for different climates. PhD Thesis, University of Aix-Marseille.

Delaygue G., Masson V., Jouzel J., Koster R. D., and Healy C. (2000a) The origin of the Antarctic precipitation: a modelling approach. *Tellus* **27**, 19–36.

Delaygue G., Jouzel J., Masson V., Koster R. D., and Bard E. (2000b) Validity of the isotopic thermometer in central Antarctica: limited impact of glacial precipitation seasonality and moisture origin. *Geophys. Res. Lett.* **27**, 2677–2680.

Delmotte M. (1997) Enregistrements climatiques à Law-Dome: variabilité pour les périodes récentes et pour la déglaciation. Thèse de doctorat, Université Joseph Fourier.

Druyan L. M. and Koster R. D. (1989) Sources of Sahel precipitation for simulated drought and rainy seasons. *J. Climate* **2**, 1348–1466.

Ehhalt D. H. and Knott K. (1965) Kinetische isotopentrennungbei der verdampfung von wasser. *Tellus* **17**, 389–397.

Ehhalt D. H., Knott K., Nagel J. F., and Vogel J. C. (1963) Deuterium and oxygen 18 in rain water. *J. Geophys. Res.* **68**, 3775–3780.

Ekaykin A. A., Lipenkov V. Y., Barkov N. I., Petit J. R., and Masson-Delmotte V. (2002) Spatial and temporal variability in isotope composition of recent snow in the vicinity of Vostok Station: implications for ice-core record interpretation. *Ann. Glaciol.* **35**, 181–186.

Epstein S. and Mayeda T. (1953) Variations of $^{18}O$ content of waters from natural sources. *Geochim. Cosmochim. Acta* **4**, 213–224.

Epstein S. and Sharp R. P. (1970) Antarctic ice sheet: stable isotope analyses of Byrd station cores and interhemispheric climatic implications. *Science* **16**, 1570–1572.

Eriksson E. (1965) Deuterium and oxygen-18 in deuterium and other natural waters: some theoretical considerations. *Tellus* **16**, 498–512.

Fawcett P. J., Agustdottir A. M., Alley R. B., and Shuman C. A. (1997) The Younger Dryas termination and north Atlantic deepwater formation: insight from climate model simulations and Greenland ice core data. *Paleocanography* **12**, 23–38.

Federer B., Brichet N., and Jouzel J. (1982) Stable isotopes in hailstones: Part I. The isotopic cloud model. *J. Atmos. Sci.* **39**, 1323–1335.

Firestone J. (1995) Resolving the Younger Dryas event through borehole thermometry. *J. Glaciol.* **41**, 39–50.

Fisher D. A. (1991) Remarks on the deuterium excess in precipitation in cold regions. *Tellus* **43B**, 401–407.

Fisher D. A. (1992) Stable isotope simulations using a regional stable isotope model coupled to a zonally averaged global model. *Cold Reg. Sci. Technol.* **21**, 61–77.

Friedman I. (1953) Deuterium content of natural waters and other substances. *Geochim. Cosmochim. Acta* **4**, 89–103.

Friedman I., Machta L., and Soller R. (1962) Water vapor exchange between a water droplet and its environment. *J. Geophys. Res.* **67**, 2761–2766.

Friedman I., Redfield A. C., Schoen B., and Harris J. (1964) The variation of the deuterium content of natural waters in the hydrologic cycle. *Rev. Geophys.* **2**, 177–224.

Gallée H., Guyomarc'h G., and Brun E. (2001) Impact of snow drift on the Antarctic ice sheet surface mass balance: possible sensitivity to snow-surface properties. *Boundary-Layer Meteorol.* **99**, 1–19.

Gat J. (1980) The isotopes of hydrogen and oxygen in precipitation. In *Handbook of Environmental Isotopes Geochemistry: Volume 1. "The Terrestrial Environment A"* (eds. P. Fritz and J. C. Fontes). Elsevier, vol. 1, pp. 22–46.

Gedzelman S. D. and Arnold R. (1994) Modeling the isotopic content of precipitation. *J. Geophys. Res.* **99**, 10455–10471.

Gonfiantini R. (1978) Standards for stable isotope measurements in natural compounds. *Nature* **271**, 534–536.

Grootes P. M., Stuiver M., White J. W. C., Johnsen S. J., and Jouzel J. (1993) Comparison of the oxygen isotope records from the GISP2 and GRIP Greenland ice cores. *Nature* **366**, 552–554.

Hageman R., Nief G., and Roth E. (1970) Absolute isotopic scale for deuterium analysis of natural waters. Absolute D/H ratio for SMOW. *Tellus* **22**, 712–715.

Hendricks M. B., De Paolo D. J., and Cohen R. C. (2000) Space and time variation of $\delta D$ and $^{18}O$ in precipitation: can paleotemperature be estimated from ice cores? *Global Biogeochem. Cycles* **14**, 851–861.

Holdsworth G. (2001) Calibration changes in the isotopic thermometer according to different climatic states. *Geophys. Res. Lett.* **28**, 2625–2628.

Hoffmann G. and Heimann M. (1993) Water tracers in the ECHAM general circulation model. In *Isotope Techniques in the Study of Past and Current Environmental Changes in the Hydrosphere and the Atmosphere.* IAEA, Vienna, pp. 3–14.

Hoffmann G. and Heimann M. (1997) Water isotope modeling in the Asian monsoon region. *Quat. Int.* **37**, 115–128.

Hoffmann G., Werner M., and Heimann M. (1998) Water isotope module of the ECHAM atmospheric general circulation model: a study on time scales from days to several years. *J. Geophys. Res.* **103**, 16871–16896.

Hoffmann G., Masson V., and Jouzel J. (2000) Stable water isotopes in atmospheric general circulation models. *Hydrol. Process.* **14**, 1385–1406.

IAEA (1981) Statistical treatment of environmental isotope data in precipitation. IAEA, Technical Report Series no. 206, Vienna.

IAEA (1992) Statistical treatment of data on environmental isotopes in precipitation. IAEA, Technical Report Series no. 331, Vienna, 720pp.

Johnsen S. J., Dansgaard W., and White J. W. (1989) The origin of Arctic precipitation under present and glacial conditions. *Tellus* **41**, 452–469.

Johnsen S. J., Clausen H. B., Dansgaard W., Fuhrer K., Gundestrup N. S., Hammer C. U., Iversenn P., Jouzel J., Stauffer B., and Steffensen J. P. (1992a) Irregular glacial interstadials recorded in a new Greenland ice core. *Nature* **359**, 311–313.

Johnsen S. J., Claussen H. B., Dansgaard W., Gundestrup N. S., Hansson N. S., Jonsson P., Steffenssen J. P., and Sveinbjörnsdottir A. E. (1992b) A "deep" ice core from East Greenland. *Medellelser om Groenland Geosci.* **29**, 1–22.

Johnsen S. J., Dahl-Jensen D., Dansgaard W., and Gundestrup N. (1995) Greenland paleotemperatures derived from GRIP bore hole temperature and ice core isotope profiles. *Tellus* **47B**, 624–629.

Johnsen S. J., Clausen H. B., Dansgard W., Gundestrup N. S., Sveinbjörnsdottir A., Jouzel J., Hammer C. U., Andersson U., Fisher D., and White J. (1997) The $\delta^{18}O$ record along the GRIP deep ice core. *J. Geophys. Res.* **102**, 26397–26410.

Johnsen S., Dahl-Jensen D., Gundestrup N., Steffenssen J. P., Clausen H. B., Miller H., Masson-Delmotte V., Sveinbjörnsdottir A. E., and White J. (2001) Oxygen isotope and palaeotemperature records from six Greenland ice-core stations: Camp Century, Dye-3, GRIP, GISP2, Renland and NorthGRIP. *J. Quat. Sci.* **16**, 299–307.

Joussaume S. and Jouzel J. (1993) Paleoclimatic tracers: an investigation using an atmospheric General Circulation Model under ice age conditions: 2. Water isotopes. *J. Geophys. Res.* **98**, 2807–2830.

Joussaume S., Jouzel J., and Sadourny R. (1984) A general circulation model of water isotope cycles in the atmosphere. *Nature* **311**, 24–29.

Joussaume S. J., Sadourny R., and Vignal C. (1986) Origin of precipitating water in a numerical simulation of the July climate. *Ocean–Air Interact.* **1**, 43–56.

Jouzel J. (1986) Isotopes in cloud physics: multisteps and multistages processes. In *Handbook of Environmental Isotopes Geochemistry: Volume 2: "The Terrestrial Environment B"*. (eds. P. Fritz and J. C. Fontes). Elsevier, vol. 2, pp. 61–112.

Jouzel J. (1999) Calibrating the isotopic paleothermometer. *Science* **286**, 910–911.

Jouzel J. and Koster R. (1996) A reconsideration of the initial conditions used for stable water isotopes models. *J. Geophys. Res.* **101**, 22933–22938.

Jouzel J. and Merlivat L. (1980) Growth regime of hailstones as deduced from simultaneous deuterium and oxygen 18 measurements. In *VIIIème International Conference on Cloud Physics*. Conference Proceedings AIMPA, Clermont-Ferrand, pp. 253–256.

Jouzel J. and Merlivat L. (1984) Deuterium and oxygen 18 in precipitation: modeling of the isotopic effects during snow formation. *J. Geophys. Res.* **89**, 11749–11757.

Jouzel J. and Souchez R. A. (1982) Melting-refreezing at the glacier sole and the isotopic composition of the ice. *J. Glaciol.* **28**, 35–42.

Jouzel J., Merlivat L., and Roth E. (1975) Isotopic study of hail. *J. Geophys. Res.* **80**, 5015–5030.

Jouzel J., Brichet N., Thalmann B., and Federer B. (1980) A numerical cloud model to interpret the isotope content of hailstones. In *VIIIème International Conference on Cloud Physics*. Conference AIMPA, Clermont-Ferrand, France, pp. 249–252.

Jouzel J., Merlivat L., Petit J. R., and Lorius C. (1983) Climatic information over the last century deduced from a detailed isotopic record in the South Pole snow. *J. Geophys. Res.* **88**, 2693–2703.

Jouzel J., Merlivat L., and Federer B. (1985) Isotopic study of hail. The $\delta D$-$\delta^{18}O$ relationship and growth history of large hailstones. *Quart. J. Roy. Meteorol. Soc.* **111**, 495–514.

Jouzel J., Russell G. L., Suozzo R. J., Koster R. D., White J. W. C., and Broecker W. S. (1987a) Simulations of the HDO and $H_2^{18}O$ atmospheric cycles using the NASA/GISS general circulation model: the seasonal cycle for present-day conditions. *J. Geophys. Res.* **92**, 14739–14760.

Jouzel J., Lorius C., Petit J. R., Genthon C., Barkov N. I., Kotlyakov V. M., and Petrov V. M. (1987b) Vostok ice core: a continuous isotope temperature record over the last climatic cycle (160,000 years). *Nature* **329**, 402–408.

Jouzel J., Koster R. D., Suozzo R. J., Russell G. L., White J. W., and Broecker W. S. (1991) Simulations of the HDO and $H_2^{18}O$ atmospheric cycles using the NASA GISS General Circulation Model: sensitivity experiments for present day conditions. *J. Geophys. Res.* **96**, 7495–7507.

Jouzel J., Barkov N. I., Barnola J. M., Bender M., Chappelaz J., Genthon C., Kotlyakov V. M., Lipenkov V., Lorius C., Petit J. R., Raynaud D., Raisbeck G., Ritz C., Sowers T., Stievenard M., Yiou F., and Yiou P. (1993) Extending the Vostok ice-core record of paleoclimate to the penultimate glacial period. *Nature* **364**, 407–412.

Jouzel J., Koster R. D., Suozzo R. J., and Russell G. L. (1994) Stable water isotope behaviour during the LGM: a GCM analysis. *J. Geophys. Res.* **99**, 25791–25801.

Jouzel J., Vaikmae R., Petit J. R., Martin M., Duclos Y., Stievenard M., Lorius C., Toots M., Mélières M. A., Burckle L. H., Barkov N. I., and Kotlyakov V. M. (1995) The two-step shape and timing of the last deglaciation in Antarctica. *Clim. Dyn.* **11**, 151–161.

Jouzel J., Waelbroeck C., Malaizé B., Bender M., Petit J. R., Barkov N. I., Barnola J. M., King T., Kotlyakov V. M., Lipenkov V., Lorius C., Raynaud D., Ritz C., and Sowers T. (1996) Climatic interpretation of the recently extended Vostok ice records. *Clim. Dyn.* **12**, 513–521.

Jouzel J., Alley R. B., Cuffey K. M., Dansgaard W., Grootes P., Hoffmann G., Johnsen S. J., Koster R. D., Peel D., Shuman C. A., Stievenard M., Stuiver M., and White J. (1997) Validity of the temperature reconstruction from ice cores. *J. Geophys. Res.* **102**, 26471–26487.

Jouzel J., Petit J. R., Souchez R., Barkov N. I., Lipenkov V. Y., Raynaud D., Stievenard M., Vassiliev N. I., Verbeke V., and Vimeux F. (1999) More than 200 m thick of lake ice above the subglacial Lake Vostok, Antarctica. *Science* 2138–2141.

Jouzel J., Hoffmann G., Koster R. D., and Masson V. (2000) Water isotopes in precipitation: data/model comparison for present-day and past climates. *Quat. Sci. Rev.* **19**, 363–379.

Jouzel J., Masson V., Cattani O., Falourd S., Stievenard M., Stenni B., Longinelli A., Johnsen S. J., Steffenssen J. P., Petit J. R., Schwander J., Souchez R., and Barkov N. I. (2001) A new 27 Ky high resolution East Antarctic climate record. *Geophys. Res. Lett.* **28**, 3199–3202.

Jouzel J., Vimeux F., Caillon N., Delaygue G., Hoffmann G., Masson V., and Parrenin F. (2003) Magnitude of the isotope/ temperature scaling for interpretation of central Antarctic ice cores. *J. Geophys. Res.* **108**(D12), doi:10.1029/2002 JD002677.

Kessler E. (1969) On the distribution and continuity of water substance in atmospheric circulation. *Meteorol. Monogr.* **10**, 84.

Koster R. D., Jouzel J., Suozzo R., Russel G., Broecker W., Rind D., and Eagleson P. S. (1986) Global sources of local precipitations as determined by the NASA/GISS GCM. *Geophys. Res. Lett.* **43**, 121–124.

Koster R. D., Jouzel J., Suozzo R. J., and Russel G. L. (1992) Origin of July Antarctic precipitation and its influence on deuterium content: a GCM analysis. *Clim. Dyn.* **7**, 195–203.

Koster R. D., De Valpine P., and Jouzel J. (1993) Continental water recycling and stable water isotope concentration. *Geophys. Res. Lett.* **20**, 2215–2218.

Krinner G., Genthon C., and Jouzel J. (1997) GCM analysis of local influences on ice core $\delta$ signals. *Geophys. Res. Lett.* **24**, 2825–2828.

Lang C., Leuenberger M., Schwander J., and Johnsen S. J. (1999) 16 °C rapid temperature variation in central Greenland 70,000 years ago. *Science* **286**, 934–937.

Legrand M., Jouzel J., and Raynaud D. (1994) Past climate and trace gas content of the atmosphere inferred from polar ice cores. In *ERCA Publication* (eds.), pp. 453–477.

Leuenberger M., Lang C., and Schwander J. (1999) $\delta^{15}N$ measurements as a calibration tool for the paleothermometer and gas-ice age differences: a case study for the 8200 B.P. event on GRIP ice. *J. Geophys. Res.* **104**(D18), 22163–22170.

Lorius C. and Merlivat L. (1977) Distribution of mean surface stable isotope values in East Antarctica. Observed changes with depth in a coastal area. In *Isotopes and Impurities in Snow and Ice. Proceedings of the Grenoble Symposium Aug./Sep. 1975* (eds. IAHS), IAHS, vol. 118, pp. 125–137.

Lorius C., Merlivat L., Jouzel J., and Pourchet M. (1979) A 30,000 yr isotope climatic record from Antarctic ice. *Nature* **280**, 644–648.

Lorius C., Jouzel J., Ritz C., Merlivat L., Barkov N. I., Korotkevitch Y. S., and Kotlyakov V. M. (1985) A 150,000-year climatic record from Antarctic ice. *Nature* **316**, 591–596.

Majoube M. (1971a) Fractionement en Oxygène 18 et en deutérium entre l'eau et sa vapeur. *J. Chim. Phys.* **10**, 1423–1436.

Majoube M. (1971b) Fractionement en Oxygène 18 et en deutérium entre la glace et sa vapeur. *J. Chim. Phys.* **68**, 635–636.

Mason B. J. (1971) *The Physics of Clouds* (ed. Oxford Press).

Mathieu R., Pollard D., Cole J., Webb R., White J. W. C., and Thompson S. (2002) Simulation of stable water isotope variations by the GENESIS GCM for present-day conditions. *J. Geophys. Res.* **107**, D4, 10.1029.

Merlivat L. (1978) Molecular diffusivities of $H_2{}^{16}O$, $HD^{16}O$, and $H_2{}^{18}O$ in gases. *J. Chem. Phys.* **69**, 2864–2871.

Merlivat L. and Jouzel J. (1979) Global climatic interpretation of the deuterium–oxygen 18 relationship for precipitation. *J. Geophys. Res.* **84**, 5029–5033.

Merlivat L. and Nief G. (1967) Fractionnement isotopique lors des changements d'état solide-vapeur et liquide-vapeur de l'eau à des températures inférieures à 0 °C. *Tellus* **19**, 122–127.

Morgan V., Delmotte M., van Ommen T., Jouzel J., Chappellaz J., Woon S., Masson-Delmotte V., and Raynaud D. (2002) The timing of events in the last deglaciation from a coastal east Antarctic core. *Science* **297**, 1862–1864.

Noone D. and Simmonds I. (2002a) Associations between $\delta^{18}O$ of water and climate parameters in a simulation of atmospheric circulation for 1979–1995. *J. Climate.* **15**(22), 3150–3169.

Noone D. and Simmonds I. (2002b) Annular variations in moisture transport mechanisms and the abundance of $\delta^{18}O$ in Antarctic snow. *J. Geophys. Res.* **107**(D24), 4742, doi: 10.1029/2002JD002262.

O'Neil (1968) Hydrogen and oxygen isotope fractionation between ice and water. *J. Chem. Phys.* **72**, 3683–3684.

Parrenin F., Jouzel J., Waelbroeck C., Ritz C., and Barnola J. M. (2001) Dating the Vostok ice core by an inverse method. *J. Geophys. Res.* **106**, 31837–31851.

Paterson W. S. B. and Clarke G. K. C. (1978) Comparison of theoretical and observed temperature profiles in Devon Island ice cap, Cananda. *Geophys. J. Roy. Astron. Soc.* **55**, 615–632.

Petit J. R., White J. W. C., Young N. W., Jouzel J., and Korotkevich Y. S. (1991) Deuterium excess in recent Antarctic snow. *J. Geophys. Res.* **96**, 5113–5122.

Petit J. R., Jouzel J., Raynaud D., Barkov N. I., Barnola J. M., Basile I., Bender M., Chappellaz J., Davis J., Delaygue G., Delmotte M., Kotyakov V. M., Legrand M., Lipenkov V. Y., Lorius C., Pépin L., Ritz C., Saltzman E., and Stievenard M. (1999) Climate and atmospheric history of the past 420000 years from the Vostok ice core, Antarctica. *Nature* **399**, 429–436.

Raisbeck G. M., Yiou F., and Jouzel J. (2002) Cosmogenic 10Be as a high resolution correlation tool for climate records. In *Goldschmidt Conference*, Conference Davos, Switzerland.

Ramirez E., Hoffmann G., Taupin J. L., Francou B., Ribstein P., Caillon N., Landais A., Petit J. R., Pouyaud B., Schotterer U., and Stiévenard M. (2003) A new Andean deep ice core from the Illimani (6,350 m), Bolivia. *Earth Planet. Sci. Lett.* **212**, 337–350.

Raynaud D. and Lorius C. (1973) Climatic implications of total gas content in ice at Camp Century. *Nature* **243**, 283–284.

Ritz C. (1992) Un modele thermo-mecanique d'evolution pour le bassin glaciaire Antarctique Vostok-Glacier Byrd: sensibilite aux valeurs des parametres mal connus. Thèse d'Etat, Univ. de Grenoble.

Robin G. d. Q. (1977) Ice cores and climatic changes. *Phil. Trans. Roy. Soc. London, Ser. B* **280**, 143–168.

Rogers R. R. (1979) *A Short Course in Clouds Physics* (ed. Pergamon).

Rommelaëre V. (1997) Trois problèmes inverses en glaciologie. PhD Thesis (in french), University Grenoble I, France.

Rozanski K., Araguas-Araguas L., and Gonfiantini R. (1992) Relation between long-term trends of oxygen-18 isotope composition of precipitation and climate. *Science* **258**, 981–985.

Rozanski K., Araguas-Araguas L., and Gonfiantini R. (1993) Isotopic pattern in modern global precipitation. In *Climate Change in Continental Isotopic Records*, Geophysical Monograph 78 (eds. P. K. Swart, K. C. L. J. McKenzie, and S. Savin). American Geophysical Union, Washington, DC, pp. 1–37.

Salamatin A. N., Lipenkov V. Y., Barkov N. I., Jouzel J., Petit J. R., and Raynaud D. (1998) Ice core age dating and paleothermometer calibration on the basis of isotopes and temperature profiles from deep boreholes at Vostok station (East Antarctica). *J. Geophys. Res.* **103**, 8963–8977.

Schwander J., Sowers T., Barnola J. M., Blunier T., Malaizé B., and Fuchs A. (1997) Age scale of the air in the summit ice: implication for glacial-interglacial temperature change. *J. Geophys. Res.* **D16**, 19483–19494.

Schwander J., Jouzel J., Hammer C. U., Petit J. R., Udisti R., and Wolff E. (2001) A tentative chronology of the EPICA Dome C ice core. *Geophys. Res. Lett.* **28**, 4243–4246.

Severinghaus J. P. and Brook E. (1999) Simultaneous tropical-Arctic abrupt climate change at the end of the last glacial period inferred from trapped air in polar ice. *Science* **286**, 930–934.

Severinghaus J. P., Brook E. J., Sowers T. and Alley R. B. (1996) Gaseous thermal diffusion as a gas-phase stratigraphic marker of abrupt warmings in ice core climate records. *EOS Supplement*, AGU Spring meeting, 157.

Severinghaus J. P., Sowers T., Brook E., Alley R. B., and Bender M. L. (1998) Timing of abrupt climate change at the end of the Younger Dryas interval from thermally fractionated gases in polar ice. *Nature* **391**, 141–146.

Severinghaus J. P., Grachev A., Luz B., and Caillon N. (2003) A method for precise measurement of argon 40/36 and krypton/argon ratios in trapped air in polar ice with applications to past firn thickness and abrupt climate change in Greenland and at Siple Dome, Antarctica. *Geochim. Cosmochim. Acta* **67**(3), 325–343.

Steig E., Brook E. J., White J. W. C., Sucher C. M., Bender M. L., Lehman S. J., Morse D. L., Waddigton E. D., and Clow G. D. (1998) Synchronous climate changes in Antarctica and the North Atlantic. *Science* **282**, 92–95.

Stenni B., Masson V., Johnsen S. J., Jouzel J., Longinelli A., Monnin E., Roethlisberger R., and Selmo E. (2001) An oceanic cold reversal during the last deglaciation. *Science* **293**, 2074–2077.

Stewart M. K. (1975) Stable isotope fractionation due to evaporation and isotopic exchange of falling water drops: application to atmospheric processes and evaporation of lakes. *J. Geophys. Res.* **80**, 1133–1146.

Taylor C. B. (1972) *The Vertical Distribution of the Isotopic Concentrations of Tropospheric Water Vapour over Continental Europe and their Relationship to Tropospheric Structure*. Report: INS-R, 107. N. Z. Dep. Sci. Ind.; Res., Inst. Nucl. Sci.

Taylor K. C., Hammer C. U., Alley R. B., Clausen H. B., Dahl-Jensen D., Gow A. J., Gundestrup N. S., Kipfstuhl J., Moore J. C., and Waddington E. D. (1993) Electrical conductivity measurements from the GISP2 and GRIP Greenland ice cores. *Nature* **366**, 549–552.

Thompson L. G., Mosley-Thompson E., Davis M. E., Lin P.-N., Henderson K. A., Cole-Dai J., Bolzan J. F., and Liu K.-B. (1995) Late Glacial stage and Holocene tropical-ice core records from Huascaran Peru. *Science* **269**, 46–50.

Thompson L. G., Davis M. E., Mosley-Thompson E., Sowers T. A., Henderson K. A., Zagorodnov V. S., Lin P.-N., Mikhalenko V. N., Campen R. K., Bolzan J. F., Cole-Dai J., and Francou B. (1998) A 25,000-tear tropical climate history from bolivian ice cores. *Science* **282**, 1858–1864.

van Ommen T. D. and Morgan V. (1997) Calibrating the ice core paleothermometer using seasonality. *J. Geophys. Res.* **102**, 9351–9357.

Vimeux F., Cuffey K., and Jouzel J. (2002) New insights into southern hemisphere temperature changes from Vostok ice cores using deuterium excess correction. *Earth. Planet. Sci. Lett.* **203**, 829–843.

Watanabe O., Jouzel J., Johnsen S., Parrenin F., Shoji H., and Yoshida (2003) Homogeneous climate variability accross East Antarctica over the last three glacial cycles. *Nature* **422**, 509–512.

Werner M., Heimann M. and Hoffmann G. (1998) Stable water isotopes in Greenland ice cores: ECHAM4 model simulations versus field measurements. In *International Symposium on Isotope Techniques in the Study of Past and Current Environmental Changes in the Hydrosphere and the Atmosphere*. IAEA Conference Vienna (Austria), pp. 603–612.

Werner M., Mikolajewicz U., Heimann M., and Hoffmann G. (2000) Borehole versus isotope temperatures on Greenland: seasonality does matter. *Geophys. Res. Lett.* **27**, 723–726.

Werner M., Heimann M., and Hoffmann G. (2001) Isotopic composition and origin of polar precipitation in present and glacial climate simulations. *Tellus* **53B**, 53–71.

Werner M. and Heimann M. (2002) Modeling interannual variability of water isotopes in Greenland and Antarctica. *J. Geophys. Res.* **107**, No. D1, p. ACL-1 1-13.

Yurtserver Y. and Gat J. (1981) Atmospheric waters. In *Stable Isotope Hydrology: Deuterium and Oxygen18 in the Water Cycle*. IAEA, Vienna, pp. 103–142.

# 4.09
# Radiocarbon

## W. S. Broecker

### Columbia University, Palisades, NY, USA

## 4.09.1 INTRODUCTION

Willard Libby's invention of the radiocarbon dating method revolutionized the fields of archeology and Quaternary geology because it brought into being a means to correlate events that occurred during the past $3.5 \times 10^4$ years on a planet-wide scale (Libby *et al.*, 1949). This contribution was recognized with the award of the Nobel Prize for Chemistry. In addition, radiocarbon measurements have been a boon to the quantification of many processes taking place in the environment, to name a few: the rate of "ventilation" of the deep ocean, the turnover time of humus in soils, the rate of growth of cave deposits, the source of carbon-bearing atmospheric particulates, the rates of gas exchange between the atmosphere and water bodies, the replacement time of carbon atoms in human tissue, and depths of bioturbation in marine sediment. Some of these applications have been greatly aided by the creation of excess $^{14}$C atoms as the result of nuclear tests conducted in the atmosphere. Since the 1960s, this so-called bomb

radiocarbon has made its way into all of the Earth's active carbon reservoirs. To date, tens of thousands of radiocarbon measurements have been made in laboratories throughout the world.

## 4.09.2 PRODUCTION AND DISTRIBUTION OF $^{14}$C

Radiocarbon atoms are produced when protons knocked loose by cosmic ray impacts encounter the nuclei of atmospheric nitrogen atoms. The reaction is as follows (Equation (1)):

$$n + {}^{14}N \rightarrow {}^{14}C + p \qquad (1)$$

The half-life of these $^{14}$C atoms is 5,730 years. Hence, they have on average 8,270 years to distribute themselves through the Earth's active carbon reservoirs. Radiocarbon atoms decay by emitting an electron, thereby converting a neutron to a proton returning the nucleus to its original $^{14}$N form.

Once produced, radiocarbon atoms become oxidized to $CO_2$ gas and join the Earth's carbon cycle (see Figure 1). $CO_2$ exchange with the inorganic carbon dissolved in the Earth's surface waters carries these atoms into the sea, lakes, and rivers. Photosynthesis moves the $^{14}C$ into both terrestrial and aquatic plants and from there to the animals that feed upon them. Radiocarbon also gets incorporated into shell and coral, and into soil and cave carbonates.

### 4.09.3  MEASUREMENTS OF RADIOCARBON

Initially all measurements were made by counting the $\beta$-particles emitted during radioactive decay of $^{14}C$. The first wave of laboratories followed Libby's lead and using screen wall Geiger counters measured the $\beta$-particles emitted by carbon black spun onto the inside of stainless steel cylinders (Figure 2). But the advent of nuclear testing created a serious problem for this method. Airborne strontium-90, caesium-137, and other fission products became absorbed onto the carbon black adding to the sample's radioactivity. Hans Suess (USGS, Washington), Hessel deVries

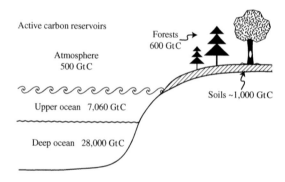

Active carbon reservoirs

Forests
600 GtC

Atmosphere
500 GtC

Soils ~1,000 GtC

Upper ocean   7,060 GtC

Deep ocean   28,000 GtC

**Figure 1**  Preanthropogenic distribution of carbon among the Earth's active reservoirs (units: gigatons of carbon or $10^{15}$ g of carbon).

(The Netherlands), and Gordon Fergusson (New Zealand) pioneered the transition to gas counting. Most laboratories filled proportional counters with highly purified $CO_2$ gas generated by burning or acidifying the sample. A few used acetylene. Other laboratories went a step further converting the sample carbon to liquid benzene and measured its radioactivity in a scintillation counter. In the early 1980s, it was realized that the $^{14}C$ atoms themselves could be measured using high-energy tandem Van de Graaff accelerators. During the latter part of the 1980s and early 1990s, accelerator mass spectrometry largely replaced the traditional decay counting techniques. The huge advantage of this new approach was that measurements required only 1 mg of carbon as opposed to 1 g for decay counting. The routine accuracy ($\pm 5$ per mil) achievable in atom counting is comparable to that obtained in decay-counting laboratories. Also comparable is the ability to measure the minute amounts of $^{14}C$ in very old (i.e., $>3.5 \times 10^4$ yr) samples. This innovation of atom counting has opened up a large number of applications previously untouchable by decay counting. The convention for expressing $^{14}C$ results is shown in Table 1.

### 4.09.4  TIMESCALE CALIBRATION

It was a biophysicist, deVries (1958), who first clearly demonstrated that the radiocarbon timescale was imperfect, i.e., that radiocarbon years were not identical to calendar years. Fascinated by the potential of Libby's new method, this brilliant Dutch scientist plunged into all its aspects. deVries devised means to reduce the background of gas counters by greatly increasing the effectiveness of the shielding from cosmic ray mesons. Not only did he strive to extend the method

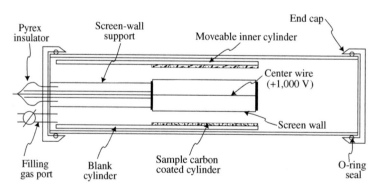

Pyrex insulator

Screen-wall support

Moveable inner cylinder

End cap

Center wire (+1,000 V)

Screen wall

Filling gas port

Blank cylinder

Sample carbon coated cylinder

O-ring seal

**Figure 2**  Diagrammatic representation of the Libby screen wall Geiger counter. Half of the moveable inner stainless steel cylinder was coated with the sample carbon and the other half left bare. By gently tipping the counter, the cylinder could be moved back and forth alternatively exposing first the sample portion and then the blank portion of the cylinder to the active (i.e., screen-wall portion) of the counter. In this way, the background count rate could be monitored and subtracted from the total sample plus background count rate.

**Table 1** The international convention for expressing radiocarbon results on contemporary materials in delta units. The standard is the age-corrected $^{14}C/C$ ratio in 1850 wood. The $^{13}C$ correction is designed to remove that part of the radiocarbon variability associated with the isotope separations occurring in nature (e.g., during photosynthesis and air–sea exchange). Also listed are some important characteristics of radiocarbon.

*Definitions:*

$$\delta^{14}C = \left( \frac{^{14}C/C \text{ sample} - {}^{14}C/C \text{ standard}}{^{14}C/C \text{ standard}} \right) 1,000$$

$$\Delta^{14}C = \delta^{14}C - 2\left(\delta^{13}C + 25\right)\left(1 + \frac{\delta^{14}C}{1,000}\right)$$

where

$$\delta^{13}C = \left( \frac{^{13}C/^{12}C \text{ sample} - {}^{13}C/^{12}C \text{ standard}}{^{13}C/^{12}C \text{ standard}} \right)$$

*Age correction:*

$$^{14}C/C_{\text{age corrected}} = {}^{14}C/C_{\text{measured}} \, e^{\lambda t}$$

where $t$ is the calendar age of the sample (referenced to 1950).

*Characteristics:*

$$^{14}C/C(\text{for } \Delta^{14}C = 0\%) = 1.18 \times 10^{-12}$$

$$t_{1/2} = 5,730 \text{ yr}$$

$$T_{\text{mean}} = 8,270 \text{ yr}$$

$$\lambda = \frac{1}{8,270} \text{ yr}^{-1}$$

$$\text{Decay rate} = \frac{1\%}{82.7 \text{ yr}}$$

A computerized calibration program (CALIB) is available.

beyond its nominal range ($3.5 \times 10^4$ yr), he also worked to improve the precision of measurements so that he could document small imperfections in the radiocarbon timescale based on measurements on materials of known calendar age. He was able to document such deviations and he postulated that they reflected changes in the $^{14}C/C$ ratio in atmospheric $CO_2$ due to perturbations in the flux of cosmic rays entering the Earth's atmosphere. Hessel deVries stands out in my mind as the great genius of this field. Had his life not been snuffed out just as he was reaching his prime, he would have dominated the radiocarbon world for decades. But, fortunately, during his short career, not only did he put his finger on the calibration problem that even today dominates much of the effort in the field, he also trained a graduate student, Minze Stuiver, who would devote much of his career to calibration studies and to measurements of $^{14}C$ in oceanic carbon.

#### 4.09.4.1 Calibration Based on Tree Rings

Nature has provided a marvelous set of calibration materials, namely, cellulose in the annual growth rings in trees. Single living trees provide annual ring series extending back 1,000 or more years. More important, tree trunks preserved in swamp muck and in riverine alluvium can be cross-dated based on ring-width "fingerprints," thus greatly extending the calibration range. Treasure troves of such material lie in Holocene sediments of the main Danube and Rhine rivers. Its flood-stage deposits constitute a major source of sand and gravel for the German construction industry. In the course of "mining" these deposits, large trunks of trees are often encountered. Over the years, the late Bernd Becker and members of his Hohenheim dendrochronology team responded to alerts hastening to the site of a new find to cut out a section to add to their ever-growing collection. A single radiocarbon measurement would tell them the approximate growth period of this tree. Measurements of the ring widths would allow them to tie this section precisely into an ever more complete and accurate master chronology. Once this chronology was established, very detailed high-precision $^{14}C$ measurements were made in the laboratory of Bernd Kromer in Heidelberg. This amazing record extends back to ~8,000 yr ago. In order to extend the growth-ring series further back in time, it was necessary for the Becker team to switch from oaks to pines. This species appeared in northern Europe soon after the abrupt warming which brought to an end the final cold punctuation of last glacial period (i.e., the Younger Dryas). Working closely with Bernd Kromer, who conducted ultra-precise $^{14}C$ measurements (to $\pm 2\%$) in his Heidelberg laboratory on both the oak and pine series, Becker and co-workers were satisfactorily able to splice the pine series onto the oak series creating a master chronology covering the last $1.1919 \times 10^4$ yr (Kromer *et al.*, 1986; Becker and Kromer, 1993; Kromer and Spurk, 1998; Friedrich *et al.*, 1999). Efforts are currently underway using tree trunk series from central Europe and south of the Alps to extend the chronology back into glacial time.

The thousands of measurements which contributed to this calibration effort were largely made in five laboratories: that of Hans Suess in La Jolla, California; that of Paul Damon in Tuscon, Arizona; that of Gordon Pearson in Belfast Northern Ireland; and, of course, those of Minze Stuiver in Seattle, Washington, and Bernd Kromer in Heidelberg, Germany. The resulting calibration curve was a boon to archeologists faced with the task of relating radiocarbon dates to calendar dates for historical events, as well as to geophysicists interested in probing the causes for the $^{14}C/C$ ratio changes. A compilation base on these tree-ring calibration measurements is shown in Figure 3.

Two features of this record stand out, a long-term trend of decreasing $^{14}C$ culminating ~1,500 yr ago and century-duration fluctuations around this trend. Minze Stuiver was the first to demonstrate that the century-duration fluctuations were, at least in part, related to sunspot activity. He did this by showing that during the course of

the so-called Maunder minimum (a period from 1645 to 1715 when no spots were observed on the Sun) the $^{14}C/C$ ratio in atmospheric $CO_2$ underwent a significant increase (Stuiver and Quay, 1980, 1981). Stuiver attributed this increase to the shutdown of the ions streaming into space from the Sun's spots (and hence also of the so-called heliomagnetic field). This magnetic field serves to partially shield the solar system from incoming cosmic rays. Similar increases in the ratio $^{14}C/C$ were established by Stuiver to have occurred between 1280 and 1345 and between 1420 and 1540 (Stuiver and Quay, 1980, 1981). By analogy to the Maunder, they have been termed the Spörer and Wolf sunspot minima (see Figure 4).

A reasonably convincing case has been made that the longer-term Holocene trend is a consequence of changes in the Earth's magnetic field which also diverts incoming cosmic rays. This case rests on measurements designed to reconstruct the strength of the Earth field at times in the past. The first such attempt was by

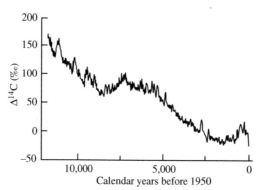

**Figure 3** Tree-ring-based reconstruction of the temporal trend of the atmosphere's $^{14}C/C$ ratio over the last $1.1854 \times 10^4$ yr calendar years. $\Delta^{14}C$ is the deviation of the atmospheric ratio from that in pre-Industrial time (~1800) (see Table 1) (source Stuiver *et al.*, 1998).

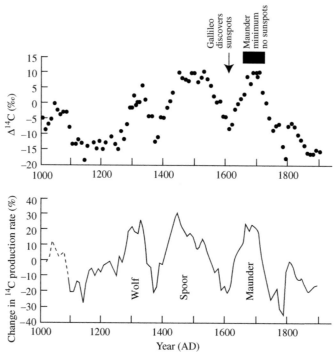

**Figure 4** Temporal fluctuations in the $^{14}C/C$ ratio (given in the $\Delta$ units defined in Table 1) as determined from measurements on Douglas fir wood from the Pacific Northwest (upper panel). The record is cut off in the year 1900 because after that time the influence of $^{14}C$ free $CO_2$ released by fossil fuel burning became significant. In the lower panel are shown the changes in the production rate of $^{14}C$ required to produce the observed temporal changes in atmospheric $^{14}C$ to C ratio. The production rates are deconvolved from the observed $\Delta^{14}C$ changes using a simplified atmosphere–ocean–terrestrial biosphere model. Because of the interchange of carbon atoms with the ocean and the terrestrial biosphere, the changes in atmospheric $^{14}C/C$ ratio lag the production-rate changes. As can be seen, the last of the peaks in production occurred during a time when no spots were observed on the Sun (i.e., during the Maunder minimum). As telescopic observations of the Sun (by Galileo) commenced in the early 1600s, the existence of the Spörer and Wolf sunspot minima is based on the $^{14}C$ results. The measurements and modeling were carried out by Stuiver and Quay (1980). Bard *et al.* (1997) showed that the observed $^{14}C$ variations can be generated from measurements of $^{10}Be$ in ice cores.

E. Thellier and O. Thellier (1959), who measured the magnetic field in ancient ceramics and then reheated them beyond the Curie point and allowed them to cool in the same fashion as originally. They then remeasured the magnetic field. These early ceramic measurements have been supplemented by reconstructions based on igneous rocks and deep-sea cores, which extend back many radiocarbon half-lives (see Bard (1998) for an excellent summary).

However, lurking in the wings is yet another mechanism by which the atmospheric $^{14}C/C$ ratio may have been changed. It involves changes in the rate of ventilation of the deep sea. Around 70% of the Earth's cosmic ray-produced $^{14}C$ atoms currently reside in the deep sea. Because $^{14}C$ is lost by radiodecay during residence in the deep sea, the $^{14}C/C$ ratio for the inorganic carbon (i.e., $HCO_3^-$, $CO_3^{2-}$, and $CO_2$) dissolved in these waters averages ~16% lower than that in the atmospheric $CO_2$. Hence, were the rate of ocean mixing to have been slower in the past than now, the contrast between the atmospheric and deep-sea $^{14}C/C$ ratios would have increased and, consequently, the ratio in the atmosphere would have been higher than now. Later in the chapter, a dramatic example of such a change will be discussed. For the Holocene (i.e., the last $1.15 \times 10^4$ yr), there is no concrete evidence that ocean mixing contributed to the $^{14}C/C$ changes. Minze Stuiver has pondered whether the changes in sunspot activity might be linked to the Holocene's small climate changes. If so, then both sunspot activity and ocean mixing may have contributed to the $^{14}C/C$ ratio fluctuations.

An extremely important contribution to this subject was made by an Indian scientist, Devendra Lal, who made calculations aimed at determining the exact dependence on the Earth's magnetic field strength of the influx of cosmic rays to the Earth's atmosphere, and hence also on the rate of production of $^{14}C$ atoms (Lal, 1988).

While resolving most of the inconsistencies between radiocarbon and historic ages, the calibration curve also reveals a fundamental limitation of the radiocarbon method. During some time periods, the century-duration changes in $^{14}C/C$ ratio created reversals in the age sequence. These reversals give rise to multiple possible calendar ages for a single radiocarbon measurement. Only by measuring several samples in stratigraphic sequence, it is possible to distinguish among these multiple possibilities (see Stuiver, 1982).

In a related manner, radiocarbon dating is powerless to aid in verifying the authenticity of art objects purported to have made in the period 1700–1950. The reason is that during this period the radiocarbon content of the atmosphere decreased at a rate closely matching the rate of $^{14}C$ decay (i.e., 1% per 83 yr). There are two reasons for this decline. First, the excess $^{14}C$ atoms produced during the Maunder minimum were being mixed into the ocean and sequestered in the terrestrial biosphere. Second, the burning of fossil fuels added $CO_2$ free of radiocarbon to the atmosphere, thereby extending this decline into the twentieth century. In the early 1950s, this downward trend was reversed by the production of $^{14}C$ during nuclear tests (see Section 4.09.6.2). Hence, based on radiocarbon measurements, a forgery made using wood or parchment dating just prior to 1950 could not be distinguished from the real thing created during the eighteenth century.

### 4.09.4.2 Calibration Based on Corals

Working at Columbia University's Lamont–Doherty Earth Observatory with French isotope geochemists Edouard Bard and Bruno Hamelin, Richard Fairbanks found a means to extend the calibration curve beyond the Holocene into late glacial time. This team adopted a new analytical method developed at Caltech to make very precise uranium and thorium isotope measurements using conventional thermal ionization mass spectrometry (Edwards *et al.*, 1986/1987) instead of the less precise alpha counting method used previously. Just as the transition from decay to atom counting revolutionized $^{14}C$ dating, this innovation revolutionized dating based on the uranium decay-series isotope $^{230}Th$ (half-life $7.5 \times 10^4$ yr). This new technique was applied to corals obtained from a series of shallow borings Fairbanks conducted off the island of Barbados. These corals formed during the last deglaciation as sea level rose in response to the melting of the glacial age icecaps. The Fairbanks team conducted both $^{230}Th$-age and $^{14}C$-age determinations on these corals, which were ideal for the task as they contained no original thorium and had never been exposed to $CO_2$-charged groundwater. The $^{230}Th$ results produced a big surprise. Rather than revealing a cyclicity in the Earth's magnetic field as a number of authors had predicted, the offset between calendar and radiocarbon ages became larger and larger as one went back in time (Bard *et al.*, 1990a,b) (see Figure 5).

### 4.09.4.3 Other Calibration Schemes

The measurements on Fairbanks' Barbados corals spurred efforts to find other means of extending the calibration curves back in time. Several tacks were taken. One obvious strategy was to count annual layers (varves) in lake and marine sediments. Stuiver, the hero of the calibration effort, had adopted this approach way back in the

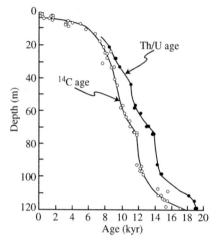

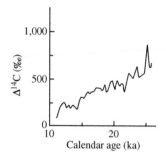

**Figure 5** $^{14}$C (open circles) and $^{230}$Th (closed circles) based age determinations on corals obtained from shallow drillings off the island of Barbados. As can be seen, the magnitude of the offset between the two ages increases back in time indicating that the $^{14}$C/C ratio in atmosphere and upper ocean carbon was declining during the period 18,000–8,000 yr ago. As the corals chosen for these measurements were shallow growing, the $^{230}$Th ages portray the rise in sea level during the last deglaciation. These results were reported in papers by Fairbanks (1989) and Bard *et al.* (1990a,b).

**Figure 6** Extension of the tree-ring-based and Barbados coral-based calibration curve back to $3.2 \times 10^4$ yr based on $^{14}$C and $^{230}$Th measurements on a stalagmite from a submerged cave in the Bahamas. The growth of the stalagmite came to a halt $\sim 1.1 \times 10^4$ yr ago when the rising sea invaded the cave. The upper portion of the record which overlaps in time the Barbados coral record was used to establish the initial $^{230}$Th to $^{232}$Th ratio and also the reservoir correction for the $^{14}$C ages (source Beck *et al.*, 2001).

1960s (Stuiver, 1970, 1971). Another approach was to put to use the annual layering in long ice cores from Greenland. As the ice itself cannot be dated using $^{14}$C, this required that the $^{18}$O (i.e., air temperature) record in the ice be correlated with ice rafting events in radiocarbon-dated sediment cores from the northern Atlantic (Voelker *et al.*, 1998). Finally, coupled $^{230}$Th and $^{14}$C measurements on cave formations have been utilized (see, e.g., Beck *et al.*, 2001). In my estimation, none of these approaches rests on an entirely firm foundation. Varve counting is often subjective, e.g., varves may be missing or doublets may represent a single year. However, Kitagawa and van der Plicht (1998) make an impressive case for a varve sequence in a Japanese lake. Greenland's ice cores provide an excellent example of the difficulty. Counts by a European team and by an American team in two long and virtually identical cores from Greenland's summit locale yield results that differ by as much as several percent (see Southon, 2002). Further, the correlation of the ice record with the marine record is fraught with subjectivity and there are large discrepancies between radiocarbon dates on coexisting planktic and benthic foraminifera. Stalagmites offer perhaps the most promise but, unlike corals, during growth they often incorporate some $^{230}$Th. Also, the initial $^{14}$C/C ratio is offset from that for atmospheric $CO_2$ because the $CO_2$ in cave waters is derived from the oxidation of organic

matter in the overlying soil. But by far the most worrisome aspect of the cave formation approach is contamination with younger $CaCO_3$ (infilling of pores, recrystallization, etc.). A $3.5 \times 10^4$-year-old sample contains only (1/64)th its original amount of $^{14}$C. This contamination problem is far more severe for $CaCO_3$ than for wood from which chemically inert cellulose can be extracted. Because of these problems, I do not trust any existing extensions of calibration curve beyond $2.5 \times 10^4$ yr. Much more work will have to be done before this task can be declared a success. And, of course, the chances of ever extending the calibration beyond $4.7 \times 10^4$ year (only one part in 256 of the original $^{14}$C remains) are indeed slim. However, the $^{14}$C and $^{230}$Th measurements of Beck *et al.* (2001) on a Bahamas stalagmite did allow the calibrations curve to be reliably extended back to $\sim 2.5 \times 10^4$ yr (see Figure 6). Clearly this effort constituted a major advance.

### 4.09.4.4 Cause of the Long-term $^{14}$C Decline

One might ask what accounts for the decline in atmospheric $^{14}$C/C ratio over the last $2.5 \times 10^4$ yr. The prime suspect is the drop to near-zero values in the Earth's magnetic field $\sim 4 \times 10^4$ yr ago (Bonhommet and Zähringer, 1969). This drop, referred to as the Laschamp event, is named after a volcanic field in France where it was first identified. The existence of this event has now been confirmed by measurements of magnetic intensity in deep-sea sediments that reveal a strong minimum close to $4 \times 10^4$ yr ago. Also, it shows up as a doubling of the abundance of $^{10}$Be (another cosmogenic isotope created in the Earth's atmosphere) at this time in the Greenland ice cores (Yiou *et al.*, 1997). Both the drop to near-zero values of magnetic intensity and the doubling of the $^{10}$Be concentration are consistent with a temporary shutdown of the Earth's field (presumably associated with a false polarity reversal). The existence of the Laschamp event has provided a strong impetus to extend the calibration curve to times before this temporary shutdown of the Earth's magnetic field. However, as already stated, this challenge will prove to be an extremely difficult one.

It is unlikely that the Laschamp event alone can explain the observed decline in the $^{14}$C/C ratio. The duration of this event (several thousand years) was too short and the magnitude of its impact on $^{14}$C production was too small to produce a large enough increase in $^{14}$C inventory to persist for $\sim 2 \times 10^4$ yr. After $2.3 \times 10^4$ yr had elapsed, the excess would have been diminished by a factor of 16. However, as shown by a number of marine sediment-based magnetic field reconstructions (see Laj *et al.*, 1996), the Laschamp event was followed by a long-term buildup in the strength of the Earth's magnetic field (see Figure 7) (see Bard, 1998, for summary). This slow buildup appears to be adequate to account for the post $4 \times 10^4$ yr decline in atmospheric $^{14}$C.

### 4.09.4.5 Change in Ocean Operation

In addition to attempts to extend the radiocarbon calibration curve back further in time, there have also been attempts to enhance its detail. One of these attempts stands out. Konrad Hughen, while a graduate student at the University of Colorado, carried out a detailed set of radiocarbon measurements on planktonic foraminifera shells contained in varved sediment from the Cariaco Basin just off Venezuela. One of the attributes of this core is that the interval correlating with the Younger Dryas cold event is marked by a distinct color change. Further, the boundaries of this interval are very sharp. Hughen used his radiocarbon measurements to splice the Cariaco varve sequence onto the Rhine Valley tree-ring sequence. He then used his varve counts to extend the calibration record several thousand years beyond that established using tree rings (i.e., through the entire Younger Dryas and the underlying Bölling Allerod warm period). His resulting reconstruction revealed a very exciting feature (see Figure 8). During the first 200 yr of the Younger Dryas, the $^{14}$C/C ratio in upper ocean dissolved inorganic carbon soared by 5‰. Then, during the remaining 1,000 yr of this cold snap, the radiocarbon content drifted back down.

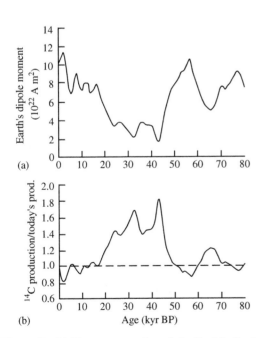

(a)

(b)

**Figure 7** (a) The reconstruction of the Earth's dipole moment (Laj *et al.*, 1996) based on magnetic measurements made on deep-sea sediments and calibrated against measurements on volcanic rocks of known age. (b) The radiocarbon production rates reconstructed from the magnetic dipole reconstruction (a). Frank *et al.* (1997) have shown that a similar reconstruction is obtained based on ice-core $^{10}$Be results.

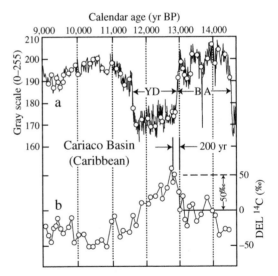

**Figure 8** Reconstruction of the radiocarbon content of atmospheric $CO_2$ during the deglaciation interval (Hughen *et al.*, 1998). This reconstruction was obtained from a varved sediment core from the Cariaco Basin. The position of the individual $^{14}$C measurements is shown by the circles. The Younger Dryas (YD) and Bolling-Allerod (BA) intervals are clearly shown in gray scale record. As can be seen, the $^{14}$C content of surface ocean carbon rose by $50 \pm 10‰$ in the first two centuries of the Younger Dryas event. After that, it slowly declined. Assuming that the air–sea difference in the $^{14}$C/C ratio remained the same as today's, these changes can be transferred to atmospheric $CO_2$.

The coincidence of the initiation of the $^{14}$C/C ratio rise with the onset of the Younger Dryas led Hughen to propose that this rise was tied to a change in the rate of oceanic mixing rather than to a change in magnetic shielding. Such a $^{14}$C rise would be a logical consequence of a shutdown in production of new deep water in the northern Atlantic Ocean. In today's ocean, roughly three quarters of the $^{14}$C atoms resupplying the deep-sea inventory enter via this route. A number of authors have shown that the sudden release of melt water stored in proglacial Lake Agazzis into the St. Lawrence drainage at the time of the onset of the Younger Dryas would have flooded the northern Atlantic with enough freshwater to shut down the Atlantic's conveyor circulation.

If Hughen's hypothesis proves to be correct then one might ask: Why not attribute other aspects of offset of radiocarbon years from calendar years to changes in ocean mixing? In particular, could the higher atmospheric $^{14}$C/C ratio during peak glacial time ($(2.3–1.5) \times 10^4$ yr ago) be attributed to more sluggish ocean circulation? Here the radiocarbon method itself provides an answer. It is based on differences between the radiocarbon age of coexisting planktonic (surface-dwelling) and benthic (bottom-dwelling) foraminifera shells picked from glacial sections of deep-sea sediments. Due to rapidity of $CO_2$ exchange with the atmosphere, the $^{14}$C/C ratio in surface ocean dissolved inorganic carbon is closely linked to that in atmospheric $CO_2$. Thus, the difference between the $^{14}$C/C ratio in surface water and that in deep water recorded by coexisting planktic and benthic foraminifera shells provides a record of how the rate of deep-sea ventilation differed between late glacial time ($(2.3–1.5) \times 10^4$ yr ago) and the Holocene ($1.1 \times 10^4$ yr to present). While these measurements suggest that ventilation was a bit slower during late glacial time, the rate of mixing was not nearly slow enough to explain the 20–30% higher atmospheric $^{14}$C/C ratio at that time (see Shackleton *et al.*, 1988; Broecker *et al.*, 1990).

Another radiocarbon-based study showed just how rapidly changes in the patterns of deep ventilation in the ocean can occur. As part of his PhD research with MIT's Ed Boyle, Jess Adkins made a study of a single benthic coralite, which grew at a depth of 1.8 km in the northwestern Atlantic Ocean during the early stages of the last deglaciation. He found an amazing thing. While the base and crest of this several centimeter-high mushroom-shaped coralite yielded nearly identical $^{230}$Th ages of $1.54 \times 10^4$ yr, the $^{14}$C age of the base was $1.385 \times 10^4$ yr and that of the crest $1.452 \times 10^4$ yr (Adkins *et al.*, 1998). As this coralite probably formed in a period of just a few decades, this difference in radiocarbon age requires that the water in which the coralite grew underwent a

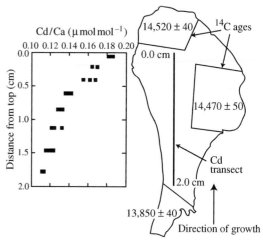

**Figure 9** Measurements of $^{14}$C/C ratio in a benthic coral from 1.8 km depth in the northern Atlantic reveal that the base has an apparent age 650 radiocarbon years younger than the top. Yet $^{230}$Th ages on the base and crest are nearly identical ($1.54 \times 10^4$ yr calendar years). Adkins *et al.* (1998) (reproduced by permission of Jess Adkins) attribute this difference to an abrupt invasion of low $^{14}$C and high cadmium content Southern Ocean water.

sudden and large drop in the $^{14}$C/C ratio. When Adkins compared his $^{14}$C to $^{230}$Th age offsets with those obtained by the Fairbanks team, he concluded that the apparent deep Atlantic ventilation age jumped from ~400 yr to ~1,100 yr. Puzzled by how this could have happened, he measured the cadmium content of the calcite along a traverse from the stem to the coralite crest and found that it increased from 0.11 μmol Cd mol Ca$^{-1}$ to 0.18 μmol Cd mol Ca$^{-1}$ (see Figure 9). Only one explanation seems to fit these observations. The production of low-cadmium content and high-radiocarbon water in the northern Atlantic must have come to an abrupt halt allowing the more dense waters with higher cadmium content and lower radiocarbon from the Southern Ocean to flood northward, displacing upward the less dense waters of the North Atlantic origin.

For all these studies, there is a necessity to establish the reservoir age for the glacial surface ocean. Several authors (Bard *et al.*, 1994; Sikes *et al.*, 2000; Waelbroeck *et al.*, 2001; Siani *et al.*, 2001) have attempted to do this and find puzzling older reservoir ages for glacial times. Clearly, more effort must go into such reconstructions.

## 4.09.5 RADIOCARBON AND SOLAR IRRADIANCE

As solar energy output and radiocarbon production are both tied to sunspot activity, it might be possible to reconstruct past irradiance variations from the perturbations in the atmosphere's $^{14}$C/C ratio reconstructed from measurements on tree rings of known calendar age. Indeed, based on radiocarbon reconstructions, Stuiver (1980) and Stuiver and Braziunas (1993) suggested that the Little Ice Age might be the result of reduced solar luminosity during periods such as the Maunder minimum when sunspot activity was shut down. However, because the records Stuiver used were quite short and since changes in solar luminosity were, at the time his paper was written, poorly documented, his proposal failed to receive wide acceptance.

The advent of satellite observations of the solar output made it clear that there is indeed a tie between energy output and sunspot activity (see Figure 10). However, as these variations are very small and the record covers only the last two decades, this finding made only a modest impact with regard to interest in the Sun as a driver of Earth climate fluctuations.

The situation changed with the publication of a paper by Bond *et al.* (2001) in which it was shown that a correlation exists between the cosmic ray flux changes required to generate the tree-ring-based reconstruction of atmospheric $^{14}$C/C ratio record for the last $1 \times 10^4$ yr and an ice-rafting index for the radiocarbon-dated northern Atlantic sediments developed by Bond *et al.* (1997, 1999) (see Figure 11). This index involves the ratio of red-coated (i.e., iron-stained) to total ice-rafted grains. During the Holocene the index has

fluctuated on a timescale on average of 750 yr (i.e., 1,500 yr for a complete cycle) between highs of ~16% and lows of only a few percent. Bond reasons that times of high index represent cold spells and those of low index, warm spells. His argument is based on the observation that the sources of red-coated grains lie poleward of sources devoid of red-coated grains, and hence, to get them to the site of his deep-sea cores requires colder surface water conditions. As a confirmation that the changes the $^{14}$C/C ratio reflect production rather than changes in ocean circulation, Bond *et al.* (2001) point to the Holocene record of $^{10}$Be in Greenland ice cores, which yield a close match in timing and in amplitude of the cosmic ray flux changes reconstructed from $^{14}$C measurements.

A confirmation of Bond's climate interpretation has been obtained from the dating of wood and peat being carried by summer melt water from beneath the retreating mountain glaciers in the European Alps (Hormes *et al.*, 1998). The forests and bogs in which these materials were generated must date from warmer times when the glaciers were even smaller than they are today. Three such times have been documented by radiocarbon dating many tens of wood and peat samples. They match quite well with three of Bond's warm intervals (see Figure 11). Further, evidence in support of Bond's temperature cycles comes from fossil-tree remains found north of the present tree

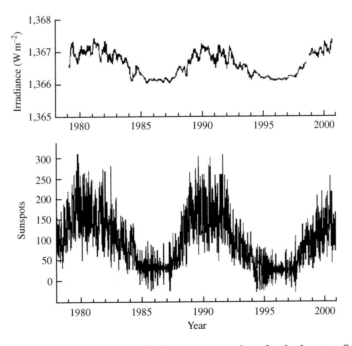

**Figure 10** Comparison of the solar irradiance with the sunspot numbers for the last two Schwabe cycles. The irradiance record is a compilation of data from different satellites. During periods of high solar activity there are more sunspots darkening a small part of the solar disk (visible in the negative excursions of the irradiance). However, the brightness of the Sun is increased at the same time, overcompensating the darkening effect of the sunspots.

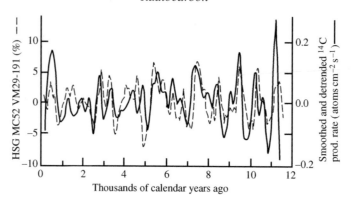

**Figure 11** Comparison between the fluctuations in the fraction of red-coated lithic grains (dashed curve) with that of cosmogenic nuclide production (smooth curve) in the Earth's atmosphere. This correspondence provides powerful evidence that the Holocene's small cyclic temperature changes were paced by changes in solar luminosity (source Bond *et al.*, 2001).

line on Russia's Kola Peninsula (Hiller *et al.*, 2001). They document the existence of a northward expansion of the forest during medieval time (1000–1300). This corresponds to the period during which the Vikings colonized southern Greenland.

### 4.09.6 THE "BOMB" $^{14}$C TRANSIENT

During the late 1950s and early 1960s, hydrogen bomb tests carried out in the atmosphere by Russia, Britain, and USA led to the production of manmade radiocarbon atoms. In this case, the neutrons that collided with atmospheric nitrogen nuclei were by-products of the fusion reactions. Large-scale anthropogenic production of $^{14}$C came to a halt with the implementation of the ban on atmospheric weapons tests on January 1, 1963. At that time, the number of manmade $^{14}$C atoms in the atmosphere was roughly equal to the number of natural radiocarbon atoms (i.e., the $^{14}$C/C ratio nearly doubled). Since then, this atmospheric excess has steadily dwindled until as of the year 2000, the bomb-produced atoms constituted only ~10% of the atmospheric radiocarbon inventory (see Figure 12). This decrease has three causes:

(i) exchange with oceanic $\sum CO_2$,

(ii) exchange with terrestrial biospheric carbon, and

(iii) continued addition of $^{14}$C free fossil-fuel-derived $CO_2$ molecules to the atmosphere.

Figure 13 shows an estimate as to how the distribution of bomb radiocarbon atoms among these active carbon reservoirs has evolved.

#### 4.09.6.1 Radiocarbon as a Tracer for Ocean Uptake of Fossil Fuel $CO_2$

While $^{14}CO_2$ is in a sense a perfect tracer for fossil fuel $CO_2$, the quite different time histories for their inputs and the large difference between the times required for the surface ocean to respond to isotopic compared with chemical transients makes the task more complicated than it might seem. Even so, the situation for the ocean turns out to be simpler than that for the land. There are three reasons for this. First, the ocean has a far greater degree of lateral homogeneity. Second, uptake by the sea is governed by basic inorganic chemistry while that by the land biota involves complex biotic cycles. Third, the documentation of the radiocarbon transient through direct measurement is far more extensive for the ocean.

#### 4.09.6.2 Ocean Uptake of $^{14}CO_2$ and $CO_2$

As of early 2000s, it has not been possible to make an accurate direct assessment of the amount of fossil fuel $CO_2$ taken up by the sea. The reason is related to the fact that even in the surface mixed layer the increase can have been no more than 2–3% (10% increase in the atmosphere's $CO_2$ partial pressure produces only a 1% increase in the surface ocean's dissolved inorganic carbon content (i.e., $CO_2 + HCO_3^- + CO_3^{2-}$)). Unfortunately, until the GEOSECS global survey (1972–1978) no measurements of the concentration of dissolved inorganic carbon (i.e., $CO_2 + HCO_3^-$, and $CO_3^{2-}$) in the ocean of sufficient accuracy were made. Further, as the GEOSECS measurements are accurate to only ±1%, even they are not quite up to the task. Only with the TTO surveys in the Atlantic during the early 1980s and the WOCE surveys in the Pacific and Indian oceans during the early 1990s was a base level established with sufficient accuracy (i.e., 0.1%). Hence the uptake must be estimated using ocean–atmosphere models rather than direct measurements. Whether these models are of the simple box-diffusion or complex three-dimensional general circulation variety, it is necessary to calibrate (in the case of box models) or constrain (in the case of general circulation

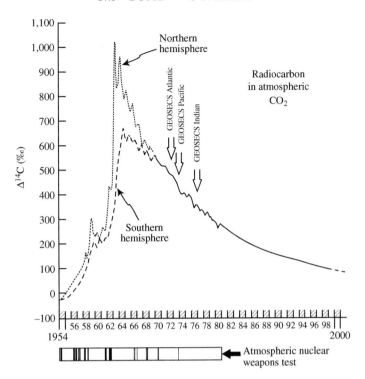

**Figure 12** $\Delta^{14}C$ for atmospheric $^{14}C$ following the onset of H-bomb testing in 1952. The buildup continued until mid-1963 and since then the $\Delta^{14}C$ to C ratio has been declining toward its prenuclear value as the excess bomb testing $^{14}C$ atoms are transferred to the ocean and terrestrial biosphere. Although China and India, who were not signatories to the test ban treaty explode, conducted atmospheric tests in the late 1960s and early 1970s, the amount of $^{14}C$ produced was negligible. Within decades, the $\Delta^{14}C$ will drop below zero due to the continued emission of $^{14}C$-free fossil fuel $CO_2$. The times of the GEOSECS ocean surveys are shown. The pre-1980 data are from Nydal and Lövseth (1983).

models) them with tracer measurements. The most powerful tracer for this purpose is the radiocarbon produced during the atmospheric testing of nuclear weapons.

Two resistances limit the rate at which bomb $^{14}CO_2$ and fossil fuel $CO_2$ are transferred from the atmosphere to the sea. The first is transfer across the air–sea interface and the second is the rate at which surface waters are mixed into the ocean's interior. Clearly, the relative strengths of these two resistances are of importance. Were the resistance posed by vertical mixing in the ocean to be much smaller than that posed by $CO_2$ exchange across the air–sea interface, then the air-to-sea difference in $CO_2$ partial pressure and in $^{14}CO_2$ partial pressure would remain large, and the rate of uptake by the sea would be dominated by the rate of air-to-sea $CO_2$ exchange. Were the opposite the case, then ocean uptake would be limited by the rate of vertical mixing within the sea. It turns out that for $CO_2$ itself the resistance posed by vertical mixing in the sea dominates. Waters in the tens-of-meters thick wind-mixed layer are able to reach 85% or so of saturation with the atmosphere's fossil fuel $CO_2$ burden before being mixed into the interior. For $^{14}CO_2$ the situation is quite different. The resistance posed by the interface is

comparable to that posed by vertical mixing within the sea. The reason for this difference between $^{14}CO_2$ and $CO_2$ has to do with the chemical speciation of carbon in surface water. The proportions of $CO_2$, $HCO_3^-$, and $CO_3^{2-}$ in the surface ocean are roughly 10 : 1,800 : 200. While isotopic equilibration requires that the $^{14}C$ in all three species be exchanged, chemical equilibration is accomplished by the reaction in Equation (2):

$$CO_2 + CO_3^{2-} + H_2O \rightarrow 2HCO_3^- \qquad (2)$$

Because of this, the time required for chemical equilibration turns out to be roughly an order of magnitude smaller (i.e., 200/1,800) than that for isotopic equilibration.

Taken together, measurements of the vertical distribution of bomb radiocarbon in the sea and the air-to-surface-sea difference in the bomb $^{14}C/C$ ratio permit both the laterally averaged rate of air–sea $CO_2$ exchange and the laterally averaged rate of vertical mixing to be estimated. As part of the GEOSECS survey, the distribution of radiocarbon throughout the entire world ocean was measured (in the laboratories of Minze Stuiver (University of Washington) and Gote Ostlund (University of Miami)). As this survey was conducted $12 \pm 3$ yr after the implementation

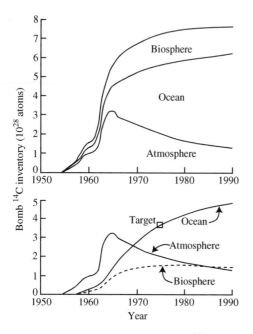

**Figure 13** Distribution of bomb testing $^{14}C$ between the atmosphere, ocean, and terrestrial biosphere as reconstructed by Broecker and Peng (1994). The ocean contribution is obtained by a model constrained by the inventory based on the GEOSECS survey. The contribution of the terrestrial biosphere is based on estimates of the biomass and turnover times for trees and active soil humus.

of the ban on atmospheric testing, the average bomb $^{14}C$ atom had a little more than a decade to enter the sea and be mixed into its interior.

In order to make use of this transient, the contributions of the natural and bomb radiocarbon had to be separated. Three sets of observations went into this separation. First, use was made of $^{14}C/C$ measurements on surface waters collected very early in the nuclear testing era (Broecker *et al.*, 1960) and also with results of measurements on prenuclear growth ring-dated corals and mollusks (Druffel and Linick, 1978; Druffel, 1981, 1989). Second, use was made of the tritium released to the atmosphere during nuclear tests. As this bomb-test tritium swamped the natural tritium present in the ocean, the vertical distribution of tritium in the sea could be used to establish the limit of penetration of bomb radiocarbon. Finally, based on the radiocarbon analyses made on thermocline waters free of bomb tritium, it was shown that there was a close correlation between the natural $^{14}C/C$ ratio and the dissolved silica content of the water (Broecker *et al.*, 1995). More recently, it was shown that there was an even better correlation between natural radiocarbon and salinity-normalized alkalinity. These relationships were essential in separating the two "brands" of $^{14}C$ in the waters from the upper portion of the main oceanic thermocline.

Taken together, the distributions of natural and bomb radiocarbon in the sea give us an important insight regarding the manner in which the ocean mixes. The steady-state distribution of natural radiocarbon constrains the replacement timescale for waters in the deep sea to be on the order of one millennia. The transient distribution of bomb radiocarbon at the time of the GEOSECS survey tells us that on a timescale of a decade, radiocarbon atoms are able to exchange with ~10% of the ocean's dissolved inorganic carbon (i.e., $CO_2 + HCO_3^- + CO_3^{2-}$). Taken together, these two constraints suggest that the fraction of the ocean volume accessed by vertical mixing increases with the square root of penetration time. This concept gave rise to the early one-dimensional box-diffusion models for fossil fuel $CO_2$ uptake by the ocean (Oeschger *et al.*, 1975; Seigenthaler *et al.*, 1980). Such models consist of a well-mixed atmosphere and a well-mixed surface ocean layer underlain by a diffusive half-space. The so-called eddy diffusivity assigned to this half-space was set by fitting the horizontally averaged vertical distribution of bomb radiocarbon.

One might conclude that the crude box-model representation of the ocean would have been soon eclipsed by ocean general circulation models capable of simulating, in three dimensions, the full suite of ocean currents and mixing processes. While certainly a critical step in the evolution of our ability to predict the split of fossil fuel $CO_2$ between the ocean and atmosphere, the task of creating models that match the distributions of not only temperature and salinity, but also those of natural and bomb radiocarbon, is a daunting one. Properly simulating the flow pathways and flow rates in the 600–1,200 m depth range dominated by the ocean's intermediate water masses has proven particularly challenging. Consequently, fossil fuel $CO_2$ uptake estimates made with these three-dimensional models are, in my estimation, as yet no more reliable than those based on the one-dimensional $^{14}C$-calibrated box-diffusion models.

Both types of models suggest that $35 \pm 5\%$ of the $CO_2$ produced to date by fossil fuel burning has been taken up in the ocean. This represents ~16% of the ocean's capacity for fossil fuel $CO_2$ uptake (i.e., of the amount of uptake were the entire volume of the ocean to be equilibrated with the atmosphere). One might ask why this percentage is higher than that based on the distribution of bomb radiocarbon. The answer is that while the $^{14}C$ atoms had at the time of the GEOSECS survey been in existence for only ~12 yr, the average existence time for fossil fuel $CO_2$ molecules is more like 30 yr. Since the square root of 30/12 is 1.6, hence, to the extent that the diffusive characterization of ocean mixing is correct, fossil fuel $CO_2$ should have been able to penetrate a 1.6 times larger volume than bomb $^{14}C$.

The challenge for ocean modelers is to achieve a match with the observed distributions of bomb $^{14}C$ not only for the time of the GEOSECS survey (the 1970s) but also for the time of the TTO (1980–1981) and WOCE (early 1990s) surveys. Only then can modelers be confident that they can predict the split in excess $CO_2$ between atmosphere and ocean for any given future fossil-fuel-use scenario. This is tricky because the uptake capacity for $CO_2$ by surface ocean waters depends on their carbonate ion concentration. Clearly, as $CO_2$ builds up in the atmosphere, the carbonate ion concentration in surface waters will decrease, but the rate of decrease will depend on the manner in which the ocean mixes. Advective overturning brings "virgin" water high in $CO_3^{2-}$ concentration to the surface. However, diffusive mixing creates an ever increasing carbonate ion gradient between "spent" surface water and "virgin" deep water. Different models yield different ratios of diffusive to advective mixing.

### 4.09.6.3  Terrestrial Uptake of $^{14}CO_2$ and $CO_2$

In recent years the response of the terrestrial biosphere to the increase in the atmosphere's $CO_2$ content has become a hot topic. Thousands of chamber experiments show that at least on the short term (a year or two) given more $CO_2$, plants grow faster. Further, global terrestrial carbon inventories based on the rate of the atmosphere's $CO_2$ increase and $O_2$ decline indicate that storage is on the increase. This increase is happening despite large-scale deforestation that, of course, drives down carbon stocks. Three factors appear to be responsible for this increase: (i) the excess $CO_2$ in the atmosphere; (ii) the dispersal of fixed nitrogen evaporated from farmland and given off in automobile exhausts; and (iii) regrowth of forests on land previously cleared for farming. Unlike the ocean, the land surface is a highly checkered mosaic of vegetation types and growth histories making the development of models capable of predicting future storage an enormously challenging task.

While radiocarbon is not nearly as valuable to this exercise as it is in the case of the ocean, it does have a role to play. More than half of the terrestrial carbon inventory is stored in soils. The humus in soils consists of a host of complex organic compounds. The evolution of storage in this reservoir will be driven by two competing impacts. Increasing planetary temperature will lead to more rapid oxidation of these humic compounds and hence will tend to drive down the planetary inventory. In contrast, increasing plant growth will lead to increased storage of new humic compounds and hence tend to drive up the inventory. Radiocarbon's role is to constrain the turnover times for the compounds making up humus.

The temporal trend of the $^{14}C/C$ ratio in soil humus (from prenuclear time to the present) makes it possible to separate the contribution of "passive" compounds that have very long soil residence times (measured in hundreds to thousands of years) and hence will be little changed by man's activities, and those "active" compounds which have relatively short chemical lifetimes (measured in decades). It also allows the characterization of the average turnover time of these "active" components. Further, by conducting such measurements on soils from different climate zones, it is possible to get a handle on how the turnover time of active compounds depends on temperature. Clearly, knowledge of this dependence is critical to the prediction of future global humus inventories.

Unfortunately, there was no GEOSECS-type survey of the world's soils during the 1970s. The decay counting method in use at that time required such large soil samples that processing was extremely cumbersome and stored samples were, by and large, far too small. Hence, interest in this subject lagged until the mid-1980s when atom counting became available. And of course, interest was further heightened when the Kyoto Accord permitted excess carbon storage in soils to be deducted from $CO_2$ emissions. However, unlike the ocean whose bomb $^{14}C$ inventory continues to rise, that in soils peaked in the 1970s and has subsequently declined. Hence, an enormous opportunity was largely missed—"largely" because, fortunately, here and there soils have been collected and stored.

As part of his PhD thesis research at Lamont–Doherty, Kevin Harrison summarized his own and published radiocarbon measurements on soils (see Figures 14 and 15). He was able to show three quite important things (Harrison *et al.* 1993a,b):

(i) In order to reconcile both the lower than modern pre-1950 $^{14}C/C$ ratios in soil humus and the time history of the bomb $^{14}C$-induced rise in the $^{14}C/C$ ratio, he concluded that about one quarter of the carbon in soils is in either inactive (i.e., turns over very slowly) or totally inert (immune to destruction) compounds.

(ii) In order to account for the temporal evolution of bomb $^{14}C$ in the soil, he concluded that the mean residence time of the active humus component was $25 \pm 10$ yr.

(iii) In order to account for the lower $^{14}C/C$ ratio found for agricultural soil he concluded that a sizable portion of the active component had been lost as the result of farming practices (tilling, harvesting, etc.) thereby enhancing the contribution of the low $^{14}C$ inactive component.

Sue Trumbore of the University of California, Irvine demonstrated that the turnover time of humic material is strongly dependent on soil temperature ranging from 100 yr at high latitudes

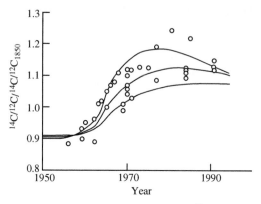

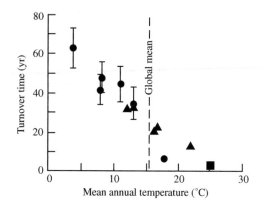

**Figure 14** Plot of measurements of the $^{14}$C/C ratio in untilled soils from a range of locales. The results are expressed as the $^{14}$C/C ratio in the bulk soil carbon to that in age-corrected 1850 wood. The curves are based on a model which assumes that 25% of the carbon is essentially inert (mean $^{14}$C age 3,700 yr) and that 75% turns over on timescales of 40 yr (lower curve), 25 yr (middle curve), and 15 yr (upper curve) (source Harrison and Broecker, 1993a).

**Figure 16** Turnover time for the active humus component of soils from the Brazilian rainforest (square), Hawaii (triangles), and the Sierra Nevada Mountains (circles). The mean global temperature is shown for reference (source Trumbore *et al.*, 1996).

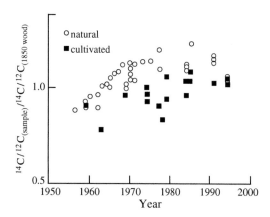

**Figure 15** Comparison of $^{14}$C to C measurements on natural soils and those on cultivated soils. The lower ratios are presumably the result of the loss of a sizable portion of the active humus component as the result of agricultural practice (source Harrison *et al.*, 1993a).

to as little as a decade in the tropics (Trumbore *et al.*, 1996) (see Figure 16).

### 4.09.7  FUTURE APPLICATIONS

The recent development of a much smaller and hence less-expensive accelerator mass spectrometer capable of state-of-the-art radiocarbon measurements will allow large users to have an in-house unit rather than having to send their samples off to one of the dozen or so existing major centers. These small units operate at $5 \times 10^5$ V rather than at several million volts as do the conventional accelerator mass spectrometers.

Also they will fit into a standard-size laboratory rather than one the size of a small airplane hanger. Further, like conventional mass spectrometers, these so-called "tandys" (i.e., small tandem Van de Graaff) can be maintained and operated by the users rather than by a team of specialists. Thus, measurements, which now cost from $300 to $600 each, will be carried out for more like $100 each. This will allow scientists to take on problems requiring hundreds of measurements rather than being budgetarily restricted to projects involving only tens of measurements.

As the bomb $^{14}$C transient demonstrated the value of radiocarbon as an environmental tracer, availability of "tandy" might lead to small-scale tracer experiments designed to study the allocation of carbon by plants in natural environments. Because of the high accuracy and high sensitivity of $^{14}$C measurements, such tracer experiments can be carried out at a few times the ambient atmospheric $^{14}$C/C ratio and hence pose no environmental hazard.

### REFERENCES

Adkins J. F., Cheng H., Boyle E. A., Druffel E. R. M., and Edwards R. L. (1998) Deep-sea coral evidence for rapid change in ventilation of the deep North Atlantic 15,400 years ago. *Science* **280**, 725–728.

Bard E. (1998) Geochemical and geophysical implications of the radiocarbon calibration. *Geochim. Cosmochim. Acta* **62**, 2025–2038.

Bard E., Hamelin B., Fairbanks R. G., and Zindler A. (1990a) Calibration of $^{14}$C timescale over the past 30,000 years using mass spectrometric U–Th ages from Barbados corals. *Nature* **345**, 405–410.

Bard E., Hamelin B., Fairbanks R. G., Zindler A., Arnold M., and Mathieu G. (1990b) U/Th and $^{14}$C ages of corals from Barbados and their use for calibrating the $^{14}$C time scale beyond 9,000 years BP. In *Proceedings of* the 5th International Conference on AMS (eds. F. Yiou and G. Raisbeck). *Nucl. Inst. Met.* **B52**, 461–468.

Bard E., Arnold M., Mangerud J., Paterne M., Labeyrie L., Duprat J., Mélières M.-A., Sønstegaard E., and Duplessy J.-C. (1994) The North Atlantic atmosphere-sea surface $^{14}C$ gradient during the Younger Dryas climatic event. *Earth Planet. Sci. Lett.* **126**, 275–287.

Bard E., Raisbeck G. M., Yiou F., and Jouzel J. (1997) Solar modulation of cosmogenic nuclide production over the last millennium: comparison between $^{14}C$ and $^{10}Be$ records. *Earth Planet. Sci. Lett.* **150**, 453–462.

Beck J. W., Richards D. A., Edwards R. L., Silverman B. W., Smart P. L., Donahue D. J., Hererra-Osterheld S., Burr G. S., Calsoyas L., Jull A. J. T., and Biddulph D. (2001) Extremely large variations of atmospheric $^{14}C$ concentration during the last glacial period. *Science* **292**, 2453–2458.

Becker B. and Kromer B. (1993) The continental tree-ring record—absolute chronology, $^{14}C$ calibration, and climatic change at 11 ka. *Palaeogeogr. Palaeoclimatol. Palaeoecol.* **103**, 67–71.

Bond G. C., Showers W., Cheseby M., Lotti R., Almasi P., deMenocal P., Priore P., Cullen H., Hajdas I., and Bonani G. (1997) A pervasive millennial-scale cycle in North Atlantic Holocene and glacial climates. *Science* **278**, 1257–1266.

Bond G. C., Showers W., Elliot M., Evans M., Lotti R., Hajdas I., Bonani G., and Johnson S. (1999) The North Atlantic's 1–2 kyr climate rhythm: relation to Heinrich events, Dansgaard/Oeschger cycles, and the Little Ice Age. In *Mechanisms of Global Climate Change at Millennial Time Scales*, Geophysical Monograph Series 112 (eds. P. Clark, R. Webb, and L. D. Keigwin). American Geophysical Union, Washington, DC, pp. 35–58.

Bond G. C., Kromer B., Beer J., Muscheler R., Evans M. N., Showers W., Hoffmann S., Lotti-Bond R., Hajdas I., and Bonani G. (2001) Persistent solar influence on North Atlantic climate during the Holocene. *Science* **294**, 2130–2152.

Bonhommet N. and Zähringer J. (1969) Paleomagnetism and potassium argon determinations of the Laschamp geomagnetic polarity event. *Earth Planet. Sci. Lett.* **6**, 43–46.

Broecker W. S. and Peng T.-H. (1994) Stratospheric contribution to the global bomb radiocarbon inventory: model versus observation. *Global Biogeochem. Cycles* **8**, 377–384.

Broecker W. S., Gerard R., Ewing M., and Heezen B. C. (1960) Natural radiocarbon in the Atlantic Ocean. *J. Geophys. Res.* **65**, 2903–2931.

Broecker W. S., Peng T.-H., Trumbore S., Bonani G., and Wolfli W. (1990) The distribution of radiocarbon in the glacial ocean. *Global Biogeochem. Cycles* **4**, 103–117.

Broecker W. S., Sutherland S., and Smethie W. (1995) Oceanic radiocarbon: separation of the natural and bomb components. *Global Biogeochem. Cycles* **9**, 263–288.

Druffel E. R. M. (1981) Radiocarbon in annual coral rings from the eastern tropical Pacific Ocean. *Geophys. Res. Lett.* **8**, 59–62.

Druffel E. R. M. (1989) Decade time scale variability of ventilation in the North Atlantic: high-precision measurements of bomb radiocarbon in banded corals. *J. Geophys. Res.* **94**, 3271–3285.

Druffel E. R. M. and Linick T. W. (1978) Radiocarbon in annual coral rings of Florida. *Geophys. Res. Lett.* **5**, 913–916.

deVries H. L. (1958) Variation in concentration of radiocarbon with time and location on Earth. *Proc. Koninkl. Ned. Akad. Wetenschap.* **61**, 94–102.

Edwards R. L., Chen J. H., and Wasserburg G. J. (1986/1987) $^{238}U$–$^{234}U$–$^{230}Th$–$^{232}Th$ systematics and the precise measurement of time over the past 500,000 years. *Earth Planet. Sci. Lett.* **81**, 175–192.

Fairbanks R. G. (1989) A 17,000-year glacio-eustatic sea level record: influence of glacial melting rates on the Younger Dryas event and deep ocean circulation. *Nature* **342**, 637–647.

Frank M., Schwarz B., Baumann S., Kubik P. W., Sater M., and Mangini A. (1997) A 200 kyr record of cosmogenic radionuclide production rate and geomagnetic field intensity from $^{10}Be$ in globally stacked deep-sea sediments. *Earth Planet. Sci. Lett.* **149**, 121–129.

Friedrich M., Kromer B., Spurk M., Hofmann J., and Kaiser K. F. (1999) Paleo-environment and radiocarbon calibration as derived from Late Glacial/Early Holocene tree-ring chronologies. *Quat. Int.* **61**, 27–39.

Harrison K. G., Broecker W., and Bonani G. (1993a) A strategy for estimating the impact of $CO_2$ fertilization on soil carbon storage. *Global Biogeochem. Cycles* **7**, 69–80.

Harrison K. G., Broecker W. S., and Bonani G. (1993b) The effect of changing land use on soil $^{14}C$. *Science* **262**, 725–726.

Hiller A., Boettger T., and Kremenetski C. (2001) Medieval climatic warming recorded by radiocarbon dated alpine tree-line shift on the Kola Peninsula, Russia. *Holocene* **11**, 491–497.

Hormes A., Schlüchter C., and Stocker T. F. (1998) Minimal extension phases of unteraar glacier (Swiss Alps) during the Holocene based on $^{14}C$ analysis of wood. *Radiocarbon* **40**, 809–817.

Hughen K. A., Overpeck J. T., Lehman S. J., Kasgarian M., Southon J., Peterson L. C., Alley R., and Sigman D. M. (1998) Deglacial changes in ocean circulation from an extended radiocarbon calibration. *Nature* **391**, 65–68.

Kitagawa H. and van der Plicht J. (1998) Atmospheric radiocarbon calibration to 45,000 yr BP: Late Glacial fluctuations and cosmogenic isotope production. *Science* **279**, 1187–1190.

Kromer B. and Spurk M. (1998) Revision and tentative extension of the tree-ring based $^{14}C$ calibration, 9,200–11,855 cal BP. *Radiocarbon* **40**, 1117–1125.

Kromer B., Rhein M., Bruns M., Schoch-Fisher H., Munnich K. O., Stuiver M., and Becker B. (1986) Radiocarbon calibration data for the sixth to the eighth millennia BC. In Calibration issue. *Radiocarbon* **28**, 954–960.

Laj C., Mazaud A., and Duplessy J. C. (1996) Geomagnetic intensity and $^{14}C$ abundance in the atmosphere and ocean during the past 50 kyr. *Geophys. Res. Lett.* **23**, 2045–2048.

Lal D. (1988) Theoretically expected variations in the terrestrial cosmic-ray production rates of isotopes. In *Solar–Terrestrial Relationships and the Earth Environment in the Last Millennia* (ed. X. C. V. Corso). Soc. Italiana de Fisica., pp. 216–233.

Libby W. F., Anderson E. C., and Arnold J. R. (1949) Age determination by radiocarbon content: worldwide assay of natural radiocarbon. *Science* **109**, 227–228.

Nydal R. and Lövseth K. (1983) Tracing bomb $^{14}C$ in the atmosphere 1962–1980. *J. Geophys. Res.* **88**, 3621–3642.

Oeschger H., Siegenthaler U., Gugelmann A., and Schotterer U. (1975) A box-diffusion model to study the carbon dioxide exchange in nature. *Tellus* **27**, 168–192.

Seigenthaler U., Heimann M., and Oeschger H. (1980) $^{14}C$ variations caused by changes in the global carbon cycle. *Radiocarbon* **22**, 177–191.

Shackleton N. J., Duplessy J.-C., Arnold M., Maurice P., Hall M. A., and Cartlidge J. (1988) Radiocarbon age of the last glacial Pacific deep water. *Nature* **335**, 708–711.

Siani G., Paterne M., Michel E., Sulpizio R., Sbrana A., Arnold M., and Haddad G. (2001) Mediterranean Sea surface radiocarbon reservoir age changes since the Last Glacial maximum. *Science* **294**, 1917–1920.

Sikes E. L., Samson C. R., Guilderson T. P., and Howard W. R. (2000) Old radiocarbon ages in the southwest Pacific Ocean during the Last Glacial period and deglaciation. *Nature* **405**, 555–559.

Southon J. (2002) A first step to reconciling the GRIP and GISP2 Ice-core chronologies, 0–14,500 yr BP. *Quat. Res.* **57**, 32–37.

Stuiver M. (1970) Long-term $^{14}C$ variations. In *Radiocarbon Variations and Absolute Chronology*, 12th Nobel Symposium (ed. I. U. Olsson). Wiley, New York, pp. 197–213.

Stuiver M. (1971) Evidence for the variation of atmospheric $^{14}$C content in the Late Quaternary. In *Late Cenozoic Glacial Ages* (ed. K. K. Turekian). Yale University Press, New Haven, CT, pp. 57–70.

Stuiver M. (1980) Solar variability and climatic change during the current millennium. *Nature* **286**, 868–871.

Stuiver M. (1982) A high-precision calibration of the AD radiocarbon time scale. *Radiocarbon* **24**, 1–26.

Stuiver M. and Braziunas T. F. (1993) Sun, ocean, climate, and atmospheric $^{14}$CO$_2$: an evaluation of causal and spectral relationships. *Holocene* **34**, 289–305.

Stuiver M. and Quay P. D. (1980) Changes in atmospheric carbon-14 attributed to a variable Sun. *Science* **207**, 11–19.

Stuiver M. and Quay P. D. (1981) Atmospheric $^{14}$C changes resulting from fossil fuel CO$_2$ release and cosmic ray flux variability. *Earth Planet. Sci. Lett.* **53**, 349–362.

Stuiver M., Reimer P. J., Bard E., Beck J. W., Burr G. S., Hughen K. A., Kromer B., McCormac G., van der Plicht J., and Spurk M. (1998) Intcal98 radiocarbon age calibration, 24,000–0 cal BP. *Radiocarbon* **40**, 1126–1159.

Thellier E. and Thellier O. (1959) Sur l'intensite du champ magnetique terrestre dans le passe historique et geologique. *Ann. Geophys.* **15**, 285–378.

Trumbore S. E., Chadwick O. A., and Amundson R. (1996) Rapid exchange between soil carbon and atmospheric carbon dioxide driven by temperature change. *Science* **272**, 393–396.

Voelker A. H. L., Sarnthein M., Grootes P. M., Erlenkeuser H., Laj C., Mazaud A., Nadeau M.-J., and Schleicher M. (1998) Correlation of marine $^{14}$C ages from the Nordic seas with the GISP2 isotope record: implications for radiocarbon calibration baeyond 25 ka BP. *Radiocarbon* **40**, 517–534.

Waelbroeck C., Duplessy J.-C., Michel E., Labeyrie L., Paillard D., and Duprat J. (2001) The timing of the last deglaciation in North Atlantic climate records. *Nature* **412**, 724–727.

Yiou F., Raisbeck G. M., Baumgartner S., Beer J., Hammer C., Johnsen S., Jouzel J., Kubik P. W., Lestringuez J., Stievenard M., Suter M., and Yiou P. (1997) Beryllium 10 in the Greenland Ice Core Project ice core at Summit, Greenland. *J. Geophys. Res.* **102**, 26783–26794.

# 4.10
# Natural Radionuclides in the Atmosphere

## K. K. Turekian and W. C. Graustein
### Yale University, New Haven, CT, USA

## 4.10.1  INTRODUCTION

Natural radioactive nuclides in the atmosphere have two principal sources—radon and its progeny derived from Earth's surface and cosmic-ray-produced nuclides. Dust from the elevation of soils can also provide secondary sources of these nuclides. Suitable accommodation for these sources must be made if only those species having gaseous precursors are to be considered.

There is a radon isotope in each of the three major natural-decay chains: $^{238}U$ ($^{222}Rn$, half-life = 3.8 d), $^{232}Th$ ($^{220}Rn$, half-life = 55 s), and $^{235}U$ ($^{219}Rn$, half-life = 3.9 s). Almost all radon in the atmosphere is produced in soils and is transported to the atmosphere by diffusion. Because its longer half-life allows for greater diffusive transport, most radon entering the atmosphere is $^{222}Rn$; its radioactive decay scheme is presented in Table 1.

An early summary of $^{222}Rn$ and its progeny in the atmosphere was provided by Turekian et al. (1977).

Table 2 lists the radioactive species produced by cosmic rays acting on the gaseous components of the atmosphere, which include nitrogen, oxygen, and all the rare gases as targets. A detailed discussion of the formation of radioactive species by cosmic-ray bombardment has been published by Lal (2001) based on his pioneering work since 1967 in a classic paper by Lal and Peters (1967). Radiocarbon ($^{14}C$) is not discussed in this chapter. Because of its central role in many Earth's surface processes, a separate chapter is found in this volume (see Chapter 4.09).

**Table 1** The decay chain for $^{222}$Rn.

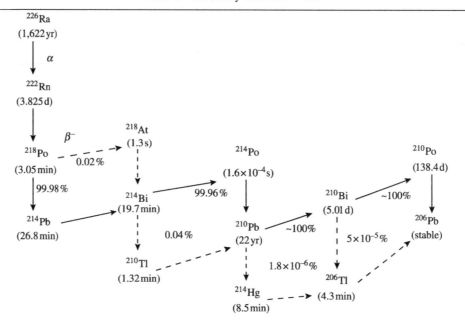

**Table 2** Production rates of several isotopes in the Earth's atmosphere; arranged in order of decreasing half-lives.

| Isotope | Half-life | Production rate (atoms cm$^{-2}$ s$^{-1}$) | | Global inventory |
|---|---|---|---|---|
| | | Troposphere | Total atmosphere | |
| $^3$He | Stable | $6.7 \times 10^{-2}$ | 0.2 | $3.2 \times 10^3$ t[a] |
| $^{10}$Be | $1.5 \times 10^5$ yr | $1.5 \times 10^{-2}$ | $4.5 \times 10^{-2}$ | 260 t |
| $^{26}$Al | $7.1 \times 10^5$ yr | $3.8 \times 10^{-5}$ | $1.4 \times 10^{-4}$ | 1.1 t |
| $^{31}$Kr[b] | $2.3 \times 10^5$ yr | $5.2 \times 10^{-7}$ | $1.2 \times 10^{-5}$ | 8.5 kg |
| $^{36}$Cl | $3.0 \times 10^5$ yr | $4 \times 10^{-4}$ | $1.1 \times 10^{-3}$ | 15 t[d] |
| $^{14}$C | 5,730 yr | 1.1 | 2.5 | 75 t |
| $^{39}$Ar[c] | 268 yr | $4.5 \times 10^{-3}$ | $1.3 \times 10^{-2}$ | 52 kg |
| $^{32}$Si | ~150 yr | $5.4 \times 10^{-5}$ | $1.6 \times 10^{-4}$ | 0.3 kg |
| $^3$H | 12.3 yr | $8.4 \times 10^{-2}$ | 0.25 | 3.5 kg |
| $^{22}$Na | 2.6 yr | $2.4 \times 10^{-5}$ | $8.6 \times 10^{-5}$ | 1.9 g |
| $^{36}$S | 87 d | $4.9 \times 10^{-4}$ | $1.4 \times 10^{-3}$ | 4.5 g |
| $^7$Be | 53 d | $2.7 \times 10^{-2}$ | $8.1 \times 10^{-2}$ | 3.2 g |
| $^{37}$Ar | 35 d | $2.8 \times 10^{-4}$ | $8.3 \times 10^{-4}$ | 1.1 g |
| $^{33}$P | 25.3 d | $2.2 \times 10^{-4}$ | $6.8 \times 10^{-4}$ | 0.6 g |
| $^{32}$P | 14.3 d | $2.7 \times 10^{-4}$ | $8.1 \times 10^{-4}$ | 0.4 g |

Source: Based on Lal and Peters (1967).
[a] The inventory of this stable nuclide is based on its atmospheric inventory, which includes an appreciable contribution from crustal degassing of $^3$He.
[b] Based on atmospheric $^{31}$Kr/Kr ratio of $(5.2 \pm 0.4) \times 10^{-13}$.   [c] Based on atmospheric $^{39}$Ar/Ar ratio of $(0.107 \pm 0.004)$ dpm$^{-1}$ l Ar (STP).
[d] Includes a rough estimate of $^{36}$Cl produced by the capture of neutrons at the Earth's surface.

There are more than 30 radionuclides produced in the atmosphere by these two natural processes. This chapter focuses on describing different ways in which the observed distributions of several of these nuclides can be used to infer patterns and rates of transport, mixing, and removal of these nuclides and of other constituents of the atmosphere.

We would like to define here several terms. "Activity" refers to the number of radioactive decays of a nuclide per unit time and is the product

of the decay constant, $\lambda$, and the number of atoms, $N$. The SI unit of activity is the Becquerel (Bq), which represents one disintegration per second. The "disintegration per minute" or dpm often appears in the literature. An older representation was the Curie (Ci), which is defined as the activity of 1 g of pure $^{226}$Ra or $3.70 \times 10^{10}$ disintegrations per second. The picocurie (pCi), which is 0.037 Bq or 2.2 dpm, is often used in reporting radon concentrations in air.

In a decay chain shown in Table 1 at "secular equilibrium" the activities of the coupled radionuclides are equal. In a closed system, this occurs after roughly five half-lives of the shorter-lived daughter have elapsed ("parent" and "daughter" refer to the first and subsequent nuclides in the decay series being considered). For example, $^{222}$Rn (half-life $= 3.8$ d) is in secular equilibrium with its parent, $^{226}$Ra (half-life $= 1,620$ yr), in $\sim 20$ d.

### 4.10.2 RADON AND ITS DAUGHTERS

#### 4.10.2.1 Flux of Radon from Soils to the Atmosphere

The escape of $^{222}$Rn from soils is the source of $\sim 99\%$ of the $^{222}$Rn in the atmosphere. Typical radon escape rates are on the order of 1 atom cm$^{-2}$ s$^{-1}$ from the land surface, which result in a radon inventory of the global atmosphere of $\sim 1.5 \times 10^{18}$ Bq. Atmospheric radon itself is a chemically inert and unscavenged, i.e., not removed from the atmosphere by physical or chemical means. Because its half-life is much less than the mixing time of the atmosphere, it is a tracer of atmospheric transport and can be used in a synoptic approach to identify air masses derived from continental boundary layers or in a climatological manner to verify the predictions of numerical models of transport.

$^{222}$Rn is lost from the atmosphere only by its radioactive decay and it is thus the source of virtually all of the $^{214}$Bi, $^{214}$Pb, $^{214}$Po, $^{210}$Pb, $^{210}$Bi, and $^{210}$Po in the atmosphere. The quantities of radon decay products on soil dust suspended by atmospheric turbulence or volatilized from lava are small by comparison on a global scale, but may be comparable locally and episodically.

The migration of $^{222}$Rn through soils into buildings can lead to indoor air concentrations of $^{222}$Rn that pose significant radiological health hazards. Knowing the flux and understanding the factors that cause it to vary spatially and temporally improve the utility of $^{222}$Rn as an atmospheric tracer and increase the ability to predict potential health hazards National Research Council (1988).

$^{238}$U is a ubiquitous trace component of rock and soil; typical activities are $\sim 30$ mBq g$^{-1}$. In most soils, there is little transport of $^{234}$U, $^{230}$Th, or $^{226}$Ra, so the activities of $^{238}$U and $^{226}$Ra are nearly equal. When an atom of $^{226}$Ra decays in a mineral grain, the energy imparted by the recoil of the alpha particle is sufficient to displace the daughter $^{222}$Rn atom by a few tens of nanometers. Although most often $^{222}$Rn atoms remain within the grain, a fraction of them emanate from the grain and become free to move through the pore space of the soil. The fraction of $^{226}$Ra decays that result in $^{222}$Rn atoms in the pore space is termed as the emanating power or fraction. Typical values for emanating fractions are $\sim 25\%$. Within the top few meters of a soil, a fraction of the radon in the pore space is transported to the surface, where it escapes to the atmosphere.

If one makes the simplifying assumptions that transport of $^{222}$Rn is by molecular diffusion alone and that soil has uniform porosity, $^{222}$Rn concentration, and emanating power, then the fraction of $^{222}$Rn that escapes to the atmosphere as a function of depth, $L_f(z)$, is given by (e.g., Clements and Wilkening, 1974)

$$L_f(z) = a\mathrm{e}^{-bz} \qquad (1)$$

where $a$ is the $^{222}$Rn emanating power (dimensionless), $b = \sqrt{\lambda \varepsilon / D}$ (cm$^{-1}$), $\lambda$ is the decay constant of $^{222}$Rn (s$^{-1}$), $\varepsilon$ is the porosity (dimensionless), $D$ is the diffusion coefficient (cm$^2$ s$^{-1}$) of $^{222}$Rn, and $z$ is the depth (cm).

Here $1/b$ has dimensions of length and is called the "mean depth" or "relaxation depth" of radon loss. Values of $1/b$ reported in the literature range from 100 cm to 218 cm (Schery and Gaeddert, 1982; Clements and Wilkening, 1974; Graustein and Turekian, 1990; Dörr and Münnich, 1990). These terms should not be confused with "half-depth" or 0.693 mean depth.

The escape rate of $^{222}$Rn, $J$, can be derived from (1) by multiplying by the $^{226}$Ra concentration in soil and integrating over depth, yielding

$$J = \frac{a\rho \, (^{226}\mathrm{Ra})}{b} \qquad (2)$$

where $\rho$ is the density (g cm$^{-3}$) and $(^{226}$Ra$)$ is the activity (Bq g$^{-1}$).

Compared to this idealized model, the actual flux of $^{222}$Rn may be diminished by the saturation of pore space by water (the mean length of $^{222}$Rn diffusion in water is on the order of a millimeter, so saturation diminishes the flux by up to a factor of 1,000) and decreases in porosity with depth. Advection of gas through soil in response to barometric pressure change, soil gas convection, and transpiration of $^{222}$Rn saturated soil solution will increase the radon escape rate. All of these processes are difficult to model accurately, so the determination of $^{222}$Rn fluxes relies on measurements.

Several methods have been developed to measure the escape rate or flux of radon to the atmosphere; measuring accumulation of $^{222}$Rn in chambers placed on the soil surface, modeling the flux from the vertical profile of $^{222}$Rn in the atmosphere, modeling the diffusive flux from measurements of $^{222}$Rn in the interstitial gas of a soil (e.g., Dörr and Münnich, 1990), and measuring the deficiency of the $^{210}$Pb daughter with respect to its parent $^{226}$Ra (Graustein and Turekian, 1990).

As shown by Equations (1) and (2), the radon loss fraction at shallow depths closely approximates the emanating power, and the escape rate is directly proportional to the emanating power. The data of Graustein and Turekian (1990) put a lower limit on the mean emanating power of US soils at 0.23. The fraction of $^{222}$Rn lost from depth between 10 cm and 60 cm for 340 cores from North America (Graustein and Turekian, 1990, in preparation) averages 0.22 and does not vary significantly over that interval. Figure 1 shows the distribution with depth. An exponential curve with a half-depth of 100 cm (mean depth = 144 cm) is plotted for reference. The value of $b$ was assigned; the value for $a$, 0.28, was obtained from a best fit to the data. With the exception of the sections from above 20 cm and those below 180 cm, the curve fits the data well. The counting errors and small number of samples from below 160 cm diminish our confidence in using this curve to extrapolate to greater depths and we regard extrapolations as representing an upper limit rather than an estimate. The samples from above 20 cm fall to the right of the reference curve, most probably due to the presence of atmospherically derived $^{210}$Pb. The extrapolation of the reference curve to the surface yields an upper limit to the emanating fraction of 0.28.

Using these data the mean $^{222}$Rn flux from US soils is at least 1.5 atoms cm$^{-2}$ s$^{-1}$. Extrapolation using the exponential curve in Figure 1 gives an upper limit of 2.0 atoms cm$^{-2}$ s$^{-1}$. This range is 50–100% greater than the commonly used estimate of 1.0 atoms cm$^{-2}$ s$^{-1}$ for global radon flux from continents, implying that there are large areas with radon fluxes less than 1 or that the global estimate of 1.0 is low. The survey of $^{222}$Rn fluxes from Australia by Schery *et al.* (1989) averaged 1.05 atoms cm$^{-2}$ s$^{-1}$ with a standard deviation of 0.24 indicating that large areas may have significantly different $^{222}$Rn fluxes. Their data were derived from the accumulation of $^{222}$Rn in accumulator chambers and are nearly instantaneous compared to the 30 yr mean values determined from the $^{210}$Pb deficiency in soil method discussed above. That the two Rn flux measurements yield the same result is justified by the similarity of a $^{222}$Rn flux from a soil sampling site in the same cornfield (Graustein and Turekian, 1990) in which the $^{222}$Rn flux was determined by accumulator (Pearson and Jones, 1966).

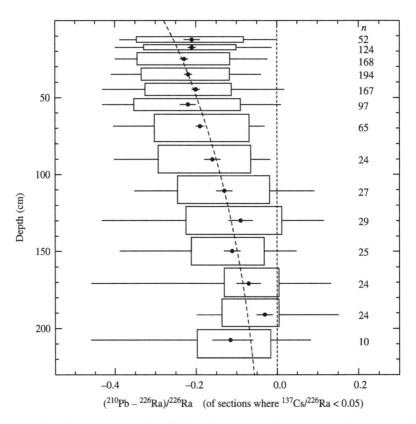

**Figure 1** Depth profile of radon loss fraction. Filled circles and error bars represent the mean and standard error of the radon loss fraction from all samples from the indicated depth for which the $^{137}$Cs/$^{226}$Ra ratio is less than 0.05. The boxes represent $1\sigma$ deviation and the whiskers the extreme values of each population. The number of samples is given in the column on the right. The curve corresponds to Equation (1) with an emanating power of 0.28 and a mean depth of 144 cm.

The spatial variability of the radon fluxes means that many measurements are needed to make precise estimates of the mean value for a region. Conen and Robertson (2002) noted in their compilation of radon fluxes determined by accumulator that there is a correlation with latitude from 30° to 50°, with a decreasing flux with increasing latitude. They ascribed this change as due primarily to the degree of water saturation of the soils, the increase toward higher latitude.

Comparisons of $^{222}$Rn flux from the upper 50 cm, based on disequilibrium measurements in soil profiles discussed above, with mean annual temperature and precipitation show only weak relations; neither $r^2$ exceeded 0.02. The major correlation is with the $^{226}$Ra activity of the soil (Figure 2).

### 4.10.2.2 Flux of Radon from the Oceans

The radium concentration in the oceans is at least a factor of a thousand less than soils; therefore, the flux is proportionally less. The radon flux from the mixed layer has been determined by Broecker and Peng (1971) and Wilkening and Clements (1975) and is as expected low and varies depending on wind stress. Its contribution to the radon burden of the atmosphere can therefore be taken as negligible. It is less than 1% of the global continental flux. A flux from shoal areas around Hawaii has been measured by Moore *et al.* (1974). Although the flux is higher than the open ocean, being driven by the higher radium concentration of the underlying rocks, on a worldwide basis it too is negligible. The major flux of radon to the atmosphere is clearly soils, mainly on the continents.

### 4.10.2.3 Distribution of Radon in the Atmosphere

The half-life of $^{222}$Rn (3.8 d) is much less than the mixing time of the atmosphere, so its concentrations are greatest near the land surface and decrease with both altitude and distance from land (e.g., Turekian *et al.*, 1977; Liu *et al.*, 1984).

Although the vertical distribution of radon over the continents is a direct consequence of supply from soils, convection upward (treated as turbulent diffusion), and radioactive decay, the pattern is different over the oceans. This difference is due to the fact that no significant source of radon exists over the oceans and the pattern is set by long-distance transport from continents. Off the northwest coast of the United States, for example, Andreae *et al.* (1988) show vertical patterns up to ~4 km ranging from constancy with elevation to increases with elevation. The concentration range from ~6 pCi m$^{-3}$ (STP) to 10 pCi m$^{-3}$ (STP) is more typical of upper troposphere air over the continents and not like the ~400 pCi m$^{-3}$ (STP) in the continental boundary layer (Figure 3).

Radon and its progeny $^{210}$Pb have been found in the stratosphere in some locations at elevated levels. Although some of $^{210}$Pb may be due to explosive volcanic penetration of the tropopause, most of the $^{210}$Pb is due to the decay of radon injected convectively mainly in the tropics (Kritz *et al.*, 1993).

Lambert *et al.* (1970) reported periodic fluctuations in the radon concentrations in boundary layer air sampled on islands and Antarctica between 40° S and 70° S. Being in the geographic zone with no large land masses, all the radon is delivered from lower latitudes by advection. The ambient concentration was found to be ~2 pCi m$^{-3}$ (STP) with occasional "radon storms" raising the concentration briefly to 10 pCi m$^{-3}$ (STP). Lambert *et al.* (1970) argued that, based on long-term observations, there is a 27–28 d cycle of radon variation in this region. The causes of changes in circulation that would yield this result are not known, but radon measurements obviously put constraints on models of hemispheric circulation.

### 4.10.2.4 Short-lived Daughters of $^{222}$Rn in the Atmosphere

As can be seen in Table 1, $^{222}$Rn decays through a number of short-lived daughters before the long-lived $^{210}$Pb is reached. The first of these radionuclides along the decay chain is $^{218}$Po with a half-life of 3 min. The 6 MeV energy of the alpha particle emitted by the $^{222}$Rn decay causes the $^{218}$Po to recoil with a velocity much greater than that of the ambient gas molecules. Collisions with gas molecules or atoms diminish

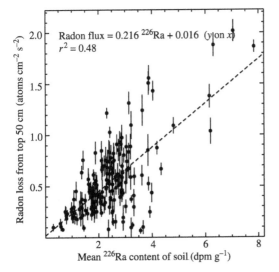

**Figure 2** Radon loss rate from the upper 50 cm of soils versus the mean radium concentration. Bars indicate calculated 1$\sigma$ error of the loss.

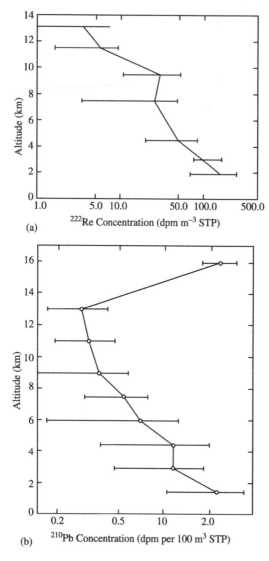

(a) $^{222}$Re Concentration (dpm m$^{-3}$ STP)

(b) $^{210}$Pb Concentration (dpm per 100 m$^3$ STP)

**Figure 3** (a) The average distribution of Rn with elevation in the North American mid-continental region based on several profiles averaged by Moore *et al.* (1973). (b) Vertical distribution of $^{210}$Pb in the atmosphere in the mid-continental region of North America (source Moore *et al.*, 1973).

its energy until it is same as that of the ambient air, at which time the $^{218}$Po diffuses through the air until it encounters a surface, typically an aerosol particle, to which it attaches.

In an enclosed space within which a large number of aerosol particles exist, such as in a mine or around cigarette smoke, such $^{222}$Rn decay derived nuclides attach to particles on the order of seconds to minutes and subsequent radioactive decay can occur on the aerosol particles. An array of short-lived energetic radioactive species in the chain add to the radioactive burden of the particle. If inhaled, these highly radioactive particles can cause health problems. This was shown in the case of uranium miners and may be a part of the cause

of the lung problems of smokers and those affected by secondary particle inhalation around smokers.

Studies of $^{214}$Bi (half-life = 19.8 min) activity relative to $^{214}$Pb (half-life = 26.8 min) in boundary layer air indicate that these are in equilibrium, indicating that the particles are not removed more rapidly than a couple of hours (Turekian *et al.*, 1999). Attaching of these short-lived daughters on aerosols, and the mean life of the aerosols being much longer than their half-lives means that the nuclide to be tracked on most atmospherically interesting timescales is $^{210}$Pb with its half-life of 22 yr.

### 4.10.2.5 $^{210}$Pb and Its Progeny

After it is produced by the decay of $^{214}$Po, $^{210}$Pb rapidly associates with aerosols in the 0.1–0.5 μm diameter size range (Knuth *et al.*, 1983). Aerosols in this size range, the so-called "accumulation mode," are large enough that their velocity due to Brownian motion is so small that diffusion is not a significant method of transport. They are also small enough that their gravitational settling velocity is much less than typical rates of vertical motion in the atmosphere. Scavenging by precipitation is the principal mechanism of removal of these aerosols. Many of the chemical species with low vapor pressure that form from gaseous precursors in that atmosphere, such as $H_2SO_4$ formed from the oxidation of $SO_2$, are also associated with accumulation-mode aerosols.

As a result, these aerosols follow the motions of the atmosphere and are responsible for the transport of much of the nonvolatile products of photochemical and oxidative reactions. $^{210}$Pb is a minor constituent of this aerosol population (one aerosol particle in $10^4$ or more will carry an atom of $^{210}$Pb) and is a useful tracer of the transport, deposition, and residence time of aerosols.

#### 4.10.2.5.1 Distribution of $^{210}$Pb in the atmosphere

Vertical profiles of Rn, $^{210}$Pb, $^{210}$Bi, and $^{210}$Po were measured over the mid-continental United States by Moore *et al.* (1973). These results for $^{222}$Rn and $^{210}$Pb are shown in Figure 3. A number of surface sites have been locations for long-term studies of $^{210}$Pb in conjunction with the measurements of radionuclides from nuclear bomb testing by the Environmental Measurements Laboratory of the Department of Energy and its precursor, the Atomic Energy Commission (Feely *et al.*, 1981). In addition, there have been measurements made throughout the United States and the islands of the Pacific and Atlantic as parts of the SEAREX and

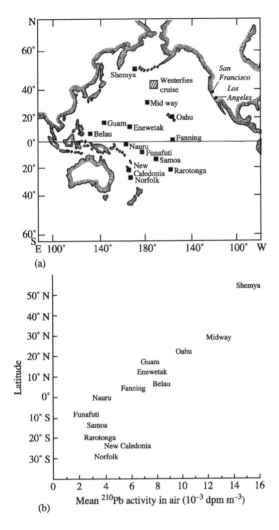

(a)

(b)

**Figure 4** (a) Location map of the sites of the SEAREX network. (b) Mean concentration of $^{210}$Pb in air in the SEAREX network. The site name is centered over the point representing the data. The error of the mean is approximated by the length of the name. Data are measured values which should be increased by ~30% to correct for filter capture efficiency (source Turekian et al., 1989).

AEROCE programs, respectively. The sampling programs complement each other and, where there are regional overlaps, the results are identical. The values across the Pacific from the SEAREX program (Turekian et al., 1989) are shown in Figure 4. There clearly is a correlation with the size of the land-mass upwind from the island sampling site. This is expected since $^{222}$Rn is the precursor to the $^{210}$Pb and is derived from soils.

### 4.10.2.5.2 *Flux of $^{210}$Pb to Earth's surface*

(i) *Methods of determinations.* The most direct method of estimating the flux of $^{210}$Pb to the Earth's surface is through bucket collection of precipitation and subsequent analysis of the water for $^{210}$Pb obtained over a specific period of time. This method has been used during the assay of precipitation for bomb-produced radionuclides such as $^{90}$Sr and $^{137}$Cs, notably by the Environmental Measurements Laboratory of the Department of Energy and its precursor, the Atomic Energy Commission. Detailed summaries, mainly along ~80° W longitude in both hemispheres plus some other selected sites in the continental US, are available from the Department of Energy as unpublished reports.

There is great value to the data obtained by such a method especially as it provides seasonal values. The method does not, however, measure the extraction of water and $^{210}$Pb from the air by "horizontal" precipitation. Horizontal precipitation is the consequence of impingement of cloud, fog, or dew droplets on leaf surfaces. The water and its constituents are then transferred to the soil by dripping or episodic vertical precipitation. Also the long-term time-averaged flux is limited by the length of time that the sampling program has been in effect.

The accumulation of atmospherically derived $^{210}$Pb in soil profiles is another method of assessing the atmospheric flux of $^{210}$Pb. This approach has already been referred to in the section on the measurement of $^{222}$Rn flux from soils to the atmosphere. Specifically, the atmospherically derived $^{210}$Pb in soils is generally found in the topmost organic-rich layer, although there is transport down into the soil profile by illuviation, bioturbation, and possibly chemical transport by chelators.

This method of assaying the long-term total atmospheric $^{210}$Pb flux has been studied extensively by Moore and Poet (1976), Nozaki et al. (1978), and Graustein and Turekian (1983, 1986, 1989). It requires that the soil profile be undisturbed for ~100 yr. In nonforested areas the directly measured atmospheric fluxes of $^{210}$Pb and the fluxes determined by soil profile method are generally in agreement, allowing for the expected short-term variability (Graustein and Turekian, 1986). The relationship of the atmospherically derived standing crop of excess $^{210}$Pb in soil profiles and the atmospheric flux is

$$\text{Flux} = (^{210}\text{Pb decay constant}) \times \text{inventory}$$

(ii) *The flux of $^{210}$Pb over the oceans.* The flux of $^{210}$Pb across the Pacific Ocean based on bucket collections has mainly been studied by Tsunogai et al. (1985) for the region around Japan, and by Turekian et al. (1989) for the rest of the Pacific as part of the SEAREX program. Figure 5 shows the flux of $^{210}$Pb over the North Pacific. The most striking pattern is the very high flux close to Japan

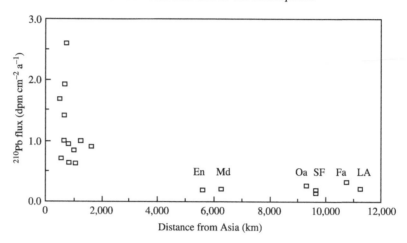

**Figure 5** $^{210}$Pb deposition flux versus distance from the Asian continent. Unlabeled points are from Tsunogai *et al.* (1985). En, Enewetak; Md, Midway; SF, San Francisco (Fuller and Hammond, 1983; Monaghan *et al.*, 1985/1986); LA, Los Angeles (Fuller and Hammond, 1983); Oa, Oahu; FA, Fanning. For the mid-ocean sites distances from Asia are measured along parallels of latitude (source Turekian *et al.*, 1989).

and a virtually constant low flux across the rest of the North Pacific including the California coast.

This pattern is different from the model expectation presented by Turekian *et al.* (1977), which predicted an exponential decrease of $^{210}$Pb flux across the North Pacific. The observations clearly indicate the efficient removal of the burden of $^{210}$Pb from the Asian continent by the time the longitude of Japan is reached. The flux of $^{210}$Pb thereafter is determined by the $^{222}$Rn concentration in the air and the scavenging efficiency by rain.

(iii) *The flux of $^{210}$Pb over the continents.* The flux of $^{210}$Pb across the United States has been studied primarily by the Environmental Measurements Laboratory on precipitation (see Feely *et al.*, 1981) and Graustein and Turekian (1986 and references therein) in soil profiles. Figure 6 shows the $^{210}$Pb flux pattern across the United States primarily from these sources but from other data as well. The major features are: (i) a general increase in flux from the Pacific coast towards the Great Plains; (ii) an approximately constant flux across the Great Plains to the Appalachians; (iii) marked orographic effects in the Sierras and the Appalachians; and (iv) a decrease along the east coast. The explanation for observation (i) is that air arriving with the prevailing westerly winds is generally depleted in $^{210}$Pb and its precursor $^{222}$Rn because of scavenging over the oceans and decay during transport, respectively, and the increase towards the center of the continent and general constancy is due to the local nature of both the source of $^{210}$Pb and its removal. The marked orographic effects (iii) are due to both the increased rainfall in the mountains and the cloud scavenging alluded to earlier. The decrease on the

east coast is due to precipitation carrying aerosols both from the $^{210}$Pb-rich continental interior and $^{210}$Pb-poor maritime air.

### 4.10.2.5.3 Residence time of $^{210}$Pb and associated species in the atmosphere

One approach to measuring the residence time of aerosols is to calculate the time required for presumed initial values in an air mass of two or more nuclides in a decay chain to evolve to the observed values. This approach yields a measure of the effect of removal processes along the path of transport of the air mass being sampled.

If we consider an air parcel as a closed system, then the activity of each nuclide in the $^{222}$Rn decay chain is described by

$$dN/dt = P - \lambda N \qquad (3)$$

where $N$ is the number of atoms per unit volume and $P$ is the production rate of the nuclide by the radioactive decay of its parent. Applying (3) to each nuclide in a decay chain and setting initial conditions leads to a set of coupled differential equations. The general form for the solution to such a set of differential equations was obtained by Bateman (1910) and is given in standard references in radiochemistry (e.g., Friedlander *et al.*, 1981).

The dotted line in Figure 7 shows the time evolution of the ratios of radon daughters in closed system in which initially $^{222}$Rn is present but its decay products are absent. We refer to this model age as the batch process time.

The atmosphere can also be modeled as a steady-state system, to which $^{222}$Rn is added at a constant rate and from which its decay products

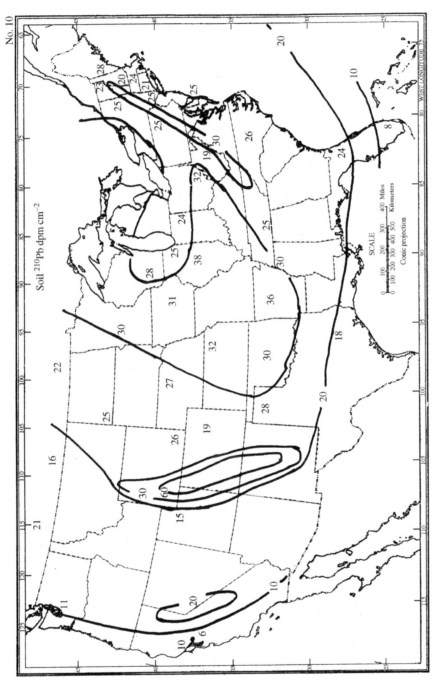

**Figure 6** $^{210}$Pb deposition flux over the United States. The cartoon is based on measurements made in the laboratory of the authors with accommodation for topography. Standing crop of 32 dpm cm$^{-2}$ equals a deposition flux of 1 dpm $^{210}$ Pb cm$^{-2}$ yr$^{-1}$.

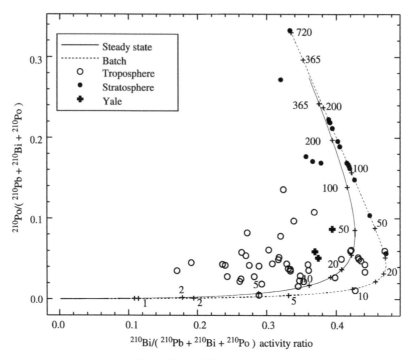

**Figure 7** Residence time implications of $^{210}$Pb–$^{210}$Bi–$^{210}$Po activities in aerosols. The lines represent the evolution of radon daughter product activities for a closed system (dotted line) or a steady-state system (solid line). The numbers indicate the age (for a closed system) or aerosol mean residence time (steady-state system) corresponding to the adjacent + mark. Data from Moore *et al.* (1973) are plotted as filled circles for samples from the stratosphere and open circles for samples from the troposphere. Yale data from the troposphere are marked by crosses.

are removed by a first-order removal process:

$$dN/dt = 0 = P - (\lambda + k)N \qquad (4)$$

where $k$, the removal constant, is the inverse of the mean residence time. It is zero for $^{222}$Rn and assumed equal for all of its decay products. The production term, $P$, for each nuclide is equal to the decay rate of its parent. The solid line in Figure 7 shows the ratio of radon daughters as a function of mean residence time.

The real atmosphere does not conform to either model assumption, but lies somewhere between them. Over the five-day timescale of $^{210}$Bi decay, there is too much mixing to allow for the identification and sampling of a parcel that has behaved as a closed system and far too little mixing and transport to allow the homogenization required by the steady-state assumptions. The curves are, however, similar enough to allow meaningful interpretation of data.

Real air samples can be thought of as a mixture of individual small parcels that are small enough to have behaved as closed systems. The plot is therefore constructed with a common denominator for both the *x*- and *y*-axes, so that a mixture of two components will plot along the line connecting the points representing the two components.

Moore *et al.* (1973) reported on a series of samples collected at various altitudes in the

troposphere and stratosphere over central North America and discussed their implications for aerosol residence times. We review their data here and add the consideration of parcel mixing. Samples obtained from the stratosphere are denoted by a filled circle. Most of these samples lie along the parcel evolution line with model ages from 40 d to 2 yr, with the bulk of the samples between 100 d and 180 d. These model ages are consistent with the meteorological understanding of the lower stratosphere, where vertical mixing is suppressed and precipitation scavenging is nil. These ages are also consistent with, though slightly smaller than estimates of stratospheric residence times derived from the behavior of fission products injected into the stratosphere by atmospheric testing of nuclear weapons. A few of the stratospheric samples fall below the evolution curves, suggesting the recent admixture of some "younger" aerosols.

By comparison, few of their tropospheric samples, represented by open circles, lie on either model evolution curve, but all lie within an envelope bounded by the batch model curve on the right. These data are consistent with mixing of aerosols with varying ages, and many of the data strongly indicate that mixing has occurred. The data that fall between the two model curves are most constrained and indicate an age between 20 d and 30 d.

The samples that plot within the envelope bounded by the steady-state evolution curve are less constrained. Although those that do not fall near the model curve show the effects of mixing of aerosols of different ages, the data do not lie along a single trend and therefore do not suggest end-members. Most of the samples are consistent with mixing of aerosols of age 1–2 d with those of up to age 50 d; a few of the samples appear to require an even more aged component.

Martell and Moore (1974) and Moore *et al.* (1973) interpreted these data to yield a mean age of aerosols in the lower troposphere of less than 4 d, increasing by a factor of 3 toward the top of the troposphere. The mixing analysis presented above does not contradict their estimate of the mean and yields information about the range of ages of the parcels that compose their samples. The combination of the approaches suggests that the troposphere is not sufficiently mixed on timescales of a week to approach homogenization, but is mixed sufficiently that a single aerosol sample almost always contains material from parcels of widely varying history.

Another approach to determining the residence time of $^{210}$Pb in the atmosphere is to divide the mean air column inventory of $^{210}$Pb by the flux of $^{210}$Pb to the surface at a given location. This quotient yields a climatological average for the removal processes at that particular site. Graustein and Turekian (1986) used the atmospheric profiles of $^{210}$Pb from Moore *et al.* (1973) and their own measured $^{210}$Pb fluxes from soil profiles and bucket collection to obtain a value of ~6 d over the central and eastern United States. As the source of $^{222}$Rn and thus $^{210}$Pb is from the ground and the major removal by precipitation is in the lower troposphere, the mean residence time is dominated by the processes of the lower troposphere. Modeling by Balkanski *et al.* (1993) shows that the mean residence time of $^{210}$Pb increases with altitude.

#### 4.10.2.5.4 *Use of $^{210}$Pb as surrogate for other atmospheric component*

(i) *Sulfate.* Of the various atmospheric components for which $^{210}$Pb can be considered a surrogate, the most applicable appears to be sulfate. The reason for this is that sulfate, like $^{210}$Pb, is derived from a gaseous precursor ($SO_2$). The conversion to an aerosol-carried species takes place at an approximately similar rate (but, as we shall see later, the oxidation rate of $SO_2$ actually is variable with season) and $^{210}$Pb and $SO_4^{2-}$ attach to similar-size aerosols.

For this reason $^{210}$Pb has been used as a tracer of the precipitation fate of $SO_4^{2-}$. Turekian *et al.* (1989) used the $SO_4^{2-}/^{210}$Pb ratio in aerosols and the flux of $^{210}$Pb measured in bucket collections to determine the $SO_4^{2-}$ flux across the Pacific Ocean. Further, they showed that the $SO_4^{2-}/^{210}$Pb in aerosols from regions of high biological productivity was higher than for normal relatively unpolluted air (Table 3) indicating a sulfate source from the oxidation of dimethyl sulfide (DMS). The measured flux of DMS from the oceans at the equator matched the biogenic flux determined from the $^{210}$Pb calculation (Table 4). (Actually, as we shall see below, this concordance is probably due to an underestimate of sulfate flux and an overestimate of the fraction of DMS oxidized to sulfate.)

A similar study for determining sulfate flux from aerosol $SO_4^{2-}/^{210}$Pb and $^{210}$Pb flux across the eastern US yielded a new insight (Graustein and Turekian, 2004a,b, in preparation). Measurements were made at the SURE sites of the Electric Power Research Institute (EPRI). The sulfate flux was measured at each site as part of the SURE sampling program and $SO_4^{2-}/^{210}$Pb was measured on composite aerosol samples covering a year's collection. The predicted $SO_4^{2-}$ flux from the $^{210}$Pb flux (measured in soil profiles, as described above) was generally less than the measured sulfate flux. The results indicate that over a period of a year or longer the increased sulfate flux must have been due to in-cloud conversion of $SO_2$ to $SO_4^{2-}$. About 12% of the precipitation flux of sulfate is by in-cloud conversion. A different approach corroborates this estimate. Tanaka *et al.* (1994), using stable isotope analyses of S in $SO_2$ and $SO_4^{2-}$ associated with rain and air, showed that homogeneous oxidation of $SO_2$ provided ~90% of the measured sulfate flux and 10% was provided by heterogeneous (in-cloud) oxidation.

**Table 3** Nonmarine $SO_4^{2-}/^{210}$Pb (in $\mu g\ dpm^{-1}$) at continental and oceanic sites as a function of season.

| Site | Summer | Winter | Source |
|------|--------|--------|--------|
| Mould Bay(76° N, 119° W) | 47 | 28 | Graustein and Barrie (unpublished) |
| Northeast United States | 306 | 216 | Graustein and Turekian (1986) |
| Shemya | 200 | 51 | Turekian *et al.* (1989) |
| Midway | 77 | 13 | Turekian *et al.* (1989) |
| Oahu | 50 | 39 | Turekian *et al.* (1989) |
| Fanning | 123 | 147 | Turekian *et al.* (1989) |

**Table 4** Comparison of biogenic sulfate deposition flux to Fanning Island (equatorial Pacific) and the DMS flux from the equatorial Pacific Ocean.

| | $SO_4^{2-}/^{210}Pb$ ($\mu$g dpm$^{-1}$) | $^{210}Pb$ flux (dpm cm$^{-2}$ yr$^{-1}$) | Biogenic $SO_4^{2-}$ flux (mmol m$^{-2}$ yr$^{-1}$) |
|---|---|---|---|
| *Sulfate deposition (from Table 3)* | | | |
| Continental air (Asia) | 50 | | |
| Fanning Island (equatorial Pacific) | | | |
| Total | 140 | | |
| Biogenic | 90[a] | 0.33 | 3.1 |
| *Dimethyl sulfide flux* | | | |
| Equatorial Pacific Ocean | | | |
| Cline and Bates (1983) | | | 2.9 |
| Andreae and Raemdonck (1983) | | | 3.3 |

[a] Biogenic = Total−continental air.

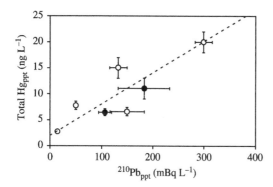

**Figure 8** Hg and $^{210}$Pb in Wisconsin (circles) and mid-Atlantic (triangles) precipitation (source Lamborg *et al.*, 2000). The filled circles represent side-by-side collections from a single event.

(ii) *Lead.* Much of the lead flux to the oceans in recent times has been from pollution sources. This lead was injected into the atmosphere primarily as a volatile compound that subsequently formed an aerosol by photolytic processes. In this sense, much of the lead transported through the atmosphere and ultimately deposited resembles the origin and fate of $^{210}$Pb. On this basis, Turekian and Cochran (1981a,b) estimated the flux of lead to Enewetak in the North Pacific. In a more detailed study, Settle *et al.* (1982) were able to calculate the flux of lead from the aerosol Pb/$^{210}$Pb at a number of diverse sites, including Pigeon Key, Florida, Tahiti, and Bermuda.

(iii) *Mercury.* The primary form of mercury in the atmosphere is as a gaseous element. Only as the mercury is oxidized to ionic species is it removed by precipitation. In this respect it resembles $^{210}$Pb, which also arises from a gas ($^{222}$Rn). Consequently, $^{210}$Pb can be used to track mercury precipitation. Lamborg *et al.* (2000) performed such a study (Figure 8) to determine the correlation of ionic mercury and $^{210}$Pb in aerosols and precipitation and to determine the flux of mercury from the $^{210}$Pb flux determined independently.

(iv) *Other components.* Although other components of aerosols may not have had gaseous precursors, the use of $^{210}$Pb as a surrogate may still be used as an approximation assuming the size fraction of the aerosol bearing both $^{210}$Pb and the component is about the same. Turekian *et al.* (1989) applied this approach to organic compounds and such elements as aluminum. Williams and Turekian (2002) extended this procedure to estimate coastal and open-ocean fluxes of osmium, a metal with isotopic signature impacting the oceans. They were able to show that a high osmium flux was evident close to shore (in New Haven, CT) that could not be characteristic of the whole oceans because of the constancy of the osmium isotope composition in seawater. An analogue to this behavior is the study of $^{210}$Pb fluxes in the SEAREX sites discussed above.

### 4.10.3 COSMOGENIC NUCLIDES

#### 4.10.3.1 Atmospheric Production of Cosmogenic Nuclides

The source of cosmogenic nuclides in the atmosphere is the interaction of galactic cosmic rays with the atoms composing the atmosphere. Cosmic rays are primarily composed of protons with energies of billions of electron volts. As these highly energetic charged particles penetrate the magnetic shield of Earth the interactions with the atoms in the atmosphere result in spallation products—fragments of the target nucleus. Some of the secondary particles, especially neutrons, undergo further reactions. This latter process is responsible, for example, for the production of $^{14}$C (radiocarbon). The major radioactive nuclides so produced in the atmosphere are given in Table 2.

The production of the radionuclides is controlled by the magnetic pattern of Earth. The production rate of the nuclides is determined by the strength of the magnetic field. The strength of the magnetic field varies as the result of both the intrinsic variations in Earth's dipole magnetic moment and the varying intensity of solar activity. At high magnetic field strengths the production rate is low and at low magnetic field strengths the production rate is high.

The period of magnetic fluctuations due to Earth's intrinsic field is not regular but has been shown to have increased over the past $2 \times 10^4$ yr. Fluctuations in the past are also recorded and clearly evident in the changing magnetic polarity of Earth over time.

The solar activity cycles are more complex. The 11 yr sunspot cycle is well known. It is accompanied with charges in the magnetic field strength of Earth. There are fluctuations in the intensity of solar activity on this timescale. As we shall see, periods of fluctuation of ~60 yr, 200 yr, and 1,500 yr have been identified through measurements of $^{10}$Be in deep-sea sediments and continental ice sheets. These results also indicate possible variations on the $10^5$ yr timescale as well.

### 4.10.3.2 $^7$Be and $^{10}$Be

The production rate of $^7$Be (half-life = 53 d) as a function of latitude and elevation by Lal and Peters (1967) is shown in Figure 9. Approximately one-third of the nuclide production rate is in the troposphere and two-thirds in the upper atmosphere (stratosphere and higher). This partitioning is valid for all radionuclides except $^{14}$C, where most is produced by secondary neutrons in the vicinity of the tropopause.

$^{10}$Be (half-life = $1.5 \times 10^6$ yr), although formed primarily in the atmosphere, is found in surface deposits because of the short residence

time of aerosols in the atmosphere and the long half-life. Although the rate of deposition is the same as the rate of production in the atmosphere ($4.5 \times 10^{-2}$ atoms cm$^{-2}$ s$^{-1}$) its distribution on Earth's surface is determined by the sites of primary stratospheric intrusion into the troposphere (~$40°$–$50°$ latitude), the focusing in the troposphere due to precipitation controls such as regional climate and orographic effects and, in the oceans, the role of concentration and transport by particles, especially those produced biologically.

Despite these controls, it is possible to use the record of $^{10}$Be accumulation at designated sites such as ice cores and certain oceanic areas as recorders of variations in the rate of supply and therefore of production with time.

As the residence time of aerosols in the stratosphere is ~2 yr and in the troposphere ~1 week, the $^7$Be/$^{10}$Be ratio of the two air masses is distinctive. Tropospheric air shows the ratio of $^7$Be relative to $^{10}$Be of 1.8, whereas stratospheric air has a ratio of 0.13. It is therefore possible to distinguish stratospheric air injected into the troposphere by considering the ratio of $^7$Be/$^{10}$Be. Of course, the stratospheric air will also be higher in $^{10}$Be than the tropospheric air. As stratospheric air will also contain ozone, the interest in this source has been strong to distinguish from pollution-based tropospheric ozone.

### 4.10.3.3 $^{35}$S and the Kinetics of SO$_2$ Oxidation and Deposition

$^{35}$S is produced by the spallation of $^{40}$Ar by cosmic rays. After formation it is quickly converted into gaseous $^{35}$SO$_2$. The subsequent oxidation and removal from the atmosphere can be tracked and the kinetics applied to terrestrial and pollution SO$_2$. For this reason it has been studied in conjunction with other cosmogenic nuclides, the most useful of which is $^7$Be. Both $^{35}$S with a half-life of 87 d and $^7$Be with a half-life of 53 d, although produced from gaseous precursors by cosmic-ray bombardment, have different initial states. Whereas $^7$Be as an ion quickly associates with aerosols, $^{35}$S as $^{35}$SO$_2$ must first be oxidized to be associated with aerosols. Alternatively, $^{35}$SO$_2$ can react with surfaces in a method called "dry deposition" to distinguish it from $^{35}$S removal in precipitation.

There is only one study using these two nuclides for deciphering the kinetics of SO$_2$ oxidation and removal from the atmosphere (Tanaka and Turekian, 1995, and earlier papers). Their study, performed in New Haven, Connecticut, applied the measurement of these nuclides in air and precipitation to the problem posed. Measurements were made for SO$_2$, $^{35}$SO$_2$, SO$_4^{2-}$, $^{35}$SO$_4^{2-}$, and $^7$Be in air samples and for the same

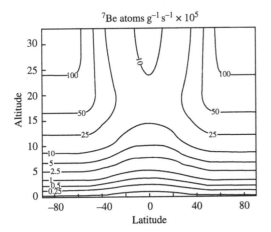

**Figure 9** Production of $^7$Be in the atmosphere as a function of latitude and elevation based on data from Lal and Peters (1967) and Lal (personal communication).

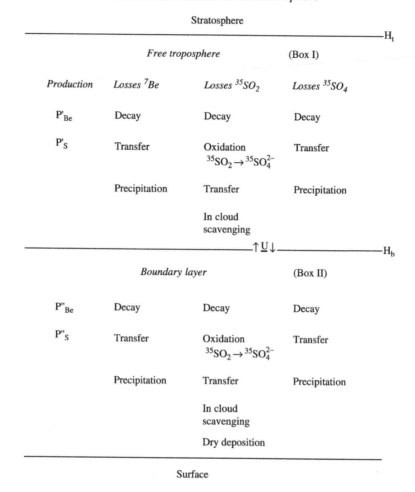

**Figure 10** Components of the box model used by Tanaka and Turekian (1995) in determining the parameters for the conversion and removal of $SO_2$ in the atmosphere based on measurements of the concentrations of cosmogenic $^{35}S$ and $^7Be$. $Ps$ are the cosmic-ray production rates of the radionuclides, $U$ is the mass transport flux between the boundary layer and the free troposphere, $H_b$ and $H_t$ are the heights of the boundary layer and the tropopause, respectively.

species in precipitation over a year's time period. Figure 10 shows the box model with the kinetic factors listed for each process modifying the composition of the air. Although the coupling of the boxes results in 12 equations with 3 unknowns, the selection of reasonable constraints results in solutions that are comparable with expectations based on experience. Figure 11 shows the variation in one parameter, $j_2$, the first-order kinetic constant for the homogeneous oxidation of $SO_2$ in the boundary layer. The value of $j_2$ is greatest in the summer months and lowest in the winter months. This prediction is compatible with the observation that $SO_2$ is the dominant species in winter air and $SO_4^{2-}$ is so in the summer.

The application to the elusive process of dry deposition of $SO_2$ was evaluated by Tanaka and Turekin (1995) using the S and $^{35}S$ data on air and precipitation samples and the model used

to construct Figure 10. They showed that in New Haven, CT the annual weighted average ratio of dry to wet precipitation was 0.26.

### 4.10.3.4 Phosphorus Isotopes

Of the isotopes listed in Table 2 the study of the cosmogenic isotopes of phosphorus $^{32}P$ (half-life = 14.3 d) and $^{33}P$ (half-life = 25.3 d) provides insights into the residence time of aerosols in upper troposphere air because of the relatively short half-lives of the nuclides. Waser and Bacon (1995) measured the concentrations of $^{32}P$ and $^{33}P$ in precipitation at Bermuda over three seasons. They concluded that, with the average activity ratio of $^{33}P$ to $^{32}P$ of 0.96 and a production activity ratio of 0.7, the average residence time of aerosols in the upper troposphere was $\sim$40 d. This increase in the

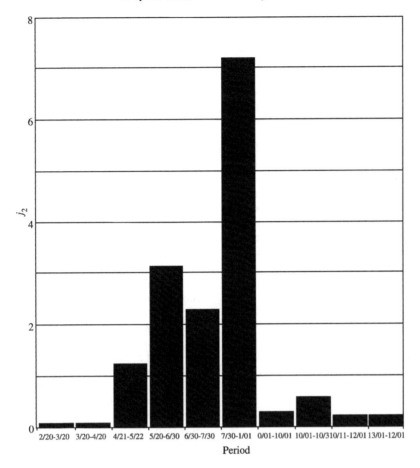

**Figure 11** The variation over the year of the homogeneous oxidation coefficient ($j_2$) of $SO_2$ in the boundary layer of Figure 10 (source Tanaka and Turekian, 1995).

residence time with height in the troposphere is compatible with the modeling results of Balkanski *et al.* (1993) based on $^{210}Pb$. When the upper-troposphere air is transported to the lower troposphere, it is subject to the more efficient removal characteristic of that layer, but with the memory of the original locus of production.

### 4.10.4 COUPLED LEAD-210 AND BERYLLIUM-7

#### 4.10.4.1 Temporal and Spatial Variation

In the absence of atmospheric motion and removal by precipitation, $^7Be$ and $^{210}Pb$ would remain where they originated—the upper troposphere and stratosphere and the lowermost meter of the atmosphere over continents and islands, respectively. In the real atmosphere, $^7Be$ is mixed downward and $^{210}Pb$ is mixed upwards and both are removed by precipitation. They are distributed through the atmosphere by eddy mixing. The residence time of aerosols is short compared to

the time required for homogenization by eddy mixing, but long compared to the life on an individual eddy. Changes in $^7Be/^{210}Pb$ with time and space reflect both vertical and horizontal transport in the atmosphere.

Because of similarity of the behavior of $^7Be$ and $^{210}Pb$ in the atmosphere, $^7Be/^{210}Pb$ is little affected by processes other than production and transport. Both $^{210}Pb$ and $^7Be$ are formed in the atmosphere as energetic single atoms. Since neither is volatile, each of them attaches to the first particles they encounter. The most abundant aerosols in terms of surface area are typically those with a diameter of $0.1-0.5$ $\mu$m, the so-called accumulation mode. This size class carries many of the chemical species in the atmosphere that have low volatility and also have gaseous precursors, such as sulfate as discussed above. Accumulation-mode aerosols are most subject to long-distance transport. Scavenging by precipitation is the principal mechanism of removal of these aerosols from the atmosphere.

The following results are from the work of Graustein and Turekian (1991).

Over continents, seasonal variation in the stability of the atmosphere has a dominant influence on the $^7$Be/$^{210}$Pb ratio. Surface heating in the summer increases convective mixing, which reduces $^{210}$Pb in surface air by mixing it through a larger volume and simultaneously increases the transport of $^7$Be to the surface. Winter stability tends to isolate surface air from the $^7$Be source and retain $^{222}$Rn and $^{210}$Pb near the surface.

The seasonal distributions of $^7$Be and $^{210}$Pb at Champaign, IL, a mid-continent site, are shown in Figure 12. The highest 24 h mean $^{210}$Pb concentrations occur principally in winter when atmospheric stability is greatest; the highest $^7$Be concentrations occur in summer when vertical mixing is strongest. The effect is pronounced, and there is little overlap between the summer and winter sets of measurements. The time series of a 45 d moving average of the $^7$Be/$^{210}$Pb ratio at Champaign, IL shows a repeating annual cycle (Figure 13) characterized by a late spring and early summer maximum and a December minimum in $^7$Be/$^{210}$Pb, and is similar to sites at the east and west coasts of North America.

The maritime pattern, determined from islands in the Atlantic, is different from the continental pattern. Island sources of $^{222}$Rn are too small to have a significant effect on $^{210}$Pb. As a result, virtually all the $^{210}$Pb observed at these sites is transported from continents. $^7$Be/$^{210}$Pb over the ocean is therefore a measure of the influence of continental sources on the local aerosol. Low values of the ratio reflect a high continental influence; high ratios indicate a relative isolation from continental sources.

At Bermuda (Figure 14) the daily data cluster more tightly than at Champaign, and there is a distinct lower bounding value of $^7$Be/$^{210}$Pb. Values that are common in continental winters are never seen in 24 h samples at Bermuda, indicating that continental boundary layer air is diluted or scavenged in transit to Bermuda in the winter.

The seasonal pattern of $^7$Be/$^{210}$Pb at island sites is six months out of phase with continental sites (Figure 15). During the summer, $^7$Be/$^{210}$Pb is nearly the same over the continent and the ocean, indicating vigorous mixing of the troposphere. In the winter, however, $^7$Be/$^{210}$Pb is higher at oceanic sites than continental sites by a factor of ~4.

Izania is located at an elevation of 2,367 m on Tenerife in the Canary Islands and is in the free troposphere. Figure 16 shows that it also exhibits a strong seasonal cycle in $^7$Be/$^{210}$Pb resembling Barbados rather than Champaign. The maximum $^7$Be/$^{210}$Pb occurs in December, the same month that the continental sites reach their minima.

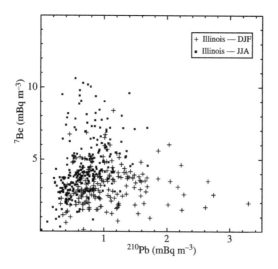

**Figure 12**   $^7$Be and $^{210}$Pb activities in 24 h air samples from Champaign, Illinois. Plus symbols indicate samples collected in December through February, and filled squares represent samples collected in June–August.

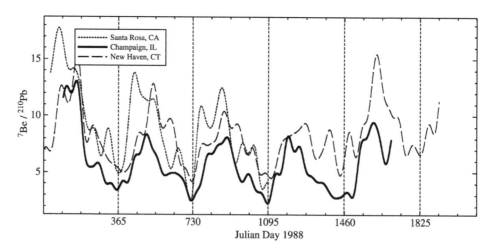

**Figure 13**   Six-year time series of $^7$Be/$^{210}$Pb in near-surface air at three continental sites. The lines represent a Gaussian weighted 45 d moving average of 24 h samples.

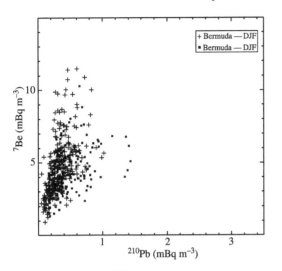

**Figure 14** $^{7}$Be and $^{210}$Pb activities in 24 h air samples from Bermuda. Plus symbols indicate samples collected in December through February and filled squares represent samples collected in June–August.

Taken together, these observations suggest that there is relatively little long-distance transport of aerosols in the marine boundary layer, and that long-distance transport is relatively efficient in the free troposphere. Wintertime stability over continents inhibits transfer of surface-source aerosols to the free troposphere and limits their transport distance. Spring and summer convective mixing over continents results in an efficient exchange between the boundary layer and free troposphere, which enhances long-range transport of aerosols and greatly increases the uniformity of aerosol composition between the continent and marine boundary layers.

### 4.10.4.2 Application of the Coupled $^{7}$Be–$^{210}$Pb System to Sources of Atmospheric Species

As $^{210}$Pb has its source in the boundary layer and $^{7}$Be has its major source in the troposphere, mainly at higher elevations, the sources of

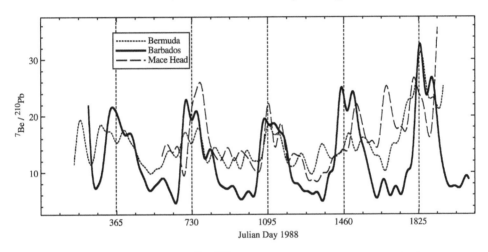

**Figure 15** Time series of $^{7}$Be/$^{210}$Pb in near-surface air at three oceanic sites.

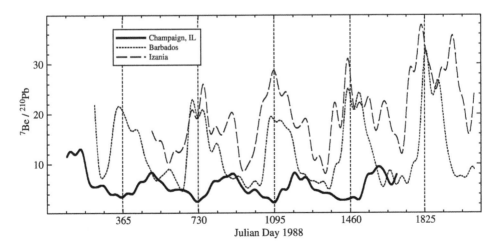

**Figure 16** Time series of $^{7}$Be/$^{210}$Pb in surface air at a continental site (Champaign), in surface air at an oceanic site (Bermuda), and in the oceanic free troposphere (Izania). The data plotted are Gaussian weighted 45 d moving averages of 24 h samples.

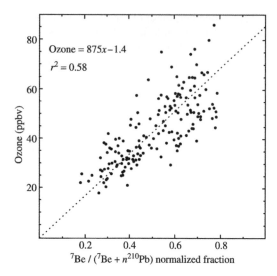

**Figure 17** Ozone versus the "normalized fraction" at Izana (Canary Islands) for the summers of 1989–1991. The "normalized fraction" is $(^7Be)/[(^7Be) + n\,(^{210}Pb)]$, where $n$ is approximated by the ratio of the standard deviation of $(^7Be)$ to the standard deviation of $(^{210}Pb)$ in the sample set. The parentheses indicate activities (source Graustein and Turekian, 1996).

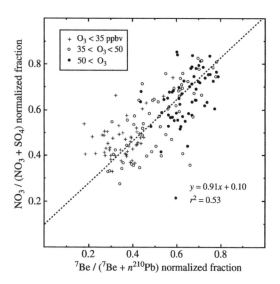

**Figure 18** "Normalized fraction" of the nitrate–sulfate system plotted against the "normalized fraction" of the $^7Be–^{210}Pb$ system as defined in the caption for Figure 17 (from unpublished analysis of AEROCE data by the authors).

chemical species in the lower free troposphere can be determined by using the ratio of the two nuclides. Such a study was performed in one of the AEROCE sites at Izania in the Canary Islands. Using a function of the $^7Be$ to $^{210}Pb$ ratio, Graustein and Turekian (1996) showed that the primary source of ozone in the lower free troposphere of the eastern Atlantic was the upper troposphere (Figure 17). An extension of the study

to nitrate using sulfate as a boundary layer index as is $^{210}Pb$ revealed that the primary source of nitrate in the lower free troposphere was also the upper free troposphere and not the boundary layer (Figure 18). Thus, both chemical species have their origin in the upper troposphere and not anthropogenic pollution source in Europe or biomass burning in Africa. Sampling in the boundary layer downwind of human activity at such sites as Bermuda, Barbados, and Mace Head (Ireland) indicates that both ozone and nitrate are assignable primarily to boundary layer sources.

In another application, Lamborg *et al.* (2000) showed that mercury is distributed homogeneously in the troposphere, as there is no correlation of elemental mercury concentration and $^7Be/^{210}Pb$.

## REFERENCES

Andreae M. O. and Raemdonck H. (1983) Dimethyl sulfide in the surface ocean and marine atmosphere: a global view. *Science* **221**, 744–747.

Andreae M. O., Berresgeim H., Andreae T. W., Kritz M. A., Bates T. S., and Merrill J. T. (1988) Vertical distribution of dimethylsulfide, sulfur dioxide, aerosol ions, and radon over the northeast Pacific Ocean. *J. Atmos. Chem.* **6**, 149–173.

Balkanski Y. J., Jacob D. J., Gardner G. M., Graustein W. C., and Turekian K. K. (1993) Transport and residence times of continental aerosols inferred from a global three-dimensional simulation of $^{210}Pb$. *J. Geophys. Res.* **98**, 20573–20586.

Bateman H. (1910) Solution of a system of differential equations occurring in the theory of radio-active transformations. *Proc. Cambridge Phil. Soc.* **15**, 423–430.

Broecker W. S. and Peng T.-H. (1971) The vertical distribution of radon in the BOMEX area. *Earth Planet. Sci. Lett.* **11**, 99–108.

Clements W. E. and Wilkening M. H. (1974) Atmospheric pressure effects on $^{222}Rn$ transport across the earth–air interface. *J. Geophys. Res.* **79**, 5025–5029.

Cline J. D. and Bates T. S. (1983) Dimethyl sulfide in the Equatorial Pacific Ocean: a natural source of sulfur to the atmosphere. *Geophys. Res. Lett.* **10**, 949–952.

Conen F. and Robertson L. B. (2002) Latitudinal distribution of radon-222 flux from continents. *Tellus* **54B**, 127–133.

Dörr H. and Münnich K. O. (1990) $^{222}Rn$ flux and soil air concentration profiles in West-Germany: soil $^{222}Rn$ as a tracer for gas transport in the unsaturated soil zone. *Tellus* **42B**, 20–28.

Feely H. W., Toonkel L., and Larsen R. (compilers) (1981) *Radionuclides and Trace Elements in Surface Air*. US Dept. Energy, Environ. Q. Rep. EML-395, appendix, New York.

Friedlander G., Kennedy J. W., Macias E. S., and Miller J. M. (1981) *Nuclear and Radiochemistry*, 3rd edn. Wiley, New York, 684pp.

Fuller C. and Hammond D. E. (1983) The fallout rate of Pb-210 on the western coast of the United States. *Geophys. Res. Lett.* **10**, 1164–1172.

Graustein W. C. and Turekian K. K. (1983) $^{210}Pb$ as a tracer of the deposition of sub-micrometer aerosols. In *Precipitation Scavenging, Dry Deposition and Resuspension* (eds. H. R. Pruppacher, R. G. Semonin, and W. G. N. Slinn). Elsevier, Amsterdam and Oxford, vol. 2, pp. 1315–1324.

Graustein W. C. and Turekian K. K. (1986) $^{210}Pb$ and $^{137}Cs$ in air and soils measure the rate and vertical distribution of aerosol scavenging. *J. Geophys. Res.* **91**, 14355–14366.

Graustein W. C. and Turekian K. K. (1989) The effects of forests and topography on the deposition of sub-micrometer aerosols measured by $^{210}$Pb and $^{137}$Cs in soils. *Agri. Forest Meteorol.* **47**, 199–220.

Graustein W. C. and Turekian K. K. (1990) Radon fluxes from soils to the atmosphere measured by $^{210}$Pb–$^{210}$Ra disequilibrium in soils. *Geophys. Res. Lett.* **17**, 841–844.

Graustein W. C. and Turekian K. K. (1991) $^{210}$Pb and $^{7}$Be trace seasonal variations in aerosol transport over North America and the North Atlantic. (Paper presented at CHEMRAWN VII Symposium, Int. Union of Pure and Appl. Chem., Baltimore, MD. Available from the authors).

Graustein W. C. and Turekian K. K. (1996) $^{7}$Be and $^{210}$Pb indicate an upper troposphere source for elevated ozone in the summertime subtropical free troposphere of the eastern North Atlantic. *Geophys. Res. Lett.* **23**, 539–542.

Graustein W. C. and Turekian K. K. (2004a) Radon flux from soils (in preparation).

Graustein W. C. and Turekian K. K. (2004b) In-cloud scavenging of $SO_2$ (in preparation).

Knuth R. H., Knutson E. O., Feely H. W., and Volchock H. L. (1983) Size distribution of atmospheric Pb and Pb-210 in rural New Jersey: implications for wet and dry deposition. In *Precipitation Scavenging, Dry Deposition and Resuspension* (eds. H. R. Pruppacher, R. G. Semonin, and W. G. N. Slinn). Elsevier, Amsterdam and Oxford, pp. 1325–1334.

Kritz M. A., Rosner S. W., Kelly K. K., Loewenstein M., and Chan K. R. (1993) Radon measurements in the lower tropical stratosphere: evidence for rapid vertical transport and dehydration of tropospheric air. *J. Geophys. Res.* **98**, 8725–8736.

Lal D. (2001) Cosmogenic isotopes. In *Encyclopedia of Ocean Sciences* (eds. J. H. Steele, S. A. Thorpe, and K. K. Turekian). Academic Press, London, pp. 550–560.

Lal D. and Peters B. (1967) Cosmic rays produced radioactivity on the earth. *Handbuch der Physik* **46**(2), 551–612.

Lambert G., Polian G., and Taupin D. (1970) Existence of periodicity in radon concentrations and in large-scale circulation at lower altitudes between 40° and 70° south. *J. Geophys. Res.* **75**, 2341–2345.

Lamborg C. H., Fitzgerald W. F., Graustein W. C., and Turekian K. K. (2000) An examination of the atmospheric chemistry of mercury using $^{210}$Pb and $^{7}$Be. *J. Atmos. Chem.* **36**, 325–338.

Liu S., McAfee J. R., and Cicerone R. J. (1984) Radon-222 and tropospheric vertical transport. *J. Geophys. Res.* **89**, 7202–7297.

Martell E. A. and Moore H. E. (1974) Tropospheric aerosol residence times: a critical review. *J. Rech. Atmos.* **8**, 903–910.

Monaghan M. C., Krishnaswami S., and Turekian K. K. (1985/1986) The global-average production rate of $^{10}$Be. *Earth Planet. Sci. Lett.* **76**, 279–287.

Moore H. E. and Poet S. E. (1976) $^{210}$Pb fluxes determined from $^{210}$Pb and $^{226}$Ra soil profiles. *J. Geophys. Res.* **81**, 1056–1058.

Moore H. E., Poet S. E., and Martell E. A. (1973) $^{222}$Rn, $^{210}$Pb, $^{210}$Bi, and $^{210}$Po profiles and aerosol residence times versus altitude. *J. Geophys. Res.* **78**, 7065–7075.

Moore H. E., Poet S. E., and Martell E. A. (1974) Origin of $^{222}$Rn and its long-lived daughters in air over Hawaii. *J. Geophys. Res.* **79**, 5019–5024.

National Research Council (1988) *Health Risks of Radon and other Internally Deposited Alpha-emitters.* National Academy Press, Washington, DC, 602pp.

Nozaki Y., DeMaster D. J., Lewis D. M., and Turekian K. K. (1978) Atmospheric Pb-210 fluxes determined from soil profiles. *J. Geophys. Res.* **83**, 4047–4051.

Pearson J. E. and Jones G. E. (1966) Soil concentrations of "emanating radium-226" and the emanation of radon-222 from soils and plants. *Tellus* **18**, 655–661.

Schery S. D. and Gaeddert D. H. (1982) Measurements of the effect of cyclic atmospheric pressure variation on the flux of $^{222}$Rn from the soil. *Geophys. Res. Lett.* **9**, 835–838.

Schery S. D., Whittlestone S., Hart K. P., and Hill S. E. (1989) The flux of radon and thoron from Australian soils. *J. Geophys. Res.* **94**, 8567–8576.

Settle D. M., Patterson C. C., Turekian K. K., and Cochran J. K. (1982) Lead precipitation fluxes at tropical oceanic sites determined from $^{210}$Pb measurements. *J. Geophys. Res.* **87**, 1239–1245.

Tanaka N. and Turekian K. K. (1995) The determination of the dry deposition flux of $SO_2$ using cosmogenic $^{35}$S and $^{7}$Be measurements. *J. Geophys. Res.* **100**, 2841–2848.

Tanaka N., Rye D. M., Xiao Y., and Lasaga A. C. (1994) Use of stable sulfur isotope systematics for evaluating oxidation pathways and in-cloud-scavenging of sulfur dioxide in the atmosphere. *Geophys. Res. Lett.* **21**, 1519–1522.

Tsunogai S., Shinagawa T., and Kurata T. (1985) Deposition of anthropogenic sulfate and Pb-210 in the western North Pacific area. *Geochem. J.* **19**, 77–90.

Turekian K. K. and Cochran J. K. (1981a) $^{210}$Pb in surface air at Enewetak and the Asian dust flux to the Pacific. *Nature* **292**, 522–524.

Turekian K. K. and Cochran J. K. (1981b) $^{210}$Pb in surface air at Enewetak and the Asian dust flux to the Pacific: a correction. *Nature* **294**, 670.

Turekian K. K., Nozaki Y., and Benninger L. K. (1977) Geochemistry of atmospheric radon and radon products. *Ann. Rev. Earth Planet. Sci.* **5**, 227–255.

Turekian K. K., Graustein W. C., and Cochran J. K. (1989) Lead-210 in the SEAREX program: an aerosol tracer across the Pacific. In *Chemical Oceanography* (ed. J. P. Riley). Academic Press, London, vol. 10, pp. 51–81.

Turekian V. C., Graustein W. C., and Turekian K. K. (1999) The $^{214}$Bi to $^{214}$Pb ratio in lower boundary layer aerosols and aerosol residence times at New Haven, Connecticut. *J. Geophys. Res.* **104**, 11593–11598.

Waser N. A. D. and Bacon M. P. (1995) Wet deposition fluxes of cosmogenic $^{32}$P and $^{33}$P and variations in the $^{33}$P/$^{32}$P ratios at Bermuda. *Earth Planet. Sci. Lett.* **133**, 71–80.

Wilkening M. H. and Clements W. E. (1975) Radon-222 from the ocean surface. *J. Geophys. Res.* **80**, 3828–3830.

Williams G. and Turekian K. K. (2002) The atmospheric supply of osmium to the oceans. *Geochim. Cosmochim. Acta* **66**, 3789–3791.

# 4.11
# The History of Planetary Degassing as Recorded by Noble Gases

D. Porcelli

*University of Oxford, UK*

and

K. K. Turekian

*Yale University, New Haven, CT, USA*

## 4.11.1 INTRODUCTION

Noble gases provide unique clues to the structure of the Earth and the degassing of volatiles into the atmosphere. Since the noble gases are highly depleted in the Earth, their isotopic compositions are prone to substantial changes due to radiogenic additions, even from scarce parent elements and low-yield nuclear processes. Therefore, noble gas isotopic signatures of major reservoirs reflect planetary differentiation processes that generate fractionations between these volatiles and parent elements. These signatures can be used to construct planetary degassing histories that have relevance to the degassing of a variety of chemical species as well.

It has long been recognized that the atmosphere is not simply a remnant of the volatiles that surrounded the forming Earth with the composition of the early solar nebula. It was also commonly thought that the atmosphere and oceans were derived from degassing of the solid Earth over time (Brown, 1949; Suess, 1949; Rubey, 1951). Subsequent improved understanding of the processes of planet formation, however, suggests that substantial volatile inventories could also have been added directly to the atmosphere. The characteristics of the atmosphere therefore reflect the acquisition of volatiles by the solid Earth during formation (see Pepin and Porcelli, 2002; Chapter 4.12), as well as the history of degassing from the mantle. The precise connection between volatiles now emanating from the Earth and the long-term evolution of the atmosphere are key subjects of modeling efforts, and are discussed below.

Major advances in understanding the behavior of terrestrial volatiles have been made based upon observations on the characteristics of noble gases that remain within the Earth. Various models have been constructed that define different components and reservoirs in the planetary interior, how materials are exchanged between them, and how the noble gases are progressively transferred to the atmosphere (see Chapter 2.06). While there remain many uncertainties, an overall process of planetary degassing can be discerned. The present chapter discusses the constraints provided by the noble gases and how these relate to the degassing of the volatile molecules formed from nitrogen, carbon, and hydrogen (see also Chapter 3.04). The evolution of particular atmospheric molecular species, such as $CO_2$, that are controlled by interaction with other crustal reservoirs and which reflect surface chemical conditions, are primarily discussed elsewhere (Chapter 8.09).

Noble gases provide the most detailed constraints on planetary degassing. A description of the available noble gas data that must be incorporated into any Earth degassing history is provided first in Section 4.11.2, and the constraints on the total extent of degassing of the terrestrial interior are provided in Section 4.11.3. Noble gas degassing models that have been used to describe and calculate degassing histories of both the mantle (Section 4.11.4) and the crust (Section 4.11.5) are then presented. These discussions then provide the context for an evaluation of major volatile cycles in the Earth (Section 4.11.6), and speculations about the degassing of the other terrestrial planets (Section 4.11.7), Mars and Venus, that are obviously based on much more limited data. The processes controlling mantle degassing are clearly related to the structure of the mantle, as discussed in Section 4.11.4. Further descriptions of mantle noble gas reservoirs and transport processes based upon multi-tracer variations in mantle-derived materials are provided in Chapter 2.06. An important aspect is the origin of planetary volatiles and whether initial incorporation was into the solid Earth or directly to the atmosphere; these issues are discussed in detail in Chapter 4.12. Basic noble gas elemental and isotopic characteristics are given in Ozima and Podosek (2001) and Porcelli *et al.* (2002). The major nuclear processes that produce noble gases within the solid Earth, and the half-lives of the major parental nuclides, are given in Table 1.

## 4.11.2 PRESENT-EARTH NOBLE GAS CHARACTERISTICS

There are various terrestrial reservoirs that have distinct volatile characteristics. Data from mid-ocean ridge basalts (MORBs) characterize the underlying convecting upper mantle, and are described here without any assumptions about the depth of this reservoir. Other mantle reservoirs are sampled by ocean island basalts (OIBs) and may represent a significant fraction of the mantle (Chapter 2.06). Note that significant krypton isotopic variations due to radiogenic additions are neither expected nor observed, and there are no isotopic fractionation observed between any terrestrial noble gas reservoirs. Therefore, no constraints on mantle degassing can be obtained from krypton, and so krypton is not discussed further. Comparison between terrestrial and solar system krypton is discussed in Chapter 4.12.

### 4.11.2.1 Surface Inventories

The atmosphere is the largest accessible terrestrial noble gas reservoir, and its composition serves as a reference for measurements of other materials. The major volatile molecules of

**Table 1** Major nuclear processes producing noble-gas isotopes in the solid earth.[a]

| Daughter | Nuclear process | Parent half-life | Yield (atoms/decay) | Comments |
|---|---|---|---|---|
| $^3$He | $^6$Li(n, $\alpha$)$^3$H($\beta$-)$^3$He | | | $^3$He/$^4$He $= 1 \times 10^{-8}$[b] |
| $^4$He | $\alpha$-decay of $^{238}$U decay series nuclides | 4.468 Ga | 8[c] | |
| $^4$He | $\alpha$-decay of $^{235}$U decay series nuclides | 0.7038 Ga | 7[c] | $^{238}$U/$^{235}$U $= 137.88$ |
| $^4$He | $\alpha$-decay of $^{232}$Th decay series nuclides | 14.01 Ga | 6[c] | Th/U $= 3.8$ in bulk Earth |
| $^{21}$Ne | $^{18}$O($\alpha$, n)$^{21}$Ne | | | $^{21}$Ne/$^4$He $= 4.5 \times 10^{-8}$[b] |
| $^{21}$Ne | $^{24}$Mg(n, $\alpha$)$^{21}$Ne | | | $^{21}$Ne/$^4$He $= 1 \times 10^{-10}$[b] |
| $^{40}$Ar | $^{40}$K $\beta^-$ decay | 1.251 Ga | 0.1048[b] | $^{40}$K $= 0.01167\%$ total K |
| $^{129}$Xe | $^{129}$I $\beta^-$ decay | 15.7 Ma | 1 | $^{129}$I/$^{127}$I $= 1.1 \times 10^{-4}$ at 4.56 Ga[d] |
| $^{136}$Xe | $^{238}$U spontaneous fission | | $4 \times 10^{-8}$[e] | |
| $^{136}$Xe | $^{244}$Pu spontaneous fission | 80.0 Ma | $7.00 \times 10^{-5}$ | $^{244}$Pu/$^{238}$U $= 6.8 \times 10^{-3}$ at 4.56 Ga[f] |

[a] From data compilations of Blum (1995), Ozima and Podosek (2001), and Pfennig *et al.* (1998). [b] Production ratio for upper crust (Ballentine and Burnard, 2002). [c] Per decay of series parent, assuming secular equilibrium for entire decay series. [d] Hohenberg *et al.* (1967). [e] Eikenberg *et al.* (1993) and Ragettli *et al.* (1994). [f] Hudson *et al.* (1989).

**Table 2** Volatile surface inventories.

| Constituent | Atmosphere (mol) | Crust (mol) |
|---|---|---|
| N | $2.760 \times 10^{20}$ | $4 \times 10^{19}$ |
| $O_2$ | $3.702 \times 10^{19}$ | |
| Ar | $1.651 \times 10^{18}$ | |
| C | $5.568 \times 10^{16}$ | $8.3 \times 10^{21}$ |
| Ne | $3.213 \times 10^{15}$ | |
| He | $9.262 \times 10^{14}$ | |
| Kr | $2.015 \times 10^{14}$ | |
| Xe | $1.537 \times 10^{13}$ | |

Note: Based on dry tropospheric air. Water generally accounts for ≤4% of air. Other chemical constituents have mixing ratios less than Xe. Data from compilation by Ozima and Podosek (2001). C atmosphere data from Keeling and Whorf (2000). Crustal C from Hunt (1972) and Ronov and Yaroshevsky (1976). Crustal N from Marty and Dauphas (2003).

**Table 3** Noble-gas and major volatile isotope composition of the atmosphere.

| Isotope | Relative abundances | Percent molar abundance |
|---|---|---|
| $^3$He | $(1.399 \pm 0.013) \times 10^{-6}$ | 0.000140 |
| $^4$He | $\equiv 1$ | 100 |
| $^{20}$Ne | $9.80 \pm 0.08$ | 90.50 |
| $^{21}$Ne | $0.0290 \pm 0.0003$ | 0.268 |
| $^{22}$Ne | $\equiv 1$ | 9.23 |
| $^{36}$Ar | $\equiv 1$ | 0.3364 |
| $^{38}$Ar | $0.1880 \pm 0.0004$ | 0.0632 |
| $^{40}$Ar | $295.5 \pm 0.5$ | 99.60 |
| $^{78}$Kr | $0.6087 \pm 0.0020$ | 0.3469 |
| $^{80}$Kr | $3.9599 \pm 0.0020$ | 2.2571 |
| $^{82}$Kr | $20.217 \pm 0.004$ | 11.523 |
| $^{83}$Kr | $20.136 \pm 0.021$ | 11.477 |
| $^{84}$Kr | $\equiv 100$ | 57.00 |
| $^{86}$Kr | $30.524 \pm 0.025$ | 17.398 |
| $^{124}$Xe | $2.337 \pm 0.008$ | 0.0951 |
| $^{126}$Xe | $2.180 \pm 0.011$ | 0.0887 |
| $^{128}$Xe | $47.15 \pm 0.07$ | 1.919 |
| $^{129}$Xe | $649.6 \pm 0.9$ | 26.44 |
| $^{130}$Xe | $\equiv 100$ | 4.070 |
| $^{131}$Xe | $521.3 \pm 0.8$ | 21.22 |
| $^{132}$Xe | $660.7 \pm 0.5$ | 26.89 |
| $^{134}$Xe | $256.3 \pm 0.4$ | 10.430 |
| $^{136}$Xe | $217.6 \pm 0.3$ | 8.857 |
| $^{14}$N | 0.0037 | 0.37 |
| $^{15}$N | $\equiv 1$ | 99.63 |
| $^{12}$C | $\equiv 1$ | 98.63 |
| $^{13}$C | 0.0113 | 1.11 |
| $^1$H | $\equiv 1$ | 99.985 |
| $^2$H | 0.00015 | 0.015 |

After Ozima and Podosek (2001) and Porcelli *et al.* (2002).

carbon, nitrogen, and hydrogen, have considerable inventories in the crust that are part of the volatile budget that has been either degassed from the mantle or initially incorporated into the atmosphere. The total surface inventory is summarized in Table 2 and includes the atmosphere, hydrosphere, and continental crust. Since helium is lost from the atmosphere, the atmospheric abundance has no significance for determining long-term evolution. Atmospheric isotopic compositions, which are generally used as standards for comparison and measurement normalization, are provided in Table 3.

### 4.11.2.2 Helium Isotopes

There are two isotopes of helium. In addition to the cosmologically produced $^4$He and $^3$He, $^4$He is produced as $\alpha$-particles during radioactive decay of various parent radionuclides, and the much less abundant $^3$He is produced from $^6$Li (Tables 1 and 2). Overall, radiogenic helium is primarily $^4$He, with a ratio of $^3$He/$^4$He $\sim 0.01$ $R_A$ (Morrison and Pine, 1955), where $R_A$ is the air value of $1.39 \times 10^{-6}$. The initial value for the Earth depends upon the origin of terrestrial noble gases (see Chapter 4.12), and is presumed to be that of the solar nebula of $^3He/^4He = 120R_A$ (Mahaffy *et al.*, 1998). The solar wind value of $330R_A$ (Benkert *et al.*, 1993) was established after deuterium burning in the Sun; if terrestrial helium was captured after significant deuterium burning, then this higher value would be the composition of the initial helium. The first clear evidence for the degassing of primordial volatiles still remaining within the solid Earth came from helium isotopes (Figure 1). MORB has an average of $8R_A$ (Clarke *et al.*, 1969; Mamyrin *et al.*, 1969; see Graham, 2002), and so is a mixture of radiogenic helium (that accounts for most of the $^4He$) with initially trapped "primordial" helium (that accounts for most of the $^3He$).

OIB has more variable $^3He/^4He$ ratios. Some are below those of MORB, probably due to radiogenic recycled components (e.g., Kurz *et al.*, 1982; Hanyu and Kaneoka, 1998). Due to their limited occurrence, these values are likely to represent only a small fraction of the total mantle. $^3He/^4He$ ratios greater than $\sim 10R_A$ provide evidence for a long-term noble gas reservoir distinct from MORB that has a time-integrated $^3He/(U + Th)$ ratio greater than that of the upper mantle (Kurz *et al.*, 1982; Allègre *et al.*, 1983) and so leads to the highest $^3He/^4He$ ratios of $32-38R_A$ found in Loihi Seamount, the youngest Hawaiian volcano (Kurz *et al.*, 1982; Rison and Craig, 1983; Honda *et al.*, 1993; Valbracht *et al.*, 1997), and Iceland (Hilton *et al.*, 1998b). A major issue has

been determining the nature of this reservoir (see Chapter 2.06), the abundances of noble gases it contains, and how it degasses (see Section 4.11.4).

Helium isotopes have a $\sim 1$ Myr residence time in the atmosphere prior to loss to space; therefore, their atmospheric abundances do not contain information about the integrated degassing history of the Earth. However, the large variations in helium isotope compositions in the Earth constrain mantle degassing models by (i) providing clear fingerprints of mantle volatile fluxes into the crust and atmosphere (see Sections 4.11.2.7 and 4.11.2.8); (ii) requiring several mantle noble gas reservoirs (see Section 4.11.4.3); (iii) indicating that the upper mantle is relatively well-mixed with respect to noble gases (see Chapter 2.06); and (iv) relating the sources of noble gases with heat production, since uranium and thorium are the dominant sources of both $^4He$ and heat in the mantle (see Section 4.11.2.7).

### 4.11.2.3 Neon Isotopes

There are three neon isotopes. The more abundant $^{20}Ne$ and $^{22}Ne$ are both essentially all primordial, as there is no significant global production of these isotopes. In contrast, $^{21}Ne$ is produced by nuclear reactions. In mantle-derived materials, measured $^{20}Ne/^{22}Ne$ ratios are greater than that of the atmosphere of 9.8, and extend toward the values of the solar wind (13.8) or

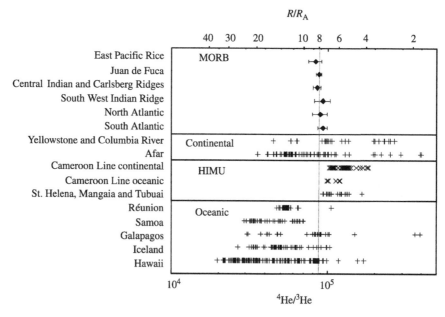

**Figure 1** Helium isotope data from various mantle-derived volcanics. The upper axis is the $^3He/^4He$ ratio ($R$) normalized to the atmospheric ratio ($R_A$). As indicated by the data for selected segments of the MORB away from ocean islands falls almost entirely within the range of $(7-9)R_A$. While there are hotspot basalts that are characterized by high U/Pb ratios and low $^3He/^4He$ ratios (HIMU), many major oceanic hotspots, as well as continental hotspots, have high $^3He/^4He$ ratios (source Porcelli and Ballentine, 2002).

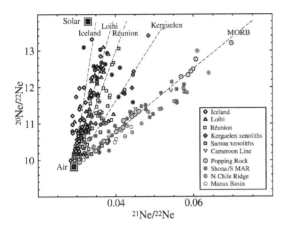

**Figure 2** Neon isotope compositions of MORB and selected OIBs. The data generally fall on correlations that extend from air contamination to higher values that characterize the trapped mantle components. Islands with high $^{3}He/^{4}He$ ratios have lower-mantle $^{21}Ne/^{22}Ne$ ratios, reflecting high $^{3}He/(U + Th)$ and $^{22}Ne/(U + Th)$ ratios (source Graham, 2002).

implanted solar wind (12.5) (Figure 2). Since these isotopes are not produced in significant quantities in the Earth, this is unequivocal evidence for storage in the Earth of at least one nonradiogenic mantle component that is distinctive from the atmosphere and has remained trapped separately since formation of the Earth. It is likely that the $^{20}Ne/^{22}Ne$ ratio of the atmosphere was originally similar to the higher values now found in the mantle, and was fractionated during losses to space (Chapter 4.12). This could only have occurred early in Earth history. Since the neon remaining in the mantle preserves the original isotopic composition, the difference from that of the atmosphere also limits the amount of mantle neon that can have subsequently degassed. The exact proportion depends upon how much fractionation of atmospheric neon originally occurred. While this is unconstrained, the observed fractionation is already considered quite extreme, and so it is unlikely that much lower $^{20}Ne/^{22}Ne$ ratios had been generated, and so only a small proportion of mantle neon is likely to have been degassed subsequently.

MORB $^{20}Ne/^{22}Ne$ and $^{21}Ne/^{22}Ne$ ratios generally are correlated (Sarda *et al.*, 1988; Moreira *et al.*, 1998), and this is likely due to mixing of variable amounts of air contamination with uniform mantle neon. The upper-mantle $^{21}Ne/^{22}Ne$ ratio of ~0.074 is higher than the solar value (0.033) due to additions of nucleogenic $^{21}Ne$ (Table 1). OIBs with high $^{3}He/^{4}He$ ratios span a similar range in $^{20}Ne/^{22}Ne$ ratios, but with lower corresponding $^{21}Ne/^{22}Ne$ ratios (Sarda *et al.*, 1988; Honda *et al.*, 1991, 1993). The OIB sources therefore have higher time-integrated

He/(U + Th) and Ne/(U + Th) ratios (Honda and McDougall, 1993).

The MORB helium and neon isotopic compositions can be used to calculate the $^{3}He/^{22}Ne$ ratio of the source region prior to any recent fractionations created during transport and eruption. Since the production ratio of $^{4}He$ to $^{21}Ne$ is fixed (Table 1), the shifts in $^{3}He/^{4}He$ and $^{21}Ne/^{22}Ne$ isotope ratios from the initial, primordial values of the Earth due to radiogenic and nucleogenic additions can be used to calculate the reservoir $^{3}He/^{22}Ne$ ratio. Using an uncontaminated MORB value of $^{21}Ne/^{22}Ne = 0.074$ and $^{21*}Ne/^{4*}He = 4.5 \times 10^{-8}$, then $^{3}He/^{22}Ne = 11$. Calculating a source value for each MORB and OIB sample individually, a mantle average of 7.7 was found (Honda and McDougall, 1998). For comparison, the solar nebula value is $^{3}He/^{22}Ne = 1.9$ (see Porcelli and Pepin, 2000). This value can be used to relate degassing of helium with the other noble gases (see Section 4.11.2.7).

Overall, neon isotopes clearly identify the noble gases presently degassing from the mantle as solar in origin (see Chapter 4.12 for a discussion of the implications for early Earth history). Taking this composition as the original value of the neon in the atmosphere, most of the atmospheric neon was degassed very early in Earth history and suffered substantial fractionating losses. The atmospheric noble gas inventory was therefore highly modified from that which was originally degassed from the solid Earth or added to the surface. Since noble gases generally behave similarly within the mantle and during degassing (see Section 4.11.2.6), the constraints on neon degassing can be applied to the other noble gases.

### 4.11.2.4 Argon Isotopes

The two minor isotopes of argon, $^{36}Ar$ and $^{38}Ar$, are essentially all primordial, with no significant radiogenic production on a global scale. The initial $^{40}Ar/^{36}Ar$ ratio of the solar system was $<10^{-3}$ (Begemann *et al.*, 1976) and orders of magnitude less than any planetary values, so essentially all $^{40}Ar$ is radiogenic. A large range in $^{40}Ar/^{36}Ar$ ratios has been measured in MORBs that is likely due to mixing of variable proportions of air argon (with $^{40}Ar/^{36}Ar = 296$) with a single, more radiogenic, mantle composition (Figure 3). The minimum value for this mantle composition is represented by the highest measured values of $2.8 \times 10^4$ (Staudacher *et al.*, 1989) to $4 \times 10^4$ (Burnard *et al.*, 1997). From correlations between $^{20}Ne/^{22}Ne$ and $^{40}Ar/^{36}Ar$ during step heating of a gas-rich MORB that was designed to separate contaminant air noble gases from those trapped within the glass, a maximum value of $^{40}Ar/^{36}Ar = 4.4 \times 10^4$ was obtained (Moreira *et al.*, 1998).

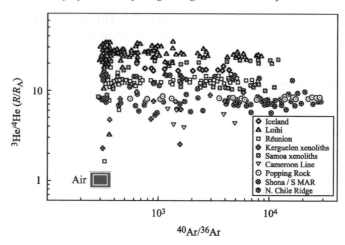

**Figure 3**  Helium and argon isotope compositions of MORB ("Popping Rock" from the Atlantic, Shona ridge section, and N. Chile ridge) and selected islands. $^{40}Ar/^{36}Ar$ ratios vary widely due to variable amounts of atmospheric contamination (source Graham, 2002).

Like terrestrial $^{20}Ne/^{22}Ne$ ratios, it might be expected that $^{38}Ar/^{36}Ar$ ratios vary due to different initial sources building the Earth or to early fractionation events. Measurements of MORB and OIB $^{38}Ar/^{36}Ar$ ratios typically are atmospheric within error, but have been of low precision due to the low abundance of these isotopes. While some high-precision analyses of MORB and OIB samples show $^{38}Ar/^{36}Ar$ ratios lower than that of the atmosphere and approaching solar values (see Pepin, 1998), but others do not (Kunz, 1999). While nonatmospheric ratios would limit the amount of argon transfer between the mantle and atmosphere, atmospheric ratios in the mantle could be explained either by early trapping of argon that had been fractionated or entrainment of atmospheric argon during subduction or melt formation (see Porcelli and Wasserburg, 1995b).

OIBs with high $^3He/^4He$ ratios reflecting high $^3He/(U + Th)$ ratios have been expected to have low $^{40}Ar/^{36}Ar$ ratios reflecting correspondingly high $^{36}Ar/K$ ratios. Measurements of $^{40}Ar/^{36}Ar$ in OIBs with $^3He/^4He > 10R_A$ are indeed consistently lower than MORB values. However, early values for Loihi glasses of $^{40}Ar/^{36}Ar < 10^3$ appear to reflect overwhelming contamination with air argon (Fisher, 1985; Patterson *et al.*, 1990). A study of basalts from Juan Fernandez (Farley *et al.*, 1993) found atmospheric contamination contained within phenocrysts introduced into the magma chamber, thus providing an explanation for the prevalence of air contamination of OIBs. Recent measurements of Loihi samples found higher $^{40}Ar/^{36}Ar$ values of 2,600–2,800 associated with high $^3He/^4He$ ratios (Hiyagon *et al.*, 1992; Valbracht *et al.*, 1997), while Trieloff *et al.* (2000) found values up to 8,000 on samples with $^3He/^4He = 24$ (and so with a helium composition midway between MORBs and the highest

OIBs). Poreda and Farley (1992) found values of $^{40}Ar/^{36}Ar \leq 1.2 \times 10^4$ in Samoan xenoliths that have intermediate $^3He/^4He$ ratios ($9–20R_A$). Kola Peninsula carbonatites with high $^{20}Ne/^{22}Ne$ ratios were used to calculate a mantle $^{40}Ar/^{36}Ar$ value of 5,000 (Marty *et al.*, 1998). Other attempts to remove the effects of air contamination have used associated neon isotopes and the debatable assumption that the contaminant Ne/Ar ratio is constant, and have also found $^{40}Ar/^{36}Ar$ values substantially lower than in MORBs (Sarda *et al.*, 2000). Overall, it appears that $^{40}Ar/^{36}Ar$ ratios in the high $^3He/^4He$ OIB source are >3,000 but probably <$10^4$, and so lower than that of the MORB source (see also Matsuda and Marty, 1995).

The unambiguous measurement of past atmospheric $^{40}Ar/^{36}Ar$ ratios would provide an important constraint on the degassing history of the atmosphere. If $^{36}Ar$ had largely degassed early (see below), then the past atmospheric $^{40}Ar/^{36}Ar$ ratio would reflect the past atmospheric abundance of $^{40}Ar$ due to subsequent $^{40}Ar$ degassing. Unfortunately, it has been difficult to find samples that have captured and retained atmospheric argon but do not contain significant amounts of either inherited radiogenic $^{40}Ar$ or potassium. Nonetheless, various studies have sought to find $^{40}Ar/^{36}Ar$ ratios that are lower than the present atmosphere, and so clearly are not dominated by either present atmosphere contamination or radiogenic $^{40}Ar$. Cadogan (1977) reported a value of $^{40}Ar/^{36}Ar = 291.0 \pm 1.5$ from the 380 Myr old Rhynie chert, while Hanes *et al.* (1985) reported $258 \pm 3$ for an old pyroxenite sill sample from the Abitibi Greenstone Belt that contains amphibole apparently produced during deuteric alteration at 2.70 Ga. Both samples are presumed to have atmospheric argon trapped at these times. Using these data, the maximum possible changes in the atmospheric $^{40}$Ar abundance is obtained by assuming that the $^{36}$Ar abundance has been constant since that time, so that the differences between these $^{40}$Ar/$^{36}$Ar ratios and the present atmospheric value reflect only lower $^{40}$Ar abundances. In this case, the Rhynie chert data suggest that the atmospheric $^{40}$Ar abundance was 1.5% lower 380 Ma ago than today. At that time, there was 1.9% less $^{40}$Ar in the Earth (assuming a bulk Earth potassium concentration of 270 ppm), and it thus appears that the same fraction of terrestrial $^{40}$Ar as today was in the atmosphere at that time. At 2.7 Ga ago, the calculated atmospheric $^{40}$Ar abundance is 12.7% lower than at present. However, there was 66% less $^{40}$Ar 2.7 Ga ago, and assuming a BSE concentration of 270 ppm K, even complete degassing of the entire Earth at that time (compared to 40% today) would not provide sufficient $^{40}$Ar for the atmosphere. The alternative that the atmospheric $^{36}$Ar abundance has doubled since 2.7 Ga ago is also unlikely if current arguments for early degassing of nonradiogenic nuclides are valid (see Sections 4.11.2.3 and 4.11.4.1). However, before pursuing further speculation, the possibility that the Abitibi sample contains either excess $^{40}$Ar trapped during formation or subsequently produced radiogenic $^{40}$Ar, and so provides an unreasonably high $^{40}$Ar/$^{36}$Ar atmospheric ratio for that time, must be discounted.

Overall, argon isotope compositions indicate that the mantle is much more radiogenic than the atmosphere, and this provides an important starting point for degassing history models. It appears that there are variations in $^{40}$Ar/$^{36}$Ar associated with different $^{3}$He/$^{4}$He ratios in the mantle, although atmospheric contamination of samples has made this difficult to quantify. It appears that the same proportion of $^{40}$Ar has been in the atmosphere over the last 380 Ma, although reliable data are required from much earlier to effectively discriminate between different degassing histories (see Section 4.11.4).

### 4.11.2.5 Xenon Isotopes

There are nine isotopes of xenon (see Table 3). On a global scale, there have been additions to $^{129}$Xe through decay of $^{129}$I ($t_{1/2} = 15.7$ Myr), which as a short-lived nuclide was only present in significant quantities early in Earth history. Additions to the heavy isotopes $^{131}$Xe, $^{132}$Xe, $^{134}$Xe, and $^{136}$Xe have also occurred by fission of $^{244}$Pu ($t_{1/2} = 80$ Myr), another short-lived nuclide, and $^{238}$U ($t_{1/2} = 4.5$ Gyr). The largest fission contributions are to $^{136}$Xe, so this isotope is usually used as an index for fissiogenic contributions. Primordial components completely account for the other isotopes and even a dominant proportion of those with later additions.

The starting point for examining xenon isotope systematics is defining the nonradiogenic isotope composition; i.e., the proportion of primordial xenon underlying the radiogenic and fissiogenic contributions. The light isotopes of atmospheric xenon ($^{124}$Xe, $^{126}$Xe, $^{128}$Xe, and $^{130}$Xe) are related to both bulk chondritic and solar xenon by very large fractionation of ~4.2% per amu (Krummenacher *et al.*, 1962), which demands a strongly fractionating planetary process (see Chapter 4.12). However, both chondritic and solar xenon, when fractionated to match the light isotopes of the atmosphere, have proportionately more $^{134}$Xe and $^{136}$Xe than presently in the atmosphere and so neither can serve as the primordial terrestrial composition (see Pepin, 2000). There is no other commonly observed solar system composition that can be used to account for atmospheric xenon by simple mass fractionation and the addition of radiogenic and fissiogenic xenon. Pepin (2000) has used isotopic correlations of chondrite data to infer the presence of a mixing component, U–Xe, that has solar light isotope ratios and when highly mass-fractionated yields the light-isotope ratios of terrestrial xenon and is relatively depleted in heavy xenon isotopes. The composition of the present atmosphere then can be obtained from fractionated U–Xe by the addition of a heavy isotope component that has the composition of $^{244}$Pu-derived fission xenon (Pepin, 2000). Further discussion of U–Xe can be found in Chapter 4.12. Therefore, the proportions of atmospheric xenon isotopes that are fissiogenic can be calculated. The budgets of radiogenic and fissiogenic xenon in the atmosphere are discussed in detail in Section 4.11.3.2 below.

MORB $^{129}$Xe/$^{130}$Xe and $^{136}$Xe/$^{130}$Xe ratios lie on a correlation extending from atmospheric ratios to higher values (Staudacher and Allègre, 1982; Kunz *et al.*, 1998), and likely reflect mixing of variable proportions of contaminant air xenon with an upper-mantle component having more radiogenic $^{129}$Xe/$^{130}$Xe and $^{136}$Xe/$^{130}$Xe ratios (Figure 4). The highest measured values thus provide lower limits for the MORB source. The MORB data demonstrate that the xenon presently in the atmosphere was in an environment with a higher Xe/I ratio than that of the xenon in the mantle, at least during the lifetime of $^{129}$I. As discussed further below, this difference can be generated either by degassing processes or by early differences in mantle reservoirs related to the formation processes of the planet.

Contributions to $^{136*}$Xe enrichments in MORBs can be from either decay of $^{238}$U over Earth history or early $^{244}$Pu decay, which in theory can be distinguished based on the spectrum of contributions to other xenon isotopes, although analyses have typically not been sufficiently precise to do so. More precise measurements can be obtained

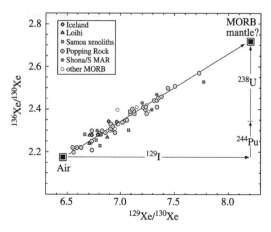

**Figure 4** Xenon isotope compositions of MORB and selected ocean islands. The excesses in $^{129}$Xe due to $^{129}$I, and in $^{136}$Xe due to $^{238}$U and $^{244}$Pu, are correlated due to mixing between mantle xenon and atmospheric contamination. The fraction of $^{136}$Xe from $^{244}$Pu calculated by Kunz *et al.* (1998) is shown for illustration (source Graham, 2002).

from the abundant xenon in some $CO_2$ well gases that have $^{129}$Xe and $^{136}$Xe enrichments similar to those found in MORBs, and are likely to be from the upper mantle (Staudacher, 1987). Precise measurements indicate that $^{244}$Pu has contributed <10–20% of the $^{136}$Xe that is in excess of the atmospheric composition (Phinney *et al.*, 1978; Caffee *et al.*, 1999). An error-weighted best fit to recent precise MORB data (Kunz *et al.*, 1998) yielded a value of 32 ± 10% for the fraction of $^{136}$Xe excesses relative to the atmospheric composition that are $^{244}$Pu-derived, although with considerable uncertainties (Marti and Mathew, 1998). The atmosphere itself, therefore, contains $^{244}$Pu-derived $^{136}$Xe. Clearly, further work is warranted on the proportion of plutonium-derived heavy xenon in the mantle, although it appears that the fissiogenic xenon is dominantly derived from uranium.

It has proven to be more difficult to characterize the xenon in OIB source regions. Xenon with atmospheric isotopic ratios in high-$^3$He/$^4$He OIB samples (e.g., Allègre *et al.*, 1983) appears to be dominated by air contamination (Patterson *et al.*, 1990; Harrison *et al.*, 1999) rather than represent mantle xenon with an air composition. Although Samoan samples with intermediate (9–20$R_A$) helium isotope ratios have been found with xenon isotopic ratios distinct from those of the atmosphere (Poreda and Farley, 1992), the xenon in these samples may have been derived largely from the MORB source. Recently, Harrison *et al.* (1999) found slight $^{129}$Xe excesses in Icelandic samples with $^{129*}$Xe/$^3$He ratios that are compatible with the ratio in a gas-rich MORB, but due to the uncertainties in the data it cannot be determined

whether there are indeed differences between the MORB and OIB sources. Trieloff *et al.* (2000) reported xenon isotope compositions in Loihi dunites and Icelandic glasses that were on the MORB correlation line and had values up to $^{129}$Xe/$^{130}$Xe = 6.9. These were accompanied by $^3$He/$^4$He ratios up to 24$R_A$, and so may contain noble gases from both MORB ($\sim$8$R_A$) and the highest $^3$He/$^4$He ratio (37$R_A$) OIB source. From these data it appears that the OIB source may have xenon that is similar to that in MORB, although there may be some differences that have not been resolvable.

The relatively imprecise measured ratios of the nonradiogenic isotopes in MORB are indistinguishable from those in the atmosphere. However, more precise measurements of mantle-derived xenon in $CO_2$ well gases have been found to have higher $^{124–128}$Xe/$^{130}$Xe ratios (Phinney *et al.*, 1978; Caffee *et al.*, 1999) that can be explained by either: (i) a mixture of $\sim$10% xenon trapped within the Earth of solar isotopic composition and $\sim$90% atmospheric xenon (subducted or added in the crust); or (ii) a mantle xenon component that has not been isotopically fractionated relative to solar xenon to the same extent as air xenon.

In sum, nonradiogenic atmospheric xenon isotopes, like those of neon, require that early degassing of noble gases occurred when fractionating losses to space were still operating. Like argon, xenon in the upper mantle is more radiogenic than the atmosphere, and this must be a feature of any reasonable degassing model. The problems of atmospheric contamination are greatest for xenon, and so there is little definitive evidence regarding the isotopic variations in the mantle.

### 4.11.2.6 Noble Gas Abundance Patterns

Noble gas abundance patterns in MORBs and OIBs scatter greatly. This is due to sample alteration as well as fractionation during noble gas partitioning between basaltic melts and a vapor phase that may then be preferentially gained or lost by the sample. Nonetheless, MORB Ne/Ar and Xe/Ar ratios that are greater than the air values are common. An example of this pattern was found in a gas-rich, relatively uncontaminated MORB sample (Figure 5) with high $^{40}$Ar/$^{36}$Ar and $^{129}$Xe/$^{130}$Xe ratios and a $^4$He/$^{40}$Ar ratio ($\sim$3) that is near that of production in the upper mantle (and so not fractionated) (Staudacher *et al.*, 1989; Moreira *et al.*, 1998). An upper-mantle pattern also can be calculated by assuming that the noble gases have been degassed from the mantle without substantial elemental fractionation, and the radiogenic nuclides are present in their production ratios. Using ratios of estimated upper-mantle

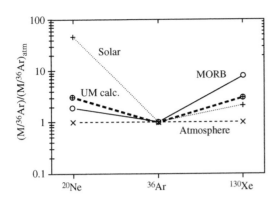

**Figure 5** The relative abundances of noble gases in gas-rich MORB (Moreira *et al.*, 1998), as calculated for the upper mantle (see text), air (Tables 2 and 3), and in the solar composition (see Ozima and Podosek, 2001). The upper mantle is enriched in neon and xenon, relative to argon, compared to the air composition.

production rates to determine the relative abundances of radiogenic $^4$He, $^{21}$Ne, $^{40}$Ar, and $^{136}$Xe, measured isotopic compositions can be used to determine ratios of nonradiogenic isotopes. For example, using the production ratio of $^{21}$Ne/$^{40}$Ar and the ratio of $^{21}$Ne/$^{22}$Ne and $^{40}$Ar/$^{36}$Ar (taking into account the amounts of non-nucleogenic $^{21}$Ne) measured in basalts that are not altered by melting and transport processes, the $^{22}$Ne/$^{36}$Ar ratio in the mantle source can be calculated. In this case, $^{22}$Ne/$^{36}$Ar$_{MORB} = 0.15 \times 10^{-4}$ and $^{130}$Xe/$^{36}$Ar$_{MORB} = 3.3 \times 10^{-4}$, significantly higher than the corresponding atmospheric values of $0.05 \times 10^{-4}$ and $1.1 \times 10^{-4}$, respectively.

It is often assumed that all the noble gases are highly incompatible during basalt genesis and so are efficiently extracted from the mantle without elemental fractionation. Experimental data for partitioning between basaltic melts and olivine are consistent with this for helium but not for the heavier noble gases, which have been found in higher concentrations in olivine than expected (Hiyagon and Ozima, 1986; Broadhurst *et al.*, 1992). However, these results may be due to experimental difficulties. Recent data (Chamorro-Perez *et al.*, 2002) indicate that the argon clinopyroxene/silicate melt partition coefficient is relatively constant and equal to ~4 × 10$^{-4}$ at pressures up to at least 80 kbar, and more recent data indicate that neon, krypton, and xenon are similarly incompatible (Brooker *et al.*, 2003). Therefore, it appears that the noble gases are all highly incompatible in the mantle, and there is no elemental fractionation between the melt and the mantle source region. However, while noble gases transported from the mantle may not be elementally fractionated, some fractionation may occur in the highly depleted melt residue due to

small differences in partition coefficients. This remains a possibility since reliable partition coefficients for most of the noble gases are unavailable.

Any fractionations that might occur during degassing of the upper mantle would be expected to be monotonic across the noble gases. Therefore, degassing alone does not explain high values for both $^{22}$Ne/$^{36}$Ar and $^{130}$Xe/$^{36}$Ar with respect to atmospheric values. If the heavy noble gases were slightly more compatible (as might be plausibly assumed), then a higher $^{130}$Xe/$^{36}$Ar value for the mantle would be generated, but accompanied by a lower $^{22}$Ne/$^{36}$Ar ratio. It is possible that the atmospheric $^{22}$Ne/$^{36}$Ar ratio was lowered instead during volatile losses to space. However, if the starting noble gas composition of the Earth was solar (see Chapter 4.12), with $^{22}$Ne/$^{36}$Ar = 37 (Geiss *et al.*, 1972), then the upper-mantle value, which is lower than the solar value, still requires explanation. The possibilities that remain are a mantle composition that was generated during Earth formation that is different from the solar value, and subduction of atmospheric argon (see Section 4.11.2.9).

### 4.11.2.7 MORB Fluxes and Upper-mantle Concentrations

Noble gas concentrations vary widely in mantle-derived volcanics due to degassing during ascent and eruption, as well as from volatile redistribution in vesicles. Therefore, these data cannot be readily used to constrain the upper-mantle concentrations. An alternative approach is to compare fluxes to the atmosphere from a region with a known rate of volcanism. The largest and most clearly defined mantle volatile flux, from mid-ocean ridges (Clarke *et al.*, 1969; Craig *et al.*, 1975), is 1,060 ± 250 mol yr$^{-1}$ $^3$He (Lupton and Craig, 1975; Farley *et al.*, 1995), and is obtained by combining seawater $^3$He concentrations in excess of dissolved air helium with seawater advection models. This value represents an average over the last 1,000 years, and while this is generally assumed to be the long-term average, this remains to be confirmed. Using the $^3$He/$^4$He in vented gas, the flux of $^4$He can be determined once the $^3$He flux is determined. The flux of $^4$He is approximately equal to the production rate for the upper mantle above 670 km (O'Nions and Oxburgh, 1983). This fact has been incorporated in some mantle degassing models, and is discussed further below. The fluxes of other noble gases from the mantle can be obtained from the $^3$He fluxes and the relative abundances in MORBs (see Section 4.11.2.6). Assuming that radiogenic $^4$He and $^{40}$Ar are degassing together, a $^{40}$Ar flux of $1.9 \times 10^{31}$ atoms a$^{-1}$ is obtained by using a radiogenic ratio of $^4He/^{40}Ar = 3$, a MORB $^3He/^4He$ ratio of $8R_A$, and noting that ~90% of mantle $^4He$ is radiogenic.

The concentration of $^3He$ in the mantle can be determined by dividing the flux of $^3He$ into the oceans by the rate of production of melt that is responsible for carrying this $^3He$ from the mantle, which is equivalent to the rate of ocean crust production of 20 km³ yr⁻¹ (Parsons, 1981). MORBs that degas quantitatively to produce a $^3He$ flux of 1,060 mol yr⁻¹ must have an average $^3He$ content of $1.96 \times 10^{-14}$ mol g⁻¹ or $4.4 \times 10^{-10}$ cm³ $^3He$(STP) g⁻¹. This $^3He$ concentration is within a factor of 2 of that obtained for the most gas-rich basalt glass of ~$10.0 \times 10^{-10}$ cm³ $^3He$(STP) g⁻¹ (Sarda *et al.*, 1988; Moreira *et al.*, 1998). Assuming that MORB is generated by an average of 10% partial melting, the source region contains $2 \times 10^{-15}$ mol g⁻¹ ($1.2 \times 10^9$ atoms $^3He$ g⁻¹). The mantle concentration of $^3He$ and $^4He$ cannot be usefully compared with the $^3He$ and $^4He$ inventories of the atmosphere since $^3He$ and $^4He$ are lost to space with a residence time of a few million years. A comparison, however, can be made with argon. Using the MORB $^3He/^{36}Ar$ ratio of 1.7, then the upper mantle has $(0.8–2.9) \times 10^9$ atoms $^{36}Ar$ g⁻¹. This is highly depleted compared to the benchmark value of $3 \times 10^{12}$ atoms $^{36}Ar$ g⁻¹ obtained by dividing the atmospheric inventory by the mass of the upper mantle. The abundances of other noble gas isotopes can be obtained similarly.

The flux of $^4He$ can be compared with that of heat, since $^4He$ is produced along with heat during radioactive decay. The Earth's global heat loss amounts to 44 TW (Pollack *et al.*, 1993). Subtracting the heat production from the continental crust (4.8–9.6 TW), and the core (3–7 TW; Buffett *et al.*, 1996) leaves 9.6–14.4 TW to be accounted for by present-day radiogenic heating and 17.8–21.8 TW as a result of secular cooling (largely from earlier heat production). O'Nions and Oxburgh (1983) pointed out that the present-day $^4He$ mantle flux was produced along with only 2.4 TW of heat, an order of magnitude less than that produced along with all the radiogenic heat presently reaching the surface, and suggested that this could be achieved by a boundary layer in the mantle through which heat could pass, but behind which helium was trapped. It has been suggested that heat and helium are transported to the surface at hotspots, where the bulk of the helium is lost, while the heat is lost subsequently at ridges (Morgan, 1998). However, evidence for such a hotspot helium flux at present or in the past, and formulation of a mantle noble gas model incorporating this suggestion, are unavailable. The possibility that the separation is due to the different mechanisms that extract heat and helium

from the mantle was investigated with a secular cooling model of the Earth (van Keken *et al.*, 2001). It was found that the ratio of the surface fluxes of $^4He$ and heat were substantially higher than presently observed except for rare excursions. However, the ratio of surface fluxes would more closely match the model results if the surface helium flux were three times greater. It is possible that since the observed helium flux represents an average over only 1,000 yr, it does not represent the flux over somewhat longer timescales (Ballentine *et al.*, 2002). However, this is difficult to assess. The alternative that remains is that some boundary is needed for separating uranium and thorium, from the upper-mantle reservoir, and allowing heat to pass more efficiently than helium, although the configuration of this boundary is unconstrained by geochemistry. The implications of this are considered further in discussions of mantle models (see Chapter 2.06). This significance for mantle degassing is that there are considerable amounts of radiogenic noble gases maintained behind such a boundary. Bercovici and Karato (2003) have proposed a model of extraction of incompatible elements at depth in the upper mantle by melting and entrainment in the down going slab.

### 4.11.2.8 Other Mantle Fluxes

The flux of $^3He$ from intraplate volcanic systems is dominantly subaerial and so it is not possible to obtain directly time-integrated flux values for even a short geological period. While the Loihi hotspot in the Pacific is submarine, calculation of $^3He$ fluxes into the ocean using ocean circulation models have not required a large flux from this location that is comparable to the $^3He$ plumes seen over ridges (Gamo *et al.*, 1987; Farley *et al.*, 1995), although recent data has seen an extensive $^3He$ plume from Loihi (Lupton, 1996).

Helium fluxes from OIB could be calculated if the rates of magmatism are known, along with the helium concentrations of the source regions or concentrations of undegassed magmas (see Porcelli and Ballentine, 2002). Estimates of the rate of intraplate magma production vary from 1% to 12% of the MORB production rate (Reymer and Schubert, 1984; Schilling *et al.*, 1978; Batiza, 1982; Crisp, 1984). The total noble gas flux from ocean islands is related to the mantle source concentrations and whether some source domains are more gas-rich than the MORB source mantle. OIBs are typically more extensively degassed than MORBs. This is readily explained for those that are erupted at shallower depths or subaerially. Also, OIBs may be more volatile-rich than MORBs and, therefore, may degas more effectively (Dixon and Stolper, 1995). High water contents of basalts lower $CO_2$ solubility, and appear to have lowered helium contents in lavas with high $^3He/^4He$ ratios along the Reykjanes Ridge (Hilton *et al.*, 2000). Estimates for the mantle source with high $^3He/^4He$ ratios have ranged from much lower than that of the MORB source, perhaps due to prior melting (Hilton *et al.*, 1997), to up to 15 times higher (Moreira and Sarda, 2000; Hilton *et al.*, 2000). It should be emphasized that in many locations, the high $^3He/^4He$ source component makes up a small fraction of the sample source and, if gas-rich, may substantially affect the helium composition but not substantially increase the helium flux. Therefore, in many cases, source concentrations may be closer to that of MORBs. In this case, the overall $^3He$ flux relative to that from MORBs is proportional to the relative melt production rate, or 1–12% that of MORBs. It is not clear to what extent more gas-rich OIBs augment this flux.

Continental settings provide a small but significant flux from the mantle. Regional groundwater systems provide time-integrated records of this flux over large areas, but are based on short timescales and dependent on the hydrogeological model used. In the Pannonian basin (4,000 km$^2$ in Hungary), for example, the flux of $^3He$ has been estimated to be from $0.8–5 \times 10^4$ $^3He$ atoms m$^{-2}$ s$^{-1}$ (Stute *et al.*, 1992) to $8 \times 10^4$ $^3He$ atoms m$^{-2}$ s$^{-1}$ (Martel *et al.*, 1989). Taking an area of $2 \times 10^{14}$ m$^2$ for continents and assuming 10% is under extension, this yields a total $^3He$ flux of 8.4–84 mol $^3He$ yr$^{-1}$ (Porcelli and Ballentine, 2002). This value compares with a mantle flux of <3 mol $^3He$ yr$^{-1}$ through the stable continental crust (O'Nions and Oxburgh, 1988).

The flux of mantle $^3He$ to the atmosphere at subduction zones from the upper mantle can be estimated from the volume of convergent zone volcanics. The estimate of Reymer and Schubert (1984) is 5% that at mid-ocean ridges. If this is largely generated by similar degrees of melting of a similar source as MORB, then the $^3He$ flux is only 5% that at the mid-ocean ridges, or 50 mol $^3He$ yr$^{-1}$ (see also Hilton *et al.*, 2002). An estimated $CO_2$ flux at convergent margins of $3 \times 10^{11}$ mol yr$^{-1}$ (Sano and Williams, 1996) can also be used to estimate the $^3He$ flux. Using a mid-ocean ridge ratio of $CO_2/^3He = 2 \times 10^9$ (Marty and Jambon, 1987) gives 150 mol $^3He$ yr$^{-1}$, although the convergent margin $CO_2/^3He$ ratio may be much greater due to the presence of recycled carbon.

In summary, the dominant noble gas flux from the mantle is from mid-ocean ridges. Other fluxes may be negligible or augment the ridge flux by up to an additional 25%.

### 4.11.2.9 Subduction Fluxes

It is possible that atmospheric noble gases that are incorporated in oceanic crust materials, including altered MORBs, hydrothermally altered crust, and sediments are subducted into the mantle and thus are mixed with primordial noble gas constituents. Available data indicate that holocrystalline MORBs and oceanic sediments contain atmospheric noble gases that are greatly enriched in the heavier noble gases. However, because concentrations vary by several orders of magnitude, an accurate average value cannot be easily determined. Measurements of pelagic sediments, with $3 \times 10^{15}$ g yr$^{-1}$ subducted (Von Huene and Scholl, 1991), have $(0.05–7) \times 10^{10}$ atoms $^{130}Xe$ g$^{-1}$ (Podosek *et al.*, 1980; Matsuda and Nagao, 1986; Staudacher and Allègre, 1988) with a mean of $6 \times 10^9$ atoms g$^{-1}$. Measurements of holocrystalline basalts found $(4–42) \times 10^7$ atoms $^{130}Xe$ g$^{-1}$ (Dymond and Hogan, 1973; Staudacher and Allègre, 1988), with a mean of $2 \times 10^8$ atoms $^{130}Xe$ g$^{-1}$. While the depth over which addition of atmospheric gases are added by alteration is unknown, it might be assumed that this occurs over the same depth as low temperature enrichment of alkaline elements of ~600 m (Hart and Staudigel, 1982). In this case, an estimated $7 \times 10^{15}$ g yr$^{-1}$ of this material is subducted. In total, these numbers result in $2 \times 10^{25}$ atoms $^{130}Xe$ yr$^{-1}$ (33 mol yr$^{-1}$) reaching subduction zones in sediments and altered basalt (Porcelli and Wasserburg, 1995a). This can be compared to the estimate of Staudacher and Allègre (1988) determined by assuming that 40–80% of the subducting flux of oceanic crust $(6.3 \times 10^{16}$ g yr$^{-1})$ is altered and contains atmosphere-derived noble gases, and that 18% of this mass is ocean sediment. In this case, a similar flux of 13.8–39 mol $^{130}Xe$ yr$^{-1}$ was obtained, along with $1.9–2.5 \times 10^6$ mol $^4He$ yr$^{-1}$, $1.9–5.9 \times 10^3$ mol $^{20}Ne$ yr$^{-1}$, $5.8–21.9 \times 10^3$ mol $^{36}Ar$ yr$^{-1}$, and $3.0–12.3 \times 10^3$ mol $^{84}Kr$ yr$^{-1}$. Subduction of noble gases at these rates over $10^9$ yr would have resulted in 1%, 90%, 110%, and 170% of the respective inventories of $^{20}Ne$, $^{36}Ar$, $^{84}Kr$, and $^{130}Xe$ to be in the upper mantle (Allègre *et al.*, 1986); however, the fraction stripped from subducting materials during deformation and arc volcanism before entering the mantle is unknown but it is likely to be substantial.

Subduction zone processing and volcanism may return much of the noble gases in the slab to the atmosphere. However, it is possible that the total amounts of noble gases reaching subduction zones are sufficiently high that subduction into the deeper mantle (i.e., beyond the zone of magma generation) of only a small fraction may have a considerable impact upon the composition of argon and xenon in the upper mantle (Porcelli and Wasserburg, 1995a,b). Staudacher and Allègre (1988) argued that subducting argon and xenon must be almost completely lost to the atmosphere during subduction zone magmatism, or the high $^{129}Xe/^{130}Xe$ and $^{136}Xe/^{130}Xe$ in the upper mantle would not have been preserved throughout Earth history. However, this conclusion is dependent upon a model of unidirectional degassing to generate the upper-mantle xenon isotope composition (see Section 4.11.4.3), and as discussed in Section 4.11.4.5 the contrary view that subducted noble gases are mixed with nonrecycled, mantle-derived xenon to produce the upper-mantle composition (Porcelli and Wasserburg, 1995a,b) is compatible with the mantle data. Note that, as discussed above, the $^{128}Xe/^{130}Xe$ ratio measured in mantle-derived xenon trapped in $CO_2$ well gases may be interpreted as a mixture of ~90% subducted xenon with 10% trapped solar xenon. Further, direct input of the subducted slab into a gas-rich deeper reservoir that has $^{40}Ar/^{36}Ar$ values that are significantly lower than in the MORB source mantle (e.g., Trieloff *et al.*, 2000) is also possible.

## 4.11.3  BULK DEGASSING OF RADIOGENIC ISOTOPES

The total fraction of a species that has been degassed to the atmosphere can only be calculated for radiogenic nuclides, which have total planetary abundances that are constrained by the parent element abundances. Most attention has focused on $^{40}Ar$, although as discussed below $^{136}Xe$ also provides valuable constraints. Similar calculations cannot be done for $^{4}He$, which does not accumulate in the atmosphere, nor for neon isotopes, since the production rate of $^{21}Ne$, as well as the amount of nonradiogenic $^{21}Ne$ in the atmosphere, are too uncertain. Also, the amount of radiogenic $^{129}Xe$ in the bulk Earth is not well constrained due to early losses from the planet.

### 4.11.3.1  The $^{40}K-^{40}Ar$ Budget

Potassium is a moderately volatile element and is depleted by a factor of ~8 in the bulk silicate Earth compared to CI chondrites, but a precise and unambiguous concentration is difficult to obtain. Estimates have been made by comparison with uranium, which like potassium is highly incompatible during melting and so is not readily fractionated between MORB and the upper mantle. There is little debate regarding the concentration of uranium, which is obtained from concentration in carbonaceous chondrites and, by assuming that refractory elements (e.g., calcium, uranium, thorium) are unfractionated from solar values

in the bulk Earth (e.g., O'Nions *et al.*, 1981). In this case, a bulk silicate Earth concentration of 21 ppb uranium is obtained (Rocholl and Jochum, 1993). If it is assumed that the MORB source value of $K/U = 1.27 \times 10^4$ (Jochum *et al.*, 1983) is the same as that of the bulk silicate Earth, then there is 270 ppm potassium that has produced a total of $2.4 \times 10^{42}$ atoms $^{40}Ar$. This is a widely accepted value. Note that the core is not a significant repository of either potassium or $^{40}Ar$ (Chabot and Drake, 1999).

The $^{40}Ar$ in the atmosphere ($9.94 \times 10^{41}$ atoms) is essentially entirely radiogenic. Assuming the crust has 0.91–2.0 wt.% K (Taylor and McLennan, 1985; Rudnick and Fountain, 1995; Wedepohl, 1995), and noting that the continents have a mean K–Ar age of $1 \times 10^9$ a (Hurley and Rand, 1969), equivalent to a ratio of $^{40}Ar/K = 9.1 \times 10^{-6}$, yields an amount of crustal $^{40}Ar$ that is only 3.1–6.8% of that in the atmosphere. Therefore, in total, 41% of the $^{40}Ar$ that has been produced is now in the atmosphere (Allègre *et al.*, 1986, 1996; Turcotte and Schubert, 1988). Thus, a significant reservoir of $^{40}Ar$ remains in the Earth. How this relates to degassing of nonradiogenic noble gas isotopes is the subject of degassing models, but identifying the fraction of radiogenic isotopes in the atmosphere provides an important overall constraint. It has sometimes been assumed that the mantle reservoir that is rich in $^{40}Ar$ is the same as that with high $^{3}He/^{4}He$ and so is also rich in $^{3}He$, although this is not the case of all mantle models (see Porcelli and Ballentine, 2002).

Uncertainty arises in the above calculation when considering that the depleted MORB-source mantle may not have a bulk silicate Earth K/U ratio, since a significant fraction of either element may have been preferentially added into the upper mantle by subduction. The bulk of the potassium and uranium originally in the upper mantle is now in the continental crust, which, therefore, may be expected to have a bulk silicate Earth K/U ratio. But this is not sufficiently well constrained due to wide variations in the potassium and uranium contents of the more differentiated continental crustal rocks. It has been suggested that the bulk Earth K/U ratio is much lower than the MORB value (Albarède, 1998; Davies, 1999). In this case, the potassium content of the Earth is much lower, and so a greater fraction of the total $^{40}Ar$ has degassed. This would make the $^{40}Ar$ budget compatible with geophysical models that have convection and mixing throughout the mantle, and so would imply that the depleted mantle that serves as the MORB source represents the bulk of the whole mantle. In such a case, there is no need to isolate and maintain a gas-rich mantle reservoir making up 60% of the mantle. However, it has been argued that the relative proportions of moderately volatile elements in the Earth lie on compositional trends defined by chondritic meteorite classes (Allègre *et al.*, 1995; Halliday and Porcelli, 2001), and trends in meteoritic Rb/Sr versus K/U are compatible with a terrestrial value of $K/U = 1.27 \times 10^4$. Only a modest reduction in the terrestrial potassium content would still be compatible with these relationships.

### 4.11.3.2 The $^{129}I$–$^{129}Xe$ and $^{244}Pu$–$^{136}Xe$ Budgets

The amount of short-lived isotope $^{129}I$ ($t_{1/2} = 15.7$ Myr) that was in the Earth or Earth-forming materials and produced $^{129*}Xe$ can be estimated from the present bulk silicate Earth concentration of stable $^{127}I$. Iodine is a volatile element and is highly depleted in the Earth relative to chondrites. Wänke *et al.* (1984) estimated a bulk silicate Earth stable $^{127}I$ concentration of 13 ppb based on the analysis of a fertile xenolith judged to represent the undepleted mantle, and this is the generally accepted estimate. McDonough and Sun (1995) arrived at a similar value of 11 ppb. Alternatively, using a bulk crust abundance of $8.6 \times 10^{18}$ g I (Muramatsu and Wedepohl, 1998) and an upper-mantle concentration of 0.8 ppb (Déruelle *et al.*, 1992), and assuming that the crust was derived from 25% of the mantle, a bulk silicate Earth value of 9 ppb is obtained (Porcelli and Ballentine, 2002). However, based on unpublished data, Déruelle *et al.* (1992) quote a concentration for the crust that is 4.3 times higher, raising the possibility that the silicate Earth concentration of iodine is substantially higher than typically estimated, although such high values have not been adopted. At 4.57 Ga ago, $(^{129}I/^{127}I)_0 = 1.1 \times 10^{-4}$ based on meteorite data (Hohenberg *et al.*, 1967; Brazzle *et al.*, 1999). For a silicate Earth value of 13 ppb $^{127}I$, $2.7 \times 10^{37}$ atoms of $^{129*}Xe$ were produced in the Earth or Earth-forming materials since formation of the solar system.

Plutonium is a highly refractory element represented naturally by a single isotope, $^{244}Pu$ ($t_{1/2} = 80$ Myr), which produces heavy xenon isotopes by fission. It is assumed that plutonium was incorporated into Earth-forming materials unfractionated relative to other refractory elements such as uranium. Meteorite data suggests that at 4.56 Ga, $(^{244}Pu/^{238}U)_0 = 6.8 \times 10^{-3}$ (Hudson *et al.*, 1989) and this is the commonly accepted value. In this case the silicate Earth, or Earth-forming materials, with a present-day bulk silicate Earth value of 21 ppb U (O'Nions *et al.*, 1981), initially had 0.29 ppb Pu. This produced $2.0 \times 10^{35}$ atoms of $^{136*}Xe$ since the start of the solar system. Other work on meteorites

(Hagee *et al.*, 1990) calculated values of $(4-7) \times 10^{-3}$, with the higher number considered more likely to represent the solar value, although a lower value remains a possibility. The amount produced by $^{238}U$ in the bulk silicate Earth ($7.5 \times 10^{33}$ atoms $^{136}Xe$) is much less, and so bulk silicate Earth (and atmospheric) $^{136*}Xe$ is dominantly plutonium-derived; even if xenon was lost over the first $10^8$ a of Earth history (see Chapter 4.12) so that half of the plutonium-derived xenon was lost, plutonium-derived $^{136}Xe$ is still 13 times more abundant in the Earth.

The greatest difficulty in constraining the global xenon budget has been in calculating the abundances of radiogenic xenon in the atmosphere (see Chapter 4.12). The composition for nonradiogenic atmospheric xenon (Section 4.11.2.5) provides ratios of $^{129}Xe/^{130}Xe = 6.053$ and $^{136}Xe/^{130}Xe = 2.075$ as the present best estimates of the isotopic composition of nonradiogenic terrestrial xenon (Pepin, 2000). Therefore, $6.8 \pm 0.30\%$ of atmospheric $^{129}Xe$ ($^{129*}Xe_{atm} = 1.7 \times 10^{35}$ atoms) and $4.65 \pm 0.5\%$ of atmospheric $^{136}Xe$ ($^{136*}Xe_{atm} = 3.81 \times 10^{34}$ atoms) are radiogenic. The $^{136*}Xe$ in the atmosphere is 20% of the total $^{136}Xe$ produced by $^{244}Pu$ in the bulk silicate Earth. However, the $^{129*}Xe_{atm}$ is only 0.8% of the total $^{129}Xe$ produced since 4.57 Ga; such a low value cannot be accounted for by incomplete degassing of the mantle nor from any uncertainties in the estimated amount of $^{129*}Xe$, and requires losses to space over an early period that is short relative to the longer time constant of $^{136}Xe$ production.

The depletion of radiogenic xenon in the atmosphere due to losses from the Earth to space must have occurred during early Earth history, when such heavy species could have been lost either from protoplanetary materials or from the growing Earth. Wetherill (1975) proposed that a "closure age" of the Earth could be calculated by assuming a two-stage history that involved essentially complete loss of $^{129*}Xe$ and $^{136*}Xe$ initially, followed by complete closure against further loss. The "closure age" also can be calculated by combining the $^{129}I$–$^{129}Xe$ and $^{244}Pu$–$^{136}Xe$ systems (Pepin and Phinney, 1976; see Chapter 4.12) to obtain a closure age of 82 Myr. If radiogenic $^{136}Xe$ was lost from the entire planet over about one half-life of $^{244}Pu$ (80 Myr), then ~40% of the $^{136}Xe$ remaining in the Earth is in the atmosphere, compatible with the fraction of $^{40}Ar$ in the atmosphere (using a silicate Earth value of 270 ppm K). Note that in the first 100 Ma, only 6% of the $^{40}Ar$ now present was produced, and losses over this time would not have significantly changed the $^{40}Ar$ budget (Davies, 1999).

The coincidence between the $^{40}K$–$^{40}Ar$ and $^{244}Pu$–$^{136}Xe$ budgets is remarkable considering that these values are the result of a series of independent, albeit somewhat uncertain, estimates. As mentioned above, in order to adjust these numbers to accommodate a greater fraction of degassing, it has been suggested that there is greater depletion of the moderately volatile potassium in the Earth (Albarède, 1998); however, the coincidence with the Pu–$^{136}$Xe budget requires a similarly lowered estimate of the amount of short-lived, refractory $^{244}$Pu in the solar nebula. Clearly, these two factors are unrelated. It might be assumed that the $^{40}$Ar budget might reflect processing of a greater fraction of the mantle, but with subduction returning a considerable fraction of the potassium over geological time, thereby creating domains in the mantle that have been degassed early but now contain a considerable budget of $^{40}$Ar by subsequent production. However, this is not possible for the $^{136}$Xe budget; all plutonium-derived $^{136}$Xe was produced early, and the present budget reflects the total processing of the mantle. Therefore, it appears that the noble gas budget requires that a considerable fraction of the mantle has not been degassed to the atmosphere.

### 4.11.4 DEGASSING OF THE MANTLE

The present value for the mid-ocean ridge flux of $^3$He to the atmosphere, if constant over 4.5 Ga, would result in a total of $4.5 \times 10^{12}$ mol $^3$He degassed. Using the mantle value of $^3$He/$^{22}$Ne = 11 (see Section 4.11.2.3), this corresponds to a total of $5 \times 10^{13}$ mol $^{22}$Ne. This is only 2% of the $^{22}$Ne presently in the atmosphere (Table 2). Therefore, present fluxes of noble gases from the mantle, applied over the history of the Earth, are insufficient to provide the inventories in the atmosphere. A degassing history that involves stronger degassing in the past is required, and a consideration of degassing models is needed to address the issue of the time dependence of the fluxes. An important factor in degassing models is the extent of noble gas recycling into the mantle by subduction. While subduction is unlikely to have a significant effect on the atmospheric inventory, it may impact the characteristics of noble gases in depleted mantle reservoirs may be significant. Therefore, models either assume that subduction of noble gases does not occur, or explicitly incorporates the effects of atmospheric inputs to the mantle.

#### 4.11.4.1 Early Earth Degassing

Models for the evolution of terrestrial noble gases must necessarily consider appropriate starting conditions. The initial incorporation of noble gases and the establishment of terrestrial characteristics are discussed in Chapter 4.12. Early degassing is likely to be very vigorous due to high accretional impact energies during Earth formation over $10^8$ yr. Loss of major volatiles from impacting materials, and noble gases, may occur after only 10% of the Earth has accreted (Ahrens *et al.*, 1989). In this case, much of the volatiles are added directly to the atmosphere rather than being degassed from the solid planet. Volatiles may also have been captured directly from the solar nebula (e.g., Porcelli *et al.*, 2001), followed by modifications due to losses to space (see Chapter 4.12). Various models for the origin of relatively volatile elements in the Earth have accounted for terrestrial volatiles by late infall of volatile-rich material (e.g., Turekian and Clark, 1969, 1975; Dreibus and Wänke, 1989; Owen *et al.*, 1992). The relative uniformity of lead and strontium isotopes in the mantle suggests that relatively volatile elements such as rubidium and lead that would also be supplied by late-accreting volatile-rich material were subsequently mixed into the deep mantle (Gast, 1960). However, loss of noble gases from impacting materials directly into the atmosphere likely inhibited their incorporation into the growing solid Earth. Therefore, noble gases supplied to the Earth in this way were unlikely to have been initially uniformly distributed in the solid Earth. It is also clear that very strong degassing of the Earth occurred during the extended period of planetary formation.

Atmospheric noble gases were also likely to have been lost to space during accretion by atmospheric erosion (Ahrens, 1993), when large impactors impart sufficient energy to the atmosphere for the constituents to reach escape velocity. In this way, each large impact during later accretion can drive away a substantial portion of the previously degassed atmosphere. Therefore, the present atmospheric abundances do not necessarily reflect the total amounts of nonradiogenic and early-produced nuclides that were degassed from accreting materials. Also, it has been suggested that the strong fractionation of neon and xenon isotopes in the atmosphere is due to hydrodynamic escape (Hunten *et al.*, 1987; Sasaki and Nakazawa, 1988; Pepin, 1991), where loss of hydrogen from the atmosphere entrains heavier species, leaving behind a fractionated residue (see Chapter 4.12). Such losses would not have affected gases within the Earth, and so would have generated isotopic contrasts between the atmosphere and internal terrestrial reservoirs. Such loss processes can account for the losses of xenon isotopes produced by short-lived $^{129}$I and $^{244}$Pu. In sum, degassing histories must include strong early degassing of the Earth as it accretes, and consider that the abundances degassed were not fully retained in the atmosphere.

### 4.11.4.2 Degassing from One Mantle Reservoir

The simplest case for atmosphere formation is unidirectional degassing from a single solid Earth reservoir, which is represented by the MORB source region. Early models focused on argon isotopes. Generally, the key assumption is that the rate of degassing at any time is directly proportional to the total amount of argon present in the mantle at that time. Also, there is no return flux from the atmosphere by subduction. Then

$$\frac{d^{36}Ar_m}{dt} = -\alpha(t)^{36}Ar_m \qquad (1)$$

and

$$\frac{d^{40}Ar_m}{dt} = -\alpha(t)^{40}Ar_m + \lambda_{40}y^{40}K_m \qquad (2)$$

where $\alpha(t)$ is the time-dependent degassing proportionality constant, $\lambda_{40} = 5.543 \times 10^{-10}\,a^{-1}$ is the total decay rate of $^{40}K$, $y = 0.1048$ is the fraction of decays of $^{40}K$ that yield $^{40}Ar$ (Table 1), and $^{36}Ar_m$, $^{40}Ar_m$, and $^{40}K_m$ are the total Earth abundances. In the simplest case, $\alpha(t)$ is a constant (Turekian, 1959; Ozima and Kudo, 1972; Fisher, 1978). This is reasonable if the mantle is well mixed and has been melted and degassed at a constant rate at mid-ocean ridges. Assuming there was no argon initially in the atmosphere, then the only free variable is $\alpha$; in this case

$$\left(\frac{^{40}Ar}{^{36}Ar}\right)_{atm} = \left(\frac{\alpha}{\alpha-\lambda}(1-e^{-\lambda t}) - \frac{\lambda}{\alpha-\lambda}\right)$$
$$\times (1-e^{-\alpha t})\left(\frac{y^{40}K_m e^{\lambda t}}{^{36}Ar_{atm}}\right)$$
$$(3)$$

Using a BSE value of 270 ppm K, a value of $\alpha = 1.82 \times 10^{-10}$ is obtained. From Equation (1), $^{36}Ar_{atm} = {}^{36}Ar_{m0}\,(1-e^{-\alpha t})$, so that the fraction of nonradiogenic $^{36}Ar$ that has degassed, $^{36}Ar_{atm}/^{36}Ar_{m0}$, is 0.62. The $^{40}Ar/^{36}Ar$ ratio of the mantle is then

$$\left(\frac{^{40}Ar}{^{36}Ar}\right)_m = \frac{\lambda_{40}}{\alpha-\lambda_{40}}(e^{(\alpha-\lambda_{40})t}-1)\frac{y^{40}K_m e^{\lambda_{40}t}}{^{36}Ar_{m0}}$$
$$(4)$$

A value of $(^{40}Ar/^{36}Ar)_{mn} = 520$ is calculated from Equation (4) for the mantle. Once higher values were measured in MORB samples, such a simple formulation no longer appeared valid. Higher ratios can be obtained if an early catastrophic degassing event occurred, removing a fraction $f$ of the $^{36}Ar$ from the mantle into the atmosphere (Ozima, 1973). In this case, the term $(1-f)^{36}Ar_{m0}$ can be substituted for $^{36}Ar_{m0}$ (see Ozima and Podosek, 1983). Then for a mantle with $^{40}Ar/^{36}Ar = 4 \times 10^4$ (see Section 4.11.2.4),

98.6% of $^{36}Ar$ was degassed initially. Alternatively, a more complicated degassing function that is steeply diminishing with time (such as $\alpha(t) = \alpha e^{-\beta t}$) can be used to match the present isotope compositions (Sarda *et al.*, 1985; Turekian, 1990), and so also involves early degassing of the bulk of the atmospheric $^{36}Ar$. Regardless of the formulation used, such early degassing is required by the high measured $^{40}Ar/^{36}Ar$ ratios and may reflect extensive devolatilization of impacting material during accretion or a greater rate of mantle melting very early in Earth history due to higher heat flow.

Another aspect of mantle degassing that can be represented in model calculations is the transfer of potassium from the upper mantle into the continental crust so that the mantle potassium content becomes time dependent. This requires including the crust as an additional model reservoir. The continents may be modeled as either attaining their complete mass very early or more gradually using some growth function (see, e.g., Ozima, 1975; Hamano and Ozima, 1978; Sarda *et al.*, 1985). However, all model formulations qualitatively agree that $^{36}Ar$ degassing dominantly occurred very early in Earth history.

A different perspective on $^{40}Ar$ degassing has been provided by Schwartzman (1973), who argued that potassium is likely to be transferred "coherently" out of the mantle with $^{40}Ar$; i.e., any $^{40}Ar$ that has been produced by potassium will be degassed when the potassium is transferred to the crust. While the $K/^{40}Ar$ ratio of magmas leaving the mantle and that of the mantle source region are thus assumed to be approximately equal, the highly depleted residue could still be fractionated, so that very radiogenic-derived $^{40}Ar/^{36}Ar$ ratios could develop. It has been pointed out that, using the budgets of potassium and $^{40}Ar$, this implies that the potassium in the crust should fully account for the $^{40}Ar$ in the atmosphere (Coltice *et al.*, 2000). This is only the case if the continental crust contains 2.0% K, which is higher than most estimates (see Section 4.11.5.1), and so additional $^{40}Ar$ was provided by additional potassium that has been recycled and which corresponds with up to ~30% of the potassium that is now in the continents (Coltice *et al.*, 2000).

Similar considerations used in the $^{40}Ar$ modeling have been applied to xenon (see Thomsen, 1980; Staudacher and Allègre, 1982; Turner, 1989). In these studies, there is greater resolution of the timing of early degassing due to the short half-lives of the parent nuclides $^{129}I$ and $^{244}Pu$. The higher $^{129}Xe/^{130}Xe$ ratio of the upper mantle is interpreted as resulting from an increase in the I/Xe ratio during the lifetime of $^{129}I$ due to degassing of xenon to the atmosphere. Regardless of the exact degassing history used, this requires that strong degassing occur very early in Earth history, compatible with the results of the argon studies. Models of degassing from a single mantle reservoir have also been applied to helium (Turekian, 1959; Tolstikhin, 1975), although there are greater degrees of freedom since the atmosphere does not preserve a record of the total abundances of helium isotopes that have been degassed.

The strong early degassing inferred in these models can be identified with the degassing that likely occurred during extended accretion of the Earth. These models have not been modified to include possible substantial losses to space. This would not affect the $^{40}$Ar budget, and would require stronger early degassing of $^{36}$Ar and modification of the total amount degassed.

### 4.11.4.3 Multiple Mantle Reservoirs

The major shortcoming of the single reservoir degassing models is that mantle heterogeneities,

especially regarding $^3$He/$^4$He ratios, cannot be explained. The second generation of mantle degassing models developed with growing evidence that, in addition to the MORB-source reservoir, another helium source was required to account for the high $^3$He/$^4$He ratios found in ocean island basalts (Hart *et al.*, 1979; Kurz *et al.*, 1982; Allègre *et al.*, 1983, 1986). These models also incorporate the degassing of a single mantle reservoir to the atmosphere. However, to explain the high OIB $^3$He/$^4$He ratios, there is an additional underlying gas-rich reservoir that is isolated from the degassing upper mantle. Therefore, these layered mantle models can be considered to incorporate two separate systems; the upper-mantle atmosphere that evolves according to the systematics described in Section 4.11.4.2, and the lower mantle (Figure 6). There is no interaction between these two systems, and the lower mantle is completely isolated, except for a minor flux to the surfaced at ocean islands that marks its existence. In order to characterize the

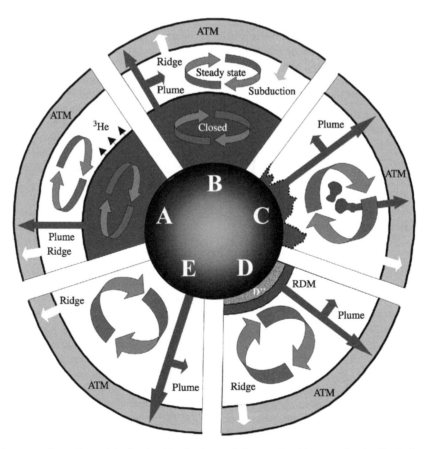

**Figure 6** A range of mantle models for the distribution and fluxes of noble gases in the Earth. Layered mantle models with the atmosphere derived from the upper mantle involve either progressive unidirectional depletion of the upper mantle (A) or an upper mantle subject to inputs from subduction and the deeper mantle, and has steady state concentrations (B). Whole mantle convection models involve degassing of the entire mantle, with helium with high $^3$He/$^4$He ratios found in OIB stored in either a deep variable-thickness layer (C), a layer of subducted material at the core–mantle boundary (D), or the core (E). The models are discussed in the text and Chapter 2.06 (source Porcelli and Ballentine, 2002).

lower-mantle reservoir, it is further assumed that the mantle was initially uniform in noble gas and parent-isotope concentrations, so that both systems had the same starting conditions.

The deep reservoir with high $^3He/^4He$ ratios is assumed to have evolved approximately as a closed system for noble gases and has bulk silicate Earth parent-nuclide concentrations. Assigning the highest OIB $^3He/^4He$ ratios to this reservoir, a comparison between the total production of $^{4*}He$ and the shift in $^3He/^4He$ from the initial terrestrial value to the present value provides an estimate of the $^3He$ concentration in this reservoir:

$$\left(\frac{^4He}{^3He}\right)_{present} - \left(\frac{^4He}{^3He}\right)_{initial} = \frac{^{4*}He}{^3He} \quad (5)$$

For a bulk silicate Earth concentration of 21 ppb U (O'Nions *et al.*, 1981) and Th/U = 3.8 (e.g., Doe and Zartman, 1979), a total of $1.02 \times 10^{15}$ atoms $^{4*}He$ g$^{-1}$ is produced over 4.5 Ga. Assigning this reservoir an Iceland value of $^3He/^4He = 37R_A$ and an initial value of $^3He/^4He = 120R_A$ (see Section 4.11.2.2), then the reservoir has $7.6 \times 10^{10}$ atoms $^3He$ g$^{-1}$. The concentration of another noble gas is required for comparison of lower-mantle noble-gas abundances with the atmosphere. Using $^3He/^{22}Ne = 11$ (see Section 4.11.2.3), a concentration in a closed system lower mantle of $7 \times 10^9$ atoms $^{22}Ne$ g$^{-1}$ is obtained. A benchmark for comparison is the atmospheric $^{22}Ne$ abundance divided by the mass of the upper mantle ($1 \times 10^{27}$ g) of $1.8 \times 10^{11}$ atoms $^{22}Ne$ g$^{-1}$, which is much higher, and might be taken to indicate that the atmosphere source reservoir was more gas-rich than any deep isolated reservoir. However, such an undegassed reservoir has a $^3He$ concentration that is still ~40 times greater than that of the MORB source (see Section 4.11.2.7), and so is still relatively gas-rich.

Nonradiogenic argon and xenon isotope concentrations of such a lower-mantle reservoir cannot be directly calculated without assuming either specific lower-mantle Ar/Ne and Xe/Ne ratios or argon and xenon isotopic compositions. Since modification of the noble gases in the atmosphere has occurred, the relative abundances of the interior of the Earth are unlikely to match those now observed in the atmosphere. For example, a closed-system lower mantle with $^{40}Ar/^{36}Ar \geq 3,000$ (see Section 4.11.2.4) and 270 ppm K has $^{40}Ar = 5.7 \times 10^{14}$ atoms g$^{-1}$ and so $^{36}Ar \leq 1.9 \times 10^{11}$ atoms g$^{-1}$. The reservoir then has a ratio of $^{22}Ne/^{36}Ar \geq 0.9$, which is much greater than the air value of 0.05. Note that some calculations have assumed that the lower-mantle concentration is equal to the atmospheric inventory divided by the mass of the upper mantle

(e.g., Hart *et al.*, 1979), which is based on the idea of an initially uniform distribution of $^{36}Ar$. This is compatible with a lower-mantle ratio of $^{40}Ar/^{36}Ar = 300$. Evidence for higher lower-mantle ratios might be taken to reflect partial degassing of the deep mantle. However, it is possible that there are high $^{40}Ar/^{36}Ar$ ratios in the lower mantle that are due not to degassing; but rather to a lower initial trapped $^{36}Ar$ concentration (Porcelli and Wasserburg, 1995b). As discussed in Sections 4.11.2.4 and 4.11.2.5, the isotopic compositions of argon and xenon that accompany the high $^3He/^4He$ ratios in the source region of ocean island basalts are still too poorly constrained to make firmer conclusions regarding the concentrations of the heavier noble gases.

Note that these closed-system considerations do not require assumptions regarding the size of the reservoir. While models have assumed that it constitutes the entire mantle below 670 km (e.g., Allègre *et al.*, 1986), it can involve a portion of a stratified mantle or material that is distributed as heterogeneities within another mantle reservoir. However, a large deep-mantle reservoir is compatible with the K–$^{40}Ar$ budget and heat-$^4He$ balance.

It should be emphasized that while an undepleted, undegassed mantle reservoir is a component in many models, there is no direct evidence that such a reservoir does indeed exist. An implication of these arguments is that the deep Earth reservoir also has not suffered removal of lithophile elements, and so contains bulk silicate Earth isotopic signatures for strontium, neodymium, hafnium, and lead. As discussed in detail in Chapter 2.06, evidence has not been found for such a component in OIBs. However, it is possible that, due to high ratios of noble gases to lithophile elements, the involvement of small amounts of material in OIB source regions imparts deep-mantle noble gas signatures but not those of other elements. Models that have attempted to describe the mantle domains with high $^3He/^4He$ ratios as being gas-poor (Graham *et al.*, 1990) have not been incorporated into a convincing model for global mantle noble gas evolution (see Porcelli and Ballentine, 2002).

### 4.11.4.4 Interacting Reservoirs

An important constraint that leads to a complete re-evaluation of the degassing models described above comes from consideration of fissiogenic $^{136}Xe$ and $^{129}Xe$ together (Ozima *et al.*, 1985). Coupled shifts in $^{136*Pu}Xe/^{130}Xe$ and $^{129}Xe/^{130}Xe$ are proportional to the $^{244}Pu/^{129}I$ ratio of the source reservoir. On a $^{129}Xe/^{130}Xe$ versus $^{136}Xe/^{130}Xe$ plot (Figure 4), the slope of the line from the values of nonradiogenic atmospheric xenon and through the composition of the present atmosphere (which has fissiogenic $^{136}$Xe largely from $^{244}$Pu, and radiogenic $^{129}$Xe from $^{129}$I) provides the $^{136*Pu}$Xe/$^{129*}$Xe ratio of the atmosphere, and so the $^{244}$Pu/$^{129}$I ratio of the source of the atmospheric xenon. If degassing occurred before these parent elements became extinct, then the source would have an increased $^{129}$I/$^{130}$Xe and $^{244}$Pu/$^{130}$Xe ratio, and so would be expected to have higher $^{136*Pu}$Xe/$^{130}$Xe and $^{129}$Xe/$^{130}$Xe ratios now than the atmosphere. Evidence from MORBs indicates that the upper mantle indeed has such higher ratios. Another feature of any xenon that has remained within this source reservoir is that it will have a higher $^{136*Pu}$Xe/$^{129*}$Xe ratio than that of air xenon; i.e., it will lie above the line that defining this ratio in Figure 4 (Ozima *et al.*, 1985). This can be most easily envisaged by considering that degassing occurred early and in a single event. At the time of xenon loss from the mantle to the atmosphere, the proportion of undecayed $^{244}$Pu will be greater than that of $^{129}$I (which has a much shorter half-life). Therefore, the remaining $^{244}$Pu and $^{129}$I will produce xenon with a higher $^{136*Pu}$Xe/$^{129*}$Xe ratio than is in the atmosphere. This is true regardless of whether or not degassing actually occurred as a single event. Note that mantle xenon will also have fissiogenic $^{136}$Xe from decay of $^{238}$U produced during later mantle evolution (see Section 4.11.2.5), and so to compare the present upper-mantle $^{136*}$Xe/$^{129*}$Xe ratio generated from short-lived parent nuclides to that of the atmosphere, the proportion of mantle $^{136*}$Xe that is plutonium-derived must be determined. As seen in Figure 7, MORB data do fall above the line through the atmosphere. However, any correction for additions from $^{238}$U will lower the $^{136}$Xe/$^{130}$Xe ratio, and unless essentially all of $^{136*}$Xe in MORB is $^{244}$Pu-derived, the corrected value for upper-mantle xenon will fall below the atmosphere line and so cannot be the residue left after atmospheric xenon degassing. The data for $CO_2$ well gases and MORB indicate that a large fraction of the $^{136*}$Xe is in fact from uranium (see Section 4.11.2.5), which is consistent with grow-in in the upper mantle due to the presently inferred $^{238}$U/$^{130}$Xe ratio (see Porcelli and Wasserburg, 1995a). Therefore, it appears inescapable that the xenon presently found in the upper mantle is not the residue from degassing of the atmosphere. Since the daughter xenon isotopes how found in the atmosphere must have been derived from the $^{244}$Pu and $^{129}$I that were in the upper mantle (see Section 4.11.3.2), the xenon now found there must have been introduced from a deeper reservoir after the atmosphere was removed.

The main uncertainty in the evaluation of xenon isotopes is the composition of terrestrial nonradiogenic xenon (see Section 4.11.2.5). However, the arguments presented above appear

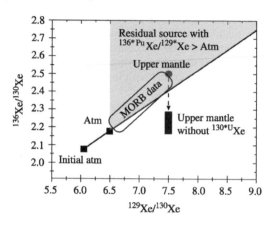

**Figure 7** The relationship between the xenon isotope compositions of the atmosphere, initial atmosphere, and the upper mantle as sampled by MORB. The line connecting the initial atmosphere (fractionated U–Xe) and the present atmosphere has a slope equal to the ratio of plutonium-derived $^{136}$Xe to radiogenic $^{129}$Xe ($^{136*}$Xe/$^{129*}$Xe) in the atmosphere. Any xenon that remained in the solid reservoir from where this was degassed must have a greater value for this ratio and so lie in the shaded region (Ozima *et al.*, 1985). While measured MORBs do so, upper-mantle compositions that have been corrected for $^{238}$U-derived $^{136}$Xe (based on MORB data of Kunz *et al.*, 1998 and on $CO_2$ well gas data of Phinney *et al.*, 1978) do not, indicating that upper-mantle xenon is not the residual left from atmosphere degassing.

to be robust when considering the possible compositions. As shown in Figure 8 and discussed in Chapter 4.12, solar wind xenon or U–Xe do not match the terrestrial light xenon isotope composition without extensive fractionation. Only U–Xe, when fractionated, provides a plausible precursor for the atmosphere (Figure 8). While the errors on this composition are small (Pepin, 2000), it is worth considering whether another composition is plausible. Due to the magnitude of the excess of $^{129}$Xe in the atmosphere (see Chapter 4.12), the amount of radiogenic $^{129}$Xe present has not been debated, so that the $^{129}$Xe/$^{130}$Xe ratio of nonradiogenic terrestrial xenon is relatively well fixed. As shown in Figure 8, the $CO_2$ well gas data can be residual of atmospheric degassing only if there is almost no plutonium-derived $^{136}$Xe in the atmosphere (i.e., the $^{136}$Xe/$^{130}$Xe ratios of the initial and present atmospheres are very similar). However, this would require a substantial downward revision of the amount of $^{244}$Pu that was present in Earth-forming materials that has been estimated completely independently and is consistent with the $^{40}$Ar budget of the Earth (Section 4.11.3.2). There appears to be no reason to make such a re-evaluation, and, as discussed in Section 4.11.4.5 below, there are mantle models that can satisfy these constraints.

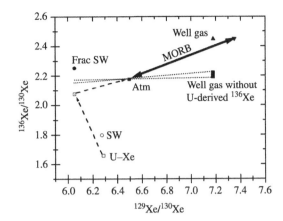

**Figure 8** The xenon isotope compositions of U–Xe and solar wind xenon, both nonfractionated and fractionated to match the light xenon isotopes, are compared to the value of the atmosphere. It is clear that fractionated solar wind xenon cannot serve as the nonradiogenic composition of the atmosphere since it has a higher $^{136}Xe/^{130}Xe$ ratio. In order for the upper mantle to have the same $^{136*}Xe/^{129*}Xe$ ratio as the atmosphere, the nonradiogenic composition of the atmosphere must lie on the dotted lines, implying that there is very little plutonium-derived $^{136}Xe$. This is contrary to the inferred $^{244}Pu$ budget of the Earth.

Another constraint on the origin of the xenon now found in the upper mantle is obtained from the composition of fissiogenic $^{136}Xe$. In a closed-system reservoir, plutonium-derived $^{136*}Xe$ will dominate over uranium-derived $^{136*}Xe$ (see Section 4.11.3.2). In a system that has been closed throughout solar system history and starting with a chondritic $^{244}Pu/^{238}U$ ratio of 0.0068, fission of $^{244}Pu$ will be 27 times that produced by $^{238}U$ (Porcelli and Wasserburg, 1995a). If xenon was lost over the first 100 Myr (about one half-life of $^{244}Pu$), plutogenic $^{136}Xe$ will still dominate. Therefore, xenon from such a reservoir that has $^{129}Xe$ excesses will have accompanying $^{136*}Xe$ that is dominantly plutonium-derived. MORB and $CO_2$ well gas data indicate that uranium-derived $^{136*}Xe$ dominates upper-mantle $^{136*}Xe$. This requires that xenon in this mantle reservoir has been in an environment with a much higher $^{238}U/^{130}Xe$ ratio after extinction of $^{244}Pu$, either due to a large decrease in Xe/U after the decay of $^{244}Pu$, or has received xenon from another reservoir. As discussed above, based on xenon isotope systematics the xenon in the upper mantle has clearly been introduced from the deep mantle, and therefore this suggests that the $^{136*Pu}Xe$ and $^{129*}Xe$ was generated in the deep mantle, and the upper mantle has served as the locale for significant late $^{136*U}Xe$ additions.

Another constraint on the relationship between atmospheric and mantle xenon is obtained by considering the nonradiogenic xenon isotopes. As discussed in Chapter 4.12, the nonradiogenic xenon isotopes presently in the atmosphere are highly fractionated with respect to the composition of xenon that is likely to have been initially trapped by the Earth. The fractionation most likely occurred during subsequent losses to space under conditions that were only present very early in Earth history. In this case, the degassing of nonradiogenic isotopes occurred early while such processes were still possible (see Pepin, 1991). In contrast, as discussed above, nonradiogenic xenon isotopes in the mantle are fractionated with respect to atmospheric xenon, with e.g., $^{128}Xe/^{130}Xe$ ratios that are intermediate between the solar and atmospheric ratio. This requires retention in the mantle of a component that was either only somewhat fractionated relative to solar xenon (due to fractionation processes before accretion by presently unknown processes), and therefore is less fractionated than atmospheric xenon, or more likely that is unfractionated solar xenon but has been mixed with subducted xenon. In the latter case, the xenon in the upper mantle is a mixture between xenon that has been stored in the deeper mantle since initial trapping with subducted xenon. Models of xenon evolution in the mantle therefore should have a return flux from the atmosphere, rather than simple unidirectional solid Earth degassing.

In sum, the xenon isotopes point to a more complicated mantle degassing history than simple unidirectional degassing into the atmosphere. It appears that first nonradiogenic noble gases were degassed to the atmosphere early. The solid reservoir volume from which these nonradiogenic gases were lost is unconstrained. This atmospheric xenon was fractionated by losses to space over $10^8$ yr, so that the atmosphere became enriched in the heavy isotopes. Radiogenic xenon was then degassed quantitatively from at least 40% of the mantle (based on the $^{136}Xe$–Pu budget). This occurred either by degassing of xenon directly from the mantle, or by transport of parent plutonium and iodine to the crust first, followed by crustal degassing. Xenon contained in another, deep-mantle reservoir and which is not residual from atmosphere removal was subsequently added to the upper mantle. The U/Xe ratio in the upper mantle was greater than that of the deeper reservoir, so that radiogenic $^{136*}Xe$ in the upper mantle became dominated by uranium-derived xenon.

An important implication of these considerations is that since the noble gases in the upper mantle are not the residual from atmosphere degassing, they provide no information regarding the degassing history of the atmosphere. Rather, the present composition of the upper mantle reflects interactions between different noble gas reservoirs. The disposition of these reservoirs, and the transfer of noble gases toward the surface, controls the overall degassing of the Earth, and these factors are considered further in the degassing models below. It should be emphasized that this is a crucial conclusion, since it renders invalid all calculations of degassing of the solid Earth based upon matching upper-mantle isotope compositions by simple degassing into the atmosphere. Rather, the relationship between mantle noble gas compositions and the degassing of the atmosphere is more complex.

### 4.11.4.5 Open-system Models

Another set of models still includes two mantle reservoirs, a MORB-source upper mantle and a deeper, gas-rich reservoir. However, these steady-state box models (Figure 6) are distinguished from the limited interaction box models discussed in Section 4.11.4.2 in being based upon the open interaction between the upper mantle and both the lower mantle (O'Nions and Oxburgh, 1983) and the atmosphere (Porcelli and Wasserburg, 1995a). Therefore, the constraints from xenon isotopes discussed above (see Section 4.11.4.4) are naturally accommodated. The steady-state model has been applied to helium isotope and heat fluxes (O'Nions and Oxburgh, 1983; Kellogg and Wasserburg, 1990), the U–Pb system (Galer and O'Nions, 1985), and the other noble gases (O'Nions and Tolstikhin, 1994; Porcelli and Wasserburg, 1995a,b). The central focus of the model is not degassing of the upper mantle to form the atmosphere, but rather mixing in the upper mantle. There are upper-mantle noble gas inputs by radiogenic production from decay of uranium, thorium, and potassium. In addition, atmospheric argon and xenon are subducted into the upper mantle, and lower-mantle noble gases are transported into the upper mantle within mass fluxes of upwelling material that carries all the deep-mantle noble gases together. The lower mantle is assumed to be an approximately closed system, in common with other mantle models (see Section 4.11.4.3). Noble gases from these sources comprise the outflows at mid-ocean ridges. The isotopic systematics of the different noble gases are linked by the assumption that transfer of noble gases from the upper mantle to the atmosphere by volcanism, as well as the transfer from the lower into the upper mantle by bulk mass flow, occurs without elemental fractionation. All model calculations assume that the upper mantle extends down to the 670 km discontinuity and so comprises 25% of the mantle. Plumes carrying lower-mantle material arise from the 670 km discontinuity.

It has been assumed that upper-mantle concentrations and isotopic compositions are in steady state, so that the inflows and outflows are equal. Since the main outflow is at mid-ocean ridges, this flux can be used to determine the upper-mantle residence times. The upper mantle is degassed to the atmosphere according to a rate constant determined by the rate of melting at ridges, and this then fixes upper-mantle residence times. The calculated values are short ($\sim$1.4 Ga) and imply that nonradiogenic primordial noble gases that cannot be supplied by subduction (including solar helium and neon) are provided from the lower mantle. An important implication of this conclusion is that the composition of the upper mantle does not carry any information regarding earlier mantle volatile history, nor how steady state conditions were reached. In this case, an extended degassing history of the mantle that includes the much earlier rigorous degassing that generated most of the atmosphere, cannot be inferred from noble gases presently found in the upper mantle.

In the open-system models, primordial non-radiogenic noble gases presently seen in the upper mantle originate in the lower mantle, and their fluxes to the atmosphere are moderated by the rates of transfer from the source region and through the upper mantle. As discussed above, the lower-mantle source region is assumed to be a closed system, and so has bulk Earth concentrations of radiogenic isotopes. However, since the lower-mantle argon and xenon isotope compositions, as represented by OIB noble gases, cannot be well constrained by available data, the nonradiogenic argon and xenon isotope concentrations are also not known. However, these can be calculated in the model from the balance of fluxes into the upper mantle. The MORB $^3$He/$^4$He ratio is a result of mixing between lower-mantle helium and production of $^4$He in the upper mantle (O'Nions and Oxburgh, 1983). Since the flux of the latter is fixed, the rate of helium transfer from the lower mantle can be calculated (Kellogg and Wasserburg, 1990), and this defines the degassing rate of the lower mantle. Neon, which follows helium, is transferred similarly, with decreases in $^3$He/$^4$He ratios accompanied by increases in $^{21}$Ne/$^{22}$Ne ratios due to coupled production of $^4$He and $^{21}$Ne. The MORB $^{40}$Ar/$^{36}$Ar ratio (when corrected for atmospheric contamination of basalts) is a mixture of lower-mantle argon (obtained from the coupling with helium fluxes), radiogenic $^{40*}$Ar produced in the upper mantle (calculated from the upper-mantle potassium concentration), and subducted air argon. When there is no subduction, the lower mantle is calculated to have ($^{40}$Ar/$^{36}$Ar) = 9,700. The MORB $^{136}$Xe/$^{130}$Xe ratio (when corrected for air contamination) is the result of mixing between lower-mantle xenon, fissiogenic $^{136*}$U Xe produced in the upper mantle, and subducted air xenon

(Porcelli and Wasserburg, 1995a). The lower-mantle ratios are established early in Earth history by decay of $^{129}I$ and $^{244}Pu$, and the $^{129}Xe/^{130}Xe$ ratio is constrained to be at least as great as that in the upper mantle. Once lower-mantle xenon is transported into the upper mantle, it is augmented by U-derived $^{136*U}Xe$. Subduction of atmospheric xenon is also possible, lowering the $^{136}Xe/^{130}Xe$ and $^{129}Xe/^{130}Xe$ ratios. Therefore, the model is consistent with the isotopic evidence that upper-mantle xenon does not have a simple direct relationship to atmospheric xenon.

The lower mantle has a $^{40}Ar/^{36}Ar$ ratio that is much higher than the atmospheric value. This can be compared to the reservoirs that supplied the atmosphere, which includes all late-accreting sources and the uppermost fraction of the mantle that has degassed $^{36}Ar$ early and $^{40}Ar$ subsequently. The higher $^{40}Ar/^{36}Ar$ ratio of the lower mantle implies the $^{36}Ar/K$ ratio of the lower mantle is much lower than the bulk ratio of the reservoirs that supplied the atmosphere. Assuming potassium was initially uniformly distributed throughout the BSE, this implies very heterogeneous incorporation of $^{36}Ar$. This contrasts with the typical *a priori* assumption of an initially uniform distribution of $^{36}Ar$. The $^{244}Pu$-derived $^{136}Xe$ and radiogenic $^{129}Xe$ in the upper mantle are derived from the lower mantle. This lower-mantle compositions, obtained by subtracting from MORB xenon the $^{238}U$-derived $^{136}Xe$ produced in the upper mantle, corresponds to closure times that are similar to that of the atmosphere, indicating that early losses occurred from the deep mantle as well (Porcelli *et al.*, 2001). These losses must have been prior to the assumed closed-system evolution.

One feature of the model is that the atmosphere generally has no *a priori* connection with other reservoirs. It simply serves as a source for subducted gases, and no assumptions are made about its origin. Since no assumption is made about the initial distribution of noble gases in the Earth, the atmospheric abundances cannot be used to derive lower-mantle concentrations. The daughter isotopes that are presently found in the atmosphere clearly were originally degassed from the upper mantle, possibly along with nonradiogenic nuclides. However, this occurred before the present character of the upper mantle was established. Therefore, the upper mantle no longer contains information regarding atmosphere formation. Further discussion of this model, and the implications for the sources of mantle-derived volcanics, is provided in Chapter 2.06.

## 4.11.4.6 Boundaries within the Mantle

The principal objection has been based on geophysical arguments advanced for greater mass exchange with the lower mantle below 670 km, and so the difficulty of maintaining a distinctive deep-mantle reservoir. The clearest indication of the scale of mantle convection comes from seismic tomography imaging of subducting slabs and mantle plumes that cross the 670 km discontinuity (e.g., Creager and Jordan, 1986; Grand, 1987, 1994; van der Hilst *et al.*, 1997). Plumes may arise from boundary layer instabilities within the mantle or at the core–mantle boundary. There is growing evidence that plumes such as those that form the Hawaiian and Icelandic hotspots have their origin at the core–mantle boundary rather than at a depth of 670 km. Ultrahigh seismic velocity zones at the boundary beneath these islands have been described as plume-induced (Helmberger *et al.*, 1998; Russell *et al.*, 1998). Recent images of the Iceland plume are compatible with a deep origin for this plume (Shen *et al.*, 1998), and a mantle structure extending to the core (Bijwaard and Spakman, 1999). A plume extending to a depth of at least 2,000 km also has been imaged beneath central Europe (Goes *et al.*, 1999), and hotspots in Hawaii, Iceland, South Pacific, and East Africa have been shown to be located above slow anomalies in the lower mantle that extend to the core (Zhao, 2001).

Although geophysical observations point to the present-day mantle convecting as a single layer, it has been argued (Allègre, 1997) that models requiring long-term mantle layering can be reconciled with geophysical observations for present-day whole mantle convection if the mode of mantle convection changed less than 1 Ga ago from layered to whole mantle convection. In this case, the mode of volatile transfer from the deep mantle, and so the rate of deep-mantle degassing, has changed dramatically. However, this idea is at odds with the thermal history of the mantle (see Chapter 2.06).

Kellogg *et al.* (1999) developed a model in which mantle below $\sim$1,700 km has a composition, and so density, that is sufficiently different from that of the shallower mantle to largely avoid being entrained and homogenized in the overlying convecting mantle. The boundary with the overlying mantle has a variable depth, and is much deeper where there is geophysical evidence for deeper slab penetration or plumes arising from the core–mantle boundary. This model preserves a region in the mantle behind which the radioelements and primitive noble gases can be preserved, while accommodating many geophysical observations. There is no geophysical evidence for such a boundary, although whether it could elude seismic detection is debated (Kellogg *et al.*, 1999; Vidale *et al.*, 2001). Also, if the overlying mantle has the composition of the MORB source, then the abyssal layer must contain a large proportion of the heat-producing elements, and it is not yet clear what this effect would have on the thermal stability of the layer or temperature contrast with the overlying mantle. Coltice and Ricard (1999) suggested an alternative model in which helium with high $^3$He/$^4$He ratios are stored in the narrow zone around the core that is composed of subducted material. This would also be consistent with the geophysical evidence for whole mantle convection, and provide a source for all primordial noble gases. However, it does not provide a reservoir for the undegassed radionuclides (Section 4.11.3), and it is unclear how such a reservoir would have high initial $^3$He concentrations (see Porcelli and Ballentine, 2002). Other reservoir configurations have also been suggested for reservoirs containing high $^3$He/$^4$He ratios, such as in small domains embedded throughout the mantle, but have not been incorporated into comprehensive models that explain all of the noble gas and geophysical observations (see Chapter 2.06; Porcelli and Ballentine, 2002).

The core has also been considered as a source of noble gases. During formation of the Earth, the core may have incorporated sufficient quantities of solar helium to now supply ocean islands with high $^3$He/$^4$He ratios (Porcelli and Halliday, 2001). In this case, such ratios trace the interaction between the core and mantle (see Chapter 2.06). However, the core cannot be used to explain the radiogenic isotope budgets (see Section 4.11.3) nor the heat–$^4$He budget, and so must be coupled with a mantle model for these features.

### 4.11.4.7 Summary

Early degassing models have described the formation of the atmosphere by progressive degassing of the upper mantle, leaving the presently observed noble gases behind to generate strongly radiogenic compositions. However, it is now clear that upper-mantle noble gases are not residual from atmosphere degassing, and have been introduced into the upper mantle from a deeper reservoir after quantitative removal of atmospheric constituents. Nonetheless, the conclusions regarding strong early degassing are compatible with the events of early Earth history. A complete model of noble gas evolution in the mantle and atmosphere remains to be constructed, although elements of earlier models may survive. It might be possible to reformulate the open-system models for a larger upper mantle, or greater mass fluxes between reservoirs, although in some cases non-steady-state upper-mantle concentrations may be required. Unfortunately, the particular configuration of the reservoirs

will remain speculative until the geophysical constraints are fully clarified.

A general outline of the history of mantle degassing can be constructed from the observations and constraints discussed above (Porcelli *et al.*, 2003):

(i) Noble gases are acquired throughout the mantle during formation of the Earth, and a substantial inventory is retained within the growing planet, although concentrations in the deeper mantle are lower.

(ii) Radiogenic nuclides produced in the first $10^8$ a are degassed from throughout the entire mantle, while isotopic differences are generated due to lower $^{130}$Xe concentrations in the deep mantle.

(iii) Strong fractionation of xenon and neon occurred during losses to space. The most plausible mechanism yet formulated involves hydrodynamic escape. Xenon is depleted by at least $10^2$. This sets the currently observed fractionation of nonradiogenic xenon in the atmosphere. This may have overlapped with the degassing of the mantle. Nonfractionating losses by atmospheric erosion may have also occurred. All daughter noble gas isotopes that had been degassed to that time were lost to space.

(iv) After losses to space has terminated at ~100 Ma, the radiogenic and fissiogenic xenon produced by the $^{244}$Pu and $^{129}$I remaining in the upper 40% of the Earth is degassed into the atmosphere, either directly from the mantle, or by transfer of the parent elements into any early crust followed by crustal degassing. This was not accompanied by a significant amount of nonradiogenic (and nonfractionated) xenon.

(v) 40% of the mantle loses $^{40}$Ar to the atmosphere and potassium to the crust, from where any further $^{40}$Ar produced is largely degassed.

(vi) Gases in 60% of the mantle are largely isolated from the atmosphere. However, a small amount of noble gases has leaked into the upper mantle, where it has mixed with radiogenic nuclides produced there, to generate the presently observed MORB compositions.

The composition of the upper mantle clearly does not provide information on the degassing history of the mantle, but rather reflects the small fluxes into this reservoir from the deep mantle, from production within the upper mantle, and possibly from subduction. The transfer of $^{40}$Ar from the crust throughout much of Earth history has been dictated by the degassing of the crust, where the potassium has been stored.

### 4.11.5 DEGASSING OF THE CRUST

The processes involved in the formation of the crust are likely to release any volatiles into the atmosphere that were present in the mantle source region, and so the crust is not expected to be a significant reservoir of primordial volatiles. However, gaseous radiogenic nuclides that are produced within the crust do not readily escape. The most extensively studied has been $^{40}$Ar.

### 4.11.5.1 Crustal Potassium and $^{40}$Ar Budget

There have been a number of studies examining the total potassium content of the continental crust, which is a key parameter for the average petrologic composition of the crust. Estimates of $K_2O$ in the crust, which has a total mass of $2.05 \times 10^{25}$ g (Cogley, 1984) range from 0.96% to 2.4% (e.g., Weaver and Tarney, 1984; Taylor and McLennan, 1985; Rudnick and Fountain, 1995; Wedepohl, 1995). Taylor and McLennan (1985) and McLennan and Taylor (1996) have emphasized that the extensive data for heat flow can be used to constrain the total abundances of the heat-producing elements. For an average crustal heat flux away from recently disturbed regions of $\sim$48 mW m$^{-2}$ (Nyblade and Pollack, 1993) and a reduced heat flux (i.e., heat conducted through the crust from the mantle) of 27 mW m$^{-2}$ (Morgan, 1984), and assuming that the heat-producing elements are in the ratios Th/U $= 3.8 \times 10^4$ and K/U $= 1.27 \times 10^4$, a value of 0.96% $K_2O$ was obtained (McLennan and Taylor, 1996). The corresponding total crustal budget is therefore $1.6 \times 10^{23}$ g K, and this produces $1.7 \times 10^{31}$ atoms $^{40}$Ar a$^{-1}$. Using the values of 21 ppb U and K/U $= 1.27 \times 10^4$ for the bulk silicate Earth, 15% of the total terrestrial potassium is in the crust. Further consideration of the mantle contribution to crustal heat flow has suggested that models with $K_2O$ up to $\sim$2% may be possible (Rudnick *et al.*, 1998), so that $\sim$30% of the terrestrial potassium may be in the crust, and producing twice as much $^{40}$Ar.

As discussed above, the K/U ratio was obtained for the MORB source, and is often taken to represent the bulk Earth. This can be applied to the crust only by assuming that there has been no fractionation of potassium and uranium during crust formation, and no preferential recycling of either element. It should be noted that using this ratio, $\sim$85% of the heat currently produced in the crust is generated by uranium and thorium (see, e.g., Turcotte and Schubert, 1982), and so heat flow is a good constraint for the total crustal uranium and thorium budgets, but a poor constraint for the potassium budget, unless the K/U ratio is known very accurately. Therefore, it is possible that the K/U ratio of the crust is somewhat different from the MORB ratio largely due to a different potassium content. In this case, the composition of the deeper mantle (not represented by MORB) may be different.

The amount of $^{40}$Ar that is in the crust can be calculated from the mean K–$^{40}$Ar age of the continental crust of 960 Ga (Hurley and Rand, 1969), which corresponds to a $^{40}$Ar/K ratio of $1.6 \times 10^{17}$ atoms $^{40}$Ar g$^{-1}$ K. This is 3–5% of the amount of $^{40}$Ar in the atmosphere.

The composition of the crust provides some interesting constraints on the composition of the mantle. If the upper mantle that supplies MORB has $\sim$50 ppm K (e.g., Hofmann, 1988), and extends down to a depth of 670 km (about one-quarter of the mantle), then it has only a small fraction of the total terrestrial potassium. Therefore, the 70–85% of the terrestrial potassium (for a bulk silicate Earth K/U $= 1.27 \times 10^4$) is in the deeper mantle. Since $\sim$40% of the terrestrial $^{40}$Ar is in the atmosphere, some of this deep-mantle material that is not depleted in potassium must have degassed. It has been argued that the upper mantle has a higher potassium concentration of $\sim$100 ppm (Korenaga and Kelemen, 2000), so that $\sim$40% of the terrestrial potassium is in the crust and upper mantle and can supply the atmosphere. In this case, there has been net transfer of potassium to the upper 25% of the Earth. Clearly, how much of the Earth has been degassed and depleted requires further constraints on the potassium budget of both the upper mantle and crust. Coltice *et al.* (2000) pointed out that more potassium is needed to supply the atmosphere than is in the crust, and since potassium and argon are both highly incompatible and so extracted together from the mantle, the additional potassium needed has been subducted.

### 4.11.5.2 Formation Time of the Crust

There has been considerable discussion about the age of the crust. Various growth curves have been proposed, from slow steady growth based on K–Ar ages (Hurley and Rand, 1969) to rapid early growth (see Armstrong, 1991), and various intermediate histories (see discussion in Taylor and McLennan, 1985). The current rate of crustal formation is $\sim$1 km$^3$ yr$^{-1}$, and so is insufficient to generate the full mass of the crust over 4.5 Ga. Crustal formation was clearly much more rapid in the past, and the general consensus is that the average crustal formation age is substantially greater than the K–$^{40}$Ar age. Therefore, it appears that $^{40}$Ar has been progressively degassed from the crust.

Considering that potassium now in the crust accounts for $\sim$30–70% of the $^{40}$Ar in atmosphere, crustal degassing may be an important component of the solid Earth degassing history (see Hamano and Ozima, 1978). If the crust formed very early, then release of $\sim$30–70% of the $^{40}$Ar in the atmosphere has been controlled by crustal processing. For slower crustal growth histories, the proportion degassed directly from the crust would depend upon the time constant for growth compared to the $1.4 \times 10^9$ a half-life of $^{40}K$.

### 4.11.5.3 Present Degassing

The present rate of continental degassing of $^{40}Ar$ into the atmosphere cannot be readily measured, and so must be inferred by considering the mechanisms of degassing. Studies of radiogenic $^4He$ in groundwaters from the Great Artesian Basin in Australia (Torgersen and Clarke, 1985; Torgersen and Ivey, 1985) concluded that the higher $^4He$ concentrations in older waters could be explained by a steady influx from production in the underlying crust. The required calculated flux was found to equal the entire production rate in the crust, and so suggested that the continental crust degasses by continuous release of all radiogenic $^4He$ from uranium- and thorium-bearing host rocks. A similar conclusion was drawn from data from another basin as well (Heaton, 1984). Additional work on $^{40}Ar$ in the Great Artesian Basin extended this conclusion to $^{40}Ar$ (Torgersen et al., 1989), implying that the continental flux to the atmosphere is $(1.7–3.3) \times 10^{31}$ atoms $^{40}Ar$ a$^{-1}$. However, there have been doubts that helium and argon can be so readily released from minerals and transported effectively across the crust. These doubts have generated discussions regarding the interpretations of groundwater flow rates and so noble gas accumulation rates (e.g., Mazor, 1995; Bethke et al., 1999).

An important factor is how radiogenic $^{40}Ar$ is released from host rocks. The release of $^{40}Ar$ from potassium-bearing minerals by diffusion has been extensively examined as part of thermal evolution studies, and the blocking temperatures of major minerals such as feldspars and biotites correspond to quantitative retention within the top ~10 km of the crust (McDougall and Harrison, 1999). The upper continental crust, down to ~10 km, is estimated to have 3.4% $K_2O$ and 2.8 ppm U (Taylor and McLennan, 1985). This is consistent with the observation that the concentration of heat-producing elements decreases with depth (e.g., Jaupart et al., 1981), although potassium may not decrease with depth as strongly as uranium and thorium (e.g., Rudnick and Fountain, 1995). Nonetheless, a dominant fraction of the potassium in the crust is below the blocking temperature of $^{40}Ar$. Further, the $^{40}Ar$ produced in the deeper crust is more likely to be transported episodically to the surface during mobilization of fluids that may also be responsible for metamorphic processes (e.g., Etheridge et al., 1984).

A process that clearly leads to widespread release of $^{40}Ar$ is weathering at the surface. In a comprehensive survey of river chemistry, Martin and Meybeck (1979) reported an average concentration of 1.4 ppm K in a total water discharge of $3.74 \times 10^{16}$ L yr$^{-1}$, leading to a total of $5.2 \times 10^{13}$ g K. This potassium is likely to be released largely from the weathering of feldspars (e.g., Wollast and Mackenzie, 1983). Assuming that this K is derived from weathering of crystalline rocks with a mean age of 960 Ma (Hurley and Rand, 1969) with $1.6 \times 10^{17}$ atoms $^{40}Ar$ g$^{-1}$ of potassium, this leads to release of $0.83 \times 10^{31}$ atoms a$^{-1}$. Mechanical weathering of the continental crust leads to discharge to the oceans of $1.6 \times 10^{16}$ g a$^{-1}$ (Milliman and Meade, 1983), which contains an average concentration of 2% K (Martin and Meybeck, 1979) or $3 \times 10^{14}$ g K. If this material is derived from cannibalization of earlier sediments (see, e.g., Veizer and Jansen, 1985) with an average age of ~500 Ma (with $0.6 \times 10^{17}$ atoms $^{40}Ar$ g$^{-1}$ of potassium), then this material contained $1.8 \times 10^{31}$ atoms $^{40}Ar$ when mobilized, which is twice that released during dissolution of potassium. Of course, the average K–Ar ages of materials that provide either detrital or dissolved constituents remains highly uncertain. The long-term rate of mechanical weathering may differ considerably from this because of recent changes in discharge due to human activities and due to the deposition within drainage basins. However, a much greater uncertainty lies in estimating how much of this material has lost $^{40}Ar$. Limited studies have found that the K–Ar ages of surface sediments often reflect those of the source rock, even where significant clay mineral formation has occurred (e.g., Hurley et al., 1961). Hurley (1966) summarized available data and suggested that detrital ages are gradually wiped out during burial. Later studies in the Gulf of Mexico confirmed that K–Ar ages decreased with depth at rates that depend upon mineral size fractions (Weaver and Wampler, 1970; Aronson and Hower, 1976), and likely reflect the redistribution of potassium, and loss of $^{40}Ar$, during diagenetic illite formation. Overall, it is clear that some poorly constrained proportion of the $^{40}Ar$ contained in sediments is released. Perhaps as much $^{40}Ar$ is released by diagenesis of sediments as derived from primary chemical weathering.

Another mechanism for degassing $^{40}Ar$ is thermal processing due to tectonic thickening of the crust in orogenic belts (see, e.g., England and Thompson, 1984). Veizer and Jansen (1979, 1985) used the distribution of basement K–Ar ages to model the thermal cycling rate of the crust, and obtained a rate of $\sim 2 \times 10^{16}$ g a$^{-1}$. It is possible that during such events, $^{40}Ar$ in the lower crust is also transported to the surface, and so degassing of the entire crustal thickness may occur together. For an average crustal composition with 1% K and $1.6 \times 10^{17}$ atoms $^{40}Ar\,g^{-1}$ of potassium (assuming an average K/Ar age of the crust of 960 Ma), $3.2 \times 10^{31}$ atoms $^{40}Ar\,a^{-1}$ are released. Given the range of possible crustal potassium contents discussed above, this value may be as much as 2 times greater.

In summary, chemical weathering and orogenic processing of the crust generate a flux of $\sim(3.8-7) \times 10^{31}$ atoms $^{40}Ar\,a^{-1}$ to the atmosphere, and this may be augmented by a presently poorly constrained flux associated with potassium in riverine sediments. Nonetheless, thermal processing of the continental crust appears to dominate the $^{40}Ar$ flux. The total flux is more than the present-day crustal production rate of $1.7-3.4 \times 10^{31}$ atoms $^{40}Ar\,a^{-1}$. If the flux does indeed exceed the present production rate, then the $^{40}Ar$ reservoir size of the continents is diminishing.

The present flux of $^{40}Ar$ from mid-ocean ridges is $1.9 \times 10^{31}$ atoms $a^{-1}$ (Section 4.11.2.7), and so substantially lower than the continental flux, indicating that at present continental processing is more important than mantle processing for $^{40}Ar$ release to the atmosphere.

## 4.11.6 MAJOR VOLATILE CYCLES

Many of the important characteristics of the atmosphere are related to the concentrations of particular molecular species, such as $O_2$. Here the absolute abundances of carbon, nitrogen, and hydrogen (as $H_2O$) as supplied from the mantle are considered in the context of models of noble gas degassing from the solid Earth. A convenient reference for comparing surface volatile inventories to mantle reservoirs is obtained by dividing the surface volatiles into the mass of the upper mantle, the minimum size of the source reservoir. However, this should not be taken to imply a particular model of degassing.

### 4.11.6.1 Carbon

*Crustal carbon.* In contrast to the noble gases, carbon near the planetary surface is concentrated in crustal rocks, and is largely divided between carbonates, with an isotopic composition of $\delta^{13}C = 0\%o$ and sedimentary organic carbon with $\delta^{13}C = -25\%o$. While Hoefs (1969) estimated a total of $2.6 \times 10^{22}$ g of C, with 66% in sedimentary rocks, Hunt (1972) reassessed the budget using sedimentary rock abundance data from Ronov and Yaroshevsky (1969) and carbon concentration data compilations to obtain a higher value of $9 \times 10^{22}$ g. Ronov and Yaroshevsky (1976) updated their earlier work with new additional data to obtain a

total budget of $1.2 \times 10^{23}$ g. A similar value ($6 \times 10^{22}$ g C) was obtained by Li (1972) when requiring that the crust have an isotopic composition similar to that of MORB (see below) of approximately $\delta^{13}C = -5$; conversely, this agreement supports the notion that the crust and mantle are isotopically similar. Therefore, considering the uncertainties involved, a value of $1.0 \times 10^{23}$ g C, or $8.3 \times 10^{21}$ mol, with perhaps a 20% uncertainty, appears to be reasonable (Table 3).

*Carbon in MORB and the upper mantle.* MORB entering the ocean crust are generally oversaturated in carbon as $CO_2$, and exhibit exsolved $CO_2$-rich vesicles. Carbon contents of MORB samples vary widely due to degassing, but a ratio of $C/^3He = 2 \times 10^9$ appears to characterize the undegassed magmas (Marty and Jambon, 1987). Combined with the initial concentration of $^3He$ obtained from the total $^3He$ ridge flux (see Section 4.11.2.7), this provides a value of 510 ppm carbon in undegassed MORB, within the range of other estimates of undegassed MORB of 900–1,800 ppm carbon (Holloway, 1998) and 400 ppm (Pineau and Javoy, 1994). Assuming carbon is incompatible, this corresponds to an upper mantle (to a depth of 670 km) concentration of 50 ppm. Therefore, the upper mantle of $1 \times 10^{27}$ g contains about half the carbon that is present in the crust. If this value applies to the entire mantle, then only about a third of the terrestrial carbon is at the surface. The carbon isotope composition of the upper mantle is $\delta^{13}C = -5\%o$ based on MORB data (e.g., Javoy and Pineau, 1991), and is similar to that of the surface. The carbon brought to the surface by MORB is quantitatively degassed during eruption and subsequent alteration, and results in a flux of $2 \times 10^{12}$ mol C $yr^{-1}$.

There may also be a significant amount of carbon in the core, with values of up to 4% possible, although this depends upon the amount available in the early Earth (Wood, 1993; Halliday and Porcelli, 2001). The amount of carbon that is supplied across the core–mantle boundary into the mantle is not known.

*Carbon at hotspots.* Ocean islands with high $^3He/^4He$ ratios clearly contain distinctive volatile components. However, there are also some data regarding the carbon flux of hotspot sources. The ratio of hotspot volcanics is an important parameter in further defining the carbon cycle. It has been argued that the $C/^3He$ ratio is equal to that of the upper mantle (Trull *et al.*, 1993). Alternatively, Poreda *et al.* (1992) argued that the ratio is higher and is $6 \times 10^9$ in Iceland, while Kingsley and Schilling (1995) argued from mixing relationships along the Mid-Atlantic Ridge that the Iceland source was $\sim 10^{10}$. Hilton *et al.* (1998a) found values of $(2-5) \times 10^9$ for

Loihi, but also argued that degassing highly modified the ratio, and concluded that the source ratio was unknown. It should be noted that plumes appear to contain contributions not only from a high $^3He/^4He$ source region, but also subducted components (e.g., Sobolev *et al.*, 2000) that may contribute carbon that increases the $C/^3He$ ratio and mask the carbon accompanying helium with high $^3He/^4He$ ratios. Measured $\delta^{13}C$ values scatter around upper-mantle values (Trull *et al.*, 1993), and a consistent composition distinctive from that of MORB has not been unambiguously documented. The flux of carbon from ocean islands depends on the accepted $C/^3He$ as well as the flux of $^3He$ relative to MORB $^3He$ flux (see above).

*Subduction.* There is a significant amount of carbon on the ocean floor that is available for subduction. The amount of carbon released by arc volcanism is $(1.6–3.1) \times 10^{12}$ mol a$^{-1}$ (see Chapter 2.06). This can be compared with estimates of the amount of subducted carbon of $(0.35–2.3) \times 10^{13}$ mol yr$^{-1}$ (Bebout, 1995; Hilton *et al.*, 2002). Therefore, while it is possible that there is no carbon return flux into the upper mantle beyond island arcs, the ranges suggest that some carbon does survive in the downgoing slabs, and this flux may even be greater than the degassing flux observed at mid-ocean ridges of $0.2 \times 10^{13}$ mol a$^{-1}$. Volcanic arcs have a range of $\sim \delta^{13}C = -2\permil$ to $-12\permil$ (Bebout, 1995), reflecting the variable proportions of subducted components.

*The global carbon cycle.* The flux of carbon brought to the surface by MORB of $2 \times 10^{12}$ molC yr$^{-1}$ can supply the entire crustal carbon inventory in 4.2 Ga; if there was stronger earlier degassing, there must have been a carbon return flux to the mantle. Much of the surface inventory indeed may have been degassed early along with noble gases (Section 4.11.4.1). It is possible that the surface inventory was initially greater than at present due to the difficulties of subducting carbon in hotter, younger slabs (Des Marais, 1985; McCulloch, 1993; Zhang and Zindler, 1993), although this effect may have been countered by increased incorporation of carbon into ocean crust by hydrothermal alteration (Sleep and Zahnle, 2001). If the carbon in the upper mantle is largely recycled, it may now be present in a steady-state abundance. This would naturally explain the high upper-mantle carbon concentration. However, since surface carbon is divided at the surface between carbonate carbon (with $\delta^{13}C = 0\permil$) and organic carbon (with $\delta^{13}C \sim -25\permil$), the subducted carbon at any location need not have the average crustal $\delta^{13}C$. Therefore, in order to reconcile the isotopic similarity between the mantle and surface, carbonate and organic carbon must be subducted in the same proportions as at the surface. Otherwise, shifts in the composition away from that of the bulk Earth will occur in both the crust and mantle. There are no obvious controls that would demand this condition. The isotopic compositions of carbon in back arc basin basalts scatter, and can be $\sim 5\permil$ lighter than that of MORB carbon (Mattey *et al.*, 1984), reflecting mixing of organic carbon with MORB carbon.

An alternative view that naturally explains the isotopic similarity between the crust and upper mantle is that the carbon degassing from the mantle is primordial carbon that was trapped during solid Earth formation, representing the continuing unidirectional upward transfer of carbon. In this case, any carbon that is subducted does not constitute a dominant fraction of carbon in the upper mantle. However, carbon is not depleted in the upper mantle to the same extent as other highly incompatible elements such as the noble gases. An indication of this is the mantle $C/^{36}Ar$ ratio, which is $10^2$ times greater than that of the surface (Marty and Jambon, 1987). This observation has been explained by having less net degassing of carbon early in Earth history, either due to higher recycling rates or more compatible behavior under more reducing conditions (Marty and Jambon, 1987). As discussed above (Section 4.11.4), the nonradiogenic noble gases may be supplied from a deeper mantle reservoir to the MORB source reservoir; if carbon is similarly supplied, then using the calculated $^3He$ content of a gas-rich reservoir (see Section 4.11.4.3) and MORB $^{36}Ar/^3He$ (see Section 4.11.2.6) and $^{36}Ar/C$ ratios as reflecting the deeper mantle, a carbon concentration of 3,000 ppm is obtained. This value is an order of magnitude less than values for carbonaceous chondrites (Kerridge, 1985), and would be even lower if some of the upper-mantle $^{36}Ar$ is from subduction.

### 4.11.6.2 Nitrogen

*Nitrogen at the surface.* Nitrogen is highly atmophile, although there is a significant fraction of surface nitrogen in the continental crust (see Table 3). Crustal nitrogen is isotopically heavier than atmospheric nitrogen, and the total nitrogen at the surface has $\delta^{15}N = 2\permil$ relative to the atmospheric value of $0\permil$ (Marty and Humbert, 1997). The total budget of $4.5 \times 10^{21}$ g, when divided into the upper mantle, yields 4.5 ppm for the bulk silicate Earth.

*Nitrogen in MORB and the upper mantle.* The concentration of $N_2$ in undegassed MORB can be determined from the value of $N_2/^{40}Ar = 120 \pm 20$ obtained from MORB data (Marty and Humbert, 1997; Marty and Zimmermann, 1999) to be 1.63 ppm. This corresponds to an upper-mantle concentration of 0.16 ppm, assuming nitrogen is incompatible during melting (Marty and Humbert, 1997). The amount of nitrogen in the upper mantle is therefore $1.6 \times 10^{20}$ g, which is only ~3% that in the crust and atmosphere. The MORB flux is equivalent to a flux of $5.0 \times 10^{10}$ mol yr$^{-1}$, or 9% of the surface $N_2$ over 4.5 Ga. Note that Javoy (1998) argued that nitrogen is relatively compatible, with an upper-mantle concentration of up to 40 ppm (Cartigny *et al.*, 2001). However, this is from mass balance calculations based upon model assumptions that volatiles in the upper mantle and on the surface are a mixture of enstatite chondrites and a late veneer of CI chondrites, and such a model has not been widely adopted.

The nitrogen isotopic composition of the upper mantle has been estimated from MORB (Marty and Humbert, 1997; Marty and Zimmermann, 1999) and diamond (Javoy *et al.*, 1984; Cartigney *et al.*, 1998; Boyd and Pillinger, 1994) data to be approximately $\delta^{15}N = -4\%_o$. Values as low as $\delta^{15}N = -25\%_o$ have been found in diamonds, indicating that there is another reservoir in the mantle.

*Nitrogen at hotspots.* $\delta^{15}N$ values from hotspots has been found to be largely positive and as high as $\delta^{15}N = +8\%_o$ as (Dauphas and Marty, 1999; Marty and Dauphas, 2003). It has been argued that these values are due to the presence of subducted components in mantle plumes (Marty and Dauphas, 2003), and is supported by the lack of correlation with helium isotope compositions.

*Subduction.* There are some data for the subduction of nitrogen. Metasedimentary rocks and organics in pelagic sediments are isotopically heavy, with organic nitrogen having $\delta^{15}N = +2\%_o$ to $+10\%_o$ in marine sediments (Peters *et al.*, 1978) and $\delta^{15}N = +2\%_o$ to $+15\%_o$ in metasediments (Haendel *et al.*, 1986; Bebout and Fogel, 1992). Metamorphosed complexes are also isotoipcally heavy (Bebout, 1995). Arc volcanics have an average of $\delta^{15}N = +7\%_o$ (Sano *et al.*, 1998). The subduction flux has been estimated from an average sediment subduction rate of $3.5 \times 10^{15}$ g yr$^{-1}$ (Ito *et al.*, 1983) and an average concentration of 125–600 ppm from the Catalina Schist to be $(0.44–2.1) \times 10^{12}$ g yr$^{-1}$ (Bebout, 1995), although N in altered seafloor basalts may also carry nitrogen (Hall, 1989). The flux to the atmosphere due to arc volcanism is $4.5 \times 10^{11}$ g a$^{-1}$ (Hilton *et al.*, 2002). Therefore, it is possible that no nitrogen is returned to the mantle, although a flux equal to that at mid-ocean ridges (and so supporting steady state mantle and surface abundances), or higher (supporting arguments for the net inflow of nitrogen (Javoy, 1998) is also possible.

*The global nitrogen cycle.* A salient feature of the nitrogen cycle is that mantle nitrogen is isotopically lighter than both the surface and subducted nitrogen. The atmosphere is either derived from another component than the mantle or has suffered preferential loss of $^{14}N$. The atmospheric composition therefore is inherited from the processes of Earth formation (see Chapter 4.12). Subsequently, the rates of mantle–crust exchange are constrained by these isotopic differences. There are limited possibilities for the source of isotopically light nitrogen in the mantle (see Chapter 4.12). Data for carbonaceous and ordinary chondrites typically show $\delta^{15}N > 0\%_o$ (Kerridge, 1985). However, E chondrites have been found to have consistently lower values, with an average of about $\delta^{15}N < -35\%_o$, and with concentrations on the order of 500 ppm (Grady *et al.*, 1986). Therefore, it has been proposed that E chondrites are the source of mantle nitrogen (Javoy, 1998). Since the meteorite concentrations are $4 \times 10^3$ times greater than presently found in the mantle, if a substantial portion of the Earth was derived from E chondrites, substantial nitrogen losses must have occurred during accretion (Javoy, 1998); conversely, only a small fraction of material needed to have been incorporated in an early protoplanet prior to sufficiently large impacts cause volatile loss (Ahrens, 1993). Alternatively, solar nitrogen may also be isotopically very light. Recent data for lunar samples indicate that solar nitrogen may be as light as $\delta^{15}N = -380\%_o$ (Hashizume *et al.*, 2000; Owen *et al.*, 2001). The consequences of incorporating solar nitrogen in the Earth has not been fully explored, although due to the high solar Ne/N ratio, incorporation of a significant amount of solar nitrogen will require strong elemental fractionation.

If nitrogen is highly incompatible along with the noble gases, then it likely is introduced into the upper mantle from another reservoir as well. Since this nitrogen may be isotopically very light, it appears that there is also a subducted, isotopically heavy, nitrogen component in the upper mantle, and this provides the dominant fraction of nitrogen. The specific proportions depend upon the isotopic compositions of each. In the case of solar nitrogen, only 3% of MORB nitrogen is solar (3 ppb), with the remainder provided by subduction. A much larger fraction of nitrogen derived from enstatite chondrites is required. If this nitrogen was introduced with $^3He$ from a deep-mantle source that also supplies plumes, then OIBs would be expected to have very light $\delta^{15}N$ values. This is not the case; rather, nitrogen appears to be dominated by subducted components. However, it is clear that both deep-mantle and subducted components are present in OIB source regions, and so primordial nitrogen may be masked. Alternatively, the composition of nitrogen subducted into the upper mantle prior to the Archean was isotopically light, and has persisted in the upper mantle over a long residence time (Marty and Dauphas, 2003).

### 4.11.6.3 Water

*Crustal $H_2O$.* The terrestrial oceans have $\delta D = 0‰$ with an inventory of $1.4 \times 10^{24}$ g $H_2O$, equivalent to 120 ppm H when divided into the mass of the upper mantle. There is an additional 20% in other surface reservoirs and crustal rocks, and the total composition is $\sim\delta D = -10‰$ to $-18‰$ (Taylor and Sheppard, 1986; Lécuyer et al., 1998).

*$H_2O$ in MORB and the upper mantle.* Water is not so readily lost by vesiculation, with measurements of unaltered MORB glasses of 0.12% to 0.33 wt.% $H_2O$ (Pineau and Javoy, 1994; Jambon, 1994; Sobolev and Chaussidon, 1996). Water behaves incompatibly during melting (see Jambon and Zimmermann, 1990), and so a concentration of 120–330 ppm is obtained for the mantle source region (see also Thompson, 1992). The upper mantle, therefore, contains at least as much water as the hydrosphere. The flux at ridges totals $(1-3) \times 10^{14}$ g a$^{-1}$. The hydrogen isotopic composition is distinctive from that of the surface, with $\delta D = -71‰$ to $-91‰$ (Craig and Lupton, 1976; Kyser and O'Neil, 1984; Poreda et al., 1986).

*$H_2O$ at hotspots.* There are some hydrogen isotope data available for hotspots. Poreda et al. (1986) found a correlation between helium and hydrogen isotopes along the Reykjanes Ridge, with values of up to $\delta D = -50‰$, significantly higher than in MORB. These samples also had an increase in water content toward Iceland, with values of up to $\sim0.35\%$. A higher water content was also found in samples from the Azores Platform, with 0.52 wt.% $H_2O$. Rison and Craig (1983) found glasses from Loihi had $\delta D = -69‰$ to $-74‰$, within the MORB range. Lighter compositions, down to $-125‰$, have been found in Hawaii and elsewhere (Deloule et al., 1991; Hauri, 2002), which suggests that isotopically light, juvenile hydrogen remains in the mantle.

*Subduction.* The flux of $H_2O$ entering subduction zones of $8.8 \times 10^{14}$ g yr$^{-1}$ or greater (Ito et al., 1983; Bebout, 1995) is substantially greater than the ridge flux, and has been interpreted as a net inflow of water (Ito et al., 1983). However, a substantial fraction is returned during arc metamorphism, dewatering, and volcanism. Data for boron isotopes and $B/H_2O$ relationships suggest that only $\sim20\%$ of the water is recycled (Chaussidon and Jambon, 1994), compatible with a steady-state mantle concentration. Due to the difficulties of obtaining total long-term values for subduction and return of water, concrete constraints on the balance of $H_2O$ will probably be impossible.

*The global $H_2O$ cycle.* The main feature of the solid Earth hydrogen cycle is the large contrast between the isotopic composition of the surface of $\delta D = -10‰$ to $-20‰$ and the upper mantle of $\delta D = -80‰$. There have been two contrasting ideas. It has been argued that upper-mantle water is derived entirely from subduction. The low $\delta D$ ratio of MORB could be controlled by those of dewatered metasedimetary and metamorphosed mafic subducted rocks that have compositions that are close to upper-mantle values, with $\delta D = -50‰$ to $-80‰$ (Magaritz and Taylor, 1976; Bebout, 1995). The average of these rocks may be somewhat heavier than MORB values, although Bell and Ihinger (2000) suggested that somewhat lighter hydrogen might be in nominally anhydrous minerals that may preferentially retain hydrogen during subduction. Alternatively, this isotopic signature has been ascribed to a primordial component established and isolated early in Earth history (e.g., Craig and Lupton, 1976; Lécuyer et al., 1998; Dauphas et al., 2000). It is possible that this is stored with $^3$He in the deep mantle. If the $H_2O$ and $^3$He are both derived from the same reservoir, then using the MORB ratio of $H_2O$ to $^3$He, and an undegassed mantle $^3$He concentration (Section 4.11.4.3), this reservoir contains 900–2,300 ppm H, which is much higher than the surface reservoir divided into the upper mantle, although somewhat below CI chondrites (Kerridge, 1985). Isotopically, CI chondrites are too heavy, and enstatite chondrites, with approximately $-460‰$, provide a better source, but contain only 50 ppm (Javoy, 1998). Overall, it appears likely that water in the upper mantle is dominated by subducted components, although hydrogen from the lower mantle may accompany $^3$He upwards, and may also account for some isotopically light mantle values.

## 4.11.7 DEGASSING OF OTHER TERRESTRIAL PLANETS

Understanding of the degassing of Mars and Venus is incomplete due to the lack of data regarding interior volatile reservoirs, and uncertainties in whether differences between the surface inventories with those on Earth are due to different degassing histories or to initial differences generated during planet formation (see Chapter 4.12). There are manifestations of volcanic activity that can be used to deduce at least a generalized history of mantle melting, but of course the precise timing and relationships to mantle structure are not well constrained. In this review, some of the information regarding the extent of planetary degassing relative to that of the Earth is considered.

### 4.11.7.1 Mars

Mars is characterized by low noble gas abundances in the atmosphere, with equivalent total planet concentrations that are 10 times less than on Earth. The amount of martian carbon is also not well known; while the atmosphere has only $7 \times 10^{18}$ g C, equivalent to 0.01 ppm for the bulk planet (Owen *et al.*, 1977), a large fraction may be stored in the polar regolith.

While the amount of water that has been observed on Mars is low, there is evidence that there was substantially more in the past. Measurements of atmospheric water vapor on Mars have found D/H values ~5 times that of the Earth and has been fractionated due to Jeans escape of hydrogen to space (Owen *et al.*, 1988). Models of hydrogen atmospheric losses (Donahue, 1995) and morphological data of features generated by surface water (Carr, 1986) both suggest that originally there may have been the equivalent of up to 500 m of water, or $\sim 7 \times 10^{22}$ g $H_2O$. The total mass of Mars is 0.11 that of the Earth, and so both planets originally may have had similar bulk water concentrations. This would also require degassing of water on both planets to similar extents.

The atmosphere of Mars has a high value of $\delta^{15}N = (+620 \pm 160)$‰ (Nier and McElroy, 1977) that may be due to fractionating losses (see section 7 in Chapter 4.12), and $(4.7 \pm 1.2) \times 10^{17}$ g N, which is equivalent to 0.7 ppb when divided into the mass of the entire planet (Owen *et al.*, 1977). Therefore, Mars appears to have $10^{-4}$ times the nitrogen on the Earth. Mathew *et al.* (1998) reported evidence for a component in a martian meteorite with $\delta^{15}N < -22$‰, suggesting that, like the Earth, the solid planet may contain nitrogen that is isotopically lighter than the atmosphere, and consistent with the modification of atmospheric nitrogen isotopes by losses.

*Radiogenic $^{40}Ar$.* The atmospheric $^{40}Ar/^{36}Ar$ ratio has been measured by Viking to be $3,000 \pm 400$ (Pepin and Carr, 1992), although a lower value of ~1,800 has been deduced from meteorite data (Pepin and Carr, 1992; Bogard, 1997; Bogard *et al.*, 2001). Based on the Viking data, there are $(7.0 \pm 1.4) \times 10^{39}$ atoms $^{40}Ar$ in the atmosphere. The Mars mantle has been estimated to have 305 ppm K (Wänke and Dreibus, 1988), so that $3.3 \times 10^{41}$ atoms $^{40}Ar$ have been produced in Mars. Therefore, only 2% of martian $^{40}Ar$ has degassed to the atmosphere, and most of the planet interior has retained the $^{40}Ar$ produced throughout its history. The history of degassing of $^{40}Ar$ from the interior has been discussed in several studies

(Volkov and Frenkel, 1993; Sasaki and Tajika, 1995; Hutchins and Jakosky, 1996; Tajika and Sasaki, 1996). There are considerable uncertainties in the history of martian volcanism and the amounts of volatiles that have been lost to space (see Chapter 4.12) that must be resolved before more definitive degassing histories can be constructed. Tajika and Sasaki (1996) argue that much of the $^{40}Ar$ has degassed from relatively recent volcanic regions, in contrast to early degassing of other volatiles, while Hutchins and Jakosky (1996) conclude that in order to account for volatiles that have been lost to space, much degassing must have occurred by processes other than volcanic outgassing.

*The radiogenic $^{129}Xe$ budget.* Martian atmospheric xenon clearly contains a considerable fraction of radiogenic $^{129}Xe$. It has been estimated that the silicate portion of Mars contains 32 ppb I (Wänke and Dreibus, 1988). Assuming that $^{129}I/^{127}I = 1.1 \times 10^{-4}$ at 4.57 Ga (Hohenberg *et al.*, 1967), then $8.44 \times 10^{36}$ atoms $^{129}Xe$ have been produced in Mars or precursor materials. Using fractionated CI chondrite xenon or solar xenon for the nonradiogenic light xenon isotope composition (see Chapter 4.12), the atmosphere is calculated to contain only 0.092% of what has been produced. Assuming there is none remaining in the planet, this corresponds to a closure age of 160 Myr. However, if only 2% has degassed (like $^{40}Ar$) to the atmosphere, then a closure age of 70 Myr is obtained. This value is similar to that of the Earth; it suggests that there may also have been losses of volatiles from Mars over the same extended period of accretion.

*The fissiogenic $^{136}Xe$ budget.* The amount of $^{136}Xe$ produced in Mars or accreting materials, assuming that the silicate portion of Mars has 16 ppb of $^{238}U$ at present (Wänke and Dreibus, 1988) and initially had $^{244}Pu/^{238}U = 0.0068$ (Hudson *et al.*, 1989), is $1.9 \times 10^{34}$ atoms $^{136}Xe$ from $^{244}Pu$ and $7.2 \times 10^{32}$ atoms $^{136}Xe$ from $^{238}U$. In contrast, there is a total of $2.8 \times 10^{33}$ atoms $^{136}Xe$ in the atmosphere. Up to ~5% of the atmospheric $^{136}Xe$ may be plutonium-derived (see Chapter 4.12); if so, and the closure age for Mars is 70 Ma, then 1–2% of the $^{244}Pu$ produced in the solid planet has degassed. This is consistent with the $^{129}Xe$ and $^{40}Ar$ budgets. It has been argued that plutogenic $^{136*}Xe$ could make up much less that 5% of the total atmospheric inventory (see Chapter 4.12), requiring even less planetary degassing and greater very early isolation of interior volatiles from the atmosphere. Reports of significant abundances of $^{244}Pu$ fission xenon in several SNC meteorites do in fact point strongly to its efficient retention in the martian crust (Marty and Marti, 2002; Mathew and Marti, 2002). Further discussion of the abundances of daughter xenon isotopes in the atmosphere is provided by Swindle and Jones (1997).

*Martian mantle noble gases.* Martian meteorites contain components other than those derived directly from the atmosphere (see detailed discussion by Swindle (2002)). Information on the relative abundances of the heavier noble gases in the mantle (Ott, 1988; Mathew and Marti, 2001) suggests that the $^{84}Kr/^{132}Xe$ ratio is at least 10 times lower than both the martian atmosphere and the solar composition. If this is truly a source feature, it indicates that heavy noble gases trapped within the planet suffered substantially different elemental fractionation than the atmosphere (see Chapter 4.12) and have not subsequently formed a dominant fraction of the atmosphere. However, it is not possible at present to conclusively determine whether the measured elemental abundance ratios reflect an interior reservoir that was initially different from atmospheric noble gases, rather than due either to planetary processing or transport and incorporation into the samples.

The noble gas isotopic composition of the martian interior is only available for xenon. Data for the martian meteorite Chassigny found xenon with little scope for radiogenic additions (Ott, 1988; Mathew and Marti, 2001), indicating that this reservoir had a high Xe/Pu ratio, at least during the lifetime of $^{244}Pu$. Data from other meteorites indicate that there are other interior martian reservoirs that contain solar xenon but with resolvable fissiogenic contributions (see Mathew and Marti, 2002; Marty and Marti, 2002), and so have had lower Xe/Pu ratios. Data for $^{129}Xe$ and $^{136}Xe$ have been used to argue that there were substantial losses of xenon from the mantle within the first 35 Myr (Marty and Marti, 2002). As discussed in Chapter 4.12, the martian mantle appears to have nonfractionated solar xenon (Ott, 1988). This contrasts with the fractionated character of atmospheric xenon, and is consistent with fractionation of the atmosphere after strong early degassing (see Chapter 4.12) and minimal recycling of xenon into the mantle.

*Martian degassing history.* The budget for radiogenic argon and xenon indicate that only a small fraction of the planet has degassed since very early in planetary history. A consequence of this is that noble gases were retained within the planet during formation from a very early stage, as indicated by the $^{129}Xe$ budget. This is supported by evidence for xenon-rich mantle domains that for a body of at least this size, accretion does not necessarily lead to strong degassing. The low atmospheric abundances of nonradiogenic noble gases may be partly due to retention in the largely undegassed planet, but the much higher $^{40}Ar/^{36}Ar$ ratio relative to the terrestrial value indicates that the initial inventory of nonradiogenic species after planet formation was

likely lower. The composition of nitrogen and water at the surface of Mars has been strongly affected by the history of losses to space, with no evidence for significant fluxes back into the planet.

The observation that the martian atmosphere has a higher $^{129}Xe/^{130}Xe$ ratio than the mantle, in contrast to the situation observed on Earth, has led to speculations about the mechanisms fractionating iodine from xenon. Both of these elements are highly incompatible, and so melting and crust formation are expected to transport them together out of the mantle. Available data suggest that iodine is more incompatible, so that the residue is expected to have a higher I/Xe ratio. In order to generate a higher atmospheric $^{129}Xe/^{130}Xe$ ratio, Musselwhite *et al.* (1991) suggested that the iodine was sequestered in the crust by hydrothermal alteration while xenon was lost to space (see Chapter 4.12). Alternatively, Musselwhite and Drake (2000) suggest that the iodine was preferentially retained within a magma ocean. These models are based upon the assumption that the mantle noble gases are the residue that is complementary to the atmosphere; this is not the case on the Earth and it is not clear whether this is true for Mars.

### 4.11.7.2  Venus

Venus is similar in size to the Earth and might be expected to have differentiated to a similar extent. However, while the early accretion history might have been similar (with the exception of the absence of a moon-forming event), silicate differentiation did not proceed according to the familiar plate tectonic mechanisms. There is, of course, no data on the interior of Venus, and so planetary degassing characteristics must be deduced from limited atmospheric data and observations of volcanic activity at the surface.

The measured $^{40}Ar/^{36}Ar$ ratio is $1.11 \pm 0.02$, substantially less radiogenic than the terrestrial atmosphere. For a mixing ratio of 21–48 ppm (Donahue and Pollack, 1983), the atmosphere contains $(1.8–4.4) \times 10^{41}$ atoms $^{40}Ar$. Divided by the mass of the planet, this corresponds to $(3.6–9.0) \times 10^{13}$ atoms $^{40}Ar\ g^{-1}$. This is 0.2–0.5 times the value for the Earth. However, Venus appears to be deficient in potassium. Data for the K/U ratio of the surface indicate that K/U = $7,220 \pm 1,220$ (Kaula, 1999), or $0.57 \pm 0.10$ times that of the value of $1.27 \times 10^4$ commonly taken for the Earth. Assuming that Venus has the same uranium concentration as the total Earth (including the core) of 14 ppb, then 12–28% of the $^{40}Ar$ produced in Venus is now in the atmosphere (see Kaula, 1999). This indicates that a substantial inventory of $^{40}Ar$ remains within the planet, possibly also accompanied by up to an equivalent fraction of nonradiogenic noble gases. In contrast, at least 40% of terrestrial $^{40}Ar$ is in the atmosphere.

Venus has about twice as much carbon (like nitrogen) at the surface than the Earth, equivalent to 26 ppm when divided into the bulk planet (von Zahn *et al.*, 1983); whether Venus is more rich in carbon therefore depends upon what volume of the mantle has degassed and how much remains in the mantle (Lécuyer *et al.*, 2000). From the radionuclide budget discussed above, it appears that Venus may have degassed to a similar extent as the Earth, but is unlikely to have been substantially more degassed. Therefore, it appears that the high abundances of carbon and nitrogen reflect greater total planetary abundances. The Venus atmosphere has ~200 ppm $H_2O$ (Hoffman *et al.*, 1980) and a D/H ratio of $(1.6 \pm 0.2) \times 10^{-2}$, i.e., ~$10^2$ times that of the Earth (Donahue *et al.*, 1982). It has been suggested that Venus originally had the same D/H value as the Earth, but has lost at least one terrestrial ocean volume of water by hydrodynamic escape, thereby generating an enrichment in deuterium (Donahue *et al.*, 1982). The ratio of water to carbon and nitrogen therefore may have been similar to that of the Earth.

Venus is also rich in nonradiogenic noble gases, with the absolute abundance of $^{36}Ar$ on Venus exceeding that on Earth by a factor >70. This is clearly due to the amount of noble gases initially supplied and retained by the planet, and is discussed further in Chapter 4.12.

Venus is similar to the Earth in mass and composition, and so might be expected to evolve similarly. However, Venus has a hot, insulating atmosphere and does not have the features of plate tectonics. These two features appear to be related. As discussed in Chapter 4.12, Venus and Earth may have started with similar atmospheres, but the Earth suffered the consequences of a late moon-forming impact. The consequence is that the atmosphere of Venus remained sufficiently insulating to maintain temperatures that prevented the formation of liquid water at the surface, and so oceans. This may have led to less effective crustal recycling, and ultimately the inhibition of plate tectonics (see Kaula, 1990). However, the styles of early Venus evolution, and the transitions between different tectonic styles, are debated. Various models have been presented for the degassing of $^{40}Ar$ from the mantle and crust and into the Venus atmosphere, but these are dependent upon the history of tectonic activity and heat loss on the planet (Sasaki and Tajika, 1995; Turcotte and Schubert, 1988; Namiki and Solomon, 1998; Kaula, 1999). Indeed, the amount of $^{40}Ar$ in the atmosphere provides a constraint on the total amount of mantle melting and transfer of potassium to the crust (Kaula, 1999). The fraction of $^{40}Ar$ that has degassed, ~12–28%, is substantially lower than that of the Earth. Kaula (1999) has

discussed various mechanisms that may account for decreased degassing of Venus. The possibility that this may reflect layering on Venus is difficult to assess, considering that relating the limited apparent degassing of the Earth to mantle structure has proven so controversial.

## 4.11.8 CONCLUSIONS

The degassing of the Earth is an integral part of the formation and thermal evolution of the planet. Degassing histories have often naturally been based on noble gas abundances and isotopic compositions. Radiogenic isotopes provide the strongest constraints on the total volume of the silicate Earth that has degassed to the atmosphere, and indicate that a large portion of the Earth remains undegassed. Nonradiogenic noble gases, incorporated during Earth formation, strongly degassed during accretion and the extreme thermal conditions on the forming Earth. The upper mantle, down to some as yet unresolved depth, is now highly degassed. Silicate differentiation producing the continental crust also likely promoted degassing, and continual processing of the continental crust has left it relatively degassed of even radiogenic daughters produced within the crust. Other volatiles have also been effectively removed from the upper mantle, although elements such as carbon may have substantial subduction fluxes that are the dominant inputs.

An outstanding question is how much of the mantle still maintains high volatile concentrations. This involves resolution of the nature of the high $^3He/^4He$ OIB-source region. Most models equate this with undepleted, undegassed mantle, although some models invoke depletion mechanisms. However, none of these has matched the end-member components seen in OIB lithophile isotope correlations. It remains to be demonstrated that a primitive component is present and so can dominate the helium and neon isotope signatures in OIB. The heavy-noble-gas characteristics in OIB must still be documented. It is not known to what extent major volatiles are stored in the deep Earth and associated with these noble gas components.

As understanding of terrestrial noble gas geochemistry has evolved, various earlier conclusions now appear to be incorrect. As discussed above, the very radiogenic argon and xenon isotope ratios of the upper mantle are not the result of early degassing of this mantle reservoir, since this is a model-dependent conclusion based on the assumption that upper-mantle noble gases are residual from atmosphere degassing. However, it is now clear that xenon isotope systematics precludes such a relationship (Ozima *et al.*, 1985). While a complete description of mantle noble gases remains to be formulated, it is clear that there are other mechanisms that can account for the observed xenon isotope variations. Nonetheless, early transfer of volatiles to the atmosphere probably did occur and was caused by impact degassing.

The interactions between the atmosphere and the mantle now appear to be more complex. The subduction of heavy noble gases may have a marked impact on mantle isotope compositions. The earlier conclusion that this was not possible was based on models of the isolation of the upper mantle or arguments about preservation of non-atmospheric $^{129}Xe/^{130}Xe$ ratios in the mantle. In fact, upper-mantle nonradiogenic xenon isotopes could be dominated by subducted xenon and admixed with very radiogenic xenon, and some models explicitly incorporate subducted xenon fluxes. Until more is conclusively known about argon and xenon isotopic variations in the mantle, subduction must be considered a potentially important process.

In devising atmosphere formation histories, it is also now clear that the present is not the key to the past. Although primordial noble gases continue to degas, their isotopic compositions do not match those of the atmosphere and limit their contribution to a small fraction of the present atmospheric inventory. Volatile species continue to be added to the atmosphere, but the dominant inputs occurred during very early Earth history.

There are, of course, many questions regarding terrestrial noble gases that remain to be explored. Some of the issues that are critical to making advances in global models of noble gas behavior include the partitioning of noble gases into the core and between mantle minerals and silicate melts. While the common assumptions regarding general behavior may very well be correct, the effects of fractionation between noble gases cannot be clearly assessed. Further refinement of planetary degassing models will come from greater resolution of the nature of the reservoirs that remain undegassed within the planet.

## ACKNOWLEDGMENT

A very useful review by N. Daup that was provided under very short notice is much appreciated.

## REFERENCES

Ahrens T. J. (1993) Impact erosion of terrestrial planetary atmospheres. *Ann. Rev. Earth Planet. Sci.* **21**, 525–555.

Ahrens T. J., O'Keefe J. D., and Lange M. A. (1989) Formation of atmospheres during accretion of the terrestrial planets. In *Origin and Evolution of Planetary and Satellite Atmospheres* (eds. S. K. Atreya, J. B. Pollack, and M. S. Matthews). University of Arizona Press, Tucson, pp. 328–385.

Albarède F. (1998) Time-dependent models of U–Th–He and K–Ar evolution and the layering of mantle convection. *Chem. Geol.* **145**, 413–429.

Allègre C. J. (1997) Limitation on the mass exchange between the upper and lower mantle: the evolving convection regime of the Earth. *Earth Planet. Sci. Lett.* **150**, 1–6.

Allègre C. J., Staudacher T., Sarda P., and Kurz M. (1983) Constraints on evolution of Earth's mantle from rare gas systematics. *Nature* **303**, 762–766.

Allègre C. J., Staudacher T., and Sarda P. (1986) Rare gas systematics: *formation* of the atmosphere, evolution and structure of the Earth's mantle. *Earth Planet. Sci. Lett.* **87**, 127–150.

Allègre C. J., Poirier J.-P., Humler E., and Hofmann A. W. (1995) The chemical composition of the Earth. *Earth Planet. Sci. Lett.* **134**, 515–526.

Allègre C. J., Hofmann A. W., and O'Nions R. K. (1996) The argon constraints on mantle structures. *Geophys. Res. Lett.* **23**, 3555–3557.

Armstrong R. L. (1991) The persistent myth of crustal growth. *Austral. J. Earth Sci.* **38**, 613–630.

Aronson J. L. and Hower J. (1976) Mechanism of burial metmorphism of argillaceous sediment: 2. Radiogenic argon evidence. *Geol. Soc. Am. Bull.* **87**, 738–744.

Ballentine C. J. and Burnard P. G. (2002) Production, release, and transport of noble gases in the continental crust. *Rev. Mineral. Geochem.* **47**, 481–538.

Ballentine C. J., van Keken P. E., Porcelli D., and Hauri E. K. (2002) Numerical models, geochemistry and the zero-paradox noble-gas mantle. *Phil. Trans. Roy. Soc. London A* **360**, 2611–2631.

Batiza R. (1982) Abundances, distribution and sizes of volcanoes in the Pacific Ocean and implications for the origin of non-hotspot volcanoes. *Earth Planet. Sci. Lett.* **60**, 195–206.

Bebout G. E. (1995) The impact of subduction-zone metamorphism on mantle-ocean chemical cycling. *Chem. Geol.* **126**, 191–218.

Bebout G. E. and Fogel M. (1992) Nitrogen isotope composition of metasedimentary rocks in the Catalina Schist, California: implications for metorphic devolitization histor. *Geochim. Cosmochim. Acta* **56**, 2839–2849.

Begemann R., Weber H. W., and Hintenberger H. (1976) On the primordial abundance of argon-40. *Astrophys. J.* **203**, L155–L157.

Bell D. R. and Ihinger P. D. (2000) The isotopic composition of hydrogen in nominally anhydrous mantle minerals. *Geochim. Cosmochim. Acta* **64**, 2109–2118.

Benkert J.-P., Baur H., Signer P., and Wieler R. (1993) He, Ne, and Ar from solar wind and solar energetic particles in lunar ilmenites and pyroxenes. *J. Geophys. Res.* **98**, 13147–13162.

Bercovici D. and Karato S.-I. (2003) Whole-mantle convection and the transition-zone water filter. *Nature* **425**, 39–44.

Bethke C. M., Zhao X., and Torgersen T. (1999) Groundwater flow and the $^4$He distribution in the Great Artesian Basin of Australia. *J. Geophys. Res.* **104**, 12999–13011.

Bijwaard H. and Spakman W. (1999) Tomographic evidence for a narrow whole mantle plume below Iceland. *Earth Planet. Sci. Lett.* **166**, 121–126.

Blum J. D. (1995) Isotopic decay data. In *Global Earth Physics: A Handbook of Physical Constants* (ed. T. J. Ahrens). American Geophysical Union, Washington, DC, pp. 271–282.

Bogard D. D. (1997) A reappraisal of the Martian Ar-36/Ar-38 ratio. *J. Geophys. Res.* **102**, 1653–1661.

Bogard D. D., Clayton R. N., Marti K., Owen T., and Turner G. (2001) Martian volatiles: isotopic composition, origin, and evolution. *Space Sci. Rev.* **96**, 425–458.

Boyd S. R. and Pillinger C. T. (1994) A preliminary study of $^{15}N/^{14}N$ in octahedral growth form diamonds. *Chem. Geol.* **116**, 43–59.

Brazzle R. H., Pravdivtseva O. V., Meshik A. P., and Hohenberg C. M. (1999) Verification and interpretation of the I–Xe chronometer. *Geochim. Cosmochim. Acta* **63**, 739–760.

Broadhurst C. L., Drake M. J., Hagee B. E., and Bernatowicz T. J. (1992) Solubility and partitioning of Ne, Ar, Kr, and Xe in minerals and synthetic basaltic melts. *Geochim. Cosmochim. Acta* **56**, 709–723.

Brooker R. A., Du Z., Blundy J. D., Kelley S. P., Allan N. L., Wood B. J., Chamorro E. M., Wartho J.-A., and Purton J. A. (2003) The "zero charge" partitioning behaviour of noble gases during mantle melting. *Nature* **423**, 738–741.

Brown H. (1949) Rare gases and the formation of the Earth's atmosphere. In *The Atmospheres of the Earth and Planets* (ed. G. P. Kuiper). University of Chicago Press, Chicago, pp. 258–266.

Buffett B. A., Huppert H. E., Lister J. R., and Woods A. W. (1996) On the thermal evolution of the Earth's core. *J. Geophys. Res.* **101**, 7989–8006.

Burnard P. G., Graham D., and Turner G. (1997) Vesicle specific noble gas analyses of Popping Rock: implications for primordial noble gases in Earth. *Science* **276**, 568–571.

Cadogan P. H. (1977) Paleoatmospheric argon in Rhynie chert. *Nature* **268**, 38–41.

Caffee M. W., Hudson G. U., Velsko C., Huss G. R., Alexander E. C., Jr., and Chivas A. R. (1999) Primordial noble cases from Earth's mantle: identification of a primitive volatile component. *Science* **285**, 2115–2118.

Carr M. H. (1986) Mars: a water-rich planet? *Icarus* **68**, 187–216.

Cartigny P., Harris J. W., and Javoy M. (1998) Eclogitic diamond formation at Jwaneng: no room for a recycled component. *Science* **280**, 1421–1424.

Cartigny P., Jendrzejewski N., Pineau F., Petit E., and Javoy M. (2001) Volatile (C, N, Ar) variability in MORB and the respective roles of mantle source heterogeneity and degassing: the case of the Southwest Indian Ridge. *Earth Planet. Sci. Lett.* **194**, 241–257.

Chabot N. L. and Drake M. J. (1999) Potassium solubility in metal: the effects of composition at 15 kbar and 1900 degrees C on partitioning between iron alloys and silicate melts. *Earth Planet. Sci. Lett.* **172**, 323–335.

Chamorro-Perez E. M., Brooker R. A., Wartho J.-A., Wood B. J., Kelley S. P., and Blundy J. D. (2002) Ar and K partitioning between clinopyroxene and silicate melt to 8 GPa. *Geochim. Cosmochim. Acta* **66**, 507–519.

Chaussidon M. and Jambon A. (1994) Boron content and isotopic composition of oceanic basalts: geochemical and cosmochemical implications. *Earth Planet. Sci. Lett.* **121**, 277–291.

Clarke W. B., Beg M. A., and Craig H. (1969) Excess $^3$He in the sea: evidence for terrestrial primordial helium. *Earth Planet. Sci. Lett.* **6**, 213–220.

Cogley J. C. (1984) Continental margins and the extent and number of the continents. *Rev. Geophys. Space Phys.* **22**, 101–122.

Coltice N. and Ricard Y. (1999) Geochemical observations and one layer mantle convection. *Earth Planet. Sci. Lett.* **174**, 125–137.

Coltice N., Albarède F., and Gillet P. (2000) $^{40}$K–$^{40}$Ar constraints on recycling continental crust into the mantle. *Science* **288**, 845–847.

Craig H. and Lupton J. E. (1976) Primordial neon, helium, and hydrogen in oceanic basalts. *Earth Planet. Sci. Lett.* **31**, 369–385.

Craig H., Clarke W. B., and Beg M. A. (1975) Excess $^3$He in deep water on the East Pacific Rise. *Earth Planet. Sci. Lett.* **2**, 125–132.

Creager K. C. and Jordan T. H. (1986) Slab penetration into the lower mantle beneath the Marianas and other island arcs of the northwest Pacific. *J. Geophys. Res.* **91**, 3573–3589.

Crisp J. A. (1984) Rates of magma emplacement and volcanic output. *J. Volcanol. Geotherm. Res.* **89**, 3031–3049.

Dauphas N. and Marty B. (1999) Heavy nitrogen in carbonatites of the Kola Peninsula: a possible signature of the deep mantle. *Science* **286**, 2488–2490.

Dauphas N., Robert F., and Marty B. (2000) The late asteroidal and cometary bombardment of Earth as recorded in water deuterium to protium ratio. *Icarus* **148**, 508–512.

Davies G. F. (1999) Geophysically constrained mantle mass flows and the Ar-40 budget: a degassed lower mantle? *Earth Planet. Sci. Lett.* **166**, 149–162.

Deloule E., Albarède F., and Sheppard S. M. F. (1991) Hydrogen isotope heterogeneities in the mantle from ion probe analysis of amphiboles from ultramafic rocks. *Earth Planet. Sci. Lett.* **105**, 543–553.

Déruelle B., Dreibus G., and Jambon A. (1992) Iodine abundances in oceanic basalts: implications for Earth dynamics. *Earth Planet. Sci. Lett.* **108**, 217–227.

Des Marais D. J. (1985) Carbon exchange between the mantle and crust and its effect upon the atmosphere: today compared to Archean time. In *The Carbon Cycle and Atmospheric $CO_2$: Natural Variations Archean to Present* (eds. E. T. Sundquist and W. S. Broecker). American Geophysical Union, Washington, DC, pp. 602–611.

Dixon J. E. and Stolper E. M. (1995) An experimental study of water and carbon dioxide solubilities in mid-ocean ridge basaltic liquids: 2. Applications to degassing. *J. Petrol.* **36**, 1633–1646.

Doe B. R. and Zartman R. E. (1979) Plumbotectonics: I. The Phanerozoic. In *Geochemistry of Hydrothermal Ore Deposits* (ed. H. L. Barnes). Wiley, New York, pp. 22–70.

Donahue T. M. (1995) Evolution of water reservoirs on Mars from D/H ratios in the atmosphere and crust. *Nature* **374**, 432–434.

Donahue T. M. and Pollack J. B. (1983) Origin and evolution of the atmosphere of Venus. In *Venus* (eds. D. Hunten, L. Colin, T. Donahue, and V. Moroz). University of Arizona Press, Tucson, pp. 1003–1036.

Donahue T. M., Hoffman J. H., Hodges R. R., Jr., and Watson A. J. (1982) Venus was wet: a measurement of the ratio of deuterium to hydrogen. *Science* **216**, 630–633.

Dreibus G. and Wänke H. (1989) Supply and loss of volatile constituents during accretion of terrestrial planets. In *Origin and Evolution of Planetary and Satellite Atmospheres* (eds. S. K. Atreya, J. B. Pollack, and M. S. Matthews). University of Arizona Press, Tucson, pp. 268–288.

Dymond J. and Hogan L. (1973) Noble gas abundance patterns in deep sea basalts—Primordial gases from the mantle. *Earth Planet. Sci. Lett.* **20**, 131–139.

Eikenberg J., Signer P., and Wieler R. (1993) U–Xe, U–Kr, and U–Pb systematics for dating uranium minerals and investigations of the production of nucleogenic neon and argon. *Geochim. Cosmochim. Acta* **57**, 1053–1069.

England P. C. and Thompson A. B. (1984) Pressure–temperature–time paths of regional metamorphism: I. Heat transfer during evolution of regions of thickened continental crust. *J. Petrol.* **25**, 894–928.

Etheridge M. A., Wall V. J., Cox S. F., and Vernon R. H. (1984) High fluid pressures during regional metamorphism and deformation: implications for mass transport and deformation mechanisms. *J. Geophys. Res.* **89**, 4344–4358.

Farley K. A., Basu A. R., and Craig H. (1993) He, Sr, and Nd isotopic variations in lavas from the Juan Fernandez archipelago. *Contrib. Mineral. Petrol.* **115**, 75–87.

Farley K. A., Maier-Reimer E., Schlosser P., and Broecker W. S. (1995) Constraints on mantle He-3 fluxes and deep-sea circulation from an oceanic general circulation model. *J. Geophys. Res.* **100**, 3829–3839.

Fisher D. E. (1978) Terrestrial potassium abundances as limits to models of atmospheric evolution. In *Terrestrial Rare Gases* (ed. M. Ozima). Japan Scientific Societies Press, Tokyo, pp. 173–183.

Fisher D. E. (1985) Noble gases from oceanic island basalts do not require an undepleted mantle source. *Nature* **316**, 716–718.

Galer S. J. G. and O'Nions R. K. (1985) Residence time of thorium, uranium, and lead in the mantle with implications for mantle convection. *Nature* **316**, 778–782.

Gamo T., Ishibashi J.-I., Sakai H., and Tilbrook B. (1987) Methane anomalies in seawater above the Loihi seamount summit area, Hawaii. *Geochim. Cosmochim. Acta* **51**, 2857–2864.

Gast P. W. (1960) Limitations on the composition of the upper mantle. *J. Geophys. Res.* **65**, 1287–1297.

Geiss J., Buehler F., Cerutti H., Eberhardt P., and Filleaux C. H. (1972) Solar wind composition experiments. *Apollo 15 Preliminary Sci. Report*, chap. 15.

Goes S., Spakman W., and Bijwaard H. (1999) A lower mantle source for central European volcanism. *Science* **286**, 1928–1931.

Grady M. M., Wright I. P., Carr L. P., and Pillinger C. T. (1986) Compositional differences in estatite chondrites based on carbon and nitrogen stable isotope measurements. *Geochim. Cosmochim. Acta* **50**, 2799–2813.

Graham D. W. (2002) Noble gases in MORB and OIB: observational constraints for the characterization of mantle source reservoirs. *Rev. Mineral. Geochem.* **47**, 247–318.

Graham D. W., Lupton F., Albarède F., and Condomines M. (1990) Extreme temporal homogeneity of helium isotopes at Piton de la Fournaise, Réunion Island. *Nature* **347**, 545–548.

Grand S. P. (1987) Tomographic inversion for shear velocity beneath the North American plate. *J. Geophys. Res.* **92**, 14065–14090.

Grand S. P. (1994) Mantle shear structure beneath the Americas and surrounding oceans. *J. Geophys. Res.* **99**, 66–78.

Haendel D., Mühle K., Nitzsche H., Stiehl G., and Wand U. (1986) Isotopic variations of the fixed nitrogen in metamorphic rocks. *Geochim. Cosmochim. Acta* **50**, 749–758.

Hagee B., Bernatowicz T. J., Podosek F. A., Johnson M. L., Burnett D. S., and Tatsumoto M. (1990) Actinide abundances in ordinary chondrites. *Geochim. Cosmochim. Acta* **54**, 2847–2858.

Hall A. (1989) Ammonium in spilitized basalts of southwest England and its implications for the recycling of nitrogen. *Geochem. J.* **23**, 19–23.

Halliday A. N. and Porcelli D. (2001) In search of lost planets—the paleocosmochemistry of the inner solar system. *Earth Planet. Sci. Lett.* **192**, 545–559.

Hamano Y. and Ozima M. (1978) Earth-atmosphere evolution model based on Ar isotopic data. In *Terrestrial Rare Gases* (ed. M. Ozima). Japan Scientific Societies Press, Tokyo, pp. 155–171.

Hanes J. A., York D., and Hall C. M. (1985) An $^{40}$Ar/$^{39}$Ar geochronological and electron microprobe investigation of an Archaean pyroxenite and its bearing on ancient atmospheric compositions. *Can. J. Earth Sci.* **22**, 947–958.

Hanyu T. and Kaneoka I. (1998) Open system behavior of helium in case of the HIMU source area. *Geophys. Res. Lett.* **25**, 687–690.

Harrison D., Burnard P., and Turner G. (1999) Noble gas behaviour and composition in the mantle: constraints from the Iceland plume. *Earth Planet. Sci. Lett.* **171**, 199–207.

Hart R., Dymond J., and Hogan L. (1979) Preferential formation of the atmosphere-sialic crust system from the upper mantle. *Nature* **278**, 156–159.

Hart S. R. and Staudigel H. (1982) The control of alkalies and uranium in seawater by ocean crust alteration. *Earth Planet. Sci. Lett.* **58**, 202–212.

Hashizume K., Chaussidon M., Marty B., and Robert F. (2000) Solar wind record on the moon: deciphering presolar from planetary nitrogen. *Science* **290**, 1142–1145.

Hauri E. (2002) SIMS analysis of volatiles in silicate glasses: 2. Isotopes and abundances in Hawaiian melt inclusions. *Chem. Geol.* **183**, 115–141.

Heaton T. H. E. (1984) Rates and sources of 4He accumulation in groundwater. *Hydrol. Sci. J.* **29**, 29–47.

Helmberger D. V., Wen L., and Ding X. (1998) Seismic evidence that the source of the Iceland hotspot lies at the core–mantle boundary. *Nature* **396**, 251–255.

Hilton D. R., McMurty G. M., and Kreulen R. (1997) Evidence for extensive degassing of the Hawaiian mantle plume from helium–carbon relationships at Kilauea volcano. *Geophys. Res. Lett.* **24**, 3065–3068.

Hilton D. R., McMurtry G. M., and Goff F. (1998a) Large variations in vent fluid $CO_2/^3$He ratios signal rapid changes in magma chemistry at Loihi seamount, Hawaii. *Nature* **396**, 359–362.

Hilton D. R., Grönvold K., Sveinbjornsdottir A. E., and Hammerschmidt K. (1998b) Helium isotope evidence for off-axis degassing of the Icelandic hotspot. *Chem. Geol.* **149**, 173–187.

Hilton D. R., Thirlwall M. F., Taylor R. N., Murton B. J., and Nichols A. (2000) Controls on magmatic degassing along the Reykjanes Ridge with implications for the helium paradox. *Earth Planet. Sci. Lett.* **183**, 43–50.

Hilton D. R., Fischer T. P., and Marty B. (2002) Noble gases in subduction zones and volatile recycling. *Rev. Mineral. Geochem.* **47**, 319–370.

Hiyagon H. and Ozima M. (1986) Partition of gases between olivine and basalt melt. *Geochim. Cosmochim. Acta* **50**, 2045–2057.

Hiyagon H., Ozima M., Marty B., Zashu S., and Sakai H. (1992) Noble gases in submarine glasses from mid-oceanic ridges and Loihi Seamount—Constraints on the early history of the Earth. *Geochim. Cosmochim. Acta* **56**, 1301–1316.

Hoefs J. (1969) Carbon. In *Handbook of Geochemistry* (ed. K. H. Wedepohl). Springer, Berlin.

Hofmann A. W. (1988) Chemical differentiation of the Earth: the relationship between mantle, continental crust, and oceanic crust. *Earth Planet. Sci. Lett.* **90**, 297–314.

Hoffman J. H., Hodges R. R., Donahue T. M., and McElroy M. B. (1980) Composition of the Venus lower atmosphere from the Pioneer Venus mass spectrometer. *J. Geophys. Res.* **85**, 7882–7890.

Hohenberg C. M., Podosek F. A., and Reynolds J. H. (1967) Xenon–iodine dating: sharp isochronism in chondrites. *Science* **156**, 233–236.

Holloway J. R. (1998) Graphite-melt equilibria during mantle melting: constraints on $CO_2$ in MORB magmas and the carbon content of the mantle. *Chem. Geol.* **147**, 89–97.

Honda M. and McDougall I. (1993) Solar noble gases in the Earth—the systematics of helium–neon isotopes in mantle-derived samples. *Lithos* **30**, 257–265.

Honda M. and McDougall I. (1998) Primordial helium and neon in the Earth—a speculation on early degassing. *Geophys. Res. Lett.* **25**, 1951–1954.

Honda M., McDougall I., Patterson D. B., Doulgeris A., and Clague D. A. (1991) Possible solar noble-gas component in Hawaiian basalts. *Nature* **349**, 149–151.

Honda M., McDougall I., Patterson D. B., Doulgeris A., and Clague D. A. (1993) Noble gases in submarine pillow basalt glasses from Loihi and Kilauea, Hawaii—a solar component in the Earth. *Geochim. Cosmochim. Acta* **57**, 859–874.

Hudson G. B., Kennedy B. M., Podosek F. A., and Hohenberg C. M. (1989) The early solar system abundance of $^{244}$Pu as inferred from the St. Severin chondrite. *Proc. 19th Lunar Planet. Sci. Conf.* 547–557.

Hunt J. M. (1972) Distribution of carbon in crust of Earth. *Bull. Am. Assoc. Petrol. Geol.* **56**, 2273–2277.

Hunten D. M., Pepin R. O., and Walker J. C. G. (1987) Mass fractionation in hydrodynamic escape. *Icarus* **69**, 532–549.

Hurley P. M. (1966) K–Ar dating of sediments. In *Potassium Argon Dating* (eds. O. A. Schaeffer and J. Zähringer). Springer, New York, pp. 134–151.

Hurley P. M. and Rand J. R. (1969) Pre-drift continental nuclei. *Science* **164**, 1229–1242.

Hurley P. M., Brookins D. G., Pinson W. H., Hart S. R., and Fairbairn H. W. (1961) K–Ar age studies of Mississippi and other river sediments. *Geol. Soc. Am. Bull.* **72**, 1807–1816.

Hutchins K. S. and Jakosky B. M. (1996) Evolution of martian atmospheric argon: implications for sources of volatiles. *J. Geophys. Res.* **101**, 14933–14949.

Ito E., Harris D. M., and Anderson A. T., Jr. (1983) Alteration of oceanic crust and geologic cycling of chlorine and water. *Geochim. Cosmochim. Acta* **47**, 1613–1624.

Jambon A. (1994) Earth degassing and large-scale geochemical cycling of volatile elements. *Rev. Mineral.* **30**, 479–517.

Jambon A. and Zimmermann J. L. (1990) Water in oceanic basalts—evidence for dehydration of recycled crust. *Earth Planet. Sci. Lett.* **101**, 323–331.

Jaupart C., Sclater J. G., and Simmons G. (1981) Heat flow studies: constraints on the distribution of uranium, thorium, and potassium in the continental crust. *Earth Planet. Sci. Lett.* **52**, 328–344.

Javoy M. (1998) The birth of the Earth's atmosphere: the behavior and fate of its major elements. *Chem. Geol.* **147**, 11–25.

Javoy M. and Pineau F. (1991) The volatiles record of a "Popping Rock" from the Mid-Atlantic Ridge at 14°N: chemical and isotopic composition of gas trapped in the vesicles. *Earth Planet. Sci. Lett.* **107**, 598–611.

Javoy M., Pineau F., and Demaiffe Dl. (1984) Nitrogen and carbon isotopic composition in the diamonds of Mbuji Mayi (Zaire). *Earth Planet. Sci. Lett.* **68**, 399–412.

Jochum K. P., Hofmann A. W., Ito E., Seufert H. M., and White W. M. (1983) K, U, and Th in mid-ocean ridge basalt glasses and heat production, K/U and K/Rb in the mantle. *Nature* **306**, 431–436.

Kaula W. M. (1990) Venus: a contrast in evolution to Earth. *Science* **247**, 1191–1196.

Kaula W. M. (1999) Constraints on Venus evolution from radiogenic argon. *Icarus* **139**, 32–39.

Keeling C. D. and Whorf T. P. (2000) Atmospheric $CO_2$ records from sites in the SIO air sampling network. In *Trends: A Compendium of Data on Global Change.* Carbon Dioxide Information Analysis Center, Oak Ridge National Laboratory, Oak Ridge, TN.

Kellogg L. H. and Wasserburg G. J. (1990) The role of plumes in mantle helium fluxes. *Earth Planet. Sci. Lett.* **99**, 276–289.

Kellogg L. H., Hager B. H., and van der Hilst R. D. (1999) Compositional stratification in the deep mantle. *Science* **283**, 1881–1884.

Kerridge J. F. (1985) Carbon, hydrogen, and nitrogen in carbonaceous chondrites: abundances and isotopic compositions in bulk samples. *Geochim. Cosmochim Acta* **49**, 1707–1714.

Kingsley R. H. and Schilling J.-G. (1995) Carbon in Mid-Atlantic Ridge basalt glasses from 28-degrees-N to 63-degrees-N—evidence for a carbon-enriched azores mantle plume. *Earth Planet. Sci. Lett.* **129**, 31–53.

Korenaga J. and Kelemen P. B. (2000) Major element heterogeneity in the mantle source of the North Atlantic igneous province. *Earth Planet. Sci. Lett.* **184**, 251–268.

Krummenacher D., Merrihue C. M., Pepin R. O., and Reynolds J. H. (1962) Meteoritic krypton and barium versus the general isotopic anomalies in meteoritic xenon. *Geochim. Cosmochim. Acta* **26**, 231–249.

Kunz J. (1999) Is there solar argon in the Earth's mantle? *Nature* **399**, 649–650.

Kunz J., Staudacher T., and Allègre C. J. (1998) Plutonium-fission xenon found in Earth's mantle. *Science* **280**, 877–880.

Kurz M. D., Jenkins W. J., and Hart S. R. (1982) Helium isotopic systematics of oceanic islands and mantle heterogeneity. *Nature* **297**, 43–46.

Kyser T. K. and O'Neil J. R. (1984) Hydrogen isotope systematics of submarine basalts. *Geochim. Cosmochim. Acta* **48**, 2123–2133.

Lécuyer C., Gillet P., and Robert F. (1998) The hydrogen isotope composition of seawater and the global water cycle. *Chem. Geol.* **45**, 249–261.

Lécuyer C., Simon L., and Guyot F. (2000) Comparison of carbon, nitrogen, and water budgets on Venus and the Earth. *Earth Planet. Sci. Lett.* **181**, 33–40.

Li Y.-H. (1972) Geochemical mass balance among lithosphere, hydrosphere, and atmosphere. *Am. J. Sci.* **272**, 119–137.

Lupton J. E. (1996) A far-field hydrothermal plume from Loihi Seamount. *Science* **272**, 976–979.

Lupton J. E. and Craig H. (1975) Excess $^3$He in oceanic basalts, evidence for terrestrial primordial helium. *Earth Planet. Sci. Lett.* **26**, 133–139.

Magaritz M. and Taylor H. P., Jr. (1976) Oxygen, hydrogen, and carbon isotope studies of the Franciscan Formation Coast Ranges, California. *Geochim. Cosmochim. Acta* **40**, 215–234.

Mahaffy P. R., Donahue T. M., Atreya S. K., Owen T. C., and Niemann H. B. (1998) Galileo probe measurements of D/H and $^3$He/$^4$He in Jupiter's atmosphere. *Space Sci. Rev.* **84**, 251–263.

Mamyrin B. A., Tolstikhin I. N., Anufriev G. S., and Kamensky I. L. (1969) Anomalous isotopic composition of helium in volcanic gases. *Dokl. Akad. Nauk SSSR* **184**, 1197–1199 (in Russian).

Marti K. and Mathew K. J. (1998) Noble-gas components in planetary atmospheres and interiors in relation to solar wind and meteorites. *Proc. Indian Acad. Sci. (Earth Planet. Sci.)* **107**, 425–431.

Martel D. J., Deak J., Dovenyi P., Horvath F., O'Nions R. K., Oxburgh E. R., Stegna L., and Stute M. (1989) Leakage of helium from the Pannonian Basin. *Nature* **432**, 908–912.

Martin J.-M. and Meybeck M. (1979) Elemental mass-balance of material carried by major world rivers. *Mar. Chem.* **7**, 173–206.

Marty B. and Dauphas N. (2003) The nitrogen record of crust–mantle interaction and mantle convection from Archean to present. *Earth Planet. Sci. Lett.* **206**, 397–410.

Marty B. and Humbert F. (1997) Nitrogen and argon isotopes in oceanic basalts. *Earth Planet. Sci. Lett.* **152**, 101–112.

Marty B. and Jambon A. (1987) C/$^3$He in volatile fluxes from the solid Earth—implications for carbon geodynamics. *Earth Planet. Sci. Lett.* **83**, 16–26.

Marty B. and Marti K. (2002) Signatures of early differentiation on Mars. *Earth Planet. Sci. Lett.* **196**, 251–263.

Marty B. and Zimmermann L. (1999) Volatiles (He, C, N, Ar) in mid-ocean ridge basalts: assessment of shallow-level fractionation and characterization of source composition. *Geochim. Cosmochim. Acta* **63**, 3619–3633.

Marty B., Tolstikhin I., Kamensky I. L., Nivin V., Balanganskaya E., and Zimmermann J.-L. (1998) Plume-derived rare gases in 380 Ma carbonatites from the Kola region (Russia) and the argon isotopic composition of the deep mantle. *Earth Planet. Sci. Lett.* **164**, 179–192.

Mathew K. J. and Marti K. (2001) Early evolution of martian volatiles: nitrogen and noble gas components in ALH84001 and Chassigny. *J. Geophys. Res.* **106**, 1401–1422.

Mathew K. J. and Marti K. (2002) Martian atmospheric and interior volatiles in the meteorite Nakhla. *Earth Planet. Sci. Lett.* **199**, 7–20.

Mathew K. J., Kim J. S., and Marti K. (1998) martian atmospheric and indigenous components of xenon and nitrogen in the Shergotty, Nakhla, and Chassigny group meteorites. *Meteorit. Planet. Sci.* **33**, 655–664.

Matsuda J.-I. and Marty B. (1995) The $^{40}$Ar/$^{36}$Ar ratio of the undepleted mantle: a re-evaluation. *Geophys. Res. Lett.* **22**, 1937–1940.

Matsuda J.-I. and Nagao K. (1986) Noble gas abundances in a deep-sea core from eastern equatorial Pacific. *Geochem. J.* **20**, 71–80.

Mattey D. P., Carr R. H., Wright I. P., and Pillinger C. T. (1984) Carbon isotopes in submarine basalts. *Earth Planet. Sci. Lett.* **70**, 196–206.

Mazor E. (1995) Stagnant aquifer concept: 1. Large scale artesian systems—Great Artesian Basin, Australia. *J. Hydrol.* **173**, 219–240.

McCulloch M. T. (1993) The role of subducted slabs in an evolving Earth. *Earth Planet. Sci. Lett.* **115**, 89–100.

McDonough W. F. and Sun S. S. (1995) The composition of the Earth. *Chem. Geol.* **120**, 223–253.

McDougall I. and Harrison T. M. (1999) *Geochronology and Thermochronology by the $^{40}Ar/^{39}Ar$ Method*. Oxford University Press, Oxford.

McLennan S. M. and Taylor S. R. (1996) Heat flow and the chemical composition of continental crust. *J. Geol.* **104**, 369–377.

Milliman J. D. and Meade R. H. (1983) World-wide delivery of river sediment to the oceans. *J. Geol.* **91**, 1–21.

Moreira M. and Sarda P. (2000) Noble gas constraints on degassing processes. *Earth Planet. Sci. Lett.* **176**, 375–386.

Moreira M., Kunz J., and Allègre C. J. (1998) Rare gas systematics in Popping Rock: isotopic and elemental compositions in the upper mantle. *Science* **279**, 1178–1181.

Morgan P. (1984) The thermal structure and thermal evolution of the continental lithosphere. *Phys. Chem. Earth* **15**, 107–193.

Morgan J. P. (1998) Thermal and rare gas evolution of the mantle. *Chem. Geol.* **145**, 431–445.

Morrison P. and Pine J. (1955) Radiogenic origin of the helium isotopes in rock. *Ann. NY Acad. Sci.* **62**, 69–92.

Muramatsu Y. W. and Wedepohl K. H. (1998) The distribution of iodine in the Earth's crust. *Chem. Geol.* **147**, 201–216.

Musselwhite D. S. and Drake M. J. (2000) Early outgassing of Mars: implications from experimentally determined solubility of iodine in silicate magmas. *Icarus* **148**, 160–175.

Musselwhite D. S., Drake M. J., and Swindle T. D. (1991) Early outgassing of Mars supported by differential water solubility of iodine and xenon. *Nature* **352**, 697–699.

Namiki N. and Solomon S. C. (1998) Volcanic degassing of argon and helium and the history of crustal production on Venus. *J. Geophys. Res.* **103**, 3655–3677.

Nier A. O. and McElroy M. B. (1977) Composition and structure of Mars' upper atmosphere: results from the neutral mass spectrometers on Viking 1 and 2. *J. Geophys. Res.* **82**, 4341–4349.

Nyblade A. A. and Pollack H. N. (1993) A global analysis of heat flow from Precambrian terrains: implications for the thermal structure of Archean and Proterozoic lithosphere. *J. Geophys. Res.* **98**, 12207–12218.

O'Nions R. K. and Oxburgh E. R. (1983) Heat and helium in the Earth. *Nature* **306**, 429–431.

O'Nions R. K. and Oxburgh E. R. (1988) Helium, volatile fluxes and the development of continental crust. *Earth Planet. Sci. Lett.* **90**, 331–347.

O'Nions R. K. and Tolstikhin I. N. (1994) Behaviour and residence times of lithophile and rare gas tracers in the upper mantle. *Earth Planet. Sci. Lett.* **124**, 131–138.

O'Nions R. K., Carter S. R., Evensen N. M., and Hamilton P. J. (1981) Upper mantle geochemistry. In *The Sea* (ed. C. Emiliani). Wiley, vol. 7, pp. 49–71.

Ott U. (1988) Noble gases in SNC meteorites: Shergotty, Nakhla, Chassigny. *Geochim. Cosmochim. Acta* **52**, 1937–1948.

Owen T., Biemann K., Rushneck D. R., Biller J. E., Howarth D. W., and Lafleur A. L. (1977) The composition of the atmosphere at the surface of Mars. *J. Geophys. Res.* **82**, 4635–4639.

Owen T., Maillard J. P., Debergh C., and Lutz B. (1988) Deuterium on Mars: the abundance of HDO and the value of D/H. *Science* **240**, 1767–1770.

Owen T., Bar Nun A., and Kleinfeld I. (1992) Possible cometary origin of heavy noble gases in the atmospheres of Venus, Earth, and Mars. *Nature* **358**, 43–46.

Owen T., Mahaffay P. R., Niemann H. B., Atreya S., and Wong M. (2001) Protosolar nitrogen. *Astrophys. J.* **553**, L77–L79.

Ozima M. (1973) Was the evolution of the atmosphere continuous or catastrophic? *Nature Phys. Sci.* **246**, 41–42.

Ozima M. (1975) Ar isotopes and Earth-atmosphere evolution models. *Geochim. Cosmochim. Acta* **39**, 1127–1134.

Ozima M. and Kudo K. (1972) Excess argon in submarine basalts and an Earth-atmosphere evolution model. *Nature Phys. Sci.* **239**, 23–24.

Ozima M. and Podosek F. A. (1983) *Noble Gas Geochemistry*. Cambridge University Press, Cambridge.

Ozima M. and Podosek F. A. (2001) *Noble Gas Geochemistry*, 2nd edn. Cambridge University Press, Cambridge.

Ozima M., Podozek F. A., and Igarashi G. (1985) Terrestrial xenon isotope constraints on the early history of the Earth. *Nature* **315**, 471–474.

Parsons B. (1981) The rates of plate creation and consumption. *Geophys. J. Roy. Astron. Soc.* **67**, 437–448.

Patterson D. B., Honda M., and McDougall I. (1990) Atmospheric contamination: a possible source for heavy noble gases basalts from Loihi Seamount, Hawaii. *Geophys. Res. Lett.* **17**, 705–708.

Pepin R. O. (1991) On the origin and early evolution of terrestrial planet atmospheres and meteoritic volatiles. *Icarus* **92**, 1–79.

Pepin R. O. (1998) Isotopic evidence for a solar argon component in the Earths mantle. *Nature* **394**, 664–667.

Pepin R. O. (2000) On the isotopic composition of primordial xenon in terrestrial planet atmospheres. *Space Sci. Rev.* **92**, 371–395.

Pepin R. O. and Carr M. (1992) Major issues and outstanding questions. In *Mars* (eds. H. H. Kieffer, B. M. Jakosky, C. W. Snyder, and M. S. Matthews). University of Arizona Press, Tucson, pp. 120–143.

Pepin R. O. and Phinney D. (1976) The formation interval of the Earth. *Lunar Sci.* **VII**, 682–684.

Pepin R. O. and Porcelli D. (2002) Origin of noble gases in the terrestrial planets. *Rev. Mineral. Geochem.* **47**, 191–246.

Peters K. E., Sweeney R. E., and Kaplan I. R. (1978) Correlation of carbon and nitrogen stable isotope ratios in sedimentary organic matter. *Limnol. Oceanogr.* **23**, 598–604.

Pfennig G., Klewe-Nebenius H., and Seelann-Eggebert W. (1998) *Karlsruhe Chart of the Nuclide*, 6th edn. (revised reprint). Institut für Instrumentelle Analytik, Karlsruhe.

Phinney D., Tennyson J., and Frick U. (1978) Xenon in $CO_2$ well gas revisited. *J. Geophys. Res.* **83**, 2313–2319.

Pineau F. and Javoy M. (1994) Strong degassing at ridge crests: the behaviour of dissolved carbon and water in basalt glasses at 14°N, Mid-Atlantic Ridge. *Earth Planet. Sci. Lett.* **123**, 179–198.

Podosek F. A., Honda M., and Ozima M. (1980) Sedimentary noble gases. *Geochim. Cosmochim. Acta* **44**, 1875–1884.

Pollack H. N., Hurter S. J., and Johnson J. R. (1993) Heat flow from the Earth's interior: analysis of the global data set. *Rev. Geophys.* **31**, 267–280.

Porcelli D. and Ballentine B. J. (2002) Models for the distribution of terrestrial noble gases and evolution of the atmosphere. *Rev. Mineral. Geochem.* **47**, 411–480.

Porcelli D. and Halliday A. N. (2001) The core as a possible source of mantle helium. *Earth Planet. Sci. Lett.* **192**, 45–56.

Porcelli D. and Pepin R. O. (2000) Rare gas constraints on early Earth history. In *Origin of the Earth and Moon* (eds. R. M. Canup and K. Righter). University of Arizona Press, Tucson, pp. 435–458.

Porcelli D. and Wasserburg G. J. (1995a) Mass transfer of xenon through a steady-state upper mantle. *Geochim. Cosmochim. Acta* **59**, 1991–2007.

Porcelli D. and Wasserburg G. J. (1995b) Mass transfer of helium, neon, argon, and xenon through a steady-state upper mantle. *Geochim. Cosmochim. Acta* **59**, 4921–4937.

Porcelli D., Woolum D., and Cassen P. (2001) Deep Earth rare gases: initial inventories, capture from the solar nebula, and losses during Moon formation. *Earth Planet. Sci. Lett.* **193**, 237–251.

Porcelli D., Ballentine C. J., and Wieler R. (2002) An introduction to noble gas geochemistry and cosmochemistry. *Rev. Mineral. Geochem.* **47**, 1–18.

Porcelli D., Pepin R. O., Ballentine C. J., and Halliday A. (2003) Xe and degassing of the Earth (in preparation).

Poreda R. J. and Farley K. A. (1992) Rare gases in Samoan xenoliths. *Earth Planet. Sci. Lett.* **113**, 129–144.

Poreda R., Schilling J. G., and Craig H. (1986) Helium and hydrogen isotopes in ocean ridge basalts north and south of Iceland. *Earth Planet. Sci. Lett.* **78**, 1–17.

Poreda R., Craig H., Arnorsson S., and Welhan J. A. (1992) Helium isotopes in Icelandic geothermal systems: 1. $^3He$, gas chemistry, and $^{13}C$ relations. *Geochim. Cosmochim. Acta* **56**, 4221–4228.

Ragettli R. A., Hebeda E. H., Signer P., and Wieler R. (1994) Uranium–xenon chronology: precise determiantion of $\alpha_{sf}*^{136}Y_{sf}$ for spontaneous fission of $^{238}U$. *Earth Planet. Sci. Lett.* **128**, 653–670.

Reymer A. and Schubert G. (1984) Phanerozoic addition rates to the continental crust. *Tectonics* **3**, 63–77.

Rison W. and Craig H. (1983) Helium isotopes and mantle volatiles in Loihi Seamount and Hawaiian Island basalts and xenoliths. *Earth Planet. Sci. Lett.* **66**, 407–426.

Rocholl A. and Jochum K. P. (1993) Th, U, and other trace elements in carbonaceous chondrites—implications for the terrestrial and solar system Th/U ratios. *Earth Planet. Sci. Lett.* **117**, 265–278.

Ronov A. B. and Yaroshevsky A. A. (1969) Chemical composition of the Earth's crust. *Am. Geophys. Union Geophys. Mon. Ser.* **13**, 37–57.

Ronov A. B. and Yaroshevsky A. A. (1976) A new model for the chemical structure of the Earth's crust. *Geochem. Int.* **13**, 89–121.

Rubey W. W. (1951) Geological history of seawater. *Bull. Geol. Soc. Am.* **62**, 1111–1148.

Rudnick R. and Fountain D. M. (1995) Nature and composition of the continental crust: a lower crustal perspective. *Rev. Geophys.* **33**, 267–309.

Rudnick R. L., McDonough W. F., and O'Connell R. J. (1998) Thermal structure, thickness, and composition of continental lithosphere. *Chem. Geol.* **145**, 395–412.

Russell S. A., Lay T., and Garnero E. J. (1998) Seismic evidence for small-scale dynamics in the lowermost mantle at the root of the Hawaiian hotspot. *Nature* **369**, 225–258.

Sano Y. and Williams S. (1996) Fluxes of mantle and subducted carbon along convergent plate boundaries. *Geophys. Res. Lett.* **23**, 2746–2752.

Sano Y., Takahata N., Nishio Y., and Marty B. (1998) Nitrogen recycling in subduction zones. *Geophys. Res. Lett.* **25**, 2289–2292.

Sarda P., Staudacher T., and Allègre C. J. (1985) $^{40}Ar/^{36}Ar$ in MORB glasses: constraints on atmosphere and mantle evolution. *Earth Planet. Sci. Lett.* **72**, 357–375.

Sarda P., Staudacher T., and Allègre C. J. (1988) Neon isotopes in submarine basalts. *Earth Planet. Sci. Lett.* **91**, 73–88.

Sarda P., Moreira M., Staudacher T., Schilling J. G., and Allègre C. J. (2000) Rare gas systematics on the southernmost Mid-Atlantic Ridge: constraints on the lower mantle and the Dupal source. *J. Geophys. Res.* **105**, 5973–5996.

Sasaki S. and Nakazawa K. (1988) Origin and isotopic fractionation of terrestrial Xe: hydrodynamic fractionation during escape of the primordial $H_2$–He atmosphere. *Earth Planet. Sci. Lett.* **89**, 323–334.

Sasaki S. and Tajika E. (1995) Degassing history and evolution of volcanic activities of terrestrial planets based on radiogenic noble gas degassing models. In *Volatiles in the Earth and Solar System*, AIP Conf. Proc. 341 (ed. K. A. Farley). AIP Press, New York, pp. 186–199.

Schilling J.-G., Unni C. K., and Bender M. L. (1978) Origin of chlorine and bromine in the oceans. *Nature* **273**, 631–636.

Schwartzman D. W. (1973) Argon degassing and the origin of the sialic crust. *Geochim. Cosmochim. Acta* **37**, 2479–2495.

Shen Y., Solomon S. C., Bjarnason I. T., and Wolfe C. J. (1998) Seismic evidence for a lower-mantle origin of the Iceland plume. *Nature* **395**, 62–65.

Sleep N. H. and Zahnle K. (2001) Carbon dioxide cycling and implications for climate on ancient Earth. *J. Geophys. Res.* **106**, 1373–1399.

Sobolev A. V. and Chaussidon M. (1996) $H_2O$ concentrations in primary melts from supra-subduction zones and mid-ocean ridges: implications for $H_2O$ storage and recycling in the mantle. *Earth Planet. Sci. Lett.* **137**, 45–55.

Sobolev A. V., Hofmann A. W., and Nikogosian I. K. (2000) Recycled oceanic crust observed in "ghost plagioclase" within the source of Mauna Loa lavas. *Nature* **404**, 986–990.

Staudacher T. (1987) Upper mantle origin for Harding County well gases. *Nature* **325**, 605–607.

Staudacher T. and Allègre C. J. (1982) Terrestrial xenology. *Earth Planet. Sci. Lett.* **60**, 389–406.

Staudacher T. and Allègre C. J. (1988) Recycling of oceanic crust and sediments: the noble gas subduction barrier. *Earth Planet. Sci. Lett.* **89**, 173–183.

Staudacher T., Sarda P., and Allègre C. J. (1989) Noble gases in basalt glasses from a Mid-Atlantic Ridge topographic high at $14°N$: geodynamic consequences. *Earth Planet. Sci. Lett.* **96**, 119–133.

Stute M., Sonntag C., Deak J., and Schlosser P. (1992) Helium in deep circulating groundwater in the Great Hungarian Plain—Glow dynamics and crustal and mantle helium fluxes. *Geochim. Cosmochim. Acta* **56**, 2051–2067.

Suess H. E. (1949) The abundance of noble gases in the Earth and the cosmos. *J. Geol.* **57**, 600–607 (in German).

Swindle T. D. (2002) Martian noble gases. *Rev. Mineral. Geochem.* **47**, 171–190.

Swindle T. D. and Jones J. H. (1997) The xenon isotopic composition of the primordial martian atmosphere: contributions from solar and fission components. *J. Geophys. Res.* **102**, 1671–1678.

Tajika E. and Sasaki S. (1996) Magma generation on Mars constrained from an $^{40}Ar$ degassing model. *J. Geophys. Res.* **101**, 7543–7554.

Taylor S. R. and McLennan S. M. (1985) *The Continental Crust: Its Composition and Evolution*. Blackwell, Oxford.

Taylor H. P., Jr. and Sheppard S. M. F. (1986) Igneous rocks: I. Processes of isotopic fractionation and isotope systematics. *Rev. Mineral.* **16**, 227–271.

Thompson A. B. (1992) Water in the Earth's upper mantle. *Nature* **358**, 295–302.

Thomsen L. (1980) $^{129}Xe$ on the outgassing of the atmosphere. *J. Geophys. Res.* **85**, 4374–4378.

Tolstikhin I. N. (1975) Helium isotopes in the Earth's interior and in the atmosphere: a degassing model of the Earth. *Earth Planet. Sci. Lett.* **26**, 88–96.

Torgersen T. and Clarke W. B. (1985) Helium accumulation in groundwater: I. An evaluation of sources and the continental flux of crustal $^4He$ in the Great Artesian Basin, Australia. *Geochim. Cosmochim. Acta* **49**, 1211–1218.

Torgersen T. and Ivey G. N. (1985) Helium accumulation in groundwater: II. A model for the accumulation of crustal $^4He$ degassing flux. *Geochim. Cosmochim. Acta* **49**, 2445–2452.

Torgersen T., Kennedy B. M., Hiyagon H., Chiou K. Y., Reynolds J. H., and Clarke W. B. (1989) Argon accumulation and the crustal degassing flux of $^{40}Ar$ in the Great Artesian Basin, Australia. *Earth Planet. Sci. Lett.* **92**, 43–56.

Trieloff M., Kunz J., Clague D. A., Harrison D., and Allègre C. J. (2000) The nature of pristine noble gases in mantle plumes. *Science* **288**, 1036–1038.

Trull T., Nadeau S., Pineau F., Polvé M., and Javoy M. (1993) C–He systematics in hotspot xenoliths: implications for mantle carbon contents and carbon recycling. *Earth Planet. Sci. Lett.* **118**, 43–64.

Turcotte D. L. and Schubert G. (1982) *Geodynamics*. Wiley, New York.

Turcotte D. L. and Schubert G. (1988) Tectonic implications of radiogenic noble gases in planetary atmospheres. *Icarus* **74**, 36–46.

Turekian K. K. (1959) The terrestrial economy of helium and argon. *Geochim. Cosmochim. Acta* **17**, 37–43.

Turekian K. K. (1990) The parameters controlling planetary degassing based on $^{40}$Ar systematics. In *From Mantle to Meteorites* (eds. K. Gopolan, V. K. Gaur, B. L. K. Somayajulu, and J. D. MacDougall). Indian Academy of Sciences, Bangalore, pp. 147–152.

Turekian K. K. and Clark S. P., Jr. (1969) Inhomogeneous accumulation of the Earth from the primitive solar nebula. *Earth Planet. Sci. Lett.* **6**, 346–348.

Turekian K. K. and Clark S. P., Jr. (1975) The non-homogeneous accumulation model for terrestrial planet formation and the consequences for the atmosphere of Venus. *J. Atmos. Sci.* **32**, 1257–1261.

Turner G. (1989) The outgassing history of the Earth's atmosphere. *J. Geol. Soc. London* **146**, 147–154.

Valbracht P. J., Staudacher T., Malahoff A., and Allègre C. J. (1997) Noble gas systematics of deep rift zone glasses from Loihi Seamount, Hawaii. *Earth Planet. Sci. Lett.* **150**, 399–411.

van der Hilst R. D., Widiyantoro S., and Engdahl E. R. (1997) Evidence for deep mantle circulation from global tomography. *Nature* **386**, 578–584.

van Keken P. E., Ballentine C. J., and Porcelli D. (2001) A dynamical investigation of the heat and helium imbalance. *Earth Planet. Sci. Lett.* **188**, 421–443.

Veizer J. and Jansen S. L. (1979) Basement and sedimentary recycling and continental evolution. *Am. J. Sci.* **87**, 341–370.

Veizer J. and Jansen S. L. (1985) Basement and sedimentary recycling: 2. Time dimension to global tectonics. *J. Geol.* **93**, 625–643.

Vidale J. E., Schubert G., and Earle P. S. (2001) Unsuccessful initial search for a mid-mantle chemical boundary with seismic arrays. *Geophys. Res. Lett.* **28**, 859–862.

Volkov V. P. and Frenkel M. Y. (1993) The modelling of Venus degassing in terms of K–Ar system. *Earth Moon Planets* **62**, 117–129.

Von Huene R. and Scholl D. W. (1991) Observations at convergent margins concerning sediment subduction, subduction erosion, and the growth of continental crust. *Rev. Geophys.* **29**, 279–316.

von Zahn U., Kumar S., Niemann H., and Prinn R. (1983) Composition of the Venus atmosphere. In *Venus* (eds. D. Hunten, L. Colin, T. Donahue, and V. Moroz). University of Arizona Press, Tucson, pp. 299–430.

Wänke H. and Dreibus G. (1988) Chemical composition and accretion history of terrestrial planets. *Phil. Trans. Roy. Soc. London* **A325**, 545–557.

Wänke H., Dreibus G., and Jagoutz E. (1984) Mantle chemistry and accretion history of the Earth. In *Archaean Geochemistry* (eds. A. Kröner, G. N. Hanson, and A. M. Goodwin). Springer, Berlin, pp. 1–24.

Weaver B. L. and Tarney J. (1984) Empirical approach to estimating the composition of the continental crust. *Nature* **310**, 575–577.

Weaver C. E. and Wampler J. M. (1970) K, Ar, Illite burial. *Geol. Soc. Am. Bull.* **81**, 3423–3430.

Wedepohl K. H. (1995) The composition of the continental crust. *Geochim. Cosmochim. Acta* **59**, 1217–1232.

Wetherill G. (1975) Radiometric chronology of the early solar system. *Ann. Rev. Nuclear Sci.* **25**, 283–328.

Wollast R. and Mackenzie F. T. (1983) The global cycle of silica. In *Silicon Geochemistry and Biogeochemistry* (ed. S. R. Aston). Academic Press, New York, pp. 39–76.

Wood B. J. (1993) Carbon in the core. *Earth Planet. Sci. Lett.* **117**, 593–607.

Zhao D. (2001) Seismic structure and origin of hotspots and mantle plumes. *Earth Planet. Sci. Lett.* **192**, 251–265.

Zhang Y. and Zindler A. (1993) Distribution and evolution of carbon and nitrogen in Earth. *Earth Planet. Sci. Lett.* **117**, 331–345.

# 4.12

# The Origin of Noble Gases and Major Volatiles in the Terrestrial Planets

D. Porcelli

*University of Oxford, UK*

and

R. O. Pepin

*University of Minnesota, Minneapolis, MN, USA*

## 4.12.1 INTRODUCTION

One of the salient characteristics of the composition of the Earth is the depletion in volatiles compared to parental solar-nebula relative abundances, and this is most pronounced in the noble gases. These chemically unreactive species, concentrated in the atmosphere, have retained many characteristics established early Earth history. A comparison between noble gases on the terrestrial planets and other solar system objects reveals significant differences in both elemental ratios and isotopic compositions and indicates that complex processes were involved in accumulating planetary volatiles from the nebula. Therefore, the atmosphere is not primary (i.e., directly acquired entirely from the solar nebula without modification). Atmophile elements have been added to the Earth in material that has contributed into the growing solid Earth and has subsequently degassed into the atmosphere, in late-accreting materials that degassed upon impact, and possibly directly to the atmosphere from the nebula as well. However, the importance of each of these sources, and the processes that modified these volatiles after initial capture, are still debated.

While considerations of the origin of planetary noble gases have been predominantly focused on those presently found in the atmosphere, noble gases still within the Earth provide further constraints about volatile trapping during planet formation. A wide range of noble-gas information for the Earth's mantle has been obtained from mantle-derived materials, and indicates that there are separate reservoirs within the Earth that have distinctive characteristics that were established early in Earth history. These must be included in comprehensive models of Earth volatile history. Also, data are now available for the atmospheres of both Venus and Mars, as well as from the interior of Mars, so that the evolution of Earth volatiles can be considered within the context of terrestrial-planet formation across the solar system.

The origins of the noble-gas features of the terrestrial-planet atmospheres and interiors have defied simple explanations and are clearly the result of a combination of acquisition and subsequent loss processes that have generated unique solar system compositions. The relevant data for noble gases in the atmospheres and interiors of the terrestrial planets, and the constraints these provide, are summarized below first. The emphasis is on the information provided by noble gases, since the signatures of volatile origin would be expected to survive most clearly in these species, although the implications for the major volatiles—nitrogen, carbon, and hydrogen—are also covered. Acquisition and loss processes are then discussed separately. Models for each planet are then described that are consistent with the available data and involve both acquisition episodes and subsequent modifications during partial loss.

The numerous debates regarding the origin of noble gases on the terrestrial planets have inevitably engendered explanations that are no longer viable, as well as more controversial viewpoints that, as of early 2000s, have not received sufficient support or substantial acceptance. These are discussed further by Pepin and Porcelli (2002).

## 4.12.2 CHARACTERISTICS OF TERRESTRIAL-PLANET VOLATILES

While the terrestrial atmosphere is readily accessible, planetary probes sent into the atmospheres of Venus and Mars have provided data to formulate model histories for these planets as well. Data from *in situ* composition measurements of the Venus atmosphere by mass spectrometers and gas chromatographs on the Pioneer Venus and Venera spacecrafts are reviewed and assessed by von Zahn *et al.* (1983); an updated summary is set out in Wieler (2002). Further data are provided by the martian Shergotty–Nakhla–Chassigny clan (SNC) meteorites. Atmophile elements found in some phases in these rocks have been found to match the probe data for the martian atmosphere and have been identified as shock-implanted atmospheric gases, most clearly seen in the glassy lithology of the SNC meteorite EET79001. These in turn have yielded more precise information about the atmosphere. Other components representing interior martian reservoirs have also been found (see review by Swindle, 2002).

Comparisons can be made between planetary atmospheres and both solar and chondritic compositions. The features of noble gases in the Sun are obtained by measurements of the solar wind. These generally represent those of the bulk solar nebula within which the planets formed, and provide the reference for consideration of terrestrial noble gases. Chondritic meteorites are rich in noble gases, and have long been considered as sources for terrestrial noble gases. While meteoritic gases have significant isotopic differences due to the inclusion of various nucleosynthetic components that escaped homogenization in the solar nebula (see Chapter 1.14), a major primordial component, designated Q gases (see Wieler, 1994), appears to have been widely distributed in the solar system. This component appears to be largely free of "exotic" components, with isotopic compositions similar to those of solar gases, and so represents noble gases trapped in fine-grained solid materials in at least some portion of the nebula. Note that helium is lost from the planetary atmospheres to space, so that early Earth characteristics have not been preserved, and is not discussed in this section.

Table 1 provides the composition of the terrestrial atmosphere, which is the standard for all noble-gas analyses, and indicates the general relative abundances of the isotopes. The masses of the rare-gas planetary reservoirs are given in Table 2.

#### 4.12.2.1 Atmospheric Noble-gas Abundance Patterns

*Noble-gas relative abundances in chondrites and terrestrial-planet atmospheres are generally highly fractionated relative to solar values, except for planetary Xe/Kr ratios. Planetary abundances*

*vary by $10^3$, from gas rich Venus to gas-poor Mars.* Rare-gas abundances in terrestrial-planetary atmospheres are listed in Table 2, and are normalized to the solar pattern in Figure 1, along with the pattern exhibited by trapped noble gases in bulk chondrites. All objects have a striking depletion in noble gases relative to solar abundances. Chondrites display a regular fractionated pattern across the noble gases, with the lightest displaying the greatest depletions. This pattern seems consistent with preferential retention of the generally more reactive heavier noble gases, and was likely established during trapping into solid grains. The Ne/Ar ratios of the terrestrial planets and chondrites are similar, with normalized total abundances in the Earth 10 times greater

**Table 1** Noble-gas isotopes in the terrestrial atmosphere.

| Isotope | Relative abundances | Principal sources of nuclides[a] |
|---------|---------------------|----------------------------------|
| $^3$He | $(1.399 \pm 0.013) \times 10^{-6}$ | Primordial; space[b] |
| $^4$He | $\equiv 1$ | $^{238}$U, $^{235}$U, and $^{232}$Th decay; space[b] |
| $^{20}$Ne | $9.80 \pm 0.08$ | Primordial |
| $^{21}$Ne | $0.0290 \pm 0.0003$ | Primordial; $^{18}$O$(\alpha, n)^{21}$Ne; $^{24}$Mg$(n, \alpha)^{21}$Ne |
| $^{22}$Ne | $\equiv 1$ | Primordial |
| $^{36}$Ar | $\equiv 1$ | Primordial |
| $^{38}$Ar | $0.1880 \pm 0.0004$ | Primordial |
| $^{40}$Ar | $295.5 \pm 0.5$ | Decay of $^{40}$K |
| $^{78}$Kr | $0.6087 \pm 0.0020$ | Primordial |
| $^{80}$Kr | $3.9599 \pm 0.0020$ | Primordial |
| $^{82}$Kr | $20.217 \pm 0.004$ | Primordial |
| $^{83}$Kr | $20.136 \pm 0.021$ | Primordial |
| $^{84}$Kr | $\equiv 100$ | Primordial |
| $^{86}$Kr | $30.524 \pm 0.025$ | Primordial |
| $^{124}$Xe | $2.337 \pm 0.008$ | Primordial |
| $^{126}$Xe | $2.180 \pm 0.011$ | Primordial |
| $^{128}$Xe | $47.15 \pm 0.07$ | Primordial |
| $^{129}$Xe | $649.6 \pm 0.9$ | Primordial; $^{129}$I decay |
| $^{130}$Xe | $\equiv 100$ | Primordial |
| $^{131}$Xe | $521.3 \pm 0.8$ | Primordial; $^{244}$Pu decay; $^{238}$U decay[c] |
| $^{132}$Xe | $660.7 \pm 0.5$ | Primordial; $^{244}$Pu decay; $^{238}$U decay[c] |
| $^{134}$Xe | $256.3 \pm 0.4$ | Primordial; $^{244}$Pu decay; $^{238}$U decay[c] |
| $^{136}$Xe | $217.6 \pm 0.3$ | Primordial; $^{244}$Pu decay; $^{238}$U decay[c] |

After Ozima and Podosek (2001).
[a] Primordial isotopes are those trapped during Earth formation and are not nucleogenic. Only globally significant other sources are included. See Ballentine and Burnard (2002) for production rates and other nuclear production mechanisms. [b] Various mechanisms supply He isotopes from space; see Torgersen (1989) for compilation. [c] The source of variations is within the solid Earth, but has not contributed significantly to the atmosphere.

**Table 2** Volatiles in terrestrial-planet atmospheres.

| Constituent | Earth | Venus | Mars |
|-------------|-------|-------|------|
| Total mass of atm (g) | $5.1 \times 10^{21}$ | $4.75 \times 10^{23}$ | $2.7 \times 10^{19}$ |
| $N_2$ (mol/g-planet) | $2.31 \times 10^{-8}$ | $1.4 \times 10^{-7}$ | $>4.2 \times 10^{-11}$ |
| $^4$He (mol/g-planet) | $1.56 \times 10^{-13}$ | $2.9 \times 10^{-11}$ | |
| $^{20}$Ne (mol/g-planet) | $4.86 \times 10^{-13}$ | $1.0 \times 10^{-11}$ | $2.5 \times 10^{-15}$ |
| $^{36}$Ar (mol/g-planet) | $9.41 \times 10^{-13}$ | $6.7 \times 10^{-11}$ | $5.8 \times 10^{-15}$ |
| $^{84}$Kr (mol/g-planet) | $1.97 \times 10^{-14}$ | $(6.7-130) \times 10^{-14}$ | $1.9 \times 10^{-16}$ |
| $^{130}$Xe (mol/g-planet) | $1.09 \times 10^{-16}$ | $<5 \times 10^{-15}$ | $3.2 \times 10^{-18}$ |
| Total mass of planet (g) | $5.98 \times 10^{27}$ | $4.87 \times 10^{27}$ | $0.64 \times 10^{27}$ |

Source: Ozima and Podosek (2001).

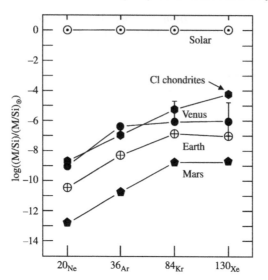

**Figure 1** Noble-gas abundances in planetary atmospheres and CI chondrites, plotted as the atom concentration relative to Si and divided by the corresponding solar ratio. Note that ranges of Kr and Xe values are shown for Venus (source Pepin, 1991).

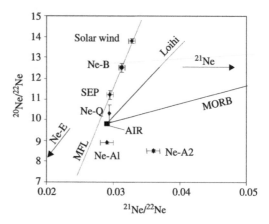

**Figure 2** The Ne-isotopic composition of the atmosphere, compared to different extraterrestrial compositions (compiled by Busemann *et al.*, 2000). MORB data fall on a correlation line extending from air values (Sarda *et al.*, 1988), and reflect mixing between air contamination of the samples with a trapped mantle component with high Ne-isotope ratios. This component is composed of radiogenic $^{21}$Ne and solar Ne (Honda *et al.*, 1993) or Ne-B (Trieloff *et al.*, 2000). OIB such as Loihi with high $^{3}$He/$^{4}$He ratios reflect mixing of air contamination with a mantle composition that has a lower $^{21}$Ne/$^{22}$Ne ratio, reflecting a higher time-integrated Ne/(U + Th) ratio (Honda *et al.*, 1991).

than on Mars, and $10^2$ times less than on Venus. In contrast, the heavier noble gases display greater variations in relative abundances. While the Kr/Ar ratio of chondrites and the Earth are similar, the terrestrial Xe/Kr ratio is much lower and similar to the solar value. This was initially thought to be due to the sequestration of terrestrial xenon in other terrestrial reservoirs. However, investigations of possible reservoirs of xenon, such as shales or glacial ice, failed to find this "missing xenon" (Bernatowicz *et al.*, 1984, 1985; Wacker and Anders, 1984; Matsuda and Matsubara, 1989). The lower terrestrial Xe/Kr ratio has since then widely been considered to be a feature of the Earth. The pattern exhibited by Mars closely follows that of the Earth, with a near-solar Xe/Kr ratio. Venus also appears to have an Xe/Kr ratio near the solar value, although uncertainties are large. However, it seems clear that Venus differs markedly from the other two terrestrial planets in having an Ar/Kr ratio that is also close to solar.

### 4.12.2.2 Atmospheric Neon

*Neon in planetary atmospheres is isotopically heavier than the solar composition, and falls within meteoritic values.* Neon-isotopic compositions for the atmosphere and solar system reservoirs are shown in Figure 2. The greatest differences between solar system bodies are seen in the proportions of $^{20}$Ne and $^{22}$Ne, which are not produced in significant quantities in large bodies. The value of $^{20}$Ne/$^{22}$Ne = 13.8 ± 0.1 (Benkert *et al.*, 1993; Pepin *et al.*, 1999) derived for the solar wind is believed to represent the initial

solar-nebula composition. Solar neon found in irradiated meteorites is a mixture, called Ne-B, of solar-wind (SW) neon and fractionated solar-energetic particle (SEP) neon (Black, 1972), and has a value of $^{20}$Ne/$^{22}$Ne ~ 12.5. Neon-isotope ratios in bulk CI chondrites scatter around an average $^{20}$Ne/$^{22}$Ne of 8.9 ± 1.3. Meteoritic Q gases have $^{20}$Ne/$^{22}$Ne = 10.1–10.7 (Busemann *et al.*, 2000). The terrestrial-atmospheric ratio of 9.8, in principle, can be derived either from mixing meteorite and solar components or by fractionation of solar neon. The difference between the solar (0.033) and atmospheric (0.029) $^{21}$Ne/$^{22}$Ne ratios is consistent with fractionation of solar neon and addition of radiogenic $^{21}$Ne (see Porcelli and Ballentine, 2002).

The $^{20}$Ne/$^{22}$Ne ratio of the martian atmosphere has been estimated to be 10.1 ± 0.7 from SNC meteorite data (Pepin, 1991). It is indistinguishable from neon in the terrestrial atmosphere, and so is substantially lower than the solar value. The Venus ratio of $^{20}$Ne/$^{22}$Ne = 11.8 ± 0.7 (Donahue and Russell, 1997) requires some fractionation relative to the solar value, although not to the same extent as neon in the Earth's atmosphere. The $^{21}$Ne/$^{22}$Ne ratios of Mars and Venus are not known sufficiently well to provide further constraints.

### 4.12.2.3 Atmospheric Argon

*Nonradiogenic argon in chondrites and planetary atmospheres is isotopically heavier than the solar composition.* The initial $^{40}Ar/^{36}Ar$ ratio in the solar system is $\sim 10^{-4}$–$10^{-3}$ (Begemann *et al.*, 1976). The atmosphere has $^{40}Ar/^{36}Ar = 296$, so that essentially all the $^{40}Ar$ has been produced by $^{40}K$ decay in the solid Earth. Regarding the two nonradiogenic isotopes, the atmospheric $^{38}Ar/^{36}Ar$ ratio of 0.188 is similar to that found in CI chondrites of $0.189 \pm 0.002$ (Mazor *et al.*, 1970) but substantially higher than some recent estimates of $0.172 \pm 0.002$ (Pepin *et al.*, 1999; Palma *et al.*, 2002) and $0.179 \pm 0.001$ (Wieler, 1998) for the solar wind. If atmospheric argon was derived from a solar source, it must have been fractionated.

The martian $^{38}Ar/^{36}Ar$ ratio is highly fractionated relative to the solar ratio; SNC meteorite analyses yield values from $0.24 \pm 0.01$ (Pepin, 1991) to 0.26–0.30 (Bogard, 1997; Garrison and Bogard, 1998). The measured Venus value of $0.180 \pm 0.019$ (Donahue and Russell, 1997) is indistinguishable from both the solar and terrestrial values.

#### 4.12.2.4 Atmospheric Krypton

*Krypton in meteorites and the Earth is isotopically fractionated relative to the solar composition, although this may not be true for krypton on Mars.* Isotopic variations due to production of krypton isotopes is not expected in bulk solar system bodies. Solar, bulk meteorite, and terrestrial krypton-isotopic compositions (Figure 3) are generally related to one another by mass fractionation, with the terrestrial atmosphere depleted in light isotopes by $\sim 0.8\%$ amu$^{-1}$ relative to the solar composition (Eugster *et al.*, 1967; Pepin *et al.*, 1995). Meteoritic krypton possibly contains an excess in $^{86}Kr$. The krypton-isotopic composition of the atmosphere of Mars (Pepin, 1991; Garrison and Bogard, 1998) is essentially indistinguishable from that of SW-Kr, but may be slightly fractionated to an isotopically lighter composition (see Garrison and Bogard, 1998; Swindle, 2002). No data are available for krypton on Venus.

#### 4.12.2.5 Terrestrial Atmospheric Xenon

*Nonradiogenic xenon on the Earth is highly fractionated, and requires early, fractionating losses to space of a nonsolar precursor. Radiogenic xenon abundances require losses to space over 100 Myr.*

##### 4.12.2.5.1 Nonradiogenic xenon

The source and composition of terrestrial nonradiogenic xenon have been difficult to constrain, because atmospheric xenon does not match any widespread solar system xenon composition.

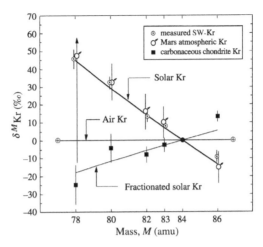

**Figure 3** Krypton isotopes in solar system volatile reservoirs, plotted as ‰ deviations of the ratio to $^{84}Kr$, and normalized to the ratio in terrestrial air (Basford *et al.*, 1973). The heavy "solar Kr" curve represents a smooth fit to the measured solar-wind isotope ratios. Measured SW-Kr from Wieler and Baur (1994) and Pepin *et al.* (1995); Mars Kr from Pepin (1991); carbonaceous chondrite Kr from Krummenacher *et al.* (1962), Eugster *et al.* (1967), and Marti (1967) (Pepin and Porcelli, 2002) (reproduced by permission of the Mineralogical Society of America from *Rev. Mineral. Geochem.* **2002**, *47*, 191–246).

The problem has been made more difficult because contributions to $^{129}Xe$ by decay of $^{129}I$, and to the heavy isotopes $^{131}Xe$ to $^{136}Xe$ by fission of $^{244}Pu$ and $^{238}U$ that cannot be independently tightly constrained. The light isotopes of atmospheric xenon are related to both bulk chondritic and solar xenon by fractionation of $\sim 4.2\%$ amu$^{-1}$ (Krummenacher *et al.*, 1962), with a clear radiogenic excess of $\sim 7\%$ in $^{129}Xe$ from decay of $^{129}I$. However, these compositions do not match the heavy xenon isotopes. As shown in Figure 4, where solar xenon is fractionated to match the light isotopes of the atmosphere, the $^{136}Xe/^{130}Xe$ ratio is well above atmospheric. Consequently, SW-Xe cannot be the primordial terrestrial composition. This exclusion also applies to the meteoritic CI-Xe and Q-Xe compositions, which are considerably richer in the heavy isotopes than SW-Xe. However, a suitable initial composition for the atmosphere has been derived using multidimensional isotopic correlations of chondrite data (Pepin, 2000). Terrestrial atmospheric xenon is then composed of the nonradiogenic composition (dubbed U-Xe) that has been fractionated, as well as radiogenic $^{129}Xe$ and fissiogenic xenon from decay of $^{244}Pu$.

U-Xe is necessarily depleted in the heaviest isotopes compared to SW-Xe, and this suggests that there is a heavy-isotope component in the Sun but not in the early Earth. This presents a conundrum, since SW-Xe, presumably reflecting the xenon composition in the accretion disk,

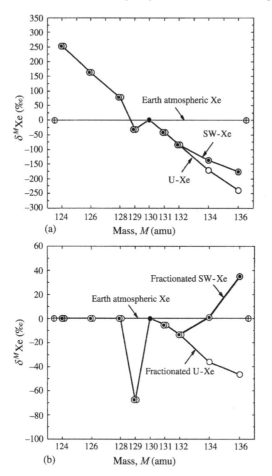

**Figure 4** Relationships of: (a) unfractionated SW-Xe and U-Xe to terrestrial atmospheric Xe, plotted as per mil differences from the air Xe composition (Table 1). (b) SW-Xe and U-Xe, after hydrodynamic escape fractionation to the extent required to match the $^{124-128}Xe/^{130}Xe$ ratios to the corresponding air Xe values. The fractionated solar-wind $^{136}Xe/^{130}Xe$ ratio is elevated above the air ratio by $\sim 10\sigma$ (source Pepin, 2000).

would arguably be the more plausible contributor to primordial planetary inventories. Note, however, that such a problematic relationship will apply to any primordial composition for the Earth. Moreover, while it is difficult to understand why U-Xe has not been widely found in other solar system materials, this also applies to any suitable composition. A possible explanation is that the xenon composition of the nebula changed over the lifetime of the disk due to changes in material supplied from the surrounding molecular cloud (Pepin *et al.*, 1995; Pepin, 2003). An important parameter of this problem is the xenon-isotope composition of the atmosphere of Jupiter, which contains gases gravitationally captured from the solar nebula. Results from the Galileo Probe mass spectrometer (Mahaffy *et al.*, 1998) have narrowed the range of possible nebular-xenon

compositions at the time of Jovian-atmospheric formation. The relative isotopic abundances fall approximately along a fractionation curve with respect to nonradiogenic-terrestrial xenon, but unfortunately are not known with sufficient precision so as to distinguish between U-Xe and SW-Xe.

#### 4.12.2.5.2 Radiogenic xenon

Heavy xenon isotopes (typically represented by $^{136}Xe$) are produced by fission of $^{244}Pu$ ($t_{1/2} = 80$ Ma) and $^{238}U$ ($t_{1/2} = 4.47$ Ga). Since all $^{244}Pu$ has decayed, and there are no stable isotopes presently representing plutonium, the original plutonium concentration in the Earth must be estimated from those of other highly refractory elements. Meteorite data suggest that at 4.57 Ga, $(^{244}Pu/^{238}U)_0 = 6.8 \times 10^{-3}$ (Hudson *et al.*, 1989; Hagee *et al.*, 1990), and so the silicate Earth, or Earth-forming materials, with 21 ppb $^{238}U$ at present, initially had 0.29 ppb $^{244}Pu$ that produced $2.0 \times 10^{35}$ atoms of $^{136*}Xe$. The amount produced by $^{238}U$ over the age of the Earth ($7.5 \times 10^{33}$ atoms $^{136}Xe$) is much less, and so bulk Earth (and atmospheric) $^{136*}Xe$ must be dominantly plutonium derived. The short-lived isotope $^{129}I$ ($t_{1/2} = 15.7$ Ma) produces $^{129*}Xe$. The bulk silicate Earth (BSE) concentration of stable $^{127}I$ is difficult to constrain. Wänke *et al.* (1984) estimated a commonly quoted BSE concentration of 13 ppb based on mantle-xenolith data, and this is consistent with other data (Déruelle *et al.*, 1992). At 4.57 Ga, $(^{129}I/^{127}I)_0 = 1.1 \times 10^{-4}$ based on meteorite data (Hohenberg *et al.*, 1967; Brazzle *et al.*, 1999). Using 13 ppb I, $2.7 \times 10^{37}$ atoms of $^{129*}Xe$ were produced in the Earth or Earth-forming materials.

Using fractionated U-Xe for the nonradiogenic composition of the atmosphere, $6.8 \pm 0.3\%$ of atmospheric $^{129}Xe$ ($^{129*}Xe_{atm} = 1.7 \times 10^{35}$ atoms) and $4.65 \pm 0.50\%$ of atmospheric $^{136}Xe$ ($^{136*}Xe_{atm} = 3.81 \times 10^{34}$ atoms) are radiogenic. The $^{136*}Xe_{atm}$ is 20% of the total $^{136}Xe$ produced by $^{244}Pu$ in the silicate Earth. In contrast, the $^{129*}Xe_{atm}$ is only 0.8% of the total $^{129}Xe$ produced; such a low value can neither be accounted for by incomplete degassing of the mantle nor from any uncertainties in the estimated amount of iodine, and requires losses to space. This must have occurred during early Earth history, when such heavy species could have been lost either from protoplanetary materials or from the growing Earth. Wetherill (1975) proposed that a "closure age" of the Earth could be calculated by assuming that essentially complete loss of $^{129*}Xe$ occurred initially, followed by complete closure against further loss. The time when this closure commenced can be calculated by

$$t = \frac{-1}{\lambda_{129}} \ln\left[ \left( \frac{^{129*}Xe_{atm}}{^{127}I} \right) \left( \frac{^{127}I}{^{129}I} \right)_0 \right] \quad (1)$$

Using an updated value for $^{129*}Xe_{atm}$, and assuming 40% of the mantle degassed to the atmosphere (see Chapter 4.11), 83 Ma is obtained. The "closure age" can also be calculated by combining the $^{129}I$–$^{129}Xe$ and $^{244}Pu$-$^{136}Xe$ systems (Pepin and Phinney, 1976):

$$t = \frac{1}{\lambda_{244} - \lambda_{129}} \ln\left[ \left( \frac{^{129*}Xe_{atm}}{^{136*}Xe_{atm}} \right) \left( \frac{^{238}U}{^{127}I} \right)_0 \right.$$
$$\left. \times \left( \frac{^{244}Pu}{^{238}U} \right)_0 \left( \frac{^{127}I}{^{129}I} \right)_0 {^{136}Y_{244}} \right] \quad (2)$$

Using a value of $7.05 \times 10^{-5}$ for the parameter $^{136}Y_{244}$ that represents the number of $^{136}Xe$ atoms produced per decay of $^{244}Pu$, a similar closure age of 96 Ma is obtained. For example, note that if atmospheric xenon loss occurred during a massive Moon-forming impact, then the closure period corresponds to the time after an instantaneous catastrophic loss event.

### 4.12.2.6 Martian Atmospheric Xenon Isotopes

*Martian atmospheric xenon is likely derived from solar xenon, with addition of radiogenic $^{129*}Xe$ and plutonium-derived heavy xenon after a closure age of ~70 Ma.* Xenon-isotope data for the martian atmosphere have been derived entirely from SNC meteorites (Figure 5). Swindle *et al.* (1986) pointed out that martian atmospheric xenon strongly resembles mass-fractionated CI-Xe, with addition of radiogenic $^{129*}Xe$ (comprising 61% of total $^{129}Xe$). This implies that there are no additional contributions to the xenon inventory from degassed fission xenon. Alternatively, Swindle and Jones (1997) demonstrated that SW-Xe could be fractionated to fall below the measured martian atmospheric $^{131-136}Xe/^{130}Xe$ ratios by amounts consistent with the presence of $^{244}Pu$-derived xenon that comprises ~5% of total $^{136}Xe$. A revised SNC database for martian atmospheric-xenon composition by Mathew *et al.* (1998) closely matches stronger hydrodynamic-escape fractionation of SW-Xe (see Section 4.12.4.2) with no $^{244}Pu$ fission-xenon contribution at all. However, a weaker SW-Xe fractionation plus a $^{244}Pu$ contribution is still allowed within the data uncertainties. An added feature of the Mathew–Marti database is that it is poorly fit by fractionated CI-Xe, and consequently points to SW-Xe rather than CI-Xe as the primordial martian composition if these data really represent the martian atmosphere.

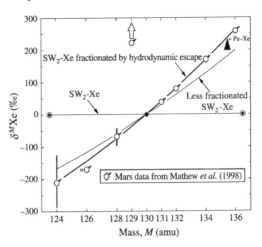

**Figure 5** The martian atmospheric composition (Mathew *et al.*, 1998) plotted relative to SW-Xe (see Pepin and Porcelli, 2002). The heavy curve demonstrates that SW-Xe in the primordial martian atmosphere can be fractionated by escape to a close fit to all atmospheric isotope ratios except $\delta^{129}Xe$ (which is off scale at +1,480‰). The light curve represents a less severely fractionated SW-Xe composition, which allows for the addition of a $^{131-136}Xe$ component with the composition of fission-produced Xe from decay of $^{244}Pu$ (as suggested by Swindle and Jones (1997)) (Pepin and Porcelli, 2002) (reproduced by permission of the Mineralogical Society of America from *Rev. Mineral. Geochem.* **2002**, *47*, 191–246).

The composition of primordial xenon on Mars and the presence or absence of $^{244}Pu$ fission xenon in the present atmosphere are important issues in the context of the provenance of accretional materials, the timing of planetary growth, and the subsequent geochemical and outgassing histories of the planet. The central question of an atmospheric $^{244}Pu$ component is currently plagued by apparent coincidences. The excellent match of fractionated SW-Xe, alone, to the atmospheric data is presumably fortuitous if Pu-Xe is present; and if it is absent, the fact that a weaker fractionation generates heavy-isotope residuals in good accord with $^{244}Pu$ fission yields must likewise be accidental. Atmospheric xenon data with higher measurement precision than is currently attainable for xenon extracted from the SNCs will eventually resolve the issue.

It has been estimated that the silicate portion of Mars contains 32 ppb I and 16 ppb of $^{238}U$ at present, equivalent to 32.3 ppb initially (Wänke and Dreibus, 1988), so that I/U = 0.54 (molar ratio) at 4.57 Ga. If the atmosphere contains plutonium-derived xenon comprising 5% of the atmospheric $^{136}Xe$, so that $(^{129*}Xe/^{136*}Xe) = 83$, then using Equation (2), a closure age of 46 Ma is obtained, which is a factor of 2 lower than that of the Earth. In this case, ~2% of the mantle has degassed radiogenic xenon, which is compatible with the $^{40}Ar$ budget (see Chapter 4.11), and so supports the identification of plutonium-derived xenon in the atmosphere. If SW-Xe is the primordial composition of the atmosphere, if there is not a clearly resolvable fission component, then the closure age is much lower. However, in this case only a very low—perhaps unreasonably low—fraction of the mantle could have degassed to the atmosphere.

### 4.12.2.7  Noble-gas Isotopes in the Terrestrial Mantle

*The mantle contains solar-derived neon and less-fractionated xenon, as well as nonatmospheric abundance patterns. There are considerable amounts of $^3He$ and other gases retained within the Earth after early losses.* Evidence for primordial noble gases in the mantle first came from helium, with $^3He/^4He$ ratios in mid-ocean ridge basalts (MORBs) that are melts of the convecting upper mantle of $\sim 8R_A$ (where $R_A$ is the atmospheric ratio) that represents a mixture of primordial $^3He$ with radiogenic helium (with $\sim 0.01R_A$). Values of up to $37R_A$ have been found in ocean islands (see Chapter 2.06).

Measured MORB $^{20}Ne/^{22}Ne$ ratios are greater than that of the atmosphere, and extend toward the solar value (Figure 2). Since these isotopes are not produced in significant quantities in the Earth, there must be trapped primordial neon in the Earth with a $^{20}Ne/^{22}Ne$ ratio at least as high as the highest measured mantle value. Samples with lower measured ratios appear to have contaminant air neon. While it has often been assumed that the trapped neon was derived directly from the solar nebula with $^{20}Ne/^{22}Ne = 13.8$ (Honda *et al.*, 1993), it has been suggested that it is the meteoritic Ne-B component with $^{20}Ne/^{22}Ne = 12.5$ (Trieloff *et al.*, 2000, 2002). While earlier measured mantle values above this (Sarda *et al.*, 1988) were insufficiently precise to firmly establish the presence of neon with a higher value, recent analyses for Icelandic samples found $^{20}Ne/^{22}Ne = 13.75 \pm 0.32$ (Harrison *et al.*, 1999), although this measurement has been debated (see Ballentine *et al.*, 2001; Trieloff *et al.*, 2000, 2001). Clearly, further work is required over this issue. High $^{21}Ne/^{22}Ne$ ratios in the mantle (Figure 2) are due to production of $^{21}Ne$ by various nuclear reactions (see Chapter 4.11).

Solar argon might be expected in the mantle as well. $^{40}Ar/^{36}Ar$ ratios of up to $4 \times 10^4$ have been found due to production of $^{40}Ar$ in the mantle (Burnard *et al.*, 1997), and so contain no information on the origin of argon. Measurements of nonradiogenic MORB and Loihi $^{38}Ar/^{36}Ar$ ratios are typically atmospheric within error, but have been of low precision due to the low abundance of these isotopes. Also, since most $^{40}Ar/^{36}Ar$ ratios are interpreted as having been

lowered due to a substantial fraction of atmospheric contamination, the $^{36}Ar$ (and $^{38}Ar$) of these samples are dominated by contamination. Two recent analyses of MORB and OIB samples have found $^{38}Ar/^{36}Ar$ ratios lower than that of the atmosphere (Valbracht *et al.*, 1997; Niedermann *et al.*, 1997), and Pepin (1998) argued that these values reflect a mixture of a solar upper-mantle composition with atmospheric argon contamination. However, new high-precision data for MORB with $^{40}Ar/^{36}Ar \sim 2.8 \times 10^4$ failed to find $^{38}Ar/^{36}Ar$ ratios that deviate from that of air (Kunz, 1999). Similar air-like $^{38}Ar/^{36}Ar$ ratios were also found in high-precision OIB analyses (Trieloff *et al.*, 2000, 2002), in contrast to the earlier measurements. Further work is certainly required, considering the importance of establishing whether trapped argon within the Earth has been fractionated relative to solar argon, and so whether the atmospheric composition was indeed generated by processes occurring in the planetary atmosphere. If solar argon is not found in the mantle, the possibility that atmospheric $^{38}Ar/^{36}Ar$ ratios reflect subduction of atmospheric argon rather than the composition of initially trapped argon also must be considered.

The measured ratios of the nonradiogenic xenon isotopes in MORBs are indistinguishable from those in the atmosphere. However, more precise measurements of mantle-derived xenon in $CO_2$ well gases have been found to have higher $^{124-128}Xe/^{130}Xe$ ratios (Phinney *et al.*, 1978; Caffee *et al.*, 1999) that can be explained by either: (i) a mixture of $\sim 10\%$ solar xenon trapped within the Earth and $\sim 90\%$ atmospheric xenon (subducted or added in the crust); or (ii) a mantle component that has not been fractionated relative to solar xenon to the same extent as air xenon. In the former case, the radiogenic composition of the nonatmospheric component can be calculated (Jacobsen and Harper, 1996); if 90% of the well-gas xenon is derived from the atmosphere (with $^{129}Xe/^{130}Xe = 6.496$), the well-gas value of $^{129}Xe/^{130}Xe = 7.2$ contains 10% of a solar component with $^{129}Xe/^{130}Xe \approx 13.5$. Note that it is not possible to distinguish whether solar Xe or U-Xe is present in the mantle.

Mantle-derived materials have a range of elemental-abundance patterns due to various fractionation processes, although the mantle pattern has been inferred from measured isotopic variations and radiogenic isotope-production ratios (see Porcelli and Ballentine, 2002). From the $^4He/^{21}Ne$ production ratio and the average coexisting shifts in $^4He/^3He$ and $^{21}Ne/^{22}Ne$ in the mantle relative to the primordial compositions, a ratio of $^3He/^{22}Ne = 11$ is obtained. This is greater than the more recent estimate of 1.9 for the solar nebula (see discussion in Porcelli and Pepin (2000)). Similarly, using the mantle $^{40}Ar/^{36}Ar$

ratio, and an estimate of the ratio of $^{21*}$Ne production in the mantle to that of $^{40*}$Ar, a ratio of $^{22}$Ne/$^{36}$Ar = 0.15 is inferred, which is substantially lower than the solar value. These values suggest that solar noble gases found in the mantle have been fractionated when the deep-Earth reservoirs were established, although it is possible that subduction of argon modified the $^{22}$Ne/$^{36}$Ar ratio, as well as changed the $^{38}$Ar/$^{36}$Ar ratio. Similarly, the calculated Xe/Ar ratio depends upon the amount of subducted xenon (Porcelli and Wasserburg, 1995b). However, a straightforward calculation is not possible without assumptions regarding where upper-mantle xenon has been stored and how much of the radiogenic noble gases have been produced in the upper mantle (see Chapter 4.11). If the nonsubducted component in MORB has $^{129}$Xe/$^{130}$Xe = 13.5 (see above) and so $^{136}$Xe/$^{130}$Xe = 4.42, and all the excess $^{136}$Xe has been produced by $^{238}$U in a reservoir over 4.5 Ga, along with all the $^{40}$Ar, with a K/U ratio of $1.27 \times 10^4$, then MORB has $^{130}$Xe/$^{36}$Ar = $8 \times 10^{-4}$. In the context of a mantle model, Porcelli and Wasserburg (1995b) obtained a ratio of $1.9 \times 10^{-3}$. These values are higher than the solar ratio of $2 \times 10^{-5}$; lower ratios are obtained if a much greater proportion of the mantle xenon is subducted so that stored mantle xenon has a higher $^{136}$Xe/$^{130}$Xe ratio.

The total amount of noble gases in the mantle provides another constraint on acquisition mechanisms. The upper-mantle $^3$He concentration has been estimated in various ways from the total $^3$He flux from mid-ocean ridges, the ratio of C/$^3$He in MORBs, and from concentrations in relatively undegassed basalts, with a range of an MORB-source mantle $^3$He concentration of (1.2–4.6) $\times 10^9$ atoms g$^{-1}$ (see Porcelli and Ballentine (2002) for discussion). However, this is a lower limit to the amount initially trapped within the mantle, since the upper mantle has clearly lost volatiles to the atmosphere (see Chapter 4.11). Evidence for a more gas rich mantle reservoir comes from ocean-island hot spots such as Hawaii and Iceland that have $^3$He/$^4$He ratios of up to 37$R_A$ compared to an MORB value of 8$R_A$ (see Graham, 2002). Clearly, these have had a higher time-integrated He/U ratio than the MORB source. Had these have been derived from a reservoir that evolved as a closed system since Earth formation with a BSE uranium concentration of 21 ppb, then this reservoir would have had $7.6 \times 10^{10}$ atom $^3$He g$^{-1}$ (e.g., Porcelli and Wasserburg, 1995a), and, with $^3$He/$^{22}$Ne = 11, it also has $6.9 \times 10^9$ atom $^{22}$Ne g$^{-1}$. For comparison, the neon inventory in the atmosphere divided into the mass of the upper mantle yields only $2 \times 10^{11}$ atom $^{22}$Ne g$^{-1}$. Other possible interpretations of high $^3$He/$^4$He ratios (e.g., Helffrich and Wood, 2001; Porcelli and Halliday, 2001)

have not been fully developed and may still require similar high initial mantle noble-gas concentrations.

As discussed above, xenon isotopes indicate that early losses of radiogenic isotopes occurred from the atmosphere or its source reservoir. It has been argued that, based on the proportion of plutonium-derived fissiogenic xenon (Kunz *et al.*, 1998), mantle data indicate that a similarly late-xenon closure age applies to the mantle as well (Porcelli *et al.*, 2001). Trapping of nonradiogenic noble gases in the Earth must have occurred early (see Section 4.12.3), and so extensive losses of radiogenic xenon also necessarily apply to these as well. Modeling of terrestrial-isotope compositions indicates that up to 99% of noble gases originally trapped in the Earth were lost, and so acquisition mechanisms must supply not only the inventories currently found, but $10^2$ times more (Porcelli *et al.*, 2001).

### 4.12.2.8 Nonradiogenic Xenon Isotopes in the Martian Mantle

*The martian mantle has high xenon concentrations and distinct abundance patterns.* Martian meteorites contain components other than those derived directly from the atmosphere (see detailed discussion by Swindle, 2002). In particular, noble gases in the dunite meteorite Chassigny appear to represent a distinct interior reservoir. The $^{84}$Kr/$^{132}$Xe ratio of 1.2 (Ott, 1988) is lower than both the martian atmosphere (20) and solar (16.9) values, but is similar to that of CI chondrites. If this is truly a source feature, it indicates that heavy noble gases trapped within the planet suffered substantially different elemental fractionation than the atmosphere. The interior $^{84}$Kr/$^{36}$Ar ratio of 0.06 is much higher than the solar value of $2.8 \times 10^{-4}$, but it is close to the atmospheric value of 0.02 and so does not display the same contrast as the Kr/Xe ratio. Unfortunately, it is not possible to determine if the measured elemental abundance ratios were modified by planetary processing or transport and incorporation into the samples.

The isotopic composition of the martian interior is only available for xenon. Data for the dunite Chassigny indicate that there is a mantle reservoir with nonradiogenic isotope ratios that appear to be indistinguishable from solar values (Ott, 1988; Mathew and Marti, 2001), and so does not exhibit the strong isotopic fractionation seen in the atmosphere. The relative abundances of $^{129}$Xe and $^{136}$Xe are also close to solar, indicating that this reservoir had a high Xe/Pu and Xe/I ratios, at least during the lifetime of $^{244}$Pu. Data from other meteorites indicate that there are other interior martian reservoirs that contain solar xenon but with resolvable fissiogenic contributions, and so have had lower Xe/Pu ratios (Marty and Marti, 2002).

### 4.12.2.9  Major Volatiles

The "surface" inventories of major volatiles that have been added directly to the Earth's surface or removed from the mantle include not only atmospheric abundances, but also those of the hydrosphere and crustal rocks. Isotopic compositions are generally reported as $\delta$ units of per mil deviations from a standard.

#### 4.12.2.9.1  Nitrogen

*Nitrogen abundances in the atmospheres of Earth and Venus are comparable, being much higher than that of Mars. Meteoritic $\delta^{15}N$ values overlap the terrestrial-atmosphere and the lower-mantle values, while that of Mars' atmosphere is much higher. Planetary $N/^{36}Ar$ ratios fall between meteorite and solar ratios.* While much of the terrestrial nitrogen at the surface is in the atmosphere with $\delta^{15}N = 0‰$ (where $\delta^{15}N = [(^{15}N/^{14}N)/(^{15}N/^{14}N)_{atm} - 1] \times 10^3)$, $\sim30\%$ is in crustal rocks (Wlotzka, 1972), and so the "surface" has $5.3 \times 10^{21}$ g N with $\delta^{15}N \approx +1‰$ (Tolstikhin and Marty, 1998). Divided into the mass of the upper mantle (the minimum degassed volume of the Earth), this is equivalent to 5 ppm N. Studies of MORBs indicate that the upper mantle contains $\sim0.16$ ppm N, assuming nitrogen is incompatible during melting (Marty and Humbert, 1997). MORBs also have a value of $\delta^{15}N \approx -3.3‰$ (e.g., Marty and Zimmermann, 1999). Overall, it appears that the bulk Earth has $\sim5$ ppm N, with $\delta^{15}N = -5‰$ to $-3‰$ (Javoy and Pineau, 1991; Marty and Humbert, 1997). However, it has been argued that there may be up to 40 ppm in the mantle, which requires that nitrogen behaves more compatibly during melting (Cartigny et al., 2001). Values as low as $\delta^{15}N = -25‰$ have been found in diamonds, indicating that there is an isotopically light nitrogen component in the mantle; since subducting materials are expected to have $\delta^{15}N > 0‰$, these values represent a maximum $\delta^{15}N$ value for nitrogen initially trapped in the mantle (Cartigny et al., 1998). Higher values in MORBs and diamonds may be either a mixture of light trapped nitrogen from the deeper mantle and subducted nitrogen, or represent mantle heterogeneities established during Earth formation.

The atmosphere of Mars has a high value of $\delta^{15}N = +620 \pm 160‰$ (Nier and McElroy, 1977) that may be due to fractionating losses (see Section 4.12.7), and $(4.7 \pm 1.2) \times 10^{17}$ g N, which is equivalent to 0.7 ppb when divided into the mass of the entire planet (Owen et al., 1977). Therefore, Mars appears to have $10^{-4}$ times the

nitrogen on the Earth. Mathew et al. (1998) reported evidence for a component in a martian meteorite with $\delta^{15}N < -22‰$, suggesting that, like the Earth, the solid planet may contain nitrogen that is isotopically lighter nitrogen than the atmosphere. The available atmospheric data for Venus is very imprecise, with $\delta^{15}N = 0 \pm 200‰$ (von Zahn et al., 1983), and there is $\sim2$ times more nitrogen ($1.1 \times 10^{22}$ g) than that on the surface of the Earth.

There has been some controversy over the solar $\delta^{15}N$ value. Nitrogen in lunar samples appears to have more than one component, and it has not been clear whether solar-wind implanted nitrogen is an isotopically light or heavy component (see Becker, 2000). While direct measurement of the solar wind by SOHO gave $\delta^{15}N = +400^{+500}_{-300}‰$ (Kallenbach et al., 1998), measurements of the Jupiter atmosphere, where solar-nebula gases have been retained, found $\delta^{15}N = -370 \pm 80‰$ (Owen et al., 2001), and a recent ion-microprobe study of lunar samples suggested that the solar wind has $\delta^{15}N < -240‰$ (Hashizume et al., 2000). A wide range of $\delta^{15}N$ values has been reported for meteorites, and the causes of the variations are not understood. The several classes that have $\delta^{15}N < 0$ are the CO and CV chondrites, with bulk concentrations of up to almost 200 ppm and $\delta^{15}N$ down to $-40‰$ (Kung and Clayton, 1978; Kerridge, 1985), and the E chondrites with $\delta^{15}N = -30‰$ to $-50‰$ and up to 800 ppm N (Kung and Clayton, 1978; Grady et al., 1986). In contrast, CM and CR chondrites are nitrogen rich with $\delta^{15}N > 0‰$, and CI chondrites have values up to 1,900 ppm and $\delta^{15}N = +50‰$ (Kung and Clayton, 1978; Kerridge, 1985). In sum, several meteorite classes and possibly solar gases can provide the light nitrogen found within the Earth. The source of nitrogen in the planetary atmospheres had isotopically lighter nitrogen than that presently observed if there were subsequent fractionating losses (see Section 4.12.4).

In considering the origin of nitrogen and the noble gases, a key parameter is the $N/^{36}Ar$ ratio, which has been used to link nitrogen and the noble gases in MORBs. The ratio for the terrestrial atmosphere is $5 \times 10^4$, compared to a value for the upper mantle of $2 \times 10^6$ (Marty, 1995). The martian ratio is $1 \times 10^4$, and may have been lowered by nitrogen losses (Fox, 1993) and also modified by $^{36}Ar$ losses. The Venus $N/^{36}Ar$ ratio is $2 \times 10^3$. Meteorites are generally much more nitrogen rich, while the solar value is much lower, with $N/^{36}Ar = 40$.

#### 4.12.2.9.2  Carbon

*Carbon on all three planets has $\delta^{13}C$ within the meteorite range, but with relatively depleted abundances.* Carbon at the terrestrial surface is largely divided between carbonates, with $\delta^{13}C = 0‰$ (where $\delta^{13}C = [(^{13}C/^{12}C)/(^{13}C/^{12}C)_{std} - 1] \times 10^3$), and organic deposits with $\delta^{13}C = -25‰$. The bulk inventory appears to have $\delta^{13}C \approx -5‰$ and a total budget of $1 \times 10^{23}$ g C. Divided into the mass of the upper mantle, this abundance is equivalent to 100 ppm. An additional 100–300 ppm may be present in the upper mantle (Trull *et al.*, 1993). The amount of carbon in the core is not known, although it may be up to several weight percent (Wood, 1993; Halliday and Porcelli, 2001), which if added back into the mantle would raise the carbon content by a factor of 10. The upper-mantle value of $\delta^{13}C = -5‰$ (see Javoy *et al.*, 1986) is similar to the bulk inventory of the crust. Venus has about twice as much carbon (like nitrogen) at the surface than the Earth, equivalent to 26 ppm when divided into the bulk planet (von Zahn *et al.*, 1983); whether Venus is more rich in carbon therefore depends upon what volume of the mantle has degassed and how much remains in the mantle (Lécuyer *et al.*, 2000). The carbon-isotopic compositions of the martian mantle appears to be as low as $\delta^{13}C = -20‰$ to $-25‰$ based on SNC data (Jakosky and Jones, 1997; Goreva *et al.*, 2003); the atmospheric composition is less well defined. The amount of martian carbon is also not well known; while the atmosphere has only $7 \times 10^{18}$ g C, equivalent to 0.01 ppm for the bulk planet (Owen *et al.*, 1977), a large fraction may be stored in the polar regolith.

The solar carbon-isotopic composition is unknown. Bulk carbonaceous chondrites have a range of values of $\delta^{13}C = 0‰$ to $-25‰$ (Kerridge, 1985), with most somewhat isotopically lighter than the Earth. Enstatite chondrites are largely within the range of $\delta^{13}C = 0‰$ to $-14‰$ (Grady *et al.*, 1986). While enstatite chondrites have <1% C, CI chondrites can have >4% (Kerridge, 1985). In sum, the Earth appears to be isotopically somewhat heavier than average meteorite compositions, but not beyond the full range of measured values. There is no shortage of carbon in most sources, although the amount on the Earth is uncertain without further constraints on the abundance in the core. It is worth considering that the Earth does not appear to have the composition of any single group or collection of meteorites, but that the bulk composition has been deduced from inter-element correlations of meteorite data (e.g., Allègre *et al.*, 2001). It has been argued that Earth-forming materials originally had greater abundances of moderately and highly volatile elements but suffered losses during accretion (Halliday and Porcelli, 2001). However, carbon may have been sequestered in the core before such losses, and based on a correlation between C/Sr and Rb/Sr in meteoritic materials may constitute 0.6–1.5% of

the core (Halliday and Porcelli, 2001). Unfortunately, in attempting to construct the Earth from meteorite abundance systematics, volatile isotope compositions do not vary as regularly, and so it is not clear how to relate $\delta^{13}C$ values between the Earth and meteorites.

### 4.12.2.9.3  Hydrogen

*The $\delta D$ value of the oceans falls within the meteoritic range but not that of measured comets. Trapped lighter hydrogen may be in the terrestrial mantle. Venus and Mars may have had similar water concentrations as Earth, but suffered losses that generated high $\delta D$ ratios.* The terrestrial oceans have $\delta D = 0‰$ with an inventory of $1.1 \times 10^{24}$ g $H_2O$, equivalent to 120 ppm H when divided into the mass of the upper mantle. The upper mantle value of $\delta D = -65‰$ to $-75‰$ may be due to subduction of crustal material that has undergone metamorphism (Margaritz and Taylor, 1976) and may have ~13–35 ppm H. However, even lighter compositions, down to $-125‰$, have been found in Hawaii and elsewhere (Deloule *et al.*, 1991; Hauri, 2002), suggesting that isotopically light, juvenile hydrogen remains in the mantle.

The Venus atmosphere has ~200 ppm $H_2O$ (Hoffman *et al.*, 1980) and a D/H ratio of $(1.6 \pm 0.2) \times 10^{-2}$ that is ~$10^2$ times that of the Earth (Donahue *et al.*, 1982). It has been suggested that Venus originally had the same D/H value as the Earth, but has lost at least one terrestrial ocean volume of water by hydro-dynamic escape, thereby generating an enrichment in deuterium (Donahue *et al.*, 1982). The ratio of water to carbon and nitrogen therefore may have been similar to that of the Earth.

Measurements of atmospheric water vapor on Mars have found D/H values ~5 times that of the Earth and have been fractionated due to Jeans escape of hydrogen to space (Owen *et al.*, 1988), and 2–5 times the terrestrial value in SNC meteorites (Watson *et al.*, 1994). Morphological data (Carr, 1986) and modeling of hydrogen atmospheric losses (Donahue, 1995) suggest that originally there may have been the equivalent of up to 500 m of water, or ~$7 \times 10^{22}$ g $H_2O$. The total mass of Mars is 0.11 times that of the Earth, and so both planets originally may have had similar bulk water concentrations.

Hydrogen was the most abundant element in the solar nebula. The D/H ratio of the solar nebula has not been preserved in the Sun due to deuterium burning to $^3He$, but has been deduced from the solar $^3He/^4He$ and He/H ratios to be $\delta D = -880‰$ (Geiss and Gloeckler, 1998). A large range of values has been measured in meteorites due to fractionation between chemical species, with bulk values largely between −200‰ and +500‰, and concentrations of 60–$10^4$ ppm H (Kerridge, 1985). Comets have long been regarded as a likely source of terrestrial water; however, recent measurements of three comets have found D/H ratios about twice that of the Earth (Balsiger *et al.*, 1995; Bockelée-Morvan *et al.*, 1998; Meier *et al.*, 1998).

## 4.12.3    ACQUISITION OF NOBLE GASES AND VOLATILES

A variety of different mechanisms have been proposed for the acquisition of noble gases. Not all of these are mutually exclusive.

### 4.12.3.1    Solar-wind Implantation

Solar noble gases are spread throughout the solar system in the solar wind, and are typically implanted by low-energy solar-wind irradiation in solar-like elemental abundance proportions in lunar and meteoritic materials. Typically, implantation extends a few nanometers into irradiated materials, so that the amount of noble gases that can be accumulated is correlated with surface area and is most efficient for dust. An available analogue is the fine material found in the lunar regolith (see, e.g., Eberhardt *et al.*, 1972). Accretion of planetesimals containing ∼25–40 wt.% of such material could account for the absolute noble-gas abundances measured in the Venus atmosphere. The very low relative abundances of solar major volatiles require another source for these species. However, the presence of substantial dust in the nebular disk prior to aggregation also greatly dampens penetration of solar wind out to much of the planet-forming region, while clearance of this dust results in larger targets. Moreover, with prolonged exposure, target materials can become saturated, and losses of helium and neon can occur from irradiated grains preferentially by diffusion during subsequent heating or even at low temperatures (Frick *et al.*, 1988). Sasaki (1991) argued that off-disk penetration of an early and intense solar-wind flux into a post-nebular environment rich in fine collisional dust could have generated an ancient reservoir of abundant irradiated dust.

Models of a solar-wind source for noble gases on the terrestrial planets have been proposed in various contexts by Wetherill (1981), Donahue *et al.* (1981), and McElroy and Prather (1981). Assuming that sufficient abundances of noble gases were accumulated in solid materials, it would be expected that due to gravitational scattering gas-bearing materials would be dispersed throughout the inner solar system and

supply Venus and Earth with similar amounts of volatiles. In this case, the present differences in noble-gas abundances may be due to subsequent loss processes (see Section 4.12.4) that are necessary to generate the presently observed elemental and isotopic fractionations. The principal problem with these hypotheses is that xenon in the atmospheres of the early Earth is not solar, but U-Xe, while that of Mars was likely SW-Xe.

Such a source has also been considered for providing the noble gases presently found within the Earth's mantle. Podosek *et al.* (2000) argued that the present concentrations of neon estimated for a deep-mantle gas rich reservoir could have been derived from irradiated, kilometer-sized planetesimals, assuming that sufficient turnover of the surfaces occurs so that the process is not limited by grain-saturation effects and that irradiation fluxes were much higher in the past. This process would not have been limited by self-shielding by solid material across the accretionary disk due to removal of dust into larger bodies, and requires that a substantial fraction of the present mass of the Earth remained as small, dispersed planetesimals until after nebula gas had dispersed. Also, the gases must be retained in growing planetesimals and ultimately into the growing Earth, without being lost due to impacts or melting. This model is dependent on the chronology of accretion and gas dispersal and the early solar-wind flux, but it remains as a possible explanation for the origin of mantle noble gases.

### 4.12.3.2    Adsorption on Accreting Materials

Another mechanism of trapping volatiles from nebular gases onto solid materials is adsorption (see Ozima and Podosek, 2001). Laboratory studies have shown that noble gases exposed to some finely divided solid materials are adsorbed on the surfaces of individual grains. Adsorption is most efficient for various forms of carbon (e.g., Frick *et al.*, 1979; Wacker, 1989), but has also been experimentally demonstrated for other minerals (e.g., Yang and Anders, 1982). Adsorbed gases on these substrates generally display elemental patterns that are fractionated relative to ambient gas-phase abundances, in which heavier elements are enriched. These elemental fractionations are remarkably uniform, considering the wide range of experimental and natural conditions under which they are produced, and are similar to planetary atmosphere patterns. However, laboratory estimates of single-stage gas/solid partition coefficients are too low by orders of magnitude to account for planetary noble-gas abundances by adsorption on free-floating nebular dust grains at nebular pressures. Also, while occasional isotopic effects have been reported in natural samples

(Phinney, 1972), these are not observed in equilibrium-adsorption experiments (Bernatowicz and Podosek, 1986), and adsorption in the nebula cannot produce the overabundance of gases that can allow subsequent fractionating losses.

### 4.12.3.3 Gravitational Capture

A number of noble-gas capture mechanisms have been suggested that involve gravitational attraction to increase local gas pressures in the surrounding nebular gases, followed by capture of nebular gases by growing protoplanetary bodies. These require the growth of protoplanets to appreciable masses (at least to about the Mercury to Mars size) prior to dissipation of the nebular gas phase, and so depend upon the relative timing of nebular dissipation versus planetary accretion. Current estimates for loss of circumstellar dust and gas are up to ~10 Ma (Podosek and Cassen, 1994). However, nebular lifetimes inferred from astronomical observation are based solely on evolution of their fine dust component, while nebular gas may still remain for longer. The abundances of gases that are captured by solid bodies, and the base pressures and temperatures of the resulting atmospheres, depend upon the sizes reached when the nebula gases are dissipated. The standard model of planetary accumulation estimates that the terrestrial planets reached full size by ~100 Ma or more (Wetherill, 1986, 1990a), although terrestrial-planet growth to ~80% of final masses may have occurred within ~20 Ma (Wetherill, 1986). Therefore, if a significant remnant of gas survived to this time, substantial gravitational capture would have occurred. Then subsequent evolutions must have involved losses that fractionated both elements and isotopes to generate the presently observed compositions.

#### 4.12.3.3.1 *Capture by planetary embryos*

While the atmospheres captured by small bodies may not provide sufficient terrestrial rare-gas abundances in themselves, sufficient quantities of gases may have accumulated within the protoplanetary bodies by gas adsorption from these atmospheres on surface materials followed by burial below the surface during continuing accretion (Pepin, 1991). The process may have played an important role in creating internal volatile reservoirs for later outgassing of secondary atmospheres on the terrestrial planets, especially for the heavy noble gases. Interaction of the atmospheric gases with the surface is governed by the pressure at the base of the atmosphere, which depends on the thermal structure of the atmosphere. This, in turn, is a sensitive function of atmospheric opacity, which

is difficult to estimate, although amplifications of surface pressure by about four to six orders of magnitude above that of the ambient nebula are likely (Pepin, 1991). Therefore, adsorption and occlusion of surface gases on and within growing planetary embryos might be a natural consequence of protoplanetary growth, in the presence of nebular gas, to bodies of up to about the Mercury size formed within <1 Myr (Wetherill, 1990b; Wetherill and Stewart, 1993). Another consideration is that the impact velocities of materials accreting to form these small bodies are generally too low to promote efficient degassing of the impactors themselves. Consequently, their volatiles also tend to be buried within the growing embryos (Tyburczy *et al.*, 1986).

If rare gases were acquired by these mechanisms, atmospheric formation would then occur by subsequent degassing and isotopic fractionation during loss to space. Rare gases trapped within the Earth and incorporated into the present deep mantle would exhibit solar isotopic compositions although conceivably accompanied by elemental fractionations. There is evidence that solar-like light rare-gas isotopic compositions exist, and the abundance pattern appears to be enriched in heavy noble gases, even though the pattern cannot be well constrained. Pepin (1991) calculated that a Mars-size terrestrial embryo could have developed the concentrations of neon that might presently be stored in a gas rich lower mantle, although not enough to account for initial deep-Earth abundances prior to accretionary losses (see Section 4.12.2.7).

#### 4.12.3.3.2 *Xenon fractionation in porous pre-planetary planetesimals*

It has been suggested (Ozima and Nakazawa, 1980; Ozima and Igarashi, 1989; Zahnle *et al.*, 1990b; Ozima and Zahnle, 1993) that fractionation of nebular xenon to produce the terrestrial composition occurred by gravitational isotopic separation in large-porous planetesimals which have now vanished from the solar system. A consequence of this mechanism is that the atmospheric rare-gas characteristics are established in accreting materials, so that rare gases presently within the deep Earth are predicted to have the same characteristics. Other noble-gas characteristics must be generated by mixing with unfractionated components and some fractionating escape to space, but it is not clear that terrestrial isotopic compositions of all three noble gases can be generated from solar compositions for any distribution of planetesimal masses accreted by the Earth.

### 4.12.3.3.3   Gravitational capture and dissolution into molten planets

If the Earth reached sufficient size in the presence of the solar nebula, a massive atmosphere of solar gases would have been gravitationally captured and supported by the luminosity provided by the growing Earth, and the underlying planet would have melted by accretional energy and the blanketing effect of the atmosphere (Hayashi *et al.*, 1979). Under these conditions, gases from this atmosphere would have been sequestered within the molten Earth by dissolution at the surface and downward mixing (Mizuno *et al.*, 1980). This mechanism can provide solar rare gases into the deep Earth with relative elemental abundances that have been fractionated according to differences in solubilities (with depletion of heavy rare gases). Initial calculations found that at least an order of magnitude more neon than presently found in the deep mantle could be dissolved into the Earth unless the atmosphere began to escape when the Earth was only partially assembled (Mizuno *et al.*, 1980; Mizuno and Wetherill, 1984; Sasaki and Nakazawa, 1990; Sasaki, 1999). As noted above, initial concentrations may have actually been $10^2$ times greater than the present abundances prior to losses at ~100 Myr after the start of the solar system, and Porcelli *et al.* (2001) and Woolum *et al.* (1999) considered the conditions required to dissolve sufficient neon to account for the initial deep-mantle inventory. If it is assumed that equilibration of the atmosphere with a thoroughly molten mantle was rapid, and uniform concentrations were maintained throughout the mantle by vigorous convection, then the initial abundances of gases retained in any mantle layer reflect surface rare-gas partial pressures when that layer solidified. The depth at which solidification occurs is determined by the surface temperature and the efficiency of convection in the molten mantle. Hence, the initial distributions of retained rare gases would be determined by the history of surface pressure and temperature during mantle cooling and solidification, i.e., the coupled cooling of Earth and atmosphere. For typical solubility coefficients (e.g., Lux, 1987), a total surface pressure ~100 atm under an atmosphere of solar composition is required to establish the initial deep-mantle neon concentration (Porcelli *et al.*, 2001), along with surface temperatures high enough to melt the deep mantle (~4,000 K). The dense atmosphere is a balance between the gravitational attraction of the nebula-derived gases and expansion due to the Earth's luminosity (energy released by accreting planetesimals and the cooling Earth). Therefore, the temperature and pressure at the base of the atmosphere evolved as the energy released by accretion declined with time once planet assembly approached completion, and as the nebular pressure declined during nebula dispersal. Woolum *et al.* (1999) demonstrated that the necessary conditions were met under a range of parameter values for both convective and radiative atmospheric structures, although many complexities remain to be resolved. It should be noted that not all situations facilitate the dissolution of atmospheric gases. At low nebular pressures and high initial luminosities, rapid magma solidification may occur without the incorporation of significant concentrations of atmospheric gases. However, in the presence of a massive atmosphere that promotes gas dissolution, the mantle cooling time is greatly extended (Tonks and Melosh, 1990). It is clear that any noble gases that were within the zone of mantle melting from earlier trapping by other mechanisms would have been incorporated into the mantle–atmosphere system and overwhelmed. Conversely, extensive melting of the Earth without sufficient surface pressures would lead to losses of these gases.

### 4.12.3.4   Accretion of Comets

Noble gases, as well as water, carbon, and nitrogen, could have been supplied to the inner planets by accretion of volatile-rich icy comets scattered inward from the outer solar system. Although noble-gas isotopic compositions in comets are unknown, it is expected that these gases, directly acquired from the solar nebula, have solar isotopic compositions. There is experimental evidence that the relative elemental abundances of heavier species (xenon, krypton, and argon) trapped in water ice at plausible comet-formation temperatures (~30 K) approximately reflect those of the ambient gas phase, and trapped noble-gas concentrations in water are substantial (Bar-Nun *et al.*, 1985; Owen *et al.*, 1991). At somewhat higher temperatures, a range of elemental fractionations is obtained with relative depletions in the lighter noble gases. In addition to physical adsorption on ice, thermodynamic modeling suggests that noble-gas incorporation in clathrates can be effective at low temperatures, and produces gases that are strongly enriched in the heavier species (Lunine and Stevenson, 1985).

The origins of volatile species on the terrestrial planets have been modeled as resulting from accretion, in variable planet-specific proportions, of rocky materials as well as three types of comets. These formed at different heliocentric distances and thus at different nebular temperatures, leading to distinctive elemental fractionation patterns in volatiles trapped in their ice from ambient nebular gases (e.g., Owen *et al.*, 1991, 1992; Owen and

Bar-Nun, 1995a,b). The different volatile relative abundances on Venus, Earth, and Mars require several different components. Owen and Bar-Nun (1995a) suggested that comets from the Jupiter region, formed under sufficiently high temperatures to be essentially devoid of noble gases and depleted in nitrogen, have supplied much of the carbon and nitrogen on the planets. Comets formed at temperatures of $\sim 50$ K then supplied heavy noble gases and established the $N/^{36}Ar$ ratios. In this way, incorporation of a few percent or less by mass of icy cometary matter into the accreting terrestrial planets could have supplied heavy noble gases. Comets formed at lower temperatures can provide the unique Venus noble-gas abundance pattern, with solar proportions of argon, krypton, and xenon but a lower neon abundance. Final modification of the terrestrial budget is required to raise the ratio of water to carbon, and it was suggested that atmospheric species were preferentially lost over water during impact erosion. Nonfractionating losses from Mars are also required to deplete volatile abundances.

Accretion of icy comet matter has long been viewed as a plausible source for Earth's water. The D/H ratio in seawater, however, is a factor of $\sim 2$ lower than that in the few comets where D/H has been measured. A significant contribution of terrestrial water by comets would still be permitted if their high D/H ratio were appropriately lowered by accretion of additional, deuterium-poor materials. Suggested possibilities for low D/H carriers include rocky planetary accretional components (Laufer *et al.*, 1999; Dauphas *et al.*, 2000), or a high influx during the heavy bombardment epoch of interplanetary dust particles heavily loaded with implanted SW-H (Pavlov *et al.*, 1999). It has also been argued that comets from different orbital distances (and so forming in different temperatures) had different D/H ratios that could have formed the bulk of the terrestrial water (Delsemme, 1999; Morbidelli *et al.*, 2000). Overall, it appears that hydrogen-isotope compositions do not exclude cometary sources, but only further define source regions.

The principal difficulty encountered by these mixing models is their inability to account for differences in nonradiogenic noble-gas isotopic distributions between Earth and Mars, and between both of these and solar compositions. Experiments specifically designed to investigate isotopic fractionation in the gas-trapping process showed maximum heavy-isotope enrichments which are too small to explain the observed offsets of martian $^{36}Ar/^{38}Ar$ and of xenon on both Mars and Earth from solar ratios (Notesco *et al.*, 1999). Also, models that rely on supply of noble gases from material trapped in the outer parts of the solar system cannot explain the abundances of

mantle noble gases, since these materials are expected to be provided as a "late veneer" when accreting bodies are supplied from a wider swathe of the nebula, and are more likely to devolatilize upon impact due to the size of the proto-Earth, rather than bury volatiles. However, comets may still have provided the major volatiles, as well as hydrogen to fuel hydrodynamic escape fractionation of noble gases.

### 4.12.3.5 Accretion of Carbonaceous Chondrites

There are various features in planetary volatiles that have suggested their derivation from accreting chondrites. As discussed above, the nonradiogenic-isotopic compositions and relative abundances of neon, argon, and krypton are similar to those found in carbonaceous chondrites (with some isotopic fractionation of krypton). However, the differences in Xe/Kr ratio and the xenon-isotope compositions appear to be irreconcilably different. In contrast, chondrites, or a mixture of chondrites classes, can be found to match nitrogen and carbon terrestrial-isotope compositions. The C/N ratios in CI and CM chondrites are similar to the planetary ratios, although CO and CV chondrites are relatively depleted in nitrogen (see compilation by Newsom, 1995). Since the ratio of carbon to noble gases in carbonaceous chondrites is much higher than in the terrestrial planets, it is possible that chondrites supplied the major elements, and such an explanation for major volatiles then requires an additional source for the noble gases. A late infall of material of chondritic composition has been hypothesized in various contexts to provide a volatile-rich oxidizing veneer (Wänke *et al.*, 1984) or source of mantle siderophiles (e.g., Chou, 1978).

A model for the total composition of the terrestrial planets involves the mixture between a highly reduced, refractory component and an oxidizing volatile-rich component with characteristics similar to CI chondrites (Dreibus and Wänke, 1987; Wänke and Dreibus, 1988). The proportions of each are determined by key elemental ratios, and initial inventories of water, noble gases, and other volatiles have been calculated that exceed present abundances, along with inferences regarding subsequent modification.

### 4.12.4 EARLY LOSSES OF NOBLE GASES TO SPACE

As discussed above, none of the acquisition mechanisms can explain the full range of volatile features observed within the terrestrial planets. However, there are various potential loss mechanisms that may have operated on the initially acquired volatile budgets, modifying isotopic compositions, elemental ratios, and absolute abundances to produce the presently observed compositions.

### 4.12.4.1 Losses During Accretion

As growth of a protoplanet proceeds with increasing accretional energy, shock-induced devolatilization of the accreting materials occurs and volatile species are transferred into the growing atmosphere, limiting the amounts of volatiles that can be buried into a planet. Data summarized by Ahrens et al. (1989) indicate that efficient loss of $CO_2$ and $H_2O$ from accreting solids on impact occurs when the planetesimal mass approaches that of Mars. Above this size, degassing would also be driven by extensive melting due to deposition of accretional energy (Safronov, 1978), and further promoted by a radiative blanketing effect if a water-rich atmosphere has accumulated (see Abe and Matsui, 1986) or a dense atmosphere has been gravitationally captured. The process will limit the burial of volatiles by accreting materials.

Loss of atmospheric gases to space can occur by impact erosion, when a sufficient transfer of energy from accreting bodies to the atmosphere occurs and a substantial portion of the proto-planetary atmosphere reaches escape velocity (see Cameron, 1983; Ahrens, 1993). For smaller accreting bodies, the maximum fraction of the atmosphere that can be expelled is $\sim 6 \times 10^{-4}$ (Vickery and Melosh, 1990), equivalent to the total above the plane tangent to the planetary surface at the impact location. However, atmospheric loss may be much greater for very large impacts by bodies exceeding lunar size (Chen and Ahrens, 1997). These impact-driven losses are not expected to generate elemental or isotopic fractionations of volatiles, and contribute only to their overall depletion.

A Moon-forming collision of an approximately Mars-sized body with Earth (Hartmann and Davis, 1975; Cameron and Ward, 1976) would clearly result in catastrophic loss of volatiles from the pre-existing atmosphere and may have caused substantial loss of deep-Earth noble gases as well. Ahrens (1990, 1993) argued that virtually complete expulsion might have occurred by direct ejection from the impacted hemisphere and by shock-induced outward ramming of the antipodal planetary surface. However, losses may have been incomplete, and this event could have been followed by additional, isotopically fractionating losses driven by thermal processes (see Section 4.12.5).

### 4.12.4.2 Hydrodynamic Escape

Thermally driven escape of atmospheric constituents to space can generate substantial isotopic fractionations in the residual atmosphere, and models involving hydrodynamic escape have been successful at reproducing planetary features (Zahnle and Kasting, 1986; Hunten et al., 1987, 1988, 1989; Sasaki and Nakazawa, 1988, 1990; Zahnle et al., 1990a; Pepin, 1991, 1994, 1997, 2000). In this process, hydrogen-rich primordial atmospheres of partially or fully accreted planets are heated at high altitudes after the nebula has dissipated, and the resulting hydrogen-escape fluxes can exert upward-drag forces on heavier atmospheric constituents sufficient to lift them out of the atmosphere. Lighter species are entrained and are lost with the outflowing hydrogen more readily than are heavier ones, leading to mass fractionation of the residual atmosphere. The energy required can be provided by intense far-ultraviolet radiation from the young sun or energy deposited by a large impact event. Hydrogen-escape fluxes high enough to sweep out and fractionate atmospheric species as massive as xenon require energy inputs that are $\sim 10^2 - 10^3$ times greater than the amount presently supplied to planetary exospheres by solar extreme ultra-violet (EUV) radiation, but are not unreasonable for the early Sun.

Various studies have developed the theory of hydrodynamic escape (Zahnle and Kasting, 1986; Hunten et al., 1987) and demonstrated that, given adequate supplies of hydrogen and energy, observed noble-gas features could be achieved. Hunten et al. (1987) and Sasaki and Nakazawa (1988) examined the derivation of terrestrial xenon from solar xenon, and Zahnle et al. (1990a) derived neon and argon compositions on Earth and Mars. Pepin (1991, 1994, 1997) examined how hydrodynamic escape could generate the full range of elemental- and isotopic-mass distributions now found in planetary atmospheres, and explored the range of suitable astrophysical and planetary conditions. The simple analytic approach (Hunten et al., 1987; Pepin, 1991, 1997; Pepin and Porcelli, 2002) assumes the presence of an isothermal atmosphere consisting of hydrogen and minor amounts of heavier components. For a given hydrogen-escape flux, $F_H$, the upward drag is sufficient to lift all constituents with masses $m_2$ less than a critical mass $m_c$ out of the atmosphere. The critical mass $m_c$ is defined as (Hunten et al., 1987)

$$m_c = m_H + \frac{kTF_H}{bgX_H} \qquad (3)$$

where $k$ is the Boltzmann constant, $T$ the atmospheric temperature, $g$ the gravitational acceleration, $X_H$ the mole fraction of H (assumed to remain ~1 throughout the escape episode), $m_H$ the mass of H, and $b$ the diffusion parameter (the product of diffusion coefficient and total number density) of mass $m_2$ in the gas. Note that the critical mass is largely dependent on the escape flux of hydrogen, with the loss of heavier species requiring a greater hydrogen flux. Values of $b$ for noble-gas diffusion in $H_2$ at various temperatures are known (Mason and Marrero, 1970; Zahnle and Kasting, 1986). These increase from xenon to neon by a factor of ~2 that varies only slightly with temperature, and are identical or nearly so for isotopes of a given element. The relative losses of minor constituents can be seen in the relationship between the ratio of the atmospheric inventory of hydrogen ($N_H$) with that of a minor constituent, $N_2$, and the escape flux $F_2$ of the minor constituent:

$$F_2 = F_H \frac{N_2}{N_H} \left[ \frac{m_c - m_2}{m_c - m_H} \right] \qquad (4)$$

when the mass $m_2$ is much smaller than the critical mass (because of a large $F_H$), the loss of this species is proportional to its inventory, and so there is no fractionation. Losses are also then maximized. For species with masses that approach $m_c$, maximum elemental and isotopic fractionation between them occurs. An important consequence is that when isotopic fractionation is occurring to a species, much lighter elements are being greatly depleted.

The energy required for escape of a particle with mass $m_1$ from its local gravitational field, at radial distance $r = r_S$ from a body of mass $M$ and radius $r_S$, is $Gm_1M/r$ erg per particle. If the global mean solar EUV input at heliocentric distance $R$ and time $t$ is $\phi(R, t)$ erg cm$^{-2}$ s$^{-1}$, the energy-limited escape flux is

$$F_H(R, t) = \frac{\phi(R, t)\varepsilon}{Gm_H M/r} \text{ particle cm}^{-2}\text{s}^{-1} \qquad (5)$$

where $\varepsilon$ is the fraction of incident EUV energy flux converted to thermal escape energy of hydrogen and $\phi(R, t)$ is the energy input at heliocentric distance $R$ and time $t$. The energy required for escape of a hydrogen atom from the surface of a body of mass $M$ and radius $r$ is $Gm_1M/r$ erg per particle. Equation (5) can be combined with Equation (3) to determine the history of the crossover mass at each planet. It can be seen here that once the history of the driving energy source for loss, i.e., the EUV flux, is known, and assumptions are made about the initial inventory and ongoing supply of the major volatile species, $H_2$, the losses of each species can be calculated.

Information on what the EUV flux might have been in the early solar system comes from astronomical observations of radiation from young solar-type stars at various stages of pre- and early main-sequence evolution. Since early solar EUV radiation could not have penetrated a full gaseous nebula to planetary distances, the applicable time dependence of stellar activity in the present model is that which follows dissipation of the dense accretion disks surrounding the classical T-Tauri stars so that solar EUV radiation could penetrate to planetary distances, at stellar ages of up to ~10 Ma (Simon *et al.*, 1985; Walter *et al.*, 1988; Strom *et al.*, 1988; Walter and Barry, 1991; Podosek and Cassen, 1994). Among present observational data, soft (~3–60 Å) X-ray fluxes are most likely to be representative of at least the short-wavelength coronal component ($\lambda < 700$ Å) of the EUV spectrum. Although there is considerable scatter from a single functional dependence of fluxes with age, most of the data between ~50 Ma and 200 Ma do indicate a decline by factors of ~5–10 from levels at ~20 Ma. Pepin (1991, 1994, 1997) assumed the EUV flux fell off exponentially, with a mean decay time of 90 Ma. Using a power-law function instead (Feigelson and Kriss, 1989) produces similar results (Pepin, 1989). The modeling assumes an energy supply, and therefore crossover masses, that decline with time through the hydrodynamic escape episode. This is plausible for both solar EUV radiation and energy initially deposited by a giant impact (Pepin, 1991, 1997). An alternative assumption of a constant-crossover mass together with a corresponding specific value for $N_H/(N_H)_0$ used in integrating Equation (4) also yields the same fractionations as those calculated for a declining crossover mass (Hunten *et al.*, 1987). However, if the crossover mass defined by the actual energy supply is more than a few hundred amu above the xenon mass region, the required value of $N_H/(N_H)_0$ is too small to be consistent with realistic initial and final hydrogen inventories on the planet.

Fractionating effects of the escape process can now be calculated analytically if specific assumptions are made about the time dependence of the hydrogen inventory; not only the history of the EUV flux, but also whether it is replenished as fast as it escapes (constant inventory model), or is lost without replenishment along with the minor atmospheric species (Rayleigh fractionation model). The inventories of the minor constituents have generally been assumed to be lost without replenishment during the escape episode. It should be emphasized that there are various adjustable modeling parameters that generate the final noble-gas patterns, although some are shared by all the planets, such as the history of the EUV supply (adjusted for distance), and possibly the initial supply of atmospheric constituents.

### 4.12.4.3 Sputtering

On Mars, it has been demonstrated that losses probably occur by sputtering, which results in fractionation of elements and isotopes in the residual atmosphere (Luhmann *et al.*, 1992; Zhang *et al.*, 1993). Oxygen atoms in the martian exosphere, ionized by solar EUV radiation and accelerated in the electric field of the solar wind, can impact species near the top of the atmosphere (the "exobase") with enough energy transfer to eject them from the planet's gravitational field. Loss rates of the dominant atmospheric constituent at the exobase ($CO_2$ on Mars) in this sputtering process depend upon the magnitudes of the EUV flux and the solar-wind velocity, and so estimates of how both electromagnetic and corpuscular radiation have evolved over solar history are needed in order to calculate sputtering losses in the past (Zhang *et al.*, 1993).

Escape fluxes of sputtered trace constituents such as noble gases and nitrogen in the atmosphere are proportional to their exobase mixing ratios with $CO_2$ (Jakosky *et al.*, 1994). These ratios are dependent upon the mass of each volatile species, so that sputtering from the exobase removes species from a fractionated "target" (the exobase), leaving the residual atmosphere enriched in heavier constituents. Depletions of lighter species are further augmented in the escape process itself since ejection efficiency from the exobase increases with decreasing atomic mass (Jakosky *et al.*, 1994). Note that there are enormous elemental fractionations at the exobase, and sputtering losses of the two heaviest noble gases are consequently extremely small. For these species, isotopic fractionation by the process has negligible influence on the composition of total atmospheric inventories.

Sputtering losses are greatly attenuated by the presence of a planetary magnetic field, most importantly because it deflects the solar wind around the planet and shields atoms photoionized in the outer atmosphere from the solar-wind electric field that would otherwise accelerate some of them downward toward the exobase (Hutchins *et al.*, 1997). For this reason the process has not been important on Earth for as long as the core dynamo has existed, and it seems unlikely that the bulk composition of the massive Venus atmosphere, with low noble-gas mixing ratios, could have been substantially affected by sputtering loss with or without the protection of a magnetic field. Escape of sputtered species is also impeded by the higher gravity of these two planets. In contrast, a thin, magnetically unshielded, and more weakly bound martian atmosphere is particularly vulnerable to sputtering erosion. Efficient operation of this fractionating loss mechanism over time on Mars is thus

linked both to atmospheric pressure history and to the timing of the disappearance of the martian paleomagnetic field (Hutchins *et al.*, 1997; Connerney *et al.*, 1999).

### 4.12.5 THE ORIGIN OF TERRESTRIAL NOBLE GASES

The most notable feature of terrestrial volatiles is the extensive fractionation of xenon isotopes, since producing this effect on such a heavy species is not generally seen elsewhere in the solar system and can only be produced under very limited circumstances. The only process that appears plausible at present is hydrodynamic escape, and so modeling the evolution of the atmosphere necessarily centers on this process. For the Earth, Hunten *et al.* (1987) and Pepin (1991) provide the first formulations, and assumed that hydrodynamic losses of primary atmospheric volatiles are driven entirely by intense EUV radiation from the young-evolving Sun. Hydrogen-outflow fluxes strong enough to enable xenon escape from Earth, and fractionation to its present isotopic composition, required atmospheric $H_2$ inventories equivalent to water abundances of up to a few weight percent of the planet's mass, and early solar EUV fluxes up to $\sim 450$ times present levels, which may be realistic if nebular dust and gas had dissipated to levels low enough for solar EUV radiation to penetrate the midplane to planetary distances within 100 Ma or so. However, energy sources other than solar EUV absorption may have powered atmospheric escape. Benz and Cameron (1990) suggested that hydrodynamic loss driven by thermal energy deposited in a giant Moon-forming impact could have generated the well-known fractionation signature in terrestrial xenon. Their model of the event calls for rapid invasion of the pre-existing primary atmosphere by extremely hot ($\sim 1.6 \times 10^4$ K) dissociated rock and iron vapor, emplacement of an orbiting rock–vapor disk with an inner edge at an altitude comparable to the atmospheric-scale height at this temperature, and longer-term heating of the top of the atmosphere by re-accretion of dissipating disk material. If a short post-impact escape episode did in fact occur, resulting in Rayleigh fractionation of whatever remnant of the primary atmosphere survived, direct and presumably nonfractionating ejection in the impact event required that atmospheric $H_2$ inventories would be reduced by at least an order of magnitude compared to models in which losses are driven only by solar EUV radiation (Pepin, 1997).

The more recent modeling (Pepin, 1991, 1997) involves evolutionary processing in two stages. In the first stage, substantial depletion of xenon from the primary atmosphere occurs, driven by deposition of atmospheric energy. This sets the xenon inventory and generates the extensive xenon-isotopic fractionation that is presently observed in the terrestrial atmosphere. However, the other, lighter noble gases are greatly over-depleted and overly fractionated isotopically at the end of the first stage. In the case where hydrodynamic escape is initially driven by a giant impact, primordial atmosphere U-Xe is fractionated to the presently observed composition while ~85% of the initial xenon is lost from the planet. Increasingly severe fractionations of the lighter noble gases from their primordial isotopic compositions are imposed during xenon escape. Residual krypton and argon are both isotopically heavy and strongly depleted relative to the present atmosphere; only 6% and 0.8%, respectively, of the initial $^{84}Kr$ and $^{36}Ar$ inventories (and 0.4% of the $^{20}Ne$) survive the event.

A second stage is required to rectify the over-fractionations in the light noble gases and reach presently observed compositions. Therefore, species degassed from the second, interior reservoir are required to mix with the fractionated atmospheric residue to compensate for the over-fractionation of krypton, argon, and neon elemental and isotopic abundances in the first stage, and produce the presently observed characteristics of these gases. While no further modifications of argon and krypton abundances are required, neon is a special case in that $^{20}Ne/^{22}Ne$ ratio in the fractionated residual atmosphere is substantially higher than the present-day value, and later addition of outgassed solar neon elevates them still more. Here an episode of solar EUV energy deposition driving hydrodynamic escape of neon at some time after GI fractionation and outgassing is needed to generate the contemporary $^{20}Ne/^{22}Ne$ ratio. Now the EUV-driven $H_2$ escape flux must be only intense enough to lift neon, but not the heavier noble gases, out of the atmosphere. The waning EUV flux (Ayres, 1997) may still have been sufficiently high (~60 times present levels) to drive neon-only escape at solar ages up to ~250 Ma, with the actual timing determined by the timescale for sufficient reduction of EUV dust-gas opacity in the nebular midplane (Prinn and Fegley, 1989).

An important aspect of the modeling results is that present atmospheric-xenon inventories are largely the fractionated relicts of the first stage of hydrodynamic escape processing of primary atmospheric xenon, while most of the krypton and lighter noble gases are products of planetary outgassing. Isotope mixing systematics impose strict upper limits on the allowed levels of "contamination" of residual primary xenon by later addition of isotopically unfractionated xenon degassed from the interior or supplied by subsequent accretion of noble-gas carrier

materials, a constraint that applies with equal force to the model in which pre-fractionated xenon is delivered to planets by porous planetesimals. Estimates of these limits for Earth and Mars (Pepin, 1991, 1994) fall well below the amounts of xenon that ordinarily would be expected to accompany the outgassed krypton components. However, it is possible that xenon was preferentially outgassed well before the bulk of the lighter noble gases following the fractional degassing mechanisms of Zhang and Zindler (1989) and Tolstikhin and O'Nions (1994), and most of it was already present in the primary atmospheres prior to xenon-fractionating hydrodynamic loss. Later post-escape isotopic evolution of atmospheric xenon is largely restricted to degassing of radiogenic $^{129*}Xe$, and of $^{131-136*}Xe$ generated primarily by $^{244}Pu$ spontaneous fission, from the upper mantle and crust.

The simple analytical theory used by Pepin (1991, 1998) assumes a hydrogen-dominated atmosphere and energy limited escape, both arguably reasonable suppositions of primordial atmospheric conditions. However, such atmospheres might also have contained substantial amounts of a heavy constituent, say $CO_2$, and in this case the escape flux of $H_2$ would have been limited by its ability to diffuse through the $CO_2$. Zahnle *et al.* (1990a) and Ozima and Zahnle (1993) have shown that only neon and some argon would be hydrodynamically lost under these conditions. However, a more complete short-term atmospheric blowoff, including heavier species, could have been driven by a very large deposition of collisional energy.

This modeled history requires that the Earth acquired two isotopically primordial volatile reservoirs during accretion, one in the planet's interior, perhaps populated by a combination of nebular gases occluded in planetary embryo materials and dissolved in molten surface materials, and the other co-accreted as a primary atmosphere degassed from impacting planetesimals or gravitationally captured from ambient nebular gases during later planetary growth. These isotopically primordial reservoirs are characterized by isotope ratios measured in the solar wind, with the important exception of xenon, which has a U-Xe precursor. The relative abundances of the noble gases for each reservoir are obtained by back-calculation through the model. The required pattern for the primary atmosphere is a progressive enrichment in the heavier noble gases relative to the solar pattern, with a much higher Ar/Ne ratio, a moderate enrichment in Kr/Ar, and an Xe/Ar ratio is enriched to a somewhat greater extent (Pepin, 1997). It has been suggested that this reflects adsorption onto planetesimals (Pepin, 1991; Section 4.12.3.3.1). The other possible sources of gases with solar isotopic compositions include implanted solar wind, which can be enriched in heavier noble gases, although through losses of the lightest species and so requiring even greater initial implanted inventories, and dissolution of gravitationally captured noble gases, which favors the more-soluble light noble gases (Porcelli *et al.*, 2001).

An important issue that remains unresolved is the relationship between the presently observed mantle reservoirs and the noble gases that degassed during formation of the atmosphere. If the composition of neon within the mantle were found to be that of Ne-B, indicating that the source was material irradiated by solar gases, then a second source must be found to supply the distinctly nonsolar primordial xenon that supplied the atmosphere. It has been argued that noble gases that were supplied to the atmosphere are not represented by those now in the mantle, which have been derived from deeper within the Earth subsequently (see Chapter 4.11), and so it is possible that different sources supplied the deep mantle and atmosphere. However, some models for the different sources are incompatible. For example, noble gases derived by gravitational capture of a hot, dense atmosphere, followed by dissolution into the Earth, would overwhelm those gases derived earlier from other sources. However, late infall of volatiles would leave those in the deep Earth unaffected.

The major volatiles clearly have not been derived from the same, solar sources as the noble gases. The carbon and nitrogen characteristics are similar to those in various meteorite classes. Javoy (1998) developed a model for construction of the Earth largely by enstatite chondrites, which naturally provides isotopically light nitrogen, as well as hydrogen, into the mantle, and can match the carbon-isotope value of the bulk Earth. The isotopically heavier surface reservoirs of nitrogen and hydrogen require some late addition from another source, perhaps CI chondrites. This model supplies an apparent overabundance of nitrogen to the surface, requiring sequestration of a considerable amount of nitrogen in the deep Earth. While there are many issues that arise in having enstatite chondrites as the dominant component of the bulk Earth, a smaller contribution may still have provided much of the major volatiles. Some limits to the total amounts of carbonaceous chondrites that can constitute the Earth are provided by moderately volatile element inventories, which are depleted relative to chondrites by an order of magnitude (see Newsom, 1995). Thus, chondrites cannot constitute >10% of the Earth. However, since the Earth is strongly depleted in major volatiles, chondrites may still provide the observed carbon and nitrogen abundances. Halliday and Porcelli (2001) used correlations of volatile and moderately

volatile elements in various meteorite classes to extrapolate to the bulk Earth composition prior to substantial accretionary losses. It was suggested that higher early carbon abundances led to higher carbon in the core than predicted from present concentrations. In this case, the similarity between the C/N ratio on Earth and Venus would either be fortuitous or reflecting late additions to the planet. Using a model for the bulk composition of the Earth that involves the supply of volatiles by an oxidizing component with CI chondrite volatile abundances, Dreibus and Wänke (1987) suggested that up to 1.3% water was initially added to the Earth but was largely converted into $H_2$ during reduction of iron, but later infall of similar material comprising 0.44% of the Earth (to account for siderophile elements in the mantle) supplied the water observed at present. In this case, the $H_2$ produced may have contributed to the atmospheric inventory fuelling hydrodynamic escape.

It should be noted that deriving the bulk composition of the Earth in general by simple models of mixing between meteoritic components has generally always found difficulties, especially with elements of greater volatility. However, the possibility remains that solid materials that contributed to the Earth, with unique major-volatile characteristics, are no longer represented in the solar system.

Hydrodynamic escape required to modify the isotopic composition of the noble gases will also fractionate nitrogen and carbon in the atmosphere, as discussed in detail by Pepin (1991). The inventories in the remaining atmosphere will be isotopically heavier, and so chondrites supplying volatiles must have had lower $\delta^{15}N$ and $\delta^{13}C$ values. A higher initial N/C ratio is also likely to have been necessary, although the fractionation between carbon and nitrogen depends upon the speciation of the species being lost. CI and CM chondrites have nitrogen that is too isotopically heavy (Kerridge, 1985) and N/C ratios that may be too low, and the CO and CV chondrites also have low N/C ratios (Mazor *et al.*, 1970). In contrast, E chondrites appear to have the necessary nitrogen and carbon compositions.

### 4.12.6  THE ORIGIN OF NOBLE GASES ON VENUS

On Venus, the noble gases do not appear to have greatly evolved from solar characteristics. The heavy rare-gas elemental abundances are similar to solar values, although this similarity does not extend to neon, since the $^{20}Ne/^{36}Ar$ ratio is low. Nonetheless, the $^{20}Ne/^{22}Ne$ ratio is closer to the solar value. Venus is also gas rich, with the absolute abundance of argon on Venus exceeding that on Earth by a factor $>70$. The pronounced differences with terrestrial atmospheric noble gases are somewhat surprising, since planets as alike in size and heliocentric distance might be expected to have acquired compositionally similar primary atmospheres from similar sources and suffered similar evolutionary processes.

The similarities with solar noble gases suggest that those in the Venus atmosphere have been derived either from solar-wind implantation of accreting materials, gravitational capture of nebular gases, or volatile-rich comets. In considering these sources, it must be noted that not only will these mechanisms supply both Venus and the Earth, but also that a strong EUV flux inferred for generating noble gas losses from the Earth and modifying initially acquired inventories would also have affected Venus. The EUV flux driving neon escape in the Earth model discussed in the preceding section must also irradiate Venus at the same time. It turns out that the relatively weak solar EUV flux needed for loss of only neon from Earth (after losses from a giant impact-fractionated xenon isotopes) is still strong enough at the orbital position of Venus to drive outflow of krypton and lighter gases from this somewhat smaller and less dense planet. However, Venusian xenon is not lost and its nonradiogenic isotopic composition is predicted to be unaltered from its primordial composition. The relationship between the loss histories of Venus and the Earth has been used to construct a model for Venusian volatile evolution (Pepin, 1991, 1997). Results of EUV-driven loss of an isotopically solar and elementally near-solar primordial atmosphere from Venus are sensitive to only one of the few remaining adjustable modeling parameters once the evolution of the Earth has been calculated; the initial $H_2$ inventory. Fractionating loss of a primary atmosphere generates by itself approximate matches to observed compositions. Thus, in contrast to the case for Earth, the presence of a component that is subsequently outgassed from the solid planet and modifies the lighter atmospheric noble-gas isotopes is not required and would comprise only modest fractions of the large present-day Venusian atmospheric inventories even if bulk-planetary concentrations were comparable to those on Earth. Using the hydrodynamic escape model and calculating back from the present noble-gas isotope compositions and relative abundances, the starting elemental ratios characterizing the preloss Venus atmosphere that are obtained fall squarely within the range of estimates calculated for Earth's pre-impact primary atmosphere (Pepin, 1997). This includes an Xe/Kr ratio somewhat above the solar ratio and due to initial trapping of solar gases. The similarity between the initial terrestrial and Venusian atmospheres is a strong indication that noble gases on both planets could have evolved, clearly in quite

different ways, from the same primordial distributions in the same types of primary planetary reservoirs.

The major volatiles on Venus appear to have been derived from a different source than the noble gases. The isotopic compositions of hydrogen and nitrogen were originally similar to those of the Earth, and nitrogen does not appear to have been derived from a solar precursor. Once the noble-gas characteristics of Venus were established, the major volatiles may have been added as a late veneer (Pepin, 1991), increasing the $N/^{36}Ar$ ratio and accounting for the similarities with the major volatiles on the Earth. As on the Earth, carbon, nitrogen, and water may have been added by comets or chondrites.

## 4.12.7  THE ORIGIN OF NOBLE GASES ON MARS

The atmosphere of Mars has several features that are distinct from that of the Earth and require a somewhat different planetary history. At likely nebular temperatures and pressures at its radial distance, Mars is too small to have condensed a dense early atmosphere from the nebula even in the limiting case of isothermal capture (Hunten, 1979; Pepin, 1991). Therefore, regardless of the plausibility of gravitational capture as a noble-gas source for primary atmospheres on Venus and Earth, some other way is needed to supply Mars. This may include solar-wind implantation or comets. An important feature is that, in contrast to Earth, martian xenon apparently did not evolve from a U-Xe progenitor, but rather from SW-Xe. This requires that accreting SW-Xe-rich materials that account for martian atmospheric xenon are from sources more localized in space or time and so have not dominated the terrestrial-atmospheric xenon budget. There are insufficient data to determineif the martian C/N ratio is like the terrestrial value, but it appears that the initial $C/H_2O$ ratio may have been. Further constraints on the sources of the major volatiles are required.

Modeling of the martian atmosphere has been reviewed by Pepin (1991, 1994, 1997) and Jakosky and Jones (1997). The recent models of martian atmospheric evolution incorporate the fractionating effects of both hydrodynamic escape and the sputtering loss mechanism proposed by Luhmann *et al.* (1992) and Zhang *et al.* (1993), and explore the consequences for elemental and isotopic fractionation of the noble gas and nitrogen in the residual atmosphere (Jakosky *et al.*, 1994; Pepin, 1994). Martian atmospheric history is divided into early and late evolutionary periods, the first characterized by an episode of hydrodynamic escape, followed by high $CO_2$ pressures and a possible greenhouse, and the second by a transition to a low-pressure environment similar to present-day conditions on the planet, perhaps initiated by abrupt polar $CO_2$ condensation ~3.7 Ga (Gierasch and Toon, 1973; Haberle *et al.*, 1992, 1994). During this second period, gas loss and fractionation occurred by sputtering, which may have been the dominant mechanism governing atmospheric $CO_2$ evolution on Mars over the past ~3–4 Ga (Luhmann *et al.*, 1992; Zhang *et al.*, 1993). Another loss process, atmospheric erosion (Melosh and Vickery, 1989), increasingly appears important (Chyba, 1990, 1991; Zahnle, 1993), and may have depleted all atmophilic species prior to the end of heavy bombardment ~3.8–3.7 Ga.

The effects of hydrodynamic escape were discussed in detail by Pepin (1991), without detailed consideration of other loss mechanisms. Such a mechanism remains as the most plausible for the fractionation of the heavy noble gases, especially xenon isotopes. The EUV-powered hydrodynamic escape episode driving neon-only loss from Earth after Giant-driven escape, and loss of krypton and lighter gases from Venus, would have been intense enough on Mars to lift all the noble gases out of its primordial atmosphere. Early in this pre-3.7 Ga epoch, xenon isotopes were therefore assumed to have been hydrodynamically fractionated to their present composition, with corresponding depletions and fractionations of lighter primordial atmospheric constituents (Pepin, 1994).

The post-3.7 Ga evolution of martian $CO_2$, $N_2$, and the noble gases has been examined by Jakosky *et al.* (1994) and Pepin (1994). The late evolutionary stage on Mars was assumed to have been trigged by atmospheric $CO_2$ pressure collapse near 3.7 Ga. Sputtering loss of an atmospheric species relative to that of $CO_2$ is directly proportional to its exobase mixing ratio with $CO_2$, and so sputtering fractionation of the atmospheric noble-gas inventory is generally modest in a pre-3.7 Ga atmosphere dominated by $CO_2$ (Jakosky *et al.*, 1994). Pressure collapse of the major atmospheric constituent abruptly increased the mixing ratios of pre-existing argon, neon, and $N_2$ at the exobase, and allowed their rapid removal by sputtering. Current abundances and isotopic compositions are entirely determined by the action of sputtering and photochemical escape on gases supplied by outgassing during the late evolutionary epoch. Since light species from the first epoch are qualitatively lost, the final distributions of the light noble gases and nitrogen are therefore decoupled from whatever their elemental and isotopic inventories might have been in the pre-3.7 Ga atmosphere. Jakosky *et al.* (1994) showed that contemporary neon, argon, and $N_2$ abundances and isotope ratios, including the uniquely low martian $^{36}Ar/^{38}Ar$ ratio, could have been generated by sputtering losses from an atmosphere that was continuously replenished by degassing of meteoritic (CI) $N_2$ and isotopically solar neon and argon, with the time-dependent rates of degassing similar to estimates of volcanic fluxes over this period. Both krypton and xenon are too massive to be significantly affected by sputtering loss and fractionation during the late evolutionary stage. The present atmospheric krypton inventory derives almost completely from solar-like krypton degassed during this period, which overwhelms any krypton fractionated earlier, while only the xenon isotopes and $\delta^{13}C$ survive as isotopic tracers of atmospheric history prior to its transition to low pressure (Pepin, 1994). The assumption that early hydrodynamic escape fractionated the nonradiogenic xenon isotopes to at least approximately their present composition severely limits subsequent additions of unfractionated xenon to the atmospheric inventory by outgassing (consistent with the low degree of planetary degassing; see Section 4.12.2.6) or late-stage veneer accretion.

Hutchins and Jakosky (1996) revisited the late evolution sputtering–degassing models to investigate in more detail the parameters controlling the evolution of neon and argon abundances and isotopes, in particular those relating to martian degassing history. Assuming that the martian mantle was not more gas rich than the bulk Earth, it was concluded that the outgassing flux of argon and neon attributable to degassing during epochs of volcanic activity would have been about one to three orders of magnitude too low to appropriately balance sputtering losses, and thus another major source of juvenile volatiles must have contributed to the atmosphere over geologic time, perhaps via input from gas-enriched hydrothermal systems. Hutchins *et al.* (1997) explored to what extent a martian paleomagnetic dipole field would have throttled sputtering losses by deflecting the solar wind around the upper atmosphere, and calculated the conditions, as functions of the time when paleomagnetic suppression of the sputtering mechanism ended, under which the combination of sputtering and degassing would still have generated present-day argon and neon distributions. The discovery by Mars Global Surveyor of large-scale remnant magnetic lineations in the old martian southern highlands (Connerney *et al.*, 1999) confirmed that an active dynamo existed, but its history is still unknown.

In the first evolutionary epoch, note that the extent of xenon fractionation from primordial to present composition is similar on both Earth and Mars despite the much smaller mass of Mars, the apparent differences (U-Xe versus SW-Xe) in their precursor xenon, the much greater overall depletion of martian noble gases, and the possibility that escape episodes were powered by distinctly different energy sources (EUV radiation on Mars versus giant impact on Earth). It is not clear if this is just coincidence, or the expression of some more-fundamental fractionating process that left similar signatures on all three of the terrestrial-planet atmospheres. In the second evolutionary epoch, the $CO_2$ pressure and isotopic history was dictated by the interplay of estimated losses to impact erosion, sputtering, and carbonate precipitation, additions by outgassing and carbonate recycling, and perhaps also by feedback stabilization under greenhouse conditions. In a subsequent model of the early martian atmosphere, Carr (1999) examined the influences of these same mechanisms in controlling $CO_2$ pressure history, and was led to similar results and conclusions. It should be stressed, however, that since almost nothing is actually known about the values of the parameters governing these various processes, models of this epoch are no more than qualitative illustrations of how they might have driven early atmospheric behavior.

## 4.12.8 CONCLUSIONS

Considerable progress has been made in the long-standing problem of understanding the sources of volatiles on and within the terrestrial planets, and the processes that modified their initial inventories down planet-specific evolutionary tracks to the amazingly divergent compositional states observed today on Earth, Mars, and Venus. However, a full volatile history of the inner solar system remains to be formulated. While various mechanisms can be evoked to explain particular features, the interconnections between capture and modifying losses require further definition, and some features remain enigmatic. Some avenues of further research include:

*Acquisition of noble gases by planetary interiors.* Further constraints are required on the characteristics of noble gases trapped within the Earth. The isotopic compositions of the heavy noble gases, the $^{20}Ne/^{22}Ne$ ratio, and the concentrations of deep Earth reservoirs are needed to further evaluate capture mechanisms. The two viable mechanisms for burying solar gases into the Earth, gravitational capture of nebular gases and solar-wind implantation of small accreting materials, require further development to formulate further tests and implications. The capture of a primordial atmosphere by gravitational attraction of nebular gas is inescapable if the growing planet reaches sufficient mass in the presence of the nebula, and so the first criterion for this mechanism is firmly establishing if such a nebular history occurred. How much gas was then trapped in the Earth is a more complex issue requiring a substantial modeling effort. Parameters that need

further consideration include the structure of the atmosphere, how long will the underlying mantle remains molten and to what depth, and how the atmosphere is affected by continuing accretion. Whether or not this supplies deep-mantle noble gases, a shallower reservoir may be more readily created which supplied noble gases to the atmosphere and may no longer be represented in the mantle. Under conditions in which melting of the underlying planet is not achieved, adsorption of noble gases, enhanced by the increased pressure of the gravitationally focused nebular gases, may be an important source of noble gases that may become buried during accretion.

The burial of material that contains solar noble gases implanted by radiation is another option for the source of mantle noble gases. This requires the opposite conditions of gravitational capture, clearance of the solar nebula prior to substantial aggregation of solids to allow penetration of solar wind to where the terrestrial planets accumulate. The solar fluxes and size of target materials must provide sufficient accumulation of noble gases so that after subsequent losses during accretion as well as during escape that caused fractionation of xenon, there are still sufficient noble gases remaining to account for the present atmospheric inventories. The choice between these mechanisms likely will be decided when there are greater constraints on the history of the solar nebula and the growth of protoplanetary material.

*Atmosphere origin and evolution.* Hydrodynamic escape models are capable of replicating details of contemporary isotopic distributions. However, the model is highly parametrized and intrinsically multistage, requiring both escape fractionation and subsequent mixing with species degassed from planetary interiors. It seems clear that some degree of hydrodynamic loss and fractionation of planetary atmospheres would have been inevitable if the required conditions for energy source, hydrogen supply, and, in the case of solar EUV-driven escape, midplane transparency to solar radiation were even partially met. However, which species are lost and the extent of fractionations generated are dependent upon various parameters that require independent substantiation. This includes the composition of the dense atmosphere and the sources and history of the energy fluxes driving hydrodynamic escape.

*Mars.* The combination of Viking *in situ* measurements and SNC meteorite data has provided a much more quantitative view of the present state and possible history of martian volatiles. Further progress requires greater precision for the atmospheric composition, more data on possible near-surface and mantle reservoirs, and further constraints on the conditions that affect the volatile evolution of all the terrestrial planets.

*Venus.* Venus is characterized only by the immensely valuable but still incomplete and relatively imprecise reconnaissance data from the Pioneer Venus and Venera spacecraft missions of the late 1970s. Additional *in situ* measurements, at precisions within the capabilities of current spacecraft instrumentation, are now necessary to refine atmospheric evolution models. Unfortunately, the possibilities of documenting the volatile inventories of the interior of the planet are more remote. A significant question that must be addressed is whether nonradiogenic xenon on Venus is compositionally closer to SW-Xe (as seen on Mars) or to the U-Xe that is seen on the Earth and so is expected to have been present within the inner solar system. Also, the extent of xenon fractionation will be an important parameter for hydrodynamic escape models; if intense solar EUV radiation drove hydrodynamic escape on the Earth, it would also impact Venus, while losses from the Earth driven by a giant impact would not be recorded there.

*Distributions in the solar system.* More data on volatiles throughout the solar system are clearly required to confidently describe the volatile acquisition history of the terrestrial planets in the proper context. There are several unknown values for the solar composition, including the nitrogen- and carbon-isotope compositions. The compositions of comets from different orbital distances are needed to assess the extent of radial transport of volatiles late in accretion history. In addition, the causes of carbon- and nitrogen-isotope variations in chondrites must be better understood. While it is clear that the Earth cannot be constructed simply by mixing of different meteorite classes, it is not yet possible to unambiguously extrapolate to the volatile compositions of protoplanetary materials.

The origin of the U-Xe that is now found in the terrestrial atmosphere, and presumably was in the nebular gases surrounding the accreting Earth, is still not adequately explained. An important parameter is the composition of xenon in the atmosphere of Jupiter, which contains gases that were captured directly from the ancient nebula and may be different from the solar wind, where the possibility of isotopic fractionation in processes transporting and releasing bulk solar xenon to and from the corona cannot be completely disregarded.

*Coupled histories of atmospheric and interior planetary volatiles.* Highly detailed models of noble-gas sources and evolution for the atmosphere and interior of the Earth have been developed separately and almost independently. However, the origin and history of atmospheric noble gases are not independent of the sources, distributions, and transport histories of noble gases within a planet—these two volatile systems must clearly be linked in nature through their primordial inventories and the processes of degassing and subduction. Indeed, many models of the terrestrial atmosphere require degassing of some portion of the planetary interior, and mantle models include degassing to the atmosphere. The degassing of Venus and Mars is much more under-constrained both by data and modeling. The question is whether the requirements implicit in each of these models for the dynamical and compositional history of the other reservoir are compatible, and, if not, what the inconsistencies are and how might they be addressed.

*Timing of solar system events.* There is an assortment of chronological information for various nebular and planet-forming processes, but further theoretical work is required to formulate a full history of planetary volatiles. The relative timing of nebular dispersal and planet growth is clearly a key parameter for controlling the acquisition of solar gases. Radiogenic xenon isotopes on Earth and Mars indicate that volatiles were lost for $\sim 50$ Ma or more. This contrasts with recent $^{182}$Hf–$^{182}$W data that suggest that the average time of separation of planetary cores (which likely occurred concomitantly with accretion), as well as Moon formation, occurred $<30$ Ma (Kleine *et al.*, 2002; Yin *et al.*, 2002). These different timescales remain to be reconciled.

## ACKNOWLEDGMENTS

The authors thank David Hilton and Richard Becker for much-appreciated comments, and Ralph Keeling for editorial handling.

## REFERENCES

Abe Y. and Matsui T. (1986) Early evolution of the Earth: accretion, atmosphere formation, and thermal history. *J. Geophys. Res.* **91**, E291–E302.

Ahrens T. J. (1990) Earth accretion. In *Origin of the Earth* (eds. H. E. Newsom and J. H. Jones). Oxford University Press, New York, pp. 211–227.

Ahrens T. J. (1993) Impact erosion of terrestrial planetary atmospheres. *Ann. Rev. Earth Planet. Sci.* **21**, 525–555.

Ahrens T. J., O'Keefe J. D., and Lange M. A. (1989) Formation of atmospheres during accretion of the terrestrial planets. In *Origin and Evolution of Planetary and Satellite Atmospheres* (eds. S. K. Atreya, J. B. Pollack, and M. S. Matthews). University of Arizona Press, Tucson, pp. 328–385.

Allègre C. J., Manhès G., and Lewin E. (2001) Chemical composition of the Earth and the volatility control on planetary genetics. *Earth Planet. Sci. Lett.* **185**, 49–69.

Ayres T. R. (1997) Evolution of the solar ionizing flux. *J. Geophys. Res.* **102**, 1641–1651.

Ballentine C. J. and Burnard P. G. (2002) Production, release, and transport of noble gases in the continental crust. *Rev. Mineral. Geochem.* **47**, 481–538.

Ballentine C. J., Porcelli D., and Wieler R. (2001) Technical comment on Trieloff *et al.* (2000). *Science* **291**, 2269a.

Balsiger H., Altwegg K., and Geiss J. (1995) D/H and $^{18}O/^{16}O$ ratio in the hydronium ion and in neutral water from *in situ* ion measurements in Comet Halley. *J. Geophys. Res.* **100**, 5827–5834.

Bar-Nun A., Herman G., Laufer D., and Rappaport M. L. (1985) Trapping and release of gases by water ice and implications for icy bodies. *Icarus* **63**, 317–332.

Basford J. R., Dragon J. C., Pepin R. O., Coscio M. R., Jr., and Murthy V. R. (1973) Krypton and xenon in lunar fines. *Proc. 4th Lunar Sci. Conf.* 1915–1955.

Becker R. H. (2000) Nitrogen on the Moon. *Science* **290**, 110–111.

Begemann F., Weber H. W., and Hintenberger H. (1976) On the primordial abundance of argon-40. *Astrophys. J.* **203**, L155–L157.

Benkert J.-P., Baur H., Signer P., and Wieler R. (1993) He, Ne, and Ar from solar wind and solar energetic particles in lunar ilmenites and pyroxenes. *J. Geophys. Res.* **98**, 13147–13162.

Benz W. and Cameron A. G. W. (1990) Terrestrial effects of the giant impact. In *Origin of the Earth* (eds. H. E. Newsom and J. H. Jones). Oxford University Press, New York, pp. 61–67.

Bernatowicz T. J. and Podosek F. A. (1986) Adsorption and isotopic fractionation of Xe. *Geochim. Cosmochim. Acta* **50**, 1503–1507.

Bernatowicz T. J., Podosek F. A., Honda M., and Kramer F. E. (1984) The atmospheric inventory of xenon and noble gases in shales: the plastic bag experiment. *J. Geophys. Res.* **89**, 4597–4611.

Bernatowicz T. J., Kennedy B. M., and Podosek F. A. (1985) Xe in glacial ice and the atmospheric inventory of noble gases. *Geochim. Cosmochim. Acta* **49**, 2561–2564.

Black D. C. (1972) On the origins of trapped helium, neon, and argon isotopic variations in meteorites: II. Carbonaceous meteorites. *Geochim. Cosmochim. Acta* **36**, 377–394.

Bockelée-Morvan D., Gautier D., Lis D. C., Young K., Keene J., Phillips T., Owen T., Crovisier J., Goldsmith P. F., Bergin E. A., Despois D., and Wootten A. (1998) Deuterated water in Comet C/1996 B2 (Hyakutake) and its implications for the origin of comets. *Icarus* **133**, 147–162.

Bogard D. D. (1997) A reappraisal of the Martian $^{36}Ar/^{38}Ar$ ratio. *J. Geophys. Res.* **102**, 1653–1661.

Brazzle R. H., Pravdivtseva O. V., Meshik A. P., and Hohenberg C. M. (1999) Verification and interpretation of the I–Xe chronometer. *Geochim. Cosmochim. Acta* **63**, 739–760.

Burnard P. G., Graham D., and Turner G. (1997) Vesicle specific noble gas analyses of popping rock: implications for primordial noble gases in Earth. *Science* **276**, 568–571.

Busemann H., Baur H., and Wieler R. (2000) Primordial noble gases in "phase Q" in carbonaceous and ordinary chondrites studied by closed-system stepped etching. *Meteorit. Planet. Sci.* **35**, 949–973.

Caffee M. W., Hudson G. U., Velsko C., Huss G. R., Alexander E. C., Jr., and Chivas A. R. (1999) Primordial noble cases from Earth's mantle: identification of a primitive volatile component. *Science* **285**, 2115–2118.

Cameron A. G. W. (1983) Origin of the atmospheres of the terrestrial planets. *Icarus* **56**, 195–201.

Cameron A. G. W. and Ward W. R. (1976) The origin of the Moon. *Lunar Sci.* **VII**, 120–122.

Carr M. H. (1986) Mars: a water-rich planet? *Icarus* **68**, 187–216.

Carr M. H. (1999) Retention of an atmosphere on early Mars. *J. Geophys. Res.* **104**, 21897–21909.

Cartigny P., Harris J. W., Phillips D., Boyd S. R., and Javoy M. (1998) Subduction-related diamonds? The evidence for a mantle-derived origin from coupled $\delta^{13}C$–$\delta^{15}N$ determinations. *Chem. Geol.* **147**, 147–159.

Cartigny P., Harris J. W., and Javoy M. (2001) Diamond genesis, mantle fractionations, and mantle nitrogen content: a study of $\delta^{13}C$–N concentrations in diamonds. *Earth Planet. Sci Lett.* **185**, 85–98.

Chen G. Q. and Ahrens T. J. (1997) Erosion of terrestrial planet atmosphere by surface motion after a large impact. *Phys. Earth Planet. Int.* **100**, 21–26.

Chou C.-L. (1978) Fractionation of siderophile elements in the Earth's upper mantle. *Proc. 9th Lunar Planet. Sci. Conf.* 219–230.

Chyba C. F. (1990) Impact delivery and erosion of planetary oceans in the early inner solar system. *Nature* **343**, 129–133.

Chyba C. F. (1991) Terrestrial mantle siderophiles and the lunar impact record. *Icarus* **92**, 217–235.

Connerney J. E. P., Acuña M. H., Wasilewski P. J., Ness N. F., Rème H., Mazelle C., Vignes D., Lin R. P., Mitchell D. L., and Cloutier P. A. (1999) Magnetic lineations in the ancient crust of Mars. *Science* **284**, 794–798.

Dauphas N., Robert F., and Marty B. (2000) The late asteroidal and cometary bombardment of Earth as recorded in water deuterium to protium ratio. *Icarus* **148**, 508–512.

Deloule E., Albarède F., and Sheppard S. M. F. (1991) Hydrogen isotope heterogeneities in the mantle from ion probe analysis of amphiboles from ultramafic rocks. *Earth Planet. Sci. Lett.* **105**, 543–553.

Delsemme A. H. (1999) The deuterium enrichment observed in recent comets is consistent with the cometary origin of seawater. *Planet. Space Sci.* **47**, 125–131.

Déruelle B., Dreibus G., and Jambon A. (1992) Iodine abundances in oceanic basalts: implications for Earth dynamics. *Earth Planet. Sci. Lett.* **108**, 217–227.

Donahue T. M. (1995) Evolution of water reservoirs on Mars from D/H ratios in the atmosphere and crust. *Nature* **374**, 432–434.

Donahue T. M. and Russell C. T. (1997) The Venus atmosphere and ionosphere and their interaction with the solar wind: an overview. In *Venus II* (eds. S. W. Bougher, D. M. Hunten, and R. J. Phillips). University of Arizona Press, Tucson, pp. 3–31.

Donahue T. M., Hoffman J. H., and Hodges R. R., Jr. (1981) Krypton and xenon in the atmosphere of Venus. *Geophys. Res. Lett.* **8**, 513–516.

Donahue T. M., Hoffman J. H., Hodges R. R., Jr., and Watson A. J. (1982) Venus was wet: a measurement of the ratio of deuterium to hydrogen. *Science* **216**, 630–633.

Dreibus G. and Wänke H. (1987) Volatiles on Earth and Mars: a comparison. *Icarus* **71**, 225–240.

Eberhardt P., Geiss J., Graf H., Grögler N., Mendia M. D., Mörgeli M., Schwaller H., and Stettler A. (1972) Trapped solar wind gases in Apollo 12 lunar fines 12001 and Apollo 11 breccia 10046. *Proc. 3rd Lunar Sci. Conf.* **2**, 1821–1856.

Eugster O., Eberhardt P., and Geiss J. (1967) The isotopic composition of krypton in unequilibrated and gas rich chondrites. *Earth Planet. Sci. Lett.* **2**, 385–393.

Feigelson E. D. and Kriss G. A. (1989) Soft X-ray observations of pre-main-sequence stars in the Chamaeleon dark cloud. *Astrophys. J.* **338**, 262–276.

Fox J. L. (1993) The production and escape of nitrogen atoms on Mars. *J. Geophys. Res.* **98**, 3297–3310.

Frick U., Mack R., and Chang S. (1979) Noble gas trapping and fractionation during synthesis of carbonaceous matter. *Proc. 10th Lunar Planet. Sci. Conf.* 1961–1973.

Frick U., Becker R. H., and Pepin R. O. (1988) Solar wind record in the lunar regolith: nitrogen and noble gases. *Proc. 18th Lunar Planet. Sci. Conf.* 87–120.

Garrison D. H. and Bogard D. D. (1998) Isotopic composition of trapped and cosmogenic noble gases in several Martian meteorites. *Meteorit. Planet. Sci.* **33**, 721–736.

Geiss J. and Gloeckler G. (1998) Abundances of deuterium and helium in the protosolar cloud. *Space Sci. Rev.* **84**, 239–250.

Gierasch P. J. and Toon O. B. (1973) Atmospheric pressure variation and the climate of Mars. *J. Atmos. Sci.* **30**, 1502–1508.

Goreva J. S., Leshin L. A., and Guan Y. (2003) Ion microprobe measurements of carbon isotopes in Martian phosphates: insights into the Martian mantle. *Lunar Planet. Sci.* **XXXIV**, #1987.

Grady M. M., Wright I. P., Carr L. P., and Pillinger C. T. (1986) Compositional differences in enstatite chondrites based on carbon and nitrogen stable isotope measurements. *Geochim. Cosmochim. Acta* **50**, 2799–2813.

Graham D. W. (2002) Noble gases in MORB and OIB: observational constraints for the characterization of mantle source reservoirs. *Rev. Mineral. Geochem.* **47**, 247–318.

Haberle R. M., Tyler D., McKay C. P., and Davis W. L. (1992) Evolution of Mars' atmosphere: where has the $CO_2$ gone? *Bull. Am. Astron. Soc.* **24**, 1015–1016.

Haberle R. M., Tyler D., McKay C. P., and Davis W. L. (1994) A model for the evolution of $CO_2$ on Mars. *Icarus* **109**, 102–120.

Hagee B., Bernatowicz T. J., Podosek F. A., Johnson M. L., Burnett D. S., and Tatsumoto M. (1990) Actinide abundances in ordinary chondrites. *Geochim. Cosmochim. Acta* **54**, 2847–2858.

Halliday A. and Porcelli D. (2001) In search of lost planets—the paleocosmochemistry of the inner solar system. *Earth Planet. Sci. Lett.* **192**, 545–559.

Harrison D., Burnard P., and Turner G. (1999) Noble gas behaviour and composition in the mantle: constraints from the Iceland Plume. *Earth Planet. Sci. Lett.* **171**, 199–207.

Hartmann W. K. and Davis D. R. (1975) Satellite-sized planetesimals and lunar origin. *Icarus* **24**, 504–515.

Hashizume K., Chaussidon M., Marty B., and Robert F. (2000) Solar wind record on the Moon: deciphering presolar from planetary nitrogen. *Science* **290**, 1142–1145.

Hauri E. (2002) SIMS analysis of volatiles in silicate glasses: 2. Isotopes and abundances in Hawaiian melt inclusions. *Chem. Geol.* **183**, 115–141.

Hayashi C., Nakazawa K., and Mizuno H. (1979) Earth's melting due to the blanketing effect of the primordial dense atmosphere. *Earth Planet. Sci. Lett.* **43**, 22–28.

Helffrich G. R. and Wood B. J. (2001) The Earth's mantle. *Nature* **412**, 501–507.

Hoffman J. H., Hodges R. R., Donahue T. M., and McElroy M. B. (1980) Composition of the Venus lower atmosphere from the Pioneer Venus mass spectrometer. *J. Geophys. Res.* **85**, 7882–7890.

Hohenberg C. M., Podosek F. A., and Reynolds J. H. (1967) Xenon–iodine dating: sharp isochronism in chondrites. *Science* **156**, 233–236.

Honda M., McDougall I., Patterson D. B., Doulgeris A., and Clague D. A. (1991) Possible solar noble-gas component in Hawaiian basalts. *Nature* **349**, 149–151.

Honda M., McDougall I., Patterson D. B., Doulgeris A., and Clague D. A. (1993) Noble gases in submarine pillow basalt glasses from Loihi and Kilauea, Hawaii—a solar component in the Earth. *Geochim. Cosmochim. Acta* **57**, 859–874.

Hudson G. B., Kennedy B. M., Podosek F. A., and Hohenberg C. M. (1989) The early solar system abundance of $^{244}Pu$ as inferred from the St. Severin chondrite. *Proc. 19th Lunar Planet. Sci. Conf.* 547–557.

Hunten D. M. (1979) Capture of Phobos and Deimos by protoatmospheric drag. *Icarus* **37**, 113–123.

Hunten D. M., Pepin R. O., and Walker J. C. G. (1987) Mass fractionation in hydrodynamic escape. *Icarus* **69**, 532–549.

Hunten D. M., Pepin R. O., and Owen T. C. (1988) Planetary atmospheres. In *Meteorites and the Early Solar System* (eds. J. F. Kerridge and M. S. Matthews). University of Arizona Press, Tucson, pp. 565–591.

Hunten D. M., Donahue T. M., Walker J. C. G., and Kasting J. F. (1989) Escape of atmospheres and loss of water. In *Origin and Evolution of Planetary and Satellite Atmospheres* (eds. S. K. Atreya, J. B. Pollack, and M. S. Matthews). University of Arizona Press, Tucson, pp. 386–422.

Hutchins K. S. and Jakosky B. M. (1996) Evolution of Martian atmospheric argon: implications for sources of volatiles. *J. Geophys. Res.* **101**, 14933–14949.

Hutchins K. S., Jakosky B. M., and Luhmann J. G. (1997) Impact of a paleomagnetic field on sputtering loss of Martian

atmospheric argon and neon. *J. Geophys. Res.* **102**, 9183–9189.

Jacobsen S. and Harper C. J. (1996) Accretion and early differentiation history of the Earth based on extinct radionuclides. In *Earth Processes: Reading the Isotopic Code*, Geophys. Monogr. 95 (eds. A. Basu and S. R. Hart). American Geophysical Union, Washington, DC, pp. 47–74.

Jakosky B. M. and Jones J. H. (1997) The history of Martian volatiles. *Rev. Geophys.* **35**, 1–16.

Jakosky B. M., Pepin R. O., Johnson R. E., and Fox J. L. (1994) Mars atmospheric loss and isotopic fractionation by solar-wind-induced sputtering and photochemical escape. *Icarus* **111**, 271–288.

Javoy M. (1998) The birth of the Earth's atmosphere: the behaviour and fate of its major elements. *Chem. Geol.* **147**, 11–25.

Javoy M. and Pineau F. (1991) The volatile record of a popping rock from the mid-Atlantic ridge at 14°N: chemical and isotopic composition of gases trapped in the vesicles. *Earth Planet. Sci. Lett.* **107**, 598–611.

Javoy M., Pineau F., and Delorme H. (1986) Carbon and nitrogen isotopes in the mantle. *Chem. Geol.* **57**, 41–62.

Kallenbach R., Geiss J., Ipavich F. M., Gloeckler G., Bochsler P., Gliem F., Hefti S., Hilchenbach M., and Hovestadt D. (1998) Isotopic composition of solar wind nitrogen: first *in situ* determination with the CELIAS/MTOF spectrometer on board SOHO. *Astrophys. J.* **507**, L185–L188.

Kerridge J. F. (1985) Carbon, hydrogen, and nitrogen in carbonaceous chondrites: abundances and isotopic compositions in bulk samples. *Geochim. Cosmochim. Acta* **49**, 1707–1714.

Kleine T., Münker C., Mezger K., and Palme H. (2002) Rapid accretion and early core formation on asteroids and the terrestrial planets from Hf–W chronometry. *Nature* **418**, 952–955.

Krummenacher D., Merrihue C. M., Pepin R. O., and Reynolds J. H. (1962) Meteoritic krypton and barium versus the general isotopic anomalies in xenon. *Geochim. Cosmochim. Acta* **26**, 231–249.

Kung C.-C. and Clayton R. N. (1978) Nitrogen abundances and isotopic compositions in stony meteorites. *Earth Planet. Sci. Lett.* **38**, 21–435.

Kunz J. (1999) Is there solar argon in the Earth's mantle? *Nature* **399**, 649–650.

Kunz J., Staudacher T., and Allègre C. J. (1998) Plutonium-fission xenon found in Earth's mantle. *Science* **280**, 877–880.

Laufer D., Notesco G., and Bar-Nun A. (1999) From the interstellar medium to Earth's oceans via comets—an isotopic study of $HDO/H_2O$. *Icarus* **140**, 446–450.

Lécuyer C., Simon L., and Guyot F. (2000) Comparison of carbon, nitrogen, and water budgets on Venus and the Earth. *Earth Planet. Sci. Lett.* **181**, 33–40.

Luhmann J. G., Johnson R. E., and Zhang M. H. G. (1992) Evolutionary impact of sputtering of the Martian atmosphere by $O^+$ pickup ions. *Geophys. Res. Lett.* **19**, 2151–2154.

Lunine J. I. and Stevenson D. J. (1985) Thermodynamics of clathrate hydrate at low and high pressures with application to the outer solar system. *Astrophys. J. Suppl. Ser.* **58**, 493–531.

Lux G. (1987) The behavior of noble gases in silicate liquids: solution, diffusion, bubbles, and surface effects, with applications to natural samples. *Geochim. Cosmochim. Acta* **51**, 1549–1560.

Magaritz M. and Taylor H. P., Jr. (1976) Oxygen, hydrogen, and carbon isotope studies of the Franciscan formation, Coast Ranges, California. *Geochim. Cosmochim. Acta* **40**, 215–234.

Mahaffy P. R., Donahue T. M., Atreya S. K., Owen T. C., and Niemann H. B. (1998) Galileo probe measurements of D/H and $^3He/^4He$ in Jupiter's atmosphere. *Space Sci. Rev.* **84**, 251–263.

Marti K. (1967) Isotopic composition of trapped krypton and xenon in chondrites. *Earth Planet. Sci. Lett.* **3**, 243–248.

Marty B. (1995) Nitrogen content of the mantle inferred from $N_2$–Ar correlation in oceanic basalts. *Nature* **377**, 326–329.

Marty B. and Humbert F. (1997) Nitrogen and argon isotopes in oceanic basalts. *Earth Planet. Sci. Lett.* **152**, 101–112.

Marty B. and Marti K. (2002) Signatures of early differentiation on Mars. *Earth Planet. Sci. Lett.* **196**, 251–263.

Marty B. and Zimmermann L. (1999) Volatiles (He, C, N, Ar) in mid-ocean ridge basalts: assessment of shallow-level fractionation and characterization of source composition. *Geochim. Cosmochim. Acta* **63**, 3619–3633.

Mason E. A. and Marrero T. R. (1970) The diffusion of atoms and molecules. *Adv. At. Mol. Phys.* **6**, 155–232.

Mathew K. J. and Marti K. (2001) Early evolution of Martian volatiles: nitrogen and noble gas components in ALH84001 and Chassigny. *J. Geophys. Res.* **106**, 1401–1422.

Mathew K. J., Kim J. S., and Marti K. (1998) Martian atmospheric and indigenous components of xenon and nitrogen in the Shergotty, Nakhla, and Chassigny group meteorites. *Meteorit. Planet. Sci.* **33**, 655–664.

Matsuda J. and Matsubara K. (1989) Noble gases in silica and their implication for the terrestrial "missing" Xe. *Geophys. Res. Lett.* **16**, 81–84.

Mazor E., Heymann D., and Anders E. (1970) Noble gases in carbonaceous chondrites. *Geochim. Cosmochim. Acta* **34**, 781–824.

McElroy M. B. and Prather M. J. (1981) Noble gases in the terrestrial planets. *Nature* **293**, 535–539.

Meier R., Owen T. C., Matthews H. E., Jewitt D. C., Bockelée-Morvan D., Biver N., Crovisier J., and Gautier D. (1998) A determination of the HDO/$H_2O$ ratio in Comet C/1995 O1 (Hale-Bopp). *Science* **279**, 842–844.

Melosh H. J. and Vickery A. M. (1989) Impact erosion of the primordial atmosphere of Mars. *Nature* **338**, 487–489.

Mizuno H. and Wetherill G. W. (1984) Grain abundance in the primordial atmosphere of the Earth. *Icarus* **59**, 74–86.

Mizuno H., Nakazawa K., and Hayashi C. (1980) Dissolution of the primordial rare gases into the molten Earth's material. *Earth Planet. Sci. Lett.* **50**, 202–210.

Morbidelli A., Chambers J., Lunine J. I., Petit J. M., Robert F., Valsecchi G. B., and Cyr K. E. (2000) Source regions and timescales for the delivery of water to the Earth. *Meteorit. Planet. Sci.* **35**, 1309–1320.

Newsom H. E. (1995) Composition of the solar system, planets, meteorites, and major terrestrial reservoirs. In *Global Earth Physics, A Handbook of Physical Constants* (ed. T. J. Ahrens). American Geophysical Union, Washington, DC, pp. 159–189.

Niedermann S., Bach W., and Erzinger J. (1997) Noble gas evidence for a lower mantle component in MORBs from the southern East Pacific Rise: decoupling of helium and neon isotope systematics. *Geochim. Cosmochim. Acta* **61**, 2697–2715.

Nier A. O. and McElroy M. B. (1977) Composition and structure of Mars' upper atmosphere: results from the neutral mass spectrometers on Viking 1 and 2. *J. Geophys. Res.* **82**, 4341–4349.

Notesco G., Laufer D., Bar-Nun A., and Owen T. (1999) An experimental study of isotopic enrichments in Ar, Kr, and Xe when trapped in water ice. *Icarus* **142**, 298–300.

Ott U. (1988) Noble gases in SNC meteorites: Shergotty, Nakhla, Chassigny. *Geochim. Cosmochim. Acta* **52**, 1937–1948.

Owen T. and Bar-Nun A. (1995a) Comets, impacts, and atmospheres. *Icarus* **116**, 215–226.

Owen T. and Bar-Nun A. (1995b) Comets, impacts, and atmospheres: II. Isotopes and noble gases. In *Volatiles in the Earth and Solar System*, AIP Conf. Proc. 341 (ed. K. A. Farley). American Institute of Physics Press, New York, pp. 123–138.

Owen T., Biemann K., Rushneck D. R., Biller J. E., Howarth D. W., and Lafleur A. L. (1977) The composition of the atmosphere at the surface of Mars. *J. Geophys. Res.* **82**, 4635–4639.

Owen T., Maillard J. P., Debergh C., and Lutz B. (1988) Deuterium on Mars: the abundance of HDO and the value of D/H. *Science* **240**, 1767–1770.

Owen T., Bar-Nun A., and Kleinfeld I. (1991) Noble gases in terrestrial planets: evidence for cometary impacts? In *Comets in the Post-Halley Era* (eds. R. L. Newburn, Jr., M. Neugebauer, and J. Rahe). Kluwer, Dordrecht, The Netherlands, vol. 1, pp. 429–437.

Owen T., Bar-Nun A., and Kleinfeld I. (1992) Possible cometary origin of heavy noble gases in the atmospheres of Venus, Earth, and Mars. *Nature* **358**, 43–46.

Owen T., Mahaffy P. R., Niemann H. B., Atreya S., and Wong M. (2001) Proto-solar nitrogen. *Astrophys. J.* **553**, L77–L79.

Ozima M. and Igarashi G. (1989) Terrestrial noble gases: constraints and implications on atmospheric evolution. In *Origin and Evolution of Planetary and Satellite Atmospheres* (eds. S. K. Atreya, J. B. Pollack, and M. S. Matthews). University of Arizona Press, Tucson, pp. 306–327.

Ozima M. and Nakazawa K. (1980) Origin of rare gases in the Earth. *Nature* **284**, 313–316.

Ozima M. and Podosek F. A. (2001) *Noble Gas Geochemistry*, 2nd edn. Cambridge University Press, Cambridge, UK.

Ozima M. and Zahnle K. (1993) Mantle degassing and atmospheric evolution: noble gas view. *Geochem. J.* **27**, 185–200.

Palma R. L., Becker R. H., Pepin R. O., and Schlutter D. J. (2002) Irradiation records in regolith materials: II. Solar-wind and solar-energetic-particle components in helium, neon, and argon extracted from single lunar mineral grains and from the Kapoeta howardite by stepwise pulse-heating. *Geochim. Cosmochim. Acta* **66**, 2929–2958.

Pavlov A. A., Pavlov A. K., and Kasting J. F. (1999) Irradiated interplanetary dust particles as a possible solution for the deuterium/hydrogen paradox of Earth's oceans. *J. Geophys. Res.* **104**, 30725–30728.

Pepin R. O. (1989) On the relationship between early solar activity and the evolution of terrestrial planet atmospheres. In *The Formation and Evolution of Planetary Systems*, Space Tel Sci. Inst. Symp. Series #3 (eds. H. A. Weaver and L. Danly). Cambridge University Press, Cambridge, UK, pp. 55–74.

Pepin R. O. (1991) On the origin and early evolution of terrestrial planet atmospheres and meteoritic volatiles. *Icarus* **92**, 2–79.

Pepin R. O. (1994) Evolution of the martian atmosphere. *Icarus* **111**, 289–304.

Pepin R. O. (1997) Evolution of Earth's noble gases: consequences of assuming hydrodynamic loss driven by giant impact. *Icarus* **126**, 148–156.

Pepin R. O. (1998) Isotopic evidence for a solar argon component in the Earth's mantle. *Nature* **394**, 664–667.

Pepin R. O. (2000) On the isotopic composition of primordial xenon in terrestrial planet atmospheres. *Space Sci. Rev.* **92**, 371–395.

Pepin R. O. (2003) On noble gas processing in the solar accretion disk. *Space Sci. Rev.* **106**, 211–230.

Pepin R. O. and Phinney D. (1976) The formation interval of the Earth. *Lunar Sci.* **VII**, 682–684.

Pepin R. O. and Porcelli D. (2002) Origin of noble gases in the terrestrial planets. *Rev. Mineral. Geochem.* **47**, 191–246.

Pepin R. O., Becker R. H., and Rider P. E. (1995) Xenon and krypton isotopes in extraterrestrial regolith soils and in the solar wind. *Geochim. Cosmochim. Acta* **59**, 4997–5022.

Pepin R. O., Becker R. H., and Schlutter D. J. (1999) Irradiation records in regolith materials: I. Isotopic compositions of solar-wind neon and argon in single lunar mineral grains. *Geochim. Cosmochim. Acta* **63**, 2145–2162.

Phinney D. (1972) $^{36}Ar$, Kr, and Xe in terrestrial materials. *Earth Planet. Sci. Lett.* **16**, 413–420.

Phinney D., Tennyson J., and Frick U. (1978) Xenon in $CO_2$ well gas revisited. *J. Geophys. Res.* **83**, 2313–2319.

Podosek F. A. and Cassen P. (1994) Theoretical, observational, and isotopic estimates of the lifetime of the solar nebula. *Meteoritics* **29**, 6–25.

Podosek F. A., Woolum D. S., Cassen P., and Nichols R. H. (2000) Solar gases in the Earth by solar wind irradiation? *10th Annual Goldschmidt Conf. Oxford.*

Porcelli D. and Ballentine C. J. (2002) Models for the distribution of terrestrial noble gases and the evolution of the atmosphere. *Rev. Mineral. Geochem.* **47**, 411–480.

Porcelli D. and Halliday A. N. (2001) The core as a possible source of mantle helium. *Earth Planet. Sci. Lett.* **192**, 45–56.

Porcelli D. and Pepin R. O. (2000) Rare gas constraints on early Earth history. In *Origin of the Earth and Moon* (eds. R. M. Canup and K. Righter). University of Arizona Press, Tucson, pp. 435–458.

Porcelli D. and Wasserburg G. J. (1995a) Mass transfer of xenon through a steady-state upper mantle. *Geochim. Cosmochim. Acta* **59**, 1991–2007.

Porcelli D. and Wasserburg G. J. (1995b) Mass transfer of helium, argon, and xenon through a steady-state upper mantle. *Geochim. Cosmochim. Acta* **59**, 4921–4937.

Porcelli D. R., Woolum D., and Cassen P. (2001) Deep Earth rare gases: initial inventories, capture from the solar nebula, and losses during Moon formation. *Earth Planet. Sci. Lett.* **193**, 237–251.

Prinn R. G. and Fegley B., Jr. (1989) Solar nebula chemistry: origin of planetary, satellite, and cometary volatiles. In *Origin and Evolution of Planetary and Satellite Atmospheres* (eds. S. K. Atreya, J. B. Pollack, and M. S. Matthews). University of Arizona Press, Tucson, pp. 435–458.

Safronov V. S. (1978) The heating of the Earth during its formation. *Icarus* **33**, 1–12.

Sarda P., Staudacher T., and Allègre C. J. (1988) Neon isotopes in submarine basalts. *Earth Planet. Sci. Lett.* **91**, 73–88.

Sasaki S. (1991) Off-disk penetration of ancient solar wind. *Icarus* **91**, 29–38.

Sasaki S. (1999) Presence of a primary solar-type atmosphere around the Earth: evidence of dissolved noble gas. *Planet. Space Sci.* **47**, 1423–1431.

Sasaki S. and Nakazawa K. (1988) Origin of isotopic fractionation of terrestrial Xe: hydrodynamic fractionation during escape of the primordial $H_2$–He atmosphere. *Earth Planet. Sci. Lett.* **89**, 323–334.

Sasaki S. and Nakazawa K. (1990) Did a primary solar-type atmosphere exist around the proto-Earth? *Icarus* **85**, 21–42.

Simon T., Herbig G., and Boesgaard A. M. (1985) The evolution of chromospheric activity and the spin-down of solar-type stars. *Astrophys. J.* **293**, 551–574.

Strom S. E., Strom K. M., and Edwards S. (1988) Energetic winds and circumstellar disks associated with low mass young stellar objects. In *Galactic and Extragalactic Star Formation* (ed. R. Pudritz). NATO Advanced Study Institute, Reidel, Dordrecht, The Netherlands, pp. 53–68.

Swindle T. D. (2002) Martian noble gases. *Rev. Mineral. Geochem.* **47**, 171–190.

Swindle T. D. and Jones J. H. (1997) The xenon isotopic composition of the primordial Martian atmosphere: contributions from solar and fission components. *J. Geophys. Res.* **102**, 1671–1678.

Swindle T. D., Caffee M. W., and Hohenberg C. M. (1986) Xenon and other noble gases in shergottites. *Geochim. Cosmochim. Acta* **50**, 1001–1015.

Tolstikhin I. N. and O'Nions R. K. (1994) The Earth's missing xenon: a combination of early degassing and of rare gas loss from the atmosphere. *Chem. Geol.* **115**, 1–6.

Tolstikhin I. N. and Marty B. (1998) The evolution of terrestrial volatiles: a view from helium, neon, argon, and nitrogen isotope modelling. *Chem. Geol.* **147**, 27–52.

Tonks W. B. and Melosh H. J. (1990) The physics of crystal settling and suspension in a turbulent magma ocean. In *Origin of the Earth* (eds. H. E. Newsom and J. H. Jones). Oxford University Press, pp. 151–171.

Torgersen T. (1989) Terrestrial helium degassing fluxes and the atmospheric helium budget: implications with respect to the degassing processes of continental crust. *Chem. Geol. (Isot. Geosci. Sect.)* **79**, 1–14.

Trieloff M., Kunz J., Clague D. A., Harrison D., and Allègre C. J. (2000) The nature of pristine noble gases in mantle plumes. *Science* **288**, 1036–1038.

Trieloff M., Kunz J., Clague D. A., Harrison D., and Allègre C. J. (2001) Reply to comment on noble gases in mantle plumes. *Science* **291**, 2269a.

Trieloff M., Kunz J., and Allègre C. J. (2002) Noble gas systematics of the Réunion mantle plume source and the origin of primordial noble gases in Earth's mantle. *Earth Planet. Sci. Lett.* **200**, 297–313.

Trull T., Nadeau S., Pineau F., Polvé M., and Javoy M. (1993) C–He systematics in hotspot xenoliths—implications for mantle carbon contents and carbon recycling. *Earth Planet. Sci. Lett.* **118**, 43–64.

Tyburczy J. A., Frisch B., and Ahrens T. J. (1986) Shock-induced volatile loss from a carbonaceous chondrite: implications for planetary accretion. *Earth Planet. Sci. Lett.* **80**, 201–207.

Valbracht P. J., Staudacher T., Malahoff A., and Allègre C. J. (1997) Noble gas systematics of deep rift zone glasses from Loihi Seamount, Hawaii. *Earth Planet. Sci. Lett.* **150**, 399–411.

Vickery A. M. and Melosh H. J. (1990) Atmospheric erosion and impactor retention in large impacts with application to mass extinctions. In *Global Catastrophes in Earth History*, Geological Society of America Special Paper 247 (eds. V. L. Sharpton and P. D. Ward). Geological Society of America, Boulder, Co, pp. 289–300.

von Zahn U., Kumar S., Niemann H., and Prinn R. (1983) Composition of the Venus atmosphere. In *Venus* (eds. D. Hunten, L. Colin, T. Donahue, and V. Moroz). University of Arizona Press, Tucson, pp. 299–430.

Wacker J. F. (1989) Laboratory simulation of meteoritic noble gases: III. Sorption of neon, argon, krypton, and xenon on carbon: elemental fractionation. *Geochim. Cosmochim. Acta* **53**, 1421–1433.

Wacker J. F. and Anders E. (1984) Trapping of xenon in ice: implications for the origin of the Earth's noble gases. *Geochim. Cosmochim. Acta* **48**, 2373–2380.

Walter F. M. and Barry D. C. (1991) Pre- and main sequence evolution of solar activity. In *The Sun in Time* (eds. C. P. Sonett, M. S. Giampapa, and M. S. Matthews). University of Arizona Press, Tucson, pp. 633–657.

Walter F. M., Brown A., Mathieu R. D., Myers P. C., and Vrba F. J. (1988) X-ray sources in regions of star formation: III. Naked T Tauri stars associated with the Taurus-Auriga complex. *Astron. J.* **96**, 297–325.

Wänke H. and Dreibus G. (1988) Chemical composition and accretion history of terrestrial planets. *Phil. Trans. Roy. Soc. London* **A325**, 545–557.

Wänke H., Dreibus G., and Jagoutz E. (1984) Mantle chemistry and accretion history of the Earth. In *Archaean Geochemistry* (eds. A. Kröner, G. N. Hanson, and A. M. Goodwin). Springer, pp. 1–24.

Watson L., Hutcheon I. D., Epstein S., and Stolper E. (1994) Water on Mars: clues from deuterium/hydrogen and water contents of hydrous phases in SNC meteorites. *Science* **265**, 86–90.

Wetherill G. W. (1975) Radiometric chronology of the early solar system. *Ann. Rev. Nuclear Sci.* **25**, 283–328.

Wetherill G. W. (1981) Solar wind origin of $^{36}$Ar on Venus. *Icarus* **46**, 70–80.

Wetherill G. W. (1986) Accumulation of the terrestrial planets and implications concerning lunar origin. In *Origin of the Moon* (eds. W. K. Hartmann, R. J. Phillips, and G. J. Taylor). Lunar and Planetary Institute, Houston, pp. 519–550.

Wetherill G. W. (1990a) Formation of the Earth. *Ann. Rev. Earth Planet. Sci.* **18**, 205–256.

Wetherill G. W. (1990b) Calculation of mass and velocity distributions of terrestrial and lunar impactors by use of theory of planetary accumulation. In *Abstracts for the International Workshop on Meteorite Impact on the Early Earth*, LPI Contr. No. 746, Lunar and Planetary Institute, Houston, pp. 54–55.

Wetherill G. W. and Stewart G. R. (1993) Formation of planetary embryos: effects of fragmentation, low relative velocity, and independent variation of eccentricity and inclination. *Icarus* **106**, 190–209.

Wieler R. (1994) "Q-gases" as "local" primordial noble gas component in primitive meteorites. In *Noble Gas Geochemistry and Cosmochemistry* (ed. J. Matsuda). Terra Scientific Publishing, Tokyo, pp. 31–41.

Wieler R. (1998) The solar noble gas record in lunar samples and meteorites. *Space Sci. Rev.* **85**, 303–314.

Wieler R. (2002) Noble gases in the solar system. *Rev. Mineral. Geochem.* **47**, 21–70.

Wieler R. and Baur H. (1994) Krypton and xenon from the solar wind and solar energetic particles in two linear ilmenites of different antiquity. *Meteoritics* **29**, 570–580.

Wlotzka F. (1972) Nitrogen. In *Handbook of Geochemistry* (ed. K. H. Wedepohl). Springer, Berlin.

Wood B. J. (1993) Carbon in the core. *Earth Planet. Sci. Lett.* **117**, 593–607.

Woolum D. S., Cassen P., Porcelli D., and Wasserburg G. J. (1999) Incorporation of solar noble gases from a nebula-derived atmosphere during magma ocean cooling. *Lunar Planet. Sci.* **XXX**, #1518 (CD-ROM). Lunar and Planetary Institute.

Yang J. and Anders E. (1982) Sorption of noble gases by solids, with reference to meteorites: III. Sulfides, spinels, and other substances: on the origin of planetary gases. *Geochim. Cosmochim. Acta* **46**, 877–892.

Yin Q., Jacobsen S. B., Yamashita K., Blichert-Toft J., Télouk P., and Albarède F. (2002) A short timescale for terrestrial planet formation from Hf–W chronometry of meteorites. *Nature* **418**, 949–952.

Zahnle K. J. (1993) Xenological constraints on the impact erosion of the early martian atmosphere. *J. Geophys. Res.* **98**, 10899–10913.

Zahnle K. J. and Kasting J. F. (1986) Mass fractionation during transonic escape and implications for loss of water from Mars and Venus. *Icarus* **68**, 462–480.

Zahnle K. J., Kasting J. F., and Pollack J. B. (1990a) Mass fractionation of noble gases in diffusion-limited hydrodynamic hydrogen escape. *Icarus* **84**, 502–527.

Zahnle K. J., Pollack J. B., and Kasting J. F. (1990b) Xenon fractionation in porous planetesimals. *Geochim. Cosmochim. Acta* **54**, 2577–2586.

Zhang M. H. G., Luhmann J. G., Bougher S. W., and Nagy A. F. (1993) The ancient oxygen exosphere of Mars: implications for atmospheric evolution. *J. Geophys. Res.* **98**, 10915–10923.

Zhang Y. and Zindler A. (1989) Noble gas constraints on the evolution of Earth's atmosphere. *J. Geophys. Res.* **94**, 13719–13737.

Printed and bound by CPI Group (UK) Ltd, Croydon, CR0 4YY

03/10/2024

01040311-0014

# Volume Subject Index

The index is in letter-by-letter order, whereby hyphens and spaces within index headings are ignored in the alphabetization (e.g. Arabian–Nubian Shield precedes Arabian Sea). Terms in parentheses are excluded from the initial alphabetization. In line with normal materials science practice, compound names are not inverted but are filed under substituent prefixes.

The index is arranged in set-out style, with a maximum of three levels of heading. Location references refer to the page number. Major discussion of a subject is indicated by bold page numbers. Page numbers suffixed by *f* or *t* refer to figures or tables.